Lecture Notes in Computer Science 7214

Commenced Publication in 1973
Founding and Former Series Editors:
Gerhard Goos, Juris Hartmanis, and Jan van Leeuwen

Editorial Board

Advanced Research in Computing and Software Science

Subline of Lectures Notes in Computer Science

Subline Series Editors

Subline Advisory Board

W0193060

Cormac Flanagan   Barbara König (Eds.)

Tools and Algorithms for the Construction and Analysis of Systems

18th International Conference, TACAS 2012
Held as Part of the European Joint Conferences
on Theory and Practice of Software, ETAPS 2012
Tallinn, Estonia, March 24 – April 1, 2012
Proceedings

 Springer

Volume Editors

Cormac Flanagan
University of California at Santa Cruz
Jack Baskin School of Engineering
Computer Science Department
1156 High Street, Santa Cruz, CA 95064, USA
E-mail: cormac@ucsc.edu

Barbara König
Universität Duisburg-Essen
Fakultät für Ingenieurwissenschaften
Abteilung für Informatik und Angewandte Kognitionswissenschaft
Lotharstraße 65, 47057 Duisburg, Germany
E-mail: barbara_koenig@uni-due.de

ISSN 0302-9743   e-ISSN 1611-3349
ISBN 978-3-642-28755-8   ISBN 978-3-642-28756-5 (eBook)
DOI 10.1007/978-3-642-28756-5
Springer Heidelberg Dordrecht London New York

Library of Congress Control Number: 2012932744

CR Subject Classification (1998): F.3, D.2, C.2, D.3, D.2.4, C.3

LNCS Sublibrary: SL 1 – Theoretical Computer Science and General Issues

Typesetting: Camera-ready by author, data conversion by Scientific Publishing Services, Chennai, India

Printed on acid-free paper

Springer is part of Springer Science+Business Media (www.springer.com)

Foreword

ETAPS 2012 is the fifteenth instance of the European Joint Conferences on Theory and Practice of Software. ETAPS is an annual federated conference that was established in 1998 by combining a number of existing and new conferences. This year it comprised six sister conferences (CC, ESOP, FASE, FOSSACS, POST, TACAS), 21 satellite workshops (ACCAT, AIPA, BX, BYTECODE, CMCS, DICE, FESCA, FICS, FIT, GRAPHITE, GT-VMT, HAS, IWIGP, LDTA, LINEARITY, MBT, MSFP, PLACES, QAPL, VSSE and WRLA), and eight invited lectures (excluding those specific to the satellite events).

The six main conferences received this year 606 submissions (including 21 tool demonstration papers), 159 of which were accepted (6 tool demos), giving an overall acceptance rate just above 26%. Congratulations therefore to all the authors who made it to the final programme! I hope that most of the other authors will still have found a way to participate in this exciting event, and that you will all continue to submit to ETAPS and contribute to making it the best conference on software science and engineering.

The events that comprise ETAPS address various aspects of the system development process, including specification, design, implementation, analysis, security and improvement. The languages, methodologies and tools that support these activities are all well within its scope. Different blends of theory and practice are represented, with an inclination towards theory with a practical motivation on the one hand and soundly based practice on the other. Many of the issues involved in software design apply to systems in general, including hardware systems, and the emphasis on software is not intended to be exclusive.

ETAPS is a confederation in which each event retains its own identity, with a separate Programme Committee and proceedings. Its format is open-ended, allowing it to grow and evolve as time goes by. Contributed talks and system demonstrations are in synchronised parallel sessions, with invited lectures in plenary sessions. Two of the invited lectures are reserved for 'unifying' talks on topics of interest to the whole range of ETAPS attendees. The aim of cramming all this activity into a single one-week meeting is to create a strong magnet for academic and industrial researchers working on topics within its scope, giving them the opportunity to learn about research in related areas, and thereby to foster new and existing links between work in areas that were formerly addressed in separate meetings.

This year, ETAPS welcomes a new main conference, *Principles of Security and Trust*, as a candidate to become a permanent member conference of ETAPS. POST is the first addition to our main programme since 1998, when the original five conferences met in Lisbon for the first ETAPS event. It combines the practically important subject matter of security and trust with strong technical connections to traditional ETAPS areas.

A step towards the consolidation of ETAPS and its institutional activities has been undertaken by the Steering Committee with the establishment of *ETAPS e.V.*, a non-profit association under German law. ETAPS e.V. was founded on April 1st, 2011 in Saarbrücken, and we are currently in the process of defining its structure, scope and strategy.

ETAPS 2012 was organised by the *Institute of Cybernetics at Tallinn University of Technology*, in cooperation with

▷ European Association for Theoretical Computer Science (EATCS)
▷ European Association for Programming Languages and Systems (EAPLS)
▷ European Association of Software Science and Technology (EASST)

and with support from the following sponsors, which we gratefully thank:

INSTITUTE OF CYBERNETICS AT TUT; TALLINN UNIVERSITY OF TECHNOLOGY (TUT); ESTONIAN CENTRE OF EXCELLENCE IN COMPUTER SCIENCE (EXCS) FUNDED BY THE EUROPEAN REGIONAL DEVELOPMENT FUND (ERDF); ESTONIAN CONVENTION BUREAU; and MICROSOFT RESEARCH.

The organising team comprised:

General Chair: *Tarmo Uustalu*

Satellite Events: *Keiko Nakata*

Organising Committee: *James Chapman, Juhan Ernits, Tiina Laasma, Monika Perkmann* and their colleagues in the *Logic and Semantics* group and *administration* of the *Institute of Cybernetics*

The ETAPS portal at http://www.etaps.org is maintained by *RWTH Aachen University*.

Overall planning for ETAPS conferences is the responsibility of its Steering Committee, whose current membership is:

Vladimiro Sassone (Southampton, Chair), Roberto Amadio (Paris 7), Gilles Barthe (IMDEA-Software), David Basin (Zürich), Lars Birkedal (Copenhagen), Michael O'Boyle (Edinburgh), Giuseppe Castagna (CNRS Paris), Vittorio Cortellessa (L'Aquila), Koen De Bosschere (Gent), Pierpaolo Degano (Pisa), Matthias Felleisen (Boston), Bernd Finkbeiner (Saarbrücken), Cormac Flanagan (Santa Cruz), Philippa Gardner (Imperial College London), Andrew D. Gordon (MSR Cambridge and Edinburgh), Daniele Gorla (Rome), Joshua Guttman (Worcester USA), Holger Hermanns (Saarbrücken), Mike Hinchey (Lero, the Irish Software Engineering Research Centre), Ranjit Jhala (San Diego), Joost-Pieter Katoen (Aachen), Paul Klint (Amsterdam), Jens Knoop (Vienna), Barbara König (Duisburg), Juan de Lara (Madrid), Gerald Lüttgen (Bamberg), Tiziana Margaria (Potsdam), Fabio Martinelli (Pisa), John Mitchell (Stanford), Catuscia Palamidessi (INRIA Paris), Frank Pfenning (Pittsburgh), Nir Piterman (Leicester), Don Sannella (Edinburgh), Helmut Seidl (TU Munich),

Scott Smolka (Stony Brook), Gabriele Taentzer (Marburg), Tarmo Uustalu (Tallinn), Dániel Varró (Budapest), Andrea Zisman (London), and Lenore Zuck (Chicago).

I would like to express my sincere gratitude to all of these people and organisations, the Programme Committee Chairs and PC members of the ETAPS conferences, the organisers of the satellite events, the speakers themselves, the many reviewers, all the participants, and Springer-Verlag for agreeing to publish the ETAPS proceedings in the ARCoSS subline.

Finally, I would like to thank the Organising Chair of ETAPS 2012, Tarmo Uustalu, and his Organising Committee, for arranging to have ETAPS in the most beautiful surroundings of Tallinn.

January 2012

Vladimiro Sassone
ETAPS SC Chair

Preface

This volume contains the proceedings of the 18th International Conference on Tools and Algorithms for the Construction and Analysis of Systems (TACAS 2012). TACAS 2012 took place in Tallinn, Estonia, March 27–30, 2012, as part of the 15th European Joint Conferences on Theory and Practice of Software (ETAPS 2012), whose aims, organization, and history are presented in the foreword of this volume by the ETAPS Steering Committee Chair, Vladimiro Sassone.

TACAS is a forum for researchers, developers, and users interested in rigorously based tools and algorithms for the construction and analysis of systems. The research areas covered by TACAS include, but are not limited to, formal methods, software and hardware verification, static analysis, programming languages, software engineering, real-time systems, communications protocols, and biological systems. The TACAS forum provides a venue for such communities at which common problems, heuristics, algorithms, data structures, and methodologies can be discussed and explored.

This year TACAS solicited four kinds of papers. Research papers are full-length papers that contain novel research on topics within the scope of the TACAS conference and relevant to tool construction. Case study papers report on case studies, describing goals of the study and the research methodologies and approach used. Regular tool papers present a new tool, a new tool component, or novel extensions to an existing tool, and focus primarily on engineering aspects. Tool demonstration papers are shorter papers that give an overview of a particular tool and its applications or evaluation.

This edition of the conference attracted a total of 147 submissions (108 research papers, 7 case study papers, 11 regular tool papers, and 21 tool demonstration papers). Each submission was evaluated by at least three reviewers, who wrote detailed evaluations and gave insightful comments. After a six-week reviewing process, we had a three-week electronic Program Committee meeting, during which we accepted 36 papers (25 research papers, 2 case study papers, 3 regular tool papers, and 6 tool demonstration papers) for presentation at the conference. In addition, the TACAS conference program was greatly enriched by the invited talk by Holger Hermanns and by the ETAPS unifying speakers Bruno Blanchet and Georg Gottlob.

One novel aspect of TACAS this year was the inclusion of a Competition on Software Verification. This volume includes an overview of the competition organization and results, presented by the TACAS Competition Chair Dirk Beyer, as well as short papers describing individual tools in the competition.

We would like to thank the authors of all submitted papers, the Program Committee members, and the external referees for their invaluable contributions. We also thank Alessandro Cimatti for serving as the TACAS Tool Chair and Dirk Beyer for organizing the Competition on Software Verification. The

EasyChair system greatly facilitated the TACAS conference submission and program selection process. Finally, we would like to express our appreciation to the ETAPS Steering Committee and especially its Chair, Vladimiro Sassone, as well as the Organizing Committee, for their efforts in making ETAPS 2012 such a successful event.

January 2012   Cormac Flanagan
                                                       Barbara König

Organization

Steering Committee

Rance Cleaveland	University of Maryland and Fraunhofer USA, USA
Joost-Pieter Katoen	RWTH Aachen, Germany
Kim G. Larsen	Aalborg University, Denmark
Bernhard Steffen	TU Dortmund, Germany
Lenore Zuck	University of Illinois at Chicago, USA

Program Committee

Rajeev Alur	University of Pennsylvania, USA
Armin Biere	Johannes Kepler University, Austria
Alessandro Cimatti	FBK-irst, Italy (Tool Chair)
Rance Cleaveland	University of Maryland and Fraunhofer USA, USA
Giorgio Delzanno	University of Genova, Italy
Javier Esparza	TU Munich, Germany
Cormac Flanagan	University of California at Santa Cruz, USA (Co-chair)
Patrice Godefroid	Microsoft Research, Redmond, USA
Susanne Graf	Université Joseph Fourier/CNRS/Verimag, France
Orna Grumberg	Technion – Israel Institute of Technology, Israel
Aarti Gupta	NEC Labs America, USA
Michael Huth	Imperial College London, UK
Ranjit Jhala	University of California at San Diego, USA
Vineet Kahlon	University of Texas at Austin, USA
Daniel Kroening	University of Oxford, UK
Marta Kwiatkowska	University of Oxford, UK
Barbara König	University of Duisburg-Essen, Germany (Co-chair)
Kim G. Larsen	Aalborg University, Denmark
Rustan Leino	Microsoft Research, Redmond, USA
Matteo Maffei	Saarland University, Germany
Ken McMillan	Cadence Berkeley Labs, USA
Doron Peled	Bar Ilan University, Israel
Anna Philippou	University of Cyprus, Cyprus
Nir Piterman	University of Leicester, UK
Arend Rensink	University of Twente, The Netherlands

Andrey Rybalchenko   TU Munich, Germany
Stefan Schwoon   ENS Cachan, France
Scott Smolka   SUNY at Stony Brook, USA
Bernhard Steffen   TU Dortmund, Germany
Serdar Tasiran   Koc University, Turkey
Lenore Zuck   University of Illinois at Chicago, USA

Program Committee of the Competition on Software Verification

Dirk Beyer   University of Passau, Germany (Chair)
Bernd Fischer   University of Southampton, UK
Vadim Mutilin   Russian Academy of Sciences, Russia
Andrey Rybalchenko   TU Munich, Germany
Carsten Sinz   Karlsruhe Institute of Technology, Germany
Michael Tautschnig   University of Oxford, UK
Helmut Veith   Vienna University of Technology, Austria
Tomas Vojnar   Brno University of Technology, Czech Republic
Georg Weissenbacher   Princeton University, USA
Philipp Wendler   University of Passau, Germany
Daniel Wonisch   University of Paderborn, Germany

Additional Reviewers

Acciai, Lucia   David, Alexandre
Aldini, Alessandro   De Mol, Maarten
Alglave, Jade   Delahaye, Benoit
Ancona, Davide   Deshmukh, Jyotirmoy
Bagherzadeh, Mehdi   Deshpande, Tushar
Ball, Thomas   Diciolla, Marco
Bartocci, Ezio   Donaldson, Alastair
Bekar, Can   Draeger, Klaus
Belinfante, Axel   Duggirala, Parasara Sridhar
Blanchette, Jasmin Christian   Duret-Lutz, Alexandre
Bouajjani, Ahmed   Edelkamp, Stefan
Bulychev, Peter   Elmas, Tayfun
Burckhardt, Sebastian   Falke, Stephan
Chen, Taolue   Feng, Lu
Cheng, Chih-Hong   Feo, Sergio
Chilton, Chris   Fernandez, Jean-Claude
Clemente, Lorenzo   Ferrara, Pietro
D'Silva, Vijay   Filiot, Emmanuel
Dalsgaard, Andreas   Flur, Shaked
Dannenberg, Frits   Forejt, Vojtech

Gaiser, Andreas
Ganty, Pierre
Garg, Pranav
Gastin, Paul
Ghamarian, Amir Hossein
Giro, Sergio
Guerrini, Giovanna
Guttman, Joshua
Haar, Stefan
Habermehl, Peter
Haddad, Serge
Haller, Leopold
Han, Tingting
He, Nannan
Heule, Marijn
Hinrichs, Timothy
Howar, Falk
Isberner, Malte
Ivancic, Franjo
Järvisalo, Matti
Kant, Gijs
Khasidashvili, Zurab
Kinder, Johannes
Komuravelli, Anvesh
Kretinsky, Jan
Kulahcioglu, Burcu
Kuru, Ismail
La Torre, Salvatore
Lal, Akash
Lamprecht, Anna-Lena
Lazic, Ranko
Leroux, Jerome
Lewis, Matt
Mandrykin, Mikhail
Matar, Hassan Salehe
Mauborgne, Laurent
Meller, Yael
Mereacre, Alexandru
Merten, Maik
Merz, Stephan
Methrayil Varghese, Praveen Thomas
Mikučionis, Marius
Moskal, Michal
Mounier, Laurent
Møller, Mikael H.

Nadel, Alexander
Naujokat, Stefan
Neubauer, Johannes
Nghiem, Truong
Nimal, Vincent
Norman, Gethin
Nyman, Ulrik
Ober, Iulian
Olesen, Mads Chr.
Palikareva, Hristina
Parker, David
Passerone, Roberto
Phan, Linh
Popeea, Corneliu
Quinton, Sophie
Rajan, Ajitha
Ramachandran, Jaideep
Raymond, Pascal
Reuss, Andreas
Ruething, Oliver
Ryvchin, Vadim
Rüthing, Oliver
Sa'Ar, Yaniv
Sangnier, Arnaud
Schewe, Sven
Seghir, Mohamed Nassim
Seidl, Martina
Sezgin, Ali
Sheinvald, Sarai
Shin, Insik
Shoham, Sharon
Simaitis, Aistis
Soliman, Sylvain
Srba, Jiri
Stoelinga, Mariëlle
Strichman, Ofer
Subasi, Omer
Tautschnig, Michael
Timmer, Mark
Tiwari, Ashish
Traonouez, Louis-Marie
Trivedi, Ashutosh
Ummels, Michael
Vaandrager, Frits
Veanes, Margus

Viswanathan, Mahesh
Vizel, Yakir
Wang, Bow-Yaw
Weissenbacher, Georg
Windmüller, Stephan

Wolf, Verena
Yrke Jørgensen, Kenneth
Zambon, Eduardo
Zhang, Lijun
Zuliani, Paolo

Table of Contents

Invited Contribution

Quantitative Models for a Not So Dumb Grid 1
    Holger Hermanns

SAT and SMT Based Methods

History-Aware Data Structure Repair Using SAT 2
    Razieh Nokhbeh Zaeem, Divya Gopinath, Sarfraz Khurshid, and
    Kathryn S. McKinley

The Guardol Language and Verification System 18
    David Hardin, Konrad Slind, Michael Whalen, and Tuan-Hung Pham

A Bit Too Precise? Bounded Verification of Quantized Digital Filters ... 33
    Arlen Cox, Sriram Sankaranarayanan, and Bor-Yuh Evan Chang

Numeric Bounds Analysis with Conflict-Driven Learning 48
    Vijay D'Silva, Leopold Haller, Daniel Kroening, and
    Michael Tautschnig

Automata

Ramsey-Based Analysis of Parity Automata 64
    Oliver Friedmann and Martin Lange

VATA: A Library for Efficient Manipulation of Non-deterministic Tree
Automata 79
    Ondřej Lengál, Jiří Šimáček, and Tomáš Vojnar

LTL to Büchi Automata Translation: Fast and More Deterministic 95
    Tomáš Babiak, Mojmír Křetínský, Vojtěch Řehák, and Jan Strejček

Model Checking

Pushdown Model Checking for Malware Detection 110
    Fu Song and Tayssir Touili

Aspect-Oriented Runtime Monitor Certification 126
    Kevin W. Hamlen, Micah M. Jones, and Meera Sridhar

Partial Model Checking Using Networks of Labelled Transition Systems
and Boolean Equation Systems 141
  Frédéric Lang and Radu Mateescu

From Under-Approximations to Over-Approximations and Back 157
  Aws Albarghouthi, Arie Gurfinkel, and Marsha Chechik

Case Studies

Automated Analysis of AODV Using UPPAAL 173
  *Ansgar Fehnker, Rob van Glabbeek, Peter Höfner,
  Annabelle McIver, Marius Portmann, and Wee Lum Tan*

Modeling and Verification of a Dual Chamber Implantable
Pacemaker 188
  *Zhihao Jiang, Miroslav Pajic, Salar Moarref, Rajeev Alur, and
  Rahul Mangharam*

Memory Models and Termination

Counter-Example Guided Fence Insertion under TSO 204
  *Parosh Aziz Abdulla, Mohamed Faouzi Atig, Yu-Fang Chen,
  Carl Leonardsson, and Ahmed Rezine*

Java Memory Model-Aware Model Checking 220
  Huafeng Jin, Tuba Yavuz-Kahveci, and Beverly A. Sanders

Compositional Termination Proofs for Multi-threaded Programs 237
  Corneliu Popeea and Andrey Rybalchenko

Deciding Conditional Termination 252
  Marius Bozga, Radu Iosif, and Filip Konečný

Internet Protocol Verification

The AVANTSSAR Platform for the Automated Validation of Trust
and Security of Service-Oriented Architectures 267
  *Alessandro Armando, Wihem Arsac, Tigran Avanesov,
  Michele Barletta, Alberto Calvi, Alessandro Cappai,
  Roberto Carbone, Yannick Chevalier, Luca Compagna,
  Jorge Cuéllar, Gabriel Erzse, Simone Frau, Marius Minea,
  Sebastian Mödersheim, David von Oheimb, Giancarlo Pellegrino,
  Serena Elisa Ponta, Marco Rocchetto, Michael Rusinowitch,
  Mohammad Torabi Dashti, Mathieu Turuani, and Luca Viganò*

Reduction-Based Formal Analysis of BGP Instances 283
  *Anduo Wang, Carolyn Talcott, Alexander J.T. Gurney,
  Boon Thau Loo, and Andre Scedrov*

Stochastic Model Checking

Minimal Critical Subsystems for Discrete-Time Markov Models 299
*Ralf Wimmer, Nils Jansen, Erika Ábrahám, Bernd Becker, and
Joost-Pieter Katoen*

Automatic Verification of Competitive Stochastic Systems............. 315
*Taolue Chen, Vojtěch Forejt, Marta Kwiatkowska,
David Parker, and Aistis Simaitis*

Coupling and Importance Sampling for Statistical Model Checking 331
Benoît Barbot, Serge Haddad, and Claudine Picaronny

Verifying pCTL Model Checking 347
Johannes Hölzl and Tobias Nipkow

Synthesis

Parameterized Synthesis 362
Swen Jacobs and Roderick Bloem

QuteRTL: Towards an Open Source Framework for RTL Design
Synthesis and Verification 377
Hu-Hsi Yeh, Cheng-Yin Wu, and Chung-Yang (Ric) Huang

Template-Based Controller Synthesis for Timed Systems 392
Bernd Finkbeiner and Hans-Jörg Peter

Provers and Analysis Techniques

Zeno: An Automated Prover for Properties of Recursive Data
Structures 407
William Sonnex, Sophia Drossopoulou, and Susan Eisenbach

A Proof Assistant for Alloy Specifications 422
*Mattias Ulbrich, Ulrich Geilmann, Aboubakr Achraf El Ghazi, and
Mana Taghdiri*

Reachability under Contextual Locking 437
Rohit Chadha, P. Madhusudan, and Mahesh Viswanathan

Bounded Phase Analysis of Message-Passing Programs................ 451
Ahmed Bouajjani and Michael Emmi

Tool Demonstrations

Demonstrating Learning of Register Automata...................... 466
*Maik Merten, Falk Howar, Bernhard Steffen, Sofia Cassel, and
Bengt Jonsson*

Symbolic Automata: The Toolkit 472
   Margus Veanes and Nikolaj Bjørner

McScM: A General Framework for the Verification of Communicating
Machines 478
   Alexander Heußner, Tristan Le Gall, and Grégoire Sutre

SLMC: A Tool for Model Checking Concurrent Systems against
Dynamical Spatial Logic Specifications 485
   Luís Caires and Hugo Torres Vieira

TAPAAL 2.0: Integrated Development Environment for Timed-Arc
Petri Nets.. 492
   *Alexandre David, Lasse Jacobsen, Morten Jacobsen,
   Kenneth Yrke Jørgensen, Mikael H. Møller, and Jiří Srba*

A Platform for High Performance Statistical Model
Checking – PLASMA.. 498
   Cyrille Jegourel, Axel Legay, and Sean Sedwards

Competition on Software Verification

Competition on Software Verification (SV-COMP) 504
   Dirk Beyer

Predicate Analysis with BLAST 2.7 (Competition Contribution) 525
   Pavel Shved, Mikhail Mandrykin, and Vadim Mutilin

CPACHECKER with Adjustable Predicate Analysis
(Competition Contribution) 528
   Stefan Löwe and Philipp Wendler

Block Abstraction Memoization for CPAchecker
(Competition Contribution) 531
   Daniel Wonisch

Context-Bounded Model Checking with ESBMC 1.17
(Competition Contribution) 534
   Lucas Cordeiro, Jeremy Morse, Denis Nicole, and Bernd Fischer

Proving Reachability Using FSHELL (Competition Contribution) 538
   *Andreas Holzer, Daniel Kroening, Christian Schallhart,
   Michael Tautschnig, and Helmut Veith*

LLBMC: A Bounded Model Checker for LLVM's Intermediate
Representation (Competition Contribution) 542
   Carsten Sinz, Florian Merz, and Stephan Falke

Predator: A Verification Tool for Programs with Dynamic Linked Data
Structures (Competition Contribution) 545
    Kamil Dudka, Petr Müller, Petr Peringer, and Tomáš Vojnar

HSF(C): A Software Verifier Based on Horn Clauses
(Competition Contribution) 549
    Sergey Grebenshchikov, Ashutosh Gupta, Nuno P. Lopes,
    Corneliu Popeea, and Andrey Rybalchenko

SatAbs: A Bit-Precise Verifier for C Programs
(Competition Contribution) 552
    Gérard Basler, Alastair Donaldson, Alexander Kaiser,
    Daniel Kroening, Michael Tautschnig, and Thomas Wahl

WOLVERINE: Battling Bugs with Interpolants
(Competition Contribution) 556
    Georg Weissenbacher, Daniel Kroening, and Sharad Malik

Author Index 559

Quantitative Models for a Not So Dumb Grid

Holger Hermanns

Dependable Systems and Software, Saarland University, Saarbrücken, Germany
http://d.cs.uni-saarland.de/hermanns/

How to dimension buffer sizes in a network on chip? What availability can be expected for the Gallileo satellite navigation system? Is it a good idea to ride a bike with a wireless brake? Can photovoltaic overproduction blow out the European electric power grid? Maybe. Maybe not. Probably? The era of power-aware, wireless and distributed systems of systems asks for strong quantitative answers to such questions.

Stochastic model checking techniques have been developed to attack these challenges [2]. They merge two well-established strands of informatics research and practice: verification of concurrent systems and performance evaluation. We review the main achievements of this research strand by painting the landscape of behavioural models for probability, time, and cost, discussing important aspects of compositional modelling and model checking techniques. Different real-life cases show how these techniques are applied in practice.

A rich spectrum of quantitative analysis challenges is posed by the 'smart grid' vision [1,4]. That vision promises a more stable, secure, and resilient power grid operation, despite increasing volatility of electric power production. It is expected to come with more decentralized and autonomous structures, and with a lot of IT put in place to manage the grid. However, that vision is in its infancy, while the reality of power production is already changing considerably in some regions of Europe. We focus on a regulation put in place by the German Federal Network Agency to increase grid stability in case of photovoltaic overproduction. We show that this regulation may in fact decrease grid stability [3]. We also propose improved and fully decentralized stabilization strategies that take inspiration from probabilistic MAC protocols. Quantitative properties of these strategies are calculated by state-of-the-art stochastic model checking tools.

References

1. Amin, M.: Smart grid: Overview, issues and opportunities: Advances and challenges in sensing, modeling, simulation, optimization and control. In: IEEE Conference on Decision and Control and European Control Conference (2011)
2. Baier, C., Haverkort, B., Hermanns, H., Katoen, J.-P.: Performance evaluation and model checking join forces. Communications of the ACM 53(9), 74–83 (2010)
3. Berrang, P., Bogdoll, J., Hahn, E.M., Hartmanns, A., Hermanns, H.: Dependability results for power grids with decentralized stabilization strategies. Reports of SFB/TR 14 AVACS - ATR 83 (2012), http://www.avacs.org
4. Hermanns, H., Wiechmann, H.: Future design challenges for electric energy supply. In: IEEE International Conference on Emerging Technologies & Factory Automation (2009)

C. Flanagan and B. König (Eds.): TACAS 2012, LNCS 7214, p. 1, 2012.
© Springer-Verlag Berlin Heidelberg 2012

History-Aware Data Structure Repair Using SAT

Razieh Nokhbeh Zaeem[1], Divya Gopinath[1], Sarfraz Khurshid[1],
and Kathryn S. McKinley[2]

[1] The University of Texas at Austin
{nokhbeh,divyagopinath}@utexas.edu, khurshid@ece.utexas.edu
[2] The University of Texas at Austin and Microsoft Research
mckinley@cs.utexas.edu

Abstract. Data structure repair corrects erroneous executions in deployed programs while they execute, eliminating costly downtime. Recent techniques show how to leverage specifications and a SAT solver to enforce specification conformance at runtime. While this powerful methodology increases the reliability of deployed programs, scalability remains a key technical challenge—satisfying a specification often results in the exploration of a huge state space. We present a novel technique, called *history-aware contract-based repair* for more efficient data structure repair using SAT. Our insight is two-fold: (1) the dynamic program trace of field writes and reads provides useful guidance to repair incorrect state mutations by a faulty program; and (2) we show how to execute SAT using unsatisfiable cores it generates, in an efficient iterative approach on successive problems with increasing state spaces, in order to utilize the history of previous runs as captured in the unsatisfiable core. We implement this approach in a new tool, called Cobbler, that repairs Java programs. Experimental results on two large applications and a library implementation of a linked list show that Cobbler significantly outperforms previous techniques for specification-based repair using SAT, and finds and repairs a previously undetected bug.

1   Introduction

Software systems are pervasive and integrated into almost every aspect of life. Software reliability is essential for life-critical, science, and business applications. Much research addresses producing reliable software in various phases of the software development life cycle before deployment, from analyzing requirements to design, implementation, and testing. However, improving the reliability of an already deployed (possibly faulty) system using error recovery is a less explored area.

In practice, systems are deployed with unknown and known unfixed bugs. When bugs cause failures, the usual tactic is to restart the program because fixing bugs and redeploying software may take months. Although the latter approach may resolve the fundamental source of the problem, system downtime is undesirable and not always feasible. Many *mission critical applications* such as operating systems, may prefer to trade slight deviations in intended functionality for system uptime. Better still, if developers annotate programs with specifications, then the runtime may restore the system state to provide its intended functionality. Continuing program execution by fixing the effect of bugs on-the-fly is called *repair*. Existing techniques for repair have not so far lived up to their full potential, because they are either not general purpose or too inefficient.

C. Flanagan and B. König (Eds.): TACAS 2012, LNCS 7214, pp. 2–17, 2012.
© Springer-Verlag Berlin Heidelberg 2012

Some critical systems include code that repairs erroneous executions on-the-fly using dedicated application specific repair routines [6,7,13,14]. Recent work introduced general purpose approaches including constraint-based repair [4,8,10] and contract-based repair [18,25], some of which utilize SAT solvers [18,25]. Constraint-based repair emphasizes data structure integrity rules and repairs the data structures when a bug leads to an invariant violation. Contract-based repair adds pre- and post-condition specifications of a method which aid in generating an accurate repair, i.e., a structure that is the same or very close to the one that a correct method would generate. General purpose repair, however, has a huge state space of possible post-states and exploring them to find a solution is currently too expensive to use in practice.

This paper seeks to make repair substantially more efficient by utilizing the *history* of code execution as well as SAT solving. Our insights are two-fold: (1) the dynamic program trace of field writes and reads provides useful guidance to identify incorrect state mutations made by a faulty program; and (2) the unsatisfiable core generated by a SAT run captures core elements of the solver's reasoning, which not only facilitates locating faults but can even be leveraged directly to optimize a successive SAT run. We utilize program traces and unsatisfiable cores in tandem to define an efficient iterative methodology where SAT is run on successive problems with increasing state spaces and each run utilizes the history of the previous run. To our knowledge, our work is the first to use the history of program execution or constraint solving in data structure repair.

History-aware repair utilizes a faulty program execution by focusing repair on fields recently modified or read by the program, thereby reducing the search space for SAT. We record program writes and reads to the key data structure with *barriers*. A barrier is a code sequence that performs an action just prior to a write or read. Barriers are widely available in commercial and research implementations of managed languages, e.g., the HotSpot and Jikes RVM Java Virtual Machines, and the .NET C# system. Our approach inserts barrier instrumentation on writes and reads or piggybacks on existing barriers.

While using the history of program execution aids in improving repair performance, its heuristic nature implies that there exist cases in which we have to perform a broader search and consider fields not included in the execution trace. In such cases, we take advantage of *UNSAT* cores, which are minimal unsatisfiable sub-formulas provided by failed SAT invocations. When SAT invocations fail, we utilize their UNSAT cores to identify faulty fields. A final SAT invocation with the list of faulty fields extracted from the UNSAT core results in a repaired data structure.

We implement repair for Java in a tool called Cobbler. Cobbler uses class invariants and method post-conditions expressed in the Alloy specification language [9]. Cobbler inserts write and read instrumentation for the specified data structures to log dynamic program behavior. When Cobbler detects a violation, it uses a SAT solver to mutate the data structure until it satisfies the specification.

We explore the efficiency and accuracy of Cobbler on microbenchmarks and two open source programs: Kodkod solver [22] and ANTLR [2,16]. We compare our history-aware contract-based repair tool, Cobbler, to contract-based repair alone using PBnJ [18] and Tarmeem [25], two repair tools which leverage user guides and heuristics along with a SAT solver. Cobbler is substantially more efficient and scalable than PBnJ and Tarmeem. We also compare Cobbler with Juzi, which uses data structure specifications for repair, but does not use method post-conditions [5]. Juzi's dedicated constraint solver is more efficient than Cobbler, but Juzi's repair is applicable to far fewer cases and Cobbler is much more accurate. Our experiments show that for small to moderate instantiations of data structures, Cobbler provides repaired data structures which are 100% to 90% similar to the correct structure in more than 90% of the cases. Cobbler also finds and repairs a previously unknown error in ANTLR.

We make the following contributions: **History-aware contract-based repair** combines the program's dynamic behavior with specifications and the current erroneous state of a program to perform repair. **Read and write barriers for repair** are an unconventional use of barriers to obtain program execution history for repair. **Minimal unsatisfiable cores** provided by SAT solvers help to reduce the search space when a field outside the execution trace should be modified. **Cobbler** is an automated portable framework for repairing Java programs that enhances real applications with repair functionality. **Evaluation** shows that Cobbler efficiently and accurately repairs text-book examples and real world programs. Cobbler's more efficient and accurate repair facilitates the use of repair in real world applications and enhances software reliability.

2   Background

This section describes data structure repair and the Alloy tool-set, which Cobbler uses.

Repair: Data structure repair corrects erroneous executions on-the-fly by enforcing data integrity constraints (also known as `repOK`) and method pre- and post-conditions (contracts). Figure 1 (a) shows the faulty output of a method, which violates the acyclicity constraint as a binary search tree. A repair tool detects the violation and fixes it by removing the dotted edge. Further fixes may be needed to enforce method contracts too.

Alloy tool-set: Alloy is a relational first order logic language [9]. An Alloy model consists of relations and constraints on them. The Alloy Analyzer performs bounded exhaustive analysis of Alloy models. A *bound* is a function which maps each relation to a set of tuples (bound: $R \rightarrow 2^T$), where each tuple consists of atoms. For each relation R, two sets are defined: a lower bound $LB(R)$, which includes all tuples that R *must* have in its instance ($inst(R)$), and an upper bound $UB(R)$, which includes all tuples that R *may* have in its instance. Therefore, $LB(R) \subseteq inst(R) \subseteq UB(R)$. Figure 1 (b) shows the relational representation of the Java object graph shown in Figure 1 (a).

We use Kodkod [22], the back-end of Alloy Analyzer, which is a SAT-based constraint solver for first order logic that supports relations, transitive closure, and partial

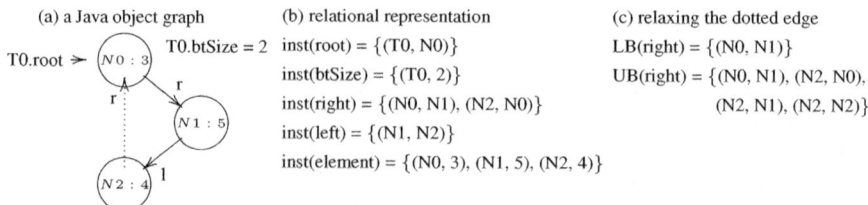

Fig. 1. Relational representation of data structures in Alloy models

models. Kodkod provides a finite model for satisfiable specifications and an UNSAT core for unsatisfiable ones. To perform repair, Kodkod suggests mutations to the data structure such that it meets the Alloy specification. Specifically, given a satisfiable relational formula and the bounds, Kodkod uses a backtracking search to find a satisfying instance. The search space is typically exponential in the number of atoms.

Kodkod allows explicit specification of upper and lower bounds for analysis, which provides partial solutions and restricts the search space. We use this functionality to specify which fields of the state can be mutated by the SAT solver to perform repair. Thus, to *relax* a field in Kodkod means to let the SAT solver suggest different values other than the one present in the faulty post-state, in order to find a satisfiable answer. Relaxing a field, which is a mutation of a field of a specific object, is done through binding a relation to suitable lower and upper bounds. For example, in Figure 1 (a) the dotted edge can be relaxed by setting the lower and upper bounds as shown in Figure 1 (c). Setting both lower and upper bounds to the same set makes it the only answer for that relation, i.e., the set becomes a partial solution for the Kodkod model.

Minimal Unsatisfiable Cores: If Kodkod cannot satisfy the constraints in a model, it produces a minimal unsatisfiable core, which is a subset of the constraints of the model. Given an unsatisfiable CNF formula X, a minimal unsatisfiable sub-formula is a subset of X's clauses that is both unsatisfiable and minimal, which means any subset of it is satisfiable. There could be many independent reasons for a formula's unsatisfiability and hence more than one minimal core. The Recycling Core Extractor algorithm, implemented as the RCE Strategy in Kodkod, returns an unsatisfiable core of specifications written in the Alloy language that is guaranteed to be sound (constraints not included in the core are irrelevant to the unsatisfiability proof) and irreducible (removal of any constraint from the set would make the remaining formula satisfiable).

3 Cobbler Framework

This section describes our history-aware contract-based repair framework.

3.1 Overview

We use class invariants and method post-conditions to detect erroneous executions. Once an error is detected, we utilize two major sources of information about the intended behavior: the specification and the dynamic trace of execution which we obtain through write and read logs. Although the post-condition specifies the expected behavior of the method, there is often a wide range of correct possibilities for a given input since there may be many ways to implement the same specification. Additionally, for a SAT-based repair framework, relaxing all fields of the data structure explodes the search space and is infeasible for real world applications.

We use the program execution to help guide our repair process. In deployed software, the program is expected to contain most of the intended logic. Furthermore, given sufficient pre-deployment testing, there should not be many bugs in the code. By observing the dynamic behavior of a faulty execution, we can substantially reduce the size of the search space and make the repair process more efficient and effective. The core idea is to focus on fields modified and/or read during the execution. To obtain the execution history, we record write and read actions performed by the program. Our implementation instruments the program, but alternatively the Java Virtual Machine could efficiently provide them [1]. We start by restricting the SAT solver to correcting written fields and values, followed by read fields during the execution, and if the SAT solver has still not found a correction, it utilizes the UNSAT core provided by the previous SAT invocations to identify and mutate faulty fields of the data structure. Hence, our technique handles both errors of *commission* when the programmer writes an incorrect assignment and errors of *omission* when she forgets to update the required fields.

While repair has various applications, it does not suit all types of software systems. For systems that cannot tolerate even slight divergences in the state of the program from the original behavior (e.g, financial systems), it is not advisable to use automatic repair routines unless complete contracts with all the required details are available.

When repair is applicable, this approach has two benefits: (1) it improves the repair performance by reducing the size of the search space, and (2) it reduces the amount of data structure perturbation introduced by the repair process by focusing on fields that a correct method conceivably would modify.

Listing 1.1 shows the repair algorithm in pseudo-code. If an assertion is violated, the repair framework initially only mutates (relaxes) fields in the write log, holding all other data structure fields constant (through providing a partial solution for the SAT solver). It then calls the SAT solver to compute correct values for the relaxed fields. If this step does not yield a structure satisfying the contracts, the next step relaxes the fields in the read and write logs. If it still is unsuccessful, it relaxes fields appearing in the UNSAT core. If the SAT solver finds no solution, there is an inconsistency in the contract itself which the repair framework reports.

```
1   if (! assertContracts ( ) ){
2       relaxSAT ( writeBarrierLog );
3       if (! assertContracts ( ) ){
4           relaxSAT ( writeBarrierLog , readBarrierLog );
5           if (! assertContracts ( ) ){
6               relaxSAT ( unsatCoreFields );
7               if (! assertContracts ( ) ){
8                   reportModelInconsistency ();}}}}
```

Listing 1.1. History-aware contract-based repair using read and write logs and unsatisfiable cores

3.2 Example

Consider a binary search tree example written in Java in Listing 1.2 and its `remove` method. In Cobbler, developers must write a specification in the Alloy relational first order logic language. Listing 1.3 shows the acyclicity and size constraints that describe a correct binary search tree in Alloy. Additional constraints include search relations on the nodes and that the elements are unique. The `repOK` method describes all method-independent constraints. The developer may also express method post-conditions, as shown in the `remove_postcondition` method. This post-condition specifies a correct `remove` with respect to the data structure and the return value from the `remove` method.

```
1  class BinarySearchTree {
2    Node root; int btSize;
3    boolean remove(int x) {
4        if (root == null) return false;
5        else {
6            boolean result;
7            if (root.element == x) {
8                Node auxRoot = new Node();
9                auxRoot.left = root;
10               result = root.remove(x, auxRoot);
11               root = auxRoot.left;
12           } else result = root.remove(x, null);
13           if (result) btSize--;//using uniqueness of elements
14           return result;}}}
15 class Node {
16    Node left, right; int element;
17    boolean remove(int x, Node parent) {
18        if (x < element) {
19            if (left != null) return left.remove(x, this);
20            else return false;
21        } else if (x > element) {
22            if (right != null) return right.remove(x, this);
23            else return false;
24        } else {
25            if (left != null && right != null) {
26                element = right.minNode().element;
27                right.remove(element, this);
28            } else if (parent.left == this) {
29                if (left != null) parent.left = left;
30                else parent.left = right;
31            } else if (parent.right == this) {
32                if (left != null) parent.right = left;
33                //to introduce bug cycle replace with left.right = parent
34                else parent.right = right;}
35        return true;}}
36    Node minNode() {...}}
```

Listing 1.2. A binary search tree implementation in Java

```
1  abstract sig BinarySearchTree {
2    root, root': lone Node,
3    btSize, btSize': one Int}
4  abstract sig Node{
5    left, left', right, right': lone Node,
6    element, element': lone Int}
7  pred repOK(t: BinarySearchTree){ //class invariant
8    //directed acyclicity
9    all n: t.root'.*(left'+right')|n !in n.^(left'+right')
10   //size OK
11   # t.root'.*(left'+right') = int t.btSize'
12   //unique elements
13   ...
14   //search property
15   ...}
16 pred remove_postcondition(This: BinarySearchTree, x: Int, removeResult: (True+
       False)){
17   repOK[This]
18   //correct remove
19   This.root.*(right+left).element − x = This.root'.*(right'+left').element'
20   //correct remove result
21   x in This.root.*(right+left).element <=> removeResult in True}
```

Listing 1.3. A binary search tree specification in Alloy

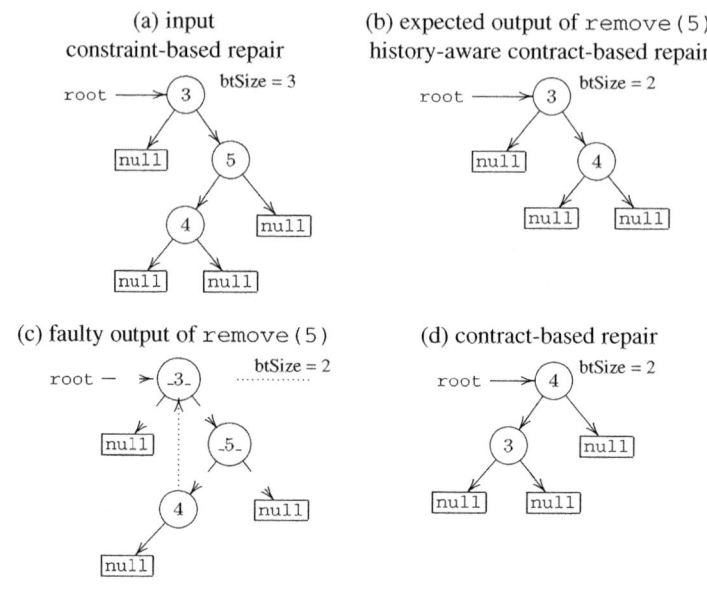

(a) input
constraint-based repair

(b) expected output of remove(5)
history-aware contract-based repair

(c) faulty output of remove(5)

(d) contract-based repair

(e) write barrier log (dotted lines in part (c)):
{[4].right, btSize}, [x] represents the node with value x before execution.
(f) read barrier log (dashed lines in part (c)):
{root, [3].element, [3].right, [5].element, [5].left, [5].right, [3].left}

Fig. 2. cycle manifested as a faulty output and the repair result

Alloy represents Java classes with signatures (e.g., sig BinarySearchTree in Listing 1.3) and field relations with a relational view. The keywords lone and one for a unary relation denote that the relation may or must not be empty, respectively. Binary relations can be defined as total or partial functions among other options (e.g., right is a partial function). We use the syntactic sugar of adding back-tick (''') to distinguish post-state Alloy relations from pre-state relations. The Alloy repOK predicate (pred) expresses data structural integrity rules. For instance, the directed acyclicity constraint specifies that for any node reachable from root by applying zero or more left or right pointers, the node cannot reach itself by following one or more left or right pointers, so it cannot traverse a cycle. * and ^ represent "zero or more" and "one or more" applications of a relation. Alloy supports membership, cardinality, and complement, in, #, and – respectively as in the acyclicity, size, and correct remove constraints.

To illustrate our repair process, consider inserting the following bug. **Bug cycle:** in Listing 1.2 line 32, replace the correct statement: parent.right = left with the incorrect: left.right = parent. For this incorrect implementation, after the method returns, checking the conjunction of repOK and the method post-condition indicates that there is an error, triggering the repair process.

To repair the erroneous output of the `cycle` faulty implementation, constraint-based repair methods [4, 8, 10] observe the cycle and remove it from Figure 2(c) to produce Figure 2(a), but fail to remove node 5. Contract-based repair techniques *without history* [18, 25] may generate Figure 2(d), which although a correct output, is different from what the program would have been generated in the absence of any bugs.

History-aware contract-based repair first invokes the SAT solver and tries to find a solution by only changing the values of the fields which the program writes into during the execution (Figure 2 (e)). These fields are distinguished by dotted lines in the faulty output. In this invocation, it does not find a solution because the program failed to update a field that needs to be modified. Our repair framework next considers changing fields read by the program (Figure 2 (f)) and shown as dashed lines. It invokes SAT to find suitable replacements for the fields written or read by the program. This invocation produces a repaired structure as shown in Figure 2 (b), which is identical to the expected output. Utilizing the barrier logs keeps us from generating Figure 2 (d) since the `left` field of node 4 is not relaxed and is held constant to be null. However, there remains a chance that a field that was not touched at all during the execution needs to be changed. Our repair framework obtains an UNSAT core from the previous SAT invocations. The UNSAT core is the conjunction of contradicting `repOK` and post-condition specifications, which were not satisfiable at the same time. In this example, if we were to proceed to the third SAT call, the UNSAT core would not include, for example, the `correct remove result` post-condition. Therefore, the final invocation of SAT would not relax the `removeResult` field.

3.3 Implementation in Cobbler

Cobbler works as follows. (1) The user provides the Java data structure class and its methods. Cobbler instruments this code with setters and getters to obtain logs of the writes and reads. Cobbler also instruments the program for our experiments to measure the repair time, edit distance and other metrics. (2) Cobbler generates a stub for the `repOK` and method post-conditions for the Java class. Cobbler extracts class-specific relations, types, and properties into the stubs, and the user enhances them with the application specific logic. (3) Cobbler then instruments the program to check the post-conditions and call the repair method when needed. (4) The user executes the Java program inside the Cobbler framework.

Figure 3 shows how the repair framework sits on top of the Java Virtual Machine and executes the Java program. The layers use shared memory to communicate. This design enhances the portability of our framework and makes it independent of JVM and the program. Alternative implementations could implement the framework inside the JVM, which would lower the overhead when pro-

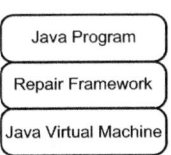

Fig. 3. The relationship between Cobbler, the Java Virtual Machine, and the program

grams are correct. When programs need to be repaired, the SAT solving time is orders of magnitude bigger than time saved by merging the repair framework into JVM.

4   Evaluation

The objectives of our evaluation are to empirically validate the hypothesis that using execution history and UNSAT cores improves the efficiency, accuracy, and scalability of contract-based repair with SAT solvers. To this end, we simulated various errors in microbenchmarks and examined two real world applications: Kodkod [22] and ANTLR [2, 16]. Cobbler discovered a previously unreported bug in the `addChild` method of ANTLR version 3.3 that resulted in a cycle in the output Tree. Our repair algorithm fixes this error accurately for a Tree with 300 nodes within 30 seconds.

Throughout the evaluation, we ran each experiment five times and reported the averages. All the experiments used a 2.50GHz Core 2 Duo processor with 4.00GB RAM running Windows 7 and Sun's Java SDK 1.6.0 JVM. All the repair frameworks used their default SAT solvers: Cobbler used MiniSat and MiniSatProver, Tarmeem used DefaultSAT4J, and PBnJ used MiniSat.

4.1   Evaluation Metrics

To evaluate the efficiency of repair, we measured: (1) **logging time:** the overhead due to logging read and write actions; (2) **check time:** the time to detect a contract violation; and (3) **repair time:** the time to search and find a repaired data structure.

To evaluable the accuracy of repair, we measure the *edit distance* between the object graphs of the repaired data structure r, and the expected data structure e that a correct implementation would produce. Note that, r satisfies the method contract but might be different from the expected output. We define edit distance as the minimum number of edge additions/deletions to change a graph to another [19, 25]. We create the correct graphs by a separate correct implementation and then measure the edit distance in set difference operations between two graphs using the relational representation discussed in Section 2. Here $inst_i(R)$ is the instance of relation R in data structure i.

Definition 1. $dist(e, r) = \sum_R(|inst_e(R) - inst_r(R)| + |inst_r(R) - inst_e(R)|).$

The lower this distance, the closer the repaired data structure is to the expected post-state data structure. We define the similarity percentage between the repaired output r and the expected output e as follows.

Definition 2. $sim(e, r) = (1 - \frac{dist(e,r)}{\sum_R |inst_e(R)|}) \times 100.$

4.2   Subject Programs

We applied Cobbler to (1) the `remove` method of Singly Linked List, (2) the `insert` method of the `Kodkod.util.ints.IntTree` class of the Kodkod solver implementation, and (3) the `deleteChild` and `addChild` methods of `BaseTree` of ANTLR.

Singly-linked list: Linked list is widely used and is a part of libraries such as `java.util.Collection`. The post-condition of the `remove(int value)` method, checks if the method has (1) deleted all nodes with elements equal to the input value, (2) maintained acyclicity, (3) inserted no new nodes, and (4) deleted no other nodes.

Red-black tree of Kodkod: Kodkod [22] is a SAT-based constraint solver for first order logic. It consists of 33,985 lines of Java code in 169 classes. The `IntTree` class with 570 lines of code and 21 methods sits at the core of the Kodkod solver, and is a generic implementation of the red-black tree data structure. Red-black tree comprises complex data structure invariants which include binary search tree invariants: every node has at most two children, key values of the left subtree are smaller and those of the right subtree are greater than the node value, and the tree is acyclic. In addition, constraints are imposed on the color of each node to keep the tree balanced: every node is either red or black, every leaf node is black, no red node has a red parent and all paths from the root to a descendant leaf contain the same number of black nodes. The `insert` method of this class comprises 58 lines of code with 67 branch statements. The post-condition of the `insert(int newKey)` method checks if an element with the new key value has been added without adding or deleting any other elements.

BaseTree of ANTLR: We use ANTLR (ANother Tool for Language Recognition) from the DaCapo 2009 benchmark suite, version 9.12 [2, 16]. ANTLR builds language parsers, interpreters, compilers, and translators from grammars. It comprises 29,710 lines of Java code, and has a download rate of about 5,000 per month. Rich tree data structures represent language grammars and are the backbone of this application. The abstract class `BaseTree` is a generic tree implementation. Each tree instance maintains a list of successor children. The `childIndex` represents its position in the list. Each child node is a tree and points back to its parent. Every node may contain a token field that represents the payload of the node. Based on the documentation and the program logic, we derived invariants for the `BaseTree` data structure such as acyclicity through children references, accurate parent-child relationships, and correct values for child indices. The `addChild(Tree node)` and `deleteChild(int childIndex)` methods are the main functions used to build and manipulate all tree structures in ANTLR. The respective post-conditions check that nodes are added or deleted without any unwarranted perturbations to the other nodes.

4.3 Errors

Table 1 enumerates all the inserted faults and, for ANTLR, a detected error. It explains the errors and displays the violated constraints. The accuracy and performance of the repair algorithm depends on which and how many fields are relaxed in each step, and the number of calls to the solver. The data structure size, size of the log, and size of violated constraint formula influence repair accuracy and efficiency. We explore these parameters with a range of errors that violate different constraints and appear in different program statements, such as incorrect field assignments, incorrect branch conditions, and errors of omission. The last column in the table indicates if the field(s) that needs to be corrected appear in the write barrier log (WB), read barrier log (RB), or neither (ALL fields).

The program logic and thus which fields Cobbler logs depends on the input structures. Faults five and six of the red-black tree `insert` method execute the same faulty code versions as that of three and four, but with a different data structure. The program writes and reads different fields on the first and second inputs and Cobbler repairs the outputs by relaxing read and written fields respectively.

Table 1. The injected faults and ANTLR addChild() fault

Method		Fault description	Violates	Error in
SLL remove	Err 1	Sets the header to null	Correct remove, Size	WB
	Err 2	Fails to update the size	Size	ALL fields
	Err 3	Deletes a node with a non-matching element	Correct remove, Size	WB
	Err 4	Introduces a cycle after performing correct remove	Acyclicity	WB
	Err 5	Breaks the list to retain only the first two nodes	Correct remove, Size	WB
	Err 6	Deletes the matching element but adds it again	Correct remove	WB
	Err 7	Fails to remove the element and updates the size incorrectly	Correct remove, Size	WB
RBT insert	Err 1	Creates a cycle of length one	Acyclicity	WB
	Err 2	Sets the color of a node to black instead of red	Color constraints	WB
	Err 3	Adds the new element as right child instead of left	Key constraints	RB
	Err 4	Violates key constraints due to a branch condition error	Key constraints	RB
	Err 5	Same as Err 3 with a different input	Key constraints	WB
	Err 6	Same as Err 4 with a different input	Key constraints	WB
	Err 7	Skips balancing of the tree after insertion	Color constraints	ALL fields
ANTLR deleteChild	Err 1	Skips deletion of the appropriate child	Correct Remove	RB
	Err 2	Skips updating children indices after deletion	Child Index constraints	ALL fields
	Err 3	Sets a wrong child index due to an incorrect branch condition in a loop	Child Index constraints	RB
	Err 4	Sets a node as its own parent	Acyclicity	WB
ANTLR addChild		Adds a node to itself as a child	Acyclicity, Child Index	WB

4.4   Subject Tools

We compare Cobbler to Juzi repair framework, which only uses structural constraints, and to Tarmeem and PBnJ, two repair frameworks that consider post-conditions too.

Juzi's assertion-based repair automatically corrects data structure violations in Java programs [5]. Upon detecting a constraint violation, Juzi searches for a repair solution based on the data structure traversal encoded in repOK [3]. Juzi further boosts its performance with symbolic execution. Since Juzi does not use a SAT solver, it is generally faster than SAT-based approaches. Juzi however does not consider method post-conditions, which causes it to miss errors that result in well-formed output. Even if repOK is violated, without the post-condition, Juzi cannot accurately correct the structure with respect to the contracts as discussed in Section 3.2. To compare Juzi and Cobbler, we manually implemented a check for the post-condition in Juzi by recording the method pre-state and the desired data structure specific post-state.

Our previous work, Tarmeem, uses Alloy contracts and a SAT solver [25]. Tarmeem repairs faulty post-states using automated and user-guided techniques, such as iterative relaxation of relations and error localization in predicates to improve the efficiency of repair. We experimented with all four of Tarmeem's heuristics and picked the best.

Samimi et al. implement a similar technique in PBnJ that executes method specifications when methods fail to produce a correct data structure [18]. They express invariants and specifications in a declarative first order relational logic. Translating them into Java methods and then invoking the methods implements program logic declaratively. This program synthesis approach leverages constraint solving technology.

4.5   Results

Figure 4 compares the performance and accuracy of repair of Cobbler, Tarmeem, Juzi, and PBnJ on the singly-linked list microbenchmark. Logging, check, and repair times

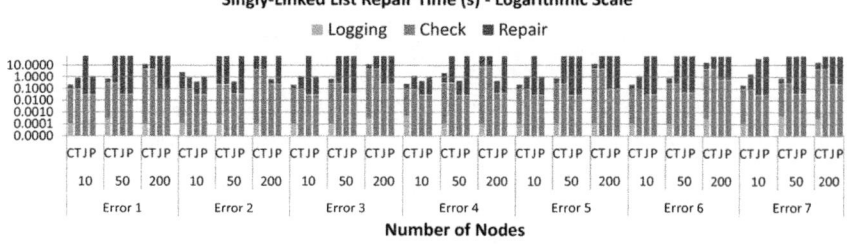

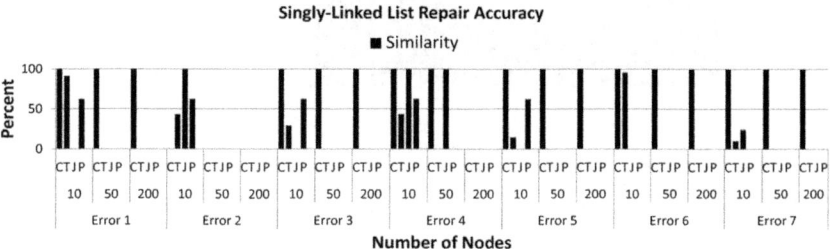

Fig. 4. Performance and accuracy: repairing singly-linked lists with Cobbler (C), Tarmeem (T), an enhanced version of Juzi (J), and PBnJ (P)

are accumulated into a single bar on a logarithmic scale. Logging time is only applicable to Cobbler and is negligible. Tarmeem and Cobbler have the same check time since they both use Kodkod evaluation (not SAT solving) to perform checks after methods execute. Juzi executes `repOK` and PBnJ translates specifications to Java assertions, which more efficiently check the data structure. Cobbler's overhead on an error-free execution includes both logging and check times. Using the approach of PBnJ to translate specifications to assertions could reduce the check time and the total overhead. We timeout after 60 seconds and report zero for accuracy upon a timeout.

Cobbler is substantially faster than all the other tools on five of the seven errors, despite the fact that Tarmeem and PBnJ receive specific user annotations to guide the repair process and Juzi performs symbolic execution. Error two skips a required update to size. Since the size field is not read or written, Cobbler does not correct it until the third call to the SAT solver, which causes its time to exceed the other repair schemes. Error four introduces a cycle. Juzi is tailored for such errors: it satisfies the constraint by breaking cycles quickly and performs better than Cobbler in this case.

Cobbler, except for one case, always produces the most accurate data structure among the four. When Cobbler does not time out, it achieves exactly the same output as expected. The edit distance between the result of a correct implementation and the repaired data structure is zero. This comparison is solely for evaluation, since in the wild, the system would not know the correct implementation.

Because Juzi solely relies on the `repOK` method instead of checking method postconditions, it does not find error six at all. Moreover, Juzi cannot access out of scope nodes that are not reachable from the header. Since Juzi does not consider the execution

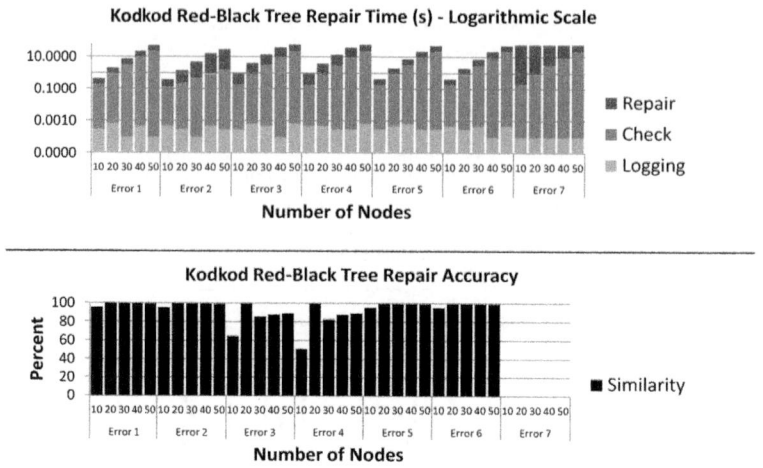

Fig. 5. Cobbler performance and accuracy: repairing Kodkod red-black trees

history, it first explores all the correct data structures nearby, but there is no guarantee that the expected output is close to the faulty one. We could enhance Juzi to work with post-conditions, as we did for evaluation of accuracy, but the original Juzi did not perform any repairs with respect to the post-conditions.

Tarmeem is not very accurate because when it invokes the SAT solver, it relaxes *all* tuples of a relation together, causing unnecessary changes. Cobbler significantly improves efficiency and accuracy over Tarmeem, especially for errors which involve incorrectly updated fields.

PBnJ's performance is similar to Tarmeem at best. The reason is that it always ignores the current faulty state and utilizes SAT to regenerate an acceptable output from scratch. It is however more accurate than Tarmeem in some cases.

Figure 5 shows the performance and accuracy of Cobbler on a faulty Kodkod red-black tree insert method. Figure 6 depicts the results of experimenting with ANTLR. We do not include the other frameworks here for brevity. Juzi always fails to repair correctly when a contract requires the addition of a node and the node is not present, because Juzi only uses those nodes currently accessible from the faulty data structure. When it does not timeout, Cobbler is very accurate on these real world applications.

The results show that the read and write field logs improve the scalability and efficiency of repair. Cobbler repairs linked lists with up to 200 nodes within 20 seconds. It performs well even on more complex data structures. For the red-black tree `remove` method, it repairs up to 50 nodes within 40 seconds and for the `deleteChild` method of ANTLR `BaseTree`, it repairs 40 nodes within 30 seconds. The size of the logs is proportional to the number of reads and writes to the data structure and was usually a few hundred bytes with a maximum of 900 bytes for error four of ANTLR.

For errors that cannot be fixed by relaxing only written and read fields, such as error two of linked list, error seven of red-black tree insert, and error two of ANTLR `deleteChild` (see Table 1), Cobbler uses the UNSAT core to identify which fields

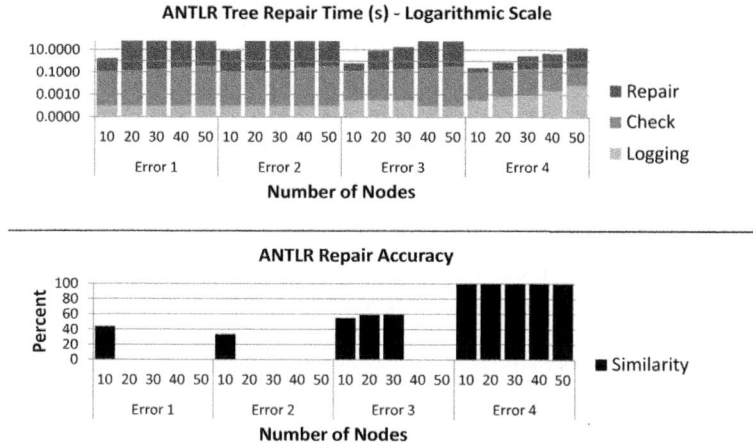

Fig. 6. Cobbler performance and accuracy: repairing ANTLR trees

to modify, and performs better than the other SAT-based tools. These cases however are challenging for Cobbler, because despite barrier logs that indicate fields of specific objects, UNSAT cores identify all fields with the same name as potentially faulty.

4.6 ANTLR `BaseTree addChild`

The public method `addChild` adds child node trees to an ANTLR `BaseTree` object. When the input tree does not have a payload (`isNil`), the method adds the children of the input tree to the children list of the current tree, otherwise, it adds the input tree itself to the children list. In the `addChild` method (v3.3), when the input tree does not have any payload, a check ensures that the current tree is not being added to itself. However, such a check is not performed for input trees with payloads. Hence, when the current tree has a payload, it may be added as a child of itself. Similarly, any tree with a payload which is already an existing child of the current tree may be added as a child again. We generated inputs that caused invariant (such as acyclicity and ascending child indices) violations. Cobbler repairs the Tree structure and restores it back to its pre-state, which is correct. This state would be the output of `addChild` if it had been implemented correctly. Cobbler repairs a tree with 300 nodes within 30 seconds.

5   Related Work

Dynamic repair aims to counteract faults at runtime and prolong system uptime. File system utilities such as fsck [6] and chkdsk [13], database check-pointing and roll-back techniques are application-specific repair routines that monitor and correct system state at runtime. DIRA [20] extends database repair with post-conditions to detect buffer overflow attacks and fix damaged data structures. Clearview [17] and Exterminator [15] also aid in repairing memory errors at runtime, but none of these techniques are suitable for repairing general purpose complex data structures. On the other hand, some commercially developed systems, such as the IBM MVS operating system [14] and the Lucent 5ESS telephone switch [7], have dedicated routines for monitoring and maintaining properties of their key data structures. These systems are tailor-made for their system structures and cannot be generalized as data structure repair tools.

Demsky and Rinard [4] pioneered constraint-based repair of data structures. Developers provide declarative constraints. The system translates the constraints into disjunctive normal form and repair solves them using an ad hoc search. The Juzi repair technique (described in Section 4.4) detects errors using user-defined repOK methods [5]. As we discussed and showed in Section 4.5, the accuracy and efficiency of Juzi suffer for errors that omit nodes and because the repair does not consider method semantics at the entry and exit. Recent improvements include Dynamic Symbolic Data Structure (DSDS) repair which builds a symbolic representation of fields and objects along the repOK executed path [8]. Whenever a predicate fails, DSDS solves the conjunction of its negation with the other path conditions. This direct generation of a satisfying result loses accuracy because it is irrespective of the exact location of corrupted object references or fields. A post-condition Java method predicate could be asserted along with the repOK to solve this problem. But as the size and complexity of properties and size of the data structure increase such techniques will not scale well.

Tarmeem [25] and PBnJ [18] (both explained in Section 4.4) overcome this limitation by using individual method pre- and post-conditions. As Section 4.5 showed, Tarmeem improves accuracy by tailoring repair to semantics, but is inefficient. PBnJ is not very efficient either, because it ignores both the faulty post-state and implementation. To improve the efficiency of PBnJ, programmers may bound the number of objects and limit changed fields, but for repairing unpredictable code errors, it does not seem feasible. Our approach instead automatically utilizes the faulty data structure and the code that produced it to prune the state space and guide repair to yield a satisfying instance as close as possible to the intended method output.

Our technique is related to, but differs substantially from, automated debugging and repair for use during testing, which focus on how to change the code rather than dynamically changing the heap [12,21,23,24]. However, as Malik et al. propose, dynamic repair actions could translate into program statements [11].

6   Conclusions

This paper introduced the idea of using program execution history for efficient and accurate contract-based data structure repair. We utilize program traces, specifically reads and writes of key fields, to direct repair of erroneous program states. Moreover, we use unsatisfiable cores provided by SAT solvers when we cannot repair the data structure by changing only read and written fields. We implemented this approach in Cobbler. Compared with previous repair techniques, our experimental results show Cobbler provides significant speedups and better accuracy, and finds and repairs a previously undetected bug in the widely used open-source ANTLR program. A promising future research avenue is to abstract concrete successful repair actions and use them to prioritize future repair actions, thus to avoid a costly search and make repair even more practical.

Acknowledgments. We thank the anonymous reviewers for their comments. This work was funded in part by the NSF under Grant Nos. CCF-0845628, IIS-0438967, CCF-1018271, CCF-0811524, and SHF-0910818, and AFOSR grant FA9550-09-1-0351.

References

1. Blackburn, S.M., Hosking, A.: Barriers: Friend or foe? In: ISMM (2004)
2. Blackburn, S.M., et al.: The DaCapo Benchmarks: Java Benchmarking Development and Analysis. In: OOPSLA (2006)
3. Boyapati, C., Khurshid, S., Marinov, D.: Korat: Automated testing based on Java predicates. In: ISSTA (2002)
4. Demsky, B., Rinard, M.: Automatic detection and repair of errors in data structures. In: OOPSLA (2003)
5. Elkarablieh, B., Garcia, I., Suen, Y.L., Khurshid, S.: Assertion-based repair of complex data structures. In: ASE (2007)
6. Ext2 fsck. manual page, http://e2fsprogs.sourceforge.net
7. Haugk, G., Lax, F., Royer, R., Williams, J.: The 5ESS(TM) switching system: Maintenance capabilities. AT&T Technical Journal 64(6 part 2), 1385–1416 (1985)
8. Hussain, I., Csallner, C.: Dynamic symbolic data structure repair. In: ICSE (2010)
9. Jackson, D.: Software Abstractions: Logic, Language, and Analysis. The MIT Press (2006)
10. Khurshid, S., García, I., Suen, Y.L.: Repairing Structurally Complex Data. In: Godefroid, P. (ed.) SPIN 2005. LNCS, vol. 3639, pp. 123–138. Springer, Heidelberg (2005)
11. Malik, M.Z., Ghori, K., Elkarablieh, B., Khurshid, S.: A case for automated debugging using data structure repair. In: ASE (2009)
12. Mayer, W., Stumptner, M.: Evaluating models for Model-Based debugging. In: ASE (2008)
13. Microsoft. chkdsk manual page, http://support.microsoft.com/kb/315265
14. Mourad, S., Andrews, D.: On the reliability of the IBM MVS/XA operating system. IEEE Transactions on Software Engineering 13(10), 1135–1139 (1987)
15. Novark, G., Berger, E.D., Zorn, B.G.: Exterminator: automatically correcting memory errors with high probability. In: PLDI (2007)
16. Parr, T., Bovet, J.: Antlr parser generator home page, http://www.antlr.org
17. Perkins, J., et al.: Automatically patching errors in deployed software. In: SOSP (2009)
18. Samimi, H., Aung, E.D., Millstein, T.: Falling Back on Executable Specifications. In: D'Hondt, T. (ed.) ECOOP 2010. LNCS, vol. 6183, pp. 552–576. Springer, Heidelberg (2010)
19. Sanfeliu, A., Fu, K.-S.: Distance measure between attributed relational graphs for pattern recognition. IEEE Trans. Systems, Man and Cybernetics 13(3), 353–362 (1983)
20. Smirnov, A., Chiueh, T.-c.: DIRA: Automatic detection, identification, and repair of control-hijacking attacks. In: NDSS (2005)
21. Staber, S., Jobstmann, B., Bloem, R.: Finding and Fixing Faults. In: Borrione, D., Paul, W. (eds.) CHARME 2005. LNCS, vol. 3725, pp. 35–49. Springer, Heidelberg (2005)
22. Torlak, E., Jackson, D.: Kodkod: A Relational Model Finder. In: Grumberg, O., Huth, M. (eds.) TACAS 2007. LNCS, vol. 4424, pp. 632–647. Springer, Heidelberg (2007)
23. Wei, Y., et al.: Automated fixing of programs with contracts. In: ISSTA (2010)
24. Weimer, W.: Patches as better bug reports. In: GPCE (2006)
25. Zaeem, R. Nokhbeh, Khurshid, S.: Contract-Based Data Structure Repair Using Alloy. In: D'Hondt, T. (ed.) ECOOP 2010. LNCS, vol. 6183, pp. 577–598. Springer, Heidelberg (2010)

The Guardol Language and Verification System

David Hardin[1], Konrad Slind[1], Michael Whalen[2], and Tuan-Hung Pham[2]

[1] Rockwell Collins Advanced Technology Center
[2] University of Minnesota

Abstract. Guardol is a domain-specific language designed to facilitate the construction of correct network guards operating over tree-shaped data. The Guardol system generates Ada code from Guardol programs and also provides specification and automated verification support. Guard programs and specifications are translated to higher order logic, then deductively transformed to a form suitable for a SMT-style decision procedure for recursive functions over tree-structured data. The result is that difficult properties of Guardol programs can be proved fully automatically.

1 Introduction

A *guard* is a device that mediates information sharing over a network between security domains according to a specified policy. Typical guard operations include reading field values in a packet, changing fields in a packet, transforming a packet by adding new fields, dropping fields from a packet, constructing audit messages, and removing a packet from a stream.

Guards are becoming prevalent, for example, in coalition forces networks, where selective sharing of data among coalition partners in real time is essential. One such guard, the Rockwell Collins Turnstile high-assurance, cross-domain guard [7], provides directional, bi-directional, and all-way guarding for up to three Ethernet connected networks. The proliferation of guards in critical applications, each with its own specialized language for specifying guarding functions, has led to the need for a portable, high-assurance guard language.

Fig. 1. Typical guard configuration

Guardol is a new, domain-specific programming language aimed at improving the creation, verification, and deployment of network guards. Guardol supports the ability to target a wide variety of guard platforms, the ability to glue together existing or mandated functionality, the generation of both implementations and formal analysis artifacts, and sound, highly automated formal analysis. Messages to be guarded, such as XML, may have recursive structure; thus a major aspect of Guardol is datatype declaration facilities similar to those available in

C. Flanagan and B. König (Eds.): TACAS 2012, LNCS 7214, pp. 18–32, 2012.

functional languages such as SML [14] or Haskell [16]. Recursive programs over such datatypes are supported by ML-style pattern-matching. However, Guardol is not simply an adaptation of a functional language to guards. In fact, much of the syntax and semantics of Guardol is similar to that of Ada: Guardol is a sequential imperative language with non-side-effecting expressions, assignment, sequencing, conditional commands, and procedures with **in/out** variables. To a first approximation **Guardol = Ada + ML**. This hybrid language supports writing complex programs over complex data structures, while also providing standard programming constructs from Ada.

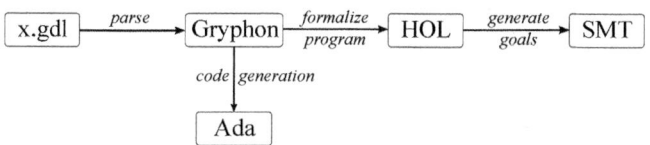

Fig. 2. Guardol system components

The Guardol system integrates several distinct components, as illustrated in Figure 2. A Guardol program in file `x.gdl` is parsed and typechecked by the Gryphon verification framework [13] developed by Rockwell Collins. Gryphon provides a collection of passes over Guardol ASTs that help simplify the program. From Gryphon, guard implementations can be generated—at present only in Ada—from Guardol descriptions. For the most part, this is conceptually simple since much of Guardol is a subset of Ada. However datatypes and pattern matching need special treatment: the former requires automatic memory management, which we have implemented via a reference-counting style garbage collection scheme, while the latter requires a phase of pattern-match compilation [18]. Since our intent in this paper is mainly to discuss the verification path, we will omit further details.

2   An Example Guard

In the following we will examine a simple guard written in Guardol. The guard applies a platform-supplied *dirty-word* operation DWO over a binary tree of messages (here identified with strings). When applied to a message, DWO can leave it unchanged, change it, or reject it with an audit string via MsgAudit.

```
type Msg        = string;
type MsgResult = {MsgOK : Msg | MsgAudit : string};
imported function DWO(Text : in Msg, Output : out MsgResult);
```

A MsgTree is a binary tree of messages. A MsgTree element can be a Leaf or a Node; the latter option is represented by a record with three fields. When the guard processes a MsgTree it either returns a new, possibly modified, tree, or it returns an audit message.

```
type MsgTree    = {Leaf | Node : [Value : Msg; Left : MsgTree; Right : MsgTree]};
type TreeResult = {TreeOK : MsgTree | TreeAudit : string};
```

The guard procedure takes its input tree in variable Input and the return value, which has type TreeResult, is placed in Output. The body uses local variables for holding the results of recursing into the left and right subtrees, as well as for holding the result of calling DWO. The guard code is written as follows:

```
function Guard (Input : in MsgTree, Output : out TreeResult) =
begin
    var ValueResult : MsgResult;
        LeftResult, RightResult : TreeResult;
    in
    match Input with
    MsgTree′Leaf ⇒ Output := TreeResult′TreeOK(MsgTree′Leaf);
    MsgTree′Node node ⇒  begin
        DWO(node.Value, ValueResult);
        match ValueResult with
        MsgResult′MsgAudit A ⇒ Output := TreeResult′TreeAudit(A);
        MsgResult′MsgOK ValueMsg ⇒  begin

           | Guard(node.Left, LeftResult); |

            match LeftResult with
            TreeResult′TreeAudit A ⇒ Output := LeftResult;
            TreeResult′TreeOK LeftTree ⇒  begin

               | Guard(node.Right, RightResult); |

                match RightResult with
                TreeResult′TreeAudit A ⇒ Output := RightResult;
                TreeResult′TreeOK RightTree ⇒
                    Output := TreeResult′TreeOK(MsgTree′Node
                                    [Value : ValueMsg, Left : LeftTree, Right : RightTree]);
            end
        end
    end
end
```

The guard processes a tree by a pattern-matching style case analysis on the Input variable. There are several cases to consider. If Input is a leaf node, processing succeeds. This is accomplished by tagging the leaf with TreeOK and assigning to Output. Otherwise, if Input is an internal node (MsgTree′Node node), the guard applies DWO to the message held at the node and recurses through the subtrees (recursive calls are marked with boxes). The only complication arises from the fact that an audit may arise from a subcomputation and must be immediately propagated. The code essentially lifts the error monad of the external operation to the error monad of the guard.

Specifying Guard Properties. Many verification systems allow programs to be annotated with assertions. Under such an approach, a program may become cluttered with assertions and assertions may involve logic constructs. Since we wanted to avoid clutter and wanted to shield programmers, as much as possible, from learning the syntax of a logic language, we decided to express specifications using Guardol programs. The key language construct facilitating verification is the *specification* declaration: it presents some code to be executed, sprinkled with assertions (which are just boolean program expressions).

Following is the specification for the guard. The code runs the guard on the tree t, putting the result in r, which is either a TreeOK element or an audit (TreeAudit). If the former, then the returned tree is named u and Guard_Stable, described below, must hold on it. On the other hand, if r is an audit, the property is vacuously true.

```
spec Guard_Correct = begin
    var t : MsgTree; r : TreeResult;
    in if (∀(M : Msg). DWO_Idempotent(M)) then  begin
        Guard(t, r);
        match r with
            TreeResult'TreeOK u ⇒ check Guard_Stable(u);
            TreeResult'TreeAudit A ⇒ skip;
    end
    else skip;
end
```

The guard code is essentially parameterized by an arbitrary policy (DWO) on how messages are treated. The correctness property simply requires that the result obeys the policy. In other words, suppose the guard is run on tree t, returning tree u. If DWO is run on every message in u, we expect to get u back unchanged, since all dirty words should have been scrubbed out in the passage from t to u. This property is a kind of idempotence, coded up in the function Guard_Stable; note that it has the shape of a *catamorphism*, which is a simple form of recursion exploited by our automatic proof component.

```
function Guard_Stable (MT : in MsgTree) returns Output : bool = begin
    var R : MsgResult;
    in
    match MT with
        MsgTree'Leaf ⇒ Output := true;
        MsgTree'Node node ⇒  begin
        DWO(node.Value, R);
        match R with
            MsgResult'MsgOK M ⇒ Output := (node.Value = M);
            MsgResult'MsgAudit A ⇒ Output := false;
        Output := Output and Guard_Stable(node.Left) and Guard_Stable(node.Right);
    end
end
```

The success of the proof depends on the assumption that the external dirty-word operation is idempotent on messages, expressed by the following program.

```
function DWO_Idempotent(M : in Msg) returns Output : bool =  begin
    var R1, R2 : MsgResult;
    in
    DWO(M, R1);
    match R1 with
        MsgResult'MsgOK M2 ⇒  begin
        DWO(M2, R2);
        match R2 with
            MsgResult'MsgOK M3 ⇒ Output := (M2 = M3);
            MsgResult'MsgAudit A ⇒ Output := false;
    end
        MsgResult'MsgAudit A ⇒ Output := true;
end
```

DWO_Idempotent calls DWO twice, and checks that the result of the second call is the same as the result of the first call, taking into account audits. If the first call returns an audit, then there is no second call, so the idempotence property is vacuously true. On the other hand, if the first call succeeds, but the second is an audit, that means that the first call somehow altered the message into one provoking an audit, so idempotence is definitely false.

3 Generating Verification Conditions by Deduction

Verification of guards is performed using HOL4 [19] and OpenSMT [2]. First, a program is mapped into a formal AST in HOL4, where the operational semantics of Guardol are encoded. One can reason directly in HOL4 about programs using the operational semantics; unfortunately, such an approach has limited applicability, requiring expertise in the use of a higher order logic theorem prover. Instead, we would like to make use of the high automation offered by SMT systems. An obstacle: current SMT systems do not understand operational semantics.[1] We surmount the problem in two steps. First, *decompilation into logic* [15] is used to deductively map properties of a program in an operational semantics to analogous properties over a mathematical function equivalent to the original program. This places us in the realm of proving properties of recursive functions operating over recursive datatypes, an undecidable setting in general. The second step is to implement a decision procedure for functional programming [20]. This procedure necessarily has syntactic limitations, but it is able to handle a wide variety of interesting programs and their properties fully automatically.

Translating Programs to Footprint Functions. First, any datatypes in the program are defined in HOL (every Guardol type can be defined as a HOL type). Thus, in our example, the types MsgTree, MsgResult, and TreeResult are translated directly to HOL datatypes. Recognizers and selectors, *e.g.*, isMsgTree_Leaf and destMsgTree_Node, for these types are automatically defined and used to translate pattern-matching statements in programs to equivalent if-then-else representations. Programs are then translated into HOL *footprint*[2] function definitions. If a program is recursive, then a recursive function is defined. (Defining recursive functions in higher order logic requires a termination proof; for our example termination is automatically proved.) Note that the HOL function for the guard (following page) is second order due to the external operator DWO: all externals are held in a record *ext* which is passed as a function argument.

3.1 Guardol Operational Semantics

The operational semantics of Guardol (see Fig. 3) describes program evaluation by an inductively defined judgement saying how statements alter the program state. The formula STEPS Γ *code* s_1 s_2 says "evaluation of statement *code* beginning in state s_1 terminates and results in state s_2". (Thus we are giving a so-called *big-step* semantics.) Note that Γ is an environment binding procedure names to procedure bodies. The semantics follows an approach taken by Norbert Schirmer [17], wherein he constructed a *generic* semantics for a large class of sequential imperative programs, and then showed how to specialize the

---

[1] Reasons for this state of affairs: there is not one operational semantics because each programming language has a different semantics; moreover, the decision problem for such theories is undecidable.

[2] Terminology lifted from the separation logic literature.

```
Guard ext Input =
if isMsgTree_Leaf Input then
|    TreeResult_TreeOK MsgTree_Leaf
else
     let ValueResult = ext.DWO ((destMsgTree_Node Input).Value, ARB)
     in if isMsgResult_MsgAudit ValueResult then
     |    TreeResult_TreeAudit (destMsgResult_MsgAudit ValueResult)
     else
          let LeftResult = | Guard ext ((destMsgTree_Node Input).Left, ARB) |
          in if isTreeResult_TreeAudit LeftResult then
          |    LeftResult
          else
               let RightResult = | Guard ext ((destMsgTree_Node Input).Right, ARB) |
               in if isTreeResult_TreeAudit RightResult then
               |    RightResult
               else
                    TreeResult_TreeOK(MsgTree_Node {Value := destMsgResult_MsgOK ValueResult;
                                                    Left   := destMsgResult_MsgOK LeftResult;
                                                    Right  := destMsgResult_MsgOK RightResult})
```

$$[Skip]\frac{}{\text{STEPS } \Gamma \text{ Skip (Normal s) (Normal s)}} \qquad [Basic]\frac{}{\text{STEPS } \Gamma \text{ (Basic f) (Normal s) (Normal (f s))}}$$

$$[Seq]\frac{\text{STEPS } \Gamma \text{ } c_1 \text{ (Normal } s_1) \text{ } s_2 \qquad \text{STEPS } \Gamma \text{ } c_2 \text{ } s_2 \text{ } s_3}{\text{STEPS } \Gamma \text{ (Seq } c_1 \text{ } c_2) \text{ (Normal } s_1) \text{ } s_3}$$

$$[withState]\frac{\text{STEPS } \Gamma \text{ (f } s_1) \text{ Normal } s_1) \text{ } s_2}{\text{STEPS } \Gamma \text{ (withState f) (Normal } s_1) \text{ } s_2}$$

$$[Cond\text{-}True]\frac{P(s_1) \qquad \text{STEPS } \Gamma \text{ } c_1 \text{ (Normal } s_1) \text{ } s_2}{\text{STEPS } \Gamma \text{ (Cond P } c_1 \text{ } c_2) \text{ (Normal } s_1) \text{ } s_2}$$

$$[Cond\text{-}False]\frac{\neg P(s_1) \qquad \text{STEPS } \Gamma \text{ } c_2 \text{ (Normal } s_1) \text{ } s_2}{\text{STEPS } \Gamma \text{ (Cond P } c_1 \text{ } c_2) \text{ (Normal } s_1) \text{ } s_2}$$

$$[Call]\frac{M.p \in \text{Dom}(\Gamma) \qquad \Gamma(M.p) = c \qquad \text{STEPS } \Gamma \text{ } c \text{ (Normal } s_1) \text{ } s_2}{\text{STEPS } \Gamma \text{ (Call M.p) (Normal } s_1) \text{ } s_2}$$

$$[Call\text{-}NotFound]\frac{M.p \notin \text{Dom}(\Gamma)}{\text{STEPS } \Gamma \text{ (Call M.p) (Normal s) Stuck}}$$

$$[Fault\text{-}Sink]\frac{}{\text{STEPS } \Gamma \text{ c (Fault f) (Fault f)}} \qquad [Stuck\text{-}Sink]\frac{}{\text{STEPS } \Gamma \text{ c Stuck Stuck}} \qquad [Abrupt\text{-}Sink]\frac{}{\text{STEPS } \Gamma \text{ c (Abrupt s) (Abrupt s)}}$$

$$[While\text{-}True]\frac{P(s_1) \qquad \text{STEPS } \Gamma \text{ c (Normal } s_1) \text{ } s_2 \qquad \text{STEPS } \Gamma \text{ (While P c) } s_2 \text{ } s_3}{\text{STEPS } \Gamma \text{ (While P c) (Normal } s_1) \text{ } s_3}$$

$$[While\text{-}False]\frac{\neg P(s)}{\text{STEPS } \Gamma \text{ (While P c) (Normal s) (Normal s)}}$$

Fig. 3. Evaluation rules

generic semantics to a particular programming language (a subset of C, for him). Similarly, Guardol is another instantiation of the generic semantics.

Evaluation is phrased in terms of a *mode* of evaluation, which describes a computation state. A computation state is either in Normal mode, or in one of a set of abnormal modes, including Abrupt, Fault, and Stuck. Usually computation is in Normal mode. However, if a Throw is evaluated, then computation proceeds in Abrupt mode. If a Guard command returns false, the computation transitions into a Fault mode. Finally, if the Stuck mode is entered, something is wrong, *e.g.*, a procedure is called but there is no binding for it in Γ.

3.2   Decompilation

The work of Myreen [15] shows how to decompile assembly programs to higher order logic functions; we do the same here for Guardol, a high-level language. A decompilation theorem

$$\vdash \forall s_1\ s_2.\ \forall x_1 \dots x_k.$$
$$s_1.proc.v_1 = x_1 \wedge \cdots \wedge s_1.proc.v_k = x_k \wedge$$
$$\mathsf{STEPS}\ \Gamma\ \boxed{code}\ (\mathsf{Normal}\ s_1)\ (\mathsf{Normal}\ s_2)$$
$$\Rightarrow$$
$$\mathtt{let}\ (o_1, ..., o_n) = \boxed{f(x_1, \dots, x_k)}$$
$$\mathtt{in}\ s_2 = s_1\ \mathtt{with}\{proc.w_1 := o_1, \dots, proc.w_n := o_n\}$$

essentially states that evaluation of *code* implements footprint function f. The antecedent $s_1.proc.v_1 = x_1 \wedge \cdots \wedge s_1.proc.v_k = x_k$ equates $x_1 \dots x_k$ to the values of program variables $v_1 \dots v_k$ in state s_1. These values form the input for the function f, which delivers the output values which are used to update s_1 to s_2.[3] Presently, the decompilation theorem only deals with code that starts evaluation in a Normal state and finishes in a Normal state.

The Decompilation Algorithm. Now we consider how to prove decompilation theorems for Guardol programs. It is important to emphasize that *decompilation is an algorithm*. It always succeeds, provided that all footprint functions coming from the Guardol program have been successfully proved to terminate.

Before specifications can be translated to goals, the decompilation theorem

$$\vdash \forall s_1\ s_2.\ \dots \mathsf{STEPS}\ \Gamma\ \boxed{\mathsf{Call}(qid)}\ (\mathsf{Normal}\ s_1)\ (\mathsf{Normal}\ s_2) \Rightarrow \cdots$$

is formally proved for each procedure *qid* in the program, relating execution of the code for procedure *qid* with the footprint function for *qid*.

Decompilation proofs are automated by forward symbolic execution of *code*, using an environment of decompilation theorems to act as summaries for procedure calls. Table 1 presents rules used in the decompilation algorithm. For the most part, the rules are straightforward. We draw attention to the Seq, withState, and Call rules. The Seq (sequential composition) rule conjoins the results of simpler commands and introduces an existential formula ($\exists t. \dots$). However, this is essentially universal since it occurs on the left of the top-level implication in the

---

[3] In our modelling, a program state is represented by a record containing all variables in the program. The notation $s.proc.v$ denotes the value of program variable v in procedure *proc* in state s. The $\mathtt{with}$-notation represents record update.

Table 1. Rewrite rules in the decompilation algorithm

Condition	Rewrite rule
$code = \mathsf{Skip}$	$\vdash \mathsf{STEPS}\ \Gamma\ \mathsf{Skip}\ s_1\ s_2 = (s_1 = s_2)$
$code = \mathsf{Basic}(f)$	$\vdash \mathsf{STEPS}\ \Gamma\ \mathsf{Basic}\ (f)\ (\mathsf{Normal}\ s_1)\ (\mathsf{Normal}\ s_2) = (s_2 = f\ s_1)$
$code = \mathsf{Seq}(c_1, c_2)$	$\vdash \mathsf{STEPS}\ \Gamma\ (\mathsf{Seq}(c_1, c_2))\ (\mathsf{Normal}\ s_1)\ (\mathsf{Normal}\ s_2) =$
	$\quad \exists t.\mathsf{STEPS}\ \Gamma\ c_1\ (\mathsf{Normal}\ s_1)\ (\mathsf{Normal}\ t)\ \wedge \mathsf{STEPS}\ \Gamma\ c_2\ (\mathsf{Normal}\ t)\ (\mathsf{Normal}\ s_2)$
$code = \mathsf{Cond}(P, c_1, c_2)$	$\vdash \mathsf{STEPS}\ \Gamma\ (\mathsf{Cond}(P, c_1, c_2))\ (\mathsf{Normal}\ s_1)\ (\mathsf{Normal}\ s_2) =$
	$\quad$ if $P\ s_1$ then $\mathsf{STEPS}\ \Gamma\ c_1\ (\mathsf{Normal}\ s_1)\ (\mathsf{Normal}\ s_2)$
	$\quad\quad$ else $\mathsf{STEPS}\ \Gamma\ c_2\ (\mathsf{Normal}\ s_1)\ (\mathsf{Normal}\ s_2)$
$code = \mathsf{withState}\ f$	$\vdash \mathsf{STEPS}\ \Gamma\ (\mathsf{withState}\ f)\ (\mathsf{Normal}\ s_1)\ s_2 = \mathsf{STEPS}\ \Gamma\ (f\ s_1)\ (\mathsf{Normal}\ s_1)\ s_2$
$code = \mathsf{Call}\ qid$	depend on whether the function is recursive or not

goal; thus it can be eliminated easily and occurrences of t can thenceforth be treated as Skolem constants. Both blocks and procedure calls in the Guardol program are encoded using withState. An application of withState stays in the current state, but replaces the current code by new code computed from the current state. Finally, there are two cases with the Call (procedure call) rule:

- The call is not recursive. In this case, the decompilation theorem for qid is fetched from the decompilation environment and instantiated, so we can derive

$$\mathtt{let}\ (o_1, ..., o_n) = f(x_1, \ldots, x_k)$$
$$\mathtt{in}\ s_2 = s_1\ \mathtt{with}\ \{qid.w_1 := o_1, \ldots, qid.w_n := o_n\}$$

  where f is the footprint function for procedure qid. We can now propagate the value of the function to derive state s_2.
- The call is recursive. In this case, an inductive hypothesis in the goal—which is a decompilation theorem for a recursive call, by virtue of our having inducted at the outset of the proof—matches the call, and is instantiated. We can again prove the antecedent of the selected inductive hypothesis, and propagate the value of the resulting functional characterization, as in the non-recursive case.

The decompilation algorithm starts by either inducting, when the procedure for which the decompilation theorem is being proved is recursive, or not (otherwise). After applying the rewrite rules, at the end of each program path, we are left with an equality between states. The proof of this equality proceeds essentially by applying rewrite rules for normalizing states (recall that states are represented by records).

Translating Specifications into Goals. A Guardol specification is intended to set up a computational context—a state—and then assert that a property holds in that state. In its simplest form, a specification looks like

```
spec name = begin
   var decls
   in
       code;
       check property;
   end
```

where *property* is a boolean expression. A specification declaration is processed as follows. First, suppose that execution of *code* starts normally in s_1 and ends normally in s_2, *i.e.*, assume STEPS Γ *code* (Normal s_1) (Normal s_2). We want to show that *property* holds in state s_2. This could be achieved by reasoning with the induction principle for STEPS, *i.e.*, by using the operational semantics; however, experience has shown that this approach is labor-intensive. We instead opt to formally leverage the decompilation theorem for *code*, which asserts that reasoning about the STEPS-behavior of *code* could just as well be accomplished by reasoning about function f. Thus, formally, we need to show

$$(\text{let } (o_1, ..., o_n) = f(x_1, \ldots, x_k)$$
$$\text{in } s_2 = s_1 \text{ with } \{name.w_1 := o_1, \ldots, name.w_n := o_n\})$$
$$\Rightarrow property \ s_2$$

Now we have a situation where the proof is essentially about how facts about f, principally its recursion equations and induction theorem, imply the property. The original goal has been freed—by sound deductive steps—from the program state and operational semantics. The import of this, as alluded to earlier, is that a wide variety of proof tools become applicable. Interactive systems exemplified by ACL2, PVS, HOL4, and Isabelle/HOL have extensive lemma libraries and reasoning packages tailored for reasoning about recursively defined mathematical functions. SMT systems are also able to reason about such functions, via universal quantification, or by decision procedures, as we discuss in Section 4.

Setting Up Complex Contexts. The form of specification above is not powerful enough to state many properties. Quite often, a collection of constraints needs to be placed on the input variables, or on external functions. To support this, specification statements allow checks sprinkled at arbitrary points in *code*:

$$\text{spec } name = \textbf{begin} \quad locals \quad \textbf{in} \quad code[\text{check } P_1, \ldots, \text{check } P_n] \quad \textbf{end}$$

We support this with a program transformation, wherein occurrences of check are changed into assignments to a boolean variable. Let V be a boolean program variable not in *locals*. The above specification is transformed into

$$\text{spec } name = \textbf{begin}$$
$$locals; \quad \text{V : bool};$$
$$\textbf{in}$$
$$\text{V} := \textbf{true}; \quad code[\text{V} := \text{V} \wedge P_1, \ldots, \text{V} := \text{V} \wedge P_n]; \quad \text{check(V)};$$
$$\textbf{end}$$

Thus V is used to accumulate the results of the checks that occur throughout the code. Every property P_i is checked in the state holding just before the occurrence of check(P_i), and all the checks must hold. This gives a flexible and concise way to express properties of programs, without embedding assertions in the source code of the program.

Recall the Guard_Correct specification. Roughly, it says *If running the guard succeeds, then running* Guard_Stable *on the result returns* true. Applying the decompiler to the code of the specification and using the resulting theorem to map from the operational semantics to the functional interpretation, we obtain the goal

$$\left(\begin{matrix} (\forall m.\ \mathsf{DWO\_Idempotent}\ ext\ m)\ \wedge \\ \mathsf{Guard}\ ext\ t = \mathsf{TreeResult\_TreeOK}\ t' \end{matrix} \right) \Rightarrow \mathsf{Guard\_Stable}\ ext\ t'$$

which has the form required by our SMT prover, namely that the catamorphism Guard_Stable is applied to the result of calling Guard. However, an SMT prover may still not prove this goal, since the following steps need to be made: (1) inducting on the recursion structure of Guard, (2) expanding (once) the definition of Guard, (3) making higher order functions into first order, and (4) elimination of universal quantification. [4]

To address the first two problems, we induct with the induction theorem for Guard, which is automatically proved by HOL4, and expand the definition of Guard one step in the resulting inductive case. Thus we stop short of using the inductive hypotheses! The SMT solver will do that. The elimination of higher order functions is simple in the case of Guardol since the function arguments (*ext* in this case) are essentially fixed constants whose behavior is constrained by hypotheses. This leaves the elimination of the universals; only the quantification on m in $\forall m.\ \mathsf{DWO\_Idempotent}\ ext\ m$ is problematic. We find all arguments of applications of ext.DWO in the body of Guard, and instantiate m to all of them (there's only one in this case), adding all instantiations as hypotheses to the goal.

4   Verification Condition Solving Using SMT

The formulas generated as verification conditions from the previous section pose a fundamental research challenge: reasoning over the structure and contents of inductive datatypes We have addressed this challenge through the use of a novel, recently-developed, decision procedure called the Suter–Dotta–Kuncak (SDK) procedure [20]. This decision procedure can be integrated into an SMT solver to solve a variety of properties over recursive datatypes. It uses catamorphisms to create abstractions of the contents of tree-structured data that can then be solved using standard SMT techniques. The benefit of this decision procedure over other techniques involving quantifiers is that it is *complete* for a large class of reasoning problems involving datatypes, as described below.

Catamorphisms. In many reasoning problems involving recursive datatypes, we are interested in *abstracting* the contents of the datatype. To do this, we

---

[4] Some of these steps are incorporated in various SMT systems, *e.g.*, many, but not all, SMT systems heuristically instantiate quantifiers. For a discussion of SMT-style induction see [10].

could define a function that maps the structure of the tree into a value. This kind of function is called a catamorphism [12] or *fold* function, which 'folds up' information about the data structure into a single value. The simplest abstraction that we can perform of a data structure is to map it into a Boolean result that describes whether it is 'valid' in some way. This approach is used in the function Guard_Stable in Section 2. We could of course create different functions to summarize the tree elements. For example, a tree can be abstracted as a number that represents the sum of all nodes, or as a tuple that describes the minimum and maximum elements within the tree. As long as the catamorphism is *sufficiently surjective* [20] and maps into a decidable theory, the procedure is theoretically *complete*. Moreover, we have found it to be *fast* in our initial experiments.

Overview of the Decision Procedure. The input of the decision procedure is a formula ϕ of literals over elements of tree terms and tree abstractions ($\mathcal{L}_C$) produced by the catamorphisms. The logic is *parametric* in the sense that we assume a datatype to be reasoned over and catamorphism used to abstract the datatype, and the existence of a decidable theory C that is the result type of the catamorphism function. The syntax of the parametric logic is depicted in Fig. 4.

The syntax of the logic ranges over datatype terms (T and S), terms of a decidable collection theory C. Tree and collection theory formulas F_T and F_C describe equalities and inequalities over terms. The collection theory describes the result of catamorphism applications. E defines terms in the element types contained within the branches of the datatypes. ϕ defines conjunctions of (restricted) formulas in the tree and collection theories. The ϕ terms are the ones solved by the SDK procedure; these can be generalized to arbitrary propositional formulas (ψ) through the use of a DPLL solver [4] which manages the other operators within the formula.

$$
\begin{array}{lll}
S &::= T \mid E & \text{Constructors' arguments} \\
T &::= t \mid \mathbb{C}_j(S_1, \ldots, S_{n_j}) \mid \mathbb{S}_{j,k_\tau}(T) & \text{Tree terms} \\
C &::= c \mid \alpha(T) \mid T_C & \mathcal{C}\text{-terms} \\
F_T &::= T = T \mid T \neq T & \text{Tree (dis)equations} \\
F_C &::= C = C \mid \mathcal{F}_C & \text{Formula of } \mathcal{L}_C \\
E &::= \text{variables of type } \mathcal{E}_k \mid \mathbb{S}_{j,k_\mathcal{E}}(T) & \text{Expression} \\
\phi &::= \bigwedge F_T \wedge \bigwedge F_C & \text{Conjunctions} \\
\psi &::= \phi \mid \neg\phi \mid \phi \vee \phi \mid \phi \wedge \phi \mid & \text{Formulas} \\
& \quad \phi \Rightarrow \phi \mid \phi \Leftrightarrow \phi &
\end{array}
$$

Fig. 4. Syntax of the parametric logic

In the procedure, we have a single datatype τ with m constructors. The j-th constructor ($1 \leq j \leq m$), $\mathbb{C}_j$, has n_j arguments ($n_j \geq 0$), whose types are either τ or $\mathcal{E}$, an element type. For each constructor $\mathbb{C}_j$, we have a list of selectors $\mathbb{S}_{j,k}$ ($1 \leq k \leq n_j$), which extracts the k-th argument of $\mathbb{C}_j$. For type safety, we may put the type of the argument to be extracted as a subscript of its selector. That is, each selector may be presented as either $\mathbb{S}_{j,k_\tau}$ or $\mathbb{S}_{j,k_\mathcal{E}}$. The decision procedure is parameterized by $\mathcal{E}$, a collection type $\mathcal{C}$, and a catamorphism function $\alpha : \tau \to \mathcal{C}$. For example, the datatype MsgTree has two constructors Leaf and Node. The former has no argument while the latter has three arguments corresponding to its Value, Left, and Right. As a result, we have three selectors for Node, including Value: MsgTree $\to$ Msg, Left : MsgTree $\to$ MsgTree, and Right : MsgTree $\to$ MsgTree. In addition, a tree can be abstracted by the catamorphism Guard_Stable : MsgTree $\to$ bool to a boolean value. In the example, $\mathcal{E}$, $\mathcal{C}$, and α are Msg, bool, and Guard_Stable, respectively.

Implementing Suter-Dotta-Kuncak in OpenSMT. We have created an implementation of the SDK decision procedure. As described earlier, the decision procedure operates over conjunctions of theory literals and generates a problem in a *base theory* (denoted C) that must be solved by another existing decision procedure. Thus, the decision procedure is not useful by itself; it must be integrated into an SMT solver that supports reasoning over complex Boolean logic formulas (rather than only conjunctions) and contains the theories necessary to solve the terms in C.

In DPLL [4], theory terms are partitioned by solver: each solver "owns" the terms in its theory. The SMT solver partitions the terms for each theory in a *purification* step. However, the SDK decision procedure as presented in [20] requires an unusual level of supervisory control over other other decision procedures within an SMT solver. That is, terms from the element theories, the collections theory, and the tree theory have to be provided to the SDK procedure where they are manipulated (possibly adding disjunctions) and eventually passed back to a SMT solver to be further processed. Thus, to implement SDK, we have re-architected a standard DPLL solver by forwarding all theory terms to the SDK procedure, letting it perform purification and formula manipulation, and passing the resulting problem instance (which no longer contains datatype terms) into an inner instance of the SMT solver to solve the (local) subproblem and return the result to the outer instance. This architecture is shown in Figure 5.

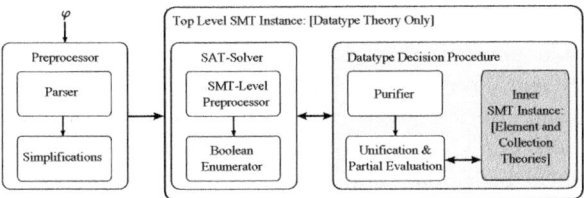

Fig. 5. Architecture for SMT solver containing SDK

We have chosen to implement SDK on OpenSMT as it supports a sufficiently rich set of theories and is freely available for extension and commercial use. In addition, OpenSMT is fast and has a simple interface between decision procedures and the underlying solver. We have also made a handful of extensions to the tool to support reasoning over mutually recursive datatypes. A drawback of the current release of

OpenSMT is that it does not support quantifiers; on the other hand, quantifiers add a source of incompleteness to the solving process that we are trying to avoid.

4.1   Experimental Results

To test our approach, we have developed a handful of small benchmark guard examples. For timings, we focus on the SMT solver which performs the interesting part of the proof search. The results are shown in Table 2, where the last guard is the running example from the paper. In our early experience, the SDK procedure allows a wide variety of interesting properties to be expressed and our initial timing experiments have been very encouraging, despite the relatively näive implementation of SDK, which we will optimize in the near future. For a point of comparison, we provide a translation of the problems to Microsoft's Z3 in which we use universal quantifiers to describe the catamorphism behavior. This approach is incomplete, so properties that are falsifiable (i.e., return SAT) often do not terminate (we mark this as 'unknown'). A better comparison would be to run our implementation against the Leon tool suite developed at EFPL [21]. Unfortunately, it is not currently possible to do so as the Leon tool suite operates directly over the Scala language rather than at the SMT level. All benchmarks were run on Windows 7 using an Intel Core I3 running at 2.13 GHz.

Table 2. Experimental results on guard examples

Test #	OpenSMT-SDK	Z3
Guard 1	5 sat (0.83s) / 1 unsat (0.1s)	5 unknown / 1 unsat (0.06 s)
Guard 2	3 unsat (0.28s)	1 unknown / 2 unsat (0.11 s)
Guard 3	3 sat (0.33s) / 4 unsat (0.46s)	1 unknown / 2 sat (0.17s) / 4 unsat (0.42 s)
DWO	1 unsat (0.34s)	1 unsat (0.11s)

5   Discussion

As a programming language with built-in verification support, Guardol seems amenable to being embedded in an IVL (Intermediate Verification Language) such as Boogie [11] or Why [3]. However, our basic aims would not be met in such a setting. The operational semantics in Section 3.1 defines Guardol program execution, which is the basis for verification. We want a strong formal connection between a program plus specifications, both expressed using that semantics, and the resulting SMT goals, which do not mention that semantics. The decompilation algorithm achieves this connection via machine-checked proof in higher order logic. This approach is simpler in some ways than an IVL, where there are two translations: one from source language to IVL and one from IVL to SMT goals. To our knowledge, IVL translations are not machine-checked, which is problematic for our applications. Our emphasis on formal models and deductive transformations should help Guardol programs pass the stringent certification requirements imposed on high-assurance guards.

Higher order logic plays a central role in our design. HOL4 implements a foundational semantic setting in which the following are formalized: program

ASTs, operational semantics, footprint functions (along with their termination proofs and induction theorems), and decompilation theorems. Decompilation extensively uses higher order features when composing footprint functions corresponding to sub-programs. Moreover, the backend verification theories of the SMT system already exist in HOL4. This offers the possibility of doing SMT proof reconstruction [1] in order to obtain higher levels of assurance. Another consequence is that, should a proof fail or take too long in the SMT system, it could be performed interactively.

As future work we plan to investigate both language and verification aspects. For example, the current Guardol language could be made more user-friendly. It is essentially a monomorphic version of ML with second order functions, owing to Guardol's external function declarations. It might be worthwhile to support polymorphic types so that the repeated declarations of instances of polymorphic types, *e.g.*, option types and list types, can be curtailed. Additionally, programs could be considerably more terse if exceptions were in the language, since explicitly threading the error monad wouldn't be needed to describe guard failures.

An interesting issue concerns specifications: guard specifications can be about *intensional* aspects of a computation, *e.g.*, its structure or sequencing, as well as its result. For example, one may want to check that data fuzzing operations always occur before encryption. However, our current framework, which translates programs to *extensional* functions, will not be able to use SMT reasoning on intensional properties. Information flow properties [5] are also intensional, since in that setting one is not only concerned with the value of a variable, but also whether particular inputs and code structures were used to produce it. Techniques similar to those in [22] could be used to annotate programs in order to allow SMT backends to reason about intensional guard properties.

Planned verification improvements include integration of string solvers and termination deferral. In the translation to SMT, strings are currently treated as an uninterpreted type and string operations as uninterpreted functions. Therefore, the system cannot reason in a complete way about guards where string manipulation is integral to correctness. We plan to integrate a string reasoner (*e.g.*, [8]) into our system to handle this problem. Finally, a weak point in the end-to-end automation of Guardol verification is termination proofs. If the footprint function for a program happens to be recursive, its termination proof may well fail, thus stopping processing. There are known techniques for defining partial functions [6,9], obtaining recursion equations and induction theorems constrained by termination requirements. These techniques remove this flaw, allowing the deferral of termination arguments while the partial correctness proof is addressed.

Acknowledgements. The TACAS reviewers did a well-informed and thorough job, kindly pointing out many mistakes and infelicities in the orginal submission.

References

1. Böhme, S., Fox, A.C.J., Sewell, T., Weber, T.: Reconstruction of Z3's Bit-Vector Proofs in HOL4 and Isabelle/HOL. In: Jouannaud, J.-P., Shao, Z. (eds.) CPP 2011. LNCS, vol. 7086, pp. 183–198. Springer, Heidelberg (2011)

2. Bruttomesso, R., Pek, E., Sharygina, N., Tsitovich, A.: The OpenSMT Solver. In: Esparza, J., Majumdar, R. (eds.) TACAS 2010. LNCS, vol. 6015, pp. 150–153. Springer, Heidelberg (2010)
3. Filliâtre, J.-C.: Deductive Program Verification. Thèse d'habilitation, Université Paris (December 11, 2011)
4. Ganzinger, H., Hagen, G., Nieuwenhuis, R., Oliveras, A., Tinelli, C.: DPLL(T): Fast Decision Procedures. In: Alur, R., Peled, D.A. (eds.) CAV 2004. LNCS, vol. 3114, pp. 175–188. Springer, Heidelberg (2004)
5. Goguen, J., Meseguer, J.: Security policies and security models. In: Proc. of IEEE Symposium on Security and Privacy, pp. 11–20. IEEE Computer Society Press (1982)
6. Greve, D.: Assuming termination. In: Proceedings of ACL2 Workshop, ACL2 2009, pp. 114–122. ACM (2009)
7. Rockwell Collins Inc. Turnstile High Assurance Guard Homepage, http://www.rockwellcollins.com/sitecore/content/Data/Products/Information_Assurance/Cross_Domain_Solutions/Turnstile_High_Assurance_Guard.aspx
8. Kiezun, A., Ganesh, V., Guo, P., Hooimeijer, P., Ernst, M.: HAMPI: A solver for string constraints. In: Proceedings of ISSTA (2009)
9. Krauss, A.: Automating recursive definitions and termination proofs in higher order logic. PhD thesis, TU Munich (2009)
10. Leino, K.R.M.: Automating Induction with an SMT Solver. In: Kuncak, V., Rybalchenko, A. (eds.) VMCAI 2012. LNCS, vol. 7148, pp. 315–331. Springer, Heidelberg (2012)
11. Leino, K.R.M., Rümmer, P.: A Polymorphic Intermediate Verification Language: Design and Logical Encoding. In: Esparza, J., Majumdar, R. (eds.) TACAS 2010. LNCS, vol. 6015, pp. 312–327. Springer, Heidelberg (2010)
12. Meijer, E., Fokkinga, M., Paterson, R.: Functional Programming with Bananas, Lenses, Envelopes, and Barbed Wire. In: Hughes, J. (ed.) FPCA 1991. LNCS, vol. 523, pp. 124–144. Springer, Heidelberg (1991)
13. Miller, S., Whalen, M., Cofer, D.: Software model checking takes off. CACM 53, 58–64 (2010)
14. Milner, R., Tofte, M., Harper, R., MacQueen, D.: The Definition of Standard ML (Revised). The MIT Press (1997)
15. Myreen, M.: Formal verification of machine-code programs. PhD thesis, University of Cambridge (2009)
16. Peyton Jones, S., et al.: The Haskell 98 language and libraries: The revised report. Journal of Functional Programming 13(1), 0–255 (2003)
17. Schirmer, N.: Verification of sequential imperative programs in Isabelle/HOL. PhD thesis, TU Munich (2006)
18. Sestoft, P.: ML Pattern Match Compilation and Partial Evaluation. In: Danvy, O., Thiemann, P., Glück, R. (eds.) Dagstuhl Seminar 1996. LNCS, vol. 1110, pp. 446–464. Springer, Heidelberg (1996)
19. Slind, K., Norrish, M.: A Brief Overview of HOL4. In: Mohamed, O.A., Muñoz, C., Tahar, S. (eds.) TPHOLs 2008. LNCS, vol. 5170, pp. 28–32. Springer, Heidelberg (2008)
20. Suter, P., Dotta, M., Kuncak, V.: Decision procedures for algebraic data types with abstractions. In: Proceedings of POPL, pp. 199–210. ACM (2010)
21. Suter, P., Köksal, A.S., Kuncak, V.: Satisfiability Modulo Recursive Programs. In: Yahav, E. (ed.) SAS 2011. LNCS, vol. 6887, pp. 298–315. Springer, Heidelberg (2011)
22. Whalen, M., Greve, D., Wagner, L.: Model checking information flow. In: Hardin, D. (ed.) Design and Verification of Microprocessor Systems for High-Assurance Applications. Springer (2010)

A Bit Too Precise? Bounded Verification of Quantized Digital Filters*

Arlen Cox, Sriram Sankaranarayanan, and Bor-Yuh Evan Chang

University of Colorado Boulder
{arlen.cox, sriram.sankaranarayanan, evan.chang}@colorado.edu

Abstract. Digital filters are simple yet ubiquitous components of a wide variety of digital processing and control systems. Errors in the filters can be catastrophic. Traditionally digital filters have been verified using methods from control theory and extensive testing. We study two alternative verification techniques: bit-precise analysis and real-valued error approximations. In this paper, we empirically evaluate several variants of these two fundamental approaches for verifying fixed-point implementations of digital filters. We design our comparison to reveal the best possible approach towards verifying real-world designs of infinite impulse response (IIR) digital filters. Our study reveals broader insights into cases where bit-reasoning is absolutely necessary and suggests efficient approaches using modern satisfiability-modulo-theories (SMT) solvers.

1 Introduction

In this paper, we present an evaluation of techniques for verification of fixed-point implementations of digital filters. Digital filters are ubiquitous in a wide variety of systems, such as control systems, analog mixed-signal (AMS) systems, and digital signal processing systems. Their applications range from automotive electronic components and medical devices to record players and musical instruments. To get them right, the design of digital filters is guided by a rich theory that includes a deep understanding of their behavior in terms of the frequency and time domain properties. Filter designers rely on a floating-point-based design and validation tools such as Matlab.

But there is a serious disconnect between filter designs and filter implementations. Implementations often use fixed-point arithmetics so that they can be implemented using special purpose digital signal processors (DSPs) or field programmable gate arrays (FPGAs) that do not support floating-point arithmetics. Meanwhile, the design tools are using floating-point arithmetics for validation. Does this disconnect between floating-point designs and fixed-point implementations matter?

The transition from floating-point to fixed-point arithmetic can lead to undesirable effects such as overflows and instabilities (e.g., limit cycles—see Section 2). They arise due to (a) the quantization of the filter coefficients, (b) input quantization, and (c) round-off errors for multiplications and additions. Thus, the fixed-point representations need

---

* This material is based upon work supported by the National Science Foundation (NSF) under Grant No. 0953941. Any opinions, findings, and conclusions or recommendations expressed in this material are those of the author(s) and do not necessarily reflect the views of NSF.

C. Flanagan and B. König (Eds.): TACAS 2012, LNCS 7214, pp. 33–47, 2012.

to be sufficiently accurate—have adequate bits to represent the integer and fraction so that undesirable effects are not observed in implementation. Naturally, an implementer faces the question whether a given design is sufficient to guarantee correctness.

Extensive testing using a large number of input signals is a minimum requirement. However, it is well-known from other types of hardware designs that testing can fall short of a full verification or an exhaustive depth-bounded search over the input space, even for relatively small depths. Therefore, the question arises whether extensive testing is good enough for filter validation or more exhaustive techniques are necessary. If we choose to perform bounded verification of fixed-point filter implementations, there are roughly two different sets of approaches. The bit-precise approach encodes the operation of the fixed-point filter to precisely capture the effect of quantization, round-offs and overflows as they happen on real hardware implementations. We then perform a bounded-depth model checking (BMC) [5] using *bit-vector* and *integer* arithmetic solvers to detect the presence of overflows and limit cycles (Section 3). An alternative approach consists of encoding the filter state using reals by over-approximating the errors conservatively. We perform an error analysis to show that such an over-approximation can be addressed using *affine* arithmetic simulations [6] or BMC using linear *real* arithmetic constraints (Section 4).

Our primary contribution is a set of experimental evaluations designed to elucidate the trade-offs between the testing and verification techniques outlined above. Specifically, we implemented the four verification approaches outlined above, as well as random testing simulators using uniform random simulation over the input signals or simulation by selecting the maximal or minimal input at each time step. We empirically compare these approaches on a set of filter implementations designed using Matlab's filter design toolbox. Overall, our experimental comparison seeks to answer four basic questions (Section 5):

1. *Is simulation sufficient to find bugs in filters?* We observe that simulation is efficient overall but seldom successful in finding subtle bugs in digital filters.
2. *Is bit-precise reasoning more precise in practice than conservative real-arithmetic reasoning?* In highly optimized filters, conservatively tracking errors produces many spurious alarms. Bit-precise reasoning seems to yield more useful results.
3. *Are bit-precise analyses usefully scalable?* We find that while less scalable than some abstract analyses, bit-precise analyses find witnesses faster than other approaches and are capable of exploring complex filters.
4. *Do bit-precise analyses allow us to address types of bugs that we could not otherwise find?* Bit-precise methods seem to be effective for discovering limit cycles (Cf. Section 2), which are hard to discover otherwise.

Motivating Digital Filter Verification. In essence, a digital filter is a function from an input signal to an output signal. A signal is a sequence of real values viewed as arriving over time. For our purposes, a digital filter is causal, that is, a value in the output signal at time t is a function of the input values at time t or before (and the previously computed output values). The construction of digital filters is typically based on a number of design templates (using specifications in the frequency domain) [16]. To design a filter, engineers select a template (e.g., "direct form" filters) and then use tools such as Matlab to compute coefficients that are used to instantiate these templates. Many templates yield linear filters (i.e., an output value is a linear combination of the preceding input values and previously computed output values). Because linear filters are so pervasive, they are an ideal target for verification tools, which have good support for linear arithmetic reasoning. Section 2 gives some basics on digital filters, but its contents are not needed to follow this example.

We used Matlab's filter design toolbox to construct a direct form I implementation of a Butterworth IIR filter with a corner frequency of 9600 Hz for a sampling frequency of 48000 Hz.[1] To the right, we compare a floating-point-based design and a fixed-point-based implementation of this filter by examining its magnitude response as a function of input frequency (top) and its impulse response (bottom). The fixed-point implementation is the result of quantizing the filter coefficients (as discussed below).[2]

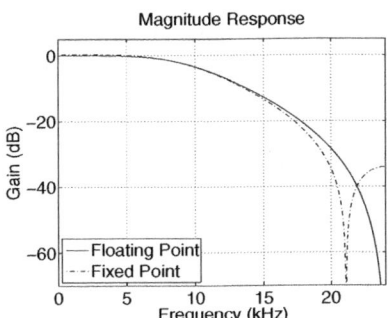

Magnitude response and impulse response are standard characterizations of filters [16]. Using these responses computed during design time the designer deduces some nice properties such as stability. Furthermore, the responses of the fixed-point implementation are often compared with the floating-point implementation. In the plots, the fixed-point implementation's response is seen to be quite "close" to the original floating-point design (certainly, where there is little attenuation—say > -20 dB). Further-

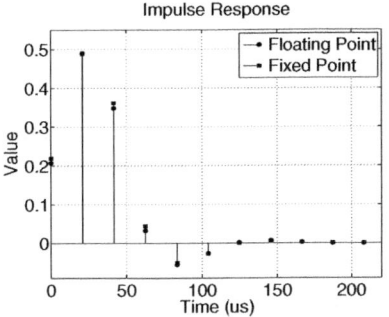

more, we see from the impulse response that the filter is stable—the output asymptotically approaches zero. Furthermore, if the inputs are bounded in the range $[-1.6, 1.6]$, the outputs will remain in the estimated range $[-2, 2]$ (Cf. Section 2). It is based on this information that the designer may choose a fixed-point representation for the implementation that uses 2 integer bits and 5 fractional bits allowing all numbers in the range $[-2, 1.96875]$ be represented with an approximation error in the range $(-0.03125, 0.03125)$; this representation leads to the quantization of the filter coefficients mentioned above.

But there are a number of problems that this popular filter design toolbox is not telling the designer, as we mention below.

Is simulation sufficient to find bugs in this filter? We estimated a range of $[-2, 2]$ for the output and our design allows for a range of $[-2, 1.96875]$. Yet, the theory used to calculate this range does not account for the presence of errors due to rounding.

---

[1] Specifically, Matlab yields coefficients $b_0 = 0.2066$, $b_1 = 0.4131$, $b_2 = 0.2066$ and $a_1 = -0.3695$, $a_2 = 0.1958$ based on floating-point calculations.

[2] Specifically, the coefficients are quantized to $b_0 = 0.21875$, $b_1 = 0.40625$, $b_2 = 0.21875$ and $a_1 = -0.375$, $a_2 = 0.1875$.

Therefore, we carried out extensive testing using a combination of uniformly random inputs vectors or randomly choosing either the maximum or minimum input value. Roughly 10^7 inputs were tested in 15 minutes. Yet, no overflows were detected.

Is bit-precise reasoning more useful in practice than conservative real-arithmetic reasoning? The conservative real-arithmetic model that tracks the range of overflow errors (Cf. Section 4) finds a spurious overflow at depth 1, yet no such overflow exists. On the other hand, bit-precise reasoning discovers an input sequence of length 5 causing an actual overflow. The solver required less than a second for each unrolling.

The difficulty of discovering this sequence through simulation or a conservative model is highlighted by the fact that small variations on this input sequence do not yield an overflow. The inset figure shows a failing input, the resulting output (fixed point) and the expected output (floating point) from the filter. We notice that there seems to be very little relation between the floating-point and the fixed-point simulations beyond $t = 100\mu s$.

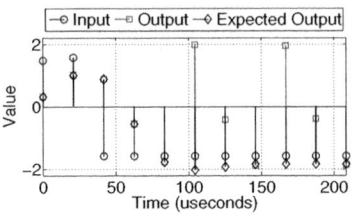

Do bit-precise analyses allow us to address types of bugs that we could not otherwise find? The quantized filter's impulse response seems to rule out the possibility of limit cycles. But then again, the impulse response did not take into account the effect of round-offs and overflows. The presence of limit cycles can potentially lead to large amplitude oscillations in the output that need further filtering. The search process for limit cycles is non-trivial and is heavily dependent on the quantization of the filter.

2   Preliminaries: Digital Filter Basics

In this section, we present some of the relevant background on filter theory. Further details on the mathematical theory of filters are discussed in standard texts [16, 19].

A discrete-time signal $x(t)$ is a function $\mathbb{Z} \mapsto \mathbb{R}$. By convention, the signal values $x(t)$ for times $t < 0$ are set to a constant default value given by $x_{<0}$.

Definition 1 (Single-Stage Digital Filter). *A single-stage digital filter is a recursive function that maps a discrete-time input signal $x(t)$ to an output discrete-time signal $y(t)$ for $t \in \mathbb{Z}$. The filter is specified in one of two direct forms. A direct form I filter is described by the tuple $\langle a, b, I, y_{<0} \rangle$, such that*

$$y(t) = \begin{cases} \sum_{i=0}^{N} b_i \, x(t-i) - \sum_{j=1}^{M} a_j \, y(t-j) & \text{if } t \geq 0 \\ y_{<0} & \text{if } t < 0 \end{cases}$$

The vectors $a \colon (a_1, \ldots, a_M) \in \mathbb{R}^M$ and $b \colon (b_0, \ldots, b_N) \in \mathbb{R}^{N+1}$ are the coefficients of the filter and describe the input-output relationship of the filter. The range $I \colon [l, u] \subseteq \mathbb{R}$ is a closed and bounded interval and is the range of the input sequence x. The constant $y_{<0} \in \mathbb{R}$ represents the initial state of the filter. Likewise, a direct form II filter is described by the tuple $\langle a, b, I, s_{<0} \rangle$, such that

$$y(t) = \sum_{i=0}^{N} b_i \, s(t-i) \qquad\qquad s(t) = \begin{cases} x(t) - \sum_{j=1}^{M} a_j \, s(t-j) & \text{if } t \geq 0 \\[2mm] s_{<0} & \text{if } t < 0 \end{cases}$$

The role of the coefficients a, b, *the input range* I, *and the initial state* $s_{<0}$ *are analogous to the corresponding components in a direct form I filter.*

A filter is said to have finite impulse response (FIR) whenever $a = 0$ and infinite impulse response (IIR), otherwise. Filters can be implemented in a single stage or multiple stages by composing individual filter stages as shown below:

Note that in a multi-stage filter implementation, the range constraint I is elided for the intermediate and final stages, but is retained just for the first input stage of the filter.

The *unit impulse* is defined by the function $\delta(t) = 1$ if $t = 0$, or $\delta(t) = 0$ if $t \neq 0$. The *impulse response* $h_F(t)$ of a digital filter F is the output produced by the unit impulse δ [16]. FIR filters have an impulse response $h_F(t) = 0$ for all $t > N$, whereas IIR filters may have an impulse response that is non-zero infinitely often.

Definition 2 (Stability). *A digital filter is* bounded-input bounded-output (BIBO) *stable if whenever the input is bounded by some interval, the output is also bounded.*

It can be easily shown that a filter F *is BIBO stable if and only if the* L_1 *norm of the impulse response* $\sum_0^\infty |h_F(t)|$ *converges.*

Let $H = \sum_0^\infty |h_F(t)|$ be the L_1 norm of the impulse response of a stable filter F. The impulse response can be used to bound the output of a filter given its input range I.

Lemma 1. *If the inputs lie in the range* $I : [-\ell, \ell]$ *then the outputs lie in the interval* $[-H\ell, H\ell]$.

Instability often manifests itself as a zero-input *limit cycle*. Given an input, the sequence of outputs forms a limit cycle if and only if there exists a number $N > 0$ and a period $\delta > 0$ wherein

$$\forall t \geq N, \quad y(t + \delta) = y(t) \text{ and } y(t) \neq 0 \text{ infinitely often and } x(t) = 0 \text{ for all time } t$$

In general, zero-input limit cycles are considered undesirable and manifest themselves as noise in the output. Further filtering may be needed to eliminate this noise.

Fixed-Point Filter Implementations. In theory, filters have real-valued coefficients and have behaviors defined over real-valued discrete-time input and output signals. In practice, implementations of these filters have to approximate the input and output signals by means of fixed- or floating-point numbers. Whereas floating-point numbers are

commonly available in general-purpose processors, most special-purpose DSP processors and/or realizations of the filters using FPGAs use fixed-point arithmetic implementations of filters.

A $\langle k, l \rangle$ fixed-point representation of a rational number consists of an integer part
represented by k binary bits and a fractional part represented by l binary bits. Given an
m-bit word $b : b_{m-1} \cdots b_0$, we can define for b its value $V(b)$ and its *two's complement*
value $V^{(2)}(b)$ as follows:

$$V(b) = \sum_{i=1}^{m-1} 2^i b_i \qquad V^{(2)}(b) = \begin{cases} V(b_{m-2} \cdots b_0) & \text{if } b_{m-1} = 0 \\ V(b_{m-2} \cdots b_0) - 2^{m-1} & \text{if } b_{m-1} = 1 \end{cases}$$

Let (b, f) be the integer and fractional words for a $\langle k, l \rangle$ fixed-point representation. The
rational represented is given by $R(b, f) = V^{(2)}(b) + \frac{V(f)}{2^l}$. The maximum value representable is given by $2^k - \frac{1}{2^l}$ and the minimum value representable is -2^k. The arithmetic
operations of addition, subtraction, multiplication and division can be carried out over
fixed-point representations, and the result approximated as long as it is guaranteed to be
within the representable range. When this constraint is violated, an overflow happens.
Overflows are handled by *saturating* wherein out-of-range values are represented by the
maximum or minimum value, or by *wrapping around*, going from either the maximum
value to the minimum, or from the minimum to the maximum upon an overflow.

A fixed-point digital filter is a digital filter where all values are represented by fixed
bit-width integer and fractional parts. In general, the implementation of a fixed-point
digital filter uses standard registers to store input and output values along with adders,
multipliers and delays. It is possible that a fixed-point implementation is unstable even
if the original filter it seeks to implement is stable.

3   Bit-Precise Encoding

In theory, bit-precise reasoning can be implemented by translating all operations at the
bit level into a propositional logic formula and solving that formula using a SAT solver.
Practically, however, there are many simplifications that can be made at the word level.
Therefore, we consider encodings of the fixed-point operations involved in a digital
filter in the theory of bit-vectors as well as linear integer arithmetic. We assume a $\langle k, l \rangle$
bit representation with k integral bits and l fractional bits. In particular, the bit-vector
representation uses the upper k-bits of a bit-vector for the integer part and the lower l-
bits for the fractional part. For the integer representation, since there is no a priori limit
to its size, an integer n is interpreted as $\frac{n}{2^l}$; then, we separately check for overflow.

Encoding Multiplication.   Fixed-point multiplication potentially doubles the number
of bits in the intermediate representation. The multiplication of two numbers with $\langle k, l \rangle$
bits produces a result of $\langle 2k, 2l \rangle$ bits. To use this result as $\langle k, l \rangle$-bit value, we must
truncate or round the number. We must remove most significant k bits of the integer
part and the l least significant bits of the fractional part.

In the theory of bit-vectors, this truncation is a bit extraction. We extract the bits in
the bit range $[k + 2l - 1 : l]$ from the intermediate result (i.e., extract the lth to the
$k + 2l - 1^{\text{st}}$ bits). In the theory of integers, we remove the lower l bits by performing an

integer division by 2^l. Because there is no size limit, we do not need to drop the upper k bits, but we perform an overflow check that simply asserts that the result fits within the permissible range at the end of each operation. That is, we check if the intermediate $\langle 2k, 2l \rangle$-bit value lies in the permissible range of the $\langle k, l \rangle$-bit representation.

Encoding Addition.   The treatment of addition is similar. Adding two fixed-point numbers with $\langle k, l \rangle$ bits produces a result of $\langle k+1, l \rangle$ bits. To use this result in as a $\langle k, l \rangle$-bit value operation, the top bit needs to be dropped.

For bit-vectors, we extract the bits in the range $[k+l-1 : 0]$. For linear integer arithmetic, we allow the overflow to happen and check using an assertion. Detecting overflow for additions involves checking whether the intermediate value using $\langle k+1, l \rangle$ bits lies inside the range of values permissible in a $\langle k, l \rangle$-bit representation.

Overflow and Wrap Around.   A subtlety lies in using wrap-around versus saturation semantics for overflow. For saturation, it is an error if any operation results in an overflow (and thus our encoding must check for it after each operation). But for wrap around, intermediate results of additions may overflow and still arrive at the correct final result, which may be in bounds. Thus, checking for overflow after each addition is incorrect in implementations that use wrap-around semantics for overflows. In terms of our encoding, if the final result of successive additions fits in the $\langle k, l \rangle$ bit range, overflows while computing intermediate results do not matter. We handle this behavior in the bit-vector encoding by allowing extra bits to represent the integer part of intermediate results (as many as $k+n$ where n is the number of additions) and checking whether the result after the last addition fits inside the range representable by a $\langle k, l \rangle$-bit representation. For the integer arithmetic representation, we simply avoid asserting the overflow condition for intermediate addition results.

Unrolling Filter Execution.   The unrolling of the filter execution takes in an argument n for the number of time steps and encodes the step-by-step execution of the filter (i.e., compute $y(0)$ up to $y(n-1)$). At each step, we assert the disjunction of the overflow conditions from the additions, multiplications, and the final output value.

Finding Limit Cycles.   To find a limit cycle of n steps, we compare a window of the output with another window of the output n steps later. The lengths of the windows are defined to be the maximum length of the coefficient vectors (i.e., the order of the filter). If these windows are equal and non-zero (for all zero inputs), then there is a limit cycle. To implement limit cycle search, we try a bounded number of values for n.

4   Real-Arithmetic Encoding

The real-valued encoding for a filter models each state variable of a fixed-point filter by a real number, while approximating the effects of quantization and round-off errors conservatively. As a result, the model includes a conservative treatment of the two sources of errors: (a) *quantization errors* due to the approximation of the filter coefficients to fit in the fixed bit-width representations and (b) *round-off errors* that happen for each multiplication and addition operation carried out for each time step.

Abstractly, a filter can be viewed as a *MIMO system* (multiple-input, multiple-output) with an internal state vector w, a control input scalar x and an output (scalar) y, wherein at each iterative step, the state is transformed as follows:

$$w(t+1) = Aw(t) + x(t)d \quad \text{and} \quad y(t+1) = c \cdot w(t+1) \,. \tag{1}$$

Note that the state vector $w(t)$ for a direct form I filter implementation includes the current and previous output values $y(t), \ldots, y(t - M)$, as well as the previous input values $x(t - 1), \ldots, x(t - N)$. The matrix A includes the computation of the output and the shifting of previous output and input values to model the delay elements. The dot-product with vector c simply selects the appropriate component in $w(t + 1)$ that represents the output at the current time.

Quantized Filter. First, we note that the quantization error in the filter coefficients is known a priori. Let $\widetilde{A}, \widetilde{d}, \widetilde{c}$ be the quantized filter coefficients. We can write the resulting filter as

$$\widetilde{w}(t+1) = \widetilde{A} \otimes \widetilde{w}(t) \oplus \widetilde{x}(t) \otimes \widetilde{d} \quad \text{and} \quad \widetilde{y}(t+1) = \widetilde{c} \otimes \widetilde{w}(t+1) \,. \tag{2}$$

Here $\otimes$ and $\oplus$ denote the multiplication and addition with possible round-off errors.

Note that since the matrix A represents the arithmetic operations with the filter coefficients as well as the action of shifting the history of inputs and outputs, the quantization error affects the non-zero and non-unit entries in the matrix A, leaving all the other entries unaltered. Likewise, the additive and multiplicative round-off errors apply only to multiplications and additions that involve constants other than 0 and 1. Comparing the original filter (1) to the quantized filter in (2), we write $\widetilde{w} = w + \Delta w$ to be the error accumulated in w. This leads to a non-deterministic iteration that jointly determines possible values of $w(t + 1)$ and $\Delta w(t + 1)$ at each time step as follows:

$$\begin{aligned}
w(t+1) &= Aw(t) + x(t)d \\
\Delta w(t+1) &\in \Delta A(w(t) + \Delta w(t)) + x(t)\Delta d + [-1,1](q|(d + \Delta d)| + r) \\
y(t+1) &= c \cdot w(t+1) \\
\Delta y(t+1) &\in \Delta c \cdot w(t+1) + (c + \Delta c) \cdot \Delta w(t+1) + [-1,1]r'
\end{aligned} \tag{3}$$

wherein q is the maximal input quantization error, and r and r' refer to the estimated maximal round off errors accumulated due to the addition and multiplication operations carried out at time step $t+1$ for each of the entries in $w(t+1)$ and $y(t+1)$, respectively. Note that $|d + \Delta d|$ refers to the vector obtained by taking the absolute value of each element in $d + \Delta d$. The round-off error for multiplication/addition of two $\langle k, l \rangle$ bit fixed point numbers is estimated to be 2^{-l}. We bound the maximum magnitude of round off errors for K arithmetic operations is $K2^{-l}$.

Our goal is to check if for a given depth bound N and bounds $[\ell, u]$ for overflow, there exist values for the input sequence $x(0), x(1), \ldots, x(N)$ such the state $\widetilde{w}(t) \notin [\ell, u]$ for some time t. Note that the values of $\Delta A, \Delta d, q, r, r'$ are available to us once the quantized coefficients and the bit-widths of the state registers, the multipliers and adders are known. As a result, the search for an input that may *potentially cause* an overflow is encoded by a linear programming problem.

Lemma 2. *Given filter coefficients (A, d, c), quantization errors $(\Delta A, \Delta d, \Delta c)$, an over-estimation of the round-off r, r' and input quantization errors q, there exists a set of linear constraints φ such that if φ is unsatisfiable then no input may cause an overflow at depth N.*

Proof. Proof consists of unrolling the iteration in Equation (3). The variables in the LP consist of inputs $x(1), \ldots, x(N)$, the state values $\boldsymbol{w}(1), \ldots, \boldsymbol{w}(N)$ and finally the outputs $y(1), \ldots, y(N)$ along with error terms $\Delta w(t)$ and $\Delta y(t)$ for $t \in [1, N]$. Note that for each step, we have a linear constraint for the state variables $\boldsymbol{w}(t + 1) = A\boldsymbol{w}(t) + x(t)\boldsymbol{d}$. Likewise, we obtain linear inequality constraints that bound the values of $\Delta \boldsymbol{w}(t + 1)$ using Equation (3). We conjoin the bounds on the input values and the overflow bounds on the outputs for each time step.

Limit Cycles. The real-arithmetic model cannot be used directly to conclude the presence or absence of limit cycles. Limit cycles in the fixed-point implementation often exist due to the presence of round-off errors and overflows that wrap around from the largest representable value to the smallest. In practice, these effects cannot be modeled using the real-arithmetic filter implementations in a straightforward manner, without introducing complex conditional expression and possibly non-linear terms.

5   Experimental Evaluation

We generated twelve filter designs in Matlab using a number of design patterns, including low-pass, band-pass and band-stop filters using Chebyshev, Butterworth, and elliptic designs. We used both multi- and single-stage designs. The designs are shown in Table 1. The nominal bit-widths of the filters were chosen such that they were the smallest that could contain the coefficients and inputs in the range $[-1, 1]$, except for lp2, whose design rationale is presented in Section 1. Our experiments also consider the effect of variations in the bit-widths.

Our experiments compare four approaches to filter verification: (a) bit-vector encoding (BV) described in Section 3, (b) the integer linear arithmetic encoding (LI) described in Section 3, (c) a real-arithmetic encoding (RA) into linear arithmetic described in Section 4, and (d) affine arithmetic [6] (AA) to track possible ranges of state and output variables conservatively. The tests were run on an Intel Core i5 750 processor with 8 GB of RAM running Ubuntu Linux. Processes were memory-limited to 1 GB and time-limited to 60 seconds for the unrolling test and 300 seconds for other tests. No processes ran out of memory.

Table 1. Benchmarks used in the experiments are designed using the Matlab Filter Design and Analysis Tool. The Type column is a choice of a function amongst **Low Pass**, **Band Stop**, and **Band Pass** and a design pattern amongst **Butterworth**, **Elliptic**, **Max Flat**, and **Chebyshev**. The Order column is the order the filter, # Stages denotes the number of stages, and the Freq. column gives the cut-off or band frequencies in kHz.

Name	Type	Order	# Stages	Freq.	Name	Type	Order	# Stages	Freq.
lp2	(LP, B)	2	1	9.6	lp10cm	(LP, MF)	2	5	0.1
lp4	(LP, B)	4	1	9.6	lp10m	(LP, MF)	10	1	0.1
lp4e	(LP, E)	4	1	9.6	bs10	(BS,C)	10	1	9.6-12
lp6	(LP, E)	6	1	9.6	bs10c	(BS,C)	2	5	9.6-12
lp6c	(LP,E)	2	3	9.6	bp8	(BP,E)	8	1	0.2-0.5
lp10c	(LP, B)	2	5	9.6	bp8c	(BP,E)	2	4	0.2-0.5

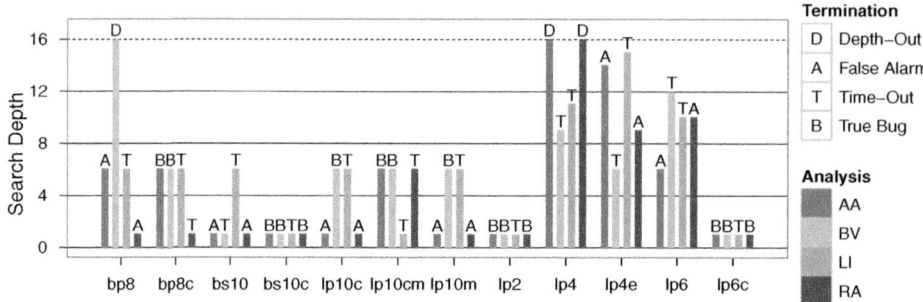

Fig. 1. Plot showing outcome for various methods on benchmarks. Timeout was set to 300 seconds and a maximum depth of 16 is shown by the dashed line.

We use the SMT solver Z3 version 3.2 [7], as it is currently the fastest known solver for both the bit-vector theory and the linear integer arithmetic theory. The framework is implemented in OCaml.

Is Simulation Sufficient to Find Bugs in Filters?   We tested all of the filters using traditional simulation based methods. To do this, we explored three possible input generation methods: (a) uniform random selection of values from the filter's input range; (b) selecting the maximum value until the output stabilized followed by the minimum value; and (c) selecting the minimum value until the output stabilized followed by the maximum value. Choices (b,c) attempt to maximize the overshoot in the filters in order to cause a potential overflow.

The filters are simulated on a fixed-point arithmetic simulator using the three input generation methods described above. The simulation was set to abort if an overflow were to be found. Each simulation was run for the standard timeout of 300 seconds. During this time filters were able to run between two and five million inputs.

There were zero overflows found by the simulations.

Is Bit-Precise Reasoning More Precise in Practice Than Conservative Real-Arithmetic Reasoning?   Figure 1 compares the outcomes of all the four techniques on our benchmarks in finding overflows. The conservative techniques, AA and RA, can yield false alarms, whereas any overflow warning raised by the bit-precise techniques, BV and LI, must be true bugs. A time-out or depth-out means no bugs were found in the allotted time or depth but of course says nothing about whether there are bugs further on. An alarm raised by the conservative techniques can be classified as being false (i.e., spurious) when a bit-precise technique is able to exceed that search depth without raising an alarm. In six out of the twelve tests (i.e., bp8, bs10, lp10c, lp10m, lp4e, lp6), both conservative approaches raised false alarms. At least one bit-precise technique was able to search deep enough to label the alarms from the conservative analyses as true (i.e., bug) or false (i.e., spurious).

Are Bit-Precise Analyses Usefully Scalable?   Figure 2 shows the performance of different methods of analysis on all twelve test filters across unrollings of 5, 8, 10 and 15. In the plot of BV vs. LI (right), we see that BV is, in general, faster than LI (above the

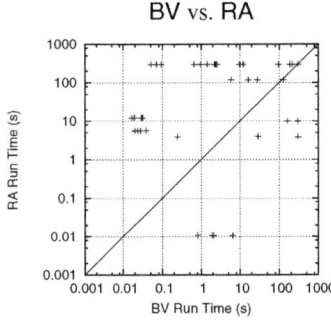
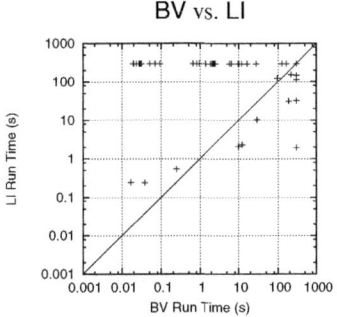

Fig. 2. Performance comparison of different analysis methods using unrollings of 5, 8, 10 and 15

line). However, the advantage is not overwhelming, suggesting that neither approach is inherently better than the other.

For both BV and LI, the unrolling depth did not have a pronounced effect on the time taken to solve benchmark instances for small unrollings. Instances wherein BV was faster at unrolling depth 5 also tended to favor BV at unrolling depth 8. Therefore, we conclude that the nature of the coefficients in the filter and its overall architecture may have a larger effect on the performance of BV and LI than the unrolling depth.

We see in the BV vs. RA plot (left), the bit-precise method BV is competitive with the conservative method RA. Whereas bit-vector theories are NP-complete, linear programs are well known to have efficient polynomial time algorithms in practice. We hypothesize that the use of an SMT solver to reason with large fractions using arbitrary precision arithmetic has a significant performance overhead. This may be a good area of application for techniques that use floating-point solvers to help obtain speedups while guaranteeing precise results [15].

The AA approximate method is very fast in comparison to all the other methods presented here. It is elided because this speed comes at a high cost in precision [18]. Furthermore, the affine arithmetic technique does not, as such, yield concrete witnesses. Therefore, it is not readily comparable to precise methods.

Effect of Unrolling Length on the Analysis. We now look deeper into the performance of encodings. We first consider how unrolling affects performance by varying the amount of unrolling from 1 to 50 on select filters.

According to Figure 3, BV, RA and LI are heavily affected by the unrolling depth. RA, even for short unrollings, times out if it does not find an error. Due to some details of implementations, the RA encoding incrementally searches for the shortest possible error unlike the BV and LI encodings. Because of this, if an error is found early, RA appears to scale well, as seen in `lp6`. AA scales well with unrolling depth, as expected. Note that the unrolling is stopped once overflow is found.

The bit-precise methods BV and LI both exhibit more unpredictable behavior. This is due to the nature of the encoding (one single monolithic encoding that searches for all paths up to a given depth limit) and the SMT solvers used. As the unrolling becomes longer, the solver is not bound to search for the shortest path first. The results from `lp2`

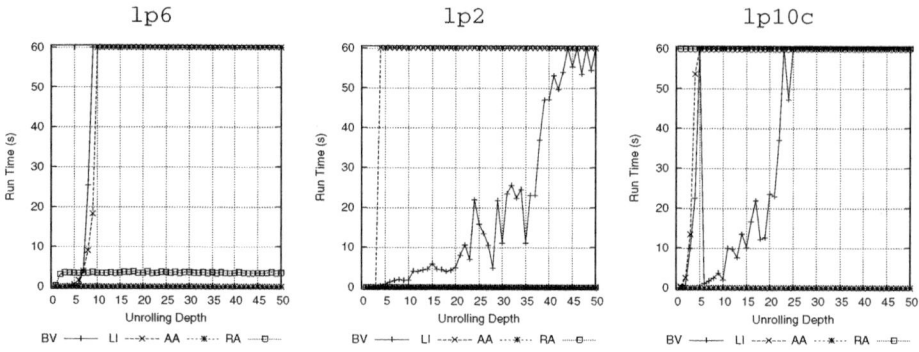

Fig. 3. Performance analysis of analysis methods as a function of unrolling depth

and `lp10c` show that longer unrollings may be faster than shorter unrollings, but there is a general trend of increasing time with unrolling depth.

Performance Impact of Bit-Widths. We also need to consider the effect that changing the precision of filters has on the analysis performance. Figure 4 shows performance for both BV and LI on two different tests across a range of bit-widths. The first test, `lp2`, is "pre-quantized" so that adding more fractional bits causes the coefficients to gain more zeros in their least significant bits. The second test, `lp6`, has large fractions in the coefficient, so meaningful bits are added when the fraction size is increased.

The first conclusion is that the total number of bits does not directly affect the time taken. Both BV and LI are faster with more integer bits. As more integer bits are added, it is possible that the abstractions used internally within the SMT solver can be coarser allowing it to come up with answers faster. As more fractional bits are added, the BV and LI approaches diverge. BV becomes much slower, and LI is not heavily affected. Once again, this behavior seems to depend critically on the coefficients in the filter.

As bit-widths are varied, the outcome typically varies from an overflow found at a low depth to unsatisfiable answers at all depths. In this case, the performance of LI is poor whenever the bit-width selected is *marginal* or nearly insufficient. If the system you are trying to analyze is marginal, but small, use BV and if it is relatively safe, but large, use LI.

Do Bit-Precise Analyses Allow Us to Find Bugs We Could Not Otherwise Find? Bit-precise analyses allow us to easily find limit cycles in

Unroll	Pass	Fail	Timeout	Mean (s)	Median (s)	Std Dev (s)
2	2	10	0	1.22	0.35	4.88
5	0	7	5	22.6	10.3	89.8
8	0	6	6	55.8	21.7	133.8

fixed-point IIR filters. Limit cycles are prevalent in fixed-point IIR filters as the inset table below shows. From our twelve test cases, the table shows the number of examples where we did not find a limit cycle (column Pass), the number where we found one (column Fail), and the remaining that timed out. The remaining columns show the mean, median, and standard deviation of the running time for limit cycle detection. Due

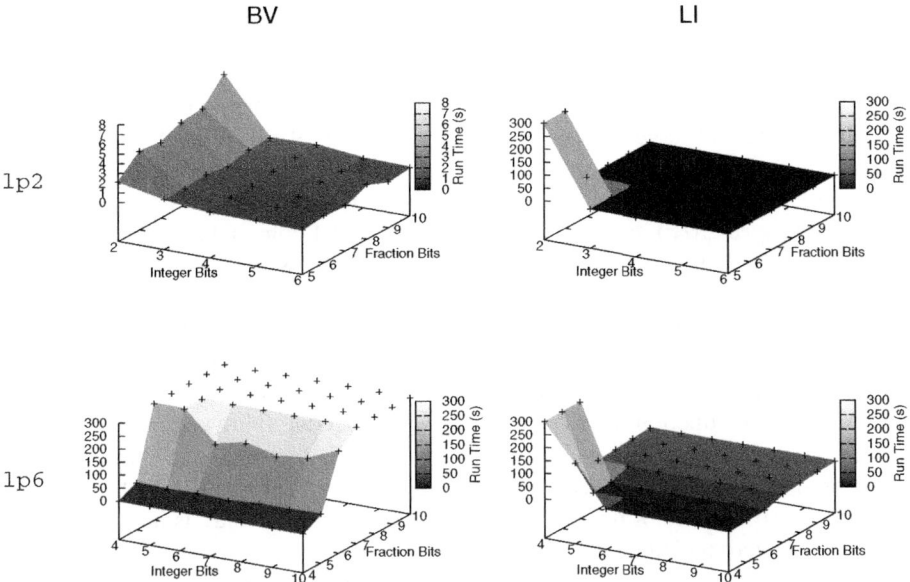

Fig. 4. Performance of bit-precise analysis methods as a function of the number of bits

to their prevalence, most limit cycles are quite easy for the SMT solver to find (using the bit-vector theory). Most limit cycles are found with short unrollings, quickly.

Because limit cycles can be detected efficiently, the designer can make informed decisions about those situations. Often designers will add extra circuitry to eliminate limit cycles, but if the designer knew the kinds of limit cycles that exist, the designer may elect to simplify the design and not add that circuitry. We have discovered limit cycles varying from small, 1-2 least significant bits, to large, oscillating from near the maximum value to near the minimum value. In the latter case, the designer may elect to design a different circuit.

6 Related Work

Verification of fixed-point digital filters has focused mostly on the problem of discovering safe bit-widths for the implementation. While verification for a specific bit-width is one method for solving this problem, other works have considered interval arithmetic, affine arithmetic [8, 13], spectral techniques [17], and combinations thereof [18].

Approaches based on SMT solvers, on the other hand, offer the promise of enhanced accuracy and exhaustive reasoning. Kinsman and Nicolici use a SMT solver to search for a precise range for each variable in fixed-point implementations of more general MIMO systems [12]. Their analysis uses the non-linear constraint solver HySAT [10] using a real-arithmetic model without modeling the errors precisely. Furthermore, since HySAT converges on an interval for each input variable, their analysis potentially lacks the ability to reason about specific values of inputs.

We have focused on comparing against some simple techniques for test input generation in this paper. Others have considered more advanced heuristics for tackling this problem [20], which may be worthy of further study.

Several researchers have tackled the difficult problem of verifying floating-point digital filters as part of larger and more complex systems [9, 14]. The static analysis approach to proving numerical properties of control systems implemented using floating point has had some notable successes [3, 11]. In particular, the analysis of digital filters has inspired specialized domains such as the ellipsoidal domain [2, 9]. While floating-point arithmetic is by no means easy to reason with, the issues faced therein are completely different from the ones considered here for fixed-point arithmetics. Whereas we focus on analyzing overflows and limit cycles, these are not significant problems for floating-point implementations. The use of bit-precise reasoning for floating-point C programs has recently been explored by Kroening et al. [4].

Yet another distinction is that of proving safety versus trying to find bugs. The approaches considered in this paper clearly focus on bug finding using bounded-depth verification. While a similar study for techniques to prove properties may be of interest, the conservative nature of the real-arithmetic model suggests that its utility in proving highly optimized implementations may also be limited.

One approach to verifying digital filters is to perform a manual proof using a theorem prover [1]. Such approaches tend to be quite general and extensible. However, they are mostly manual and often unsuitable for use by DSP designers, who may be unfamiliar with these tools.

7   Conclusion

Our results show that fixed-point digital filters designed using industry standard tools may sometimes suffer from overflow problems. Commonly used frequency-domain design techniques and extensive simulations are insufficient for finding overflows. In this work, we have compared different formal verification techniques based on bounded-model checking using SMT solvers.

We have shown that error approximation using real-arithmetic can alert designers to otherwise unknown issues in filters. These alarms are often spurious and may lead the designer to draw false conclusions about their designs. Secondly, in spite of fundamental complexity considerations, the real-arithmetic solvers can often be slower than bit-precise approaches, possibly due to the need for arbitrary precision arithmetic. The use of floating-point simplex in conjunction with arbitrary precision numbers may be a promising remedy [15].

Finally, we demonstrated that bit-precise verification is possible and efficient using modern SMT solvers. Also, bit-precise verification is able to find situations where error approximations would have otherwise prevented a designer from shrinking a filter by one more bit. We also saw that both integer and bit-vector based methods are required to achieve maximum performance.

References

[1] Akbarpour, B., Tahar, S.: Error analysis of digital filters using HOL theorem proving. Journal of Applied Logic 5(4), 651–666 (2007)

[2] Alegre, F., Feron, E., Pande, S.: Using ellipsoidal domains to analyze control systems software. CoRR, abs/0909.1977 (2009)

[3] Blanchet, B., Cousot, P., Cousot, R., Feret, J., Mauborgne, L., Miné, A., Monniaux, D., Rival, X.: Design and Implementation of a Special-Purpose Static Program Analyzer for Safety-Critical Real-Time Embedded Software (Invited Chapter). In: Mogensen, T.Æ., Schmidt, D.A., Sudborough, I.H. (eds.) The Essence of Computation. LNCS, vol. 2566, pp. 85–108. Springer, Heidelberg (2002)

[4] Brillout, A., Kroening, D., Wahl, T.: Mixed abstractions for floating-point arithmetic. In: Formal Methods in Computer Aided Design (FMCAD), pp. 69–76 (2009)

[5] Clarke, E., Biere, A., Raimi, R., Zhu, Y.: Bounded model checking using satisfiability solving. Formal Methods in System Design 19(1), 7–34 (2001)

[6] de Figueiredo, L.H., Stolfi, J.: Self-validated numerical methods and applications. In: Brazilian Mathematics Colloquium Monograph, IMPA, Rio de Janeiro (1997)

[7] de Moura, L., Bjørner, N.: Z3: An Efficient SMT Solver. In: Ramakrishnan, C.R., Rehof, J. (eds.) TACAS 2008. LNCS, vol. 4963, pp. 337–340. Springer, Heidelberg (2008)

[8] Fang, C., Rutenbar, R., Chen, T.: Fast, accurate static analysis for fixed-point finite-precision effects in DSP designs. In: International Conference on Computer-Aided Design (ICCAD), pp. 275–282 (2003)

[9] Feret, J.: Static Analysis of Digital Filters. In: Schmidt, D. (ed.) ESOP 2004. LNCS, vol. 2986, pp. 33–48. Springer, Heidelberg (2004)

[10] Fränzle, M., Herde, C., Ratschan, S., Schubert, T., Teige, T.: Efficient solving of large nonlinear arithmetic constraint systems with complex Boolean structure. JSAT 1(3-4), 209–236 (2007)

[11] Goubault, E., Putot, S.: Static Analysis of Finite Precision Computations. In: Jhala, R., Schmidt, D. (eds.) VMCAI 2011. LNCS, vol. 6538, pp. 232–247. Springer, Heidelberg (2011)

[12] Kinsman, A.B., Nicolici, N.: Finite precision bit-width allocation using SAT-modulo theory. In: Design, Automation and Test in Europe (DATE), pp. 1106–1111 (2009)

[13] Lee, D., Gaffar, A., Cheung, R., Mencer, O., Luk, W., Constantinides, G.: Accuracy-guaranteed bit-width optimization. IEEE Trans. on CAD of Integrated Circuits and Systems 25(10), 1990–2000 (2006)

[14] Monniaux, D.: Compositional Analysis of Floating-Point Linear Numerical Filters. In: Etessami, K., Rajamani, S.K. (eds.) CAV 2005. LNCS, vol. 3576, pp. 199–212. Springer, Heidelberg (2005)

[15] Monniaux, D.: On Using Floating-Point Computations to Help an Exact Linear Arithmetic Decision Procedure. In: Bouajjani, A., Maler, O. (eds.) CAV 2009. LNCS, vol. 5643, pp. 570–583. Springer, Heidelberg (2009)

[16] Oppenheim, A.V., Willsky, A.S., Nawab, S.H.: Signals & Systems, 2nd edn. Prentice Hall (1997)

[17] Pang, Y., Radecka, K., Zilic, Z.: Optimization of imprecise circuits represented by taylor series and real-valued polynomials. IEEE Trans. on CAD of Integrated Circuits and Systems 29(8), 1177–1190 (2010)

[18] Pang, Y., Radecka, K., Zilic, Z.: An efficient hybrid engine to perform range analysis and allocate integer bit-widths for arithmetic circuits. In: Asia South Pacific Design Automation Conference (ASP-DAC), pp. 455–460 (2011)

[19] Smith, J.: Introduction to Digital Filters: With Audio Applications. W3K Publishing (2007)

[20] Sung, W., Kum, K.: Simulation-based word-length optimization method for fixed-point digital signal processing systems. IEEE Transactions on Signal Processing 43(12), 3087–3090 (1995)

Numeric Bounds Analysis with Conflict-Driven Learning*

Vijay D'Silva**, Leopold Haller, Daniel Kroening, and Michael Tautschnig

Computer Science Department, University of Oxford
firstname.surname@cs.ox.ac.uk

Abstract. This paper presents a sound and complete analysis for determining the range of floating-point variables in control software. Existing approaches to bounds analysis either use convex abstract domains and are efficient but imprecise, or use floating-point decision procedures, and are precise but do not scale. We present a new analysis that elevates the architecture of a modern SAT solver to operate over floating-point intervals. In experiments, our analyser is consistently more precise than a state-of-the-art static analyser and significantly outperforms floating-point decision procedures.

1 Introduction

Automotive and avionic control software has a special structure. Few programming language constructs are used, pointers are avoided and loop iterations are often bounded by constants. Nonetheless, such software performs complex tasks, computing vehicle trajectories and approximating non-linear functions. Control software verification involves proving that IEEE 754 floating-point operations in programs are free of overflows and approximation errors. We present a new, sound and complete analysis for this problem, and demonstrate empirically that the analysis is more efficient and precise than the state of the art.

Bounds checking is the problem of determining if the value of a numeric variable lies in a given range. *Interval analysis*, a classic approach to bounds checking, propagates intervals through a program. Intervals analysis is extremely fast but woefully imprecise, producing proofs on only 17 of 33 of our safe benchmarks (see § 5). Another shortcoming of interval analysis is that imprecision cannot be distinguished from errors. An alternative approach to bounds checking is bounded model checking (BMC) with an IEEE 754 decision procedure. BMC is precise but does not scale: of 57 benchmarks, only 23 can be solved by BMC within a minute, whereas interval analysis usually requires less than a second. Another problem is that unbounded loops cannot be handled directly.

We present *Conflict Driven Fixed Point Learning* (CDFL), a new program analysis that embeds an abstract domain inside the Conflict Driven Clause Learning

---

* Supported by the Toyota Motor Corporation, EPSRC project EP/H017585/1 and ERC project 280053.
** Supported by a Microsoft Research European PhD Scholarship.

C. Flanagan and B. König (Eds.): TACAS 2012, LNCS 7214, pp. 48–63, 2012.

(CDCL) algorithm of modern SAT solvers. A SAT solver uses constraint propagation, decisions, backtracking, a conflict graph, and clause learning to decide satisfiability. We develop abstract domain analogues of these ideas: Constraint propagation uses fixed point iteration, decisions restrict the range of intervals, the conflict graph is labelled with intervals, and learning generates program analysis constraints in place of propositional clauses.

CDFL is both a static analyser and a decision procedure. From a static analysis perspective, CDFL is an abstract interpreter that uses decisions and learning to increase transformer precision. From a decision procedure perspective, CDFL is a SAT solver for program analysis constraints. CDFL is a strict generalisation of propositional CDCL in that, on acyclic programs with only Boolean variables, our analyser is a clause-learning SAT solver. Elucidating this connection is beyond the scope of this paper.

Contribution and Contents. Our new interval analysis builds on the following contributions to combine the strength of static analysis and BMC.

- A novel account of program safety as satisfiability of a set-constraint formula. Unlike the standard formulation of static analysis, which focuses on invariants, our formulation is based on error traces.
- A new interval analysis that exploits the efficiency of the interval domain while being path-sensitive and bit-level accurate.
- A tool that can prove correctness of non-linear, IEEE 754 floating-point computations using *only the interval abstraction*. Our experiments reveal that existing techniques are either imprecise or slow on such programs.

The rest of this section illustrates our approach and discusses related work. The new formulation of safety as satisfiability is in § 3 and our procedure for deciding satisfiability is in § 4. Implementation and benchmarks are discussed in § 5.

1.1 Overview

A program, as in Figure 1(a), is an acyclic control flow graph (ACFG) Edges can be labelled with loops, so this representation is not limiting. The variables x and y are mathematical integers, $[y = 0]$ is a test, and $*$ denotes non-deterministic choice. We wish to determine if the error location $\frac{1}{2}$ is reachable.

The analysis associates an interval with each location and variable. The intervals for x and y at n_1 are $[-\infty, \infty]$. The condition $[y \neq 0]$ cannot be modelled by an interval, so the interval for y at n_2 is $[-\infty, \infty]$. The intervals for x at n_4 and n_5 are $[-\infty, \infty]$, so the analysis cannot prove safety.

The analysis is refined using (for now, arbitrary) constraints on intervals. First, x is constrained to be in $[0, \infty]$ at n_4. Interval analysis concludes that x is in $[0, \infty]$ at n_5 but cannot prove safety. A second decision constrains y to $[-\infty, -124]$ at n_1. Interval analysis shows x to be in $[-\infty, -124]$ at n_5, so P is safe assuming x is in $[0, \infty]$ at n_4 and y is in $[-\infty, -124]$ at n_1. A proof by cases would repeat the analysis, once with x in $[-\infty, -1]$ at n_4 and once with y in $[-123, \infty]$ at n_1. *We do not do a proof by cases.*

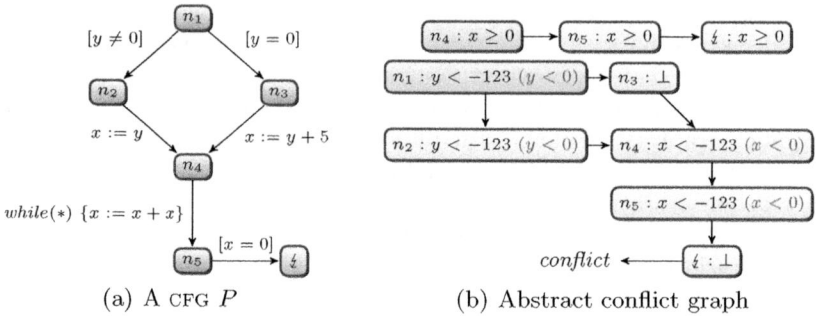

Fig. 1. A control flow graph and conflict graph

Learning is used to avoid enumerating cases. Deductions made during fixed point iteration are represented by the abstract conflict graph in Figure 1(b). For instance, the fact $x \geq 0$ at n_4 implies that the same holds at n_5 and $\frac{1}{2}$. The decisions are $(n_4 : x \geq 0)$ and $(n_1 : y < -123)$, but only the latter is connected to $(\frac{1}{2} : \bot)$, and suffices to prove safety. Conflict analysis in SAT solvers is similar. The next step is new and does not exist in SAT solvers. Constraints in the graph are generalised to the labels in parentheses. In Figure 1(b), $(n_1 : y < -123)$ generalises to $(n_1 : y < 0)$; we *learn* that all error traces must satisfy $y \geq 0$ at n_1. Our analysis backtracks, discarding all assumptions. Interval analysis is run with the learnt constraint and can prove safety.

We emphasise that learning and case-based reasoning are different. A case-based method does a proof under each assumption. Learning, however, generates constraints that preserve error reachability. These constraints are not assumptions. The procedure is simply rerun after learning.

1.2 Related Work

CDFL can be understood from the perspective of both static analysis and SAT solving. As a static analysis, CDFL is an automated refinement technique. Possibly the best-known refinement technique is CEGAR [5], which uses spurious counterexamples to refine an abstract domain and synthesise transformers. CDFL restricts transformers to eliminate sources of imprecision. The expensive operations of domain refinement and transformer synthesis are avoided.

Other work that refines an analysis without modifying the domain eliminates joins [11], constrains widening operators [23,22] or transformers [10]. In [10], a counterexample DAG is analysed and interpolants are used to constrain the analysis. Our work is similar, but a conflict graph is analysed and decisions and clauses are used to constrain the analysis. All these techniques can be viewed as instances of *trace partitioning* [19]. In technical terms, CDFL discovers program- and property-specific trace partitions. Other methods to discover trace partitions use CFG rewriting [20], and regular languages [2,12,20].

Let us discuss decision procedures. The DPLL(T) architecture leverages the efficiency of SAT solvers to reason about richer theories but separates propositional and theory reasoning. An alternative is to lift CDCL directly to a first-order theory as in generalised DPLL [17], conflict resolution [15], natural-domain SMT [7], and the cutting-planes extension [14]. Our work has similar motivations to these. First order formulae are replaced by abstract domain elements, transformers do theory propagation and set constraints are learnt instead of lemmas. We believe the abstract conflict graph and the related generalisation are new. Moreover, set-constraints allow us to handle loops.

Finally, CDFL is one of many efforts to combine static analysis and decision procedures. *Shallow integration* methods facilitate communication between engines treated as black boxes. Satisfiability Modulo Path Programs (SMPP) [13] lifts DPLL(T) to programs, using a SAT solver to guide an abstract interpreter. *Deep integration* techniques embed program analysis in satisfiability architectures. Examples are YOGI [3] and lazy annotation [16], both of which require quantifier-free first order theories. We use abstract domains and transformers.

2   Program Model and Domain

This section introduces the program model and identifies properties of the interval domain that enable conflict driven learning.

Programs. Consider sets of expressions *Exp* and Boolean expressions *BExp* over variables *Var*. We focus on numeric, machine data types, so variables take values in a finite, totally ordered set *Val* with minimum *min* and maximum *max*. IEEE 754 values are ordered by setting NaN greater than numeric values and using the arithmetic order. A statement, as below, is an assignment, conditional, sequential concatenation, non-deterministic choice or a loop.

$$s ::= x := exp \mid [b] \mid s_1; s_2 \mid choose\{s_1, s_2\} \mid loop\{s\}$$

An *acyclic control flow graph* (ACFG) is an acyclic graph $(Loc, E, stmt)$, with locations *Loc*, edges E and a function *stmt* labelling edges with statements. *Loc* contains unique initial and error locations *init* and $\frac{1}{2}$. A *loop-free* ACFG has no *loop* statements.

Concrete Semantics. Statement semantics is defined over environments $Env = Var \rightarrow Val$. A statement s defines a function $post_s : \wp(Env) \rightarrow \wp(Env)$, called a transformer. Assignments and tests have their expected semantics; concatenation is composition, choice is union and the semantics of a loop is a fixed point.

$$post_{s_1;s_2} = post_{s_2} \circ post_{s_1} \quad post_{choose\{s_1,s_2\}} = \lambda X.post_{s_1}(X) \cup post_{s_2}(X)$$
$$post_{loop\{s\}} = \lambda X.\mathsf{lfp}Y.X \cup post_s(Y)$$

We abbreviate $post_{stmt(n,m)}$ to $post_{(n_i, n_{i+1})}$. A *state* is a location with an environment. A *trace* is a sequence of states $(n_0, \varepsilon_0), \ldots, (n_k, \varepsilon_k)$ such that for all $0 \le i < k$, (n_i, n_{i+1}) is an ACFG edge and $\varepsilon_{i+1} \in post_{(n_i, n_{i+1})}(\{\varepsilon_i\})$. A program is *safe* if there is no trace as above with $n_0 = init$ and $n_k = \not4$. The *concrete domain* is the lattice of environments $\wp(Var \to Val)$ with a transformer $post_s$ for statements. Abstract interpretation with intervals is illustrated next.

Example 1. We use abstract interpretation to show that, if x is in the range $[-3, 3]$ and the statement $s = z := x; y := x * z$ is executed, y is non-negative. Let $a = \langle x \mapsto [-3, 3] \rangle$ denote that x has the range shown and other variables are unconstrained. Abstract transformers, denoted by $\hat{post}$, produce the facts below.

$$\hat{post}_{z:=x}(a) = b = \langle x \mapsto [-3, 3], z \mapsto [-3, 3] \rangle$$
$$\hat{post}_{y:=x*z}(b) = \langle x \mapsto [-3, 3], y \mapsto [-9, 9], z \mapsto [-3, 3] \rangle$$

Intervals lose the information that x and z are equal. Let m be $\langle x \mapsto [0, max] \rangle$ and $\sim m$ be $\langle x \mapsto [min, -1] \rangle$. The two intervals cover all values of x. Precision is regained by rerunning the analysis with the restrictions below.

$$\hat{post}_s(a \sqcap m) \sqcup \hat{post}_s(a \sqcap \sim m) = \langle x \mapsto [-3, 3], y \mapsto [0, 9], z \mapsto [-3, 3] \rangle$$

The transformer restriction increases precision without losing soundness. $\lhd$

Interval Abstraction. The set *Itv* of intervals over *Val* contains pairs $[l, u]$ with $l \le u$. The partial order on *Itv* is: $a \sqsubseteq b$ if b contains a. The *interval environments* domain is the lattice $IEnv = ((Var \to Itv) \cup \{\bot\}, \sqsubseteq, \sqcup, \sqcap)$ with abstract transformers $\hat{post}_s$. The least element is $\bot$ and the greatest element $\top$ maps all variables to $[min, max]$. The interval environment that maps x_1 to $[l_1, u_1]$, x_2 to $[l_2, u_2]$ and all other variables to $[min, max]$ is denoted $\langle x_1 \mapsto [l_1, u_1], x_2 \mapsto [l_2, u_2] \rangle$. Further, $\langle x \mapsto [min, c] \rangle$ is written $\langle x \le c \rangle$. The well known Galois connection $\wp(Env) \xleftrightarrow[\alpha]{\gamma} IEnv$ between environments and interval environments is recalled below.

$$\alpha(\emptyset) = \bot \qquad \alpha(X) = \left\{ x \mapsto \left[\inf_{\varepsilon \in X} \{\varepsilon(x)\}, \sup_{\varepsilon \in X} \{\varepsilon(x)\} \right] \mid x \in Var \right\}$$
$$\gamma(\bot) = \emptyset \qquad \gamma(a) = \{\varepsilon \mid \varepsilon(x) \text{ is in } a(x) \text{ for all } x\}$$

Atoms and Meet Irreducibles. The properties of elements used for decisions and clause learning are identified next. Fix a lattice $(A, \sqsubseteq, \sqcup, \sqcap)$ with elements $\bot$ and $\top$. An *atom* x of A is a least element strictly above $\bot$: $\bot \sqsubset x$ and no y satisfies $\bot \sqsubset y \sqsubset x$. The set of atoms is $Atoms(A)$. An atom of $\wp(Env)$ is a singleton. An atom of $IEnv$ maps variables to singleton intervals. An abstract transformer is *precise on atoms* if the equality $post_s \circ \gamma = \gamma \circ \hat{post}_s$ holds. We assume abstract transformers for loop-free statements have this property.

An element a is *meet irreducible* if, for all $X \subseteq A$, $\sqcap X = a$ implies a is in X. The meet irreducibles of A are $Irred_\sqcap(A)$. Meet irreducibles of $\wp(Env)$

are complements of singletons, and those of *IEnv* have the form $\langle x \leq c \rangle$ or $\langle x \geq c \rangle$. A *meet decomposition function* $decomp : A \to \wp(Irred_\sqcap(A))$ satisfies that $\sqcap decomp(a) = a$ for all a.

Example 2. The element $m = \langle x \geq 0 \rangle$ in Example 1 is meet irreducible. The complement of m in the concrete, $Env \setminus \gamma(m)$, has a precise representation in the abstract as $\sim m = \langle x < 0 \rangle$. Interval environments lack complements but can be decomposed into meet-irreducibles that have complements.

$$decomp(\langle x \mapsto [0, max], y \mapsto [1, 4] \rangle) = \{\langle x \geq 0 \rangle, \langle y \geq 1 \rangle, \langle y < 3 \rangle\}$$

Complementable meet irreducibles are required for CDCL.   ◁

An interval environment a is *precisely complementable* if there is an abstract element $\sim a$ satisfying $\gamma(a) = \wp(Env) \setminus \gamma(\sim a)$. Precise complementation differs from the standard notion of a complement in a lattice. Interval environments have the property that all meet irreducibles are precisely complementable.

3   Static Analysis as Second-Order Constraint Solving

A standard approach to analyse programs is to solve a set of equations generated by a control-flow graph. Program analysis with set-constraints makes the language explicit [8,1]. A constraint language and its satisfiability problem are defined next. Fix an ACFG $P = (Loc, E, stmt)$, a set $CVar = \{X_n | n \in Loc\}$ of *constraint variables* indexed by locations and a concrete domain $\wp(Env)$.

Second-Order Constraints. *Terms* and *constraints* are of the form below, with d ranging over concrete domain elements, and m and n over locations.

terms	$t ::= d \mid X_n \mid post_{(m,n)}(t) \mid t \cup t \mid t \cap t$
constraints	$c ::= X_n \subseteq t \mid X_n \supseteq t \mid X_n \cap t \supset \emptyset$

Let a be an interval environment. We abuse notation and write $X_l \cap a$ for $X_l \cap \gamma(a)$. A *clause* is a disjunction of constraints and a *formula* is a conjunction of clauses. Following standard convention, a clause is represented as a set of constraints and a formula as a set of clauses.

A *valuation* $v : Loc \to \wp(Env)$ maps constraint variables to sets of environments. Valuations form a lattice $(CVals, \sqsubseteq, \sqcup, \sqcap)$. The order, join and meet are lifted pointwise from $\wp(Env)$. That is, $v_1 \sqsubseteq v_2$ if $v_1(l) \sqsubseteq v_2(l)$ for all locations l, and $v_1 \oplus v_2 = \lambda l.\ v_1(l) \oplus v_2(l)$ for $\oplus$ in $\{\sqcap, \sqcup\}$. An *atomic valuation* maps every location to at most one environment. The semantics $[\![t]\!]_v$ of a term t under a valuation v is inductively defined below.

$$[\![d]\!]_v = d \qquad [\![X_n]\!]_v = v(n) \qquad [\![t_1 \cup t_2]\!]_v = [\![t_1]\!]_v \cup [\![t_2]\!]_v$$

$$[\![post_{(m,n)}(t)]\!]_v = post_{(m,n)}([\![t]\!]_v) \qquad [\![t_1 \cap t_2]\!]_v = [\![t_1]\!]_v \cap [\![t_2]\!]_v$$

A valuation v satisfies a constraint $t_1 \bowtie t_2$ if $[\![t_1]\!]_v \bowtie [\![t_2]\!]_v$ holds for $\bowtie$ in $\{\subseteq, \subset\}$. A valuation satisfies a clause if it satisfies at least one constraint in the clause and satisfies a formula if every clause in the formula is satisfied. A valuation satisfying a formula is a *solution*. A formula is *satisfiable* if it has a solution.

3.1   Safety as Satisfiability

The standard approach to checking program safety is to compute an invariant. Formally, an invariant is a solution to the formula $Inv(P)$ below.

$$Inv(P) = X_{init} \supseteq Env \wedge \bigwedge_{n \in Loc} \left\{ X_n \supseteq \bigcup_{(m,n) \in E} post_{(m,n)}(X_m) \right\}$$

The error is unreachable if an invariant also satisfies the formula $X_\notin \subseteq \emptyset$. Standard static analysis for safety can be viewed as a sound but incomplete SAT procedure for the formula $Safe(P) = Inv(P) \wedge X_\notin \subseteq \emptyset$. An alternative we propose is to search for an error – a solution to the formula below.

$$Exec(P) = X_{init} \subseteq Env \wedge \bigwedge_{n \in Loc} \left\{ \bigvee_{(m,n) \in E} X_n \subseteq post_{(m,n)}(X_m) \right\}$$

A program contains an error if a solution to $Exec(P)$ also satisfies $X_\notin \supset \emptyset$. BMC can be viewed as a SAT procedure for the formula $Err(P) = Exec(P) \wedge X_\notin \supset \emptyset$. Solutions to $Safe(P)$ and $Err(P)$ are quite different as demonstrated next.

Example 3. Revisit the ACFG P in Figure 1. An environment ε is written as $(\varepsilon(x), \varepsilon(y))$. The valuation v_1 that maps all locations to Env is an invariant and satisfies $Inv(P)$, as does $v_2 = \{n_1 \mapsto Env, n_2 \mapsto \{(i,j)|j \neq 0\}, n_3 \mapsto \{(i,j)|j = 0\}, n_4 \mapsto \{(i,j)|i \neq 0\}, n_5 \mapsto \{(i,j)|i \neq 0\}, \notin \mapsto \emptyset\}$. Only v_2 satisfies $Safe(P)$ and is strong enough to prove safety.

The condition $X_\notin \supset \emptyset$ prevents v_2 from satisfying $Err(P)$. The constraint $X_\notin \subseteq post_{[x=0]}(X_4)$ is not satisfied by v_1 so neither is $Err(P)$. In fact, $Err(P)$ is unsatisfiable. Let P' be the ACFG with the test $[y = 0]$ modified to $[y \leq 0]$. The valuation $v_3 = \{n_1 \mapsto \{(1, -5), (3, -7)\}, n_2 \mapsto \emptyset, n_3 \mapsto \{(1, -5)\}, n_4 \mapsto \{(0, -5)\}, n_5 \mapsto \{(0, -5)\}, \notin \mapsto \{(0, -5)\}\}$ contains states on an error trace. This valuation does not satisfy $Inv(P')$ or $Safe(P')$ but satisfies $Err(P')$.   ◁

To prove safety of P, we can either find an invariant satisfying $Safe(P)$ or show that $Err(P)$ is unsatisfiable.

Lemma 1. *The following conditions are equivalent for an* ACFG P. *(1) P is safe. (2) $Safe(P)$ is satisfiable. (3) $Err(P)$ is unsatisfiable.*

In propositional SAT solvers, a partial assignment represents a set of potential solutions to a formula. Decisions and constraint propagation refine this set. Over programs, we represent potential sets of errors and use transformer restriction and fixed point iteration to refine the set.

Abstract Valuations. An abstract valuation is figuratively an envelope containing potential solutions to $Err(P)$. An *abstract valuation* maps constraint variables to interval environments. An abstract valuation is *atomic* if it maps each constraint variable to an atom or to $\bot$. The abstract semantics $\|t\|_v$ of a term t with respect to an abstract valuation v is defined as expected, with abstract transformers, join and meet replacing concrete ones. An abstract valuation $\hat{v}$ *abstractly satisfies* a formula if there is a concrete solution v for which the inequality $v(X) \subseteq \gamma \circ \hat{v}(X)$ holds for all constraint variables. If the formula $Err(P)$ cannot be abstractly satisfied, the program is safe.

Example 4. Consider the ACFG $init \xrightarrow{[x<0]} n \xrightarrow{[x=4]} \lightning$ generating the formula below.

$$Err(P) = X_{init} \subseteq Env \wedge X_n \subseteq post_{[x<0]}(X_{init}) \wedge X_{\lightning} \subseteq post_{[x=4]}(X_n)$$

Standard static analysis can be viewed as refining an abstract valuation as below.

$$\hat{v}_0 = \{X_{init} \mapsto \top, X_n \mapsto \top, X_{\lightning} \mapsto \top\}$$
$$\hat{v}_1 = \{X_{init} \mapsto \top, X_n \mapsto \langle x < 0\rangle, X_{\lightning} \mapsto \top\}$$
$$\hat{v}_2 = \{X_{init} \mapsto \top, X_n \mapsto \langle x < 0\rangle, X_{\lightning} \mapsto \bot\}$$

As $X_{\lightning}$ maps to $\bot$, $Err(P)$ is not abstractly satisfied, so $\lightning$ is unreachable. $\lhd$

4   Conflict Driven Fixed Point Learning

We now present CDFL, a procedure that lifts propositional CDCL to abstract domains and program analysis constraints. Example 4 showed that standard static analysis can be viewed as a process that applies transformers to refine an abstract valuation. CDFL extends standard static analysis by using decisions, deduction, learning and backtracking to search the space of abstract valuations. Decisions restrict abstract domain elements, deduction uses transformers and set-constraint clauses and learning infers set-constraint clauses. For simplicity, heuristics for learning, decision making, and backtracking are abstracted away as non-deterministic choices. Common heuristics described in the SAT literature such as first-UIP learning, non-chronological backtracking, restarts and activity-based decision heuristics can be used to resolve this non-determinism.

4.1   Overview of CDFL

The technique is shown in Procedure CDFL. It begins with a formula $Err(P)$ and the abstract valuation $v = \lambda X.\top$. The call deduce() refines this valuation to the result of standard fixed point iteration in an abstract domain. If static analysis shows that the program is safe, our procedure terminates. Otherwise, static analysis was not precise enough and the solver enters the main loop. Thus, CDFL never does extra work if standard static analysis is sufficiently precise.

The current valuation is refined using an interval meet irreducible in the call to decide(). Consequences of this decision are inferred by a call to deduce(). Decisions and deduction alternate until one of two scenarios. If atomic(v) returns true, the valuation v cannot be refined and is returned. Either v contains an error

```
 1  (v, F) ← (λl.⊤, Err(P))
 2  deduce()
 3  if v(↯) = ⊥ then  return safe
 4  while true do
 5  │    if atomic(v) then  return (v, fail)
 6  │    decide()
 7  │    deduce()
 8  │    while (v, F) in conflict do
 9  │    │    learn()
10  │    │    if backtrack() = fail then return safe
11  │    │    deduce(v, F)
```

Procedure CDFL: Conflict Driven Fixed Point Learning

trace or the current abstraction is insufficient to prove safety. The second scenario is a conflict; the valuation v does not abstractly satisfy the formula. Learning is used to generate a reason for the conflict. Technically, learning adds a clause C to the current formula, so that $F \wedge C$ is equi-satisfiable to F. The backtracking step backtrack() then returns the solver to an earlier state that does not conflict with C. If this is not possible, $Err(P)$ is unsatisfiable and P is safe.

4.2   Data Structure and Phases of CDFL

Internally, SAT solvers use a stack to track the sequence of variable assignments of the form (x, v) where x is a propositional variable, and v is either true or false. In our procedure, the stack contains elements of the form (l, a), where l is a location and a is a meet-irreducible or is $\bot$.

A *labelled restriction* (l, a, z) consists of a location l, a meet-irreducible a and the label $z = \mathsf{d}$ if (l, a) is a decision, or $z = \mathsf{i}$ if (l, a) was inferred by deduction. The set of labelled restrictions is $\mathcal{L} = Loc \times (Irred_\sqcap(IEnv) \cup \{\bot\}) \times \{\mathsf{d}, \mathsf{i}\}$. A *stack* is a sequence of labelled restrictions, where the empty stack is ϵ, and UV denotes concatenation. A stack S defines an abstract valuation $\lfloor S \rfloor$ where $\lfloor \epsilon \rfloor = \top$ and $\lfloor S(l, a, z) \rfloor = \lfloor S \rfloor \sqcap \langle l \mapsto a \rangle$. An interval meet-irreducible a at location l *refines the stack*, denoted refines$(S, (l, a))$, if the condition $\lfloor S \rfloor(X_l) \sqcap a \sqsubset \lfloor S \rfloor(X_l)$ holds.

A *solver state* (S, F) consists of a stack of labelled restrictions S and a formula F and the *current valuation* is $\lfloor S \rfloor$. The solver is *in conflict* if some clause in F is not abstractly satisfied by $\lfloor S \rfloor$. We present the components of CDFL as state transitions made by the solver, inspired by the presentation of CDCL in [18].

Deduction. Deduction uses two rules to transform the solver state. The rule tprop applies transformers to abstract valuations, and is comparable to theory propagation in SMT solvers. The rule uprop generalises the unit rule in propositional solvers to set-constraint clauses. If deduction refines the current valuation, the new information is added to the stack. These rules are illustrated below.

Example 5. Consider the formula $Err(P)$ for the ACFG in Figure 1. The initial valuation is $v_0 = \lambda X.\top$. This valuation is refined using the clause $\{X_3 \subseteq post_{[y=0]}(X_1)\}$, and the transformer $\hat{post}_{[y=0]}$ to v_1 that maps X_3 to $\langle y = 0 \rangle$. Next, consider the clause $\{X_4 \subseteq post_{[x:=y+5]}(X_3), X_4 \subseteq post_{[x:=y]}(X_2)\}$. The right side of the first constraint evaluates to $a_1 = \langle y \mapsto [0,0], x \mapsto [5,5]\rangle$, and the second to $a_2 = \top$. One of these constraints must hold, so the weaker condition $X_4 \subseteq \gamma(a_1) \cup \gamma(a_2)$ must as well. But $a_1 \sqcup a_2 = \top$, which does not refine the current valuation of X_4, so the stack is not modified.

To illustrate the unit rule, continue with the valuation obtained above and assume that there is a clause $\{X_1 \cap \langle y < 0 \rangle \supset \emptyset, X_4 \cap \langle x > 10 \rangle \supset \emptyset\}$. The valuation v_1 does not satisfy the first constraint, so every solution must satisfy the second constraint. Every solution satisfying this constraint must also satisfy $X_4 \subseteq \langle x > 10 \rangle$, so the valuation v_1 is refined, mapping X_4 to $\langle x > 10 \rangle$.   ◁

The deduction rules are defined below.

tprop : $(S, F) \to (S(l, a, \mathsf{i}), F)$ if refines$(S, (l, a))$, $\{X_l \subseteq t_1, \ldots, X_l \subseteq t_k\} \in F$

$$\text{where } a \in decomp(\bigsqcup_{1 \leq i \leq k} \|t_i\|_{\lfloor S \rfloor} \sqcap \lfloor S \rfloor(X_l))$$

uprop : $(S, F) \to (S(l, a, \mathsf{i}), F)$ if refines$(S, (l, a))$ and $(\{X_l \cap t \supset \emptyset\} \cup C) \in F$

where C is not abstractly satisfied by $\lfloor S \rfloor$ and

where $a \in decomp(\|t\|_{\lfloor S \rfloor})$

Both of these rules are sound in the sense that if $\lfloor S \rfloor$ contains a solution to F, $\lfloor S(l, a, \mathsf{i}) \rfloor$ will also contain a solution. The function deduce applies these rules exhaustively until the valuation becomes atomic or until the solver is in conflict.

Decisions. A decision picks a location l and program variable x and constrains it with a meet irreducible. Additionally, decisions must be chosen such that they do not put the solver in conflict.

decide : $(S, F) \longrightarrow (S(l, a, \mathsf{d}), F)$ if refines$(S, (l, a))$ and

$(S(l, a, \mathsf{d}), F)$ is not in conflict

Example 6. A valid decision for the CFG P in Figure 1, for the valuation v that maps X_ℓ to $\langle x \mapsto [0,0]\rangle$ and all other locations to $\top$ is $(X_1, \langle y > 0 \rangle, \mathsf{d})$. The restriction $(\ell, \langle x > 0 \rangle, \mathsf{d})$ is *not* a valid decision because $v(X_\ell) \sqcap \langle x > 0 \rangle$ is bottom, causing a conflict.   ◁

Learning and Backtracking. Learning identifies sufficient reasons for conflicts and adds a clause that expresses the negation of that reason. For an abstract valuation v, we define the *clause complement clcomp(v)* as the clause $\{X_l \cap \sim a \supset \emptyset \mid a \in decomp(v(l))\}$. This formula is a complement in the sense that a concrete valuation is a solution of $clcomp(v)$ exactly if it is not contained in the concretisation of v.

Example 7. Let $Loc = \{1, 2\}$ and let v be an abstract valuation with $v(1) = \langle x < 0 \rangle$ and $v(2) = \langle y \mapsto [0, 10] \rangle$. Then $clcomp(v)$ contains the three constraints $X_1 \cap \langle x \geq 0 \rangle \supset \emptyset$, $X_2 \cap \langle y < 0 \rangle \supset \emptyset$ and $X_2 \cap \langle y > 10 \rangle \supset \emptyset\}$.   ◁

Backtracking is used to remove a suffix of the stack to restore the solver to a consistent state after a conflict has been encountered. Backtracking may only jump back to decision elements on the stack. Abstract rules for learning and backtracking can be stated as follows.

$$\text{learn}: \quad (S, F) \to (S, F \wedge clcomp(R)) \quad \text{if } clcomp(R) \notin F \text{ and } \gamma_V(R)$$
$$\text{contains no solutions of } F$$
$$\text{backtrack}: \quad (S_1(l, a, d)S_2, F) \to (S_1, F) \quad \text{if } (S_1, F) \text{ is not in conflict}$$

Soundness and Completeness. We denote the CDFL procedure over the interval domain as CDFL(*IEnv*). The CDFL(*IEnv*) procedure is sound, and, under certain conditions, also complete.

Theorem 1. *If P is a loop-free program and the set of values Val is finite, then* CDFL(*IEnv*) *is a sound and complete decision procedure to check safety of P.*

4.3   Abstract Conflict Graphs

In order to instantiate the learning step, heuristics for finding conflict reasons are needed. Like propositional solvers, we record deductions in a data structure called *conflict graph*, which is incrementally built by recording decisions and deductions. The nodes of the conflict graph are labelled restrictions. An example is provided in Figure 1. The two nodes without predecessors are decision nodes, all other nodes are implication nodes. The predecessors of each node n in the graph are sufficient to deduce n. Once a conflict is reached, the graph is analysed to determine a sufficient reason for unsatisfiability. A *cut* of a conflict graph (R, I) is a set $L \subseteq R$ such that any path from a decision node to the conflict node goes through L. Every cut of a conflict graph provides a conflict reason that can be used in learning. In contrast to propositional SAT, we can obtain stronger learnt clauses by generalising the nodes of the implication graph itself before obtaining a cut. Generalisation is performed by computing maximal sufficient pre-conditions in the domain of intervals, which we handle in our implementation using binary search on bounds.

Example 8. Consider the conflict graph in Figure 1. The following two sets are different cuts of the graph, and hence sufficient reasons for a conflict, $R_1 = \{n_5 : x < -123\}$, $R_2 = \{n_2 : y < -123, n_3 : \bot\}$. In learning, these cuts produce the following clauses, $C_1 = \{X_5 \cap \langle x \geq -123 \rangle \supset \emptyset\}$, $C_2 = \{X_2 \cap \langle y \geq -123 \rangle \supset \emptyset, X_3 \supset \emptyset\}$. Stronger clauses can be obtained by applying generalisation first. Consider the node $n_5 : x < -123$ in the conflict graph of Figure 1. This node is used to deduce $\frac{l}{2} : \bot$ using the conditional statement $s = [x = 0]$. The weakest pre-condition of $\frac{l}{2} : \bot$ w.r.t. s is $x < 0 \vee x > 0$, but this is not expressible as an interval element. Instead, we choose the maximal generalisation of $x < -123$ that is sufficient to prove $\frac{l}{l} : \bot$, and obtain $x < 0$. Cutting now yields the stronger clause $\{X_5 \cap \langle x \geq 0 \rangle \supset \emptyset\}$. $\lhd$

5   Implementation and Experiments

We have implemented CDFL for ANSI-C programs. The domains used are intervals over IEEE 754 floating-point numbers and machine integers. This section will show that our approach is able to efficiently prove correctness of several programs where a standard interval analysis yields a false alarm. In case our procedure fails to prove correctness it returns a concrete environment at the initial control flow node to constants. This assignment either leads to an error, or helps localise the imprecision of the abstract analysis by providing a maximal restriction that cannot be proved correct using intervals. We apply our analysis to verify properties on floating-point programs from various sources, and show that, in many cases, our analysis is as efficient as static analysis, but provides the precision of a floating-point decision procedure.

We compare our tool to the static analyser Astrée [4], which uses interval analysis, and to the bounded model checker CBMC [6], which uses a bit-precise floating-point decision procedure based on propositional encoding. Our benchmarks show highly non-linear behaviour. Astrée is not optimised for the kinds of programs we consider and introduces a high degree of imprecision. (Astrée offers simple trace partitioning heuristics for Booleans and machine integers, but not floating-point programs.) CBMC translates the floating-point arithmetic to large propositional circuits which are hard for SAT solvers. As benchmarks we use ANSI-C code originating from (a) controller code auto-generated from a Simulink model with varying loop bounds; (b) examples from the Functional Equivalence Verification Suite [21]; (c) benchmarks presented at the 2010 Workshop on Numerical Software Verification; (d) code presented by Goubault and Putot [9]; (e) hand-crafted instances that implement Taylor expansions of sine and square functions, as well as Newton-Raphson approximation. In order to allow comparison to bounded model checking, only benchmark programs with bounded loops were chosen, which were completely unrolled prior to analysis. All our 57 benchmarks, more detailed benchmark results, together with the prototype tool, are available online[1].

We discuss the following results: (1) our analysis is as precise as a full floating-point decision procedure while still being orders of magnitudes faster; (2) learning and the choice of decision heuristic yield a speed-up of more than an order of magnitude; (3) dynamic precision adjustment is observed frequently.

Efficient and Precise Analysis. In Figure 2, we show execution times for Astrée, CBMC, and our analysis (CDFL). To highlight wrong verification results or out-of-memory errors, the time for such failures was set to the timeout of

---

[1] http://www.cprover.org/cdfpl/

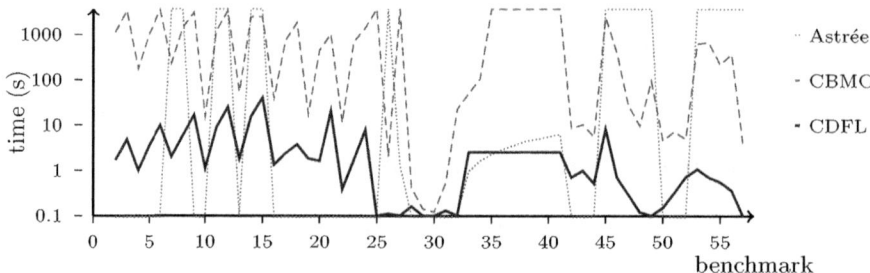

Fig. 2. Execution times of Astrée, CBMC, and CDFL; wrong results set to 3600s

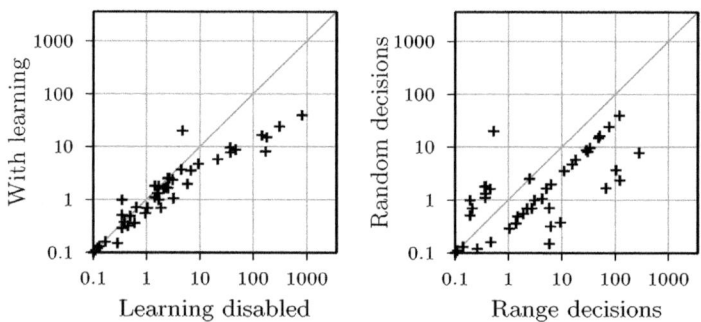

Fig. 3. Effects of learning and decision heuristics

3600 seconds. We make several observations: on average, our analysis is at least 264 times faster than CBMC. The figure 264 is a lower bound, since some runs of CBMC were aborted due to timeouts or errors. The maximum speed-up is a factor of 1595. Although Astrée is often faster than our prototype, its precision is insufficient in many cases – we obtained 16 false alerts for the 33 safe benchmarks.

Decision Heuristics and Learning. Figure 3 visualises the effects of learning and decision heuristics. Learning has a significant influence on runtime, as does the choice of a decision heuristic. We compare a random heuristic, which picks a restriction over a random variable, with a range-based one, which always aims to restrict the least restricted variable. Random decision making outperforms range-based. Activity-based heuristics common in SAT may work as well in our case.

Dynamic Precision Adjustment. One unique feature of our procedure is property-dependent refinement. The precision of the analysis dynamically adapts to match the precision required by the property. This is illustrated in Figure 4 where we check bounds on the result of computing a sine approximation under the input range $[-\frac{\pi}{2}, \frac{\pi}{2}]$. The input value is shown on the x-axis, the result of

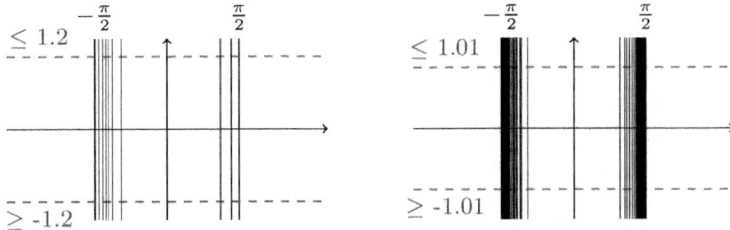

Fig. 4. Partitions explored during bounds check

the computation on the y-axis. The bound we check against is depicted as two dashed horizontal lines, boundaries of explored partitions are shown as black vertical lines. The actual maximum of the function lies at about 1.00921. As the checked bound (Figure 4 shows bounds 1.2 and 1.01) approaches this value, our procedure dynamically increases the precision of the analysis by exploring more partitions.

Limitations of CDFL(*IEnv*). Our procedure is instantiated over the domain of intervals. There are simple programs that are not amenable to interval analysis, even when additional partition-based refinement is used. Consider for example the one-line program $x := y$ together with the relational property $x = y$. Intervals are non-relational, hence CDFL(*IEnv*) would enumerate all singleton intervals over y. Similar enumeration behaviour can be found in propositional SAT solvers, which may perform badly when applied to certain, highly relational problems. This can be fixed by instantiating CDCL(A) using a richer base domain. Further, our implementation is a prototype and restricts learning to the initial control flow node, which limits performance on deep programs.

6   Conclusion

We presented a novel approach for bounds analysis that instantiates a CDCL architecture over abstract domains. In the absence of loops and for finite value domains we obtain a sound and complete analysis. Our prototype implementation witnesses the potential of this approach: our analysis is substantially more precise than a state-of-the-art static analyser and outperforms a SAT based IEEE 754 floating-point decision procedure by several orders of magnitude on small, non-linear programs.

Much research in program analysis attempts to leverage the efficiency of SAT solvers. The efficiency is the result of an intensive, community effort to discover efficient engineering techniques, and decision and learning heuristics in modern solvers. This paper has demonstrated how to lift the architecture of a modern rsat solver to program analyzers. Similarly, our approach could benefit greatly by studying heuristics and efficient engineering techniques.

The formal framework in this paper is by no means limited to bounds analysis with intervals. A number of domains, numeric or otherwise, have the complementation properties necessary for instantiations of CDFL. Examples are given by numeric domains such as octagons and polyhedra, or by equality domains. Non-numeric examples include some pointer analyses, or trace-based abstractions, for example those based on control-flow unwindings, where decisions would correspond to refinements of control structure imprecision. The instantiation of CDFL with other domains is the focus of our current work. We believe that extending our technique to new domains can yield a new class of general purpose verification tools that dynamically combine the efficiency provided by abstraction with the precision of a SAT solver.

Acknowledgments. We thank Antoine Miné for providing experimental results using Astrée. One anonymous reviewer summarised our paper better than we did and another pointed out technical loose ends.

References

1. Aiken, A.: Introduction to set constraint-based program analysis. Science of Computer Programming 35, 79–111 (1999)
2. Balakrishnan, G., Sankaranarayanan, S., Ivančić, F., Gupta, A.: Refining the control structure of loops using static analysis. In: Proc. of the Intl. Conf. on Embedded Software, pp. 49–58. ACM Press (2009)
3. Beckman, N.E., Nori, A.V., Rajamani, S.K., Simmons, R.J.: Proofs from tests. In: Proc. of Software Testing and Analysis, pp. 3–14. ACM Press (2008)
4. Blanchet, B., Cousot, P., Cousot, R., Feret, J., Mauborgne, L., Miné, A., Monniaux, D., Rival, X.: Design and Implementation of a Special-Purpose Static Program Analyzer for Safety-Critical Real-Time Embedded Software. In: Mogensen, T.Æ., Schmidt, D.A., Sudborough, I.H. (eds.) The Essence of Computation. LNCS, vol. 2566, pp. 85–108. Springer, Heidelberg (2002)
5. Clarke, E., Grumberg, O., Jha, S., Lu, Y., Veith, H.: Counterexample-Guided Abstraction Refinement. In: Emerson, E.A., Sistla, A.P. (eds.) CAV 2000. LNCS, vol. 1855, pp. 154–169. Springer, Heidelberg (2000)
6. Clarke, E., Kroning, D., Lerda, F.: A Tool for Checking ANSI-C Programs. In: Jensen, K., Podelski, A. (eds.) TACAS 2004. LNCS, vol. 2988, pp. 168–176. Springer, Heidelberg (2004)
7. Cotton, S.: Natural Domain SMT: A Preliminary Assessment. In: Chatterjee, K., Henzinger, T.A. (eds.) FORMATS 2010. LNCS, vol. 6246, pp. 77–91. Springer, Heidelberg (2010)
8. Cousot, P., Cousot, R.: Formal language, grammar and set-constraint-based program analysis by abstract interpretation. In: Proc. of Functional Programming Languages and Computer Architecture, pp. 170–181. ACM Press (June 1995)
9. Goubault, É., Putot, S.: Static Analysis of Numerical Algorithms. In: Yi, K. (ed.) SAS 2006. LNCS, vol. 4134, pp. 18–34. Springer, Heidelberg (2006)
10. Gulavani, B.S., Chakraborty, S., Nori, A.V., Rajamani, S.K.: Automatically Refining Abstract Interpretations. In: Ramakrishnan, C.R., Rehof, J. (eds.) TACAS 2008. LNCS, vol. 4963, pp. 443–458. Springer, Heidelberg (2008)

11. Gulavani, B.S., Rajamani, S.K.: Counterexample Driven Refinement for Abstract Interpretation. In: Hermanns, H. (ed.) TACAS 2006. LNCS, vol. 3920, pp. 474–488. Springer, Heidelberg (2006)
12. Gulwani, B.S., Jain, S., Koskinen, E.: Control-flow refinement and progress invariants for bound analysis. In: Proc. of Programming Language Design and Implementation, pp. 375–385. ACM Press (June 2009)
13. Harris, W.R., Sankaranarayanan, S., Ivančić, F., Gupta, A.: Program analysis via satisfiability modulo path programs. In: Proc. of Principles of Programming Languages, pp. 71–82. ACM Press (2010)
14. Jovanović, D., de Moura, L.: Cutting to the Chase Solving Linear Integer Arithmetic. In: Bjørner, N., Sofronie-Stokkermans, V. (eds.) CADE 2011. LNCS, vol. 6803, pp. 338–353. Springer, Heidelberg (2011)
15. Korovin, K., Tsiskaridze, N., Voronkov, A.: Conflict Resolution. In: Gent, I.P. (ed.) CP 2009. LNCS, vol. 5732, pp. 509–523. Springer, Heidelberg (2009)
16. McMillan, K.L.: Lazy Annotation for Program Testing and Verification. In: Touili, T., Cook, B., Jackson, P. (eds.) CAV 2010. LNCS, vol. 6174, pp. 104–118. Springer, Heidelberg (2010)
17. McMillan, K.L., Kuehlmann, A., Sagiv, M.: Generalizing DPLL to Richer Logics. In: Bouajjani, A., Maler, O. (eds.) CAV 2009. LNCS, vol. 5643, pp. 462–476. Springer, Heidelberg (2009)
18. Nieuwenhuis, R., Oliveras, A., Tinelli, C.: Solving SAT and SAT modulo theories: From an abstract Davis–Putnam–Logemann–Loveland procedure to DPLL(T). Journal of the ACM 53(6), 937–977 (2006)
19. Rival, X., Mauborgne, L.: The trace partitioning abstract domain. ACM Transactions on Programming Languages and Systems 29(5), 26 (2007)
20. Sankaranarayanan, S., Ivančić, F., Shlyakhter, I., Gupta, A.: Static Analysis in Disjunctive Numerical Domains. In: Yi, K. (ed.) SAS 2006. LNCS, vol. 4134, pp. 3–17. Springer, Heidelberg (2006)
21. Siegel, S.F., Zirkel, T.K.: A functional equivalence verification suite for high-performance scientific computing. Technical Report UDEL-CIS-2011/02, Department of Computer and Information Sciences, University of Delaware (2011)
22. Simon, A., King, A.: Widening polyhedra with landmarks. In: Proc. of the Asian Symposium on Programming Languages and Systems, pp. 166–182 (2006)
23. Wang, C., Yang, Z., Gupta, A., Ivančić, F.: Using Counterexamples for Improving the Precision of Reachability Computation with Polyhedra. In: Damm, W., Hermanns, H. (eds.) CAV 2007. LNCS, vol. 4590, pp. 352–365. Springer, Heidelberg (2007)

Ramsey-Based Analysis of Parity Automata

Oliver Friedmann[1] and Martin Lange[2]

[1] Dept. of Computer Science, Ludwig-Maximilians-University of Munich, Germany
[2] School of Electr. Eng. and Computer Science, University of Kassel, Germany

Abstract. Parity automata are a generalisation of Büchi automata that have some interesting advantages over the latter, e.g. determinisability, succinctness and the ability to express certain acceptance conditions like the intersection of a Büchi and a co-Büchi condition directly as a parity condition. Decision problems like universality and inclusion for such automata are PSPACE-complete and have originally been tackled via explicit complementation only. Ramsey-based methods are a later development that avoids explicit complementation but relies on an application of Ramsey's Theorem for its correctness. In this paper we develop new and explicit Ramsey-based algorithms for the universality and inclusion problem for nondeterministic parity automata. We compare them to Ramsey-based algorithms which are obtained from translating parity automata into Büchi automata first and then applying the known Ramsey-based analysis procedures to the resulting automata. We show that the speed-up in the asymptotic worst-case gained through the new and direct methods is exponential in the number of priorities in the parity automata. We also show that the new algorithms are much more efficient in practice.

1 Introduction

Nondeterministic Büchi automata (NBA) are the most well-known type of finite automata that work on infinite words. Much of their popularity is owed to two facts. (1) Their acceptance condition is conceptually very simple: a run is accepting iff it visits a certain subset of states infinitely often. (2) Despite this simplicity they form an expressively complete specification formalism with respect to Monadic Second-Order Logic [4], i.e. they accept exactly the regular languages of infinite words.

A lot of attention has been paid to the algorithmic treatment of fundamental decision and computation problems for regular languages represented by NBA. The complementation problem is combinatorially much more difficult than that for NFA. The fundamental difference is the fact that determinisation for NBA is provably impossible, and particularly a simple procedure like the powerset construction for NFA fails for finite automata on infinite words equipped with any reasonable acceptance condition, not just the Büchi condition. This has brought out numerous work on the complementation (and also determinisation) problem for NBA [4,13,8,14,9].

C. Flanagan and B. König (Eds.): TACAS 2012, LNCS 7214, pp. 64–78, 2012.

Clearly, other problems that generalise complementation in some way – e.g. universality, inclusion, equivalence – are also combinatorially challenging. For instance, in order to check whether $L(\mathcal{A}) \subseteq L(\mathcal{B})$ holds for two NBA $\mathcal{A}$ und $\mathcal{B}$, one would complement $\mathcal{B}$ to some $\overline{\mathcal{B}}$, build an automaton that accepts its language intersected with $L(\mathcal{A})$ – which is relatively simple using a small enhancement of the usual product construction for NFA – and check the result for emptiness. In the most complex steps of this procedure, one can choose among several complementation procedures. This choice was often made on the basis of worst-case analysis or modern aspects like symbolic implementability. For instance, Klarlund's procedure [8] runs in optimal time of $2^{\mathcal{O}(n \log n)}$ while Kupferman and Vardi's [9] can be made to work with BDDs at the expense of running in time $2^{\mathcal{O}(n^2)}$. Büchi's procedure [4] was not seen as practical which may also be caused by the fact that the literature falsely accused it of having doubly exponential running time whereas a careful analysis shows that it can be made to run in time $2^{\mathcal{O}(n^2)}$ as well.

Büchi's proof of correctness for his complementation procedure uses Ramsey's Theorem [12]. For a long time this has been regarded as a tool to handle the combinatorial difficulty in the correctness proof without much algorithmic value for the complementation problem, hence the focus on other procedures for practical applications. It was Ben-Amram, Jones and Lee [10] who first suggested to use this principle for a practical application in termination analysis called size-change termination which could have also been solved using Büchi complementation. They state that "for practical usage [. . .] the simple algorithm seems more promising than [. . .] the solution based on ω-automata". The term "simple algorithm" refers to a procedure that builds a set of finite graphs through a composition operation and searches for an idempotent graph with certain properties in it. This is basically a direct usage of the computational content of Ramsey's Theorem for this particular decision problem. Henceforth, such simple algorithms will be said to be Ramsey-based.

Next, Dax et al. [5] introduced this Ramsey-based method to the domain of temporal logic: they gave an algorithm checking validity for a formula of the linear-time μ-calculus (μ-TL) [3], a temporal fixpoint logic extending the standard LTL [11]. These problems had – until then – solely been approached using automata-theoretic machinery, i.e. explicitly using the complementation problem for NBA [16]. Dax et al. showed that the Ramsey-based method can outperform those using automata explictly. Since μ-TL is also expressively complete for regular languages, and there is a linear translation from NBA to μ-TL – mapping universality to validity – this also defines a Ramsey-based method for the universality problem for NBA. After that, Fogarty and Vardi have made this connection explicit and also investigated its practical use for the NBA universality problem in general [6,7]. The apparent use has then inspired work on further optimisations of this Ramsey-based approach to NBA universality and NBA inclusion [1].

Büchi automata are the simplest but not the only type of automaton on infinite words. Notably, the literature also considers the syntactically more general

Muller, Rabin, Streett and parity automata, all of them expressively complete w.r.t. ω-regular languages. Here we consider nondeterministic parity automata (NPA) which are computationally most elegant among those models. There are several reasons for considering NPA as a generalisation of NBA.

1. *Succinctness.* Many properties can be expressed more succinctly with NPA than with NBA. Consider, for instance the language L_0 of all words over the alphabet $\{a, b, c\}$ that also contains infinitely many b's when they contain infinitely many a's. This can be accepted by an NPA with three states in which they signal the last letter that has been seen. A b is then signaled with priority 2, an a with priority 1 and a c with priority 0. A similarly straightforward construction of an NBA for this language results in 5 states, and it does not look like a 3-state NBA for this language exists.
2. *Determinisability.* L_0 can be accepted by a deterministic parity automaton (DPA) but not by a deterministic Büchi automaton. In general, DPA are expressively complete whilst deterministic DBA are not.
3. *Expressiveness.* Certain acceptance conditions can be formulated as a parity condition but not as a Büchi condition, i.e. certain automata can be regarded as an NPA but not as an NBA.[1] For instance, the intersection of a Büchi condition with a co-Büchi condition stating that certain states should be seen infinitely often while others should only be seen finitely often, can easily be encoded by a parity condition with priorities $\{1, 2, 3\}$, compare this to L_0 above.

In this paper we develop Ramsey-based algorithms for NPA. These extend corresponding methods for NBA. The benefit of this extension is empirically shown: it is known that NPA can be translated to NBA at a moderate blow-up. We compare the new methods for NPA with the old methods for NBA obtained under this translation showing that the new methods are not only asymptotically faster but also behave better in practice. Furthermore, there is a reduction from the inclusion problem to universality that can be made to work for various types of automata including NBA and NPA. We show that this does not alleviate the use of a direct method for NPA inclusion: performing the reduction to universality and then applying the universality method for NPA is again asymptotically and practically worse. In essence, the Ramsey-based methods developed in this paper are justified by their superiority over reductions to existing methods both in theory and in practice.

2   Preliminaries

As usual, for a finite alphabet Σ we write $\Sigma^* / \Sigma^+ / \Sigma^\omega$ to denote the set of all finite / finite non-empty / infinite words over Σ. An infinite word $w \in \Sigma^\omega$ is *regular* if there are $u \in \Sigma^*$ and $v \in \Sigma^+$ s.t. $w = uv^\omega$. If w is a finite word then $|w|$ denotes its length.

---

[1] Note that their *language* is still NBA-recognisable, but this may require a different underlying automaton.

A *nondeterministic parity automaton* (NPA) is a $\mathcal{A} = (Q, \Sigma, q_0, \delta, \Omega)$ where Q is a finite set of states, Σ is a finite alphabet, $q_0 \in Q$ is a designated starting state, $\delta \subseteq Q \times \Sigma \times Q$ is the transition relation, and $\Omega : Q \to \mathbb{N}$ is the priority function.

A *run* of $\mathcal{A}$ on a word $w = a_0 a_1 a_2 \ldots \in \Sigma^\omega$ is a sequence $\rho = q_0, q_1, \ldots$ s.t. q_0 is the designated starting state and for all $i \in \mathbb{N}$ we have: $(q_i, a_i, q_{i+1}) \in \delta$. It is *accepting* if $\max\{\Omega(q) \mid \forall i \in \mathbb{N}. \exists j \geq 1. q_j = q\}$ is even. The *language* of $\mathcal{A}$ is $L(\mathcal{A}) = \{w \mid \text{there is an accepting run of } \mathcal{A} \text{ on } w\}$.

We introduce two important complexity measures for an NPA. The *size* of $\mathcal{A}$ is $|\mathcal{A}| = |Q|$. The *index* of $\mathcal{A}$ is $idx(\mathcal{A}) = |\{\Omega(q) \mid q \in Q\}|$, i.e. the number of distinct priorities used in $\mathcal{A}$. Clearly we always have $idx(\mathcal{A}) \leq |\mathcal{A}|$.

NPA are a natural generalisation of the well-known nondeterministic Büchi automata which are traditionally defined using the concept of acceptance state rather than a priority function. An accepting run is one that visists the acceptance set infinitely often. The definition used in the following is easily seen to be equivalent to that. A *nondeterministic Büchi automaton* (NBA) is a special kind of an NPA of index 2, s.t. $\Omega(q) \in \{1, 2\}$ for all $q \in Q$.

An *ω-regular language* or just *regular language* for short is a language that can be accepted by an NBA. It is known that NPA, despite being a generalisation, do not accept more than the regular languages.

Proposition 1. *For every NPA $\mathcal{A}$ of size n and index k there is an NBA $\mathcal{B}$ of size $\leq n \cdot c$ s.t. $L(\mathcal{A}) = L(\mathcal{B})$ where $c = \frac{k}{2} + 1$ if k is even and $c = \lceil \frac{k}{2} \rceil$ if k is odd.*

We quickly sketch the idea behind this construction because it is used in the comparison of the direct Ramsey-based methods for NPA with those for NBA. It is based on the fact that a run $\rho = q_0, q_1, \ldots$ of an NPA on a word is accepting iff there is an $i \in \mathbb{N}$ and an even priority p s.t. for all $j \geq i$ we have $\Omega(q_j) \leq p$, and $\Omega(q_j) = p$ for infinitely many j. Thus, the required NBA can be constructed as follows. It contains a copy of $\mathcal{A}$ with the starting state and no final states. It also has, for every even priority p, a copy of $\mathcal{A}$ in which only states with priorities at most p are preserved, and those with priority p are final. The transitions in each copy are as they are in $\mathcal{A}$. Also, there are transitions from every state in the original copy to its successors in the additional copies if they exist. This way, the NBA can mimick an accepting run of the NPA by staying in the original non-final copy until no greater priorities than the one causing acceptance are seen, and then it changes into the respective copy verifying that this priority is being seen infinitely often, and no greater one is being seen anymore.

We are particularly interested in the following decision problems for NPA.

- UNIV^P: Given an NPA $\mathcal{A}$, decide whether or not $L(\mathcal{A}) = \Sigma^\omega$.
- INCL^P: Given NPA $\mathcal{A}$ and $\mathcal{B}$, decide whether or not $L(\mathcal{A}) \subseteq L(\mathcal{B})$.

The complexity of these problems for NBA is well known; they are PSPACE-complete [14]. Together with Prop. 1 and the fact that every NBA is an NPA we immediately obtain the following.

Proposition 2. UNIV^P *and* INCL^P *are PSPACE-complete.*

3   The Ramsey-Based Method for Parity Automata

In this section we describe how to decide universality and inclusion for NPA directly using a Ramsey-based method. We compare the results with the obvious method of translating NPA into NBA first and then using the Ramsey-based methods for NBA. We focus on universality first; inclusion proves to be just a small extension of this. The completeness proofs rely on the following theorem. For any linear order $(A, <)$ let $A^2_< := \{(a, b) \mid a, b \in A, a < b\}$.

Theorem 1 (Ramsey, 1928). *Let F be a finite set and $c : \mathbb{N}^2_< \to F$. Then there is an $M \subseteq \mathbb{N}$ and an $f \in F$ such that $|M| = \infty$ and $c(i, j) = f$ for all $i, j \in M$ with $i < j$.*

3.1   Universality for NPA

For the remainder of this section fix an NPA $\mathcal{A} = (Q, \Sigma, q_0, \delta, \Omega)$ as well as $n := |\mathcal{A}|$ and $k := idx(\mathcal{A})$. Let $P = \{\Omega(q) \mid q \in Q\}$ be the set of all $\mathcal{A}$-priorities.

Words as Partial Functions from State Pairs to Priorities. We will use two total orders on the extension of $\mathbb{N}$ by one element $\dagger$. The first one is denoted $\leq$ and is the ordinary total order of type $\omega + 1$. Thus, we have $0 < 1 < \ldots < \dagger$. The *reward ordering* $\preceq$ is defined by $\dagger \ldots 3 \prec 1 \prec 0 \prec 2 \prec 4 \prec \ldots$ This reward ordering reflects the intuition of how valuable a priority of an NPA's state is for acceptance: even priorities are generally better than odd ones, and the bigger an even one the better, while small odd priorities are better than bigger ones because it is easier to subsume them in a run with an even priority elsewhere. Note that $\dagger$ is maximal for $\leq$ but minimal for $\preceq$.

Definition 1. A *box*[2] is a partial function of type $Q \times Q \dashrightarrow P$. We will sometimes write $f(q, q') = \dagger$ to denote that the value of the box f on the argument pair (q, q') is undefined.

Let f, g be two boxes. Its *composition* $f \circ g$ is the box defined by

$$(f \circ g)(q, q') := \max_{\preceq}\{\max_{\leq}\{f(q, q''), g(q'', q')\} \mid q'' \in Q\}$$

Note that the maxima are taken with respect to the two different total orders.

We will associate with every finite word $w \in \Sigma^*$ a box $[w] : Q \times Q \dashrightarrow P$ by induction on the length of w. The base cases for words of length 0 and 1 are the following.

$$[\epsilon](q, q') = \begin{cases} \Omega(q) & , \text{ if } q = q' \\ \dagger & , \text{ otherwise} \end{cases}$$

$$[a](q, q') = \begin{cases} \max_{\leq}\{\Omega(q), \Omega(q')\} & , \text{ if } q' \in \delta(q, a) \\ \dagger & , \text{ otherwise} \end{cases}$$

---

[2] See their graphical representation in Fig. 1 for an idea about the choice of this name.

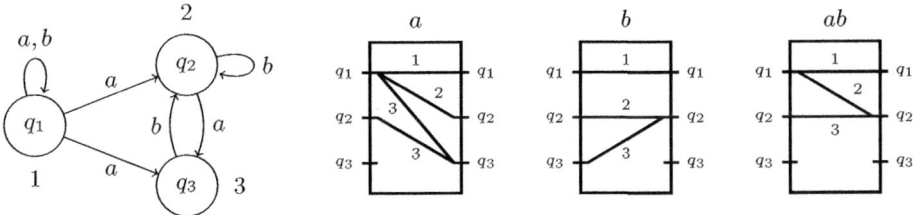

Fig. 1. An NPA with three of its boxes: $[a]$, $[b]$, and $[ab]$

We then use box composition in order to lift this to arbitrary words: $[av] := [a] \circ [v]$.

Associativity of the composition operation is not hard to establish.

Lemma 1. *For all boxes f, g, h we have $f \circ (g \circ h) = (f \circ g) \circ h$.*

With $\mathfrak{B}_A$ we denote the set of all boxes defined by any word w.r.t. the NPA A: $\mathfrak{B}_A = \{[w] \mid w \in \Sigma^*\}$. It will be used as a search space in order to decide universality of A.

Example 1. Boxes and the concept of composition can be visualised greatly. One regards the states in Q as in- and out-ports of a connector. There can be connections between in-ports and out-ports, and these connections are labeled with a priority.

Fig. 1 shows an NPA with three states having the priorities $1, 2, 3$. It also shows the boxes $[a]$ and $[b]$ w.r.t. to this NPA as well as their composition $[ab]$ which, intuitively, is obtained by merging the out-ports of $[a]$ with the in-ports of $[b]$. The priority of a merged connection is the maximum w.r.t. $\leq$ of the two connections that are being merged. Note that this may result in more than one connection, for instance between q_1 and q_2 which can go via q_2 or q_3. In the composition, only the one with the maximum w.r.t. $\preceq$ survives, i.e. the one with priority 2 rather than 3.

Definition 2. A box $[w]$ is *idempotent* if $[w] \circ [w] = [w]$, where this equality is to be understood as equality of partial functions.

Let $[w]$ be a box and $q \in Q$. We write $q[w]$ for the set of all q' that are connected to q in this box, i.e. $q[w] = \{q' \mid [w](q, q') \neq \dagger\}$. Furthermore for some $Q' \subseteq Q$ let $Q'[w] := \bigcup_{q' \in Q'} q'[w]$.

A box $[w]$ is called *bad* w.r.t. some $Q' \subseteq Q$ if for all $q \in Q'[w]$ we have: $[w](q, q)$ is either undefined or odd. A *good* box is a box that is not bad. In other words, in a bad box w.r.t. some Q' one considers first all connections from any input q' to any output q, and then all the connections from such a q to itself. We will consider bad boxes only in the context of idempotent boxes. Hence, this considers all connections in an infinite iteration of $[w]$ that are reachable from some $q' \in Q'$.

Example 2. It is easy to verify that $[ab]$ in Fig. 1 is idempotent. Note that any run in the NPA of Fig. 1 under the word $abab$ is also possible under ab already, which is essentially what idempotency of $[ab]$ means.

Furthermore, $[ab]$ is bad for any subset of $\{q_1, q_2, q_3\}$, since following connections from any of them in $[ab]$ can only lead to a subset of $\{q_1, q_2\}$, and both connections from such a state to itself are labeled with an odd priority. Note that the word $(ab)^\omega$ cannot be accepted from any state in the NPA in Fig. 1 which is essentially what badness and idempotency of $[ab]$ means.

In order to prove correctness of the universality check to be presented, we need to relate boxes to runs. We write $q \xrightarrow{w}_p q'$ if state q' is reachable from q in the transition graph of $\mathcal{A}$ on a path whose labels compose to w s.t. the highest (w.r.t. $\leq$) priority seen on this path is p. A proof of the following lemma is given in the appendix.

Lemma 2. *Let $w \in \Sigma^*$, $q, q' \in Q$, and $p \in P$. If $[w](q, q') = p$ then $q \xrightarrow{w}_p q'$.*

The converse direction is not true. Suppose that $q \xrightarrow{w}_p q'$ holds. Then we need not necessarily have $[w](q, q') = p$. simply because there may be different paths from q to q' in $\mathcal{A}$ under w, and p may only be the maximal priority on one of them. However, $[w]$ only stores one maximal priority over all such paths, namely the greatest one w.r.t. $\preceq$. Thus, we have a statement that is weaker than the converse of Lemma 2 but still sufficient to prove correctness of the search procedures in the next section. Its proof is also given in the appendix.

Lemma 3. *Let $w \in \Sigma^*$, $q, q' \in Q$, and $p \in P$. If $q \xrightarrow{w}_p q'$ then there is a p' s.t. $p \preceq p'$ and $[w](q, q') = p'$.*

Non-Universality Via Relation Testing. The following characterises (non)-universality of an NPA $\mathcal{A}$ in terms of the elements of $\mathfrak{B}_\mathcal{A}$.

Theorem 2. *$L(\mathcal{A}) \neq \Sigma^\omega$ iff there are boxes $[u]$ and $[v]$ s.t. $[v]$ is idempotent and bad w.r.t. $q_0[u]$.*

Proof. "$\Leftarrow$" Suppose that $[v]$ is idempotent and bad w.r.t. $q_0[u]$. We claim that $uv^\omega \notin L(\mathcal{A})$. For the sake of contradiction assume that $uv^\omega \in L(\mathcal{A})$. Let $q_0, q_1, \ldots$ be an accepting run of $\mathcal{A}$ on uv^ω. Let $l = |u|$ and $k = |v|$. Clearly, we have $q_0 \xrightarrow{u}_p q_l$ for some $p \in P$. According to Lemma 3 we also have $[u](q_0, q_l) = p'$ for some $p' \in P$ with $p \preceq p'$.

Since there are only finitely many states, there must be some state q that appears infinitely often in the sequence $q_l, q_{l+k}, q_{l+2k}, \ldots$. Let $i_0, i_1, \ldots$ denote the infinite ascending sequence of indices s.t. $q_{l+i_j k} = q$ for all j. Let $p_j = \max_{\leq}\{\Omega(q_{l+i_j k}), \ldots, \Omega(q_{l+i_{j+1} k})\}$. By assumption the run is accepting, hence, infinitely many p_j must be even.

Let now j be arbitrary s.t. p_j is even. It follows that $q \xrightarrow{v^{i_{j+1}-i_j}}_{p_j} q$, and, by Lemma 3, that $[v^{i_{j+1}-i_j}](q, q) = p$ for some $p \succeq p_j$. Since p_j is even, p must be even, too. Hence, $[v^{i_{j+1}-i_j}]$ is good for q. But since $[v]$ is idempotent, it follows that $[v^{i_{j+1}-i_j}] = [v]$, and this contradicts the assumption that $[v]$ is bad w.r.t. $q_0[u]$.

"$\Rightarrow$" Suppose that $L(\mathcal{A}) \neq \Sigma^\omega$, i.e. there is a word $w = a_0 a_1 \ldots \notin L(\mathcal{A})$. Consider the following colouring $c : \mathbb{N}^2 \to \mathfrak{B}_\mathcal{A}$ where $\mathbb{N}^2 := \{(i,j) \mid i \in \mathbb{N}, j \in \mathbb{N}, i < j\}$, defined by $c(i,j) = [a_i \ldots a_{j-1}]$. Since $|\mathfrak{B}_\mathcal{A}| < \infty$, Thm. 1 yields an infinite sequence $i_0, i_1, \ldots$ of indices and an $f \in \mathfrak{B}_\mathcal{A}$ s.t. $c(i_j, i_h) = f$ for all j, h with $j < h$.

First define $u := a_0 \ldots a_{i_0-1}$. According to Lemma 2, for every $q \in q_0[u]$ we have $q_0 \xrightarrow{u}_p q$ for some p. Next, note that f is idempotent because

$$f \circ f = c(i_0, i_1) \circ c(i_1, i_2) = [a_{i_0} \ldots a_{i_1-1}] \circ [a_{i_1} \ldots a_{i_2-1}] = [a_{i_0} \ldots a_{i_2-1}]$$
$$= c(i_0, i_2) = f$$

according to Lemma 1. Then define $v_j := a_{i_j} \ldots a_{i_{j+1}-1}$ for every $j \in \mathbb{N}$. Note that $w = u v_0 v_1 v_2 \ldots$, and that $f = [v_j]$ for every $j \in \mathbb{N}$.

It remains to be seen that f is bad w.r.t. $q_0[u]$. Suppose that this was not the case, i.e. it was good. Then there would be a $q \in q'[v]$ for some $q' \in q_0[u]$ s.t. $p = [v](q, q)$ is even. According to Lemma 2 we would have $q_0 \xrightarrow{u}_{p'} q'$ for some p', $q' \xrightarrow{v}_q q''$ for some p'' and $q \xrightarrow{v}_p q$. This can be iterated to form an infinite run $q_0, \ldots, q', \ldots, q, \ldots, q, \ldots$ on $u v_0 v_1 \ldots$ s.t. p is the greatest priority (w.r.t. $\geq$) that is seen infinitely often on this run, because it is the greatest (w.r.t. $\geq$) that occurs on the parts from q to itself. Since p is even, this run would be accepting, contradicting the assumption that $w \notin L(\mathcal{A})$.   □

Thm. 2 can then be used to decide (non-)universality as follows, see Algorithm 1. We keep generating boxes $[u]$, $[v]$ to see whether some $[v]$ is idempotent and bad w.r.t. to $q_0[u]$. If this is the case then uv^ω cannot be accepted by the NPA. Finiteness of the space of all boxes guarantees termination. Algorithm UP uses a set V in order to store such boxes $[v]$ whereas boxes $[u]$ need not be stored explicitly. It suffices to store all states which can be reached from the initial state under some $u \in \Sigma^*$. A set R is maintained in order to track all states that are reachable from the initial state with corresponding witnessing words, since in a non-universality check they all need to be tested for non-extendability with a loop.

Note that using sets of boxes is not necessarily the most clever way of implementing this algorithm. One can use priority lists etc. in order to avoid testing the same pair of boxes multiple times in line 7. Also, the step in line 14 is meant to remove pairs (u, Q') from R only for as long as there is still another $(v, Q'') \in R$.

At last, we analyse the asymptotic complexity of testing NPA universality this way.

Theorem 3. *For an NPA $\mathcal{A}$ with $|\mathcal{A}| = n$ and $idx(\mathcal{A}) = k$, algorithm UP tests universality in time $\mathcal{O}(2^{2((n^2 \log k)+n)})$.*

Proof. First observe that the set of boxes V can only increase (or stay the same) in an iteration of the algorithm, hence there can be at most $(k+1)^{n^2}$ many

---

Algorithm 1. UP for UNIV^{P}

1: $R \leftarrow \{(\epsilon, \{q_0\})\}$
2: $V \leftarrow \{[a] \mid a \in \Sigma\}$
3: $R' \leftarrow \emptyset$
4: $V' \leftarrow \emptyset$
5: **while** $R \neq R'$ or $V \neq V'$ **do**
6:     **for** $[v] \in V$ that are idempotent **do**
7:         **if** $\exists (u, Q') \in R$ s.t. $[v]$ bad w.r.t. Q' **then**
8:             **return** "$L(\mathcal{A})$ is not universal: $uv^\omega \notin L(\mathcal{A})$"
9:         **end if**
10:     **end for**
11:     $R' \leftarrow R$
12:     $V' \leftarrow V$
13:     $R \leftarrow R \cup \{(uv, Q'[v]) \mid (u, Q') \in R, [v] \in V\}$
14:     reduce R s.t. it contains at most one (u, Q') for every $Q' \subseteq Q$
15:     $V \leftarrow V \cup V \circ V$
16: **end while**
17: **return** "$L(\mathcal{A})$ is universal"

---

different sets V in a run. Second, observe that the set of reachable state sets R can only increase modulo inclusion in an iteration of the algorithm, hence, there cannot be more than 2^n many different sets R in a run. It follows that the number of iterations is bounded by $(k+1)^{n^2} + 2^n = \mathcal{O}(2^{n^2 \log k})$.

The number of iterations of the inner loop is bounded by $|V| \cdot |R|$, which is $(k+1)^{n^2} \cdot 2^n = \mathcal{O}(2^{(n^2 \log k)+n})$ in the worst case. The other operations in the outer loop can be easily bounded by the same term.   □

Remember that an NBA is just an NPA with priorities $1, 2$. Hence, algorithm UP can be restricted to NBA as input as well. In order to refer to it later, we call this restriction UB. It cannot be distinguished from UP in terms of pseudo code. The difference is that the search space is only of size $2^{n^2 \log 3}$. We also remark that UB coincides with the previously known Ramsey-based universality test for NBA [7].

3.2   Inclusion for NPA

There is a conceptually simple but evidently not well-known reduction from the inclusion problem to the universality problem for finite automata which can also be made to work for NPA. A proof sketch is given in the appendix.

Proposition 3. *Let $\mathcal{A}_1$ and $\mathcal{A}_2$ be two NPA over some alphabet Σ, s.t. $\mathcal{A}_i$ has size n_i states, e_i transitions and index k_i for $i \in \{1, 2\}$. There is an NPA $\mathcal{A}$ with $4 + n_1 + k_1 + n_2$ states and index $\max\{2, k_1, k_2\}$ over some alphabet Δ with $|\Delta| = e_1$ s.t. $L(\mathcal{A}) = \Sigma^\omega$ iff $L(\mathcal{A}_1) \subseteq L(\mathcal{A}_2)$.*

This construction, when applied to two NBA, does not necessarily yield an NBA. With a minor modification though it can also be used to reduce INCL^{B} to UNIV^{B}. The resulting automaton would have $3n_2 + n_1 + 1$ states.

This, together with Thm. 3 clearly yields a Ramsey-based algorithm for the inclusion problem for NPA. However, there is also a direct method which is asymptotically better. For the remainder of this section fix two NPA $\mathcal{A} = (Q^{\mathcal{A}}, \Sigma, q_0^{\mathcal{A}}, \delta^{\mathcal{A}}, \Omega^{\mathcal{A}})$ and $\mathcal{B} = (Q^{\mathcal{B}}, \Sigma, q_0^{\mathcal{B}}, \delta^{\mathcal{B}}, \Omega^{\mathcal{B}})$. We are interested to know whether or not $L(\mathcal{A}) \subseteq L(\mathcal{B})$ holds. Let $P^{\mathcal{A}} := \{\Omega^{\mathcal{A}}(q) \mid q \in Q^{\mathcal{A}}\}$ be the set of all priorities occurring in $\mathcal{A}$ and $P^{\mathcal{B}}$ be defined likewise.

Remember that in the previous section we associated to every word $w \in \Sigma^*$ a unique box $[w]$. This is not possible anymore; a word can be associated with several objects of type

$$Q^{\mathcal{A}} \times P^{\mathcal{A}} \times Q^{\mathcal{A}} \times \left(Q^{\mathcal{B}} \times Q^{\mathcal{B}} \to P^{\mathcal{B}}\right) .$$

Thus, such an object is obtained by extending a box for $\mathcal{B}$—as defined in the previous section—with two states and a priority of $\mathcal{A}$. We call these objects *typed boxes* because the two states of $Q^{\mathcal{A}}$ act as input and output types for the composition on them. A typed box of the form $(q, p, q', [w])$ is written $^q[w]_p^{q'}$.

A typed box for the empty word ϵ is $^q[\epsilon]_{\Omega(q)}^q$ for any $q \in Q^{\mathcal{A}}$, a typed box for words a of length 1 is $^q[a]_{\max\{\Omega(q), \Omega(q')\}}^{q'}$ for any $(q, a, q') \in \delta^{\mathcal{A}}$ and the composition of two typed boxes extends the composition of boxes in the following way:

$$^q[u]_p^{q'} \circ {}^{q'}[v]_{p'}^{q''} := {}^q[uv]_m^{q''}$$

where $m := \max_{\leq}\{p, p'\}$. Note that this composition is only defined if the output type of the left component equals the input type of the right component. We write $\mathfrak{B}_{\mathcal{A},\mathcal{B}}$ for the space of all typed boxes for the pair $\mathcal{A}, \mathcal{B}$ of NPA. A proof of the following lemma is given in the appendix.

Lemma 4. *Let* $w \in \Sigma^*$, $q, q' \in Q^{\mathcal{A}}$, $p \in P^{\mathcal{A}}$, $s, s' \in Q^{\mathcal{B}}$, *and* $p' \in P^{\mathcal{B}}$. *If* $^q[w]_p^{q'}(s, s') = p'$ *then* $q \xrightarrow{w}_p q'$ *and* $s \xrightarrow{w}_{p'} s'$.

An *idempotent* typed box is, as usual, a $^q[w]_p^{q'}$ s.t. $^q[w]_p^{q'} \circ {}^q[w]_p^{q'} = {}^q[w]_p^{q'}$. Note that this necessarily requires $q = q'$. A *bad* typed box w.r.t. some $Q' \subseteq Q^{\mathcal{B}}$ is a $^q[w]_p^{q'}$ s.t. p is even and the underlying untyped box $[w]$ is bad w.r.t. Q' in the sense of the previous section, i.e. there is a $q \in Q^{\mathcal{B}}$ s.t. $[w](q, q)$ is undefined or odd.

Theorem 4. *We have* $L(\mathcal{A}) \not\subseteq L(\mathcal{B})$ *iff there are* $u \in \Sigma^*$, $v \in \Sigma^+$ *and typed boxes* $^{q_0^{\mathcal{A}}}[u]_p^q$ *and* $^q[v]_{p'}^q$, *s.t.* $^q[v]_{p'}^q$ *is idempotent and bad w.r.t.* $q_0^{\mathcal{B}}[u]$.

Proof. "$\Leftarrow$" Suppose there are $^{q_0^{\mathcal{A}}}[u]_p^q$ and $^q[v]_{p'}^q$, s.t. $^q[v]_{p'}^q$ is idempotent and bad w.r.t. $q_0^{\mathcal{B}}[u]$. Using Lemma 4 we get a run $q_0^{\mathcal{A}}, \ldots, q, \ldots, q, \ldots$ of $\mathcal{A}$ on uv^ω, s.t. the maximal priority occurring infinitely often in this run is p'. This p' must be even for otherwise $^q[v]_{p'}^q$ would not be bad. Hence, we have $uv^\omega \in L(\mathcal{A})$. It remains to be seen that $uv^\omega \notin L(\mathcal{B})$.

Suppose $q_0, q_1, \ldots$ was an accepting run of $\mathcal{B}$ on uv^ω. Note that if $^q[v]_{p'}^q$ is idempotent and bad w.r.t. some set Q', then so is its underlying untyped box $[v]$. Thus, $uv^\omega \notin L(\mathcal{B})$ can be proved as in the "$\Leftarrow$"-part of the proof of Thm. 2.

"$\Rightarrow$" Suppose there is a $w = a_0 a_1 \ldots \in L(\mathcal{A}) \cap \overline{L(\mathcal{B})}$. Take an accepting run $q_0, q_1, \ldots$ of $\mathcal{A}$ on w. Here we consider the colouring $c : \mathbb{N}^2 \to \mathfrak{B}_{\mathcal{A}, \mathcal{B}}$ defined by $c(i, j) = {}^{q_i}[a_i \ldots a_j]_p^{q_{j+1}}$ where p is the maximal (w.r.t. $\leq$) priority occuring in the sequence $\Omega^{\mathcal{A}}(q_i), \ldots, \Omega^{\mathcal{A}}(j + 1)$. Since $\mathfrak{B}_{\mathcal{A}, \mathcal{B}}$ is finite, Thm. 1 yields words $u \in \Sigma^*, v \in \Sigma^+$ s.t. ${}^{q_{|u|}}[v]_{p'}^{q_{|u|}}$ is idempotent. In a way analogous to the "$\Rightarrow$"-part of the proof of Thm. 2 one can show that ${}^{q_{|u|}}[v]_{p'}^{q_{|u|}}$ is bad w.r.t. $q_0[u]$. $\qquad\square$

Inclusion can then be tested again by searching for an idempotent bad box w.r.t. to some set of reachable states. Here we maintain a set R of triples (u, q, Q') s.t. q is reachable in $\mathcal{A}$ from $q_0^{\mathcal{A}}$ under u, and all $q' \in Q'$ are reachable from $q_0^{\mathcal{B}}$ under u in $\mathcal{B}$.

Algorithm 2. IP for $\textsc{Incl}^{\textsc{P}}$

1: $R \leftarrow \{(\epsilon, q_0^{\mathcal{A}}, \{q_0^{\mathcal{B}}\})\}$
2: $V \leftarrow \{{}^q[a]_p^{q'} \mid (q, a, q') \in \delta^{\mathcal{A}}, p = \Omega(q)\}$
3: $R' \leftarrow \emptyset$
4: $V' \leftarrow \emptyset$
5: **while** $R \neq R'$ **or** $V \neq V'$ **do**
6: $\quad$ **for** ${}^q[v]_p^q \in V$ that are idempotent **do**
7: $\quad\quad$ **if** $\exists (u, q, Q') \in R$ s.t. ${}^q[v]_p^q$ bad w.r.t. Q' **then**
8: $\quad\quad\quad$ **return** "$L(\mathcal{A}) \not\subseteq L(\mathcal{B})$: $uv^{\omega} \in L(\mathcal{A}) \cap \overline{L(\mathcal{B})}$"
9: $\quad\quad$ **end if**
10: $\quad$ **end for**
11: $\quad$ $R' \leftarrow R$
12: $\quad$ $V' \leftarrow V$
13: $\quad$ $R \leftarrow R \cup \{(uv, q', Q'[v]) \mid {}^q[v]_p^{q'} \in V, (u, q, Q') \in R\}$
14: $\quad$ reduce R s.t. it contains at most one (u, q, Q') for every pair q, Q'
15: $\quad$ $V \leftarrow V \cup V \circ V$
16: **end while**
17: **return** "$L(\mathcal{A}) \subseteq L(\mathcal{B})$"

Theorem 5. *Algorithm 2 tests inclusion in time* $\mathcal{O}(2^{2((n^2 \log k) + \log m + \log k')})$ *for NPA $\mathcal{A}$ and $\mathcal{B}$ with m resp. n states and index k' resp. k.*

Proof. As in Thm. 3, it is not hard to see that the number of outer iterations is bounded by the maximal number of typed boxes and reachability sets. Since the number of typed boxes is an upper bound on the number of reachability sets, we observe that the number of iterations can be bounded by $\mathcal{O}(2^{(n^2 \log k) + \log m + \log k'})$. The runtime of the inner loop as well as the other operations can also be bounded by $\mathcal{O}(2^{(n^2 \log k) + \log m + \log k'})$. $\qquad\square$

We call IB the restriction of IP to NBA only. Its search space is decreased to $m \cdot 2^{(n^2 \log 3) + \log 3}$.

3.3 Comparing Direct and Indirect Methods

Regarding the universality problem for NPA we consider two different approaches:
(1) using algorithm UP directly, and (2) translating an NPA into an NBA and then
using UB, the restriction of UP to NBA. The asymptotic worst-case runtimes for
these two approaches compare as follows. As usual, n denotes the number of states
of the input NPA, k denotes its index.

	reductions	algorithm	runtime
(1)	—	UP	$2^{\mathcal{O}(n^2 \log k)}$
(2)	$\text{UNIV}^\text{P} \mapsto \text{UNIV}^\text{B}$	UB	$2^{\mathcal{O}(n^2 k^2)}$

This shows that the direct method devised here is asymptotically much better than the one that can be obtained from previously known reductions and
methods for NBA: (1) is polynomial in the number of occurring priorities, (2)
is exponential in the square of this number. Thus, one should expect the direct method to perform much better on NPA with more than just a very small
number of priorities.

Now consider the inclusion problem between an NPA with m states and index
k' and an NPA with n states and index k. There are even more Ramsey-based
approaches available:

1. the direct method using algorithm IP;
2. translating both NPA into NBA (Prop. 1), then using IB;
3. the reduction from inclusion to universality on the NPA side (Prop. 3), then
   using algorithm UP;
4. reducing the inclusion problem for NPA to the universality problem (Prop. 3),
   then translating the resulting single NPA into an NBA (Prop. 1), and testing
   it for universality with algorithm UB;
5. first translating the NPA into NBA (Prop. 1), then performing the reduction
   from inclusion to universality on the NBA side (Prop. 3), and finally using
   algorithm UB as well.

The asymptotic worst-case runtimes are as follows.

	reductions	alg.	runtime
(1)	—	IP	$2^{\mathcal{O}(n^2 \log k + \log(mk'))}$
(2)	$\text{INCL}^\text{P} \mapsto \text{INCL}^\text{B}$	IB	$2^{\mathcal{O}((nk)^2 + \log(mk'))}$
(3)	$\text{INCL}^\text{P} \mapsto \text{UNIV}^\text{P}$	UP	$2^{\mathcal{O}((n+k+m)^2 + \log k')}$
(4)	$\text{INCL}^\text{P} \mapsto \text{UNIV}^\text{P} \mapsto \text{UNIV}^\text{B}$	UB	$2^{\mathcal{O}((n+k+m)^2 \cdot (\max\{k,k'\})^2)}$
(5)	$\text{INCL}^\text{P} \mapsto \text{INCL}^\text{B} \mapsto \text{UNIV}^\text{B}$	UB	$2^{\mathcal{O}((nk+mk')^2)}$

One can vaguely say that the more reductions one uses, the worse the asymptotic runtime gets. Again, only the direct method devised here is polynomial in
the number of involved priorities whereas using any of the four other methods
involving a reduction of some kind results in a runtime that is exponential in at
least the number of different priorities in one of the involved automata.

4   Experimental Evaluation

The previous section argues that the direct Ramsey-based methods for parity automata devised in this paper are asymptotically, i.e. in theory, better than the methods one can obtain through reductions to Ramsey-based methods for Büchi automata. In this section, we show that this is also the case in practice. Due to space restrictions we restrict ourselves to the universality problem. Preliminary tests with the inclusion problem also show that the direct methods outperform those obtained by reductions.

4.1   A Random Model of Parity Automata

We extend the Tabakov-Vardi random model for NBA [15] to one for NPA. It is parameterized by two natural numbers $n > 0$ and $p > 0$ that result in the following automata scheme for an NPA $(Q, \Sigma, 1, \delta, \Omega)$ where $Q = \{1, \ldots, n\}$, $\Sigma = \{a, b\}$, and δ and Ω are chosen arbitrarly at random by the following distribution:

- $q' \in \delta(q, s)$ with probability $\frac{2}{n}$ for every $1 \leq q, q' \leq n$ and every $s = a, b$, and

- $\Omega(q) = \begin{cases} 2p' + 1 & \text{with probability } \frac{1}{2p} \text{ and } 0 \leq p' < p \\ 2p' + 2 & \text{with probability } \frac{1}{2p} \text{ and } 0 \leq p' < p \end{cases}$

In other words, an NPA of this model has n states, an alphabet of size two, an expected transition density of two outgoing edges per state and symbol and a priority assignment that maps every state to a priority based on a uniform distribution of $1, \ldots, 2p$. Experimentally it can be seen that this results in an NPA accepting the universal language with probability of approx. 50%.

4.2   Comparison in Practice

All tests have been carried out on a 64-bit machine with four quad-core Opteron™ CPUs. The implementation does not support parallel computations, hence, each test is run on one core only. The following tables feature the average time to decide universality over 1000 automata of a certain parameterization of the random model. They also show the average rounded number of boxes that have been created during these tests.

The first benchmark measures the effect of the number of states on the runtime. Thus, it fixes $p = 2$, i.e. the only priorities occurring are $1, \ldots, 4$. The results are presented in the first table in Fig. 2. The average runtimes distinguish the two cases of NPA accepting the universal language and a non-universal language because non-universality is much easier to establish than universality. Note that the latter requires the creation of all boxes while the former only needs to find a counterexample. This benchmark shows very clearly that the direct Ramsey-based method UP for NPA is much faster in practice than the method UB on NBA that have been obtained by translating NPA into NBA.

Benchmark 1								
	universal				non-universal			
	UP		UB		UP		UB	
states	time	boxes	time	boxes	time	boxes	time	boxes
5	0.00s	21	0.01s	23	0.00s	5	0.00s	6
10	0.08s	190	0.95s	583	0.01s	53	0.04s	64
15	1.23s	817	70.39s	6,388	0.07s	145	0.57s	272
20	3.90s	1,497	1,555.04s	40,776	0.46s	401	2.98s	811
25	19.70s	3,877	1,867.92s	43,728	0.90s	648	2.46s	846
30	72.62s	6,486	—	—	2.70s	1,106	49.18s	4,780
35	154.67s	8,868	—	—	5.11s	1,489	59.44s	5,901
40	221.01s	11,318	—	—	10.93s	2,112	70.26s	6,601

Benchmark 2								
	universal				non-universal			
	UP		UB		UP		UB	
priorities	time	boxes	time	boxes	time	boxes	time	boxes
2	0.97s	745	1.14s	677	0.05s	114	0.08s	111
4	2.74s	1,370	29.57s	5,294	0.15s	200	0.92s	238
6	2.89s	1,479	797.65s	13,049	0.17s	255	1.79s	332
8	5.28s	2,297	1,158.08s	28,261	0.22s	327	2.02s	511
10	3.56s	2,226	—	—	0.34s	400	6.39s	939
12	4.03s	2,120	—	—	0.33s	477	8.37s	1,498
14	4.13s	1,766	—	—	0.23s	374	10.69s	1,450
16	4.36s	2,755	—	—	0.31s	450	31.11s	1,402

Fig. 2. Average runtimes and number of created boxes in the benchmarks

The second benchmark measures the effect that the number of different priorites has on the runtime. It fixes $n = 16$, i.e. every automaton has exactly 16 states. See the second table in Fig. 2 for the results. Again, the direct method of algorithm UP outperforms the indirect method of algorithm UB by far.

5   Conclusion and Further Work

We have presented direct Ramsey-based methods that solve the universality and inclusion problem for nondeterministic parity automata. These direct methods turn out to be more efficient than indirect methods obtained from translating parity into Büchi automata and then performing the corresponding Ramsey-based analysis on these. Also, the reduction from inclusion to universality is equally non-viable in this context.

The work presented here can be continued into several directions. It remains to be seen whether optimisations for Ramsey-based methods as they can be done for NBA [1,2] can be lifted to yield similar speed-ups in the Ramsey-based methods for NPA.

References

1. Abdulla, P.A., Chen, Y.-F., Clemente, L., Holík, L., Hong, C.-D., Mayr, R., Vojnar, T.: Simulation Subsumption in Ramsey-Based Büchi Automata Universality and Inclusion Testing. In: Touili, T., Cook, B., Jackson, P. (eds.) CAV 2010. LNCS, vol. 6174, pp. 132–147. Springer, Heidelberg (2010)
2. Abdulla, P.A., Chen, Y.-F., Clemente, L., Holík, L., Hong, C.-D., Mayr, R., Vojnar, T.: Advanced Ramsey-Based Büchi Automata Inclusion Testing. In: Katoen, J.-P., König, B. (eds.) CONCUR 2011. LNCS, vol. 6901, pp. 187–202. Springer, Heidelberg (2011)
3. Banieqbal, B., Barringer, H.: Temporal Logic with Fixed Points. In: Banieqbal, B., Pnueli, A., Barringer, H. (eds.) Temporal Logic in Specification. LNCS, vol. 398, pp. 62–73. Springer, Heidelberg (1989)
4. Büchi, J.R.: On a decision method in restricted second order arithmetic. In: Proc. Congress on Logic, Method, and Philosophy of Science, pp. 1–12. Stanford University Press, Stanford (1962)
5. Dax, C., Hofmann, M.O., Lange, M.: A Proof System for the Linear Time μ-Calculus. In: Arun-Kumar, S., Garg, N. (eds.) FSTTCS 2006. LNCS, vol. 4337, pp. 273–284. Springer, Heidelberg (2006)
6. Fogarty, S., Vardi, M.Y.: Büchi Complementation and Size-Change Termination. In: Kowalewski, S., Philippou, A. (eds.) TACAS 2009. LNCS, vol. 5505, pp. 16–30. Springer, Heidelberg (2009)
7. Fogarty, S., Vardi, M.Y.: Efficient Büchi Universality Checking. In: Esparza, J., Majumdar, R. (eds.) TACAS 2010. LNCS, vol. 6015, pp. 205–220. Springer, Heidelberg (2010)
8. Klarlund, N.: Progress measures for complementation of ω-automata with applications to temporal logic. In: Proc. 32nd Annual Symp. on Foundations of Computer Science, FOCS 1991, pp. 358–367. IEEE (1991)
9. Kupferman, O., Vardi, M.Y.: Weak alternating automata are not that weak. ACM Transactions on Computational Logic 2(3), 408–429 (2001)
10. Lee, C.S., Jones, N.D., Ben-Amram, A.M.: The size-change principle for program termination. ACM SIGPLAN Notices 36(3), 81–92 (2001)
11. Pnueli, A.: The temporal logic of programs. In: Proc. 18th Symp. on Foundations of Computer Science, FOCS 1977, pp. 46–57. IEEE, Providence (1977)
12. Ramsey, F.P.: On a problem of formal logic. Proc. London Mathematical Society, Series 2 30(4), 338–384 (1928)
13. Safra, S.: On the complexity of ω-automata. In: Proc. 29th Symp. on Foundations of Computer Science, FOCS 1988, pp. 319–327. IEEE (1988)
14. Sistla, A.P., Vardi, M.Y., Wolper, P.: The complementation problem for Büchi automata with applications to temporal logic. TCS 49(2-3), 217–237 (1987)
15. Tabakov, D., Vardi, M.Y.: Experimental Evaluation of Classical Automata Constructions. In: Sutcliffe, G., Voronkov, A. (eds.) LPAR 2005. LNCS (LNAI), vol. 3835, pp. 396–411. Springer, Heidelberg (2005)
16. Vardi, M.Y.: A temporal fixpoint calculus. In: ACM (ed.) Proc. Conf. on Principles of Programming Languages, POPL 1988, pp. 250–259. ACM, NY (1988)

VATA: A Library for Efficient Manipulation of Non-deterministic Tree Automata[*]

Ondřej Lengál[1], Jiří Šimáček[1,2], and Tomáš Vojnar[1]

[1] FIT, Brno University of Technology, IT4Innovations Centre of Excellence, Czech Republic
[2] VERIMAG, UJF/CNRS/INPG, Gières, France

Abstract. In this paper, we present VATA, a versatile and efficient open-source tree automata library applicable, e.g., in formal verification. The library supports both explicit and semi-symbolic encoding of non-deterministic finite tree automata and provides efficient implementation of standard operations on both. The semi-symbolic encoding is intended for tree automata with large alphabets. For storing their transition functions, a newly implemented MTBDD library is used. In order to enable the widest possible range of applications of the library even for the semi-symbolic encoding, we provide both bottom-up and top-down semi-symbolic representations. The library implements several highly optimised reduction algorithms based on downward and upward simulations as well as algorithms for testing automata inclusion based on upward and downward antichains and simulations. We compare the performance of the algorithms on a set of test cases and we also compare the performance of VATA with our previous implementations of tree automata.

1 Introduction

Several current formal verification techniques are based on *finite tree automata* (TA). Some of these techniques are: (abstract) regular tree model checking [3,5] applied, e.g., for verification of programs with complex dynamic data structures [6,11], implementation of decision procedures of several logics, such as MSO or WSkS [17], or verification of programs manipulating heap structures with data [18]. The success of these techniques often depends on the performance of the underlying implementation of TA.

Currently, there exist several available tree automata libraries, they are, however, mostly written in OCaml (e.g., Timbuk/Taml [10]) or Java (e.g., LETHAL [9]) and they do not always use the most advanced algorithms known to date. Therefore, they are not suitable for tasks which require the available processing power be utilised as efficiently as possible. An exception from these libraries is MONA [17] implementing decision procedures over WS1S/WS2S, which contains a highly optimised TA package written in C, but, alas, it supports only binary deterministic tree automata. At the same time, it turns out that determinisation is often a very significant bottleneck of using TA, and a lot

---

[*] This work was supported by the Czech Science Foundation within projects No. P103/10/0306 and 102/09/H042, the Czech Ministry of Education within projects COST OC10009 and MSM 0021630528, and the EU/Czech IT4Innovations Centre of Excellence project CZ.1.05/1.1.00/02.0070.

C. Flanagan and B. König (Eds.): TACAS 2012, LNCS 7214, pp. 79–94, 2012.

of effort has therefore been invested into developing efficient algorithms for handling non-deterministic tree automata without a need to ever determinise them.

In order to allow researchers focus on developing verification techniques rather than reimplementing and optimising a TA package, we provide VATA[1], an easy-to-use open-source library for efficient manipulation of non-deterministic TA. VATA supports many of the operations commonly used in automata-based formal verification techniques over two complementary encodings: explicit and semi-symbolic. The *explicit* encoding is suitable for most applications that do not need to use alphabets with a large number of symbols. However, some formal verification approaches make use of such alphabets, e.g., the approach for verification of programs with complex dynamic data structures [5] or decision procedures of the MSO or WSkS logics [17]. Therefore, in order to address this issue, we also provide the *semi-symbolic* encoding of TA, which uses *multi-terminal binary decision diagrams* [8] (MTBDDs), an extension of reduced ordered binary decision diagrams [7] (BDDs), to store the transition function of a TA. In order to enable the widest possible range of applications of the library even for the semi-symbolic encoding, we provide both bottom-up and top-down semi-symbolic representations.

At the present time, the main application of the structures and algorithms implemented in VATA for handling explicitly encoded TA is the Forester tool for verification of programs with complex dynamic data structures [11]. The semi-symbolic encoding of TA has so far been used mainly for experiments with various newly proposed algorithms for handling TA.

In this paper, we do not present all exact details of the algorithms implemented in the library as they can be found in the referenced literature. Rather, we give an overview of the algorithms available, while mentioning various interesting optimisations that we used when implementing them. Based on experimental evidence, we argue that these optimisations are crucial for the performance of the library.

2   Preliminaries

A *ranked alphabet* Σ is a finite set of symbols together with a ranking function $\# : \Sigma \to \mathbb{N}$. For $a \in \Sigma$, the value $\#a$ is called the *rank* of a. For any $n \geq 0$, we denote by Σ_n the set of all symbols of rank n from Σ. Let ε denote the empty sequence. A *tree* t over a ranked alphabet Σ is a partial mapping $t : \mathbb{N}^* \to \Sigma$ that satisfies the following conditions: (1) the domain of t, $dom(t)$, is a finite prefix-closed subset of $\mathbb{N}^*$ and (2) for each $v \in dom(t)$, if $\#t(v) = n \geq 0$, then $\{i \mid vi \in dom(t)\} = \{1, \ldots, n\}$. Each sequence $v \in dom(t)$ is called a *node* of t. For a node v, we define the ith *child* of v to be the node vi, and the ith *subtree* of v to be the tree t' such that $t'(v') = t(viv')$ for all $v' \in \mathbb{N}^*$. A *leaf* of t is a node v which does not have any children, i.e., there is no $i \in \mathbb{N}$ with $vi \in dom(t)$. We denote by T_Σ the set of all trees over the alphabet Σ.

A (finite, non-deterministic) *tree automaton* (abbreviated sometimes as TA in the following) is a quadruple $\mathcal{A} = (Q, \Sigma, \Delta, F)$ where Q is a finite set of states, $F \subseteq Q$ is a set of final states, Σ is a ranked alphabet, and Δ is a set of transition rules. Each transition rule is a triple of the form $((q_1, \ldots, q_n), a, q)$ where $q_1, \ldots, q_n, q \in Q, a \in \Sigma$,

---

[1] http://www.fit.vutbr.cz/research/groups/verifit/tools/libvata/

and $\#a = n$. We use equivalently $(q_1,\ldots,q_n) \xrightarrow{a} q$ and $q \xrightarrow{a} (q_1,\ldots,q_n)$ to denote that $((q_1,\ldots,q_n),a,q) \in \Delta$. The two notations correspond to the *bottom-up* and *top-down* representation of tree automata, respectively. Note that we can afford to work interchangeably with both of them since we work with non-deterministic tree automata, which are known to have an equal expressive power in their bottom-up and top-down representations. In the special case when $n = 0$, we speak about the so-called *leaf rules*, which we sometimes abbreviate as $\xrightarrow{a} q$ or $q \xrightarrow{a}$.

Let $\mathcal{A} = (Q,\Sigma,\Delta,F)$ be a TA. A *run* of $\mathcal{A}$ over a tree $t \in T_\Sigma$ is a mapping $\pi : dom(t) \to Q$ such that, for each node $v \in dom(t)$ of rank $\#t(v) = n$ where $q = \pi(v)$, if $q_i = \pi(vi)$ for $1 \leq i \leq n$, then Δ has a rule $(q_1,\ldots,q_n) \xrightarrow{t(v)} q$. We write $t \xRightarrow{\pi} q$ to denote that π is a run of $\mathcal{A}$ over t such that $\pi(\varepsilon) = q$. We use $t \Longrightarrow q$ to denote that $t \xRightarrow{\pi} q$ for some run π. The *language* accepted by a state q is defined by $L_{\mathcal{A}}(q) = \{t \mid t \Longrightarrow q\}$, while the language of a set of states $S \subseteq Q$ is defined as $L_{\mathcal{A}}(S) = \bigcup_{q \in S} L_{\mathcal{A}}(q)$. When it is clear which TA $\mathcal{A}$ we refer to, we only write $L(q)$ or $L(S)$. The language of $\mathcal{A}$ is defined as $L(\mathcal{A}) = L_{\mathcal{A}}(F)$.

A *downward simulation* on TA $\mathcal{A} = (Q,\Sigma,\Delta,F)$ is a preorder relation $\preceq_D \subseteq Q \times Q$ such that if $q \preceq_D p$ and $(q_1,\ldots,q_n) \xrightarrow{a} q$, then there are states $p_1,\ldots,p_n$ such that $(p_1,\ldots,p_n) \xrightarrow{a} p$ and $q_i \preceq_D p_i$ for each $1 \leq i \leq n$. Given a TA $\mathcal{A} = (Q,\Sigma,\Delta,F)$ and a downward simulation $\preceq_D$, an *upward simulation* $\preceq_U \subseteq Q \times Q$ induced by $\preceq_D$ is a relation such that if $q \preceq_U p$ and $(q_1,\ldots,q_n) \xrightarrow{a} q'$ with $q_i = q$ for some $1 \leq i \leq n$, then there are states $p_1,\ldots,p_n,p'$ such that $(p_1,\ldots,p_n) \xrightarrow{a} p'$ where $p_i = p$, $q' \preceq_U p'$, and $q_j \preceq_D p_j$ for each j such that $1 \leq j \neq i \leq n$.

3   Design of the Library

The library is designed in a modular way (see Fig. 1). The user can choose a module encapsulating her preferred automata encoding and its corresponding operations. Various encodings share the same general interface so it is easy to swap one encoding for another, unless encoding-specific functions or operations are taken advantage of.

Thanks to the modular design of the library, it is easy to provide an own encoding of tree (or word) automata and effectively exploit the remaining parts of the infrastructure, such as parsers and serializers from/to different formats, the unit testing framework, performance tests, etc.

The VATA library is implemented in C++ using the Boost C++ libraries. In order to avoid expensive look-ups of entry points of virtual methods in the *virtual-method table* of an object and to fully exploit compiler's capabilities of code inlining and optimisation of code according to static analysis, the library heavily exploits polymorphism using C++ function templates instead of using virtual methods for core functions. We are convinced that this is the main reason why the performance of the optimised code (the -O3 flag of gcc) is up to 10 times better than the performance of the non-optimised code (the -O0 flag of gcc).

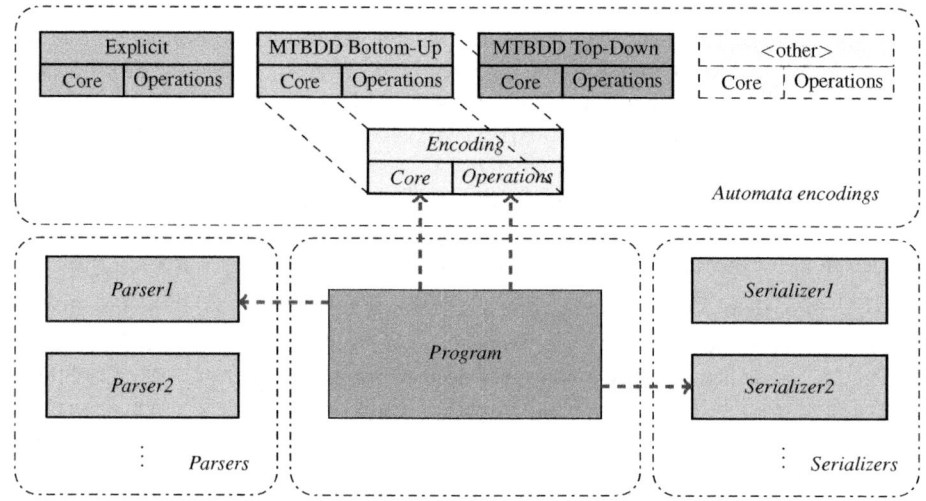

Fig. 1. The architecture of the VATA library

3.1 Explicit Encoding

In the explicit representation of TA used in VATA, top-down transitions having the form $q \xrightarrow{a} (q_1,\ldots,q_n)$ are stored in a *hierarchical data structure similar to a hash table*. More precisely, the top-level lookup table maps states to *transition clusters*. Each such cluster is itself a lookup table that maps alphabet symbols to a set of pointers to tuples of states. The set of pointers to tuples of states is represented using a red-black tree. The tuples of states are stored in a designated hash table to further reduce the required amount of space (by not storing the same tuples of states multiple times). An example of the encoding is depicted in Fig. 2.

Hence, in order to insert the transition $q \xrightarrow{a} (q_1,\ldots,q_n)$ into the transition table, one proceeds using the following algorithm:

1. Find a transition cluster which corresponds to the state q in the top-level lookup table. If such a cluster does not exist, create one.
2. In the given cluster, find a set of pointers to tuples of states reachable from q over a. If the set does not exist, create one.
3. Obtain the pointer to the tuple $(q_1,\ldots,q_n)$ from the tuple lookup table and insert it into the set of pointers.

If one ignores the worst-case time complexity of the underlying data structures (which, according to our experience, has usually a negligible real impact only), then inserting a single transition into the transition table requires a constant number of steps only. Yet the representation provides a more efficient encoding than a plain list of transitions because some transitions share the space required to store the parent states (e.g., state q in the transition $q \xrightarrow{a} (q_1,\ldots,q_n)$). Moreover, some transitions also share the alphabet symbol and each tuple of states appearing in the set of transitions is stored only

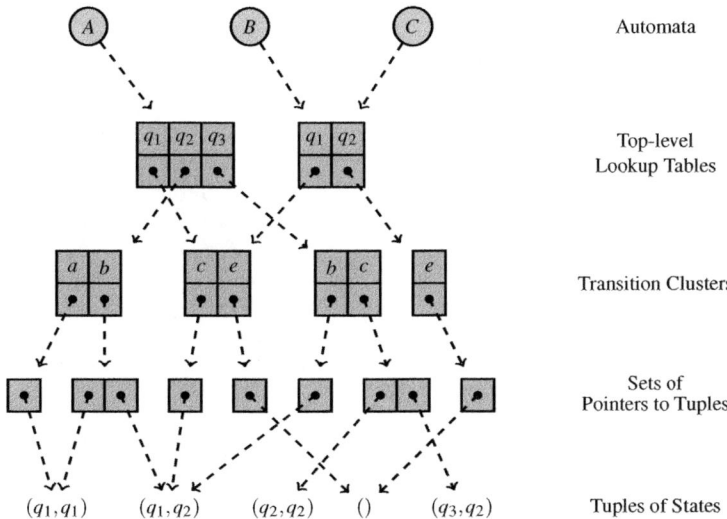

Fig. 2. An example of the VATA's explicit encoding of transition functions of three automata A, B, C. In particular, one can see that A contains a transition $q_1 \xrightarrow{c} (q_1, q_2)$: it suffices to follow the corresponding arrows. Moreover, B also contains the same transition (and the corresponding part of the transition table is shared with A). Finally, C has the same transitions as B.

once. Additionally, the encoding allows us to easily perform certain critical operations, such as finding a set of transitions $q \xrightarrow{a} (q_1, \ldots, q_n)$ for a given state q. This is useful, e.g., during the elimination of (top-down) unreachable states or during the top-down inclusion checking.

In some situations, one needs to manipulate many tree automata at the same time. As an example, we can mention the method for verifying programs with dynamic linked data structures introduced in [11] where (in theory) one needs to store one automaton representing a content of the heap for each reachable state of the program. To improve the performance of our library in such scenarios, we adapt the *copy-on-write* principle. Every time one needs to create a copy of an automaton A to be subsequently modified, it is enough to create a new automaton A' which obtains a pointer to the transition table of A (which requires constant time). Subsequently, as more transitions are inserted into A' (or A), only the part of the shared transition table which gets modified is copied (Fig. 2 provides an illustration of this feature).

3.2   Semi-symbolic Encoding

The semi-symbolic encoding uses *multi-terminal binary decision diagrams* (MTBDDs) to encode transition functions of tree automata. MTBDDs are an extension of *binary decision diagrams* (BDDs), a popular data structure for compact encoding and manipulation with Boolean formulae. In contrast to BDDs that are used to represent a function $b : \mathbb{B}^n \to \mathbb{B}$ for some $n \in \mathbb{N}$ and $\mathbb{B} = \{0, 1\}$, MTBDDs extend the co-domain to an arbitrary set S, i.e., they represent a function $m : \mathbb{B}^n \to S$.

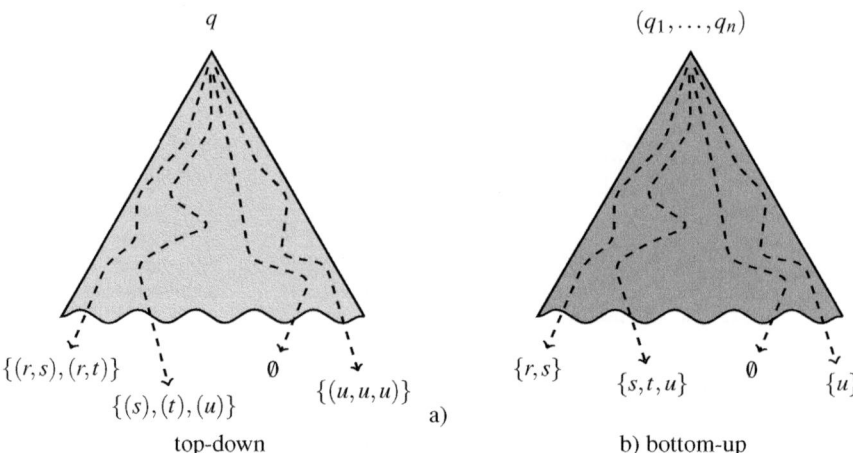

Fig. 3. The (a) top-down and (b) bottom-up semi-symbolic encodings of transition functions. Paths in the MTBDD correspond to symbols.

We support two representations of semi-symbolic automata: top-down and bottom-up. The *top-down* representation (see Fig. 3a) maintains for each state q of a tree automaton an MTBDD that maps the binary representation of each symbol f concatenated with the binary representation of its arity n onto a set of tuples of states $T = \{(q_1, \ldots, q_n), \ldots\}$ such that for all $(q_1, \ldots, q_n) \in T$ there exist the transition $q \xrightarrow{f} (q_1, \ldots, q_n)$ in the automaton. The arity is encoded in the MTBDD as a part of the symbol in order to be able to distinguish between several instances of the same symbol with different arity. The library thus supports a slight extension of tree automata in which a symbol does not have a fixed arity.

The *bottom-up* representation (see Fig. 3b), on the other hand, maintains for each tuple $(q_1, \ldots, q_n) \in Q^*$ an MTBDD that maps the binary representation of each symbol f onto a set of states $S = \{q, \ldots\}$ such that, for all $q \in S$, it holds that the transition $(q_1, \ldots, q_n) \xrightarrow{f} q$ is in the automaton. Note that the bottom-up representation does not need to encode the arity of the symbol f into the MTBDD as it is given by the arity of the tuple for which the MTBDD is maintained. It is easy to see that the two presented encodings are mutually convertible (see [13] for the algorithm).

MTBDD Package. Our previous implementation of semi-symbolically represented tree automata used a customisation of the CUDD [20] library for manipulating MTBDDs. The experiments in [12] and profiling of the code showed that the overhead of the customised library is too large. Moreover, the customisation of CUDD did not provide an easy and transparent way of manipulating MTBDDs. These two facts showed that VATA would greatly benefit from a major redesign of the MTBDD backend. Therefore, we created our own generic implementation of MTBDDs with a clean and simple-to-use interface.

The new MTBDD package uses *shared* MTBDDs for each domain, which means that all MTBDDs for the given domain are connected in a single *directed acyclic graph* (DAG), and an MTBDD corresponds to a pointer to a node in the DAG. In order to prevent memory leaks, each node of the MTBDD contains a reference counter of other nodes or variables pointing to it. In case the counter reaches zero, the node is deleted from the memory. Because of these implementation choices, copying an MTBDD can be easily done by simply copying the pointer to the root node of the copied MTBDD and incrementing its reference counter.

There are two types of nodes of the MTBDD: *internal nodes* and *leaf nodes*. A leaf node contains a value from the domain of the MTBDD, while an internal node contains a variable name and pointers to the *low* and *high* children of the node. In addition, nodes of both types also contain the aforementioned reference counter. The nodes are manipulated using pointers to them only, and the distinction between a leaf node and an internal node is done according to the least significant bit of the pointer (the compiler aligns these data structures to addresses which are multiples of 4, this bit can therefore be neglected and when accessing the value of a node pointer simply masked out).

For our use, we implemented unary, binary, and ternary *Apply* operations, which are operations that, given a unary, binary, or ternary function and one, two, or three MTBDDs, respectively, generate a new MTBDD the leaves of which correspond to the values of the given function applied to the provided MTBDDs. Note that the provided function does not need to be a pure function but my also have a side-effect. Further, we also provide *VoidApply* operations which are Apply operations that do not build a new MTBDD but that have a side-effect only. For operations that do not need to build new MTBDDs but rather, e.g., only collect data from the leaf nodes, using *VoidApply* saves a considerable and unnecessary overhead. During the execution of an *Apply* operation, both internal and leaf nodes are cached using hash tables.

The newly implemented MTBDD package does not support MTBDD reordering so far, yet the library performs better when compared to our original implementation of a semi-symbolic encoding that used customised CUDD.

4 Supported Operations

As we described in the previous section, the VATA library allows a user to choose one of three available encodings: the explicit top-down, the semi-symbolic top-down, and the semi-symbolic bottom-up. Depending on the choice, certain TA operations may or may not be available. The following operations are supported by at least one of the representations: union, intersection, elimination of (bottom-up, top-down) unreachable states, inclusion checking (bottom-up, top-down), computation of (maximum) simulation relations (downward, upward), and language preserving size reduction based on simulation equivalence. In some cases, multiple implementations of an operation are available, which is especially the case for language inclusion. This is because the different implementations are based on different heuristics that may work better for different applications as witnessed also by our experiments described in Section 5.

Below, we do not discuss the relatively straightforward implementation of the most basic operations on TA and we comment on the more advanced operations only.

4.1   Removing Unreachable States

As the performance of many operations on automata depends on the size of the automaton (in the sense of the size of the state set and the size of the transition table), it is often desirable to remove both bottom-up and top-down unreachable states. Indeed, such states are useless: bottom-up unreachable states cannot be used to generate a finite tree and although top-down unreachable states can generate a finite tree, this tree cannot be a subtree of any tree accepted by the automaton.

Removing both bottom-up *unreachable states* for the bottom-up representation and top-down unreachable states for the top-down representation can be easily done by a single traversal through the automaton. Nevertheless, sometimes, e.g., when checking language inclusion of the automata, it is useful to also remove states unreachable in the opposite direction.

The procedure for removing top-down unreachable states from a tree automaton represented in a bottom-up semi-symbolic way generates a directed graph (Q, E) where Q is the state set of the input automaton and $(q, r) \in E$ if $\exists a \in \Sigma : q \xrightarrow{a} (q_1, \ldots, q_n), \exists 1 \leq i \leq n : r = q_i$. When the graph is created, the states that are backward unreachable from the final states are removed from the automaton in a simple traversal.

Removing bottom-up unreachable states for the top-down semi-symbolic representation is more complex. First, the automaton is traversed in the top-down manner while creating an *And-Or* graph $(N_\forall, N_\exists, E)$ where $N_\forall = Q$, Q is the state set of the input automaton and represents the *And* nodes of the graph, and $N_\exists \subseteq Q^*$ represents the *Or* nodes. The set of edges E contains the edge $(q, (q_1, \ldots, q_n))$ if there exists the transition $q \xrightarrow{a} (q_1, \ldots, q_n)$ for some $a \in \Sigma$ in the automaton, and the edge $((q_1, \ldots, q_n), q)$ if $\exists 1 \leq i \leq n : q_i = q$. The algorithm starts by marking the node labelled by $()$ (which is an Or node) and proceeds by marking the nodes of the graph using the following rules: an Or node o is marked if there exists a marked node a such that $(o, a) \in E$, and an And node a is marked if all nodes o such that $(a, o) \in E$ are marked. When no new nodes can be marked, the states of the automaton are reduced to only those that correspond to marked And nodes in the graph.

4.2   Downward and Upward Simulation

Downward simulation relations can be computed over two tree automata representations in VATA: the explicit top-down and the semi-symbolic top-down encoding. The explicit variant first translates a tree automaton into a labelled transition system (LTS) as described in [1]. Then the simulation relation for this system is computed using an implementation of the state-of-the-art algorithms for computing simulations on LTSs [19,14] with some further optimisations mentioned in Section 4.6. Finally, the result is projected back to the set of states of the original automaton.

The semi-symbolic variant uses a simpler simulation algorithm based on a generalisation of [16] to trees.

Upward simulation can currently be computed over the explicit representation only. The computation is again performed via a translation to an LTS (the details are in [1]), and the relation is computed using the engine for computing simulation relations on LTSs as above.

4.3   Simulation-Based Size Reduction

In a typical setting, one often wants to use a representation of tree automata that is as small as possible in order to reduce the memory consumption and/or speed up operations on the automata (especially the potentially costly ones, such as inclusion testing). To achieve that, the classical approach is to use determinisation and minimisation. However, the minimal deterministic tree automata can still be much bigger than the original non-deterministic ones. Therefore, VATA offers a possibility to reduce the size of tree automata without determinisation by their quotienting w.r.t. an equivalence relation—currently, only the downward simulation equivalence is supported.

The procedure works as follows: first, the downward simulation relation $\preceq_D$ is computed for the automaton. Then, the symmetric fragment of $\preceq_D$ (which is an equivalence) is extracted, and each state appearing within the transition function is replaced by a representative of the corresponding equivalence class. A further reduction is then based on the following observation: if an automaton contains a transition $q \xrightarrow{a} (q_1, \ldots, q_n)$, any additional transition $q \xrightarrow{a} (r_1, \ldots, r_n)$ where $r_i \preceq_D q_i$ can be omitted since it does not contribute to the language of the result (recall that, for the downward simulation preorder $\preceq_D$, it holds that $q \preceq_D r \implies L(q) \subseteq L(r)$).

4.4   Bottom-Up Inclusion

Bottom-up inclusion testing is implemented for the explicit top-down and the semi-symbolic bottom-up representation in VATA. As its name suggests, the algorithm naturally proceeds in the bottom-up way, therefore the top-down encoding is not very suitable here. In the case of the explicit representation, however, one can afford to build a temporary bottom-up encoding since the overhead of such a translation is negligible compared to the complexity of following operations.

Both the explicit and semi-symbolic version of the bottom-up inclusion algorithm are based on the approach introduced in [4]. Here, the main principle used for checking whether $L(\mathcal{A}) \subseteq L(\mathcal{B})$ is to search for a tree which is accepted by $\mathcal{A}$ and not by $\mathcal{B}$ (thus being a witness for $L(\mathcal{A}) \not\subseteq L(\mathcal{B})$). This is done by simultaneously traversing both $\mathcal{A}$ and $\mathcal{B}$ from their leaf rules while generating pairs $(p_{\mathcal{A}}, P_{\mathcal{B}}) \in Q_{\mathcal{A}} \times 2^{Q_{\mathcal{B}}}$ where $p_{\mathcal{A}}$ represents a state into which $\mathcal{A}$ can get on some input tree and $P_{\mathcal{B}}$ is the set of *all* states into which $\mathcal{B}$ can get over the same tree. The inclusion then does clearly not hold iff it is possible to generate a pair consisting of an accepting state of $\mathcal{A}$ and of exclusively non-accepting states of $\mathcal{B}$.

The algorithm collects the so far generated pairs $(p_{\mathcal{A}}, P_{\mathcal{B}})$ in a set called *Visited*. Another set called *Next* is used to store the generated pairs whose successors are still to be explored. One can then observe that whenever one can reach a counterexample to inclusion from $(p_{\mathcal{A}}, P_{\mathcal{B}})$, one can also reach a counterexample from any $(p_{\mathcal{A}}, P'_{\mathcal{B}} \subseteq P_{\mathcal{B}})$

as $P'_\mathcal{B}$ allows less runs than $P_\mathcal{B}$. Using this observation, both mentioned sets can be represented using antichains. In particular, one does not need to store and further explore any two elements comparable w.r.t. $(=,\subseteq)$, i.e., by equality on the first component and inclusion on the other component.

Clearly, the running time of the above algorithm strongly depends on the total number of pairs $(p_\mathcal{A},P_\mathcal{B})$ taken from *Next* for further processing. Indeed, this is one of the reasons why the antichain-based optimisations helps. According to our experience, the number of pairs which needs to be processed can further be reduced when processing the pairs stored in *Next* in a suitable order. Our experimental results have shown that we can achieve a very good improvement by preferring those pairs $(p_\mathcal{A},P_\mathcal{B})$ which have smaller (w.r.t. the size of the set) second component.

Yet another way that we found useful when improving the above algorithm is to optimise the way the algorithm computes the successors of a pair from *Next*. The original algorithm picks a pair $(p_\mathcal{A},P_\mathcal{B})$ from *Next* and puts it into *Visited*. Then, it finds all transitions of the form $(p_{\mathcal{A},1},\ldots,p_{\mathcal{A},n}) \xrightarrow{a} p$ in $\mathcal{A}$ such that $(p_{\mathcal{A},i},P_{\mathcal{B},i}) \in Visited$ for all $1 \leq i \leq n$ and $(p_{\mathcal{A},j},P_{\mathcal{B},j}) = (p_\mathcal{A},P_\mathcal{B})$ for some $1 \leq j \leq n$. For each such transition, it finds all transitions of the form $(q_1,\ldots,q_n) \xrightarrow{a} q$ in $\mathcal{B}$ such that $q_i \in P_{\mathcal{B},i}$ for all $1 \leq i \leq n$. Here, the process of finding the needed $\mathcal{B}$ transitions is especially costly. In order to speed it up, we cache for each alphabet symbol a, each position i, and each set $P_{\mathcal{B},i}$, the set of transitions $\{(q_1,\ldots,q_n) \xrightarrow{a} q \in \Delta_\mathcal{B} : q_i \in P_{\mathcal{B},i}\}$ at the first time it is used in the computation of successors. Then, whenever we need to find all transitions of the form $(q_1,\ldots,q_n) \xrightarrow{a} q$ in $\mathcal{B}$ such that $q_i \in P_{\mathcal{B},i}$ for all $1 \leq i \leq n$, we find them simply by intersecting the sets of transitions cached for each $(P_{\mathcal{B},i},i,a)$.

Next, we propose another modification of the algorithm which aims to improve the performance especially in those cases where finding a counterexample to inclusion requires us to build representatives of trees with higher depths or in the cases where the inclusion holds. Unlike the original approach which moves only one pair $(p_\mathcal{A},P_\mathcal{B})$ from *Next* to *Visited* at the beginning of each iteration of the main loop, we add the newly created pairs $(p_\mathcal{A},P_\mathcal{B})$ into *Next* and *Visited* at the same time (immediately after they are generated). This, according to our experiments, allows *Visited* to converge faster towards the fixpoint.

Finally, another optimisation of the algorithm presented in [4] appeared in [2]. This optimisation maintains the sets *Visited* and *Next* as antichains w.r.t. $(\preceq_U, \succeq_U^{\exists\forall})^2$. Hence, more pairs can be discarded from these sets. Moreover, for pairs that cannot be discarded, one can at least reduce the sets on their right-hand side by removing states that are simulated by some other state in these sets (this is based on the observation that any tree accepted from an upward-simulation-smaller state is accepted from an upward-simulation-bigger state too). Finally, one can also use upward simulations between states of the two automata being compared. Then, one can discard any pair $(p_\mathcal{A},P_\mathcal{B})$ such that there is some $p_\mathcal{B} \in P_\mathcal{B}$ that upward-simulates $p_\mathcal{A}$ because it is then clear that no tree can be accepted from $p_\mathcal{A}$ that could not be accepted from $p_\mathcal{B}$. All these opti-

---

[2] One says that $P \preceq_U^{\exists\forall} Q$ holds iff $\forall p \in P \, \exists q \in Q : p \preceq_U q$. Note also that the upward simulation must be parameterised by the identity in this case [2].

misations are also available in VATA and can optionally be used—they are not used by default since the computation of the upward simulation can be quite costly.

4.5 Top-Down Inclusion

Top-down inclusion checking is supported by the explicit top-down and semi-symbolic top-down representations in VATA. Note that when one tries to solve inclusion of TA languages top-down in a naïve way, using a plain subset-construction-like approach, one immediately hits a problem due to the top-down successors of particular states are *tuples* of states. Hence, after one step of the construction, one needs to check inclusion on tuples of states, then tuples of tuples of states, etc. However, there is a way how to get out of this trap as shown in [15,12]. Very roughly said, the main idea of the approach resembles a conversion from the *disjunctive normal form* (DNF) to the *conjunctive normal form* (CNF) taking into account that top-down transitions of tree automata form a kind of and-or graphs (the disjunctions are between top-down transitions and conjunctions among the successors within particular transitions).

VATA contains an implementation of the top-down inclusion checking algorithm of [12]. This algorithm uses several optimisations, e.g., caching of results of auxiliary language inclusion queries between states of the automata whose languages are being compared. More precisely, when checking whether $L(\mathcal{A}) \subseteq L(\mathcal{B})$ holds for two tree automata $\mathcal{A}$ and $\mathcal{B}$, the algorithm stores a set of pairs $(p_{\mathcal{A}}, P_{\mathcal{B}}) \in Q_{\mathcal{A}} \times 2^{Q_{\mathcal{B}}}$ for which the language inclusion $L(p_{\mathcal{A}}) \subseteq L(P_{\mathcal{B}})$ has been shown *not* to hold. As a further optimisation, the set is stored as an antichain based on comparing the states w.r.t. the downward simulation preorder. The use of the downward simulation is one of the main advantages of this approach compared with the bottom-up inclusion checking since this preorder is cheaper to compute and usually richer than the upward simulation. Indeed, [12] shows that top-down inclusion checking is often—though not always—superior to bottom-up inclusion checking.

Moreover, VATA has recently been extended by a new version of the top-down inclusion checking algorithm that extends the original version by caching even the pairs $(p_{\mathcal{A}}, P_{\mathcal{B}}) \in Q_{\mathcal{A}} \times 2^{Q_{\mathcal{B}}}$ for which the language inclusion $L(p_{\mathcal{A}}) \subseteq L(P_{\mathcal{B}})$ has been shown to hold. This extension is far from trivial since the caching must be done very carefully in order to avoid a sort of circular reasoning when answering the various auxiliary language inclusion queries. A precise description of this rather involved algorithm is beyond the scope of this article, and so we refer an interested reader to [13]. As our experiments show, the new kind of caching comes with some overhead, which does not allow it to always win over the previous algorithm, but there are still many cases in which it performs significantly better.

4.6 Computing Simulation over LTS

The explicit part of VATA uses a highly optimised LTS simulation algorithm proposed in [19] and greatly improved in [14]. The main idea of the algorithm is to start with an overapproximation of the simulation preorder (a possible initial approximation is the relation $Q \times Q$) which is then iteratively pruned whenever it is discovered that the simulation relation cannot hold for certain pairs of states. For a better efficiency, the algorithm represents the current approximation R of the simulation being computed using a so-called *partition-relation pair*. The partition splits the set of states into subsets (called *blocks*) whose elements are equivalent w.r.t. R, and a relation obtained by lifting R to blocks.

In order to be able to deal with the partition-relation pair efficiently, the algorithm needs to record for each block a matrix of counters of size $|Q||\Sigma|$ where, for the given LTS, Q is the set of states and Σ is the set of labels. The counters are used to count how many transitions going from the given state via a given symbol a lead to states in the given block (or blocks currently considered to be bigger w.r.t. the simulation). This information is then used to optimise re-computation of the partition-relation pair when pruning the current approximation of the simulation relation being computed (for details see, e.g., [19]). Since the number of blocks can (and often does) reach the number of states, the naïve solution requires $|Q|^2|\Sigma|$ counters in the worst case. It turns out that this is one of the main barriers which prevents the algorithm from scaling to systems with large alphabets and/or large sets of states.

Working towards a remedy for the above problem, one can observe that the mentioned algorithm actually works in several phases. At the beginning, it creates an initial estimation of the partition-relation pair which typically contains large equivalence classes. Then it initialises the counters for each element of the partition. Finally, it starts the iterative partition splitting. During this last phase, the counters are only decremented or copied to the newly created blocks. Moreover, the splitting of some block is itself triggered by decrementing some set of counters to 0. In practice, late phases of the iteration typically witness a lot of small equivalence classes having very sparsely populated counters with 0 being the most abundant value.

This suggests that one could use sparse matrices containing only the non-zero elements. Unfortunately, according to our experience, this turns out to be the worst possible solution which strongly degrades the performance. The reason is that the algorithm accesses the counters very frequently (it either increments them by one or decrements them by one), hence any data structure with non-constant time access causes the computation to stall. A somewhat better solution is to record the non-zero counters using a hash table, but the memory requirements of such representation are not yet reasonable.

Instead, we are currently experimenting with storing the counters in blocks, using a copy-on-write approach and a zeroed-block deallocation. In short, we divide the matrix of counters into a list of blocks of some fixed size. Each block contains an additional counter (a block-level counter) which sums up all the elements within the block. As soon as a block contains a single non-zero counter only, it can safely be deallocated—the content of the non-zero counter is then recorded in the block-level counter.

Our initial experiments show that, using the above approach, one can easily reduce the memory consumption by the factor of 5 for very large instances of the problem compared to the array-based representation used in [14]. The best value to be used as the size of blocks of counters is still to be studied—after some initial experiments, we are currently using blocks of size $\sqrt{|Q|}$.

5   Experimental Evaluation of VATA

In order to illustrate the level of optimisation that has been achieved in VATA and that can be exploited in its applications (like the Forester tool [11]), we compared its performance against Timbuk and the prototype library considered in [12], which—despite its prototype status—already contained a quite efficient TA implementation.

The comparison of performance of VATA (using the explicit encoding) and Timbuk was done for union and intersection of more than 3,000 pairs of TA. On average, VATA was over 20,000 times faster on union and over 100,000 times faster on intersection.

When comparing VATA with the prototype library of [12], we concentrated on language inclusion testing which is one of the most costly operations on non-deterministic TA. In particular, we conducted a set of experiments evaluating the performance of the VATA's optimised TA language inclusion algorithms on pairs of TA obtained from *abstract regular tree model checking* of the algorithm for rebalancing red-black trees after insertion or deletion of a leaf node (which is the same test set that was used in [12]).

5.1   Experiments with the Explicit Encoding

For the explicit encoding, we measured for each inclusion method the fraction of cases in which the method was the fastest among the evaluated methods on the set of almost 2000 tree automata pairs. The results of this experiment are given in Table 1. The columns are labelled as follows: column `expldown` is for pure downward inclusion checking, column `expldown+s` is for downward inclusion using downward simulation, `expldown-opt` is a column for pure downward inclusion checking with the optimisation proposed in Section 4.5, and column `expldown-opt+s` is downward inclusion checking with simulation using the same optimisation. Columns `explup` and `explup+s` give the results for pure upward inclusion checking and upward inclusion checking with simulation respectively. The timeout was set to 30 s.

Table 1. Experiments with inclusion for the explicit encoding

	expldown	expldown+s	expldown-opt	expldown-opt+s	explup	explup+s
Winner	36.35 %	4.15 %	32.20 %	3.15 %	24.14 %	0.00 %
Timeouts	32.51 %	18.27 %	32.51 %	18.27 %	0.00 %	0.00 %

Table 2. Experiments with the explicit encoding for cases when inclusion does not hold

	expldown	expldown+s	expldown-opt	expldown-opt+s	explup	explup+s
Winner	39.85 %	0.00 %	35.30 %	0.00 %	24.84 %	0.00 %
Timeouts	26.01 %	20.31 %	26.01 %	20.31 %	0.00 %	0.00 %

Table 3. Experiments with the explicit encoding for cases when inclusion holds

	expldown	expldown+s	expldown-opt	expldown-opt+s	explup	explup+s
Winner	0.00 %	47.28 %	0.00 %	35.87 %	16.85 %	0.00 %
Timeouts	90.80 %	0.00 %	90.80 %	0.00 %	0.00 %	0.00 %

We also checked the performance of the algorithms for cases when inclusion either *does* or *does not* hold in order to explore the ability of the algorithms to either find a counterexample in the case when inclusion does not hold, or prove the inclusion in case it does. These results are given in Table 2 and Table 3.

When compared to our previous implementation, VATA performed almost always better. The average speed-up was even as high as 200 times for pure downward inclusion checking. The old implementation was faster in about 2.5 % of the cases, and the difference was not significant.

5.2   Experiments with the Semi-symbolic Encoding

We performed a set of similar experiments for the semi-symbolic encoding, the results of which are given in Table 4. The columns are labelled as follows: column symdown is for pure downward inclusion checking, column symdown+s is for downward inclusion using downward simulation, symdown-opt is a column for pure downward inclusion checking with the optimisation proposed in Section 4.5 and column symdown-opt+s is downward inclusion checking with simulation using the same optimisation. Column symup gives the results for pure upward inclusion checking. The timeout was again set to 30 s.

As in the experiments for the explicit encoding, we also checked the performance of the algorithms for cases when inclusion either *does* or *does not* hold. These results are given in Table 5 and Table 6.

When compared to our previous implementation, VATA again performs significantly better, with the pure upward inclusion being on average over 300 times faster and the pure downward inclusion being even over 3000 times faster.

Table 4. Experiments with inclusion for the semi-symbolic encoding

	symdown	symdown+s	symdown-opt	symdown-opt+s	symup
Winner	44.02 %	0.00 %	31.73 %	0.00 %	24.25 %
Timeouts	5.87 %	77.93 %	5.87 %	78.00 %	22.26 %

Table 5. Experiments with the semi-symbolic encoding for cases when inclusion does not hold

	symdown	symdown+s	symdown-opt	symdown-opt+s	symup
Winner	45.03 %	0.00 %	33.06 %	0.00 %	21.91 %
Timeouts	2.48 %	80.03 %	2.48 %	80.09 %	23.39 %

Table 6. Experiments with the semi-symbolic encoding for cases when inclusion holds

	symdown	symdown+s	symdown-opt	symdown-opt+s	symup
Winner	19.74 %	0.00 %	0.00 %	0.00 %	80.26 %
Timeouts	72.37 %	36.84 %	72.37 %	36.84 %	0.00 %

6 Conclusion

This paper introduced and described a new efficient and open-source non-deterministic tree automata library that supports both explicit and semi-symbolic encoding of the tree automata transition function. The semi-symbolic encoding makes use of our own MTBDD package instead of the previously used customisation of the CUDD library.

We wish to continue in this work by attempting to implement a simulation-aware symbolic encoding of antichains using BDDs. Further, we wish to implement other operations, such as determinisation (which, however, is generally desired to be avoided), or complementation (which we so far do not know how to compute without first determinising the automaton).

Finally, we hope that a public release of our library will attract more people to use it and even better contribute to the code base. Indeed, we believe that the library is written in a clean and understandable way that should make such contributions possible.

References

1. Abdulla, P.A., Bouajjani, A., Holík, L., Kaati, L., Vojnar, T.: Composed Bisimulation for Tree Automata. In: Ibarra, O.H., Ravikumar, B. (eds.) CIAA 2008. LNCS, vol. 5148, pp. 212–222. Springer, Heidelberg (2008)
2. Abdulla, P.A., Chen, Y.-F., Holík, L., Mayr, R., Vojnar, T.: When Simulation Meets Antichains (On Checking Language Inclusion of Nondeterministic Finite (Tree) Automata). In: Esparza, J., Majumdar, R. (eds.) TACAS 2010. LNCS, vol. 6015, pp. 158–174. Springer, Heidelberg (2010)
3. Abdulla, P.A., Jonsson, B., Mahata, P., d'Orso, J.: Regular Tree Model Checking. In: Brinksma, E., Larsen, K.G. (eds.) CAV 2002. LNCS, vol. 2404, pp. 555–568. Springer, Heidelberg (2002)
4. Bouajjani, A., Habermehl, P., Holík, L., Touili, T., Vojnar, T.: Antichain-Based Universality and Inclusion Testing over Nondeterministic Finite Tree Automata. In: Ibarra, O.H., Ravikumar, B. (eds.) CIAA 2008. LNCS, vol. 5148, pp. 57–67. Springer, Heidelberg (2008)
5. Bouajjani, A., Habermehl, P., Rogalewicz, A., Vojnar, T.: Abstract Regular Tree Model Checking. In: ENTCS, vol. 149. Elsevier (2006)
6. Bouajjani, A., Habermehl, P., Rogalewicz, A., Vojnar, T.: Abstract Regular Tree Model Checking of Complex Dynamic Data Structures. In: Yi, K. (ed.) SAS 2006. LNCS, vol. 4134, pp. 52–70. Springer, Heidelberg (2006)
7. Bryant, R.E.: Graph-based Algorithms for Boolean Function Manipulation. IEEE Trans. Computers (1986)
8. Clarke, E.M., McMillan, K.L., Zhao, X., Fujita, M., Yang, J.: Spectral Transforms for Large Boolean Functions with Applications to Technology Mapping. In: FMSD, vol. 10. Springer (1997)
9. Claves, P., Jansen, D., Holtrup, S.J., Mohr, M., Reis, A., Schatz, M., Thesing, I.: The LETHAL Library (2009), http://lethal.sourceforge.net/
10. Genet, T.: Timbuk/Taml: A Tree Automata Library (2003), http://www.irisa.fr/lande/genet/timbuk
11. Habermehl, P., Holík, L., Rogalewicz, A., Šimáček, J., Vojnar, T.: Forest Automata for Verification of Heap Manipulation. In: Gopalakrishnan, G., Qadeer, S. (eds.) CAV 2011. LNCS, vol. 6806, pp. 424–440. Springer, Heidelberg (2011)
12. Holík, L., Lengál, O., Šimáček, J., Vojnar, T.: Efficient Inclusion Checking on Explicit and Semi-Symbolic Tree Automata. In: Bultan, T., Hsiung, P.-A. (eds.) ATVA 2011. LNCS, vol. 6996, pp. 243–258. Springer, Heidelberg (2011)

13. Holík, L., Lengál, O., Šimáček, J., Vojnar, T.: Efficient Inclusion Checking on Explicit and Semi-Symbolic Tree Automata. Tech. rep. FIT-TR-2011-04, FIT BUT, Czech Rep. (2011)
14. Holík, L., Šimáček, J.: Optimizing an LTS-Simulation Algorithm. In: Proc. of MEMICS 2009, Znojmo, CZ, FI MU, pp. 93–101 (2009) ISBN 978-80-87342-04-6
15. Hosoya, H., Vouillon, J., Pierce, B.C.: Regular Expression Types for XML. ACM Trans. Program. Lang. Syst. 27 (2005)
16. Ilie, L., Navarro, G., Yu, S.: On NFA Reductions. In: Karhumäki, J., Maurer, H., Păun, G., Rozenberg, G. (eds.) Theory Is Forever. LNCS, vol. 3113, pp. 112–124. Springer, Heidelberg (2004)
17. Klarlund, N., Møller, A., Schwartzbach, M.I.: MONA Implementation Secrets. International Journal of Foundations of Computer Science 13(4) (2002)
18. Madhusudan, P., Parlato, G., Qiu, X.: Decidable Logics Combining Heap Structures and Data. SIGPLAN Not. 46 (2011)
19. Ranzato, F., Tapparo, F.: A New Efficient Simulation Equivalence Algorithm. In: Proc. of LICS 2007. IEEE CS (2007)
20. Somenzi, F.: CUDD: CU Decision Diagram Package Release 2.4.2 (May 2011)

LTL to Büchi Automata Translation: Fast and More Deterministic*

Tomáš Babiak, Mojmír Křetínský, Vojtěch Řehák, and Jan Strejček

Faculty of Informatics, Masaryk University
Botanická 68a, 60200 Brno, Czech Republic
{xbabiak, kretinsky, rehak, strejcek}@fi.muni.cz

Abstract. We introduce improvements in the algorithm by Gastin and Oddoux translating LTL formulae into Büchi automata via very weak alternating co-Büchi automata and generalized Büchi automata. Several improvements are based on specific properties of any formula where each branch of its syntax tree contains at least one *eventually* operator and at least one *always* operator. These changes usually result in faster translations and smaller automata. Other improvements reduce non-determinism in the produced automata. In fact, we modified all the steps of the original algorithm and its implementation known as LTL2BA. Experimental results show that our modifications are real improvements. Their implementations within an LTL2BA translation made LTL2BA very competitive with the current version of SPOT, sometimes outperforming it substantially.

1 Introduction

A translation of LTL formulae into equivalent Büchi automata plays an important role in many algorithms for LTL model checking, LTL satisfiability checking etc. For a long time, researchers aimed to find fast translations producing Büchi automata with a small number of states. This goal has led to the developments of several translation algorithms and many heuristics and optimizations including input formula reductions and optimizations of produced Büchi automata, see e.g. [3,4,9,18,11,12,10,7].

As the time goes, the translation objectives and their importance are changing. In particular, [17] demonstrates that for higher performance of the subsequent steps of the model checking process, it is more important to minimize the number of states with nondeterministic choice than the number of all states in resulting automata. Note that there are LTL formulae, e.g. FGa, for which no equivalent deterministic Büchi automaton exists. Further, model checking practice shows that one LTL formula is usually used in many different model checking tasks. Hence, it pays to invest enough computation time to get high quality (more

---

* The authors are supported by The Czech Science Foundation, grants 102/09/H042 (Babiak), 201/09/1389 (Křetínský), P202/10/1469 (Řehák, Strejček), P202/12/G061 (Křetínský, Řehák, Strejček), and P202/12/P612 (Řehák).

C. Flanagan and B. König (Eds.): TACAS 2012, LNCS 7214, pp. 95–109, 2012.

deterministic and/or minimal) automata as it may reduce computation time of many model checking tasks.

The new objectives lead to the developments of algorithms focusing on quality of produced automata. For example, [5] presents an effective algorithm translating LTL formulae of the fragment called *obligation* (see [14]) into *weak deterministic Büchi automata (WDBA)*. Moreover, WDBA can be minimized by the algorithm of [13]. There is also a SAT-based algorithm searching for minimal (nondeterministic) Büchi automata [8]. The main disadvantage of all the mentioned determinization and minimization algorithms is their long running time which limits their use.

Our research returns to the roots: we focus on a fast translation producing a relatively good output. This approach is justified by the following facts:

- The mentioned algorithms producing high quality automata often need, for a given LTL formula, some equivalent automaton as an input.
- The mentioned algorithms are usually feasible for short formulae only or for formulae with a simple structure.
- Given a fresh LTL formula, it can be useful to run vacuity checks, namely satisfiability of the formula and its negation, to detect bugs in the formula. In these checks, time of the LTL to automata translation can be more significant than time needed for subsequent computations (see [16]). Hence, we need a fast translator to support an early detection of bugs in formulae.

Considering the speed of an LTL to Büchi automata translation, LTL2BA [11] and SPOT [7] are two leading tools. Based on extensive experiments on LTL satisfiability checking, [16] even states:

> *The difference in performance between SPOT and LTL2BA, on one hand, and the rest of explicit tools is quite dramatic.*

Each of the two tools is based on different algorithms.

In LTL2BA, the translation proceeds in three basic steps:

1. A given LTL formula is translated into a *very weak alternating automaton (VWAA)* with a co-Büchi accepting condition.
2. The alternating automaton is then translated into a *transition-based generalized Büchi automaton (TGBA)*, i.e. a generalized Büchi automaton with sets of accepting transitions instead of accepting states.
3. The generalized automaton is transformed (degeneralized) into a Büchi automaton (BA).

Each of the three automata is simplified during the translation.

SPOT translates a given LTL formula to a TGBA using a tableau method presented in [3]. The TGBA is then translated to a BA. Note that the model checking algorithm natively implemented in SPOT works directly with TGBAs. Prior to a translation, both LTL2BA and SPOT try to decrease the number of temporal operators in a given input formula by applications of reduction rules.

While the LTL to automata translation in SPOT is under the gradual development following the current trends (see [6] for improvements made in the last four years), LTL2BA underwent only one minor update in 2007 since its creation in 2001. In particular, SPOT reflects the changes in objectives. Therefore, SPOT usually produces more deterministic and smaller automata than LTL2BA, while LTL2BA is often a bit faster.

Our Contribution. We introduce several modifications of LTL2BA on both algorithmic and implementation levels. We suggest changes in all the steps of the translation algorithm. Our experimental results indicate that each modified step has a mostly positive effect on the translation. The new translator, called LTL3BA, is usually faster than the original LTL2BA and it produces smaller and more deterministic automata. Moreover, comparison of LTL3BA and the current version of SPOT (run without WDBA minimization that is very slow) shows that the produced automata are of similar quality and LTL3BA is usually faster.

Some modifications employ an observation that each LTL formula containing at least one *always* operator and at least one *eventually* operator on each branch of its syntax tree (with possible exceptions of branches going to the left subformula of any *until* or *release* operator) is prefix invariant. We call them *alternating* formulae. Indeed, validity of each alternating formula on a given word u depends purely on a suffix of u. In other words, it is not affected by any finite prefix of u. We apply this observation to construct new rules for formula reductions. Further, the observation justifies some changes in constructions of VWAA and TGBA. Intuitively, a state of a VWAA corresponds to a subformula that has to be satisfied by the rest of an accepted word. If the corresponding subformula is an alternating formula, then the state can be temporarily suspended for finitely many steps of the automaton.

Other changes in a VWAA construction are designed to lower nondeterminism. This is also a motivation for new simplification rules applied on intermediate automata. These rules remove some transitions of the automaton and hence reduce the number of nondeterministic choices in produced automata. The original simplification rules can be seen as special cases of the new rules. An effective implementation of this simplification required to change representation of transitions. Further, we add one ad-hoc modification speeding up the translation of selected (sub)formulae. Finally, we modify a simplification rule merging some states of resulting BA.

The rest of the paper is organized as follows. The next section recalls the definitions of LTL, VWAA, and TGBA, as presented in [11]. Section 3 focuses on alternating formulae and its properties. Sections 4, 5, 6, and 7 present new rules for formula reductions, modified translation of LTL to VWAA (including generalized simplification of VWAA), modified translation of VWAA to TGBA, and modified rule for simplification of BA, respectively. Finally, Section 8 is devoted to experimental results. The last section summarizes the achieved improvements.

2   Preliminaries

Linear Temporal Logic (LTL). The syntax of LTL [15] is defined as follows

$$\varphi ::= tt \mid a \mid \neg\varphi \mid \varphi \vee \varphi \mid \varphi \wedge \varphi \mid \mathsf{X}\varphi \mid \varphi \mathsf{U} \varphi,$$

where tt stands for *true*, a ranges over a countable set AP of *atomic propositions*, X and U are temporal operators called *next* and *until*, respectively. The logic is interpreted over infinite words over the alphabet $\Sigma = 2^{AP'}$, where $AP' \subseteq AP$ is a finite subset. Given a word $u = u(0)u(1)u(2)\ldots \in \Sigma^\omega$, by u_i we denote the ith suffix of u, i.e. $u_i = u(i)u(i+1)\ldots$.

The semantics of LTL formulae is defined inductively as follows:

$$
\begin{aligned}
&u \models tt \\
&u \models a &&\text{iff } a \in u(0) \\
&u \models \neg\varphi &&\text{iff } u \not\models \varphi \\
&u \models \varphi_1 \vee \varphi_2 &&\text{iff } u \models \varphi_1 \text{ or } u \models \varphi_2 \\
&u \models \varphi_1 \wedge \varphi_2 &&\text{iff } u \models \varphi_1 \text{ and } u \models \varphi_2 \\
&u \models \mathsf{X}\varphi &&\text{iff } u_1 \models \varphi \\
&u \models \varphi_1 \mathsf{U} \varphi_2 &&\text{iff } \exists i \geq 0 \, . \, (\, u_i \models \varphi_2 \text{ and } \forall 0 \leq j < i \, . \, u_j \models \varphi_1 \,)
\end{aligned}
$$

We say that a word u *satisfies* φ whenever $u \models \varphi$. Two formulae φ, ψ are *equivalent*, written $\varphi \equiv \psi$, if for each alphabet Σ and each $u \in \Sigma^\omega$ it holds $u \models \varphi \iff u \models \psi$. Given an alphabet Σ, a formula φ defines the language $L^\Sigma(\varphi) = \{u \in \Sigma^\omega \mid u \models \varphi\}$. We often write $L(\varphi)$ instead of $L^{2^{AP(\varphi)}}(\varphi)$, where $AP(\varphi)$ denotes the set of atomic propositions occurring in the formula φ.

We extend the LTL with derived temporal operators:

- $\mathsf{F}\varphi$ called *eventually* and equivalent to $tt \, \mathsf{U} \, \varphi$,
- $\mathsf{G}\varphi$ called *always* and equivalent to $\neg\mathsf{F}\neg\varphi$, and
- $\varphi \, \mathsf{R} \, \psi$ called *release* and equivalent to $\neg(\neg\varphi \, \mathsf{U} \, \neg\psi)$.

In the following, *temporal formula* is a formula where the topmost operator is neither conjunction, nor disjunction. A formula without any temporal operator is called *state formula*. Note that a and tt are both temporal and state formulae. An LTL formula is in *positive normal form* if no operator occurs in the scope of any negation. Each LTL formula can be easily transformed to positive normal form using De Morgan's laws for operators $\vee$ and $\wedge$, equivalences for derived operators, and the following equivalences:

$$\neg(\varphi_1 \, \mathsf{U} \, \varphi_2) \equiv \neg\varphi_1 \, \mathsf{R} \, \neg\varphi_2 \qquad \neg(\varphi_1 \, \mathsf{R} \, \varphi_2) \equiv \neg\varphi_1 \, \mathsf{U} \, \neg\varphi_2 \qquad \neg\mathsf{X}\varphi \equiv \mathsf{X}\neg\varphi$$

Very Weak Alternating co-Büchi Automata (VWAA). A VWAA is a tuple $\mathcal{A} = (Q, \Sigma, \delta, I, F)$, where

- Q is a finite set of *states*, and we let $Q' = 2^Q$,
- Σ is a finite *alphabet*, and we let $\Sigma' = 2^\Sigma$,
- $\delta : Q \to 2^{\Sigma' \times Q'}$ is a *transition function*,

- $I \subseteq Q'$ is a set of *initial states,*
- $F \subseteq Q$ is a set of *accepting states,* and
- there exists a partial order on Q such that, for each state $q \in Q$, all the states occurring in $\delta(q)$ are lower or equal to q.

Note that the transition function δ uses Σ' instead of Σ. This enables to merge transitions that differ only by action labels. We sometimes use a propositional formula α over AP to describe the element $\{a \in \Sigma \mid a$ satisfies $\alpha\}$ of Σ'.

A *run* σ of VWAA $\mathcal{A}$ over a word $w = w(0)w(1)w(2)\ldots \in \Sigma^\omega$ is a labelled directed acyclic graph (V, E, λ) such that:

- V is partitioned into $\bigcup\limits_{i=0}^{\infty} V_i$ with $E \subseteq \bigcup\limits_{i=0}^{\infty} V_i \times V_{i+1}$,
- $\lambda : V \to Q$ is a labelling function,
- $\{\lambda(x) \mid x \in V_0\} \in I$, and
- for each $x \in V_i$, there exist $\alpha \in \Sigma'$, $q \in Q$ and $O \in Q'$ such that $w(i) \in \alpha$, $q = \lambda(x)$, $O = \{\lambda(y) \mid (x, y) \in E\}$, and $(\alpha, O) \in \delta(q)$.

A run σ is *accepting* if each branch in σ contains only finitely many nodes labelled by accepting states (co-Büchi acceptance condition). A word w is *accepted* if there is an accepting run over w.

Transition Based Generalized Büchi Automata (TGBA). A TGBA is a tuple $\mathcal{G} = (Q, \Sigma, \delta, I, \mathcal{F})$, where

- Q is a finite set of *states,*
- Σ is a finite *alphabet,* and we let $\Sigma' = 2^\Sigma$
- $\delta : Q \to 2^{\Sigma' \times Q}$ is a total *transition function,*
- $I \subseteq Q$ is a set of *initial states,* and
- $\mathcal{T} = \{T_1, T_2, \ldots, T_m\}$ where $T_j \subseteq Q \times \Sigma' \times Q$ are sets of *accepting transitions.*

A run ρ of TGBA $\mathcal{G}$ over a word $w = w(0)w(1)w(2)\ldots \in \Sigma^\omega$ is a sequence of states $\rho = q_0 q_1 q_2 \ldots$, where $q_0 \in I$ is an initial state and, for each $i \geq 0$, there exists $\alpha \in \Sigma'$ such that $w(i) \in \alpha$ and $(\alpha, q_{i+1}) \in \delta(q_i)$. A run ρ is *accepting* if for each $1 \leq j \leq m$ it uses infinitely many transitions from T_j. A word w is *accepted* if there is an accepting run over w.

3   Alternating Formulae

We define the class of *alternating formulae* together with the classes of *pure eventuality* and *pure universality* formulae introduced in [9]. Let φ ranges over general LTL formulae. The classes of *pure eventuality* formulae μ, *pure universality* formulae ν, and *alternating* formulae ξ are defined as:

$$\mu ::= \mathsf{F}\varphi \mid \mu \vee \mu \mid \mu \wedge \mu \mid \mathsf{X}\mu \mid \varphi \mathsf{U} \mu \mid \mu \mathsf{R} \mu \mid \mathsf{G}\mu$$
$$\nu ::= \mathsf{G}\varphi \mid \nu \vee \nu \mid \nu \wedge \nu \mid \mathsf{X}\nu \mid \nu \mathsf{U} \nu \mid \varphi \mathsf{R} \nu \mid \mathsf{F}\nu$$
$$\xi ::= \mathsf{G}\mu \mid \mathsf{F}\nu \mid \xi \vee \xi \mid \xi \wedge \xi \mid \mathsf{X}\xi \mid \varphi \mathsf{U} \xi \mid \varphi \mathsf{R} \xi \mid \mathsf{F}\xi \mid \mathsf{G}\xi$$

Note that there are alternating formulae, e.g. $(a \, U \, (GFb)) \wedge (c \, R \, (GFd))$, that are neither pure eventuality formulae, nor pure universality formulae. Properties of the respective classes of formulae are summarized in the following lemmata.

Lemma 1. *[9] Let μ be a pure eventuality formula and ν be a pure universality formula. For all $u \in \Sigma^*$, $w \in \Sigma^\omega$ it holds:*

$$uw \models \mu \impliedby w \models \mu$$

$$uw \models \nu \implies w \models \nu$$

Lemma 2. *Let ξ be an alternating formula. For all $u \in \Sigma^*$, $w \in \Sigma^\omega$ it holds:*

$$uw \models \xi \iff w \models \xi$$

In other words, pure eventuality formulae define *left-append closed* languages, pure universality formulae define *suffix closed* languages, and alternating formulae define *prefix-invariant* languages. The proof of Lemma 2 can be found in the full version of this paper [1].

Corollary 1. *Every alternating formula ξ satisfies $\xi \equiv X\xi$.*

Hence, in order to check whether w satisfies ξ it is possible to skip an arbitrary long finite prefix of the word w.

We use this property in new rule for formula reduction. Further, it has brought us to the notion of alternating formulae *suspension* during the translation of LTL to Büchi automata. We employ suspension on two different levels of the translation: the construction of a VWAA from an input LTL formula and the transformation of a VWAA into a TGBA.

4   Improvements in Reduction of LTL Formulae

Many rules reducing the number of temporal operators in an LTL formula have been presented in [18] and [9]. In this section we present some new reduction rules. For the rest of this section, φ, ψ range over LTL formulae and γ ranges over alternating ones.

$$X\varphi \, R \, X\psi \equiv X(\varphi \, R \, \psi) \qquad \varphi \, U \, \gamma \equiv \gamma \qquad F\gamma \equiv \gamma \qquad X\gamma \equiv \gamma$$
$$X\varphi \vee X\psi \equiv X(\varphi \vee \psi) \qquad \varphi \, R \, \gamma \equiv \gamma \qquad G\gamma \equiv \gamma$$

The following equivalences are valid only on assumption that φ implies ψ.

$$\psi \, U \, (\varphi \, U \, \gamma) \equiv \psi \, U \, \gamma \qquad\qquad \varphi \wedge (\psi \wedge \gamma) \equiv (\varphi \wedge \gamma)$$
$$(\psi \, R \, \gamma) \, R \, \varphi \equiv \gamma \, R \, \varphi \qquad\qquad \psi \vee (\varphi \vee \gamma) \equiv (\psi \vee \gamma)$$
$$\varphi \, U \, (\gamma \, R \, (\psi \, U \, \rho)) \equiv \gamma \, R \, (\psi \, U \, \rho)$$

Further, we have extended the set of rules deriving implications of the form $\varphi \Rightarrow \psi$. The upper formula is a precondition, the lower one is a conclusion.

$$\frac{G\varphi \Rightarrow \psi}{G\varphi \Rightarrow X\psi} \qquad \frac{\varphi \Rightarrow F\psi}{X\varphi \Rightarrow F\psi} \qquad \frac{\varphi \Rightarrow \psi}{X\varphi \Rightarrow X\psi}$$

5   Improvements in LTL to VWAA Translation

First, we recall the original translation of LTL to VWAA according to [11]. The translation utilizes two auxiliary operators:

- Let $\Sigma' = 2^\Sigma$, and let $Q' = 2^Q$. Given $J_1, J_2 \in 2^{\Sigma' \times Q'}$, we define

$$J_1 \otimes J_2 = \{(\alpha_1 \cap \alpha_2, O_1 \cup O_2) \mid (\alpha_1, O_1) \in J_1 \text{ and } (\alpha_2, O_2) \in J_2\}.$$

- Let ψ be an LTL formula in positive normal form. We define $\overline{\psi}$ by:
  - $\overline{\psi} = \{\{\psi\}\}$ if ψ is a temporal formula,
  - $\overline{\psi_1 \wedge \psi_2} = \{O_1 \cup O_2 \mid O_1 \in \overline{\psi_1} \text{ and } O_2 \in \overline{\psi_2}\}$,
  - $\overline{\psi_1 \vee \psi_2} = \overline{\psi_1} \cup \overline{\psi_2}$.

Let φ be an LTL formula in positive normal form. An equivalent VWAA with a co-Büchi acceptance condition is constructed as $\mathcal{A}_\varphi = (Q, \Sigma, \delta, I, F)$, where Q is the set of temporal subformulae of φ, $\Sigma = 2^{AP(\varphi)}$, $I = \overline{\varphi}$, F is the set of all U-subformulae of φ, i.e formulae of the type $\psi_1 \cup \psi_2$, and δ is defined as follows:

$$\delta(tt) = \{(\Sigma, \emptyset)\}$$
$$\delta(p) = \{(\Sigma_p, \emptyset)\} \text{ where } \Sigma_p = \{a \in \Sigma \mid p \in a\}$$
$$\delta(\neg p) = \{(\Sigma_{\neg p}, \emptyset)\} \text{ where } \Sigma_{\neg p} = \Sigma \setminus \Sigma_p$$
$$\delta(X\psi) = \{(\Sigma, O) \mid O \in \overline{\psi}\}$$
$$\delta(\psi_1 \cup \psi_2) = \Delta(\psi_2) \cup \left(\Delta(\psi_1) \otimes \{(\Sigma, \{\psi_1 \cup \psi_2\})\}\right)$$
$$\delta(\psi_1 R \psi_2) = \Delta(\psi_2) \otimes \left(\Delta(\psi_1) \cup \{(\Sigma, \{\psi_1 R \psi_2\})\}\right)$$

$$\Delta(\psi) = \delta(\psi) \text{ if } \psi \text{ is a temporal formula}$$
$$\Delta(\psi_1 \vee \psi_2) = \Delta(\psi_1) \cup \Delta(\psi_2)$$
$$\Delta(\psi_1 \wedge \psi_2) = \Delta(\psi_1) \otimes \Delta(\psi_2)$$

Using the partial order "is a subformula of" on states of $\mathcal{A}_\varphi$, one can easily prove that $\mathcal{A}_\varphi$ is very weak.

Improved Translation. In order to implement the suspension of alternating formulae, we modify the way the transition function δ handles the binary operators U, R, $\vee$, and $\wedge$. The original transition function δ reflects the following identities:

$$\varphi_1 \cup \varphi_2 \equiv \varphi_2 \vee (\varphi_1 \wedge X(\varphi_1 \cup \varphi_2)) \qquad \varphi_1 R \varphi_2 \equiv \varphi_2 \wedge (\varphi_1 \vee X(\varphi_1 R \varphi_2))$$

However, if φ_1 is an alternating formula we apply the relation $\varphi_1 \equiv X\varphi_1$ to obtain the following identities:

$$\varphi_1 \cup \varphi_2 \equiv \varphi_2 \vee (X\varphi_1 \wedge X(\varphi_1 \cup \varphi_2)) \qquad \varphi_1 R \varphi_2 \equiv \varphi_2 \wedge (X\varphi_1 \vee X(\varphi_1 R \varphi_2))$$

Using these identities, the formula φ_1 is effectively suspended and checked one step later. Similarly, in the case of disjunction or conjunction, each disjunct or conjunct corresponding to an alternating formula is suspended for one step as well. Correctness of these changes clearly follows from properties of alternating formulae. Note that δ is defined over formulae in positive normal form only. The translation treats each formula $\mathsf{F}\psi$ as $tt\,\mathsf{U}\,\psi$ and each formula $\mathsf{G}\psi$ as $(\neg tt)\,\mathsf{R}\,\psi$.

We introduce further changes to the transition function δ in order to generate automata which exhibits more determinism. In particular, we build a VWAA with only one initial state. Similarly, each state corresponding to a formula of a type $\mathsf{X}\varphi$ generates only one successor corresponding to φ. These changes can add an extra initial state and an extra state for each X-subformula comparing to the original construction. However, this drawback is often suppressed due to the consecutive optimizations during the construction of a TGBA.

Now we present a modified construction of VWAA. Given an input LTL formula φ in positive normal form, an equivalent VWAA with a co-Büchi acceptance condition is constructed as $\mathcal{A}_\varphi = (Q, \Sigma, \delta, I, F)$, where Q is the set of all subformulae of φ, Σ and F are defined as in the original construction, $I = \{\varphi\}$, and δ is defined as follows:

$$\delta(tt) = \{(\Sigma, \emptyset)\}$$
$$\delta(p) = \{(\Sigma_p, \emptyset)\} \text{ where } \Sigma_p = \{a \in \Sigma \mid p \in a\}$$
$$\delta(\neg p) = \{(\Sigma_{\neg p}, \emptyset)\} \text{ where } \Sigma_{\neg p} = \Sigma \backslash \Sigma_p$$
$$\delta(\mathsf{X}\psi) = \{(\Sigma, \{\psi\})\}$$
$$\delta(\psi_1 \vee \psi_2) = \Delta(\psi_1) \cup \Delta(\psi_2)$$
$$\delta(\psi_1 \wedge \psi_2) = \Delta(\psi_1) \otimes \Delta(\psi_2)$$
$$\delta(\psi_1 \,\mathsf{U}\, \psi_2) = \begin{cases} \Delta(\psi_2) \cup (\{(\Sigma, \{\psi_1\})\} \otimes \{(\Sigma, \{\psi_1 \,\mathsf{U}\, \psi_2\})\}) & \text{if } \psi_1 \text{ is alternating,} \\ \Delta(\psi_2) \cup (\Delta(\psi_1) \otimes \{(\Sigma, \{\psi_1 \,\mathsf{U}\, \psi_2\})\}) & \text{otherwise.} \end{cases}$$
$$\delta(\psi_1 \,\mathsf{R}\, \psi_2) = \begin{cases} \Delta(\psi_2) \otimes (\{(\Sigma, \{\psi_1\}), (\Sigma, \{\psi_1 \,\mathsf{R}\, \psi_2\})\}) & \text{if } \psi_1 \text{ is alternating,} \\ \Delta(\psi_2) \otimes (\Delta(\psi_1) \cup \{(\Sigma, \{\psi_1 \,\mathsf{R}\, \psi_2\})\}) & \text{otherwise.} \end{cases}$$

$$\Delta(\psi) = \begin{cases} \{(\Sigma, \{\psi\})\} & \text{if } \psi \text{ is a temporal alternating formula,} \\ \delta(\psi) & \text{if } \psi \text{ is a temporal formula that is not alternating.} \end{cases}$$
$$\Delta(\psi_1 \vee \psi_2) = \Delta(\psi_1) \cup \Delta(\psi_2)$$
$$\Delta(\psi_1 \wedge \psi_2) = \Delta(\psi_1) \otimes \Delta(\psi_2)$$

Motivation for our changes in the translation can be found in Figures 1 and 2. Each figure contains (a) the VWAA constructed by the original translation and (b) the VWAA constructed by our translation with suspension. Figure 1 shows the effect of suspension of alternating subformula $\mathsf{GF}a$ in computation of transitions leading from the initial state. It can be easily proved that whenever one start with a formula reduced according to Section 4, then each suspension of an alternating temporal subformula leads just to reduction of transitions in the resulting VWAA, i.e., no state is added. On the other hand, if an alternating non-temporal subformula ψ is suspended or the new definition of $\delta(\mathsf{X}\psi)$ is used, then the resulting VWAA can contain one more reachable state corresponding to the formula ψ. However, other states may become unreachable and, in particular, the automaton can also have more deterministic states as illustrated by Figure 2.

Optimization of VWAA. In the original algorithm, the VWAA is optimized before it is translated to a TGBA. In particular, if there are two transitions

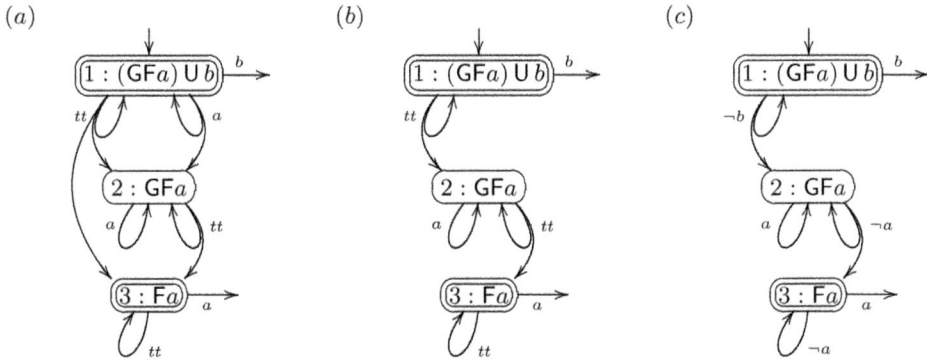

Fig. 1. VWAA for $(GFa) \cup b$ generated by (a) the translation of [11], (b) our translation with suspension, and (c) our translation with suspension and further determinization

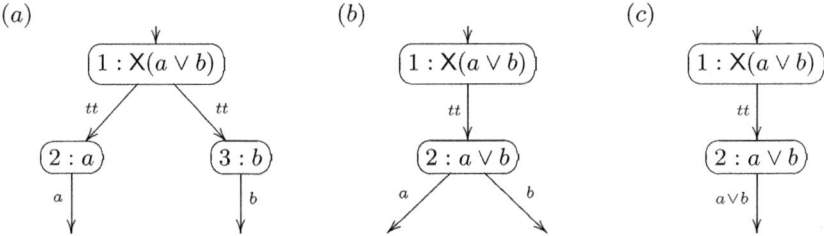

Fig. 2. VWAA for $X(a \vee b)$ generated by (a) the translation of [11], (b) our translation with suspension, and (c) the translation with suspension and further determinization

$t_1 = (q, \alpha_1, O_1)$ and $t_2 = (q, \alpha_2, O_2)$ satisfying $\alpha_2 \subseteq \alpha_1$ and $O_1 \subseteq O_2$, then t_2 is removed as it is implied by t_1.

We suggest a generalization of this principle: if $O_1 \subsetneq O_2$ then replace the label α_2 in t_2 by $\alpha_2 \wedge \neg \alpha_1$. If $O_1 = O_2$, replace both transitions by the transition $(q, \alpha_1 \vee \alpha_2, O_1)$. Note that if $\alpha_2 \Rightarrow \alpha_1$, i.e. $\alpha_2 \subseteq \alpha_1$, then $\alpha_2 \wedge \neg \alpha_1 \equiv \neg tt$ and transition t_2 can be removed as before. Our generalized optimization rule increase determinism of the produced VWAA as illustrated by automata (c) of Figures 1 and 2.

6   Improvements in VWAA to TGBA Translation

First, we recall the translation of VWAA to TGBA introduced in [11]. Let $\mathcal{A}_\varphi = (Q, \Sigma, \delta, I, F)$ be a VWAA with a co-Büchi acceptance condition. We define $\mathcal{G}_\mathcal{A} = (Q', \Sigma, \delta', I, \mathcal{T})$ to be a TGBA where:

- $Q' = 2^Q$, i.e. a state is a set of states of $\mathcal{A}_\varphi$ and represents their conjunction,
- $\delta''(\{q_1, q_2, \ldots, q_n\}) = \bigotimes\limits_{i=1}^{n} \delta(q_i)$ is the non-optimized transition function,
- δ' is the optimized transition function defined as the set of $\preccurlyeq$-minimal transitions of δ'' where the relation $\preccurlyeq$ is defined by $t_1 \preccurlyeq t_2$ iff $t_1 = (O, \alpha_1, O_1)$, $t_2 = (O, \alpha_2, O_2)$, $\alpha_2 \subseteq \alpha_1$, $O_1 \subseteq O_2$, and $\forall T_f \in \mathcal{T}$, $t_2 \in T_f \Rightarrow t_1 \in T_f$, and
- $\mathcal{T} = \{T_f \mid f \in F\}$ where
  $T_f = \{(O, \alpha, O') \mid f \notin O' \text{ or } \exists (\beta, O'') \in \delta(f), \alpha \subseteq \beta \text{ and } f \notin O'' \subseteq O'\}$.

Improved Translation. Our algorithm for a VWAA to TGBA translation differs from the original one only in definition of δ, where we also integrate the idea of suspension of alternating formulae. Recall that each state q_i of a VWAA is a subformula of an input LTL formula and each state of a TGBA is identified with a conjunction of states of a VWAA. Let $O = \{q_1, \ldots, q_n\}$ be a state of a TGBA. Then transitions leading from O in a TGBA correspond to combinations of transitions leading from $q_1, \ldots, q_n$ in a VWAA. If q_i is an alternating formula and thus it satisfies $q_i \equiv \mathsf{X} q_i$, we can effectively decrease the number of transition combinations that need to be considered during computation of $\delta'(O)$ provided we suspend a full processing of q_i to the succeeding states of the TGBA. More precisely, for the purpose of computation of $\delta'(O)$, we set $\delta(q_i) = \{(\Sigma, \{q_i\})\}$. To construct a TGBA equivalent to the VWAA, we have to ensure that q_i will not be suspended forever during any accepting run of the TGBA. Hence, we enable suspension only in the states that are not on any accepting cycle in a TGBA.

Let M be the minimal set containing all VWAA states of the form $\psi \mathsf{R} \rho$ and all subformulae of their right operands ρ. One can easily observe each TGBA state lying on some accepting cycle is a subset of M. The VWAA states outside M, called *progress formulae*, push TGBA computations towards accepting cycles. Suspension is enabled in a TGBA state only if it contains a progress formula. However, if all progress formulae in a TGBA state are alternating, their suspension is not allowed (as suspended progress formulae would not enforce any progress).

Formally, for each TGBA state $O = \{q_1, \ldots, q_n\}$ we define $\delta''(O)$ as follows:

$$\delta''(O) = \bigotimes_{i=1}^{n} \delta_O(q_i), \text{ where}$$

$$\delta_O(q_i) = \begin{cases} \{(\Sigma, \{q_i\})\} & \text{if } O \text{ contains a progress non-alternating formula} \\ & \text{and } q_i \text{ is an alternating formula,} \\ & \text{or } O \text{ contains a progress formula} \\ & \text{and } q_i \text{ is an alternating non-progress formula,} \\ \delta(q_i) & \text{otherwise.} \end{cases}$$

We have obtained better results when we restrict the definition of progress formulae to temporal progress formulae.

Note that the original translation of VWAA to TGBA uses a correct but nonstandard definition of accepting sets T_f. In fact, our modification is correct

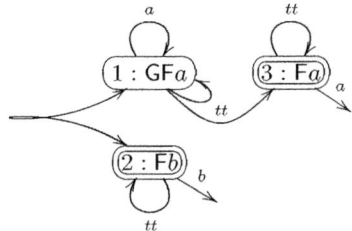

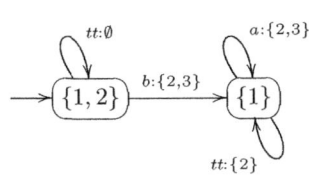

Fig. 3. A VWAA $\mathcal{A}_\psi$ corresponding to GFa $\wedge$ Fb

Fig. 4. A TGBA $\mathcal{G}_\psi$ corresponding to the VWAA of Figure 3

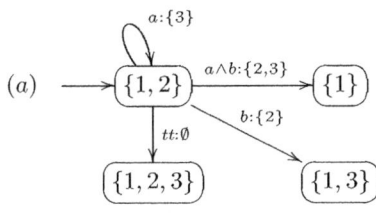

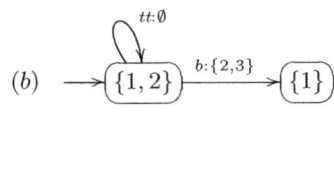

Fig. 5. Transitions leading from state $\{1, 2\}$ in the TGBA constructed from the VWAA of Figure 3 by (a) the translation of [11] and by (b) our translation with suspension

only if we change the definition of these sets to the natural one (see [1] for a explanation). Intuitively, for each accepting state f of the VWAA with a co-Büchi acceptance, we compute a set T_f of all TGBA transitions that do not contain any VWAA transition looping in f. Formally, $\mathcal{T} = \{T_f \mid f \in F\}$ where

$$T_f = \{(O, \alpha, O') \mid f \notin O \text{ or } (\exists(\beta, O'') \in \delta(f), \exists(\gamma, O''') \in \bigotimes_{f' \in O \smallsetminus \{f\}} \delta(f')$$
$$\text{such that } f \notin O'', \alpha = \beta \wedge \gamma, \text{ and } O' = O'' \cup O''')\}.$$

To demonstrate the effect of suspension during the construction of a TGBA, consider the VWAA $\mathcal{A}_\psi$ for the formula $\psi = $ GFa $\wedge$ Fb depicted in Figure 3. The construction of an equivalent TGBA $\mathcal{G}_\psi$ starts in the initial state $\{1, 2\}$ that corresponds to a conjunction of states 1 and 2 of $\mathcal{A}_\psi$. Figure 5 depicts the transitions of $\mathcal{G}_\psi$ leading from the initial state when constructed by (a) the original translation of [11] and by (b) our translation with suspension. Note that the state 1 corresponding to the alternating formula GFa is suspended in the TGBA state $\{1, 2\}$ as the state 2 corresponds to a non-alternating progress formula Fb. In both cases, the TGBA has two sets of accepting transitions, T_2 and T_3. Each transition in the TGBA is labelled by a propositional formula over AP and by a subset of $\{2, 3\}$ indicating to which sets of T_2, T_3 the transition belongs.

Comparing to the original VWAA to TGBA translation without any optimizations, the application of suspension leads to automata with fewer states.

However, if we enable the optimizations suggested in [11], the original translation often constructs automata with the same number of states as our translation with suspension. For example, in the TGBA constructed from the VWAA of Figure 3, the optimizations merge states $\{1,2,3\}$ and $\{1,3\}$ with $\{1,2\}$ and $\{1\}$, respectively. In this particular case, both approaches lead to the same automaton $\mathcal{G}_\psi$ as shown in Figure 4. However, this is not the case in general. Using suspension, automata with either more or less states can be constructed. However, the translation with suspension is usually slightly faster.

In addition, we detect that both the original and the improved algorithms spend a lot of time when computing transitions of TGBA states equivalent to a formula of the form $\rho = G\alpha_0 \wedge \bigwedge_{1 \le i \le n} GF\alpha_i$ where $n \ge 0$ and $\alpha_0, \alpha_1, \ldots, \alpha_n$ are formulae without any temporal operator. As such TGBA states are very frequent in practice, we use an optimization that detects these TGBA states and directly constructs the optimal transitions.

7   Optimization of BA

We slightly modify one optimization rule suggested in [11]. It is applied on a resulting BA. The rule says that states q_1 and q_2 of a BA can be merged if $\delta(q_1) = \delta(q_2)$ and $q_1 \in F \iff q_2 \in F$. This rule typically fails to merge the states with a self loop. We suggest to add a new rule where the condition $\delta(q_1) = \delta(q_2)$ is replaced by $\delta(q_1)[q_1/r] = \delta(q_2)[q_2/r]$, where r is a fresh artificial state and $\delta(q)[q/r]$ is a $\delta(q)$ with all occurrences of q as a target node replaced by r.

8   Implementation and Experimental Result

We have implemented all the modifications suggested in the previous sections (and formula reduction rules suggested in [9]) in order to evaluate their effect. The implementation is based on LTL2BA and therefore called *LTL3BA*. Besides the changed algorithms, we also made some other, implementation related changes. In particular, we represent transition labels by BDDs and transitions are represented by C++ STL containers.

In this section, we compare LTL3BA with LTL2BA (v1.1) and SPOT (v0.7.1). For the comparison of results, we use lbtt testbench tool [19] to measure, for each translator, the number of states and transitions[1] of resulting automata, and the time of the computation. Further, we extend lbtt to count the number of produced deterministic automata. To be able to compare the results, we set SPOT (option -N) to output automata in the form of never claim for SPIN as that is the output of LTL2BA as well. All experiments were done on a server with 8 processors Intel® Xeon® X7560, 448 GiB RAM and a 64-bit version of GNU/Linux.

---

[1] To solve the problem with different representation of transitions in automata produced by different tools, we count all transitions leading from a state q to a state r as one.

Table 1. Comparison of translators on two sets of random formulae. Time is in seconds, 'det. BA' is the number of deterministic automata produced by the translator. Note that, using WDBA minimization, SPOT failed to translate 6 formulae of Benchmark2 within the one hour limit. In order to see the effect of WDBA minimization to other formulae, we state in braces the original results increased by the values obtained when these 6 formulae were translated withut WDBA minimization.

Translator	Benchmark1				Benchmark2			
	States	Trans.	Time	det. BA	States	Trans.	Time	det. BA
SPOT	1 561	5 729	7.47	55	14 697	95 645	68.46	221
SPOT+WDBA	1 587	5 880	10.81	88	13 097	77 346	5 916.45	373
					(14 408)	(94 248)	(5 919.43)	(373)
LTL2BA	2 118	9 000	0.81	25	24 648	232 400	18.57	84
LTL3BA(1)	1 621	5 865	1.26	27	17 107	129 774	22.25	92
LTL3BA(1,2)	1 631	6 094	1.41	54	15 936	115 624	9.04	237
LTL3BA(1,2,3)	1 565	5 615	1.41	54	14 113	91 159	8.53	240
LTL3BA(1,2,3,4)	1 507	5 348	1.38	54	13 244	85 511	8.30	240

First we compare the translators on two sets, Benchmark1 and Benchmark2, of random formulae generated by lbtt. Benchmark1 contains 100 formulae of the length 15–20 and their negations. Benchmark2 contains 500 formulae of the length 15–30 and their negations. The exact lbtt parameters used to generate the formulae are in [1]. Table 1 presents the cumulative results of translations of all formulae in the two sets. The table also illustrates the gradual effect of modifications of each step of the translation (1,2,3,4 refers to modifications introduced in Sections 4, 5, 6, and 7 in the respective order; e.g. LTL3BA(1) uses the original algorithm with our formula reduction while LTL3BA(1,2,3,4) refers to the translation with all the suggested modifications). Finally, the table contains the results for SPOT with WDBA minimization, which has the longest running time but provides the best results. The automata produced by LTL3BA are in sum slightly better than the automata produced by SPOT. Further, LTL3BA seems to be much faster.

Further, we compare the execution time of translators running on parametric formulae from [11] and [16]. We use SPOT with the recommended option -r4, i.e. with the input formula reduction as the only optimization. To get a comparable settings of LTL3BA, we switched off the generalized optimization of VWAA. We gradually increase the parameter of the formulae until a translator fails to finish the translation in one hour limit. The results are partly depicted in Figure 6 (the rest is in [1]).

The graphs show that, in general, LTL3BA is slightly slower than LTL2BA and faster than SPOT on small formulae. With increasing parameter, LTL3BA outperforms LTL2BA (with exception of $S(n)$ where LTL2BA fails before its running time reaches the limit), while SPOT sometimes remains slower, but sometimes eventually outperform LTL3BA.

For more experimental results (including the benchmark of [2]) see [1].

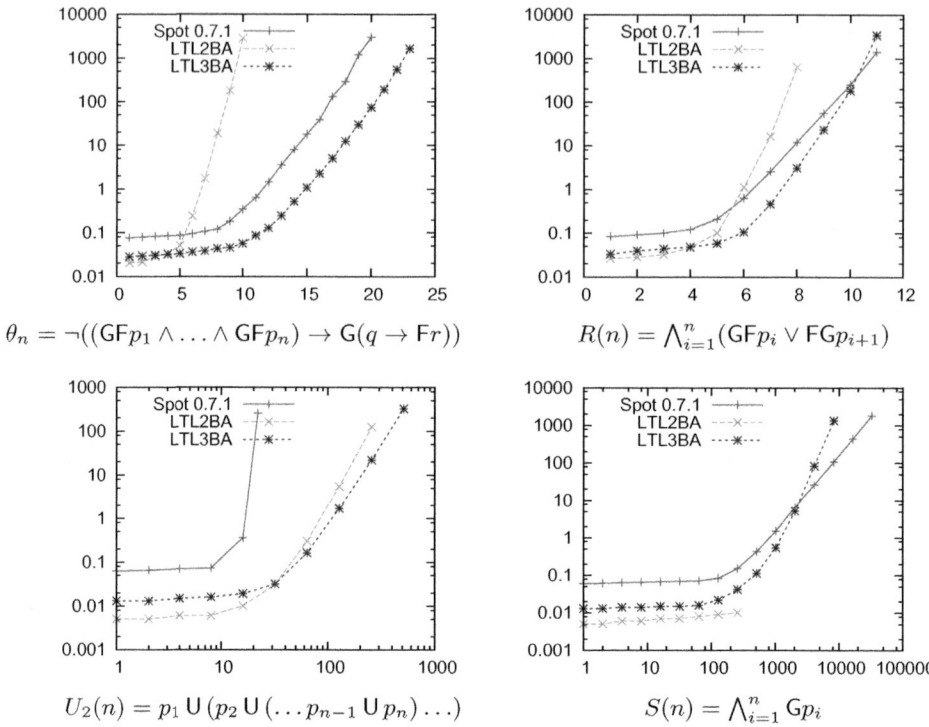

Fig. 6. Time consumption for parametric formulae constructed within an hour (the vertical axes are logarithmic and represent time in seconds, while the horizontal axes are linear or logarithmic and represent the parameter n)

9   Conclusion

We have focused on LTL to BA translations with the stress on their speed-up while maintaining outputs of a good quality. We have introduced several modifications of LTL2BA on both algorithmic and implementation levels. Among others, we have identified an LTL subclass of "alternating" formulae, validity of which does not depends on any finite prefix of the word.

Our experimental results indicate that our modifications have a mostly positive effect on each step of the translation. The new translator called LTL3BA is usually faster than the original LTL2BA and it produces smaller and more deterministic automata. Moreover, comparison of LTL3BA and the current version of SPOT (run without WDBA minimization that is very slow) shows that the produced automata are of similar quality and LTL3BA is usually faster.

LTL3BA has served as an experimental tool to demonstrate that our modifications are improvements and their applicability to other LTL to BA translations is a subject of further research.

Acknowledgments. The authors would like to thank three anonymous refrees and Alexandre Duret-Lutz for valuable comments.

References

1. Babiak, T., Křetínský, M., Řehák, V., Strejček, J.: LTL to Büchi Automata Translation: Fast and More Deterministic. CoRR, abs/1201.0682 (2012)
2. Cichoń, J., Czubak, A., Jasiński, A.: Minimal Büchi automata for certain classes of LTL formulas. In: DEPCOS-RELCOMEX 2009, pp. 17–24. IEEE (2009)
3. Couvreur, J.-M.: On-the-Fly Verification of Linear Temporal Logic. In: Wing, J.M., Woodcock, J. (eds.) FM 1999. LNCS, vol. 1708, pp. 253–271. Springer, Heidelberg (1999)
4. Daniele, M., Giunchiglia, F., Vardi, M.Y.: Improved Automata Generation for Linear Temporal Logic. In: Halbwachs, N., Peled, D.A. (eds.) CAV 1999. LNCS, vol. 1633, pp. 249–260. Springer, Heidelberg (1999)
5. Dax, C., Eisinger, J., Klaedtke, F.: Mechanizing the Powerset Construction for Restricted Classes of ω-Automata. In: Namjoshi, K.S., Yoneda, T., Higashino, T., Okamura, Y. (eds.) ATVA 2007. LNCS, vol. 4762, pp. 223–236. Springer, Heidelberg (2007)
6. Duret-Lutz, A.: LTL translation improvements in Spot. In: VECoS 2011, eWiC. British Computer Society (2011) (to appear)
7. Duret-Lutz, A., Poitrenaud, D.: SPOT: An extensible model checking library using transition-based generalized Büchi automata. In: MASCOTS 2004, pp. 76–83. IEEE (2004)
8. Ehlers, R., Finkbeiner, B.: On the Virtue of Patience: Minimizing Büchi Automata. In: van de Pol, J., Weber, M. (eds.) Model Checking Software. LNCS, vol. 6349, pp. 129–145. Springer, Heidelberg (2010)
9. Etessami, K., Holzmann, G.J.: Optimizing Büchi Automata. In: Palamidessi, C. (ed.) CONCUR 2000. LNCS, vol. 1877, pp. 153–167. Springer, Heidelberg (2000)
10. Fritz, C.: Constructing Büchi Automata from Linear Temporal Logic Using Simulation Relations for Alternating Büchi Automata. In: Ibarra, O.H., Dang, Z. (eds.) CIAA 2003. LNCS, vol. 2759, pp. 35–48. Springer, Heidelberg (2003)
11. Gastin, P., Oddoux, D.: Fast LTL to Büchi Automata Translation. In: Berry, G., Comon, H., Finkel, A. (eds.) CAV 2001. LNCS, vol. 2102, pp. 53–65. Springer, Heidelberg (2001)
12. Giannakopoulou, D., Lerda, F.: From States to Transitions: Improving Translation of LTL Formulae to Büchi Automata. In: Peled, D.A., Vardi, M.Y. (eds.) FORTE 2002. LNCS, vol. 2529, pp. 308–326. Springer, Heidelberg (2002)
13. Löding, C.: Efficient minimization of deterministic weak omega-automata. Information Processing Letters 79(3), 105–109 (2001)
14. Manna, Z., Pnueli, A.: A hierarchy of temporal properties. In: PODC 1990, pp. 377–410. ACM Press (1990)
15. Pnueli, A.: The temporal logic of programs. In: FOCS 1977, pp. 46–57. IEEE (1977)
16. Rozier, K.Y., Vardi, M.Y.: LTL Satisfiability Checking. In: Bošnački, D., Edelkamp, S. (eds.) SPIN 2007. LNCS, vol. 4595, pp. 149–167. Springer, Heidelberg (2007)
17. Sebastiani, R., Tonetta, S.: "More Deterministic" vs. "Smaller" Büchi Automata for Efficient LTL Model Checking. In: Geist, D., Tronci, E. (eds.) CHARME 2003. LNCS, vol. 2860, pp. 126–140. Springer, Heidelberg (2003)
18. Somenzi, F., Bloem, R.: Efficient Büchi Automata from LTL Formulae. In: Emerson, E.A., Sistla, A.P. (eds.) CAV 2000. LNCS, vol. 1855, pp. 248–263. Springer, Heidelberg (2000)
19. Tauriainen, H., Heljanko, K.: Testing LTL formula translation into Büchi automata. STTT 4(1), 57–70 (2002)

Pushdown Model Checking for Malware Detection[*]

Fu Song and Tayssir Touili

LIAFA, CNRS and Univ. Paris Diderot, France
{song,touili}@liafa.jussieu.fr

Abstract. The number of malware is growing extraordinarily fast. Therefore, it is important to have efficient malware detectors. Malware writers try to obfuscate their code by different techniques. Many of these well-known obfuscation techniques rely on operations on the stack such as inserting dead code by adding useless push and pop instructions, or hiding calls to the operating system, etc. Thus, it is important for malware detectors to be able to deal with the program's stack. In this paper we propose a new model-checking approach for malware detection that takes into account the behavior of the stack. Our approach consists in : (1) Modeling the program using a Pushdown System (PDS). (2) Introducing a new logic, called SCTPL, to represent the malicious behavior. SCTPL can be seen as an extension of the branching-time temporal logic CTL with variables, quantifiers, and predicates over the stack. (3) Reducing the malware detection problem to the model-checking problem of PDSs against SCTPL formulas. We show how our new logic can be used to precisely express malicious behaviors that could not be specified by existing specification formalisms. We then consider the model-checking problem of PDSs against SCTPL specifications. We reduce this problem to emptiness checking in Symbolic Alternating Büchi Pushdown Systems, and we provide an algorithm to solve this problem. We implemented our techniques in a tool, and we applied it to detect several viruses. Our results are encouraging.

1 Introduction

To identify viruses, existing antivirus systems use either code emulation or signature (pattern) detection. These techniques have some limitations. Indeed, emulation based techniques can only check the program's behavior in a limited time interval, whereas signature based systems are easy to get around. To sidestep these limitations, instead of executing the program or making a syntactic check over it, virus detectors need to use analysis techniques that check the *behavior* (not the syntax) of the program in a *static* way, i.e. without executing it. Towards this aim, we propose in this paper to use *model-checking* for virus detection. Model-checking has already been used for virus detection in [6,20,9,11,16,15,17]. However, these works model the program as a finite state graph (automaton). Thus, they are not able to model the stack of the programs, and cannot track the effects of the push, pop and call instructions. However, as decribed in [19], many obfuscation techniques rely on operations over the stack. Indeed, many antivirus systems determine whether a program is malicious by checking the calls it makes to

---

[*] Work partially funded by ANR grant ANR-08-SEGI-006.

C. Flanagan and B. König (Eds.): TACAS 2012, LNCS 7214, pp. 110–125, 2012.

the operating system. Hence, several virus writers try to hide these calls by replacing them by push and return instructions [19]. Therefore, it is important to have analysis techniques that can deal with the program stack.

We propose in this paper a novel model-checking technique for malware detection that takes into account the behavior of the stack. Our approach consists in modeling the program using a *pushdown system* (PDS), and defining a new logic, called *SCTPL*, to express the malicious behavior.

Using pushdown systems as program model allows to consider the program stack. In our modeling, the PDS control locations correspond to the program's control points, and the PDS stack mimics the program's execution stack. This allows the PDS to mimic the behavior of the program. This is different from standard program translations to PDSs where the control points of the program are stored in the stack [13,5]. These standard translations assume that the program follows a standard compilation model, where the return addresses are never modified. We do not make such assumptions since behaviors where the return addresses are modified can occur in malicious code. We only make the assumption that pushes and pops can be done only using *push*, *pop*, *call*, and *return* operations, not by manipulating the stack pointer.

The logic SCTPL that we introduce is an extension of the CTPL logic that allows to use predicates over the stack. CTPL was introduced in [16,15,17]. It can be seen as an extension of CTL with variables and quantifiers. In CTPL, propositions can be predicates of the form $p(x_1, \ldots, x_n)$, where the x_i's are free variables or constants. Free variables can get their values from a finite domain. Variables can be universally or existentially quantified. CTPL is as expressive as CTL, but it allows a more succinct specification of the malicious behavior. For example, consider the statement "The value *data* is assigned to some register, and later, the content of this register is pushed onto the stack." This statement can be expressed in CTL as a large formula enumerating all the possible registers:

$$EF \ (mov(eax, data) \wedge AF \ push(eax)) \vee$$
$$EF \ (mov(ebx, data) \wedge AF \ push(ebx)) \vee$$
$$EF \ (mov(ecx, data) \wedge AF \ push(ecx)) \ \vee \ ...$$

where every instruction is regarded as a predicate, i.e., *mov(eax, data)* is a predicate. However, the CTL formula is large for such a simple statement. Using CTPL, this can be expressed by the CTPL formula $\exists r \ EF \ (mov(r, data) \wedge AF \ push(r))$ which expresses in a succinct way that there exists a *register* r such that the above holds. [16,15,17] show how this logic is adequate to specify some malicious behaviors. However, CTPL does not allow to specify properties about the stack (which is important for malicious code detection as explained above).

For example, consider Figure 1(a). It corresponds to a critical fragment of the Email-worm Avron [14] that shows the typical behavior of an email worm: it calls an API function *GetModuleHandleA* with 0 as its parameter. This allows to get the entry address of its own executable so that later, it can infect other files by copying this executable into them. (Parameters to a function in assembly are passed by pushing them onto the stack before a call to the function is made. The code in the called function later

	l'_1: mov eax,0
	l'_2: push eax
l_1: mov eax,0	l'_3: push ebx
l_2: push eax	l'_4: pop ebx
l_3: call ds:GetModuleHandleA	l'_5: call ds:GetModuleHandleA
(a)	(b)

Fig. 1. (a) Worm fragment; (b) Obfuscated fragment

retrieves these parameters from the stack.) Using CTPL, we can specify this malicious behavior by the following formula:

$$\exists\, r_1\ \mathbf{EF}\left(mov(r_1,0) \wedge \mathbf{EX\ E}[\neg\exists r_2\ mov(r_1,r_2)\ \mathbf{U}\ (push(r_1)\wedge\right.$$

$$\mathbf{EX\ E}[\neg\exists\, r_3\ (push(r_3) \vee pop(r_3))\ \mathbf{U}\ call(GetModuleHandleA)])])\big). \tag{1}$$

This formula states that there exists a register r_1 assigned by 0 such that the value of r_1 is not modified until it is pushed onto the stack. Later the stack is not changed until function *GetModuleHandleA* is called. This specification can detect the fragment in Figure 1(a). However, a worm writer can easily use some obfuscation techniques in order to escape this specification. For example, let us introduce one push followed by one pop after *push eax* at line l_2 as done in Figure 1(b). By doing so, this fragment keeps the same malicious behavior than the fragment in Figure 1(a). However, it cannot be detected by the above CTPL formula. Since the number of pushes and pops that can be added by the worm writer can be arbitrarily large, it is always possible for worm developers to change their code in order to escape a given CTPL formula.

To overcome this problem, we introduce the SCTPL logic which extends CTPL by predicates over the stack. Such predicates are given by regular expressions over the stack alphabet and some free variables (which can also be existantially and universally quantified). Using our new logic SCTPL, the malicious behavior of Figures 1(a) and (b) can be specified as follows:

$$\psi = \exists r_1\ \mathbf{EF}\left(mov(r_1,0) \wedge \mathbf{EX\ E}[\neg\exists r_2\ mov(r_1,r_2)\mathbf{U}\ (push(r_1) \wedge \mathbf{EX\ E}[\neg\big(push(r_1)\right.$$

$$\vee(\exists r_3(pop(r_3) \wedge r_1\Gamma^*)))\mathbf{U}\ (call(GetModuleHandleA) \wedge r_1\Gamma^*)])]\big) \tag{2}$$

where $r_1\Gamma^*$ is a regular predicate expressing that the topmost symbol of the stack is r_1. The SCTPL formula ψ states that there exists a register r_1 assigned by 0 such that the value of r_1 is not changed until it is pushed onto the stack. Then, r_1 is never pushed onto the stack again nor popped from it until the function *GetModuleHandleA* is called. When this call is made, the topmost symbol of the stack has to be r_1. This ensures that *GetModuleHandleA* is called with 0 as parameter. This specification can detect both fragments in Figure 1, because it allows to specify the content of the stack when *GetModuleHandleA* is called. Note that it is important to use pushdown systems as model in order to have specifications with predicates over the stack.

The main contributions of this paper are:

1. We present a new technique to translate a binary program into a pushdown system that mimics the program's behavior (a malicious program is usually an executable,

i.e., a binary program). Our translation is different from standard program translations to PDSs that need to assume that the program follows a standard compilation
model, where the return addresses are never modified. Our translation does not need
to make this assumption since malicious code may have a non standard form.

2. We introduce the SCTPL logic and show how it can be used to efficiently and
precisely characterize malicious behaviors.

3. We propose an algorithm for model checking pushdown systems against SCTPL
specifications. We reduce this problem to checking emptiness in Symbolic Alternating Büchi Pushdown Systems (SABPDS) and we propose an algorithm to solve
this emptiness problem.

4. We implemented our techniques in a tool that we successfully applied to detect
several viruses.

Related Work. Model-checking and static analysis techniques have been applied to
detect malicious behaviors e.g. in [6,20,9,11,16,15,17]. However, all these works are
based on modeling the program as a finite-state system, and thus, they miss the behavior
of the stack. As we have seen, being able to track the stack is important for many
malicious behaviors. [7] use tree automata to represent a set of malicious behaviors.
However, [7] cannot specify predicates over the stack content.

[19] keeps track of the stack by computing an abstract stack graph which finitely represents the infinite set of all the possible stacks for every control point of the program.
Their technique can detect only obfuscated calls and obfuscated returns. Using SCTPL,
we are able to detect more malicious behaviors.

[18] performs context-sensitive analysis of *call* and *ret* obfuscated binaries. They use
abstract interpretation to compute an abstraction of the stack. We believe that our techniques are more precise since we do not abstract the stack. Moreover, the techniques of
[18] were only tried on toy examples, they have not been applied for malware detection.

[5] uses pushdown systems for binary code analysis. However, [5] has not been applied for malware detection. Moreover, the translation from programs to PDSs in [5]
assumes that the program follows a standard compilation model where calls and returns
match. Several malicious behaviors do not follow this model. Our translation from a
control flow graph to a PDS does not make this assumption.

[10] defines a language for specifying malicious behavior in terms of dependences
between system calls. Compared to SCTPL, the specification language of [10] does not
take the stack into account and is only able to express safety properties (no CTL like
properties), whereas SCTPL does. On the other hand, [10] is able to automatically derive the malicious specifications by comparing the execution behavior of a known malware against the execution behaviors of a set of benign programs. It would be interesting
to see if their techniques can be extended to automatically derive SCTPL specifications
of malicious behaviors.

LTL or CTL model-checking with regular predicates over the stack was considered
in [12,21]. These works do not consider variables and quantifiers.

Outline. We give our formal model in Section 2. In Section 3, we introduce our SCTPL
logic. Our SCTPL model checking algorithm for pushdown systems is given in Section
4. The experiments we made for malware detection are reported in Section 5.

2   Formal Model: Pushdown Systems

We model a binary code by a pushdown system (PDS). In our modeling, the PDS control locations correspond to the program's control points, and the PDS stack mimics the program's execution stack. This is different from standard program translations to PDSs where the control points of the program are stored in the stack [13,5]. These standard translations assume that the program follows a standard compilation model, where the return addresses are never modified. We do not make such assumptions since behaviors where the return addresses are modified can occur in malicious code. We only make the assumption that pushes and pops can be done only using *push*, *pop*, *call*, and *return* operations, not by manipulating the stack pointer.

Formally, a *Pushdown System* (PDS) is a tuple $\mathcal{P} = (P, \Gamma, \Delta, \sharp)$, where P is a finite set of control locations, Γ is the stack alphabet, $\Delta \subseteq (P \times \Gamma) \times (P \times \Gamma^*)$ is a finite set of transition rules, and $\sharp \in \Gamma$ is the bottom stack symbol. A configuration of $\mathcal{P}$ is $\langle p, \omega \rangle$, where $p \in P$ and $\omega \in \Gamma^*$. If $((p, \gamma), (q, \omega)) \in \Delta$, we write $\langle p, \gamma \rangle \hookrightarrow \langle q, \omega \rangle$. For technical reasons, we assume that the bottom stack symbol $\sharp$ is never popped from the stack, i.e., there is no transition rule of the form $\langle p, \sharp \rangle \hookrightarrow \langle q, \omega \rangle \in \Delta$.

The successor relation $\leadsto_{\mathcal{P}} \subseteq (P \times \Gamma^*) \times (P \times \Gamma^*)$ is defined as follows: if $\langle p, \gamma \rangle \hookrightarrow \langle q, \omega \rangle$, then $\langle p, \gamma \omega' \rangle \leadsto_{\mathcal{P}} \langle q, \omega \omega' \rangle$ for every $\omega' \in \Gamma^*$. For every configuration $c, c' \in P \times \Gamma^*$, c is a successor of c' iff $c \leadsto_{\mathcal{P}} c'$. A path is a sequence of configurations $c_0, c_1, \ldots$ s.t. $c_i \leadsto_{\mathcal{P}} c_{i+1}$ for every $i \geq 0$.

3   Malicious Behavior Specification

In this section, we introduce the Stack Computation Tree Predicate Logic (SCTPL), the formalism we use to specify malicious behavior.

3.1   Environments, Predicates and Regular Expressions

From now on, we fix the following notations. Let $X = \{x_1, x_2, \ldots\}$ be a finite set of variables ranging over a finite domain $\mathcal{D}$. Let $B : X \cup \mathcal{D} \longrightarrow \mathcal{D}$ be an environment function that assigns a value $c \in \mathcal{D}$ to each variable $x \in X$, and such that $B(c) = c$ for every $c \in \mathcal{D}$. $B[x \leftarrow c]$ denotes the environment function such that $B[x \leftarrow c](x) = c$ and $B[x \leftarrow c](y) = B(y)$ for every $y \neq x$. $Abs_x(B)$ is the set of all the environments B' s.t. for every $y \neq x$, $B'(y) = B(y)$. Let $\mathcal{B}$ be the set of all the environment functions.

Let $AP = \{a, b, c, \ldots\}$ be a finite set of atomic propositions, AP_X be a finite set of atomic predicates of the form $b(\alpha_1, \ldots, \alpha_m)$ such that $b \in AP$, $\alpha_i \in X \cup \mathcal{D}$ for every i, $1 \leq i \leq m$, and $AP_{\mathcal{D}}$ be a finite set of atomic predicates of the form $b(\alpha_1, \ldots, \alpha_m)$ such that $b \in AP$ and $\alpha_i \in \mathcal{D}$ for every i, $1 \leq i \leq m$.

Let $\mathcal{P} = (P, \Gamma, \Delta, \sharp)$ be a PDS s.t. $\Gamma \subseteq \mathcal{D}$. Let $\mathcal{R}$ be a finite set of regular variable expressions e over $X \cup \Gamma$ defined by:

$$e ::= \emptyset \mid \epsilon \mid a \in X \cup \Gamma \mid e + e \mid e \cdot e \mid e^*$$

The language $L(e)$ of a regular variable expression e is a subset of $P \times \Gamma^* \times \mathcal{B}$ defined inductively as follows: $L(\emptyset) = \emptyset$; $L(\epsilon) = \{(\langle p, \epsilon \rangle, B) \mid p \in P, B \in \mathcal{B}\}$; $L(x)$, where $x \in X$

is the set $\{(\langle p, \gamma \rangle, B) \mid p \in P, \gamma \in \Gamma, B \in \mathcal{B} : B(x) = \gamma\}$; $L(\gamma)$, where $\gamma \in \Gamma$ is the set $\{(\langle p, \gamma \rangle, B) \mid p \in P, B \in \mathcal{B}\}$; $L(e_1 + e_2) = L(e_1) \cup L(e_1)$; $L(e_1 \cdot e_2) = \{(\langle p, \omega_1 \omega_2 \rangle, B) \mid (\langle p, \omega_1 \rangle, B) \in L(e_1); (\langle p, \omega_2 \rangle, B) \in L(e_2)\}$; and $L(e^*) = \{(\langle p, \omega^* \rangle, B) \mid (\langle p, \omega \rangle, B) \in L(e)\}$. E.g., $(\langle p, \gamma_1 \gamma_1 \gamma_2 \rangle, B)$ is an element of $L(x^* \gamma_2)$ when $B(x) = \gamma_1$.

3.2 Stack Computation Tree Predicate Logic

We are now ready to define our new logic SCTPL. Intuitively, a SCTPL formula is a CTL formula where predicates and regular variable expressions are used as atomic propositions. Using regular variable expressions allows to express predicates on the stack content of the PDS. Moreover, since predicates and regular variable expressions contain variables, we allow quantifiers over variables. For technical reasons, we suppose w.l.o.g. that formulas are given in positive normal form, i.e., negations are applied only to atomic propositions. Indeed, each CTL formula can be written in positive normal form by pushing the negations inside. Moreover, we use the operator $\tilde{U}$ as a dual of the until operator for which the stop condition is not required to occur. Then, standard CTL operators can be expressed as follows: $EF\psi = E[trueU\psi]$, $AF\psi = A[trueU\psi]$, $EG\psi = E[false\tilde{U}\psi]$ and $AG\psi = A[false\tilde{U}\psi]$.

More precisely, the set of *SCTPL formulas* is given by (where $x \in X$, $a(x_1, ..., x_n) \in AP_X$ and $e \in \mathcal{R}$):

$$\varphi ::= a(x_1, ..., x_n) \mid \neg a(x_1, ..., x_n) \mid e \mid \neg e \mid \varphi \wedge \varphi \mid \varphi \vee \varphi \mid \forall x \varphi$$
$$\mid \exists x \varphi \mid AX\varphi \mid EX\varphi \mid A[\varphi U\varphi] \mid E[\varphi U\varphi] \mid A[\varphi \tilde{U}\varphi] \mid E[\varphi \tilde{U}\varphi]$$

Let φ be a SCTPL formula. The closure $cl(\varphi)$ denotes the set of all the subformulas of φ including φ. The size $|\varphi|$ of φ is the number of elements of $cl(\varphi)$. Let $AP^+(\varphi) = \{a(x_1, ..., x_n) \in AP_X \mid a(x_1, ..., x_n) \in cl(\varphi)\}$, $AP^-(\varphi) = \{a(x_1, ..., x_n) \in AP_X \mid \neg a(x_1, ..., x_n) \in cl(\varphi)\}$, $Reg^+(\varphi) = \{e \in \mathcal{R} \mid e \in cl(\varphi)\}$, $Reg^-(\varphi) = \{e \in \mathcal{R} \mid \neg e \in cl(\varphi)\}$, and $cl_{\tilde{U}}(\varphi)$ be the set of formulas of $cl(\varphi)$ in the form of $E[\varphi_1 \tilde{U}\varphi_2]$ or $A[\varphi_1 \tilde{U}\varphi_2]$.

Given a PDS $\mathcal{P} = (P, \Gamma, \Delta, \sharp)$ s.t. $\Gamma \subseteq \mathcal{D}$, let $\lambda : AP_{\mathcal{D}} \rightarrow 2^P$ be a labeling function that assigns a set of control locations to a predicate. Let $c = \langle p, w \rangle$ be a configuration of $\mathcal{P}$. $\mathcal{P}$ satisfies a SCTPL formula ψ in c, denoted by $c \models_\lambda \psi$, iff there exists an environment $B \in \mathcal{B}$ s.t. $c \models_\lambda^B \psi$, where $c \models_\lambda^B \psi$ is defined by induction as follows:

- $c \models_\lambda^B a(x_1, ..., x_n)$ iff $p \in \lambda\big(a(B(x_1), ..., B(x_n))\big)$.
- $c \models_\lambda^B \neg a(x_1, ..., x_n)$ iff $p \notin \lambda\big(a(B(x_1), ..., B(x_n))\big)$.
- $c \models_\lambda^B e$ iff $(c, B) \in L(e)$.
- $c \models_\lambda^B \neg e$ iff $(c, B) \notin L(e)$.
- $c \models_\lambda^B \psi_1 \wedge \psi_2$ iff $c \models_\lambda^B \psi_1$ and $c \models_\lambda^B \psi_2$.
- $c \models_\lambda^B \psi_1 \vee \psi_2$ iff $c \models_\lambda^B \psi_1$ or $c \models_\lambda^B \psi_2$.
- $c \models_\lambda^B \forall x \psi$ iff $\forall v \in \mathcal{D}$, $c \models_\lambda^{B[x \leftarrow v]} \psi$.
- $c \models_\lambda^B \exists x \psi$ iff $\exists v \in \mathcal{D}$ s.t. $c \models_\lambda^{B[x \leftarrow v]} \psi$.
- $c \models_\lambda^B AX \psi$ iff $c' \models_\lambda^B \psi$ for every successor c' of c.
- $c \models_\lambda^B EX \psi$ iff there exists a successor c' of c s.t. $c' \models_\lambda^B \psi$.
- $c \models_\lambda^B A[\psi_1 U \psi_2]$ iff for every path $\pi = c_0, c_1, ...,$ of $\mathcal{P}$ with $c_0 = c$, $\exists i \geq 0$ s.t. $c_i \models_\lambda^B \psi_2$ and $\forall 0 \leq j < i : c_j \models_\lambda^B \psi_1$.

- $c \models^B_\lambda E[\psi_1 U \psi_2]$ iff there exists a path $\pi = c_0, c_1, \ldots,$ of $\mathcal{P}$ with $c_0 = c$ s.t. $\exists i \geq 0$, $c_i \models^B_\lambda \psi_2$ and $\forall 0 \leq j < i, c_j \models^B_\lambda \psi_1$.
- $c \models^B_\lambda A[\psi_1 \tilde{U} \psi_2]$ iff for every path $\pi = c_0, c_1, \ldots,$ of $\mathcal{P}$ with $c_0 = c$, $\forall i \geq 0$ s.t. $c_i \not\models^B_\lambda \psi_2$, $\exists 0 \leq j < i$ s.t. $c_j \models^B_\lambda \psi_1$.
- $c \models^B_\lambda E[\psi_1 \tilde{U} \psi_2]$ iff there exists a path $\pi = c_0, c_1, \ldots,$ of $\mathcal{P}$ with $c_0 = c$ s.t. $\forall i \geq 0$ s.t. $c_i \not\models^B_\lambda \psi_2$, $\exists 0 \leq j < i$ s.t. $c_j \models^B_\lambda \psi_1$.

Intuitively, $c \models^B_\lambda \psi$ holds iff the configuration c satisfies the formula ψ under the environment B. Note that a path π satisfies $\psi_1 \tilde{U} \psi_2$ iff either ψ_2 holds everywhere in π, or the first occurrence in the path where ψ_2 does not hold must be preceeded by a position where ψ_1 holds.

3.3   Modeling Malicious Behaviors Using SCTPL

SCTPL can be used to precisely specify several malicious behaviors. We needed stack predicates to express most of the specifications. Thus, SCTPL is necessary to specify these behaviors, CTPL is not sufficient. We describe here how e.g., email worms can be specified using SCTPL. The typical behavior of an email worm can be summarized as follows: the worm will first call the API *GetModuleFileNameA* in order to get the name of its executable. For this, the worm needs to call this function with 0 and m as parameters (m corresponds to the address of a memory location), i.e., with $0m$ on the top of the stack since parameters to a function in assembly are passed through the stack. *GetModuleFileNameA* will then write the name of the worm executable on the address m. Then, the worm will copy its file (whose name is at the address m) to other locations using the function *CopyFileA*. It needs to call *CopyFileA* with m as parameter, i.e., with m on the top of the stack. Figure 2(a) shows a disassembled fragment of a code corresponding to this typical behavior. This behavior can be expressed by the SCTPL formula of Figure 2(b). In this formula, Line 2 expresses that there exists a register r_0 such that the address of the memory location m is assigned to r_0, and such that the value of r_0 does not change until it is pushed onto the stack (subformula $\neg \exists v(mov(r_0, v) \vee lea(r_0, v)) \, U push(r_0)$). Line 3 guarantees that r_0 is not pushed nor popped from the stack until *GetModuleFileNameA* is called, and $0r_0$ is on the top of the stack (the predicate $0r_0 \Gamma^*$ ensures this). This guarantees that when *GetModuleFileNameA* is called, r_0 still contains the address of m. Thus, the name of the worm file returned by *GetModuleFileNameA* will be put at the address m. Line 4 is similar to Line 2. It expresses that there exists a register r_1 such that the address of the memory location m is assigned to r_1, and such that the value of r_1 does not change until it is pushed onto the stack. This guarantees that when r_1 is pushed to the stack, it contains the address of m. Line 5 expresses that r_1 is not pushed nor popped from the stack until *CopyFileA* is called, and r_1 is on the top of the stack (the predicate $r_1 \Gamma^*$ ensures this). This guarantees that when *CopyFileA* is called, the value of r_1 is still m. Thus, *CopyFileA* will copy the file whose name is at the address m. Note that we need predicates over the stack to express in a precise manner this specification.

4   SCTPL Model-Checking for Pushdown Systems

In this section, we give an efficient SCTPL model checking algorithm for Pushdown systems. Our procedure works as follows: we reduce this model checking problem to the

$$\begin{aligned}
&\dots\\
&lea\ eax,\ [ebp + ExistingFileName]\\
&push\ eax\\
&push\ 0\\
&call\ ds : GetModuleFileNameA\\
&\dots\\
&lea\ eax,\ [ebp + ExistingFileName]\\
&push\ eax\\
&call\ ds : CopyFileA\\
&\dots
\end{aligned}$$

(a)

1.　$\psi_{ew} = \exists m \Big(\exists r_0 \Big($

2.　$\mathbf{EF}\big(lea(r_0, m) \wedge \mathbf{EX}\ \mathbf{E}\big[\neg \exists v(mov(r_0, v) \vee lea(r_0, v))\mathbf{U}\big(push(r_0)$

3.　$\wedge\ \mathbf{EX}\ \mathbf{E}[\neg(push(r_0) \vee \exists v(pop(v) \wedge r_0\Gamma^*))\mathbf{U}(call(GetModuleFileNameA) \wedge 0\ r_0\Gamma^*$

4.　$\wedge\ \exists r_1 \big(\mathbf{EF}(lea(r_1, m) \wedge \mathbf{EX}\ \mathbf{E}[\neg \exists v(mov(r_1, v) \vee lea(r_1, v))\mathbf{U}(push(r_1)$

5.　$\wedge\ \mathbf{EX}\ \mathbf{E}[\neg(push(r_1) \vee \exists v(pop(v) \wedge r_1\Gamma^*))\mathbf{U}call(CopyFileA) \wedge r_1\Gamma^*])])))])\Big)\Big)\Big)$

(b)

Fig. 2. (a) Email worm (b) Specification of Email worm

emptiness problem in Symbolic Alternating Büchi Pushdown Systems (SABPDS), and we give an algorithm to solve this emptiness problem. To achieve this reduction, we use *variable automata* to represent regular variable expressions. This section is structured as follows. First, we introduce variable automata. Then, we define Symbolic Alternating Büchi Pushdown Systems. Next, we show how SCTPL model checking for PDSs can be reduced to emptiness checking of SABPDSs.

In the remainder of this section, we let X be a finite set of variables ranging over a finite domain $\mathcal{D}$, and $\mathcal{B}$ be the set of all the environment functions $B : X \cup \mathcal{D} \longrightarrow \mathcal{D}$.

4.1 Variable Automata

Given a PDS $\mathcal{P} = (P, \Gamma, \Delta, \sharp)$ s.t. $\Gamma \subseteq \mathcal{D}$, a *Variable Automaton* (VA) is a tuple $M = (Q, \Gamma, \delta, q_0, A)$, where Q is a finite set of states; Γ is the input alphabet; $q_0 \subseteq Q$ is an initial state; $A \subseteq Q$ is a finite set of accepting states; and δ is a finite set of transition rules of the form: $p \xrightarrow{\alpha} \{q_1, ..., q_n\}$ where α can be x, $\neg x$, or γ, for any $x \in X$ and $\gamma \in \Gamma$.

Let $B \in \mathcal{B}$. A run of VA on a word $\gamma_1, ..., \gamma_m$ under B is a tree of height m whose root is labelled by the initial state q_0, and each node at depth k labelled by a state q has h children labelled by $p_1, ..., p_h$, respectively, such that: either $q \xrightarrow{\gamma_k} \{p_1, ..., p_h\} \in \delta$ and $\gamma_k \in \Gamma$; or $q \xrightarrow{x} \{p_1, ..., p_h\} \in \delta$, $x \in X$ and $B(x) = \gamma_k$; or $q \xrightarrow{\neg x} \{p_1, ..., p_h\} \in \delta$, $x \in X$ and $B(x) \neq \gamma_k$. A branch of the tree is accepting iff the leaf of the branch is an accepting state. A run is accepting iff all its branches are accepting. A word $\omega \in \Gamma^*$ is accepted by a VA under an environment $B \in \mathcal{B}$ iff the VA has an accepting run on the word ω

under the environment B. The language of a VA M, denoted by $L(M)$, is a subset of $(P \times \Gamma^*) \times \mathcal{B}$. $(\langle p, \omega \rangle, B) \in L(M)$ iff M accepts the word ω under the environment B. We can show that:

Theorem 1. *VAs are effectively closed under boolean operations.*

Theorem 2. *For every regular expression $e \in \mathcal{R}$, one can effectively compute in polynomial time a VA M such that $L(M) = L(e)$.*

4.2   Symbolic Alternating Büchi Pushdown Systems

Definition 1. *A Symbolic Alternating Büchi Pushdown System (SABPDS) is a tuple $\mathcal{BP} = (P, \Gamma, \Delta, F)$, where P is a finite set of control locations; $\Gamma \subseteq \mathcal{D}$ is the stack alphabet; $F \subseteq P \times 2^{\mathcal{B}}$ is a set of accepting states; Δ is a finite set of transitions of the form $\langle p, \gamma \rangle \overset{\mathfrak{R}}{\hookrightarrow} [\langle p_1, \omega_1 \rangle, ..., \langle p_n, \omega_n \rangle]$ where $p \in P$, $\gamma \in \Gamma$, for every i, $1 \le i \le n$: $p_i \in P$, $\omega_i \in \Gamma^*$, and $\mathfrak{R} : (\mathcal{B})^n \longrightarrow 2^{\mathcal{B}}$ is a function that maps a tuple of environments to a set of environments.*

A configuration of a SABPDS is a tuple $\langle [p, B], \omega \rangle$, where $p \in P$ is a control location, $B \in \mathcal{B}$ is an environment and $\omega \in \Gamma^*$ is the stack content. $[p, B] \in P \times \mathcal{B}$ is an accepting state iff $\exists [p, \beta] \in F$ s.t. $B \in \beta$. Let $t = \langle p, \gamma \rangle \overset{\mathfrak{R}}{\hookrightarrow} [\langle p_1, \omega_1 \rangle, ..., \langle p_n, \omega_n \rangle] \in \Delta$ be a transition, n is the width of the transition t. For every $\omega \in \Gamma^*$, $B, B_1, ..., B_n \in \mathcal{B}$, if $B \in \mathfrak{R}(B_1, ..., B_n)$, then the configuration $\langle [p, B], \gamma \omega \rangle$ (resp. $\{ \langle [p_1, B_1], \omega_1 \omega \rangle, ..., \langle [p_n, B_n], \omega_n \omega \rangle \}$) is an immediate predecessor (resp. immediate successor) of $\{ \langle [p_1, B_1], \omega_1 \omega \rangle, ..., \langle [p_n, B_n], \omega_n \omega \rangle \}$ (resp. $\langle [p, B], \gamma \omega \rangle$). A run ρ of $\mathcal{BP}$ from an initial configuration $\langle [p_0, B_0], \omega_0 \rangle$ is a tree in which the root is labeled by $\langle [p_0, B_0], \omega_0 \rangle$, and the other nodes are labeled by elements of $(P \times \mathcal{B}) \times \Gamma^*$. If a node of ρ labeled by $\langle [p, B], \omega \rangle$ has n children labeled by $\langle [p_1 B_1], \omega_1 \rangle, ..., \langle [p_n, B_n], \omega_n \rangle$, respectively, then, necessarily, $\langle [p, B], \omega \rangle$ is an immediate predecessor of $\{ \langle [p_1, B_1], \omega_1 \rangle, ..., \langle [p_n, B_n], \omega_n \rangle \}$ in $\mathcal{BP}$.

A path $c_0 c_1 ...$ of a run ρ is an *infinite* sequence of configurations where c_0 is the root of ρ and for every $i \ge 0$, c_{i+1} is one of the children of the node c_i in ρ. The path is accepting iff it visits infinitely often configurations with accepting states. A run ρ is accepting iff all its paths are accepting. Note that an accepting run has only *infinite* paths. A configuration c is accepted (or recognized) by $\mathcal{BP}$ iff $\mathcal{BP}$ has an accepting run starting from c. The language of $\mathcal{BP}$, denoted by $\mathcal{L}(\mathcal{BP})$, is the set of configurations accepted by $\mathcal{BP}$.

The predecessor functions $Pre_{\mathcal{BP}}$, $Pre^*_{\mathcal{BP}}$ and $Pre^+_{\mathcal{BP}}$: $2^{(P \times \mathcal{B}) \times \Gamma^*} \longrightarrow 2^{(P \times \mathcal{B}) \times \Gamma^*}$ are defined as follows: $Pre_{\mathcal{BP}}(C) = \{ c \in (P \times \mathcal{B}) \times \Gamma^* \mid$ some immediate successor of c is a subset of C$\}$, $Pre^*_{\mathcal{BP}}$ is the reflexive and transitive closure of $Pre_{\mathcal{BP}}$, $Pre_{\mathcal{BP}} \circ Pre^*_{\mathcal{BP}}$ is denoted by $Pre^+_{\mathcal{BP}}$.

SABPDS vs. ABPDS. An *Alternating Büchi Pushdown System* (ABPDS for short) [21] can be seen as a SABPDS:

Lemma 1. *Given a SABPDS $\mathcal{BP} = (P, \Gamma, \Delta, F)$, one can compute an equivalent ABPDS $\mathcal{BP}'$ that simulates $\mathcal{BP}$ in $O(|\Delta| \cdot |\mathcal{B}|^{k+1})$ time, where k is the maximum of the widths of the transition rules in Δ and $|\mathcal{B}| = |D|^{|X|}$.*

Symbolic Alternating Multi-Automata. To finitely represent infinite sets of configurations of SABPDSs, we use Symbolic Alternating Multi-Automata.

Let $\mathcal{BP} = (P, \Gamma, \Delta, F)$ be a SABPDS, a *Symbolic Alternating Multi-Automaton* (SAMA) is a tuple $\mathcal{A} = (Q, \Gamma, \delta, I, Q_f)$, where Q is a finite set of states, Γ is the input alphabet, $\delta \subseteq (Q \times \Gamma) \times 2^Q$ is a finite set of transition rules, $I \subseteq P \times 2^{\mathcal{B}}$ is a finite set of initial states, $Q_f \subseteq Q$ is a finite set of final states. An *Alternating Multi-Automaton* (AMA) is a SAMA such that $I \subseteq P \times \{\emptyset\}$.

We define the reflexive and transitive transition relation $\longrightarrow_\delta \subseteq (Q \times \Gamma^*) \times 2^Q$ as follows: (1) $q \xrightarrow{\epsilon}_\delta \{q\}$ for every $q \in Q$, where ϵ is the empty word, (2) if $q \xrightarrow{\gamma} \{q_1, ..., q_n\} \in \delta$ and $q_i \xrightarrow{\omega}_\delta Q_i$ for every $1 \leq i \leq n$, then $q \xrightarrow{\gamma\omega}_\delta \bigcup_{i=1}^n Q_i$. The automaton $\mathcal{A}$ recognizes a configuration $\langle [p, B], \omega \rangle$ iff there exist $Q' \subseteq Q_f$ and $\beta \subseteq \mathcal{B}$ s.t. $B \in \beta$, $[p, \beta] \in I$ and $[p, \beta] \xrightarrow{\omega}_\delta Q'$. The language of $\mathcal{A}$, denoted by $L(\mathcal{A})$, is the set of configurations recognized by $\mathcal{A}$. A set of configurations is regular if it can be recognized by a SAMA. Similarly, AMAs can also be used to recognize (infinite) regular sets of configurations for ABPDSs.

Proposition 1. *Let $\mathcal{A} = (Q, \Gamma, \delta, I, Q_f)$ be a SAMA. Then, deciding whether a configuration $\langle [p, B], \omega \rangle$ is accepted by $\mathcal{A}$ can be done in $O(|Q| \cdot |\delta| \cdot |\omega| + \tau)$ time, where τ denotes the time used to check whether $B \in \beta$ for some $B \in \mathcal{B}, \beta \subseteq \mathcal{B}$.*

Remark 1. The time τ used to check whether $B \in \beta$ depends on the representation of B and β. In particular, if we use BDDs to represent sets of environment functions, checking whether $B \in \beta$ can be done in $\tau = O(\lceil log|\mathcal{D}| \rceil \cdot |X|)$ [8].

Computing the Language of an SABPDS. We can extend the algorithm of [21] that computes an AMA that recognizes the language of an ABPDS to obtain an algorithm that computes the language of an SABPDS. More precisely:

Theorem 3. *Let $\mathcal{BP} = (P, \Gamma, \Delta, F)$ be a SABPDS, then we can compute a SAMA $\mathcal{A}$ that recognizes $\mathcal{L}(\mathcal{BP})$ in $O\left(|P|^2 \cdot 2^{2|\mathcal{B}|} \cdot |\Gamma| \cdot |\Delta| \cdot 2^{5|P| \cdot 2^{|\mathcal{B}|}}\right)$ time.*

Remark 2. Note that another way to compute $\mathcal{L}(\mathcal{BP})$ is to apply Lemma 1 and produce an equivalent ABPDS $\mathcal{BP}'$ that simulates $\mathcal{BP}$, and then apply the algorithm of [21] to compute an AMA that recognizes $\mathcal{L}(\mathcal{BP}')$. In practice, in the symbolic case (for SABPDS), the sets of environments β's can be compactly represented using BDDs for example, whereas in the explicit case (for ABPDS), all the environments B's have to be considered. Thus, the algorithm behind Theorem 3 will behave better in practice. This is confirmed by the experiments we run where, in the majority of cases, this algorithm terminates in few seconds, whereas if we compute an equivalent ABPDS and apply the algorithm of [21], we run out of memory.

Examples of Functions $\mathcal{R}$. We give some examples of functions $\mathcal{R}$ that will be used.

$$- \; equal(B_1, ..., B_n) = \begin{cases} \{B_1\} & \text{if } B_i = B_j \text{ for every } 1 \leq i, j \leq n, \text{ or } n = 1 \\ \emptyset & \text{otherwise.} \end{cases}$$

This function checks that all the B_i's are equal and returns $\{B_1\}$ (which is equal to $\{B_i\}$ for any i) if this is the case and the emptyset otherwise.

- $meet^x_{\{c_1,...,c_n\}}(B_1, ..., B_n) = \begin{cases} Abs_x(B_1) \text{ if } B_i(x) = c_i \text{ and } B_i(y) = B_j(y) \text{ for } y \neq x, \\ \qquad \text{for every } 1 \leq i, j \leq n, \\ \emptyset \qquad \text{otherwise.} \end{cases}$

This function checks whether $B_i(x) = c_i$ for every i, $1 \leq i \leq n$, and for every $y \neq x$ and every i, j, $1 \leq i, j \leq n$ $B_i(y) = B_j(y)$. It returns $Abs_x(B_1)$ (which is equal to $Abs_x(B_i)$ for any i) if this is the case and the emptyset otherwise.

- $join^x_c(B_1, ..., B_n) = \begin{cases} \{B_1\} \text{ if } B_i = B_j \text{ and } B_i(x) = c, \text{ for every } 1 \leq i, j \leq n, \\ \emptyset \qquad \text{otherwise.} \end{cases}$

This function checks whether $B_i(x) = c$ for every i. If this is the case, it returns $equal(B_1, ..., B_n)$, otherwise, it returns the emptyset.

- $join^{\neg x}_c(B_1, ..., B_n) = \begin{cases} \{B_1\} \text{ if } B_i = B_j \text{ and } B_i(x) \neq c, \text{ for every } 1 \leq i, j \leq n, \\ \emptyset \qquad \text{otherwise.} \end{cases}$

This function checks whether $B_i(x) \neq c$ for every i. If this is the case, it returns $equal(B_1, ..., B_n)$, otherwise, it returns the emptyset.

4.3 From SCTPL Model Checking for PDSs to Emptiness of SABPDS

Let $\mathcal{P} = (P, \Gamma, \Delta, \sharp)$, $\lambda : AP_\mathcal{D} \to 2^P$ be a labeling function, and φ be a SCTPL formula. For every configuration $\langle p, \omega \rangle$, our goal is to determine whether $\langle p, \omega \rangle \models_\lambda \varphi$, i.e., whether there exists an environment $B \in \mathcal{B}$ s.t. $\langle p, \omega \rangle \models^B_\lambda \varphi$. We proceed as follows: we compute a symbolic alternating Büchi pushdown system $\mathcal{BP}$ s.t. $\langle p, \omega \rangle \models^B_\lambda \varphi$ iff $\langle [\langle p, \varphi \rangle, B], \omega \rangle \in \mathcal{L}(\mathcal{BP})$. Then, $\langle p, \omega \rangle \models_\lambda \varphi$ iff there exists $B \in \mathcal{B}$ such that $\langle p, \omega \rangle \models^B_\lambda \varphi$.

Let $Reg^+(\varphi) = \{e_1, ..., e_k\}$ and $Reg^-(\varphi) = \{e_{k+1}, ..., e_m\}$ be the two sets of regular variable expressions[1] that occur in φ. As shown in Theorems 2 and 1, for every i, $1 \leq i \leq k$ we can construct VAs $M_{e_i} = (Q_{e_i}, \Gamma, \delta_{e_i}, s_{e_i}, A_{e_i})$ such that $L(M_{e_i}) = L(e_i)$; and for every j, $k < j \leq m$ we can construct VAs $M_{\neg e_j} = (Q_{\neg e_j}, \Gamma, \delta_{\neg e_j}, s_{\neg e_j}, A_{\neg e_j})$ such that $L(M_{\neg e_j}) = (P \times \Gamma^*) \times \mathcal{B} \setminus L(e_j)$. We suppose w.l.o.g. that the states of these automata are distinct. Let M be the union of all these automata, $\mathcal{F}$ be the union of all the final states of these automata A_{e_i}'s and $A_{\neg e_j}$'s and S be the union of all the states of these automata Q_{e_i}'s and $Q_{\neg e_j}$'s.

Let $\mathcal{BP}_\varphi = (P', \Gamma, \Delta', F)$ be the SABPDS defined as follows: $P' = P \times cl(\varphi) \cup S$; $F = F_1 \cup F_2 \cup F_3 \cup F_4$, where $F_1 = \{[(p, a(x_1, ..., x_n)), \beta] \mid a(x_1, ..., x_n) \in AP^+(\varphi)$ and $\beta = \{B \in \mathcal{B} \mid p \in \lambda(a(B(x_1), ..., B(x_n)))\}\}$; $F_2 = \{[(p, \neg a(x_1, ..., x_n)), \beta] \mid \neg a(x_1, ..., x_n) \in AP^-(\varphi)$ and $\beta = \{B \in \mathcal{B} \mid p \notin \lambda(a(B(x_1), ..., B(x_n)))\}\}$; $F_3 = P \times cl_U(\varphi) \times \{\mathcal{B}\}$; and $F_4 = \mathcal{F} \times \{\mathcal{B}\}$.

Δ' is the smallest set of transition rules that satisfy the following. For every control location $p \in P$, every subformula $\psi \in cl(\varphi)$, and every $\gamma \in \Gamma$:

1. if $\psi = a(x_1, ..., x_n)$ or $\psi = \neg a(x_1, ..., x_n)$; $\langle (p, \psi), \gamma \rangle \overset{equal}{\hookrightarrow} \langle (p, \psi), \gamma \rangle \in \Delta'$;

2. if $\psi = \psi_1 \wedge \psi_2$; $\langle (p, \psi), \gamma \rangle \overset{equal}{\hookrightarrow} [\langle (p, \psi_1), \gamma \rangle, \langle (p, \psi_2), \gamma \rangle] \in \Delta'$;

---

[1] $AP^+(\varphi)$, $AP^-(\varphi)$, $Reg^+(\varphi)$ and $Reg^-(\varphi)$ are as defined in Section 3.2.

3. if $\psi = \psi_1 \vee \psi_2$; $\langle (p, \psi), \gamma \rangle \overset{equal}{\hookrightarrow} \langle (p, \psi_1), \gamma \rangle \in \Delta'$ and $\langle (p, \psi), \gamma \rangle \overset{equal}{\hookrightarrow} \langle (p, \psi_2), \gamma \rangle \in \Delta'$;

4. if $\psi = \exists x \, \psi_1$; $\langle (p, \psi), \gamma \rangle \overset{meet^x_{[c]}}{\longrightarrow} \langle (p, \psi_1), \gamma \rangle \in \Delta'$, for every $c \in \mathcal{D}$;

5. if $\psi = \forall x \, \psi_1$; $\langle (p, \psi), \gamma \rangle \overset{meet^x_{\mathcal{D}}}{\longrightarrow} [\langle (p, \psi_1), \gamma \rangle, \cdots, \langle (p, \psi_1), \gamma \rangle] \in \Delta'$, where $\langle (p, \psi_1), \gamma \rangle$ is repeated m times in $[\langle (p, \psi_1), \gamma \rangle, \cdots, \langle (p, \psi_1), \gamma \rangle]$, where m is the number of elements in $\mathcal{D}$;

6. if $\psi = EX\psi_1$; $\langle (p, \psi), \gamma \rangle \overset{equal}{\hookrightarrow} \langle (p', \psi_1), \omega \rangle \in \Delta'$ for every $\langle p, \gamma \rangle \hookrightarrow \langle p', \omega \rangle \in \Delta$;

7. if $\psi = AX\psi_1$; $\langle (p, \psi), \gamma \rangle \overset{equal}{\hookrightarrow} [\langle (p_1, \psi_1), \omega_1 \rangle, ..., (p_l, \psi_1), \omega_l \rangle] \in \Delta'$ such that for every i, $1 \le i \le l$, $\langle p, \gamma \rangle \hookrightarrow \langle p_i, \omega_i \rangle \in \Delta$ and these transitions are all the transitions of Δ that have $\langle p, \gamma \rangle$ as left hand side;

8. if $\psi = E[\psi_1 U \psi_2]$; $\langle (p, \psi), \gamma \rangle \overset{equal}{\hookrightarrow} [\langle (p, \psi_1), \gamma \rangle, \langle (p', \psi), \omega \rangle] \in \Delta'$ for every rule $\langle p, \gamma \rangle \hookrightarrow \langle p', \omega \rangle \in \Delta$, and $\langle (p, \psi), \gamma \rangle \overset{equal}{\hookrightarrow} \langle (p, \psi_2), \gamma \rangle \in \Delta'$;

9. if $\psi = A[\psi_1 U \psi_2]$; $\langle (p, \psi), \gamma \rangle \overset{equal}{\hookrightarrow} [\langle (p, \psi_1), \gamma \rangle, \langle (p_1, \psi), \omega_1 \rangle, ..., \langle (p_l, \psi), \omega_l \rangle] \in \Delta'$ such that for every i, $1 \le i \le l$, $\langle p, \gamma \rangle \hookrightarrow \langle p_i, \omega_i \rangle \in \Delta$ and these transitions are all the transitions of Δ that have $\langle p, \gamma \rangle$ as left hand side, and $\langle (p, \psi), \gamma \rangle \overset{equal}{\hookrightarrow} \langle (p, \psi_2), \gamma \rangle \in \Delta'$;

10. if $\psi = E[\psi_1 \tilde{U} \psi_2]$; $\langle (p, \psi), \gamma \rangle \overset{equal}{\hookrightarrow} [\langle (p, \psi_2), \gamma \rangle, \langle (p', \psi), \omega \rangle] \in \Delta'$ for every $\langle p, \gamma \rangle \hookrightarrow \langle p', \omega \rangle \in \Delta$, and $\langle (p, \psi), \gamma \rangle \overset{equal}{\hookrightarrow} [\langle (p, \psi_2), \gamma \rangle, \langle (p, \psi_1), \gamma \rangle] \in \Delta'$;

11. if $\psi = A[\psi_1 \tilde{U} \psi_2]$; $\langle (p, \psi), \gamma \rangle \overset{equal}{\hookrightarrow} [\langle (p, \psi_2), \gamma \rangle, \langle (p_1, \psi), \omega_1 \rangle, ..., \langle (p_l, \psi), \omega_l \rangle] \in \Delta'$ such that for every i, $1 \le i \le l$, $\langle p, \gamma \rangle \hookrightarrow \langle p_i, \omega_i \rangle \in \Delta$ and these transitions are all the transitions of Δ that have $\langle p, \gamma \rangle$ as left hand side, and $\langle (p, \psi), \gamma \rangle \overset{equal}{\longrightarrow} [\langle (p, \psi_1), \gamma \rangle, \langle (p, \psi_2), \gamma \rangle] \in \Delta'$;

12. if $\psi = e$: $\langle (p, \psi), \gamma \rangle \overset{equal}{\hookrightarrow} \langle s_e, \gamma \rangle \in \Delta'$, where s_e is the initial state of M_e,

13. if $\psi = \neg e$: $\langle (p, \psi), \gamma \rangle \overset{equal}{\hookrightarrow} \langle s_{\neg e}, \gamma \rangle \in \Delta'$, where $s_{\neg e}$ is the initial state of $M_{\neg e}$,

14. for every transition $q \overset{\alpha}{\to} \{q_1, ..., q_n\}$ in M; $\langle q, \gamma \rangle \overset{\mathfrak{R}}{\hookrightarrow} \{\langle q_1, \epsilon \rangle, ..., \langle q_n, \epsilon \rangle\} \in \Delta'$, where
    (a) $\mathfrak{R} = equal$ if $\alpha = \gamma$,
    (b) $\mathfrak{R} = join^x_\gamma$ if $\alpha = x \in \mathcal{X}$,
    (c) $\mathfrak{R} = join^{\neg x}_\gamma$ if $\alpha = \neg x$ and $x \in \mathcal{X}$.

15. for every $q \in \mathcal{F}$; $\langle q, \sharp \rangle \overset{equal}{\hookrightarrow} \langle q, \sharp \rangle \in \Delta'$.

Roughly speaking, $\mathcal{BP}_\varphi$ could be seen as the product of $\mathcal{P}$ and φ. $\mathcal{BP}_\varphi$ recognizes all the configurations $\langle [(p, \psi), B], \omega \rangle$ s.t. $\langle p, \omega \rangle$ satisfies ψ under B. Thus $\mathcal{BP}_\varphi$ has an accepting run from $\langle [(p, \psi), B], \omega \rangle$ if and only if the configuration $\langle p, \omega \rangle$ satisfies ψ under B. Due to lack of space, we only explain the case $\psi = e$. In this case, the SABPDS $\mathcal{BP}_\varphi$ accepts $\langle [(p, \psi), B], \omega \rangle$ iff $(\langle p, \omega \rangle, B) \in L(M_e)$. To check this, $\mathcal{BP}_\varphi$ first goes to state $[s_e, B]$ by Item 12, where s_e is the initial state of M_e, then it continues to check whether ω is accepted by M_e under the environment B. This is ensured by Items 14. Item 14 allows $\mathcal{BP}_\varphi$ to mimic a run of M_e on ω under the environment B: if $\mathcal{BP}_\varphi$ is in state $[q, B]$ and the topmost symbol of its stack is γ, then:

– Item 14(a) deals with the case where $q \overset{\gamma}{\longrightarrow} \{q_1, ..., q_2\}$ is a transition in δ_e. In this case, $\mathcal{BP}_\varphi$ moves to the next states $[q_1, B], ..., [q_n, B]$ while popping γ from the stack. Popping γ allows $\mathcal{BP}_\varphi$ to check the rest of the word. The function *equal* guarantees that all the environments are the same.

- Item 14(b) deals with the case where $q \xrightarrow{x} \{q_1, ..., q_2\}$, $x \in X$ is a transition in δ_e. In this case, $\mathcal{BP}_\varphi$ can continue to mimic a run of M_e under the environment B only if $B(x) = \gamma$. If this holds, $\mathcal{BP}_\varphi$ moves to the next states $[q_1, B], ..., [q_n, B]$ and pops γ from the stack, which allows $\mathcal{BP}_\varphi$ to check the rest content of the stack. The function $join_\gamma^x$ ensures that all the environments are the same and the value of $B(x)$ is γ.

- Similarly, Item 14(c) deals with the case where $q \xrightarrow{\neg x} \{q_1, ..., q_2\}$ is in δ_e.

Thus, $(\langle p, \omega \rangle, B) \in L(M_e)$ iff M_e reaches final states $f_1, ..., f_n$ of M_e after reading the word ω, i.e., iff $\mathcal{BP}_\varphi$ reaches a set of states $[f_1, B], ..., [f_n, B]$ with an empty stack (a stack containing only the bottom stack symbol $\sharp$). This is why F_4 is a set of accepting states. Moreover, since all the accepting paths are infinite, Item 15 adds a loop on every configuration $\langle [f, B], \sharp \rangle$ where f is a final state of M and $\sharp$ is the stack symbol (this makes the paths of $\mathcal{BP}_\varphi$ that reach a state $\langle [f, B], \sharp \rangle$ accepting). Formally, we can show:

Theorem 4. *Given a PDS $\mathcal{P} = (P, \Gamma, \Delta, \sharp)$, a function $\lambda : \mathrm{AP}_\mathcal{D} \longrightarrow 2^P$, a SCTPL formula φ, and a configuration $\langle p, \omega \rangle$ of $\mathcal{P}$, we have: for every $B \in \mathcal{B}$, $\langle p, \omega \rangle \models_\lambda^B \varphi$ iff $\mathcal{BP}_\varphi$ has an accepting run from the configuration $\langle [(p, \varphi)], B], \omega \rangle$.*

4.4   SCTPL Model-Checking for PDSs

Given a PDS $\mathcal{P} = (P, \Gamma, \Delta, \sharp)$, a labeling function λ, and a SCTPL formula φ, thanks to Theorems 4 and 3, and due to the fact that $\mathcal{BP}_\varphi$ has $O(|P| \cdot |\varphi| + k)$ states and $O((|P| \cdot |\Gamma| + |\Delta|) \cdot |\varphi| + d)$ transitions, where k and d are the number of states and the number of transitions of the union M of the Variable Automata involved in φ; we get the following:

Corollary 1. *Given a PDS $\mathcal{P} = (P, \Gamma, \Delta, \sharp)$, a SCTPL formula φ and a labeling function λ, we can effectively compute a SAMA $\mathcal{A}$ in time $O\big((|P||\varphi| + k)^2 \cdot 2^{2|\mathcal{B}|} \cdot |\Gamma| \cdot ((|P||\Gamma| + |\Delta|)|\varphi| + d) \cdot 2^{5(|P||\varphi|+k) \cdot 2^{|\mathcal{B}|}}\big)$, where k is the number of states of M and d is the number of transition rules of M such that for every configuration $\langle p, \omega \rangle$ of $\mathcal{P}$:*

1. *$\langle p, \omega \rangle \models_\lambda \varphi$ iff there exists a $B \in \mathcal{B}$ s.t. $\mathcal{A}$ recognizes $\langle [(p, \varphi)], B], \omega \rangle$.*
2. *for every $B \in \mathcal{B}$: $\langle p, \omega \rangle \models_\lambda^B \varphi$ iff $\mathcal{A}$ recognizes $\langle [(p, \varphi)], B], \omega \rangle$.*

Thus, thanks to this corollary and to Proposition 1, it follows that it is possible to determine whether a PDS configuration satisfies a SCTPL formula:

Corollary 2. *It is possible to decide whether a PDS configuration satisfies a SCTPL formula.*

Remark 3. We can transform every SCTPL formula ψ to an equivalent CTL with regular valuations formula ψ'. Then, applying [21], we can construct an AMA recognizing all the configurations which satisfy ψ'. However, in practice, thanks to the compact representation of the sets of environments β's using BDDs, model-checking SCTPL using our symbolic techniques behaves much better than reducing SCTPL to CTL with regular valuations and then applying [21]. Indeed, the experiments we run show that in most of the cases, our symbolic algorithm for SCTPL model-checking terminates in few seconds, whereas translating the SCTPL formula to CTL with regular valuations and then applying [21] would run out of memory.

Table 1. Detection of real malwares

Examples	|P|	Our techniques		SABPDS→ABPDS		SCTPL→CTLr		Result
		Time(s)	Mem(Mb)	Time(s)	Mem(Mb)	Time(s)	Mem(Mb)	
Klez.a	42	1.62	10.8	-	MemOut	-	MemOut	Y
Klez.b	45	1.55	10.8	-	MemOut	-	MemOut	Y
Klez.c	41	1.27	8.9	-	MemOut	-	MemOut	Y
Klez.d	51	1.47	10.3	-	MemOut	-	MemOut	Y
Klez.e	52	0.77	7.0	-	MemOut	-	MemOut	Y
Klez.f	50	0.76	7.0	-	MemOut	-	MemOut	Y
Klez.g	47	0.75	7.0	-	MemOut	-	MemOut	Y
Klez.i	49	0.74	7.0	-	MemOut	-	MemOut	Y
Klez.j	55	0.74	7.0	-	MemOut	-	MemOut	Y
Mydoom.c	210	145.20	322.8	-	MemOut	-	MemOut	Y
Mydoom.e	288	123.22	267.5	-	MemOut	-	MemOut	Y
Mydoom.g	256	117.50	256.7	-	MemOut	-	MemOut	Y
Predec.j	25	0.23	0.81	-	MemOut	56.14	36.16	Y
Netsky.a	69	2.73	14.5	-	MemOut	-	MemOut	Y
Akez	42	0.22	0.3	-	MemOut	0.44	2.49	Y
Netsky.b	80	2.73	14.5	-	MemOut	-	MemOut	Y
Netsky.c	78	2.73	14.5	-	MemOut	-	MemOut	Y
Netsky.d	72	2.73	14.5	-	MemOut	-	MemOut	Y
Alcaul.h	48	0.83	0.9	-	MemOut	1.14	6.88	Y
Uedit32	180	92.58	100.94	-	MemOut	-	MemOut	N
Alcaul.l	2	0.30	0.7	-	MemOut	0.86	3.96	Y
Cygwin32	212	23.72	123.31	-	MemOut	-	MemOut	N
cmd.exe	202	1.44	25.52	-	MemOut	-	MemOut	N
Alcaul.o	68	0.20	0.6	-	MemOut	0.83	3.37	Y
Mydoor.ar	256	113.2	227.4	-	MemOut	-	MemOut	Y
Adson.1559	52	0.22	2.1	-	MemOut	-	MemOut	Y
Adson.1651	54	0.23	2.1	-	MemOut	-	MemOut	Y
Adson.1703	55	0.25	2.1	-	MemOut	-	MemOut	Y
Adson.1734	54	0.31	2.6	-	MemOut	-	MemOut	Y
Alcaul.d	62	0.20	0.8	-	MemOut	47.70	51	Y
Alcaul.i	88	4.38	0.28	-	MemOut	159.88	169.64	Y
Alcaul.j	79	0.30	2.1	-	MemOut	218.25	198.71	Y
Oroch.3982	89	3.70	7.72	-	MemOut	-	MemOut	Y
KME	145	999.31	20.04	-	MemOut	-	MemOut	Y
Anar.a	41	1.16	1.60	885.33	343.24	54.92	34.12	Y
Anar.b	47	1.49	1.60	891.42	348.54	56.14	36.16	Y
Atak.b	126	762.34	18.15	-	MemOut	-	MemOut	Y
Alcaul.c	33	0.12	0.3	-	MemOut	0.41	2.19	Y
Bagle.d	88	652.23	16.96	-	MemOut	-	MemOut	Y
Alcaul.f	52	0.09	0.3	-	MemOut	0.53	2.23	Y
Alcaul.b	50	0.06	0.2	-	MemOut	0.28	1.18	Y
Alcaul.e	49	0.49	0.9	-	MemOut	1.03	5.28	Y
Alcaul.g	53	0.31	0.7	-	MemOut	0.97	4.45	Y
Evol.a	102	9.58	3.22	-	MemOut	-	MemOut	Y
Alcaul.k	52	0.26	0.6	-	MemOut	0.76	3.65	Y
Alcaul.m	53	0.20	0.6	-	MemOut	0.88	3.37	Y
Alcaul.n	34	0.12	0.3	-	MemOut	0.44	2.28	Y
Klinge	78	237.50	4.49	-	MemOut	0.83	3.37	Y
Atak.f	220	23.4	139.1	-	MemOut	-	MemOut	Y
Mydoor.ay	328	124.2	232.5	-	MemOut	-	MemOut	Y

5   Experiments

We implemented our techniques in a tool for malware detection. We use IDAPro [3] as disassembler. We use BDDs to represent sets of environments. We carried out different experiments. We obtained interesting results. In particular, our tool was able to detect several viruses taken from [14]. Our results are reported in Table 1. **Column** $|P|$ gives the number of control locations of the PDS model. Every program is checked against several malicious behaviors. A program is declared as a potential virus if it satisfies one of the specifications. **Column** *time(s)* and *mem(Mb)* give the time (in seconds) and the memory (in Mb). The last **Column** *result* is Y is the program contains the malicious behaviors given in **Column** *Formula*, and N if not. We also compared our techniques against translating SABPDS to ABPDS (**Columns** "SABPDS→ABPDS"), or translating SCTPL to CTL with regular valuations (**Columns** "SCTPL→CTLr"). We were able to detect all the viruses that we considered, whereas applying the translation from SABPDS to ABPDS or from SCTPL to CTL with regular valuations would run out of memory in most of the cases, and thus cannot detect the viruses. Our tool was also able to deduce that some benign programs are not viruses. E.g. we tried the following benign programs: *Uedit32*, a fragment of Ultra Edit Text Editor software by IDM Computer Solutions; *Cygwin32* a fragment of the Setup software of Cygwin, a Linux-like environment for Windows. *cmd.exe* is the Microsoft-supplied command-line interpreter.

Table 2. Detection of obfuscated Viruses

Obfuscation	Our techniques detection rate	Avira antivirus detection rate	Qihoo 360 antivirus detection rate	Avast antivirus detection rate
nop-insertion	**100**%	65%	55%	60%
code-reordering	**100**%	40%	35%	45%
register-renaming	**100**%	25%	25%	30%
stack-operation	**100**%	20%	25%	20%
procedure-split	**100**%	5%	5%	5%

Moreover, we run several experiments to check how robust are our techniques in virus detection in case the virus writers use obfuscation techniques. To this aim, we considered some of the viruses of Table 1, and we added several obfuscations manually such as: instruction reordering (reordering the instructions inside the code and using jump instructions so that the control flow is not changed), dead code insertion, register renaming, splitting the code into several procedures, adding useless stack operations, etc. We tested 5 variants for each type of obfuscation of the viruses Mydoom.g, Netsky.a, Bagle.d, Adson.1734 and Akez. The results are reported in Table 2. Our techniques were able to detect all these variations, whereas the three well known and widely used free antiviruses *Avira* [2], *Qihoo 360* [4] and *Avast* [1] were not able to detect several of these virus variations.

References

1. Avast antivirus, free version, http://www.avast.com
2. Avira antivirus, free version, http://www.avira.com
3. IDA Pro, http://www.hex-rays.com/idapro/
4. Qihoo 360 antivirus, http://www.360.cn
5. Balakrishnan, G., Reps, T., Kidd, N., Lal, A., Lim, J., Melski, D., Gruian, R., Yong, S., Chen, C.-H., Teitelbaum, T.: Model Checking x86 Executables with CodeSurfer/x86 and WPDS++. In: Etessami, K., Rajamani, S.K. (eds.) CAV 2005. LNCS, vol. 3576, pp. 158–163. Springer, Heidelberg (2005)
6. Bergeron, J., Debbabi, M., Desharnais, J., Erhioui, M., Lavoie, Y., Tawbi, N.: Static detection of malicious code in executable programs. In: SREIS (2001)
7. Bonfante, G., Kaczmarek, M., Marion, J.-Y.: Architecture of a Morphological Malware Detector. Journal in Computer Virology 5, 263–270 (2009)
8. Bryant, R.E.: Symbolic boolean manipulation with ordered binary-decision diagrams. ACM Comput. Surv. 24(3) (1992)
9. Christodorescu, M., Jha, S.: Static analysis of executables to detect malicious patterns. In: 12th USENIX Security Symposium (2003)
10. Christodorescu, M., Jha, S., Kruegel, C.: Mining specifications of malicious behavior. In: ISEC (2008)
11. Christodorescu, M., Jha, S., Seshia, S.A., Song, D.X., Bryant, R.E.: Semantics-aware malware detection. In: IEEE Symposium on Security and Privacy (2005)
12. Esparza, J., Kucera, A., Schwoon, S.: Model checking LTL with regular valuations for pushdown systems. Inf. Comput. 186(2) (2003)
13. Esparza, J., Schwoon, S.: A BDD-Based Model Checker for Recursive Programs. In: Berry, G., Comon, H., Finkel, A. (eds.) CAV 2001. LNCS, vol. 2102, pp. 324–336. Springer, Heidelberg (2001)
14. Heavens, V.: http://vx.netlux.org
15. Holzer, A., Kinder, J., Veith, H.: Using Verification Technology to Specify and Detect Malware. In: Moreno Díaz, R., Pichler, F., Quesada Arencibia, A. (eds.) EUROCAST 2007. LNCS, vol. 4739, pp. 497–504. Springer, Heidelberg (2007)
16. Kinder, J., Katzenbeisser, S., Schallhart, C., Veith, H.: Detecting Malicious Code by Model Checking. In: Julisch, K., Krügel, C. (eds.) DIMVA 2005. LNCS, vol. 3548, pp. 174–187. Springer, Heidelberg (2005)
17. Kinder, J., Katzenbeisser, S., Schallhart, C., Veith, H.: Proactive detection of computer worms using model checking. IEEE Transactions on Dependable and Secure Computing 7(4) (2010)
18. Lakhotia, A., Boccardo, D.R., Singh, A., Manacero, A.: Context-sensitive analysis of obfuscated x86 executables. In: PEPM (2010)
19. Lakhotia, A., Kumar, E.U., Venable, M.: A method for detecting obfuscated calls in malicious binaries. IEEE Trans. Software Eng. 31(11) (2005)
20. Singh, P.K., Lakhotia, A.: Static verification of worm and virus behavior in binary executables using model checking. In: IAW (2003)
21. Song, F., Touili, T.: Efficient CTL Model-Checking for Pushdown Systems. In: Katoen, J.-P., König, B. (eds.) CONCUR 2011. LNCS, vol. 6901, pp. 434–449. Springer, Heidelberg (2011)

Aspect-Oriented Runtime Monitor Certification*

Kevin W. Hamlen, Micah M. Jones, and Meera Sridhar

University of Texas at Dallas
{hamlen,micah.jones1,meera.sridhar}@utdallas.edu

Abstract. In-lining runtime monitors into untrusted binary programs
via aspect-weaving is an increasingly popular technique for efficiently
and flexibly securing untrusted mobile code. However, the complexity of
the monitor implementation and in-lining process in these frameworks
can lead to vulnerabilities and low assurance for code-consumers. This
paper presents a machine-verification technique for aspect-oriented in-
lined reference monitors based on abstract interpretation and model-
checking. Rather than relying upon trusted advice, the system verifies
semantic properties expressed in a purely declarative policy specification language. Experiments on a variety of real-world policies and Java
applications demonstrate that the approach is practical and effective.

Keywords: Abstract interpretation, in-lined reference monitors, model-
checking, security.

1   Introduction

Software security systems that employ purely static analyses to detect and reject
malicious code are limited to enforcing decidable security properties. Unfortunately, most useful program properties, such as safety and liveness properties,
are not generally decidable and can therefore only be approximated by purely
static analyses. For example, signature-based antivirus products accept or reject programs based on their syntax rather than their runtime behavior, and
therefore suffer from dangerous false negatives, inconvenient false positives, or
both (cf., [16]). This has shifted software security research increasingly toward
more powerful dynamic analyses, but these dynamic systems are often far more
difficult to formally verify than provably sound static analyses.

An increasingly important family of such dynamic analyses are those that
modify untrusted binary code prior to its execution. *In-lined reference monitors*
(IRMs) instrument untrusted code with new operations that perform runtime
security checks before potentially dangerous operations [27]. The approach is motivated by improved efficiency (since IRMs require fewer context switches than
external monitors), deployment flexibility (since in-lining avoids modifying the
VM or OS), and precision (since IRMs can monitor internal program operations

* Supported by AFOSR award FA9550-08-1-0044 and NSF award NSF-1065216. Any
views expressed do not necessarily reflect those of the NSF or AFOSR.

C. Flanagan and B. König (Eds.): TACAS 2012, LNCS 7214, pp. 126–140, 2012.

not readily visible to an external monitor). Most modern IRM systems are implemented using some form of *aspect-oriented programming* (AOP) [32,28,7,8,14]. Such IRMs are implemented as *pointcut-advice* pairs: pointcuts identify security-relevant program operations and *advice* prescribes local code transformations sufficient to guard such operations. This suffices to enforce safety policies [27,18] and some liveness policies [26].

To provide exceptionally high assurance guarantees, recent work has sought to reduce the (potentially large) trusted computing bases (TCBs) of IRM frameworks by separately machine-verifying the *self-monitoring* code they produce [17,1,30,31]. For example, the S^3MS project uses a contract-based verifier [1] to avoid trusting the the much larger in-liner (over 900K lines of Java code if one includes the underlying AspectJ system [22]) that generates the IRMs.

However, TCB-minimization of large IRM systems has been frustrated by the inevitable inclusion of significant, trusted code within the AOP-style policy specifications themselves. Verifiers for these systems can prove that the IRM system has correctly in-lined the policy-prescribed advice code but not that this advice actually enforces the desired policy. Past case studies have demonstrated that such advice is extremely difficult to write correctly, especially when the policy is intended to apply to large classes of untrusted programs rather than individual applications [21]. Moreover, in many domains, such as web ad security, policy specifications change rapidly as new attacks and vulnerabilities are discovered (cf., [23,29,30]). Thus, the considerable effort that might be devoted to formally verifying one particular aspect implementation quickly becomes obsolete when the aspect is revised in response to a new threat.

To address this open challenge, we present Cheko✓: the first IRM-certification framework that verifies full, AOP-style IRMs against purely declarative policy specifications without trusting the code that implements the IRM. Cheko✓ uses light-weight model-checking and abstract interpretation to verify untrusted (but verifiably type-safe) Java bytecode binaries against trusted policy specifications that lack advice. Policies declaratively specify how security-relevant program operations affect an abstract system security state. Unlike contracts, which denote code transformations, policies in our system therefore denote pure code properties. Such properties can be enforced by untrusted aspects that dynamically detect impending policy violations and take corrective action. The woven aspects are verified (along with the rest of the self-monitoring code) against the trusted policy specification prior to its execution.

Cheko✓ was inspired by our prior work on model-checking IRMs [30,29,9], but includes numerous substantial theoretic and pragmatic leaps beyond those earlier works. These include:

- support for a full-scale Java IRM framework (the SPoX IRM system [14,20]) that includes *stateful* (history-based) policies, event detection by pointcut-matching, and IRM implementations that combine (untrusted) before- and after-advice insertions;
- a novel approach to dynamic pointcut verification using Constraint Logic Programming (CLP) [19]; and

- proofs of correctness based on Cousot's abstract interpretation framework [5] that link the denotational semantics of SPoX policies to the operational semantics of the abstract interpreter.

Section 2 begins with related work, followed by an overview of the SPoX policy language and the rewriter in §3. Section 4 presents a high-level description of the verification algorithm. (A more detailed treatment with proofs is available in the companion technical report [15].) Section 5 presents in-depth case-studies of several security policy classes that we enforced on numerous real-world applications, and discusses challenges faced in implementing and verifying these policies. Finally, §6 concludes with recommendations for future work.

2   Related Work

IRMs were first formalized in the PoET/PSLang/SASI systems [11,27,10], which implement IRMs for Java bytecode and GNU assembly code. IRM systems have subsequently been developed for many architectures (cf., [24,4]). Most of these express security policies in an AOP or AOP-like language with pointcut expressions for identifying security-relevant binary program operations, and code fragments (advice) that specify actions for detecting and prohibiting impending policy violations. A hallmark of these systems is their ability to enforce history-based, stateful policies that permit or prohibit each event based on the history of past events exhibited by the program. This is typically achieved by expressing the security policy as an automaton [27,25] whose state is *reified* into the untrusted program as a protected global variable. The IRM tracks the current security state at runtime by consulting and updating the variable as events occur.

Machine-certification of IRMs was first proposed as type-checking [33]—an idea that was later extended and implemented in the Mobile system [17]. Mobile transforms Microsoft .NET bytecode binaries into safe binaries with typing annotations in an effect-based type system. The annotations constitute a proof of safety that a type-checker can separately verify to prove that the transformed code is safe. Type-based IRM certification is efficient and elegant but does not currently support dynamic pointcut matching. It has therefore not been applied to AOP-style IRMs to our knowledge.

ConSpec [2,1] adopts a security-by-contract approach to AOP IRM certification. Its certifier performs a static analysis that verifies that contract-specified guard code appears at each security-relevant code point. While certification-via-contract facilitates natural expression of policies as AOP programs, it has the disadvantage of including the potentially complex advice code in the TCB.

Our prior work [30] is the first to adopt a model-checking approach to verify such IRMs without trusted guard code. The prototype IRM certifier in [30] supports reified security state, but it does not support dynamic pointcuts and its support for advice is limited to a very constrained form of before-advice. It therefore does not support real-world IRM systems or their policies, which regularly employ dynamic pointcuts and after-advice.

In contrast, the verifier presented in this work targets SPoX [14,20], a fully featured, purely declarative AOP IRM system for Java bytecode. SPoX policies are advice-free; any advice that implements the IRM remains untrusted and must therefore undergo verification. Policy specifications consist of pointcuts and declarative specifications of how pointcut-matching events affect the security state. The abstract security state-changes specified by SPoX policies are significantly higher-level and simpler than the arbitrary advice code admitted by non-declarative AOP languages. Thus, SPoX policies are a significant TCB reduction over AOP contracts that implement them.

3   Policy Language and Rewriter

As an example of how software security policies are specified in SPoX, Fig. 1 specifies a policy that permits applications to send at most 10 email messages per run. The policy says that `Mail.send` API calls increment security state s up to 10, but an 11th call triggers a policy violation. Such a policy could be useful for preventing spam.

```
1  (state name="s")
2  (forall "i" from 0 to 9
3    (edge name="count"
4      (call "Mail.send")
5      (nodes "s" i, i + 1)))
6  (edge name="10emails"
7    (call "Mail.send")
8    (nodes "s" 10,#))
```

Fig. 1. A policy permitting at most 10 email-send events

More formally, SPoX policies denote *security automata* [3]—finite- or infinite-state machines that accept languages of permissible *event sequences*. Sets of edges in the security automaton are described by **edge** structures, each of which consists of a pointcut expression (Lines 4 and 7) and at least one **nodes** declaration (Lines 5 and 8). The pointcut expression defines a common label for the edges in the set, while each **nodes** declaration imposes a transition pre-condition and post-condition for a particular state variable. The pre-condition constrains the set of source states to which the edge applies, and the post-condition describes how the state changes when an event satisfying the pointcut expression and all pre-conditions is exhibited. Events that satisfy none of the outgoing edge labels of the current security state leave the security state unchanged. Policy-violations are explicitly identified with the reserved post-condition "#".

SPoX derives its pointcut language from AspectJ, allowing policy writers to develop policies that regard static and dynamic method calls and their arguments, object pointers, and lexical contexts, among other properties. In order to remain fully declarative, SPoX omits explicit, imperative advice. Instead, policies declaratively specify how security-relevant events change the current security automaton state. Rewriters then synthesize their own advice in order to enforce the prescribed policy. The use of declarative state-transitions instead of imperative advice facilitates formal, automated reasoning about policies without the need to reason about arbitrary code [21].

The SPoX rewriter takes as input a Java binary archive (JAR) and a SPoX policy, and outputs a new application in-lined with an IRM that enforces the policy. The high-level in-lining approach is essentially the same as the other IRM systems discussed in §2. Each method body is scanned for potentially security-relevant instructions, and sequences of guard instructions are in-lined around those to detect and preclude policy-violations at runtime.

In-lined guard code must track event histories if the policy is stateful. To do so, the rewriter reifies abstract security state variables into the untrusted code as static, private class fields. The guard code then tracks the abstract security state by consulting and updating the corresponding fields. The runtime guards must also evaluate any statically undecidable portions of pointcut expressions to decide whether impending events are actually security-relevant. For example, to evaluate pointcut (argval 1 (intgt 2)), it might dynamically test whether $x > 2$, where x is the impending operation's first argument.

$(A{=}S \wedge A{=}T)$	0.1
`1 if (Policy.s >= 0 && Policy.s <= 9)`	
	$(A{=}S \wedge A{=}T \wedge S{\geq}0 \wedge S{\leq}9)$ 1.1
`2     Policy.temp_s := Policy.s+1;`	
	$(A{=}S \wedge A{=}T' \wedge S{\geq}0 \wedge S{\leq}9 \wedge T{=}S{+}1)$ 2.1
	$(A{=}S \wedge A{=}T \wedge (S{<}0 \vee S{>}9))$ 2.2
`3 if (Policy.s == 10)`	
	$(A{=}S \wedge A{=}T' \wedge S{\geq}0 \wedge S{\leq}9 \wedge T{=}S{+}1 \wedge S{=}10)$ 3.1
	$(A{=}S \wedge \overline{A{=}T} \wedge \overline{(S{<}0 \vee S{>}9)} \wedge S{=}10)$ 3.2
`4     call System.exit(1);`	
	$(A{=}S \wedge A{=}T' \wedge S{\geq}0 \wedge S{\leq}9 \wedge T{=}S{+}1 \wedge S{\neq}10)$ 4.1
	$(A{=}S \wedge A{=}T \wedge (S{<}0 \vee S{>}9) \wedge S{\neq}10)$ 4.2
`5 Policy.s := Policy.temp_s;`	
	$(A{=}S' \wedge A{=}T' \wedge S'{\geq}0 \wedge S'{\leq}9 \wedge T{=}S'{+}1 \wedge S'{\neq}10 \wedge S{=}T)$ 5.1
	$(A{=}S' \wedge A{=}T \wedge (S'{<}0 \vee S'{>}9) \wedge S'{\neq}10 \wedge S{=}T)$ 5.2

`6 call Mail.send();`

$(A'{=}S' \wedge A'{=}T' \wedge S'{\geq}0 \wedge S'{\leq}9 \wedge T{=}S'{+}1 \wedge S'{\neq}10 \wedge S{=}T \wedge A'{=}I \wedge I{\geq}0 \wedge I{\leq}9 \wedge A{=}I{+}1)$ 6.1

$(A'{=}S' \wedge A'{=}T' \wedge S'{\geq}0 \wedge S'{\leq}9 \wedge T{=}S'{+}1 \wedge S'{\neq}10 \wedge S{=}T \wedge A'{=}10 \wedge A{=}\#)$ 6.2

$(A'{=}S' \wedge A'{=}T' \wedge S'{\geq}0 \wedge S'{\leq}9 \wedge T{=}S'{+}1 \wedge \overline{S'}{\neq}10 \wedge S{=}T \wedge (A'{<}0 \vee A'{>}9) \wedge A'{\neq}10 \wedge A{=}A')$ 6.3

$(A'{=}S' \wedge A'{=}T \wedge (S'{<}0 \vee S'{>}9) \wedge S'{\neq}10 \wedge S{=}T \wedge A'{=}I \wedge I{\geq}0 \wedge I{\leq}9 \wedge A{=}I{+}1)$ 6.4

$(A'{=}S' \wedge A'{=}T \wedge (S'{<}0 \vee S'{>}9) \wedge S'{\neq}10 \wedge S{=}T \wedge A'{=}10 \wedge A{=}\#)$ 6.5

$(A'{=}S' \wedge A'{=}T \wedge (S'{<}0 \vee S'{>}9) \wedge S'{\neq}10 \wedge S{=}T \wedge (A'{<}0 \vee A'{>}9) \wedge A'{\neq}10 \wedge A{=}A')$ 6.6

Fig. 2. An abstract interpretation of instrumented pseudocode

The left column of Fig. 2 gives pseudocode for an IRM that enforces the policy in Fig. 1. For each call to method `Mail.send`, the IRM tests two possible preconditions: (1) $0 \leq s \leq 9$ and (2) $s = 10$. In the first case, it increments s; in the second, it aborts the process. Observe that in this example security state s has been reified as two separate fields of class `Policy` (s and `temp_s`) in order to prevent join point conflicts. This reflects a reality that any given policy has a variety of IRM implementations, many of which contain unexpected quirks that address non-obvious, low-level enforcement details.

4   Verifier

Our verifier takes as input (1) a SPoX security policy, (2) an instrumented, type-safe Java bytecode program, and (3) some optional, untrusted hints from the rewriter (detailed shortly). It either accepts the program as provably policy-satisfying or rejects it as potentially policy-violating. Type-safety is checked by the JVM, allowing our verifier to safely assume that all bytecode operations obey standard Java memory-safety and well-formedness. This keeps tractable the task of reliably identifying security relevant operations and field accesses.

The main verifier engine uses abstract interpretation to non-deterministically explore all control-flow paths of the untrusted code, inferring an abstract program state at each code point. A model-checker then proves that each abstract state is policy-adherent, thereby verifying that no execution of the code enters a policy-violating program state. Policy-violations are modeled as *stuck states* in the operational semantics of the verifier—that is, abstract interpretation cannot continue when the current abstract state fails the model-checking step. This results in conservative rejection of the untrusted code. The verifier is expressed as a bisimulation of the program and the security automaton. Abstract states in the analysis conservatively approximate not only the possible contents of memory (e.g., stack and heap contents) but also the possible security states of the system at each code point.

The heart of the verification algorithm involves inferring and verifying relationships between the abstract program state and the abstract security state. When policies are stateful, this involves verifying relationships between the abstract security state and the corresponding reified security state(s). These relationships are complicated by the fact that although the reified state often precisely encodes the actual security state, there are also extended periods during which the reified and abstract security states are not synchronized at runtime. For example, guard code may preemptively update the reified state to reflect a future security state that will only be reached after subsequent security-relevant events, or it may retroactively update the reified state only after numerous operations that change the security state have occurred. These two scenarios correspond to the insertion of before- and after-advice in AOP IRM implementations. The verification algorithm must be powerful enough to automatically track these relationships and verify that guard code implemented by the IRM suffices to prevent policy violations.

To aid the verifier in this task, we modified the SPoX rewriter to export two forms of untrusted hints along with the rewritten code: (1) a relation $\sim$ that associates policy-specified security state variables s with their reifications r, and (2) marks that identify code regions where related abstract and reified states might not be *synchronized* according to the following definition:

Definition 1 (Synchronization Point). *A synchronization point (SYNC) is an abstract program state with constraints ζ such that proposition $\zeta \wedge (\bigvee_{r \sim s}(r \neq s))$ is unsatisfiable.*

Chekov✓ uses these hints (without trusting them) to guide the verification process and to avoid state-space explosions that might lead to conservative rejection of safe code. In particular, it verifies that all non-marked instructions are *SYNC*-preserving, and each outgoing control-flow from a marked region is *SYNC*-restoring. This modularizes the verification task by allowing separate verification of marked regions, and controls state-space explosions by reducing the abstract state to *SYNC* throughout the majority of binary code which is not security-relevant. Providing incorrect hints causes Chekov✓ to reject (e.g., when it discovers that an unmarked code point is potentially security-relevant) or converge more slowly (e.g., when security-irrelevant regions are marked and therefore undergo unnecessary extra analysis), but it never leads to unsound certification of unsafe code.

A Verification Example. Figure 2 demonstrates a verification example step-by-step. The pseudocode constitutes a marked region in the target program, and the verifier requires that the abstract interpreter is in the *SYNC* state immediately before and after. At each code point, the verifier infers an abstract program state that includes one or more conjunctions of constraints on the abstract and reified security state variables. These constraints track the relationships between the reified and abstract security state. Here, variable A represents the abstract state variable s from the policy in Fig. 1. Reifications `Policy.s` and `Policy.temp_s` are written as S and T, respectively, with $S \sim A$ and $T \sim A$. Thus, state *SYNC* is given by constraint expression $(A = S \wedge A = T)$ in this example.

The analysis begins in the *SYNC* state, as shown in constraint list 0.1. Line 1 is a conditional, and thus spawns two new constraint lists, one for each branch. The positive branch (1.1) incorporates the conditional expression $(S \geq 0 \wedge S \leq 9)$ in Line 2, whereas the negative branch (2.2) incorporates the negation of the same conditional. The assignment in Line 2 is modeled by alpha-converting T to T' and conjoining constraint $S = T + 1$; this yields constraint list 2.1.[1]

Unsatisfiable constraint lists are opportunistically pruned to reduce the state space. For example, list 3.1 shows the result of applying the conditional of Line 3 to 2.1. Conditionals 1 and 3 are mutually exclusive, resulting in contradictory expressions $S \leq 9$ and $S = 10$; therefore, 3.1 is dropped. Similarly, 3.2 is dropped because no control-flows exit Line 4.

To interpret a security-relevant event such as the one in Line 6, the verifier simulates the traversal of all edges in the security automaton. In typical policies, any given instruction fails to match a majority of the pointcut labels in the policy, so most are immediately dropped. The remaining edges are simulated by conjoining each edge's pre-conditions to the current constraint list and modeling the edge's post-condition as a direct assignment to A. For example, edge `count` in Fig. 1 imposes pre-condition $(0 \leq I \leq 9) \wedge (A = I)$, and its post-condition can be modeled as assignment $A := I + 1$. Applying these to list 5.1 yields list 6.1. Likewise, 6.2 is the result of applying edge `10emails` to 5.1, and 6.4 and 6.5 are the results of applying the two edges (respectively) to 5.2.

---

[1] The $+$ operation here denotes modular addition to model arithmetic overflows.

Constraints 6.3 and 6.6 model the possibility that no explicit edge matches, and therefore the security state remains unchanged. They are obtained by conjoining the negations of all of the edge pre-conditions to states 5.1 and 5.2, respectively. Thus, security-relevant events have a multiplicative effect on the state space, expanding n abstract states into at worst $n(m+1)$ states, where m is the number of potential pointcut matches.

If any constraint list is satisfiable and contains the expression $A = \#$, the verifier cannot disprove the possibility of a policy violation and therefore conservatively rejects. Constraints 6.2 and 6.5 both contain this expression, but they are unsatisfiable, proving that a violation cannot occur. Observe that the IRM guard at Line 3 is critical for proving the safety of this code because it introduces constraint $S' \neq 10$ that makes these two lists unsatisfiable. If Lines 3–4 were not included, the verifier would reject at this point because constraints 6.2 and 6.5 are satisfiable with $A = \#$ without clause $S' \neq 10$.

At all control-flows from marked to unmarked regions, the verifier requires a constraint list that implies $SYNC$. In this example, constraints 6.1 and 6.6 are the only remaining lists that are satisfiable, and conjoining them with the negation of $SYNC$ expression $(A = S) \wedge (A = T)$ yields an unsatisfiable list. Thus, this code is accepted as policy-adherent.

Dynamically Decided Pointcuts. Verification of events corresponding to statically undecidable pointcuts (such as `argval`) requires analysis of dynamic checks inserted by the rewriter, which consider the contents of the stack and local variables at runtime. Numeric comparisons are translated directly into constraint expressions; for example, the instruction `if(x>2)` introduces clause $X > 2$ for the positive branch and $X \leq 2$ for the negative branch. Non-numeric dynamic pointcuts (e.g., `streq` pointcut expressions) are modeled by reducing them to equivalent integer encodings. For example, to support dynamic string regexp-matching, Chekov✓ introduces a boolean-valued variable X_{re} for each string-typed program variable x and policy regexp re. Program operations that test x against re introduce constraint $X_{re} = 1$ in their positive branches and $X_{re} = 0$ in their negative branches. An in-depth verification example involving dynamically decidable pointcuts is provided in the companion technical report [15].

Limitations. Our verifier supports most forms of Java reflection, but in order to safely track write-accesses to reified security state fields, the verifier requires such fields to be static, private class members, and it conservatively rejects programs that contain reflective field-write operations within classes that contain reified state. Thus, in order to pass verification, rewriters must implement reified state fields within classes that do not perform write-reflection. This is standard practice for most IRM systems including SPoX, so did not limit any of our tests. Instrumented programs may detect and respond to the presence of the IRM through read-reflection, but not in a way that violates the policy.

Our system supports IRMs that maintain a global invariant whose preservation across the majority of the rewritten code suffices to prove safety for small sections of security-relevant code, followed by restoration of the invariant. Our experience with existing IRM systems indicates that most IRMs do maintain such an invariant (*SYNC*) as a way to avoid reasoning about large portions of security-irrelevant code in the original binary. However, IRMs that maintain no such invariant, or that maintain an invariant inexpressible in our constraint language, cannot be verified by our system. For example, an IRM that stores object security states in a hash table cannot be certified because our constraint language is not sufficiently powerful to express collision properties of hash functions and prove that a correct mapping from security-relevant objects to their security states is maintained by the IRM.

To keep the rewriter's annotation burden small, our certifier also uses this same invariant as a loop-invariant for all cycles in the control-flow graph. This includes recursive cycles in the call graph as well as control-flow cycles within method bodies. Most IRM frameworks do not introduce such loops to non-synchronized regions. However, this limitation could become problematic for frameworks wishing to implement code-motion optimizations that separate security-relevant operations from their guards by an intervening loop boundary. Allowing the rewriter to suggest different invariants for different loops would lift the limitation, but taking advantage of this capability would require the development of rewriters that infer and express suitable loop invariants for the IRMs they produce. To our knowledge, no existing IRM systems yet do this.

While our certifier is provably convergent (since it arrives at a fixpoint for every loop through enforcing *SYNC* on loop back-edges), it can experience state-space explosions that are exponential in the size of each contiguous, unsynchronized code region. Typical IRMs limit such regions to relatively small, separate code blocks scattered throughout the rewritten code; therefore, we have not observed this to be a significant limitation in practice. However, such state-space explosions could be controlled without conservative rejection by applying the same solution above. That is, rewriters could suggest state abstractions for arbitrary code points, allowing the certifier to forget information that is unnecessary for proving safety and that leads to a state-space explosion. Again, the challenge here is developing rewriters that can actually generate such abstractions.

Our current implementation and theoretical analysis are for purely serial programs; concurrency support is reserved for future work. Analysis, enforcement, and certification of multithreaded IRMs is an ongoing subject of current research with several interesting open problems (cf., [6]).

Soundness. Our certifier forms the centerpiece of the TCB of the system, allowing the monitor and monitor-producing tools to remain untrusted. An unsound certifier (i.e., one that fails to reject some policy-violating programs) can lead to system compromise and potential damage. It is therefore important to establish exceptionally high assurance for the certification algorithm. We proved the soundness of our approach using Cousot's abstract interpretation framework [5].

The proof models the verification algorithm as the small-step operational semantics of an abstract machine. A corresponding concrete operational semantics models the Java VM's interpretation of bytecode instructions. For brevity, the concrete and abstract operational semantics concern a small, relevant core subset of Java bytecode instructions rather than the full bytecode language. The core language is semantically connected to full Java bytecode through Classic-Java [13,14]. Bisimulation of the abstract and concrete machines provably satisfies a soundness property that relates abstract states to the concrete states they abstract. This is proved via the following progress and preservation lemmas.

Lemma 1 (Progress). *If abstract machine state $\hat{\chi}$ is a sound abstraction of concrete machine state χ, and $\hat{\chi}$ takes a step (i.e., the verifier does not reject), then χ takes a step (i.e., the concrete machine does not exhibit a policy violation).*

Lemma 2 (Preservation). *If abstract machine state $\hat{\chi}$ soundly abstracts concrete machine state χ, and χ steps to χ', then $\hat{\chi}$ steps to some state $\hat{\chi}'$ that is a sound abstraction of χ'.*

The preservation lemma proves that a bisimulation of the abstract and concrete machines preserves the soundness relation, while the progress lemma proves that as long as the soundness relation is preserved, the abstract machine anticipates all policy violations of the concrete machine. Both proofs are standard (but lengthy) structural inductions over the respective operational semantic derivations. Together, these two lemmas dovetail to form an induction over arbitrary length execution sequences, proving that programs accepted by the verifier will not violate the policy. Detailed operational semantics and proofs can be found in the companion technical report [15].

5   Case Studies

Our prototype verifier implementation consists of 5200 lines of Prolog and 9100 lines of Java. The Prolog code runs under 32-bit SWI-Prolog 5.10.4, which communicates with Java via the JPL interface. The Java side parses SPoX policies and Java bytecode, and compares bytecode instructions to the policy to recognize security-relevant events. The Prolog code forms the core of the verifier, and handles control-flow analysis, model-checking, and linear constraint analysis using CLP. Model-checking is only applied to code that the rewriter has marked as security-relevant. Unmarked code is subjected to a linear scan that ensures that it lacks security-relevant instructions and reified security state modifications.

We have used our prototype implementation to rewrite and then successfully verify several Java applications, discussed throughout the remainder of the section. Statistics are summarized in Table 1. All tests were performed on a Dell Studio XPS notebook computer running Windows 7 64-bit with an Intel i7-Q720M quad core processor, a Samsung PM800 solid state drive, and 4 GB of memory. A more detailed description of each application can be found in [15].

In Table 1, file sizes are expressed in three parts: the original size of the main program before rewriting, the size after rewriting, and the size of system libraries that needed to be verified (but not rewritten). Verification of system library code is required to verify the safety of control-flows that pass through them. Likewise, each cell in the classes column has two parts: the number of classes in the main program and the number of classes in the libraries.

Table 1. Experimental Results

Program	Policy	File Sizes (KB) old / new / libs			# Classes old / libs		Rewrite Time (s)	# Evts.	Total Verif. Time (s)	Model Check Time (s)
EJE	NoExecSaves	439/	439/	0	147/	0	6.1	1	202.8	16.3
RText		1264/	1266/	835	448/	680	52.1	7	2797.5	54.5
JSesh		1923/	1924/	20878	863/	1849	57.8	1	5488.1	196.0
vrenamer	NoExecRename	924/	927/	0	583/	0	50.1	9	1956.8	41.0
jconsole	NoUnsafeDel	35/	36/	0	33/	0	0.6	2	115.7	15.1
jWeather	NoSndsAftrRds	288/	294/	0	186/	0	12.3	46	308.2	156.7
YTDownload		279/	281/	0	148/	0	17.8	20	219.0	53.6
jfilecrypt	NoGui	303/	303/	0	164/	0	9.7	1	642.2	2.8
jknight	OnlySSH	166/	166/	4753	146/	2675	4.5	1	650.1	3.0
Multivalent	EncrpytPDF	1115/	1116/	0	559/	0	129.9	7	3567.0	26.9
tn5250j	PortRestrict	646/	646/	0	416/	0	85.4	2	2598.2	23.6
jrdesktop	SafePort	343/	343/	0	163/	0	8.3	5	483.0	17.8
JVMail	TenMails	24/	25/	0	21/	0	1.6	2	35.1	8.0
JackMail		165/	166/	369	30/	269	2.5	1	626.7	8.9
Jeti	CapLgnAttmpts	484/	484/	0	422/	0	15.3	1	524.3	8.8
ChangeDB	CapMembers	82/	83/	404	63/	286	4.3	2	995.3	12.0
projtimer	CapFileCreat	34/	34/	0	25/	0	15.3	1	56.2	6.1
xnap	NoFreeRide	1250/	1251/	0	878/	0	24.8	4	1496.2	56.4
Phex		4586/	4586/	3799	1353/	830	69.4	2	5947.0	172.7
Webgoat	NoSqlXss	429/	431/	6338	159/	3579	16.7	2	10876.0	120.0
OpenMRS	NoSQLInject	1781/	1783/	24279	932/	17185	78.7	6	2897.0	37.3
SQuirreL	SafeSQL	1788/	1789/	1003	1328/	626	140.2	1	3352.1	37.3
JVMail	LogEncrypt	25/	26/	0	22/	0	1.8	6	71.3	43.2
jvs-vfs	CheckDeletion	277/	277/	0	127/	0	4.4	2	193.9	6.3
sshwebproxy	EncryptPayload	36/	37/	389	19/	16	1.1	5	66.7	7.0

Six of the rewritten applications listed in Table 1 (vrenamer, jWeather, jrdesktop, Phex, Webgoat, and SQuirreL) were initially rejected by our verifier due to a subtle security flaw that our verifier uncovered in the SPoX rewriter. For each of those cases, a bytecode analysis revealed that the original code contained a form of generic exception handler that can potentially hijack control-flows within IRM guard code. This could cause the abstract and reified security state to become desynchronized, breaking soundness. We corrected this by manually editing the rewritten bytecode to exclude guard code from the scope of the outer exception handler. This resulted in successful verification. Our fix could be automated by in-lining inner exception handlers for guard code to protect them from interception by an outer handler.

The following discussion groups the case-studies into four policy classes. SPoX policies are provided in a generalized form representative of the various instantiations of the policy that we used for specific applications. The real policies substitute the simple pointcut expressions in each sample with more complex, application-specific pointcuts that are here omitted for space reasons.

Filename Guards. Our NoExecSaves policy (generalized below) prevents file-creation operations from specifying a file name with an executable extension. Such a policy could be used to prevent malware propagation.

```
1 (edge name="saveToExe"
2   (nodes "s" 0,#)
3   (and (call "java.io.FileWriter.new")
4     (argval 1 (streq ".*\.(exe|bat|...)"))
5     (withincode "FileSystem.saveFile")))
```

The regular expression in Line 4 matches any string that ends in an executable file extension. There are many file extensions that are considered to be executable on Windows; we included all listed at [12]. This policy was enforced on three applications: EJE, a Java code editor; RText, a text editor; and JSesh, a heiroglyphics editor for use by archaeologists. After rewriting, each program halted when we tried to save a file with a prohibited extension.

Another policy that prevents deletion of policy-specified file directories (not shown) was enforced on jconsole. The policy monitors directory-removal system API calls for arguments that match a regular expression specifying names of protected directories. For vrenamer, a mass file-renaming application, we prohibited files being renamed to include executable extensions.

Event ordering. A canonical information flow policy in the IRM literature prohibits all network-send operations after a secret file has been read. The following NoSndsAftrRds policy prevents calls to Socket.getOutputStream after any call to java.io.File where the first argument refers to the Windows directory.

```
1 (edge name="FileRead"
2   (nodes "s" 0,1)
3   (and (call "java.io.File.*")
4     (argval 1 (streq "[A-Za-z]*:\\Windows\\.*"))))
5 (edge name="NetworkSend"
6   (nodes "s" 1,#)
7   (call "java.net.Socket.getOutputStream"))
```

We enforced this policy on jWeather, a weather widget application, and YouTube Downloader (YTDownload in the table), which downloads videos from YouTube. Neither program violated the policy, so no change in behavior occurred. However, both programs access many files and sockets, so SPoX instrumented both programs with a large number of security checks.

For multivalent, a document browsing utility, we enforced a policy that disallows saving a PDF document until a call has first been made to its built-in encryption method. The two-state policy is similar to the one shown above.

Malicious SQL and XSS protection. SPoX's use of string regular expressions facilitates natural specifications of policies that protect against SQL injection and cross-site scripting attacks. One such policy is NoSqlXss, a policy that uses whitelisting to exclude potentially dangerous input characters. We enforced NoSqlXss on Webgoat.

One edge definition in the policy contained a large number of dynamic argval pointcuts (twelve); nevertheless, verification time remained roughly linear in the size of the rewritten code because the verifier was able to significantly prune the search space by combining redundant constraints and control-flows during model-checking and abstract interpretation.

A similar policy was used to prevent SQL injection attacks on a search function in OpenMRS. The library portion of this application is extremely large but contains no security-relevant events; thus, our non-stateful verification approach for unmarked code regions was crucial for avoiding state-space explosions.

We also enforced a blacklisting policy (not shown) on the database access client SQuirreL, preventing SQL commands which drop, alter, or rename tables or databases. The policy used a regular expression guard to disallow all SQL commands that implement these operations.

Ensuring advice execution. Most aspectual policy languages (e.g., [4,2,10,26]) allow policies to include explicit advice code that implements IRM guards and interventions. Such systems can be applied to create custom implementations of SPoX policies, such as those that perform custom actions when impending violations are detected. Chekov✓ can then take the SPoX policy as input and verify that the implementation correctly enforces the policy.

To simulate this, we manually added encryption and logging calls immediately prior to email-send events in JVMail. Each email is therefore encrypted, then logged, then sent. The SPoX policy LogEncrypt requires these events occur in that order. After inserting the advice, we used the verifier to prove that the rewritten JVMail application satisfies the policy. A similar policy was applied to the Java Virtual File System (jvs-vfs), only allowing file deletion after execution of advice code that consults the user. Finally, we enforced a policy on sshwebproxy that requires the proxy to encrypt messages before sending.

6   Conclusion and Future Work

IRMs provide a more powerful alternative to purely static analysis, allowing precise enforcement of a much larger and sophisticated class of security policies. Combining this power with a purely static analysis that independently checks the instrumented, self-monitoring code results in an effective, provably sound, and flexible hybrid enforcement framework. Additionally, an independent certifier allows for the removal of the larger and less general rewriter from the TCB.

We developed Chekov✓—the first automated, model-checking-based certifier for an aspect-oriented, real-world IRM system [14]. Chekov✓ uses a flexible and semantic static code analysis, and supports difficult features such as reified security state, event detection by pointcut-matching, combinations of untrusted before- and after-advice, and pointcuts that are not statically decidable. Strong formal guarantees are provided through proofs of soundness and convergence based on Cousot's abstract interpretation framework. Since Chekov✓ performs independent certification of instrumented binaries, it is flexible enough to accommodate a variety of IRM instrumentation systems, as long as they provide (untrusted) hints about reified state variables and locations of security-relevant events. Such hints are easy for typical rewriter implementations to provide, since they typically correspond to in-lined state variables and guard code, respectively.

Our focus was on presenting main design features of the verification algorithm, and an extensive practical study using a prototype implementation of the tool. Experiments revealed at least one security vulnerability in the SPoX IRM system, indicating that automated verification is important and necessary for high assurance in these frameworks.

In future work we intend to turn our development toward improving efficiency and memory management of the tool. Much of the overhead we observed in experiments was traceable to engineering details, such as expensive context-switches between the separate parser, abstract interpreter, and model-checking modules. These tended to eclipse more interesting overheads related to the abstract interpretation and model-checking algorithms. We also intend to examine more powerful rewriter-supplied hints that express richer invariants. Such advances will provide greater flexibility for alternative IRM implementations of stateful policies.

References

1. Aktug, I., Dam, M., Gurov, D.: Provably Correct Runtime Monitoring. In: Cuellar, J., Sere, K. (eds.) FM 2008. LNCS, vol. 5014, pp. 262–277. Springer, Heidelberg (2008)
2. Aktug, I., Naliuka, K.: ConSpec - a formal language for policy specification. Science of Comput. Prog. 74, 2–12 (2008)
3. Alpern, B., Schneider, F.B.: Recognizing safety and liveness. Distributed Computing 2, 117–126 (1986)
4. Chen, F., Roşu, G.: Java-MOP: A Monitoring Oriented Programming Environment for Java. In: Halbwachs, N., Zuck, L.D. (eds.) TACAS 2005. LNCS, vol. 3440, pp. 546–550. Springer, Heidelberg (2005)
5. Cousot, P., Cousot, R.: Abstract interpretation: A unified lattice model for static analysis of programs by construction or approximation of fixpoints. In: Proc. Sym. on Principles of Prog. Lang., pp. 234–252 (1977)
6. Dam, M., Jacobs, B., Lundblad, A., Piessens, F.: Security Monitor Inlining for Multithreaded Java. In: Drossopoulou, S. (ed.) ECOOP 2009. LNCS, vol. 5653, pp. 546–569. Springer, Heidelberg (2009)
7. Dantas, D.S., Walker, D.: Harmless advice. In: Proc. ACM Sym. on Principles of Prog. Lang. (POPL), pp. 383–396 (2006)
8. Dantas, D.S., Walker, D., Washburn, G., Weirich, S.: AspectML: A polymorphic aspect-oriented functional programming language. ACM Trans. Prog. Lang. and Systems 30(3) (2008)
9. DeVries, B.W., Gupta, G., Hamlen, K.W., Moore, S., Sridhar, M.: ActionScript bytecode verification with co-logic programming. In: Proc. ACM Workshop on Prog. Lang. and Analysis for Security (PLAS), pp. 9–15 (2009)
10. Erlingsson, Ú.: The Inlined Reference Monitor Approach to Security Policy Enforcement. Ph.D. thesis, Cornell University, Ithaca, New York (2004)
11. Erlingsson, Ú., Schneider, F.B.: SASI enforcement of security policies: A retrospective. In: Proc. New Security Paradigms Workshop (NSPW), pp. 87–95 (1999)
12. FileInfo.com: Executable file types (2011),
    http://www.fileinfo.com/filetypes/executable

13. Flatt, M., Krishnamurthi, S., Felleisen, M.: Classes and mixins. In: Proc. ACM Sym. on Principles of Prog. Lang. (POPL), pp. 171–183 (1998)
14. Hamlen, K.W., Jones, M.: Aspect-oriented in-lined reference monitors. In: Proc. ACM Workshop on Prog. Lang. and Analysis for Security (PLAS), pp. 11–20 (2008)
15. Hamlen, K.W., Jones, M.M., Sridhar, M.: Chekov: Aspect-oriented runtime monitor certification via model-checking (extended version). Tech. rep., Dept. of Comput. Science, U. Texas at Dallas (May 2011)
16. Hamlen, K.W., Mohan, V., Masud, M.M., Khan, L., Thuraisingham, B.: Exploiting an antivirus interface. Comput. Standards & Interfaces J. 31(6), 1182–1189 (2009)
17. Hamlen, K.W., Morrisett, G., Schneider, F.B.: Certified in-lined reference monitoring on. NET. In: Proc. ACM Workshop on Prog. Lang. and Analysis for Security (PLAS), pp. 7–16 (2006)
18. Hamlen, K.W., Morrisett, G., Schneider, F.B.: Computability classes for enforcement mechanisms. ACM Trans. Prog. Lang. and Systems 28(1), 175–205 (2006)
19. Jaffar, J., Maher, M.J.: Constraint logic programming: A survey. J. Log. Program., 503–581 (1994)
20. Jones, M., Hamlen, K.W.: Enforcing IRM security policies: Two case studies. In: Proc. IEEE Intelligence and Security Informatics (ISI) Conf., pp. 214–216 (2009)
21. Jones, M., Hamlen, K.W.: Disambiguating aspect-oriented policies. In: Proc. Int. Conf. on Aspect-Oriented Software Development (AOSD), pp. 193–204 (2010)
22. Kiczales, G., Hilsdale, E., Hugunin, J., Kersten, M., Palm, J., Griswold, W.G.: An Overview of AspectJ. In: Lee, S.H. (ed.) ECOOP 2001. LNCS, vol. 2072, pp. 327–353. Springer, Heidelberg (2001)
23. Li, Z., Wang, X.: FIRM: Capability-based inline mediation of Flash behaviors. In: Proc. Annual Comput. Security Applications Conf. (ACSAC), pp. 181–190 (2010)
24. Ligatti, J.A.: Policy Enforcement via Program Monitoring. Ph.D. thesis, Princeton University, Princeton, New Jersey (2006)
25. Ligatti, J., Bauer, L., Walker, D.: Edit automata: Enforcement mechanisms for run-time security policies. Int. J. Information Security 4(1-2), 2–16 (2005)
26. Ligatti, J., Bauer, L., Walker, D.: Run-time enforcement of nonsafety policies. ACM Trans. Information and Systems Security 12(3) (2009)
27. Schneider, F.B.: Enforceable security policies. ACM Trans. Information and Systems Security 3(1), 30–50 (2000)
28. Shah, V., Hill, F.: An aspect-oriented security framework. In: Proc. DARPA Information Survivability Conf. and Exposition, vol. 2 (2003)
29. Sridhar, M., Hamlen, K.W.: ActionScript In-Lined Reference Monitoring in Prolog. In: Carro, M., Peña, R. (eds.) PADL 2010. LNCS, vol. 5937, pp. 149–151. Springer, Heidelberg (2010)
30. Sridhar, M., Hamlen, K.W.: Model-Checking In-Lined Reference Monitors. In: Barthe, G., Hermenegildo, M. (eds.) VMCAI 2010. LNCS, vol. 5944, pp. 312–327. Springer, Heidelberg (2010)
31. Sridhar, M., Hamlen, K.W.: Flexible in-lined reference monitor certification: Challenges and future directions. In: Proc. ACM Workshop on Prog. Lang. meets Program Verification (PLPV), pp. 55–60 (2011)
32. Viega, J., Bloch, J.T., Chandra, P.: Applying aspect-oriented programming to security. Cutter IT J. 14(2) (2001)
33. Walker, D.: A type system for expressive security policies. In: Proc. of ACM Sym. on Principles of Prog. Lang. (POPL) (2000)

Partial Model Checking Using Networks of Labelled Transition Systems and Boolean Equation Systems

Frédéric Lang and Radu Mateescu

VASY Project Team, INRIA Grenoble Rhône-Alpes/LIG, Montbonnot, France
{Frederic.Lang,Radu.Mateescu}@inria.fr

Abstract. Partial model checking was proposed by Andersen in 1995 to verify a temporal logic formula compositionally on a composition of processes. It consists in incrementally incorporating into the formula the behavioural information taken from one process — an operation called quotienting — to obtain a new formula that can be verified on a smaller composition from which the incorporated process has been removed. Simplifications of the formula must be applied at each step, so as to maintain the formula at a tractable size. In this paper, we revisit partial model checking. First, we extend quotienting to the network of labelled transition systems model, which subsumes most parallel composition operators, including m among n synchronisation and parallel composition using synchronisation interfaces, available in the E-LOTOS standard. Second, we reformulate quotienting in terms of a simple synchronous product between a graph representation of the formula (called formula graph) and a process, thus enabling quotienting to be implemented efficiently and easily, by reusing existing tools dedicated to graph compositions. Third, we propose simplifications of the formula as a combination of bisimulations and reductions using Boolean equation systems applied directly to the formula graph, thus enabling formula simplifications also to be implemented easily and efficiently. Finally, we describe an implementation in the CADP (*Construction and Analysis of Distributed Processes*) toolbox and present some experimental results in which partial model checking uses hundreds of times less memory than on-the-fly model checking.

1 Introduction

Concurrent safety critical systems can be verified using *model checking* [14], i.e., automatic evaluation of a temporal property against a model of the system. Although successful in many applications, model checking may face state explosion, particularly when the number of concurrent processes grows.

State explosion can be tackled by *divide-and-conquer* approaches regrouped under the vocable *compositional verification*, which take advantage of the compositional structure of the concurrent system. One such approach, which we call *compositional model generation* in this paper, consists in building the model of the system — usually an LTS (*Labelled Transition System*) — in a stepwise manner, by successive compositions and minimisations modulo equivalence relations,

C. Flanagan and B. König (Eds.): TACAS 2012, LNCS 7214, pp. 141–156, 2012.

possibly using *interface constraints* [23,27] to avoid explosion of intermediate compositions. Tools using this approach [19,28,29,16] are available in the CADP (*Construction and Analysis of Distributed Processes*) [20] toolbox.

In this paper, we explore a dual approach named *partial model checking*, proposed by Andersen [2,4] for concurrent processes running asynchronously and composed using CCS parallel composition and restriction operators. For a modal μ-calculus [26] formula φ and a process composition $P_1 \| \ldots \| P_n$, Andersen uses an operation $\varphi /\!/ P_1$ called *quotienting* of the formula φ w.r.t. the process P_1, so that $P_1 \| \ldots \| P_n$ satisfies φ if and only if the smaller composition $P_2 \| \ldots \| P_n$ satisfies $\varphi /\!/ P_1$. In addition, simplifications can (must) be applied to $\varphi /\!/ P_1$ to reduce its size. Partial model checking is the incremental application of quotienting and simplifications, so that state explosion is avoided if the size of intermediate formulas can be kept sufficiently small.

Partial model checking has been adapted and used successfully in various contexts, such as state-based models [5,6], synchronous state/event systems [10], and timed systems [9,12,31,32,33]. It has also been specialised for security properties [34]. More recently, it has been generalised to the full CCS process algebra, with an application to the verification of parameterised systems [8].

In this paper, we focus on partial model checking of the modal μ-calculus applied to (untimed) concurrent asynchronous processes. By considering only binary associative composition operators, previous works [2,4,8] are not directly applicable to more general operators, such as m among n synchronisation and parallel composition by synchronisation interfaces [21], present in the E-LOTOS standard and variants [13,25]. Our first contribution in this paper is thus a generalisation of partial model checking to networks of LTSs [28], a general model that subsumes parallel composition, hiding, cutting, and renaming operators of standard process languages (CCS, CSP, μCRL, LOTOS, E-LOTOS, etc.).

In realistic cases, partial model checking handles huge formulas and processes, thus requiring efficient implementations. Our second contribution is a reformulation of quotienting as a simple synchronous product, which can itself be represented in the network model, between a graph representing the formula (called a *formula graph*) and the behaviour graph of a process, thus enabling efficient implementation using existing tools dedicated to graph manipulations. Our third contribution is the reformulation of formula simplifications as a combination of graph reductions and partial evaluation of the formula graph using a BES (*Boolean Equation System*) [1]. Verifying modal μ-calculus formulas of arbitrary alternation depth is generally exponential in the size of the process graph, while verifying the alternation-free fragment remains of linear complexity. Our fourth contribution is a specialisation of the technique to alternation-free μ-calculus formulas. Finally, we present an implementation in CADP and a case-study that illustrates the complementarity between partial and on-the-fly model checking.

Paper Overview. The modal μ-calculus is presented in Sect. 2, networks of LTSs in Sect. 3, the generalisation of quotienting to networks and its reformulation as a synchronous product in Sect. 4, simplification rules in Sect. 5, rules specific to alternation-free μ-calculus formulas in Sect. 6, our implementation in Sect. 7, a case study in Sect. 8, and concluding remarks in Sect. 9.

2   The Modal μ-Calculus

An LTS (*Labelled Transition System*) is a tuple $(\Sigma, A, \longrightarrow, s_0)$, with Σ a set of states, A a set of labels, $\longrightarrow \subseteq \Sigma \times A \times \Sigma$ the (labelled) transition relation, and $s_0 \in \Sigma$ the initial state. Properties of LTSs can be expressed in the modal μ-calculus [26], whose syntax and semantics are defined in the table below.

$$
\begin{array}{ll}
\varphi ::= \mathbf{ff} & [\![\mathbf{ff}]\!]\,\rho = \emptyset \\[2pt]
\mid \neg\varphi_0 & [\![\neg\varphi_0]\!]\,\rho = \Sigma \setminus [\![\varphi_0]\!]\,\rho \\[2pt]
\mid \varphi_1 \vee \varphi_2 & [\![\varphi_1 \vee \varphi_2]\!]\,\rho = [\![\varphi_1]\!]\,\rho \cup [\![\varphi_2]\!]\,\rho \\[2pt]
\mid \langle a \rangle\,\varphi_0 & [\![\langle a \rangle\,\varphi_0]\!]\,\rho = \{s \in \Sigma \mid s \xrightarrow{a} s' \wedge s' \in [\![\varphi_0]\!]\,\rho\} \\[2pt]
\mid X & [\![X]\!]\,\rho = \rho(X) \\[2pt]
\mid \mu X.\varphi_0 & [\![\mu X.\varphi_0]\!]\,\rho = \bigcap\{U \subseteq \Sigma \mid [\![\varphi_0]\!]\,(\rho \oslash [U/X]) \subseteq U\}
\end{array}
$$

Formulas (φ) are built from Boolean connectors, the possibility modality ($\langle \_ \rangle$), and the minimal fix-point operator (μ) over propositional variables X. We write $\mathsf{fv}\,(\varphi)$ (resp. $\mathsf{bv}\,(\varphi)$) for the set of variables free (resp. bound) in φ and call a *closed formula* any formula φ s.t. $\mathsf{fv}\,(\varphi) = \emptyset$. We assume that all bound variables have distinct names, and for $X \in \mathsf{bv}\,(\varphi)$, we write $\varphi[X]$ for the (unique) sub-formula of φ of the form $\mu X.\varphi_0$. Given φ_1 and φ_2, we write $\varphi_1[\varphi_2/X]$ for substituting all free occurrences of X in φ_1 by φ_2. Derived operators are defined as usual: $\mathbf{tt} = \neg\mathbf{ff}$, $\varphi_1 \wedge \varphi_2 = \neg(\neg\varphi_1 \vee \neg\varphi_2)$, $[a]\,\varphi_0 = \neg\,\langle a \rangle\,\neg\varphi_0$ (necessity modality), and $\nu X.\varphi_0 = \neg\mu X.\neg\varphi_0[\neg X/X]$ (maximal fix-point operator).

A propositional context ρ is a partial function mapping propositional variables to sets of states and $\rho \oslash [U/X]$ stands for a propositional context identical to ρ except that X is mapped to U. The interpretation $[\![\varphi]\!]\,\rho$ (also written $[\![\varphi]\!]$ if ρ is empty) of a state formula on an LTS in a propositional context ρ (which maps each variable free in φ to a set of states) denotes the subset of states satisfying φ in that context. The Boolean connectors are interpreted as usual in terms of set operations. The possibility modality $\langle a \rangle\,\varphi_0$ (resp. the necessity modality $[a]\,\varphi_0$) denotes the states for which some (resp. all) of their outgoing transitions labelled by a lead to states satisfying φ_0. The minimal fix-point operator $\mu X.\varphi_0$ (resp. the maximal fix-point operator $\nu X.\varphi_0$) denotes the least (resp. greatest) solution of the equation $X = \varphi_0$ interpreted over the complete lattice $\langle 2^\Sigma, \emptyset, \Sigma, \cap, \cup, \subseteq \rangle$. A state s satisfies a closed formula φ if and only if $s \in [\![\varphi]\!]$.

To ensure a proper definition of fix-point operators, it suffices that formulas φ are *syntactically monotonic* [26], i.e., have an even number of negations on every path between a variable occurrence X and the μ or ν operator that binds X. Negations can then be eliminated from formulas using the identities defining the derived operators. We write $\hat{\varphi}$ the formula obtained after eliminating all negations in φ. A formula φ is *alternation-free* if there is no sub-formula of $\hat{\varphi}$ of the form $\mu X.\varphi_1$ (resp. $\nu X.\varphi_1$) containing a sub-formula of the form $\nu Y.\varphi_2$ (resp. $\mu Y.\varphi_2$) such that $X \in \mathsf{fv}(\varphi_2)$. The *fix-point sign* of a variable X in φ is μ (resp. ν) if $\hat{\varphi}[X]$ has the form $\mu X.\varphi$ (resp. $\nu X.\varphi$).

In this paper, we consider *block-labelled* formulas φ in which each propositional variable X is labelled by a unique natural number k, called its *block number*. Initially, we require that in every sub-formula of $\hat{\varphi}$ of the form $\mu X^k.\varphi_0$ (resp. $\nu X^k.\varphi_0$), every sub-formula $\mu Y^{k'}.\varphi_1$ (resp. $\nu Y^{k'}.\varphi_1$) satisfies $k' \geq k$, and every sub-formula $\nu Y^{k'}.\varphi_1$ (resp. $\mu Y^{k'}.\varphi_1$) satisfies $k' > k$. In addition, variables bound in disjoint sub-formulas may have the same block number only if they have the same fix-point sign, and by convention, block number 0 must be a μ-block (so that $k > 0$ in any formula $\nu X^k.\varphi$). We write $\mathsf{blocks}(\varphi)$ the set of block numbers occurring in φ. A block-labelled formula φ is *alternation-free* if $k' \geq k$ for all $X^k \in \mathsf{bv}(\varphi)$ and all $Y^{k'} \in \mathsf{fv}(\varphi[X^k])$. Any unlabelled formula is alternation-free if and only if it can be block-labelled to satisfy that constraint.

In the remainder of this paper, we will consider block-labelled formulas φ in *disjunctive form*, i.e., built only using the operators shown in the table above.

3   Networks of LTSs

Networks of LTSs (or *networks* for short) are inspired from the MEC [7] and FC2 [11] synchronisation vectors and were introduced in [28] as an intermediate model to represent compositions of LTSs using various operators.

Background. We write $n..m$ for the set of integers ranging from n to m, or the empty set if $n > m$. A *vector* $\mathbf{v}$ of size n is a total function on $1..n$. For $i \in 1..n$, we write $\mathbf{v}[i]$ for $\mathbf{v}$ applied to i, denoting the element of $\mathbf{v}$ stored at index i. We write $(e_1, \ldots, e_n)$ for the vector $\mathbf{v}$ of size n such that $(\forall i \in 1..n)\ \mathbf{v}[i] = e_i$. In particular, $()$ is a vector of size 0. Given $n \geq 1$ and $i \in 1..n$, $\mathbf{v}_{\backslash i}$ denotes the projection of $\mathbf{v}$ on to the set of indices $1..n \setminus \{i\}$, defined as the vector of size $n-1$ such that $(\forall j \in 1..i-1)\ \mathbf{v}_{\backslash i}[j] = \mathbf{v}[j]$ and $(\forall j \in i..n-1)\ \mathbf{v}_{\backslash i}[j] = \mathbf{v}[j+1]$.

A *network of LTSs* N of size n is a pair $(\mathbf{S}, V)$, where $\mathbf{S}$ is a vector of LTSs (called *individual LTSs*) of size n, and V is a set of *synchronisation rules*, each rule having the form $(\mathbf{t}, a)$ with a a label and $\mathbf{t}$ a vector of size n, called the *synchronisation vector*, of labels and occurrences of a special symbol $\bullet$ distinct from any label. We write Σ_i, A_i, $\longrightarrow_i$, and s_i^0 for the sets of states and labels, the transition relation, and the initial state of $\mathbf{S}[i]$. N can be associated to a (global) LTS $\mathsf{lts}(N)$ which is the parallel composition of individual LTSs. Each $(\mathbf{t}, a) \in V$ defines transitions labelled by a, obtained either by synchronisation (if more than one index i is such that $\mathbf{t}[i] \neq \bullet$) or by interleaving (otherwise) of individual LTS transitions. Formally, $\mathsf{lts}(N) = (\Sigma, A, \longrightarrow, \mathbf{s}_0)$, where $\Sigma = \Sigma_1 \times \ldots \times \Sigma_n$, $A = \{a \mid (\mathbf{t}, a) \in V\}$, $\mathbf{s}_0 = (s_1^0, \ldots, s_n^0)$, and $\longrightarrow$ is the smallest relation satisfying:

$$(\mathbf{t}, a) \in V \wedge (\forall i \in 1..n) \left(\begin{array}{l} (\mathbf{t}[i] = \bullet \wedge \mathbf{s}'[i] = \mathbf{s}[i]) \vee \\ (\mathbf{t}[i] \neq \bullet \wedge \mathbf{s}[i] \xrightarrow{\mathbf{t}[i]}_i \mathbf{s}'[i]) \end{array} \right) \Rightarrow \mathbf{s} \xrightarrow{a} \mathbf{s}'$$

$A(\mathbf{t})$ denotes the set of *active* LTS (indices), defined by $\{i \mid i \in 1..n \wedge \mathbf{t}[i] \neq \bullet\}$.

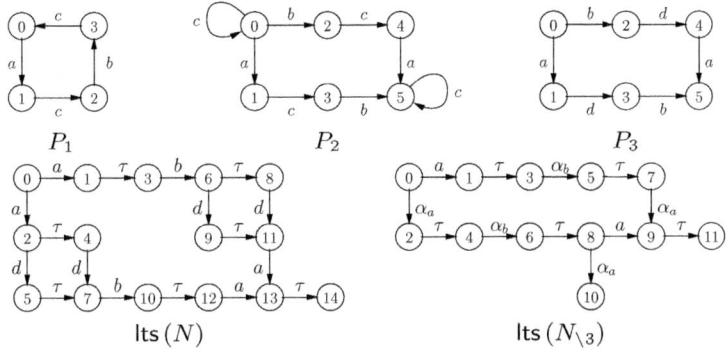

Fig. 1. Labelled Transition Systems for N defined in Ex. 1

Example 1. Let a, b, c, and d be labels, and P_1, P_2, and P_3 be the processes defined in Fig. 1 (top), where 0's denote initial states. Let $N = ((P_1, P_2, P_3), V)$ with $V = \{((a, a, \bullet), a), ((a, \bullet, a), a), ((b, b, b), b), ((c, c, \bullet), \tau), ((\bullet, \bullet, d), d)\}$, whose global LTS is in Fig. 1 (bottom left). The first two rules express a nondeterministic synchronisation on a between either P_1, P_2 or P_1, P_3. The third rule expresses a multiway synchronisation on b. The fourth rule yields an internal (τ) transition. The fifth rule expresses full interleaving of transitions labelled by d.

The network of LTSs model subsumes most hiding, renaming, cutting, and parallel composition operators present in process algebras (CCS, CSP, LOTOS, μCRL, etc.), but also more expressive operators, such as m among n synchronisation and parallel composition using synchronisation interfaces [21] present in E-LOTOS [25] and LOTOS NT [13]. For instance, the rules $\{((a, a, \bullet), a), ((a, \bullet, a), a), ((\bullet, a, a), a)\}$ realize 2 among 3 synchronisation on a.

Sub-network extraction. Computing the interactions of a process P_i with its environment in a composition of processes $\|_{j \in 1..n} P_j$ is easy when $\|$ is a binary and associative parallel composition operator, since $\|_{j \in 1..n} P_j = P_i \| (\|_{j \in 1..n \setminus \{i\}} P_j)$. However, as argued in [21], binary and associative parallel composition operators are of limited use when considering, e.g., m among n synchronisation. A more involved operation named *sub-network extraction* is necessary for networks. $N = (\mathbf{S}, V)$ being a network of size n, we assume a function $\alpha(\mathbf{t}, a)$ that assigns an unused label to each $(\mathbf{t}, a) \in V$. Given $i \in 1..n$, we define $N_{\setminus i} = (\mathbf{S}_{\setminus i}, V_{\setminus i})$ the sub-network of N modeling the environment of $\mathbf{S}[i]$ in N, where $V_{\setminus i} = \{(\mathbf{t}_{\setminus i}, a) \mid (\mathbf{t}, a) \in V \wedge i \notin A(\mathbf{t})\} \cup \{(\mathbf{t}_{\setminus i}, \alpha(\mathbf{t}, a)) \mid (\mathbf{t}, a) \in V \wedge \{i\} \subset A(\mathbf{t})\}$. N is semantically equivalent to the network $((\mathbf{S}[i], \mathsf{lts}(N_{\setminus i})), V')$ with V' the following set of rules, which define the interactions between $\mathbf{S}[i]$ and $N_{\setminus i}$:

$$\begin{aligned}
&\{ ((\bullet, \quad a), \qquad a) \mid (\mathbf{t}, a) \in V \wedge i \notin A(\mathbf{t}) \quad \} \cup \\
&\{ ((\mathbf{t}[i], \alpha(\mathbf{t}, a)), a) \mid (\mathbf{t}, a) \in V \wedge \{i\} \subset A(\mathbf{t}) \} \cup \\
&\{ ((a, \quad \bullet), \qquad a) \mid (\mathbf{t}, a) \in V \wedge \{i\} = A(\mathbf{t}) \}
\end{aligned}$$

Each $\alpha(\mathbf{t}, a)$ is a unique interaction label between $\mathbf{S}[i]$ and $N_{\backslash i}$, which aims at avoiding erroneous interactions in case of nondeterministic synchronisation.

Example 2. N being defined in Ex. 1, $N_{\backslash 3}$ has vector of LTSs (P_1, P_2) and rules $\{((a, a), a), ((a, \bullet), \alpha_a), ((b, b), \alpha_b), ((c, c), \tau)\}$ with $\alpha_a = \alpha((a, \bullet, a), a)$ and $\alpha_b = \alpha((b, b, b), b)$; $\mathsf{lts}(N_{\backslash 3})$ is depicted in Fig. 1 (bottom right); Composing it with P_3 using $\{((\bullet, a), a), ((a, \alpha_a), a), ((b, \alpha_b), b), ((\bullet, \tau), \tau), ((d, \bullet), d)\}$ yields $\mathsf{lts}(N)$.

Note that if a had been used instead of α_a in the above synchronisation rules, then the composition of $N_{\backslash 3}$ with P_3 would have enabled, in addition to the (correct) binary synchronisations on a between P_1 and P_2 and between P_1 and P_3, the (incorrect) multiway synchronisation on a between the three of P_1, P_2, and P_3. Indeed, the label a resulting from the synchronisation between P_1 and P_2 in $N_{\backslash 3}$ — rule $((a, a), a)$ in $N_{\backslash 3}$ — could synchronise with the label a in P_3 — rule $((a, a), a)$ in the composition between $N_{\backslash 3}$ and P_3. Note however that $\mathbf{t}[i]$ can be used instead of $\alpha(\mathbf{t}, a)$ when the network does not have nondeterministic synchronisation on $\mathbf{t}[i]$, as is the case for b and α_b in this example. In this paper we use $\alpha(\mathbf{t}, a)$ uniformly to avoid complications.

4   Quotienting for Networks Using Networks

To check a closed formula φ on a network $N = (\mathbf{S}, V)$, one can choose an LTS $\mathbf{S}[i]$, compute the quotient of the formula φ with respect to $\mathbf{S}[i]$, and check the resulting quotient formula on the smaller (at least in number of individual LTSs, but also hopefully in global LTS size) network $N_{\backslash i}$. The quotient formula is written $\varphi \mathbin{/\!\!/}_i^{\emptyset} s_0^i$ and defined as follows for formulas in disjunctive form:

$$\mathbf{ff} \mathbin{/\!\!/}_i^B s = \mathbf{ff} \qquad\qquad X^k \mathbin{/\!\!/}_i^B s = \varphi[X^k] \mathbin{/\!\!/}_i^B s$$

$$(\neg\varphi_0) \mathbin{/\!\!/}_i^B s = \neg(\varphi_0 \mathbin{/\!\!/}_i^B s) \qquad (\varphi_1 \vee \varphi_2) \mathbin{/\!\!/}_i^B s = (\varphi_1 \mathbin{/\!\!/}_i^B s) \vee (\varphi_2 \mathbin{/\!\!/}_i^B s)$$

$$(\mu X^k.\varphi_0) \mathbin{/\!\!/}_i^B s = \begin{cases} X_s^k & \text{if } X_s^k \in B \\ \mu X_s^k.(\varphi_0 \mathbin{/\!\!/}_i^{B \cup \{X_s^k\}} s) & \text{otherwise} \end{cases}$$

$$(\langle a \rangle \varphi_0) \mathbin{/\!\!/}_i^B s = \bigvee_{(\mathbf{t},a) \in V} \begin{pmatrix} (\quad i \notin A(\mathbf{t}) \wedge & \langle a \rangle (\varphi_0 \mathbin{/\!\!/}_i^B s) \ ) \ \vee \\ (\{i\} \subset A(\mathbf{t}) \wedge \bigvee_{s \xrightarrow{\mathbf{t}[i]}_i s'} & \langle \alpha(\mathbf{t}, a) \rangle (\varphi_0 \mathbin{/\!\!/}_i^B s')) \ \vee \\ (\{i\} = A(\mathbf{t}) \wedge \bigvee_{s \xrightarrow{\mathbf{t}[i]}_i s'} & (\varphi_0 \mathbin{/\!\!/}_i^B s')) \end{pmatrix}$$

This definition generalises Andersen's [2], specialised for CCS, to networks. The major difference is the definition of $(\langle a \rangle \varphi_0) \mathbin{/\!\!/}_i^B s$, CCS composition corresponding to vectors $((a, \bullet), a)$, $((\bullet, a), a)$, or $((a, \bar{a}), \tau)$, a and $\bar{a}$ being an action and its *co-action*, making the use of special labels $\alpha(\mathbf{t}, a)$ not necessary. A slightly minor difference is that we use μ-calculus terms instead of equations. Any sub-formula produced by quotienting has the same block number as the original sub-formula, reflecting the order of equation blocks in Andersen's work. The set B keeps track of new variables already introduced in the quotient formula. Quotienting is well-defined, because formulas are finite, every $\varphi[X^k]$ has the form $\mu X^k.\varphi_0$, and the size of the set B is bounded by $|\,\mathsf{bv}(\varphi)| \times |\Sigma_i|$.

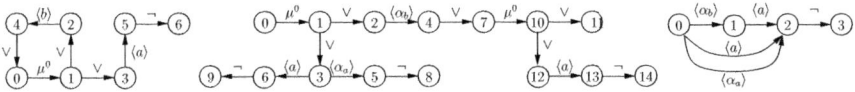

Fig. 2. Examples of formula graphs

Example 3. The μ-calculus formula $\mu X^0.\langle a\rangle \mathbf{tt} \vee \langle b\rangle X^0$ (existence of a path of zero or more b leading to an a) can be rewritten to disjunctive form as $\mu X^0.\langle a\rangle\neg\mathbf{ff} \vee \langle b\rangle X^0$. Quotienting of this formula with respect to P_3 in the network N introduced in Ex. 1 yields the formula $\mu X_0^0.\langle a\rangle\neg\mathbf{ff} \vee \langle \alpha_a\rangle\neg\mathbf{ff} \vee \langle \alpha_b\rangle\mu X_2^0.\langle a\rangle\neg\mathbf{ff} \vee \mathbf{ff}$.

We now show that quotienting can be implemented as a network that realises a product between an LTS encoding the formula (called a *formula graph*) and an individual LTS of the network under verification. The formula graph corresponding to a formula φ in disjunctive form is an LTS whose states are identified with sub-formulas of φ and whose transitions are labelled by $\vee$, $\neg$, μ^k (k being a block number), and $\langle a\rangle$ (a being any action of the network under verification). The initial state of the formula graph is φ, $\mathbf{ff}$ is a deadlock state, and each sub-formula has transitions as follows:

$$X^k \xrightarrow{\vee} \varphi[X^k] \qquad \neg\varphi_0 \xrightarrow{\neg} \varphi_0 \qquad \langle a\rangle\varphi_0 \xrightarrow{\langle a\rangle} \varphi_0$$
$$\varphi_1 \vee \varphi_2 \xrightarrow{\vee} \varphi_1 \qquad \varphi_1 \vee \varphi_2 \xrightarrow{\vee} \varphi_2 \qquad \mu X^k.\varphi_0 \xrightarrow{\mu^k} \varphi_0$$

Formula graphs are finite, connected, and every circular path (i.e., from one state to itself) contains at least one transition that is labelled by μ^k. We write $\mathrm{enc}(\varphi)$ the formula graph of φ. Conversely, every formula graph $P = (S, A, \rightarrow, s_0)$ can be decoded into the closed formula $\mathrm{dec}(P, s_0, \emptyset)$ as follows, where E is a mapping of the form $\{s \mapsto k \mid s \in \Sigma \wedge k \in \mathbb{N}\}$:

$$\mathrm{dec}(P, s, E) = \begin{cases} X_s^k & \text{if } s \mapsto k \in E \\ \bigvee_{s \xrightarrow{\sigma} s'} \delta_\sigma^s(P, s', E) & \text{otherwise} \end{cases} \quad \text{where}$$
$$\delta_\vee^s(P, s', E) = \mathrm{dec}(P, s', E) \qquad \delta_\neg^s(P, s', E) = \neg\,\mathrm{dec}(P, s', E)$$
$$\delta_{\langle a\rangle}^s(P, s', E) = \langle a\rangle\,\mathrm{dec}(P, s', E) \qquad \delta_{\mu^k}^s(P, s', E) = \mu X_s^k.\,\mathrm{dec}(P, s', E \cup \{s \mapsto k\})$$

This definition implies that a deadlock state decodes as $\mathbf{ff}$ (empty disjunction). dec is well-defined, the mapping E ensuring termination. Although the states of a formula graph are identified by formulas, only the transition labels are required for decoding. In figures, states will be be simply identified by numbers.

Example 4. The formula graph corresponding to the formula $\mu X^0.(\langle a\rangle\mathbf{tt}) \vee \langle b\rangle X^0$ introduced in Ex. 3 is depicted in Fig. 2 (left), where 0 denotes the initial state.

Proposition 1. *If φ is a closed formula, then $\mathrm{dec}(\mathrm{enc}(\varphi), \varphi, \emptyset) = \varphi$, modulo commutativity ($\phi_1 \vee \phi_2 = \phi_2 \vee \phi_1$), idempotence ($\phi \vee \phi = \phi$), and renaming of each propositional variable $X^k \in \mathrm{bv}(\varphi)$ into $X_{\varphi[X^k]}^k$.*

Proof. This is a corollary of the more general property stating that for every sub-formula ϕ of φ, if $\{\varphi[Y^k] \mapsto k \mid Y^k \in \mathsf{fv}(\phi)\} \subseteq E$ and $E \cap \{\varphi[Y^k] \mapsto k \mid Y^k \in \mathsf{bv}(\phi)\} = \emptyset$, then $\mathsf{dec}(\mathsf{enc}(\varphi), \phi, E) = \phi$ (structural induction on ϕ).

Using this encoding, the quotienting of a formula φ with respect to the ith LTS of a network $N = (\mathbf{S}, V)$ can be realised as a synchronous product, using the network $((\mathsf{enc}(\varphi), \mathbf{S}[i]), V /\!/_i)$, where $V /\!/_i$ denotes the following set of rules:

$$
\begin{array}{lll}
\{ ((\sigma, \ \bullet), \quad \sigma) & \mid \sigma \in \{\neg, \vee\} \cup \{\mu^k \mid k \in \mathsf{blocks}(\varphi)\} \ \} & \cup \\
\{ (((\langle a \rangle, \ \bullet), \quad \langle a \rangle) & \mid (\mathbf{t}, a) \in V \wedge i \notin A(\mathbf{t}) \} & \cup \\
\{ (((\langle a \rangle, \mathbf{t}[i]), \langle \alpha(\mathbf{t}, a) \rangle)) \mid (\mathbf{t}, a) \in V \wedge \{i\} \subset A(\mathbf{t}) \} & & \cup \\
\{ (((\langle a \rangle, \mathbf{t}[i]), \vee) & \mid (\mathbf{t}, a) \in V \wedge \{i\} = A(\mathbf{t}) \} &
\end{array}
$$

Proposition 2. *If* $P = \mathsf{lts}((\mathsf{enc}(\varphi), \mathbf{S}[i]), V /\!/_i)$ *then* $\mathsf{dec}(P, (\varphi, s_0^i), \emptyset) = \varphi /\!/_i^\emptyset s_0^i$, *modulo commutativity, idempotence, and renaming of each propositional variable* $Y_t^k \in \mathsf{bv}(\varphi /\!/_i^B s_0^i)$ *into* $\mathsf{X}_{(\varphi[Y^k], t)}^k$

Proof. A state of P has the form (ϕ, s), where ϕ is a sub-formula of φ and s is a state of $\mathbf{S}[i]$. The proof uses a slighty more general lemma: if $E = \{(\varphi[Y^k], t) \mapsto k \mid Y_t^k \in B\}$ then $\mathsf{dec}(P, (\phi, s), E) = \phi /\!/_i^B s$ (structural induction on $\phi /\!/_i^B s$).

Example 5. Consider the network N of Ex. 1 and the formula of Ex. 4. Quotienting of the formula with respect to P_3 involves the following set of rules:
$$\{((\neg, \bullet), \neg), ((\vee, \bullet), \vee), ((\mu^0, \bullet), \mu^0), (((\langle a \rangle, \bullet), \langle a \rangle), (((\langle a \rangle, a), \langle \alpha_a \rangle), (((\langle b \rangle, b), \langle \alpha_b \rangle)\}$$
It yields the formula graph depicted in Fig. 2 (middle). This graph encodes as expected the quotient formula of Ex. 3, which can be evaluated on $N_{\backslash 3}$.

Working with formulas in disjunctive form is crucial: branches in the formula graph denote disjunctions between sub-formulas (*or-nodes*). During composition between the formula graph and an individual LTS, the impossibility to synchronise on a modality $\langle a \rangle$ (no transition labelled by $\mathbf{t}[i]$ in the current state of the individual LTS) denotes invalidation of the corresponding sub-formula, which merely disappears, in conformance with the equality $\mathbf{ff} \vee \varphi_0 = \varphi_0$.

5   Formula Graph Simplifications

The size (number of states) of a formula graph of size n quotiented with respect to an LTS of size m is bounded by $n \times m$. Hence, as observed by Andersen [2], simplifications are needed to keep intermediate quotiented formulas at a reasonable size. We present in Fig. 3 several simplifications applying to formula graphs, as conditional rules of the form "$l \rightsquigarrow r \ (cond)$" where l and r are subsets of transition relations, such that every variable representing a state (written $s, s_1, s_2, \dots$) or a label (written $\sigma, \sigma_1, \sigma_2, \dots$) in r or in the condition $cond$ must occur in l. It means that all transitions matching the left-hand side so that $cond$ is satisfied can be replaced by the transitions of the right-hand side.
Elimination of $\vee$-*transitions* (1). This rule is essential to eliminate the transitions labelled by $\vee$ introduced by synchronisation rules of the form $(((\langle a \rangle, \mathbf{t}[i]), \vee)$

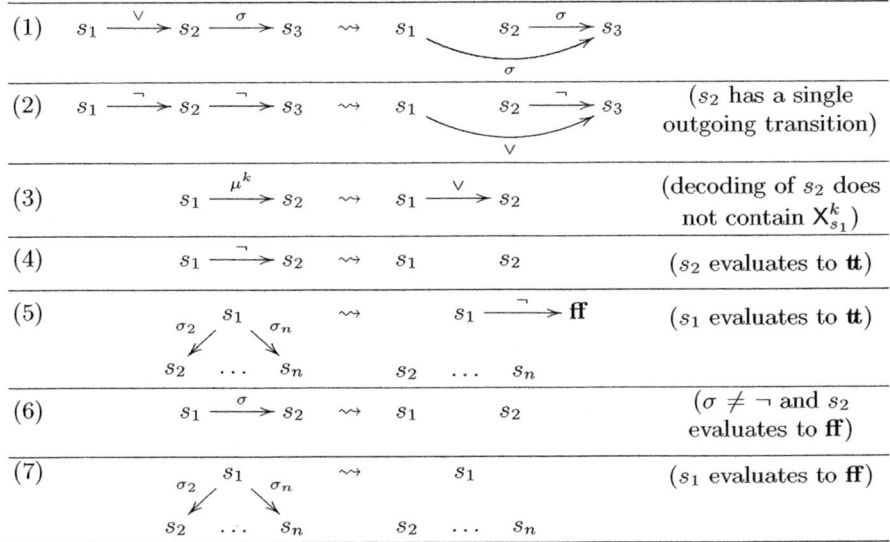

Fig. 3. Simplification rules applying to formula graphs

during quotienting. It can be achieved efficiently by applying reduction modulo $\tau^*.a$ equivalence [17], $\vee$-transitions being interpreted as internal (τ) transitions.

Elimination of double-negations (2). This rule can be used after the previous one to simplify formulas of the form $\neg\neg\varphi$, which may appear, e.g., in the quotienting of $\neg\langle a\rangle\neg\varphi'$ with an LTS that offers an action synchronising with a.

Elimination of μ-transitions (3). The transition from s_1 to s_2 denotes a propositional variable $X_{s_1}^k$, which does not occur free in the formula if at least one of the following sufficient (and checkable in linear time) conditions holds: (i) s_1 and s_2 are not in the same strongly connected component; (ii) s_1 satisfies the recursive condition "s_1 has a single predecessor p, distinct from the initial state, and either p has a single μ-transition to s_1 or p satisfies this condition, recursively". This condition is well-founded as long as it is applied to reachable states.

Evaluation of constant sub-formulas (4–7). To decide whether a state denotes a sub-formula that evaluates to a constant in any context, we consider the following BES, consisting in blocks T^k and F^k ($k \in 0..n$) of respective signs μ and ν, n being the greatest block number in the formula graph. Blocks are ordered so that $k < k'$ implies T^k (resp. F^k) is before $T^{k'}$ (resp. $F^{k'}$):

$$T^k : \Big\{\ T_s^k =_\mu \bigvee\nolimits_{s \xrightarrow{\vee} s'} T_{s'}^k \vee \bigvee\nolimits_{s \xrightarrow{\neg} s'} F_{s'}^k \vee \bigvee\nolimits_{s \xrightarrow{\mu^{k'}} s'} T_{s'}^{k'}\ \Big\}_{s \in \Sigma}$$

$$F^k : \Big\{\ F_s^k =_\nu \bigwedge\nolimits_{s \xrightarrow{\vee} s'} F_{s'}^k \wedge \bigwedge\nolimits_{s \xrightarrow{\langle\beta\rangle} s'} F_{s'}^k \wedge \bigwedge\nolimits_{s \xrightarrow{\neg} s'} T_{s'}^k \wedge \bigwedge\nolimits_{s \xrightarrow{\mu^{k'}} s'} F_{s'}^{k'}\ \Big\}_{s \in \Sigma}$$

We consider only the variables reachable from $T_{s_0}^0$ or $F_{s_0}^0$, s_0 being the initial state of the formula graph. A state s denotes $\mathbf{tt}$ (resp. $\mathbf{ff}$) if the Boolean variables T_s^k (resp. F_s^k) evaluate to $\mathbf{tt}$ in all (reachable) blocks k. Due to the presence of modalities, there may be states s and blocks k such that T_s^k and F_s^k are both false, indicating that the corresponding sub-formula is not constant. Intuitively, T_s^k expresses that s evaluates to $\mathbf{tt}$ in block k if one of its successors following a transition labelled by $\vee$ or $\mu^{k'}$ evaluates to $\mathbf{tt}$, or one of its successors following a transition labelled by $\neg$ evaluates to $\mathbf{ff}$. Variable F_s^k expresses that state s evaluates to $\mathbf{ff}$ in block k if all its successors following transitions labelled by $\vee$, $\mu^{k'}$, or modalities (by applying the identity $\langle a \rangle \mathbf{ff} = \mathbf{ff}$) evaluate to $\mathbf{ff}$ and all its successors following transitions labelled by $\neg$ evaluate to $\mathbf{tt}$. Regarding fix-point signs, observe that $F_{\mu X^k.X^k}^k =_\nu F_{\mu X^k.X^k}^k$ and $T_{\mu X^k.X^k}^k =_\mu T_{\mu X^k.X^k}^k$ respectively evaluate to $\mathbf{tt}$ and $\mathbf{ff}$, reflecting that $\mu X^k.X^k$ evaluates to $\mathbf{ff}$ as expected.

Repeated applications of quotienting progressively eliminate modalities, until none of them remains in the formula graph, which then necessarily evaluates to a constant equal to the result of evaluating the formula on the whole network. *Sharing of equivalent sub-formulas.* In addition to the above rules, reducing a formula graph modulo strong bisimulation does not change its decoding, modulo idempotence, renaming of propositional variables, and unification of equivalent variables defined in the same block. Strong bisimulation reduction can thus decrease the size of intermediate formula graphs. The reader may note that the heuristic to determine that two variables denote equivalent sub-formulas given in Andersen's work [2] is similar to the definition of strong bisimulation on LTSs.

A careful comparison between the simplifications proposed by Andersen [2] and ours would be useful and is left for further work.

Example 6. After applying the above simplifications to the formula graph of Ex. 5, we obtain the (smaller) formula graph depicted in Fig. 2 (right), which corresponds to the formula $(\langle a \rangle \mathbf{tt}) \vee (\langle \alpha_a \rangle \mathbf{tt}) \vee (\langle \alpha_b \rangle \langle a \rangle \mathbf{tt})$.

Example 7. The graph corresponding to $\mu X^0.(\langle a \rangle \mu Y^0.\langle b \rangle X^0) \vee \langle c \rangle X^0$ reduces as expected to a deadlock state representing the constant $\mathbf{ff}$ (left as an exercise).

6   Simplification of Alternation-Free Formula Graphs

Simplifications apply to μ-calculus formulas of arbitrary alternation depth. We focus here on the alternation-free μ-calculus fragment, which has a linear-time model checking complexity [15] and is therefore more suitable for scaling up to large LTSs. We propose a variant of constant sub-formula evaluation specialised for alternation-free formulas, using alternation-free BESs [1].

Even in the case of alternation-free formulas, the above BES is not alternation-free due to the cyclic dependency between T^k and F^k, e.g., when evaluating sequences of $\neg$-transitions. In Fig. 4, we propose a refinement of this BES, which splits each variable T_s^k of sign μ into two variables T_s^{+k} of sign μ and F_s^{-k} of sign ν, which evaluate to true iff the sub-formula corresponding to state s is preceded by an even (for T_s^{+k}) or odd (for F_s^{-k}) number of negations and evaluates to true. Variable F_s^k is split similarly. This BES is a generalisation, for formula graphs containing negations and modalities, of the BES characterising the solution of alternation-free Boolean graphs outlined in [35].

$$T^k : \begin{cases} T_s^{+k} =_\mu \bigvee_{s \xrightarrow{\vee} s'} T_{s'}^{+k} \vee \bigvee_{s \xrightarrow{\neg} s'} T_{s'}^{-k} \vee \bigvee_{s \xrightarrow{\mu k'} s'} T_{s'}^{+k'} \\ T_s^{-k} =_\mu \bigwedge_{s \xrightarrow{\vee} s'} T_{s'}^{-k} \wedge \bigwedge_{s \xrightarrow{\langle \beta \rangle} s'} T_{s'}^{-k} \wedge \bigwedge_{s \xrightarrow{\neg} s'} T_{s'}^{+k} \wedge \bigwedge_{s \xrightarrow{\mu k'} s'} F_{s'}^{+k'} \end{cases}_{s \in \Sigma}$$

$$F^k : \begin{cases} F_s^{+k} =_\nu \bigwedge_{s \xrightarrow{\vee} s'} F_{s'}^{+k} \wedge \bigwedge_{s \xrightarrow{\langle \beta \rangle} s'} F_{s'}^{+k} \wedge \bigwedge_{s \xrightarrow{\neg} s'} F_{s'}^{-k} \wedge \bigwedge_{s \xrightarrow{\mu k'} s'} F_{s'}^{+k'} \\ F_s^{-k} =_\nu \bigvee_{s \xrightarrow{\vee} s'} F_{s'}^{-k} \vee \bigvee_{s \xrightarrow{\neg} s'} F_{s'}^{+k} \vee \bigvee_{s \xrightarrow{\mu k'} s'} T_{s'}^{+k'} \end{cases}_{s \in \Sigma}$$

Fig. 4. BES for the evaluation of constant alternation-free formulas

For general formulas, this BES is not alternation-free due to the cyclic dependencies between T^k and $F^{k'}$, of different fix-point signs. Yet, for alternation-free block-labelled formulas, it is alternation-free, since each dependency from T^k to $F^{k'}$ (or from F^k to $T^{k'}$) always traverses a μ-transition preceded by an odd number of negations, which switches to a different block number $k' > k$.

7 Implementation

We have implemented partial model checking of alternation-free μ-calculus formulas using CADP, which provided much of what was needed:

- Individual processes can be described in the language LOTOS [24], or in the LOTOS NT variant of E-LOTOS [25], among others, for which CADP contains tools to generate LTSs automatically.
- Process compositions can be described in the EXP.OPEN 2.0 language [28], which provides various parallel composition operators, such as synchronisation vectors [7], process algebra operators (LOTOS, CCS, CSP, μCRL), and the generalised parallel composition operator of E-LOTOS [21]. It also provides generalised operators for hiding, renaming, and cutting labels based on a representation of label sets using regular expressions. The EXP.OPEN 2.0 tool compiles its input into a network of LTSs. It then generates C code for representing the transition relation [18], so that the LTS can be either generated or traversed on-the-fly using various libraries. For partial model checking, the EXP.OPEN 2.0 tool has been slightly extended both to implement sub-network extraction and to generate the network representing the parallel composition between the formula graph and a chosen individual LTS.
- Alternation-free μ-calculus formulas can be handled by the EVALUATOR 3.5 on-the-fly model checker [38], in which an option has been added for compiling a formula into a formula graph.
- Reductions modulo $\tau^*.a$ equivalence and strong bisimulation are achieved using respectively the REDUCTOR and BCG_MIN tools of CADP.

Elimination of double-negations, of μ-transitions, and evaluation of constant formulas have been implemented in a new prototype tool (1,000 lines of C code), which relies on the CAESAR_SOLVE library [37] for solving alternation-free BES. Finally, selection of the LTS w.r.t. which the formula is quotiented at each step is done using the principles described in [16] for networks of LTSs.

8   Experimentation

We have used partial model checking in a case-study in avionics, namely the verification of a communication protocol between a plane and the ground, based on TFTP (*Trivial File Transfer Protocol*)/UDP (*User Datagram Protocol*) [22].

The system consists in two instances of the TFTP connected by UDP using a FIFO buffer. We considered five scenarios, named A to E, depending whether each instance may write and/or read a file. We also checked the (alternation-free) μ-calculus (branching-time) properties named $A01$ to $A28$, studied in [22], both using the well-established on-the-fly model checker EVALUATOR 3.5 [38] of CADP and using the partial model checking approach described in this paper. These experiments were done on a 64-bit computer with 148 gigabytes of memory.

The results summarized in Tab. 1 give, for each scenario, the LTS size in kilo-states (ks), and for each property, the peak of memory in megabytes (MB) used by on-the-fly model checking (column fly) and partial model checking (column pmc). Some properties being irrelevant to some scenarios (e.g., they concern a read or write operation absent in the corresponding scenario), they have not been checked, explaining the shaded cells. The symbol "⋆" corresponds to unfinished verifications that used too much memory. For lack of space, times are not

Table 1. Experimental results for the TFTP/UDP case study

Prop	Scenario A 1,963 ks fly	pmc	Scenario B 867 ks fly	pmc	Scenario C 35,024 ks fly	pmc	Scenario D 40,856 ks fly	pmc	Scenario E 19,436 ks fly	pmc
A01	199	6	89	6	2,947	24	3,351	27	1,530	23
A02	207	6	93	6	3,156	25	3,631	28	1,612	10
A03	182	6	80	6	2,737	6	3,162	6	1,386	6
A04	199	6	89	6	2,947	6	3,351	29	1,530	7
A05	10	6	7	6	7	6	7	6	10	10
A06	187	6	85	6	2,808	6	3,249	7	1,428	6
A07	187	6	85	6	2,808	6	3,249	6	1,428	6
A08	186	6	80	6	2,745	6	3,170	6	1,390	6
A09a							3,290	28	1,488	6
A09b					2,955	6				
A10					3,354	6			1,674	6
A11					3,206	6	4,444	7	1,711	6
A12					620	⋆	133	⋆	101	⋆
A13							4,499	⋆	2,094	⋆
A14	267	6			3,988	23			2,107	15
A15			118	15	521	⋆	156	⋆	1,524	59
A16									186	8
A17					667	⋆	569	2,702		
A18			85	6	476	11	255	6	1,391	6
A19			207	6	6,352	90	8,753	13	3,104	55
A20	31	9			837	21			261	25
A21	374	6			4,958	25			2,817	25
A22			35	7			427	1,271	191	650
A23			170	6			6,909	9	3,039	40
A24	41	9			427	1,786				
A25	391	6			5,480	40				
A26	195	6			2,857	15			1,477	10
A27	228	6			3,534	6			1,871	6
A28			102	6	3,654	22	4,032	6	1,821	6

reported but each partial model checking experiment that used less than 100 MB of memory took from a few seconds to less than a minute. Note that the major part of time and memory are used by formula simplifications, as compared to the low complexity of the synchronous product operation used for quotienting.

These results confirm that partial model checking may be much more efficient (up to 600 times less memory in this example) than on-the-fly model checking. For several properties, we observe that partial model checking sometimes allows complete evaluation of formulas before they have been quotiented with respect to all individual LTSs, because the truth value of the formula is independent of some individual LTS. However, in a few cases, partial model checking leads to combinatorial explosion (properties $A12$, $A13$, $A15$, and $A17$) while on-the-fly model checking is efficient. This is inherent to the structure of the system, intermediate quotients needing to capture a large part of the behaviour before the truth value of the formula can be computed. This shows that both approaches are complementary and worthy of being used concurrently.

9   Conclusion

The original contributions of this paper are the following: (1) Partial model checking has been generalised to the network model, which subsumes many parallel composition operators. (2) An efficient implementation of quotienting with respect to an individual LTS has been proposed, using a simple synchronous product between this LTS and a graph representation of the formula. A key is the representation of the formula in a disjunctive form (using negations), which turns every node of the formula graph into an *or-node*. (3) An efficient implementation of formula simplifications has also been proposed, using a combination of existing algorithms (such as reductions modulo equivalence relations), simple transformations, and traversals of the formula graph using a BES. Using a graph equivalence relation to simplify the formula was already proposed in [8], where the formula was translated into an *and-or-graph* and then reduced modulo strong bisimulation. We use a weaker relation ($\tau^*.a$ equivalence) that enables more reduction of the formula graph, and we apply it directly on simple LTSs, thus allowing efficient LTS reduction tools to be used without any modification. Our simplifications integrate smoothly in the approach, both quotienting and simplifications applying to the same graph representation, without encoding and decoding formulas back and forth. (4) A specialisation to the case of alternation-free formulas (using alternation-free BES) has also been presented, showing that partial model checking may result in much better performance than complementary approaches, such as on-the-fly model checking. Only small software developments were required, thanks to the wealth of functionalities available in CADP. The approach would be also applicable to formulas of arbitrary alternation depth using a solver for BES of arbitrary alternation depth.

The implementation of quotienting as a synchronous product opens the way for combining partial model checking with techniques originating from compositional model generation, such as (compositional) τ-confluence reduction [30,36,40], or restriction using interface constraints following the approach developed in [23] and refined in [19,27,29]. Note also that partial model checking and compositional model generation are complementary. Although it is difficult in general to know which of them will be most efficient, a reasonable methodology is to try compositional model generation first (because one then obtains a single model on which all formulas of interest can be evaluated). In case of failure, partial model checking can then be used for each formula.

As future work, we also plan to study partial model checking of certain μ-calculus formulas of alternation depth 2 describing the existence of complex cycles (e.g., $\nu X.\mu Y.(\langle b \rangle X \vee \langle a \rangle Y)$, expressing the infinite repetition of sequences belonging to the regular language $a^*.b$), which can still be checked in linear-time using specialised BES resolution algorithms [39] generalising the detection of accepting cycles in Büchi automata.

References

1. Andersen, H.R.: Model checking and Boolean graphs. Theoretical Computer Science 126(1), 3–30 (1994)
2. Andersen, H.R.: Partial Model Checking. In: Proc. of Logic in Computer Science LICS. IEEE Computer Society Press (1995)
3. Andersen, H.R., Lind-Nielsen, J.: MuDiv: A Tool for Partial Model Checking. In: Proc. of CONCUR (1996)
4. Andersen, H.R., Lind-Nielsen, J.: Partial Model Checking of Modal Equations: A Survey. STTT 2, 242–259 (1999)
5. Andersen, H.R., Staunstrup, J., Maretti, N.: Partial Model Checking with ROB-DDs. In: Brinksma, E. (ed.) TACAS 1997. LNCS, vol. 1217, pp. 35–49. Springer, Heidelberg (1997)
6. Andersen, H.R., Staunstrup, J., Maretti, N.: A Comparison of Modular Verification. In: Bidoit, M., Dauchet, M. (eds.) CAAP 1997, FASE 1997, and TAPSOFT 1997. LNCS, vol. 1214, Springer, Heidelberg (1997)
7. Arnold, A.: MEC: A System for Constructing and Analysing Transition Systems. In: Sifakis, J. (ed.) CAV 1989. LNCS, vol. 407, pp. 117–132. Springer, Heidelberg (1990)
8. Basu, S., Ramakrishnan, C.R.: Compositional Analysis for Verification of Parameterized Systems. In: Garavel, H., Hatcliff, J. (eds.) TACAS 2003. LNCS, vol. 2619, pp. 315–330. Springer, Heidelberg (2003)
9. Berard, B., Laroussinie, F.: Verification compositionnelle des p-automates. Tech. Report Lot 4.1, RNTL, projet AVERROES (2003)
10. Bodentien, N., Vestergaard, J., Friis, J., Kristoffersen, K., Larsen, K.: Verification of State/Event Systems by Quotienting. Tech. Report RS-99-41, BRICS (1999)
11. Bouali, A., Ressouche, A., Roy, V., de Simone, R.: The Fc2Tools Set: a Toolset for the Verification of Concurrent Systems. In: Alur, R., Henzinger, T.A. (eds.) CAV 1996. LNCS, vol. 1102, pp. 441–445. Springer, Heidelberg (1996)
12. Cassez, F., Laroussinie, F.: Model-Checking for Hybrid Systems by Quotienting and Constraints Solving. In: Emerson, E.A., Sistla, A.P. (eds.) CAV 2000. LNCS, vol. 1855, pp. 373–388. Springer, Heidelberg (2000)
13. Champelovier, D., Clerc, X., Garavel, H., Guerte, Y., Lang, F., McKinty, C., Powazny, V., Serwe, W., Smeding, G.: Reference Manual of the LOTOS NT to LOTOS Translator (Version 5.4). INRIA/VASY (2011)

14. Clarke, E., Grumberg, O., Peled, D.: Model Checking. MIT Press (2000)
15. Cleaveland, R., Steffen, B.: A Linear-Time Model-Checking Algorithm for the Alternation-Free Modal Mu-Calculus. FMSD 2(2), 121–147 (1993)
16. Crouzen, P., Lang, F.: Smart Reduction. In: Giannakopoulou, D., Orejas, F. (eds.) FASE 2011. LNCS, vol. 6603, pp. 111–126. Springer, Heidelberg (2011)
17. Fernandez, J.-C., Mounier, L.: "On the Fly" Verification of Behavioural Equivalences and Preorders. In: Larsen, K.G., Skou, A. (eds.) CAV 1991. LNCS, vol. 575, pp. 181–191. Springer, Heidelberg (1992)
18. Garavel, H.: OPEN/CAESAR: An Open Software Architecture for Verification, Simulation, and Testing. In: Steffen, B. (ed.) TACAS 1998. LNCS, vol. 1384, pp. 68–84. Springer, Heidelberg (1998)
19. Garavel, H., Lang, F.: SVL: a Scripting Language for Compositional Verification. In: Proc. of FORTE. IFIP. Kluwer Academic Publishers (2001)
20. Garavel, H., Lang, F., Mateescu, R., Serwe, W.: CADP 2010: A Toolbox for the Construction and Analysis of Distributed Processes. In: Abdulla, P.A., Leino, K.R.M. (eds.) TACAS 2011. LNCS, vol. 6605, pp. 372–387. Springer, Heidelberg (2011)
21. Garavel, H., Sighireanu, M.: A Graphical Parallel Composition Operator for Process Algebras. In: Proc. of FORTE/PSTV. IFIP. Kluwer (1999)
22. Garavel, H., Thivolle, D.: Verification of GALS Systems by Combining Synchronous Languages and Process Calculi. In: Păsăreanu, C.S. (ed.) Model Checking Software. LNCS, vol. 5578, pp. 241–260. Springer, Heidelberg (2009)
23. Graf, S., Steffen, B.: Compositional Minimization of Finite State Systems. In: Clarke, E., Kurshan, R.P. (eds.) CAV 1990. LNCS, vol. 531, pp. 186–196. Springer, Heidelberg (1991)
24. ISO/IEC. LOTOS — A Formal Description Technique Based on the Temporal Ordering of Observational Behaviour. ISO International Standard 8807 (1989)
25. ISO/IEC. Enhancements to LOTOS (E-LOTOS). ISO International Standard 15437 (2001)
26. Kozen, D.: Results on the Propositional μ-calculus. TCS 27, 333–354 (1983)
27. Krimm, J.-P., Mounier, L.: Compositional State Space Generation from LOTOS Programs. In: Brinksma, E. (ed.) TACAS 1997. LNCS, vol. 1217, pp. 239–258. Springer, Heidelberg (1997)
28. Lang, F.: Exp.Open 2.0: A Flexible Tool Integrating Partial Order, Compositional, and On-The-Fly Verification Methods. In: Romijn, J.M.T., Smith, G.P., van de Pol, J. (eds.) IFM 2005. LNCS, vol. 3771, pp. 70–88. Springer, Heidelberg (2005)
29. Lang, F.: Refined Interfaces for Compositional Verification. In: Najm, E., Pradat-Peyre, J.-F., Donzeau-Gouge, V.V. (eds.) FORTE 2006. LNCS, vol. 4229, pp. 159–174. Springer, Heidelberg (2006)
30. Lang, F., Mateescu, R.: Partial Order Reductions Using Compositional Confluence Detection. In: Cavalcanti, A., Dams, D.R. (eds.) FM 2009. LNCS, vol. 5850, pp. 157–172. Springer, Heidelberg (2009)
31. Laroussinie, F., Larsen, K.: Compositional Model Checking of Real Time Systems. In: Lee, I., Smolka, S.A. (eds.) CONCUR 1995. LNCS, vol. 962, pp. 27–41. Springer, Heidelberg (1995)
32. Laroussinie, F., Larsen, K.: CMC: A Tool for Compositional Model Checking of Real-Time Systems. In: Proc. of FORTE (1998)
33. Larsen, K., Pettersson, P., Yi, W.: Compositional and Symbolic Model Checking of Real-Time Systems. In: Proc. of the IEEE Real-Time Symposium (1995)

34. Martinelli, F.: Symbolic Partial Model Checking for Security Analysis. In: Gorodet-sky, V., Popyack, L.J., Skormin, V.A. (eds.) MMM-ACNS 2003. LNCS, vol. 2776, pp. 122–134. Springer, Heidelberg (2003)

35. Mateescu, R.: Efficient Diagnostic Generation for Boolean Equation Systems. In: Graf, S. (ed.) TACAS 2000. LNCS, vol. 1785, pp. 251–265. Springer, Heidelberg (2000)

36. Mateescu, R.: On-the-fly State Space Reductions for Weak Equivalences. In: Proc. of FMICS. ERCIM. ACM Computer Society Press (2005)

37. Mateescu, R.: CAESAR_SOLVE: A Generic Library for On-the-Fly Resolution of Alternation-Free Boolean Equation Systems. STTT 8(1), 37–56 (2006)

38. Mateescu, R., Sighireanu, M.: Efficient On-the-Fly Model-Checking for Regular Alternation-Free Mu-Calculus. SCP 46(3), 255–281 (2003)

39. Mateescu, R., Thivolle, D.: A Model Checking Language for Concurrent Value-Passing Systems. In: Cuellar, J., Sere, K. (eds.) FM 2008. LNCS, vol. 5014, pp. 148–164. Springer, Heidelberg (2008)

40. Pace, G.J., Lang, F., Mateescu, R.: Calculating τ-Confluence Compositionally. In: Hunt Jr., W.A., Somenzi, F. (eds.) CAV 2003. LNCS, vol. 2725, pp. 446–459. Springer, Heidelberg (2003)

From Under-Approximations to Over-Approximations and Back

Aws Albarghouthi[1], Arie Gurfinkel[2], and Marsha Chechik[1]

[1] Department of Computer Science, University of Toronto, Canada
[2] Software Engineering Institute, Carnegie Mellon University, USA

Abstract. Current approaches to software model checking can be divided into over-approximation-driven (OD) and under-approximation-driven (UD). OD approaches maintain an abstraction of the transition relation of a program and use abstract reachability to build an inductive invariant (or find a counterexample). At the other extreme, UD approaches attempt to construct inductive invariants by generalizing from finite paths through the control-flow graph of the program.

In this paper, we present UFO, an algorithm that unifies OD and UD approaches in order to leverage both of their advantages. UFO is parameterized by the degree to which over- and under-approximations drive the analysis. At one extreme, UFO is a novel interpolation-based (UD) algorithm that generates interpolants to label (refine) multiple program paths using a single SMT solver query. At the other extreme, UFO uses an abstract domain to drive the analysis, while using interpolants to strengthen the abstraction.

We have implemented UFO in LLVM and applied it to programs from the Competition on Software Verification. Our experimental results demonstrate the utility of our algorithm and the benefits of combining UD and OD approaches.

1 Introduction

In recent years, we have witnessed a divergence in software model checking techniques. Traditionally, as promoted by the SLAM project [3], software model checkers implemented a variant of the counterexample-guided abstraction refinement (CEGAR) [11] loop, where over-approximating abstractions of programs are computed. In cases where spurious counterexamples are introduced by over-approximation, refinement is used to eliminate them. We henceforth categorize such techniques as *over-approximation-driven* (OD). OD techniques mainly rely on *predicate abstraction* [16] for computing an abstract post operator. In the refinement stage, new predicates (facts) are added to build more precise abstractions. OD techniques can supply us with efficient safety proofs, when relatively coarse abstractions are sufficient to prove correctness. Unfortunately, it is often the case that a large number of predicates is required to reach a deep error or to compute an inductive invariant, causing the abstraction step to be very expensive.

On the other hand, *under-approximation-driven* (UD) software model checking techniques are becoming more popular. Such techniques attempt to construct a

C. Flanagan and B. König (Eds.): TACAS 2012, LNCS 7214, pp. 157–172, 2012.

program invariant by generalizing from finite program paths. For example, in [24], McMillan uses *Craig Interpolants*, derived from proofs of unsatisfiability of paths to an error location, to compute program invariants. In our previous work [2], we used predicate abstraction to generalize symbolic program executions. Note that SYNERGY [17] and DASH [4] are considered under-approximation-driven according to our categorization, as they use weakest-precondition computations along infeasible symbolic paths to refine a partition of the state space with the goal of computing an invariant. Testing in these techniques only acts as a way of choosing which symbolic paths to examine. UD techniques avoid the expensive step of computing an abstract post operator, giving them an advantage over OD techniques. Unfortunately, due to the fact that they are not driven by an abstract domain, they may have to examine a large number of program paths to compute an inductive invariant or find an erroneous execution.

In this paper, our goal is to resolve the disconnect between OD and UD approaches to software model checking. Specifically, we present UFO, a software model checking algorithm for sequential programs that is parameterized by the degree to which over- and under-approximations drive the analysis. UFO makes two contributions: (1) it combines UD and OD approaches, and (2) at the UD extreme, it is a novel interpolation-based algorithm that generates interpolants to label (refine) multiple program paths using a single SMT solver query. This allows UFO to exploit an SMT solver's ability to enumerate program executions, giving it an advantage over other interpolation-based algorithms, e.g., [24,23], that explicitly enumerate program paths.

We have implemented UFO in the LLVM compiler infrastructure [22] and experimented with various instantiations of it on benchmarks from the Competition on Software Verification [5]. Our experimental results show the utility of our interpolation-based algorithm. Moreover, they show that augmenting UFO with an abstract domain (e.g., predicate abstraction) often outperforms both the OD and UD extremes.

The rest of the paper is organized as follows: In Sec. 2, we illustrate the operation of UFO on an example. In Sec. 3, we provide the definitions and notation used in the paper. In Sec. 4, we present the UFO algorithm. In Sec. 5, we present the refinement procedure. In Sec. 6, we describe our UFO implementation and present our experimental evaluation. In Sec. 7, we place UFO in the context of related work. Finally, in Sec. 8, we conclude the paper and outline directions for future work.

2   Overview

The core of UFO is a UD algorithm parameterized by an abstract POST operator and a novel interpolation-based refinement procedure. In this section, we illustrate the novel parts of UFO by instantiating POST to always return *true*, the weakest admissible POST. In practice, we also instantiate POST with Boolean and Cartesian predicate abstractions.

Consider function foo shown in Fig. 1(a), which takes n as a parameter. We want to prove that location 8 with label ERROR is unreachable for any value of n.

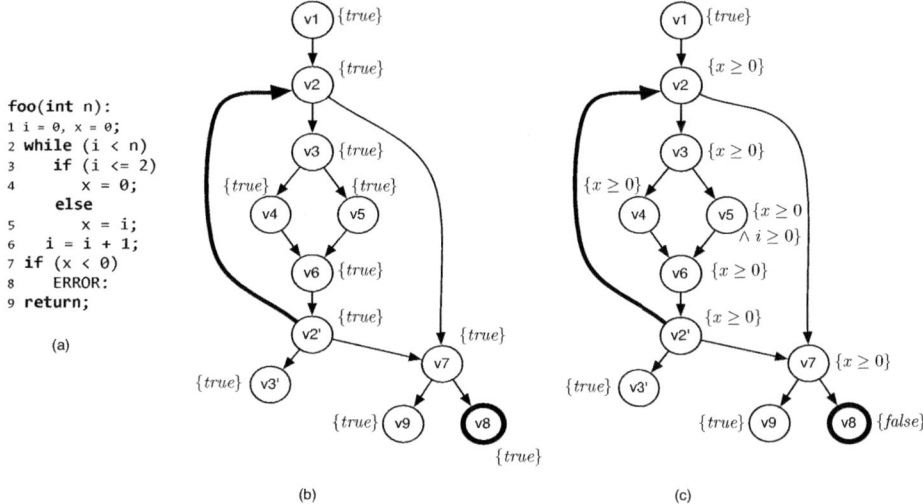

```
foo(int n):
1  i = 0, x = 0;
2  while (i < n)
3     if (i <= 2)
4        x = 0;
   else
5        x = i;
6     i = i + 1;
7  if (x < 0)
8     ERROR:
9  return;
```

(a)

Fig. 1. (a) Safe function foo. ARGs for foo (b) 1st iteration; (c) after refinement

Constructing the ARG. UFO starts by constructing an *Abstract Reachability Graph* (ARG) for the program. The ARG for foo is shown in Fig. 1(b). Each node in the ARG relates to a location in foo. For example, nodes v_2 and v_2' represent location 2. UFO associates a label $\{\varphi_i\}$ for each node v_i, where φ_i is an over-approximation of the set of reachable states at v_i.

UFO expands the ARG using the *recursive iteration strategy* [8]. That is, the innermost loop is unrolled first. In our example, UFO starts by creating nodes v_1 and v_2. Following the recursive iteration strategy, it enters the loop, creating nodes v_3, v_4, v_5, v_6, and v_2'. All nodes are initially labelled with *true*, the weakest possible over-approximation.

Upon reaching node v_2', UFO adds v_3' and v_7 as children of v_2'. At this point, the label of v_2' is subsumed by the one of v_2. We say that v_2' is *covered* by v_2 and show it as the bold back-edge in Fig. 1(b). Therefore, UFO exits the loop and goes on to process node v_7, adding v_8 (ERROR node) and v_9.

Refining the ARG. Once the ARG has been completely expanded, UFO checks if the label on the ERROR node v_8 is UNSAT. The label is *true*, which means that there is a potential execution to location 8 that either does not enter the loop at v_2 or takes one iteration before exiting through v_2'. To check if such an execution exists, UFO constructs a formula representing all executions in the ARG: for each node v_i that can reach v_8, it creates the following formula μ_i:

$$\mu_1 : c_{v_1} \Rightarrow (i_0 = 0 \wedge x_0 = 0 \wedge c_{v_2})$$
$$\mu_2 : c_{v_2} \Rightarrow ((i_0 < n \wedge c_{v_3}) \vee (i_0 \geq n \wedge x_4 = x_0 \wedge c_{v_7}))$$
$$\mu_3 : c_{v_3} \Rightarrow ((i_0 \leq 2 \wedge c_{v_4}) \vee (i_0 > 2 \wedge c_{v_5}))$$
$$\mu_4 : c_{v_4} \Rightarrow (x_1 = 0 \wedge x_3 = x_1 \wedge c_{v_6})$$

$$\mu_5 : c_{v_5} \Rightarrow (x_2 = i_0 \wedge x_3 = x_2 \wedge c_{v_6})$$
$$\mu_6 : c_{v_6} \Rightarrow (i_1 = i_0 + 1 \wedge c_{v_2'})$$
$$\mu_2' : c_{v_2'} \Rightarrow (i_1 \geq n \wedge x_4 = x_3 \wedge c_{v_7})$$
$$\mu_7 : c_{v_7} \Rightarrow (x_4 \geq 0 \wedge c_{v_8})$$

For example, μ_2 specifies that if control reaches node v_2 (represented by the Boolean *control variable* c_{v_2}), then either $i_0 < n$ and control goes to v_3, or

$i_0 \geq n$ and control goes to v_7. To avoid naming conflicts, each time a variable appears on the left hand side of an assignment, it is given a fresh subscript (e.g., x becomes x_1 in μ_4).

The formula $c_{v_1} \wedge \mu_1 \wedge \cdots \wedge \mu_7$ is UNSAT. Hence, there is no feasible execution in the ARG that can reach v_8. At this point, *Craig interpolants* [13] are used to relabel the ARG. Given a pair of formulas (A, B) s.t. $A \wedge B$ is UNSAT, an interpolant for (A, B) is a formula I s.t. $A \Rightarrow I$, $I \Rightarrow \neg B$, and I is over the variables shared between A and B. To derive a new label for v_7, UFO sets $A = c_{v_1} \wedge \mu_1 \wedge \cdots \wedge \mu'_2$ and $B = \mu_7$. A possible interpolant I for (A, B) is $c_{v_7} \wedge x_4 \geq 0$. To remove the instrumentation variable c_{v_7}, UFO sets it to *true*. It also removes the subscript from x_4 to arrive at a formula $x \geq 0$ over program variables. The new labels generated by interpolants are shown in Fig. 1(c). In Sec. 5, we formalize the process of deriving labels from interpolants and prove that for any edge (v_i, v_j) in the ARG, the resulting labels for v_i and v_j form a Hoare triple with respect to the program statement on the edge.

Note that in Fig. 1(c), v_8 is labelled with $\{false\}$ and v'_2 is still covered, since its label $x \geq 0$ is subsumed by the label on v_2. Therefore, UFO terminates execution declaring foo safe. When applied to this example, the algorithm in [24] requires at least two refinements, as the control-flow graph (CFG) is unrolled into a tree, thus creating two paths to ERROR through each branch of the conditional statement (location 3). UFO, on the other hand, unrolls the CFG into a DAG and exploits the power of SMT solvers for enumerating paths.

3   Abstract Reachability Graphs

Here, we present the notation and definitions used in the rest of the paper.

Programs. A *program* P is a tuple $(\mathcal{L}, \Delta, \text{en}, \text{err}, \text{Var})$, where $\mathcal{L}$ is a finite set of control locations, Δ is a finite set of actions, $\text{en} \in \mathcal{L}$ is the entry location of P, $\text{err} \in \mathcal{L}$ is the error location, and Var is the set of variables of program P. An *action* $(\ell_1, T, \ell_2) \in \Delta$ represents an edge in the control flow graph of P, where $\ell_1, \ell_2 \in \mathcal{L}$ and T is a program statement. We assume that there does not exist an action $(\ell_1, T, \ell_2) \in \Delta$ s.t. $\ell_1 = \text{err}$.

A program statement is either an assume statement $\text{assume}(Q)$, where Q is a Boolean expression over Var, or an assignment statement $\text{x} = E$, where x is a variable in Var and E is an expression over the variables in Var. We use the notation $[\![T]\!]$ to denote the standard semantics of a program statement T. For example, for an assignment statement $\text{x} = \text{x} + 1$, $[\![\text{x} = \text{x} + 1]\!]$ is $x' = x + 1 \wedge \forall y \in \text{Var} \cdot y \neq x \Rightarrow y' = y$. For a formula ϕ, we use ϕ' to denote ϕ with all variables replaced by their primed versions.

We say that a program P is *safe* iff there does not exist a feasible execution that starts in en and reaches err through the actions in Δ.

Weak Topological Ordering. A *Weak Topological Ordering* (WTO) [8] of a directed graph $G = (V, E)$ is a well-parenthesized total-order, denoted $\prec$, of V without two consecutive "(" s.t. for every edge $(u, v) \in E$:

$$(u \prec v \wedge v \notin \omega(u)) \vee (v \preceq u \wedge v \in \omega(u)),$$

where elements between two matching parentheses are called a *component*, the first element of a component is called a *head*, and $\omega(v)$ is the set of heads of all components containing v.

Let $v \in V$, and U be the innermost component that contains v in the WTO. We write WTONEXT(v) for an element $u \in U$ that immediately follows v, if it exists, and for the head of U otherwise.

Let U_s be a component with head v. Suppose that U_s is a subcomponent of some component U. If there exists a $u \in U$ s.t. $u \notin U_s$ and u is the first element in the total-order s.t. $v \prec u$, then WTOEXIT(v) = u. Otherwise, WTOEXIT(v) = w, where w is the head of U. Now suppose that U_s is not a subcomponent of any other component, then WTOEXIT(v) = u, where u is the first element in the total-order s.t. $u \notin U_s$ and $v \prec u$. Intuitively, if the WTO represented program locations, then WTOEXIT(v) is the first control location visited after exiting the loop headed by v. For example, for function foo in Fig. 1(d), a WTO of the control locations is 1 (2 3 4 5 6) 7 8 9, where 2 is the head of the component comprising the while loop. WTONEXT(2) = 3, WTONEXT(6) = 2, and WTOEXIT(2) = 7. Note that WTONEXT and WTOEXIT are partial functions and we only use them where they have been defined.

Abstract Reachability Graphs (ARGs). Let $P = (\mathcal{L}, \Delta, \text{en}, \text{err}, \text{Var})$ be a program. A *Reachability Graph* (RG) of P is a tuple $(V, E, v_{\text{en}}, \nu, \tau)$, where (V, E, v_{en}) represents a directed acyclic graph (DAG) rooted at the *entry node* $v_{\text{en}} \in V$, $\nu : V \to \mathcal{L}$ is a map from nodes to control locations of P, where $\nu(v_{\text{en}}) = \text{en}$, and $\tau : E \to \Delta$ is a map from edges to actions of P s.t. for every edge $(u, v) \in E$, there exists an action $(\nu(u), \tau(u, v), \nu(v)) \in \Delta$.

An *Abstract Reachability Graph* (ARG) $\mathcal{A}$ of P is a tuple $(U, \psi, \sqsubseteq, \sqsubseteq_t)$, where $U = (V, E, v_{\text{en}}, \nu, \tau)$ is an RG of P, ψ is a map from nodes V to formulas over Var, $\sqsubseteq$ is the ancestor relation over the nodes of U, and $\sqsubseteq_t$ is a fixed linearization of the topological ordering of the nodes of U. A node v s.t. $\nu(v) = \text{err}$ is called an *error node*.

A node $v \in V$ is *covered* iff there exists a node $u \in V$ that *dominates* v and there exists a set of nodes $X \subset V$, where $\psi(u) \Rightarrow \bigvee_{x \in X} \psi(x)$ and $\forall x \in X \cdot x \sqsubseteq u \wedge \nu(x) = \nu(u)$. A node u *dominates* v iff all paths from v_{en} to v pass through u. Every node v dominates itself.

Definition 1 (Well-labeledness of ARGs). *An ARG $\mathcal{A} = (U, \psi, \sqsubseteq, \sqsubseteq_t)$, where $U = (V, E, v_{\text{en}}, \nu, \tau)$, for a program $P = (\mathcal{L}, \Delta, \text{en}, \text{err}, \text{Var})$ is well-labelled iff (1) $\psi(v_{\text{en}}) \equiv$ true; and (2) $\forall (u, v) \in E, \psi(u) \wedge [\![\tau(u, v)]\!] \Rightarrow \psi(v)'$.*

An ARG is *safe* iff for all $v \in V$ s.t. $\nu(v) = \text{err}$, $\psi(v) \equiv false$. An ARG is *complete* iff for all uncovered nodes u, for all $(\nu(u), T, \ell)$, there exists an edge (u, v) s.t. $\nu(v) = \ell$ and $\tau(u, v) = T$.

Theorem 1 (Program Safety). *If there exists a safe, complete, and well-labelled ARG for a program P, then P is safe.*

The proof of this theorem follows from Theorem 1 in [24].

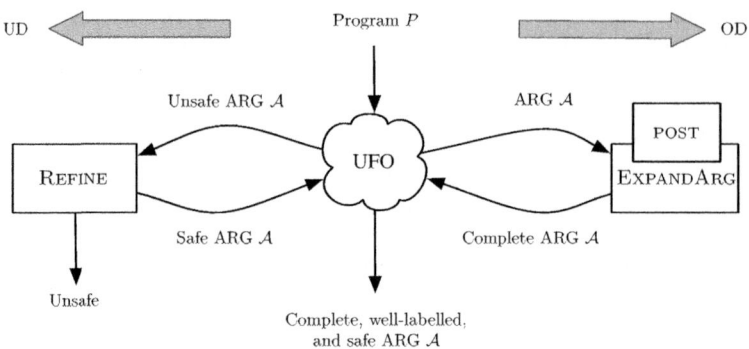

Fig. 2. High level description of UFO

```
 1: func UFOMAIN (Program P) :          29: func EXPANDARG () :
 2:    create node v_en                  30:    v ← v_en
 3:    ψ(v_en) ← true, ν(v_en) ← en      31:    while true do
 4:    marked(v_en) ← true               32:       EXPANDNODE(v)
 5:    labels ← ∅                        33:       if marked(v) then
 6:    while true do                     34:          marked(v) ← false
 7:       EXPANDARG()                    35:          ψ(v) ← ⋁_{(u,v)∈E} POST(u,v)
 8:       if ψ(v_err) is UNSAT then      36:          for all (v,w) ∈ E do marked(w) ← true
 9:          return SAFE                 37:       else if labels(v) bound then
10:       labels ← REFINE()              38:          ψ(v) ← labels(v)
11:       if labels = ∅ then             39:          for all {(v,w) ∈ E | labels(w) unbound} do
12:          return UNSAFE               40:             marked(w) ← true
13:       clear AH and FN                41:       if v = v_err then break
                                         42:       if ν(v) is head of a component then
14: func GETFUTURENODE (ℓ ∈ L) :         43:          if ψ(v) ⇒ ⋁_{u∈AH(ν(v))} ψ(u) then
15:    if FN(ℓ) exists then              44:             erase AH(ν(v)) and FN(ν(v))
16:       return FN(ℓ)                   45:             l ← WTOEXIT(ν(v))
17:    create node v                     46:             v ← FN(l); erase FN(l)
18:    ψ(v) ← true; ν(v) ← ℓ             47:             for all {(v,w) ∈ E | ∄u ≠ v·(u,w) ∈ E} do
19:    FN(l) ← v                         48:                erase FN(ν(w))
20:    return v                          49:             continue
                                         50:          add v to AH(ν(v))
21: func EXPANDNODE (v ∈ V) :            51:       l ← WTONEXT(ν(v))
22:    if v has children then            52:       v ← FN(l); erase FN(l)
23:       for all (v,w) ∈ E do
24:          FN(ν(w)) ← w
25:    else
26:       for all (ν(v), T, ℓ) ∈ Δ do
27:          w ← GETFUTURENODE(ℓ)
28:          E ← E ∪ {(v,w)}; τ(v,w) ← T
```

Fig. 3. The UFO Algorithm. Implementation of REFINE is presented in Sec. 5

4   The UFO Algorithm

In this section, we describe our verification algorithm UFO that takes a program
P with a designated error location v_{err} and determines whether v_{err} is reachable.
The output of the algorithm is either an execution of P that ends in v_{err}, or
a complete, well-labeled, and safe ARG of P. The novelty of UFO lies in its

combination of UD and OD techniques. Fig. 2 illustrates the two main states of UFO: (1) exploring (OD), and (2) generalizing (UD). Exploring an ARG is done by unwinding the CFG while *computing* node labels using an abstract post operator POST. Generalizing is done by *guessing* (typically using interpolants) a safe labelling of the current ARG from a proof of infeasibility of unsafe executions in $\mathcal{A}$.

The pseudo-code for the algorithm is given in Fig. 3. Function EXPANDARG (line 29) is responsible for the exploration, and REFINE (line 10) is used for generalization. Note that UFO is parameterized by POST (line 35) – more precise POST makes it more OD-like and less precise POST – more UD-like. We present the main parts of the algorithm in this section, and an implementation of REFINE in Sec. 5.

Main Loop. UFOMAIN is the main function of UFO. It receives a program $P = (\mathcal{L}, \Delta, \mathsf{en}, \mathsf{err}, \mathsf{Var})$ as input and attempts to prove that P is safe (or unsafe) by constructing a complete, well-labelled, and safe ARG for P (or by finding an execution to err). The function EXPANDARG is used to construct an ARG $\mathcal{A} = (U, \psi, \sqsubseteq, \sqsubseteq_T)$ for P. By definition, it always constructs a complete, well-labelled ARG. Line 8 of UFOMAIN checks if the result of EXPANDARG is a safe ARG by checking whether the label on the node v_{err} is satisfiable (by construction, v_{err} is the only node in $\mathcal{A}$ s.t. $\nu(v_{\mathsf{err}}) = \mathsf{err}$). If $\psi(v_{\mathsf{err}})$ is UNSAT, then $\mathcal{A}$ is safe, and UFO terminates by declaring the program safe (following Theorem 1). Otherwise, REFINE is used to compute new labels. In Definition 2, we provide a specification of REFINE that maintains the soundness of UFO. In Sec. 5, we present a refinement algorithm satisfying Definition 2.

Definition 2 (Specification of Refine). *If there exists a feasible execution to v_{err} in $\mathcal{A}$, then* REFINE *returns an empty map (*labels $= \emptyset$*). Otherwise, it returns a map from nodes to labels s.t.* labels$(v_{\mathsf{err}}) \equiv$ false, labels$(v_{\mathsf{en}}) \equiv$ true, *and* $\forall(u, v) \in E' \cdot$ labels$(u) \wedge [\![\tau(u, v)]\!] \Rightarrow$ labels$(v)'$*, where E' is E restricted to edges along paths to v_{err}. That is, the labeling precludes erroneous executions and maintains well-labelledness of $\mathcal{A}$ (per Definition 1).*

Constructing the ARG. EXPANDARG adopts a standard *recursive iteration strategy* [8] for unrolling a CFG into an ARG. To do so, it makes use of a *weak topological ordering* (WTO) [8] of program locations. A recursive iteration strategy starts by unrolling the innermost loops until "stabilization", i.e., until a loop head is covered, before exiting to the outermost loops. We assume that the first location in the WTO is en and the last one is err.

EXPANDARG maintains two global maps: AH (active heads) and FN (future nodes). For a loop head l, AH(l) is the set of nodes $V_\ell \subseteq V$ for location l that are heads of the component being unrolled. When a loop head is covered (line 43), all active heads belonging to its location are removed from AH (line 44). FN maps a location to a single node and is used as a worklist, i.e., it maintains the next node to be explored for a given location. Example 1 demonstrates the operation of EXPANDARG.

Example 1. Consider the process of constructing the ARG in Fig. 1(b) for function foo in Fig. 1(a). When EXPANDARG processes node v_2' (i.e., when $v = v_2'$ at line 31), $\mathsf{AH}(2) = \{v_2\}$, since the component (2 3 4 5 6) representing the loop is being unrolled and v_2 is the only node for location 2 that has been processed. When UFO covers v_2' (line 43), it sets $\mathsf{AH}(2) = \emptyset$ (line 44) since the component has stabilized and UFO has to exit it. Here, WTOEXIT(2) = 7, so UFO continues processing from node $v_7 = \mathsf{FN}(7)$ (the node for the first location after the loop).

Suppose REFINE returned a new label for node v. When EXPANDARG updates $\psi(v)$ (line 38), it *marks* all of its children that do not have labels in *labels*. This is used to strengthen the labels of v's children w.r.t the refined over-approximation of reachable states at v, using the operator POST (line 35). EXPANDARG only attempts to cover nodes that are loop heads. It does so by checking if the label on a node v is subsumed by the labels on $\mathsf{AH}(\nu(v))$ (line 43). If v is covered, UFO exits the loop (line 45); otherwise, it adds v to $\mathsf{AH}(\nu(v))$.

Post Operator. UFO is parameterized by the abstract operator POST. For sound implementations of UFO, POST should take an edge (u, v) as input and return a formula ϕ s.t. $\psi(u) \wedge [\![\tau(u, v)]\!] \Rightarrow \phi'$, thus maintaining well-labelledness of the ARG. In the UD case, POST always returns *true*, the weakest possible abstraction. In the combined UD+OD case, POST is driven by an abstract domain, e.g., based on predicate abstraction.

Theorem 2 (Soundness). *Given a program P, if UFO run on P terminates with SAFE, the resulting ARG $\mathcal{A}$ is safe, complete, and well-labelled. If UFO terminates with UNSAFE, then there exists an execution that reaches err in P.*

5   Refinement

In this section, we present our refinement procedure REFINE. It begins by computing a formula φ, called an *ARG condition*, representing all executions in a given ARG. If φ is unsatisfiable, REFINE invokes an interpolation-based algorithm is to compute new labels for the ARG.

ARG Condition. Given an ARG $\mathcal{A}$ with an entry and an error nodes v_{en}, v_{err}, respectively, we define ARGCOND as follows:

$$\text{ARGCOND}(v_{err}) \triangleq c_{u_1} \wedge \mu_1 \wedge \cdots \wedge \mu_n, \tag{1}$$

$$\text{where } \mu_i = (c_{u_i} \Rightarrow \bigvee_{(u_i, w) \in E} (c_w \wedge encode(u_i, w))), \tag{2}$$

$u_1 = v_{en}$, and $u_1, \ldots, u_n$ is the sequence of all nodes, excluding v_{err}, that can reach v_{err} in $\mathcal{A}$, ordered by $\sqsubseteq_t$, and c_{u_i} is a fresh Boolean *control variable* representing the node u_i.

If $encode(\cdot, \cdot)$ is *true*, then ARGCOND(v_{err}) is satisfiable iff there exists a path from v_{en} to v_{err} in $\mathcal{A}$. $encode(u, v)$ is a formula describing the semantics of an edge (u, v) that is used to restrict satisfying assignments of ARGCOND(v_{err}) to feasible executions. For example, for function foo in Fig. 1(a), we encode the statement on edge (v_4, v_6) as follows: $encode(v_4, v_6) = (x_1 = 0 \land x_3 = x_1)$. where x_1 is a fresh name for variable x, and $x_3 = x_1$ is used to equate the name of x at node v_6 (which is x_3) with the value of x after executing this edge.

For the purpose of presentation, we provide a simplified definition of *encode*. In practice, we use the SSA-based encoding defined in [18]. Let $\mathsf{SVar} = \{x_v \mid x \in \mathsf{Var} \land v \in V\}$ be the set of variables that can appear in $encode(\cdot, \cdot)$. That is, for each variable $x \in \mathsf{Var}$ and node $v \in V$, we create a *symbolic variable* $x_v \in \mathsf{SVar}$. The map $\mathsf{SMap} : \mathsf{SVar} \to \mathsf{Var}$ associates each x_v with its program variable x. The predicate $\mathsf{inScope} : \mathsf{SVar} \times V$ is defined so that $\mathsf{inScope}(x_u, v)$ holds iff $u = v$. If $\mathsf{inScope}(x_u, v)$ holds, we say that x_u is *in-scope* at node v; otherwise, it is *out-of-scope* at v.

Definition 3 (encode). *For an edge $(u, v) \in E$: If $\tau(u, v)$ is an assignment statement $x = E$, then* $\mathrm{encode}(u, v) = (x_v = E[x \leftarrow x_u]) \land \forall y \in \mathsf{Var} \cdot y \neq x \Rightarrow y_v = y_u$. *If $\tau(u, v)$ is an assume statement $\mathtt{assume}(Q)$, then* $\mathrm{encode}(u, v) = Q[x \leftarrow x_u \mid x \in \mathrm{var}(Q)] \land \forall y \in \mathsf{Var} \cdot y_v = y_u$, *where $\mathrm{var}(Q)$ is the set of variables appearing in Q.*

For example, for an edge $(u, v) \in E$ s.t. $\tau(u, v)$ is x = x + 1, $encode(u, v) = x_v = x_u + 1 \land y_v = y_u$, assuming $\mathsf{Var} = \{x, y\}$.

Lemma 1. *Given an ARG $\mathcal{A}$, there exists a total onto map from satisfying assignments of $\mathrm{ArgCond}(v_{\mathrm{err}})$ to feasible program executions from v_{en} to v_{err}.*

Labels from Interpolants. Given an ARG $\mathcal{A}$ with error node v_{err}, when $\mathrm{ArgCond}(v_{\mathrm{err}})$ is unsatisfiable, REFINE must return a set of labels for the nodes of the ARG that satisfy well-labelledness conditions and the specification of RE-FINE (Definitions 1 and 2). We now show how to extract such labels from an interpolant sequence of $\mathrm{ArgCond}(v_{\mathrm{err}})$. For a sequence of formulas $A_1, \ldots, A_n$ s.t. $\bigwedge_{i \in [1,n]} A_i$ is UNSAT, an *interpolant sequence* [24,10] $I_1, \ldots, I_{n+1}$ is defined as follows: (1) $I_1 \equiv true$, (2) $\forall i \in [1, n] \cdot I_i \land A_i \Rightarrow I_{i+1}$, (3) I_i is over the variables shared between $A_1, \ldots, A_{i-1}$ and $A_i, \ldots, A_n$, and (4) $I_{n+1} \equiv false$.

Let $I_1, \ldots, I_{n+1}$ be an interpolant sequence for the sequence of formulas $(c_{u_1} \land \mu_1) \land \mu_2 \land \cdots \land \mu_n$ constituting $\mathrm{ArgCond}(v_{\mathrm{err}})$. By definition, an interpolant I_i is an over-approximation of the set of states at nodes in $u_i, \ldots, u_n$ that are directly reachable from states at nodes in $u_1, \ldots, u_{i-1}$. For node u_i, this includes all states reachable at u_i since all incoming edges to u_i are from nodes that are topologically before it, i.e., $u_1, \ldots, u_{i-1}$.

Example 2. Consider node v_2' in Figure 1(c). An interpolant for v_2' is $I_2' = (c_{v_2'} \land x_3 \geq 0) \lor (c_{v_7} \land x_4 \geq 0)$. Informally, I_2' specifies that either execution reaches v_2' with $x_4 \geq 0$, or it reaches v_7 with with $x_4 \geq 0$. v_7 states appear in the formula because v_7 is directly reachable from node v_2 which comes before v_2' in the topological order. $\qquad\Box$

For each node u_i, our goal is to extract the set of reachable states at u_i from the interpolant I_i. For instance, for node v_2' from Example 2, we want to extract

$x_3 \geq 0$, the set of reachable states at v'_2, from the interpolant I'_2. To do so, we use the following transformation:

$$\text{CLEAN}(I_i) \triangleq$$
$$\forall\{x \mid x \in var(I_i) \wedge \neg inScope(x, u_i)\} \cdot \forall\{c_{u_j} \mid u_j \in V\} \cdot I[c_{u_i} \leftarrow \top], \quad (3)$$

where $var(I_i)$ is the set of variables appearing in I_i.

Example 3. Continuing Example 2, $\text{CLEAN}(I'_2) = \forall x_4, c_{v_7} \cdot I[c_{v'_2} \leftarrow \top] = x_3 \geq 0$. x_4 is quantified out since it is out-of-scope at v'_2. By replacing each variable y in the resulting formula with $\text{SMap}(y)$, we get the label $\{x \geq 0\}$ for v'_2, as shown in Figure 1(c).

By definition, $\text{CLEAN}(I_i)$ is a formula over the variables in-scope at u_i. Theorem 3 states that the labels produced by CLEAN result in a safe ARG and satisfy well-labelledness properties. That is, for any two nodes u_i, u_j, where there is an edge (u_i, u_j), the labels produced for u_i and u_j form a Hoare triple w.r.t the statement $\tau(u_i, u_j)$ encoded as $encode(u_i, u_j)$.

Theorem 3. *Let* $I'_k = \text{CLEAN}(I_k)$. *(a) If* $k = 1$, *then* $I'_k \equiv$ *true, and if* $k = n$, *then* $I'_k \equiv$ *false. (b) For any two nodes* $u_i, u_j \in V$ *s.t.* $(u_i, u_j) \in E$, $I'_i \wedge encode(u_i, u_j) \Rightarrow I'_j$, *where* $I'_j = \text{CLEAN}(I_j)$.

Proof. Part (a) follows from the definition of an interpolant sequence. Part (b):

$$I_i \wedge \mu_i \wedge \cdots \wedge \mu_{j-1} \Rightarrow I_j$$
(set c_{u_i} to $\top$ and logic)
$$\Rightarrow I_i[c_{u_i} \leftarrow \top] \wedge c_{u_j} \wedge encode(u_i, u_j) \wedge \mu_{i+1} \wedge \cdots \wedge \mu_{j-1} \Rightarrow I_j$$
(let $\Pi = \{c_{u_{i+1}}, \ldots, c_{u_{j-1}}\}$)
$$\Rightarrow I_i[c_{u_i} \leftarrow \top, \Pi \leftarrow \bot] \wedge c_{u_j} \wedge encode(u_i, u_j) \Rightarrow I_j$$
(set c_{u_j} to $\top$)
$$\Rightarrow I_i[\Pi \leftarrow \bot, c_{u_i} \leftarrow \top, c_{u_j} \leftarrow \top] \wedge encode(u_i, u_j) \Rightarrow I_j[c_{u_j} \leftarrow \top]$$
(use $(\forall x.f) \Rightarrow f$)
$$\Rightarrow I'_i \wedge encode(u_i, u_j) \Rightarrow I_j[c_{u_j} \leftarrow \top]$$
(out-of-scope variables of u_j are not in the antecedent)
$$\Rightarrow I'_i \wedge encode(u_i, u_j) \Rightarrow I'_j \qquad \square$$

Finally, REFINE returns the labeling map $\{u_i \mapsto \text{CLEAN}(I_i)' \mid i \in [1, n+1]\}$, where $u_{n+1} = v_{\text{err}}$ and $\text{CLEAN}(I_i)' = \text{CLEAN}(I_i)[x \leftarrow \text{SMap}(x) \mid x \in \text{SVar}]$.

In summary, our refinement technique uses a *single* SMT query φ to decide feasibility of *all* unsafe executions of an ARG $\mathcal{A}$. When φ is unsatisfiable, it extracts a new labeling for $\mathcal{A}$ that rules out all infeasible unsafe executions from an interpolant sequence of φ .

6   Implementation and Evaluation

Implementation. We have implemented UFO in the LLVM compiler infrastructure [22] and used it to verify properties of C programs from the 2012

Competition on Software Verification [5]. We used MATHSAT4 [9] for SMT-checking and interpolation, and Z3 [25] for quantifier elimination. Our implementation, benchmarks, and complete experimental results are available at http://www.cs.toronto.edu/~aws/ufo.

We used LLVM to heavily optimize all input programs prior to analysis. Because the benchmarks are meant for verification tools, these optimizations might be unsound with respect to the intended verification semantics. However, in all but one case (pipeline), our verification results are as expected: we find a bug in buggy programs, and prove safety of safe ones. Furthermore, we have implemented a proof and a counterexample checker that verify that the results produced by UFO are sound with respect to our semantics of C. All results discussed here have been validated by an appropriate checker.

Evaluation. For the evaluation, we used the ntdrivers-simplified, ssh-simplified, and systemc benchmarks from [5], and the pacemaker benchmarks from [1]. Overall, we had 105 C programs: 48 safe and 57 buggy. All experiments were conducted on an Intel Xeon 2.66GHz processor running a 64-bit Linux, with a 300 second time and 4GB memory limits per program, respectively.

We have evaluated 5 configurations of UFO: (1) a pure UD, called UUFO, where POST always returns *true*; (2) with Cartesian predicate abstraction, called CPUFO; (3) with Boolean predicate abstraction, called BPUFO; (4) a pure OD with Cartesian predicate abstraction, called CP, and a pure OD with Boolean predicate abstraction, called BP. Note that Boolean predicate abstraction is more precise, but is exponentially more expensive than Cartesian abstraction.

The results are summarized in Table 1. For each configuration, we show the number of instances solved (#SOLVED), number of safe (#SAFE) and unsafe (#UNSAFE) instances solved, number of unsound results (#UNSOUND), where a result is unsound if it does not agree with the benchmark categorization in [5], and the total time.

On these benchmarks, CPUFO performs significantly better than all other configurations, both in total time and number of instances solved. The UUFO configuration is a close second. We have also compared our results against the UD tool WOLVERINE [21] that implements a version of IMPACT [24] algorithm. All configurations of UFO perform significantly better than WOLVERINE.

Furthermore, we compared our tool against the results of the extensive study reported in [7] for the state-of-the-art OD tools CPACHECKER [7], BLAST [6], and SATABS [12]. Both UUFO and CPUFO configurations are able to solve all buggy transmitter examples. However, according to [7], CPACHECKER, BLAST, and SATABS are unable to solve most of these examples, even though they are run on a faster processor with a 900s time limit and 16GB of memory. Additionally, on the ntdrivers-simplified, UUFO, CPUFO and BPUFO perform significantly better than all of the aforementioned tools.

Table 2 presents a detailed comparison between different configurations of UFO on 32 (out of 105) programs. In the table, we show time, number of iterations (#ITER), and time spent in interpolation (#ITIME) and post (#PTIME), respectively. Times taken by other parts of the algorithm (such as CLEAN) were

Table 1. Summary of results on 105 C programs

Algorithm	#Solved	#Safe	#Unsafe	#Unsound	Total Time (s)
uUfo	78	22	56	0	8,289
cpUfo	79	22	57	1	7,838
bpUfo	69	17	52	1	11,260
Cp	49	10	39	0	15,363
Bp	71	19	52	1	10,018
Wolverine	38	18	20	5	19,753

insignificant and are omitted. Cp configuration was not able to solve all but one of these examples, and is omitted as well.

In this sample, cpUfo is best overall, however, it is often not the fastest approach on any given example. This is representative of its performance over the whole benchmark. As expected, both uUfo and cpUfo spend most of their time in computing interpolants, while bpUfo and Bp spend most of their time in predicate abstraction.

The results show that there is clearly a synergy between UD and OD-driven parts of the analysis. For example, in toy1_BUG and s3_srvr_1a, predicate abstraction decreases the number of required iterations. Several of the buggy examples from the token_ring family cannot be solved by a UD-only uUfo configuration alone. However, there are also some interactions. For many of the safe cases that require a few iterations, uUfo performs better than other combinations. For many unsafe cases that bpUfo can solve, it performs much better alone than in a combination.

In summary, our results show that the novel UD-driven algorithm that underlies Ufo (uUfo configuration) is very effective compared to the state-of-the-art approaches. Furthermore, there is a clear synergy in combining UD and OD approaches, with cpUfo performing the best overall. However, there are also some interactions where the combination does not result in the best of the individual approaches. Managing these interactions effectively is the subject of future work.

7   Related Work

In this section, we place Ufo in the context of related work. Specifically, we compare it with the most related UD and OD verification techniques.

Ufo is based on a novel interpolation-driven verification algorithm. It extends Impact [24], by unrolling the program into a DAG instead of a tree and by using a single SMT query to both discharge all infeasible unsafe executions and to compute new labels. In effect, Ufo uses the SMT solver to enumerate acyclic program paths, whereas Impact enumerates those paths explicitly. Furthermore, Ufo extends Impact by using an abstract post operator during exploration. As we show in our experiments, this can lead to fewer iterations and faster verification.

We have recently developed an inter-procedural extension of Impact, called Whale [1]. Whale works on loop-free recursive programs and uses interpolants

Table 2. Results of running Ufo on 33 programs from the benchmarks. All times are in seconds.

Program	uUfo Time	Iter	iTime	cpUfo Time	#Iter	iTime	pTime	bpUfo Time	Iter	iTime	pTime	Bp Time	Iter	iTime	pTime
						Unsafe Programs									
kundu1	-	-	-	24.22	4	20.3	1.84	122.88	4	56.9	54.66	33.39	3	20.23	10.95
kundu2	1.24	2	1.16	2.74	2	2.08	0.6	8.15	2	1.2	5.66	8.6	2	3.49	4.3
s3_srvr_11	1.91	4	1.67	2.78	4	1.58	0.89	118.41	4	1.72	112.76	4.25	3	2.6	1.37
s3_srvr_12	4.17	4	3.85	5.07	3	3.44	1.36	5.36	3	3.58	1.44	8.19	3	5.85	1.91
token_ring.08	12.34	4	11.84	13.5	4	11.07	1.91	19.64	3	3.7	14.62	14.15	3	1.85	11.12
token_ring.09	12.54	4	11.98	22.66	4	19.72	2.35	-	-	-	-	167.49	3	3.85	157.58
token_ring.10	15.6	4	15.05	14.02	3	11.99	1.69	-	-	-	-	-	-	-	-
token_ring.11	29.69	4	29.08	22.47	4	18.52	3.19	156.76	3	4.57	145.99	66.59	3	4.21	58.68
token_ring.12	26.94	4	26.31	13.98	3	11.45	2.15	-	-	-	-	-	-	-	-
token_ring.13	36.56	4	35.76	34.17	4	29.38	4.02	-	-	-	-	-	-	-	-
token_ring.14	10.3	3	9.99	33.49	4	29.17	3.59	-	-	-	-	-	-	-	-
token_ring.15	51.79	4	51.11	34.19	4	29.18	4.17	-	-	-	-	-	-	-	-
toy1	96.49	10	89.08	79	9	68.04	6.98	13.54	3	4.77	7.96	-	-	-	-
toy2	12.83	5	12.24	60.73	8	50.71	6.14	-	-	-	-	-	-	-	-
ddd3	0.66	4	0.5	0.19	2	0.04	0.11	0.18	2	0.03	0.11	0.27	2	0.05	0.2
						Safe Programs									
pc_sfifo_1	-	-	-	-	-	-	-	3.51	3	2.24	0.79	-	-	-	-
s3_clnt_1	11.03	10	8.18	15.68	10	8.2	4.5	-	-	-	-	14.52	5	1.92	8.21
s3_clnt_2	16	10	11.35	20.02	10	10.86	4.67	-	-	-	-	-	-	-	-
s3_clnt_3org	28.87	11	17.35	37.08	11	17.6	8.17	-	-	-	-	-	-	-	-
s3_clnt_3	13.02	10	9.14	17.42	10	9.01	4.45	-	-	-	-	-	-	-	-
s3_clnt_4	13.4	10	9.62	17.45	10	9.16	4.58	-	-	-	-	-	-	-	-
s3_srvr_1a	5.2	10	2.95	5.16	8	2.32	1.07	0.76	4	0.17	0.39	0.43	3	0.07	0.26
s3_srvr_1b	1.37	7	1	2.9	7	1.71	0.69	0.89	5	0.47	0.28	-	-	-	-
s3_srvr_2	171.15	17	116.82	184.01	17	112.65	18.65	-	-	-	-	-	-	-	-
s3_srvr_3	133.07	17	99.96	147.55	17	98.69	16.02	-	-	-	-	33.71	5	1.07	21.18
s3_srvr_4	-	-	-	-	-	-	-	-	-	-	-	8	4	0.74	5.36
s3_srvr_8	101.4	14	76.6	115.08	14	73.9	17.62	-	-	-	-	-	-	-	-
token_ring.01	98.18	18	81.58	23.64	10	17.72	1.78	0.69	4	0.27	0.23	0.69	4	0.19	0.31
token_ring.02	-	-	-	-	-	-	-	2.15	4	0.71	0.7	2.63	4	1.06	0.59
token_ring.03	-	-	-	-	-	-	-	76.18	4	4.74	37.66	-	-	-	-
token_ring.04	-	-	-	-	-	-	-	-	-	-	-	152.62	4	10.82	2.45
token_ring.05	-	-	-	-	-	-	-	-	-	-	-	149.35	4	8.25	97.48

derived from DAG encodings of a procedure to compute procedure summaries. The intra-procedural technique presented here is orthogonal to that of Whale, and the interpolation-based refinement presented here is new. It would be interesting to see if the combination of UD and OD in Ufo can be adapted to the inter-procedural setting of Whale.

Dash [4] uses weakest-precondition (WP) over infeasible program paths to partition (i.e., refine) a program's state space. In contrast, Ufo refines multiple program paths at the same time. Moreover, Ufo uses interpolants for refinement, an approach that has been shown to provide more relevant predicates than WP-based refinement [20]. We believe that our multi-path refinement strategy can be easily implemented in Dash to generate test-cases for multiple frontiers or split different regions at the same time.

At a high level, Ufo is similar to the abstract algorithm Smash [15], in the sense that it combines over- and under-approximations. In [15], the only instantiation of Smash that is experimented with is an under-approximation-driven algorithm based on Dash [4], where no abstract domain is used. In this paper, we have experimented with multiple instances of Ufo, ranging from UD to OD. Other differences between SMASH and UFO include the fact that UFO refines multiple paths at the same time, whereas SMASH considers a single path at a time.

Lazy abstraction [19] is the closest OD algorithm to UFO. UFO can be seen as extending lazy abstraction in two directions. First, UFO unrolls a program into a DAG (and not a tree). Second, it uses all the labels produced by interpolation, and only applies predicate abstraction to the "frontier" nodes that are not known to reach an error location.

We are not the first to apply interpolation to multiple program paths. In [14], Esparza et al. use interpolants to find predicates that eliminate multiple spurious counterexamples simultaneously. Their algorithm uses an eager abstraction-refinement loop, and a BDD-based interpolation procedure. In contrast, the refinement in UFO uses an SMT-solvers-based interpolation procedure. It is not clear whether BDD-based techniques of [14] can be efficiently adapted to the SMT-based setting.

8   Conclusion

Software model checkers can be divided into over-approximation-driven (OD) (e.g., SLAM [3]) and under-approximation-driven (UD) (e.g., IMPACT [24]). An OD software model-checker maintains an abstraction of the transition relation of a program and uses abstract reachability to build an inductive invariant or find a counterexample. A UD model checker avoids the (potentially expensive) abstraction step, and instead attempts to guess an inductive invariant by generalizing from finite paths through the control-flow graph of the program. Until now, combinations of these techniques have not been explored.

In this paper, we presented UFO – a model checking algorithm that tightly couples UD and OD approaches. At the core of UFO is a UD algorithm that is parameterized by an abstract POST operator, and a novel interpolation-based refinement procedure. The refinement procedure uses a *single* SMT query to decide feasibility of *all* unsafe executions in an unrolling of a program's CFG.

We have implemented UFO within LLVM [22], and experimented with two variants of POST based on Boolean and Cartesian predicate abstractions. We have evaluated our implementation on benchmarks from the Competition on Software Verification [5]. Our results show that UFO is very competitive compared to the state-of-the-art. There is a clear synergy in combining UD and OD approaches. However, there are also undesirable interactions. We believe that this work opens new avenues for exploring combinations of UD- and OD-based approaches to verification, a direction we hope to explore in the future.

References

1. Albarghouthi, A., Gurfinkel, A., Chechik, M.: Whale: An Interpolation-Based Algorithm for Inter-procedural Verification. In: Kuncak, V., Rybalchenko, A. (eds.) VMCAI 2012. LNCS, vol. 7148, pp. 39–55. Springer, Heidelberg (2012)

2. Albarghouthi, A., Gurfinkel, A., Wei, O., Chechik, M.: Abstract Analysis of Symbolic Executions. In: Touili, T., Cook, B., Jackson, P. (eds.) CAV 2010. LNCS, vol. 6174, pp. 495–510. Springer, Heidelberg (2010)
3. Ball, T., Rajamani, S.K.: The SLAM Toolkit. In: Berry, G., Comon, H., Finkel, A. (eds.) CAV 2001. LNCS, vol. 2102, pp. 260–264. Springer, Heidelberg (2001)
4. Beckman, N.E., Nori, A.V., Rajamani, S.K., Simmons, R.J.: "Proofs from Tests". In: Proc. of ISSTA 2008, pp. 3–14 (2008)
5. Beyer, D.: Competition On Software Verification (2012), http://sv-comp.sosy-lab.org/
6. Beyer, D., Henzinger, T.A., Jhala, R., Majumdar, R.: The Software Model Checker BLAST. STTT 9(5-6), 505–525 (2007)
7. Beyer, D., Keremoglu, M.E.: CPAchecker: A Tool for Configurable Software Verification. In: Gopalakrishnan, G., Qadeer, S. (eds.) CAV 2011. LNCS, vol. 6806, pp. 184–190. Springer, Heidelberg (2011)
8. Bourdoncle, F.: Efficient Chaotic Iteration Strategies with Widenings. In: Pottosin, I.V., Bjorner, D., Broy, M. (eds.) FMP&TA 1993. LNCS, vol. 735, pp. 128–141. Springer, Heidelberg (1993)
9. Bruttomesso, R., Cimatti, A., Franzén, A., Griggio, A., Sebastiani, R.: The MathSAT 4 SMT Solver. In: Gupta, A., Malik, S. (eds.) CAV 2008. LNCS, vol. 5123, pp. 299–303. Springer, Heidelberg (2008)
10. Cimatti, A., Griggio, A., Sebastiani, R.: Efficient Generation of Craig Interpolants in Satisfiability Modulo Theories. ACM Trans. Comput. Log. 12(1), 7 (2010)
11. Clarke, E., Grumberg, O., Jha, S., Lu, Y., Veith, H.: Counterexample-Guided Abstraction Refinement. In: Emerson, E.A., Sistla, A.P. (eds.) CAV 2000. LNCS, vol. 1855, pp. 154–169. Springer, Heidelberg (2000)
12. Clarke, E., Kroning, D., Sharygina, N., Yorav, K.: SATABS: SAT-Based Predicate Abstraction for ANSI-C. In: Halbwachs, N., Zuck, L.D. (eds.) TACAS 2005. LNCS, vol. 3440, pp. 570–574. Springer, Heidelberg (2005)
13. Craig, W.: Three Uses of the Herbrand-Gentzen Theorem in Relating Model Theory and Proof Theory. J. of Symbolic Logic 22(3), 269–285 (1957)
14. Esparza, J., Kiefer, S., Schwoon, S.: Abstraction Refinement with Craig Interpolation and Symbolic Pushdown Systems. In: Hermanns, H. (ed.) TACAS 2006. LNCS, vol. 3920, pp. 489–503. Springer, Heidelberg (2006)
15. Godefroid, P., Nori, A.V., Rajamani, S.K., Tetali, S.D.: Compositional May-Must Program Analysis: Unleashing the Power of Alternation. In: Proc. of POPL 2010, pp. 43–56 (2010)
16. Graf, S., Saïdi, H.: Construction of Abstract State Graphs with PVS. In: Grumberg, O. (ed.) CAV 1997. LNCS, vol. 1254, pp. 72–83. Springer, Heidelberg (1997)
17. Gulavani, B., Henzinger, T., Kannan, Y., Nori, A., Rajamani, S.: SYNERGY: a New Algorithm for Property Checking. In: FSE 2006, pp. 117–127 (2006)
18. Gurfinkel, A., Chaki, S., Sapra, S.: Efficient Predicate Abstraction of Program Summaries. In: Bobaru, M., Havelund, K., Holzmann, G.J., Joshi, R. (eds.) NFM 2011. LNCS, vol. 6617, pp. 131–145. Springer, Heidelberg (2011)
19. Henzinger, T., Jhala, R., Majumdar, R., Sutre, G.: Lazy Abstraction. In: Proc. of POPL 2002, pp. 58–70 (2002)
20. Henzinger, T.A., Jhala, R., Majumdar, R., McMillan, K.L.: Abstractions from Proofs. In: Proc. of POPL 2004, pp. 232–244 (2004)
21. Kroening, D., Weissenbacher, G.: Interpolation-Based Software Verification with WOLVERINE. In: Gopalakrishnan, G., Qadeer, S. (eds.) CAV 2011. LNCS, vol. 6806, pp. 573–578. Springer, Heidelberg (2011)

22. Lattner, C., Adve, V.: LLVM: A Compilation Framework for Lifelong Program Analysis & Transformation. In: CGO 2004 (2004)
23. McMillan, K.L.: Lazy Annotation for Program Testing and Verification. In: Touili, T., Cook, B., Jackson, P. (eds.) CAV 2010. LNCS, vol. 6174, pp. 104–118. Springer, Heidelberg (2010)
24. McMillan, K.L.: Lazy Abstraction with Interpolants. In: Ball, T., Jones, R.B. (eds.) CAV 2006. LNCS, vol. 4144, pp. 123–136. Springer, Heidelberg (2006)
25. de Moura, L., Bjørner, N.: Z3: An Efficient SMT Solver. In: Ramakrishnan, C.R., Rehof, J. (eds.) TACAS 2008. LNCS, vol. 4963, pp. 337–340. Springer, Heidelberg (2008)

Automated Analysis of AODV Using UPPAAL[*]

Ansgar Fehnker[1,4], Rob van Glabbeek[1,4], Peter Höfner[1,4], Annabelle McIver[1,2], Marius Portmann[1,3], and Wee Lum Tan[1,3]

[1] NICTA
[2] Department of Computing, Macquarie University
[3] School of ITEE, The University of Queensland
[4] Computer Science and Engineering, University of New South Wales

Abstract. This paper describes an automated, formal and rigorous analysis of the Ad hoc On-Demand Distance Vector (AODV) routing protocol, a popular protocol used in wireless mesh networks.

We give a brief overview of a model of AODV implemented in the UPPAAL model checker. It is derived from a process-algebraic model which reflects precisely the intention of AODV and accurately captures the protocol specification. Furthermore, we describe experiments carried out to explore AODV's behaviour in all network topologies up to 5 nodes. We were able to automatically locate problematic and undesirable behaviours. This is in particular useful to discover protocol limitations and to develop improved variants. This use of model checking as a diagnostic tool complements other formal-methods-based protocol modelling and verification techniques, such as process algebra.

1 Introduction

Route finding and maintenance are critical for the performance of networked systems, particularly when mobility can lead to highly dynamic and unpredictable environments; such operating contexts are typical in wireless mesh networks (WMNs). Hence correctness and good performance are strong requirements of routing algorithms. The Ad hoc On-Demand Distance Vector (AODV) routing protocol [12] is a widely used routing protocol designed for WMNs and mobile ad hoc networks (MANETs). It is one of the four protocols defined in an RFC (Request for Comments) document by the IETF MANET working group. AODV also forms the basis of new WMN routing protocols, like the upcoming IEEE 802.11s wireless mesh network standard [8].

Usually, routing protocols are optimised to achieve key objectives such as providing self-organising capability, overall reliability and performance in typical network scenarios. Additionally, it is important to guarantee protocol properties such as loop freedom for *all* scenarios, including non-typical, unanticipated ones. This is particularly relevant for highly dynamic MANETs and WMNs.

The traditional approaches for the analysis of MANET and WMN routing protocols are simulation and test-bed experiments. While these are important

---

[*] First steps towards this analysis appeared in [6].

C. Flanagan and B. König (Eds.): TACAS 2012, LNCS 7214, pp. 173–187, 2012.

and valid methods for protocol evaluation, there are limitations: they are re-source intensive and time-consuming. The challenges of extensive experimental evaluation are illustrated by recent discoveries of limitations of protocols that have been under intense scrutiny over many years. An example is [10].

We believe that formal methods in general and model checking in particular can help in this regard. Model checking is a powerful method that can be used to validate key correctness properties in finite representations of a formal system model. In the case that a property is found not to hold, the model checker produces evidence for the fault in the form of a "counter-example" summarising the circumstances leading to it. Such diagnostic information provides important insights into the cause and correction of these failures.

In [5], we specified the AODV routing protocol in the process algebra AWN. The specification follows well-known programming constructs and lends itself well for comparison with the original specification of the protocol in English. Based on such a comparison we believe that the AWN model provides a com-plete and accurate formal specification of the core functionality of AODV. In developing the formal specification, we discovered a number of ambiguities in the IETF RFC [12]. Our process algebraic formalisation captures these by sev-eral interpretations, each with slightly different AWN code.

In this paper we follow an interpretation of the RFC, which we believe to be the closest to the spirit of the AODV routing protocol. We show how to obtain executable versions of this AWN specification, in the language of the UPPAAL model checker [1,9]. By deriving the UPPAAL model from the AWN model, the accuracy of the AWN model is transferred to the UPPAAL model.

The executable UPPAAL model is used to confirm and discover the presence of undesirable behaviour. We check important properties against all topologies of up to 5 nodes, which also includes dynamic topologies with one link going up or down. This exhaustive search confirmed known and revealed new problems of AODV, and let us quantify in how many topologies a particular error can occur. Subsequently, the same experiments for modifications of AODV showed the pro-posed modifications can all but eliminate certain problems for static topologies, and significantly reduce them for dynamic topologies. The automated analysis of routing protocols presented in this paper combined with formal reasoning in AWN provides a powerful tool for the development and rigorous evaluation of new protocols and variations, and improvements of existing ones.

2   Ad Hoc On-Demand Distance Vector Routing Protocol

2.1   The Basic Routine

AODV [12] is a widely used routing protocol designed for WMNs and MANETs. It is a reactive routing protocol, where the route between a source and a desti-nation node is established on an on-demand basis. A route discovery process is initiated when a source node s has data to send to a destination node d, but has no valid corresponding routing table entry. In this case, node s broadcasts a route request (RREQ) message in the network. The RREQ message is re-broadcast

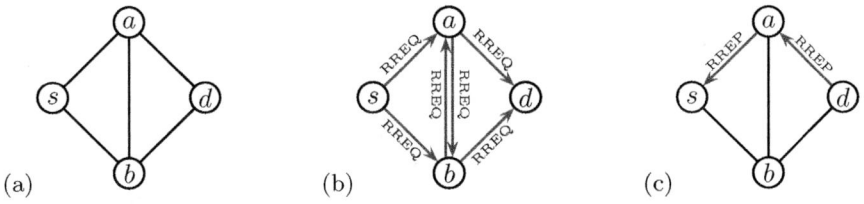

Fig. 1. Example network topology

and forwarded by other intermediate nodes in the network, until it reaches the destination node d (or an intermediate node that has a valid route to node d). Every node that receives the RREQ message will create a routing table entry to establish a *reverse route* back to node s. In response to the RREQ message, the destination node d (or an intermediate node that has a valid route to node d) unicasts a route reply (RREP) message back along the previously established reverse route. At the end of this route discovery process, an end-to-end route between the source node s and destination node d is established. Usually, all nodes on this route have a routing table entry to both the source node s and destination node d. An example topology, indicating which nodes are in transmission range of each other, as well as the flow of RREQ and RREP messages, is given in Figure 1. In the event of link and route breaks, AODV uses route error (RERR) messages to inform affected nodes. Sequence numbers are another important aspect of AODV, and are used to indicate the freshness of routing table entries for the purpose of preventing routing loops.

2.2   Process Algebraic Model of AODV

The process algebra AWN [4,5] has been developed specifically for modelling WMN routing protocols. It is designed in a way to be easily readable and treats three necessary features of WMNs protocols: *data structures*, *local broadcast*, and *conditional unicast*. Data structures are used to model routing tables etc.; local broadcast models message sending to *all* directly connected nodes; and conditional unicast models the message sending to one particular node and chooses a continuation process dependent on whether the message is successfully delivered.

In AWN, delivery of broadcast messages is "guaranteed", i.e., they are received by any neighbour that is directly connected. The abstraction to a guaranteed broadcast enables us to interpret a failure of message delivery (under assumptions on the network topology) as an imperfection in the protocol, rather than as a consequence of unreliable communication. Section 4.3, for example, describes a simple network topology and a scenario for which AODV fails to discover a route, even if broadcast is guaranteed. The failure is a shortcoming of the protocol itself, and cannot be excused by unreliable communication.

Conditional unicast models an abstraction of an acknowledgment-of-receipt mechanism that is typical for unicast communication but absent in broadcast communication, as implemented by the link layer of relevant wireless standards such as IEEE 802.11. The AWN model captures the bifurcation depending on the success of the unicast, while abstracting from all implementation details.

In [5], we used AWN to model AODV according to the IETF RFC [12]. The model captures all core functionalities as well as the interface to higher protocol layers via the injection and delivery of application layer data, and the forwarding of data packets at intermediate nodes. Although the latter is not part of the AODV protocol specification, it is necessary for a practical model of any reactive routing protocol where protocol activity is triggered via the sending and forwarding of data packets. In addition, our model contains neither ambiguities nor contradictions, both of which are often present in specifications written in natural languages, such as in the RFC3561 (see e.g. [5]).

The AWN model of AODV contains a main process, called AODV, for every node of the network, which handles messages received and calls the appropriate process to handle them. The process also handles the forwarding of any queued data packet if a valid route to its destination is known. Four other processes handle one particular message type each, like RREQ. The network as a whole is modelled as a parallel composition of these processes. Special primitives allow us to express whether two nodes are connected. Full details of the process algebra description on which our UPPAAL model is based can be found in [5].

3   Modelling AODV in UPPAAL

UPPAAL [1,9] is an established model checker for *networks of timed automata*, used in particular for protocol verification. We use UPPAAL for the following reasons: (1) UPPAAL provides two synchronisation mechanisms—binary and broadcast synchronisation, which translate to uni- and broadcast communication; (2) it provides common data structures, such as arrays and structs, and a C-like programming language to define updates on these data structures; (3) in the future, AWN (and therefore also our models) will be extended with time and probability—UPPAAL provides mechanisms and tools for both.

Our process-algebraic model of AODV has been used to prove essential properties, such as loop freedom for popular interpretations of [12]—independent of a particular topology. The UPPAAL model is derived from the AWN specification that comes closest to the spirit of the AODV routing protocol.

Section 3.2 explains the translation and the simplifying assumptions in detail.

3.1   UPPAAL Automata

Since our models do not yet use time (or probabilities) they are simply *networks of automata* with guards. The state of the system is determined, in part, by the values of data variables that can be either shared between automata, or local. We assume a data structure with several types, variables ranging over these types, operators and predicates. Common Boolean and arithmetic expressions are used to denote data values and statements about them.

Each automaton is a graph, with locations, and edges between locations. Every edge has a guard, optionally a synchronisation label, and an update. Synchronisation occurs via so-called channels; for each channel a there is one label $a!$ to denote the sender, and $a?$ to denote the receiver. Transitions without labels are internal; all other transitions use one of two types of synchronisation.

In *binary handshake* synchronisation, one automaton having an edge with a label that has the suffix ! synchronises with another automaton with an edge having the same label that has a ?-suffix. These two transitions synchronise when both guards are true in the current state, and only then. When the transition is taken both locations change, and the updates will be applied to the state variables; first the updates on the !-edge, then the updates on the ?-edge. If there is more than one possible pair, then the transition is selected non-deterministically.

In *broadcast* synchronisation, one automaton with a !-labelled edge synchronises with a set of other automata that all have an edge with a matching ?-label. The initiating automaton can change its location, and apply its update, if the guard on its edge evaluates to true. It does not require a second synchronising automaton. Automata with a matching ?-labelled edge have to synchronise if their guard is currently true. They change their location and update the state. The automaton with the !-edge will update the state first, followed by the other automata in some lexicographic order. If more than one automaton can initiate a transition on an !-edge, the choice will be made non-deterministically.

3.2   From AWN to UPPAAL

Every node in the network is modelled as a single automaton, each having its own data structures such as a routing table and message buffer. The implementation of the data structure defined in AWN is straightforward, since both AWN and UPPAAL allow C-style data structures. A routing table `rt` for example is an array of entries, one entry for every node. An entry is given by the data type

```
typedef struct
{ SQN dsn;       //destination sequence number
  bool flag;     //validity of a routing table entry
  int hops;      //distance (hop count) to the destination
  IP nhop;       //next hop (is 0 if no route)
} rtentry;
```

where `SQN` denotes a data type for sequence numbers and `IP` denotes one for all IP address. In our model, these types are mapped to integers.

The local message buffer is modelled as an array `msglocal`. UPPAAL will warn if during model checking an out-of-bounds error occurs, i.e., if the array was too small. Each message is a struct with fields `msgtype` which can take values PKT, RREQ, RREP, or RERR, integer `hops` for the distance from the originator of the message, sequence number `rreqid` to identify a route request, a destination IP `dip`, a destination sequence number `dsn`, an originator IP `oip`, an originator sequence number `osn`, and a sender IP `sip`. The model contains functions `addmsg`, `deletemsg` and `nextmsg`, to add a message, delete a message, or to return the type of the next message in the buffer.

Table 1. Excerpt of AWN spec for AODV. A few cases for RREQ handling.

AODV(ip,sn,rt,rreqs,store) $\stackrel{def}{=}$

1. /*depending on the message on top of the message queue, the node calls different processes*/
2. . . .
3. [msg = rreq(hops, rreqid, dip, dsn, oip, osn, sip) $\wedge$ (oip, rreqid) $\in$ rreqs]
4.   /*silently ignore RREQ, i.e. do nothing, except update the entry for the sender*/
5. $[\![$rt := update(rt, (sip, 0, val, 1, sip))$]\!]$. /*update the route to sip*/
6.   AODV(ip,sn,rt,rreqs,store)
7. + [msg = rreq(hops, rreqid, dip, dsn, oip, osn, sip) $\wedge$ (oip, rreqid) $\notin$ rreqs $\wedge$ dip = ip]
8.   /*answer the RREQ with a RREP*/
9. $[\![$rt := update(rt, (oip, osn, val, hops + 1, sip))$]\!]$ /*update the routing table*/
10. $[\![$rreqs := rreqs $\cup$ {(oip, rreqid)}$]\!]$ /*update the array of already seen RREQ*/
11. $[\![$sn := max(sn, dsn)$]\!]$ /*update the sqn of ip*/
12. $[\![$rt := update(rt, (sip, 0, val, 1, sip))$]\!]$ /*update the route to sip*/
13.   **unicast**(nhop(rt,oip),rrep(0,dip,sn,oip,ip)) .
14.   AODV(ip,sn,rt,rreqs,store)
15. + [msg = rreq(hops, rreqid, dip, dsn, oip, osn, sip) $\wedge$(oip, rreqid) $\notin$ rreqs $\wedge$ dip $\neq$ ip $\wedge$
        (dip $\notin$ vD(rt) $\vee$ sqn(rt,dip) $<$ dsn $\vee$ sqnf(rt,dip) = unk)]
16.   /*forward RREQ*/
17. $[\![$rt := update(rt, (oip, osn, val, hops + 1, sip))$]\!]$ /*update routing table*/
18. $[\![$rreqs := rreqs $\cup$ {(oip, rreqid)}$]\!]$ /*update the array of already seen RREQ*/
19. $[\![$rt := update(rt, (sip, 0, val, 1, sip))$]\!]$ /*update the route to the sender*/
20.   **broadcast**(rreq(hops + 1,rreqid,dip,max(sqn(rt, dip), dsn),oip,osn,ip)) .
21.   AODV(ip,sn,rt,rreqs,store)
22. + [rreq(hops, rreqid, dip, dsn, oip, osn, sip) $\wedge$ . . .]
23.   . . .

Connections between nodes are determined by a *connectivity graph*, which is specified by a Boolean-valued function `isconnected`. This graph presents one particular topology and is not derived from our AWN specification, since the specification is valid for *all* topologies. Communication is modelled as an atomic synchronised transition between a sender, on an !-edge, with a receiver, on a matching ?-edge. The guard of the sender depends on local data, e.g. buffer and routing table, while the guard of the receiver is `isconnected`. This means that in broadcast communication the sender will take the transition regardless of `isconnected`, while disconnected nodes will not synchronise. In unicast communication the transition is blocked if the intended recipient is not connected, but there is a matching broadcast transition that sends an error message in this case. When the transition is taken, the sender copies its message to a global variable `msgglobal`, and the receiver copies it subsequently to its local buffer `msglocal`.

AODV uses unicast for RREP and PKT messages, and broadcast for RERR and RREQ messages. To model unicast, the UPPAAL model has one binary handshake channel for every pair of nodes. For example, `rrep[i][j]` is used for transitions modelling the sending of a route reply from node i to j. To model broadcast, we use one broadcast channel for every node. For example, `rreq[i]` is used for the route requests of node i. To model new packets from i to j, generated by the user layer, the model contains a channel `newpkt[i][j]`.

The AWN model of Table 1 is an excerpt of the AODV specification presented in [5]—the full specification and a detailed explanation can be found there. The excerpt presented here differs slightly from the original model:[1] (1) we abstract

---

[1] It can be shown that the model presented here behaves identical to the AWN model in [4]; in other words, they are behaviourally equivalent.

Table 2. Excerpt of UPPAAL model. A few cases for RREQ handling.

```
 1. ...
 2. aodv -> aodv {
 3. guard nextmsg()==RREQ && rreqs[msglocal[0].oip][msglocal[0].rreqid];
 4. sync  tau[ip]?;
 5. assign sipupdate(), deletemsg();  },
 6. aodv -> aodv {
 7. guard nextmsg()==RREQ&&!rreqs[msglocal[0].oip][msglocal[0].rreqid]&&msglocal[0].dip==ip;
 8. sync  rrep[ip][oipnhop()]!;
 9. assign updatert(msglocal[0].oip,msglocal[0].osn,1,msglocal[0].hops+1,msglocal[0].sip),
10.        rreqs[msglocal[0].oip][msglocal[0].rreqid]=1,
11.        sn=max(sn,msglocal[0].dsn),
12.        sipupdate(),
13.        msgglobal=createrep(0,msglocal[0].dip,sn,msglocal[0].oip,ip), deletemsg();  },
14. aodv -> aodv {
15. guard nextmsg()==RREQ&&!rreqs[msglocal[0].oip][msglocal[0].rreqid]&&msglocal[0].dip!=ip
           && (!rt[msglocal[0].dip].flag || msglocal[0].dsn>rt[msglocal[0].dip].dsn
           || rt[msglocal[0].dip].dsn==0);
16. sync  rreq[ip]!;
17. assign updatert(msglocal[0].oip,msglocal[0].osn,1,msglocal[0].hops+1,msglocal[0].sip),
18.        rreqs[msglocal[0].oip][msglocal[0].rreqid]=1,
19.        sipupdate(),
20.        msgglobal=createreq(msglocal[0].hops+1,msglocal[0].rreqid,msglocal[0].dip,
           max(msglocal[0].dsn, rt[msglocal[0].dip].dsn),msglocal[0].oip,msglocal[0].osn,ip),
21.        deletemsg();  },
22. ...
```

from *precursors*, an additional data structure that is maintained by AODV (2) the model in [5] uses 6 different processes; here processes are inlined into the body of the main AODV process. This reduces the number of processes to one and yields an automaton with one control location; (3) the model in [5] uses nesting of conditions and updates, while this model has been flattened to correspond more closely with the limitations of the UPPAAL syntax—in UPPAAL the *guards* are evaluated before any update, AWN has no such restriction.

Table 1 depicts three of the cases in the AWN model for handling route requests. In each, a condition is checked, the routing tables and local data are updated, and it returns to the main AODV process AODV(ip, sn, rt, rreqs, store). Table 2 shows the corresponding edges from the UPPAAL model, one edge for every case. Like the AWN model, which goes from the process AODV to AODV, the UPPAAL model will go from control location aodv to itself (Lines 2, 6 and 14).

Each edge evaluates a guard in Lines 3, 7 and 15 in Table 2. These line numbers, and the line numbers mentioned in the remainder of this section correspond to the same line number in Table 1. Whenever the AWN specification uses set membership ((oip, rreqid) $\in$ rreqs), the UPPAAL model uses a 2-dimensional Boolean array rreqs to encode membership; whenever the AWN model uses a flag to denote a *known* sequence number (sqnf(rt,dip) = unk), the UPPAAL model compares with a distinguished value (rt[msglocal[0].dip].dsn==0).

Depending on whether a case requires no transmission, unicast, or broadcast, the UPPAAL model synchronises on a tau, a binary, or a broadcast channel (Lines 4, 8 and 16). The tau channel for internal transitions allows for optimisations; it could have been left empty. We discuss this later in this section.

After synchronisation the state is updated. For all route request messages we update the routing table for the sender sip (Lines 5, 12 and 19). The fact that the message was received means that sender `sip` is one hop away. Except for the first case (Lines 4) the routing table is updated (Lines 9 and 17), and the route request is added to the set of processed route requests (Lines 10 and 18). In case that a node receives a request, and it is the destination, it increments its sequence number, if necessary (Line 11), before it sends a route reply.

The last two steps in the UPPAAL model that complete a transmission first create a new message and copy it to the global variable `msgglobal` (Lines 13 and 20), and then delete the first element of the local message buffer. In the AWN model, these steps are part of the communication primitives.

The full UPPAAL models a node by an automaton with one control location and 26 edges: 19 cases for processing the different routing messages, four cases for receiving routing messages—one case for each type—two cases for sending data packets, and one case for handling new data packets. The case distinction is complete, i.e at least one transition is enabled and process messages if the buffers and queues are not empty.

Both the UPPAAL and the AWN model maintain a FIFO buffer for incoming messages. Any newly generated message only depends on the content of messages previously received. This implies that the timing of internal transitions that discard incoming messages is not relevant for route discovery. The UPPAAL model exploits this fact and assigns a higher priority to internal transitions. To implement priorities we labelled those transitions `tau`. This is is an effective measure to reduce the state space, at the expense that UPPAAL is now unable to check liveness properties; for this paper this is not a limitation, as all properties can be expressed as safety properties.

4   Experiments

Our automated analysis of AODV considers 3 properties that relate to "route discovery" for all topologies up to 5 nodes, with up to one topology change, and scenarios with two new data packets.

4.1   Scenarios and Topologies

The experiments consider scenarios with two initial data packets in networks with up to 5 nodes. Initially all routing tables and buffers are empty. The originator and the destination of the data packets are identified as nodes A, B, or C. The new data packets may arrive as depicted in Figure 2. In the first scenario a packet from A to B is followed by a packet from A to C; in the second a packet from B to A by a packet from C to A; in the third a packet from A to B by a packet from B to C; and in the final scenario a packet from B to C by a packet from A to B. The originator of the first new packet initiates a route discovery process first, the originator of the second non-deterministically after the first. The different scenarios are implemented by a simple automaton, `tester`. Since the different topologies cover all possible permutations, these four

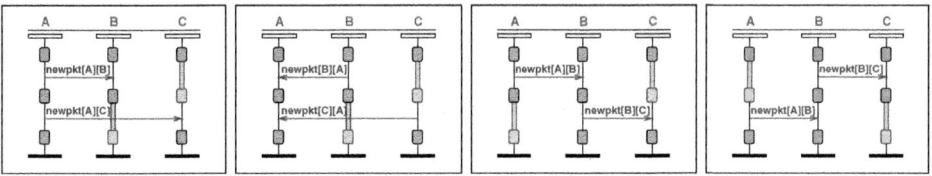

Fig. 2. Sequence charts illustrating four scenarios for initiating two route requests

scenarios cover all scenarios for injecting two new packets with either different originators or different destinations.

Additional to A, B and C, we add up to two nodes that may relay messages, but do not create data packets themselves. We consider only topologies in which nodes A, B and C are connected, either directly, or indirectly. This ensures that the route discovery is at least theoretically possible. If it fails, then it won't be because the nodes are not connected, but due to failure of the protocol.

We consider three classes of topologies. The first class are static topologies. Given the constraints that node A, B and C are connected, and that there are at most 5 nodes, this gives 444 topologies, after topologies that are identical up to symmetries are removed. The second class considers pairs of topologies from the first class, in which the second topology can be obtained by adding a new link. This models a dynamic topology in which a link is added. There are 1978 such pairs. The third class considers the same pairs, but now moves from the second topology to the first. This models a link break. Note that after deletion, nodes A, B and C are still connected. In our UPPAAL model a change of topology is modelled by another automaton. It may add or remove a link exactly once, non-deterministically, after the first route request arrives at the destination.

4.2 Properties

This paper considers three desirable properties of any routing protocol such as AODV. The first property is that once all routing messages have been processed a route from the originator to the destination has been found. In UPPAAL syntax this safety property can be expressed as:

$$A[]((tester.final \&\& emptybuffers()) \text{ imply} \atop (node(OIP).rt[DIP].nhop!=0)) \tag{1}$$

The CTL formula $A[]\phi$ is satisfied if ϕ holds on all states along all paths. The variable `node(OIP).rt` models the routing table of the originator node OIP, and the field `node(OIP).rt[DIP].nhop` represents the next hop for destination DIP. All initiated requests will have been made, iff automaton `tester` is in location `final`, the message buffers are empty iff function `emptybuffers` returns *true*, and the originator OIP has a route to node DIP iff `node(OIP).rt[DIP].nhop!=0`.

The second property is related, namely that once all messages are processed, then no sub-optimal route has been found. Here, sub-optimal means that the number of hops is greater than the shortest path. In case that the topology changes, we take the greater distance. In UPPAAL this can be expressed as

$$
\begin{aligned}
&\texttt{A[]((tester.final \&\& emptybuffers()) imply}\\
&\quad\texttt{(node(OIP).rt[DIP].hops<=distance[OIP][DIP]))}
\end{aligned}
\tag{2}
$$

Here, the array `distance` encodes the distance matrix. Note, that this fails if the route at the end is sub-optimal. It does not fail if at the end, either an optimal, or no route has been found. If the first two properties are satisfied, it means that it is guaranteed that an optimal route will be found when all messages have been processed. Note that it is known that AODV does not guarantee that optimal routes will be found. Nevertheless, an implementation or modification of AODV can be said to perform better if this property fails for fewer topologies.

The third property is even stronger than the second, namely that no sub-optimal routes will be found at all. It does not hold if a better optimal route replaces a sub-optimal route that was found first.

$$
\texttt{A[](node(OIP).rt[DIP].hops<=distance[OIP][DIP])}
\tag{3}
$$

If the third property holds, then the second must hold as well. In the experiments we will check all three properties for both originator-destination pairs at once.

4.3   Modifications

The basic UPPAAL model is based on the process algebraic AWN model, which reflects a common interpretation of the RFC with all ambiguities resolved. It is known that AODV does not guarantee that optimal routes will be found, or even any routes at all [6,10].[2] Our experiments quantify how many topologies are affected by these problems, and also what impact slight modifications of the protocol have. We will refer to the basic model as *model 1*, and discuss three proposed variants of AODV.

Forwarding All Route Replies. It is a known problem that nodes drop route reply messages under certain conditions.[3] During our experiments we found this problem even in the smallest topology, a static linear topology with only three nodes, and only two links: node A is connected to node B and B to node C. Both node B and C initiate a route request to A. For this topology and scenario, UPPAAL finds a counterexample for Property (1), i.e., it is possible that no route will been found when all messages have been processed.

Fig. 3 depicts a message sequence chart of the relevant part of the counterexample. Initially, both B and C initiate a route request for A. We refer to the first request as BA-request, and to the second as CA-request. First, node B sends the BA-request to A and C (Step 1 in Fig. 3), then node C its CA-request to

---

[2] AODV proposes to repeat the route discovery process if the first discovery process fails. However, this solution does not solve the problems entirely (see [5]).

[3] This problem has already been raised on the MANET mailing list in Oct 2004 (http://www.ietf.org/mail-archive/web/manet/current/msg05702.html).

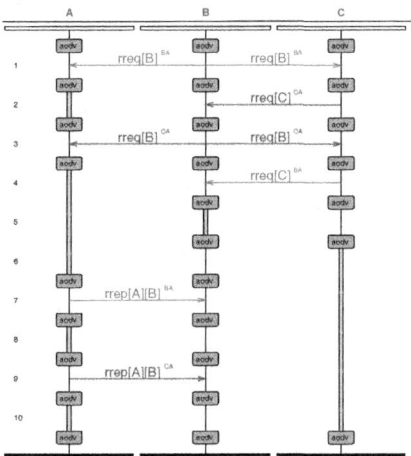

Fig. 3. Message sequence chart illustrating failed route discovery. Wide vertical lines mean that local states do not change in this transition. The superscripts indicate the corresponding originator and destination of the route discovery process.

B (Step 2). Node B forwards the CA-request (Step 3), node C the BA-request (Step 4). Node C will correctly ignore the CA-request that it received from B, since it is the originator (Step 5). Similarly, B will ignore the BA-request (Step 6). Node A will then reply to the BA-request (Step 7), and node B will update its routing table (Step 8) to include a route to A. Node A will also reply to the CA-request (Step 9), but B will ignore this message (Step 10), since it does not contain new information for B. Node A's reply to the CA-request will not arrive at C.

The discarding of the RREP message happens according to the RFC specification of AODV [12]. It states that an intermediate node only forwards the RREP message if it is not the originator node *and* it uses the RREP to update its route entry to the destination. In this case, node B is not the originator, but it also did not use the route reply to update its route. It already had an optimal route, as a result of the BA-request. This type of problem can arise whenever one node has to relay multiple route requests for the same destination.

A possible solution would be to forward every reply received by a node. Our *model 2* implements this change. Obviously, this increases the number of control messages generated during route discovery. However, this is compensated by the reduced need to repeat sending the route request in case no route has been found, the solution proposed by AODV.[4] In the experiment section we will see that this modification effectively addresses the problem.

---

[4] Moreover, a repeated route request need not be any more successful than the first.

Replying to Improving Requests. Counterexamples found by UPPAAL show that a source for sub-optimal routes is the property of AODV to only reply to the first route request. All subsequent requests with the same request ID (`rreqid`) will be ignored (Line 3 of Tables 1 and 2), even if the subsequent requests arrived via a shorter route. *Model 3* modifies the rule for the handling of route requests. It will not only reply to the first request, but also to a subsequent request (with the same request ID) with an improved hop count.

Recovering from Failed Replies. Analysis of UPPAAL's counterexamples show that a main reason for failed route discovery is that a node marks a request as having been replied to, even if the node detected the reply failed due to the link being broken in the time between the received request and the sent reply. The node will ignore other requests with the same request ID that may arrive later. *Model 4* introduces two changes: it does not mark a request as seen if the reply fails, and it replies to other requests in the same route discovery process.

This change should be considered with care, since it changes the rules with respect to sequence numbers. These numbers are an essential part of AODV being loop free, and there is currently no guarantee that this change will not violate some essential invariants of the proof [5]. We included the results nevertheless, as they show that there is still significant potential to improve AODV.

4.4 Experimental Results

The experimental results tell for how many topologies UPPAAL could show the absence of counterexamples, and thus allow quantification of the impact of improvements. However, the analysis uses a non-deterministic model, rather than a probabilistic model. For each topology it is reported whether a counterexample exists, but not how likely it is to occur. Neither can we assume that the topologies themselves are randomly distributed. Depending on the application only certain types of topologies might occur in practice. Nevertheless, it is fair to assume that a modification that leads to fewer topologies with counterexamples constitutes an improvement w.r.t. the considered property.

Table 3 presents the results of the experiments. Most relevant for all classes of topologies are Property (1), a route is found, Property (2), no sub-optimal route is found in the end, and the combination of these, i.e., an optimal route is found.

The results demonstrate that the problem of ignoring route replies as described in Figure 3 occurs even for about 50% of all static topologies. *Model 1* satisfies Property (2) only for half of all static topologies. The proposed modification solves this problem entirely for static topologies. The other modifications further improve the quality of the routes; in 99.1% of static topologies Property (2) holds, i.e., the route was in the end always optimal. The slight drop in Property (3) is explained by the fact that in a few cases, where no route was found at all for *model 1*, a sub-optimal route was found in the other models.

The results for static topologies are roughly repeated if we consider topologies in which a link is added. There were a few surprising instances though, in which adding a link was instrumental in finding a sub-optimal route.

Table 3. Model checking result for the four models and three classes of topologies. It gives the percentage of topologies for which there exists no counterexample.

		Property (1)	Property (2)	Property (3)	Property (1) & (2)	all properties
static	model 1	52.7%	93.2%	50.7%	50.0%	13.5%
	model 2	100.0%	93.2%	47.5%	93.2%	47.5%
	model 3	100.0%	99.1%	47.5%	99.1%	47.5%
	model 4	100.0%	99.1%	47.5%	99.1%	47.5%

		Property (1)	Property (2)	Property (3)	Property (1) & (2)	all properties
add link	model 1	57.5%	90.8%	49.1%	53.3%	18.1%
	model 2	100.0%	90.6%	46.2%	90.6%	46.2%
	model 3	100.0%	97.8%	46.2%	97.8%	46.2%
	model 4	100.0%	96.3%	46.2%	96.3%	46.2%

		Property (1)	Property (2)	Property (3)	Property (1) & (2)	all properties
remove link	model 1	26.7%	90.5%	59.7%	26.2%	6.0%
	model 2	53.0%	89.4%	57.1%	51.2%	28.9%
	model 3	53.0%	93.1%	57.1%	52.8%	28.9%
	model 4	75.4%	94.0%	54.0%	73.8%	41.0%

The results are, as expected, not quite as positive if a link gets removed. For the baseline model it is only guaranteed for one quarter of all topologies that a route will be found. Relaying all route replies, and not marking requests if the reply fails, improves this result. For three quarters of all topologies in which a link was removed it was shown that an optimal route will be found.

The main reason of the failures that remain is that a route reply might get lost because of some intermediate link break on the path back to the destination. A possible solution to this problem could be to maintain a set of back-up routes, or to implement different error responses. However, this requires a significant change and fundamentally changes the characteristics of AODV.

For the experiments we used an Intel Core2 CPU 2.13GHz processor with 2GB internal memory, running Ubuntu 11.04. We used UPPAAL 4.0.13. Of each of the models described in this section, we checked 17600 instances, altogether 70400 instances. As indication of the state space and runtimes, we checked an invariant on all instances of *model 4* for a topology in which a link is removed. These instances have larger state spaces than others, since these scenario have also to trigger the transitions for error handling. The models have an average of 9400 states, the largest model has 475000 states, and the median is 2700. Exploring these state spaces took on average 1.73 seconds user time, at most 81 seconds, and the median was 0.57. These run times show that an automated, systematic and rigorous analysis of reasonable rich routing protocols is feasible.

5   Related Work

Other researchers have used formal specification and analysis techniques to investigate the correctness and performance of AODV; we survey the sample related to model checking.

Bhargavan et al. [2] were amongst the first to use model checking on a draft of AODV, demonstrating the feasibility and value of automated verification of routing protocols. Their investigations using the SPIN model checker revealed that in some circumstances routes containing loops can be created. The proposed variation which guarantees loop freedom were not included in the current standard.

Musuvathi et al. [11] introduced the CMC model checker primarily to search for coding errors in implementations of protocols written in C. They use AODV as an example and, as well as discovering a number of errors, they also found a problem with the specification itself which has since been corrected.

Chiyangwa and Kwiatkowska [3] use the timing features of UPPAAL to study the relationship between the timing parameters and the performance of route discovery. They established a dependence between the lifetime of a route and the size of the network, although their study only considered the initiation of a single route discovery process, and a static linear topology. In [6], we confirmed some of the problems they discovered, and show their independence of time.

Other researchers have used model checking to analyse other routing protocols. Wibling et al. [14] for example used SPIN and UPPAAL to verify aspects of the LUNAR protocol, which is also used in ad hoc routing for wireless networks. In particular the timing feature of UPPAAL was used to check upper and lower bounds on route finding and packet delivery times. The scenarios considered included a limited number of topology changes where problems were suspected.

De Renesse and Aghvami [13] used SPIN to study the WARP protocol. To reduce the overhead on model checking, various simplifications were imposed on a five-node network, including a single source and destination and limitations on the degree that the network can change.

Fehnker et al. have used the model checker UPPAAL to analyse a TDMA time synchronisation protocol [7]. Similarly to our approach they considered all topologies in a certain class, but did not cover dynamic topologies.

Our approach is in line with these related works. However, it is unique in the sense that our UPPAAL model complements our process-algebraic specification of AODV. As mentioned before, these two approaches to formal protocol modelling, specification and evaluation, if used together, can provide a powerful tool for the development and rigorous evaluation of new protocols and variations, and improvements of existing ones. Currently, our UPPAAL model is derived by hand directly from the AWN specification, but an automatic translation from AWN in the style of Musuvathi et al. [11] is possible, and remains as future work.

6   Conclusions and Outlook

The aim of this ongoing work is to complement by model checking a process algebraic description of WMN routing protocols in general, and AODV in particular. The used description of AODV described in [5] is amongst the first detailed formal models. Having the ability of automatically deriving an UPPAAL model from an AWN specification and thus model checking formal specifications allows the confirmation and detailed diagnostics of suspected errors. The availability of an executable model becomes especially useful in the evaluation of proposed improvements to AODV, as we have shown.

We have sketched possible modifications of AODV, which have been evaluated by formal and rigorous analysis by means of model checking. An analysis of these modifications by means of process algebra is part of future work. We have set up an environment where we can test a whole bunch of different topologies in a systematic manner. This will allow us to do a fast comparison between standard AODV and proposed variations in contexts known to be problematic.

References

1. Behrmann, G., David, A., Larsen, K.G.: A Tutorial on UPPAAL. In: Bernardo, M., Corradini, F. (eds.) SFM-RT 2004. LNCS, vol. 3185, pp. 200–236. Springer, Heidelberg (2004)
2. Bhargavan, K., Obradovic, D., Gunter, C.A.: Formal verification of standards for distance vector routing protocols. J. ACM 49(4), 538–576 (2002)
3. Chiyangwa, S., Kwiatkowska, M.: A Timing Analysis of AODV. In: Steffen, M., Zavattaro, G. (eds.) FMOODS 2005. LNCS, vol. 3535, pp. 306–321. Springer, Heidelberg (2005)
4. Fehnker, A., van Glabbeek, R.J., Höfner, P., McIver, A., Portmann, M., Tan, W.L.: A process algebra for wireless mesh networks. In: Seidl, H. (ed.) European Symposium on Programming (ESOP 2012). Springer, Heidelberg (in press, 2012)
5. Fehnker, A., van Glabbeek, R.J., Höfner, P., McIver, A., Portmann, M., Tan, W.L.: A process algebra for wireless mesh networks used for modelling, verifying and analysing AODV. Tech. Rep. 5513, NICTA (2012), http://www.nicta.com.au/pub?id=5513
6. Fehnker, A., van Glabbeek, R.J., Höfner, P., McIver, A.K., Portmann, M., Tan, W.L.: Modelling and analysis of AODV in UPPAAL. In: 1st International Workshop on Rigorous Protocol Engineering (2011)
7. Fehnker, A., van Hoesel, L., Mader, A.: Modelling and Verification of the LMAC Protocol for Wireless Sensor Networks. In: Davies, J., Gibbons, J. (eds.) IFM 2007. LNCS, vol. 4591, pp. 253–272. Springer, Heidelberg (2007)
8. Hiertz, G.R., Denteneer, D., Max, S., Taori, R., Cardona, J., Berlemann, L., Walke, B.: IEEE 802.11s: the WLAN mesh standard. IEEE Wireless Communications 17(1), 104–111 (2010)
9. Larsen, K.G., Pettersson, P., Yi, W.: UPPAAL in a nutshell. International Journal of Software Tools for Technology Transfer 1(1-2), 134–152 (1997)
10. Miskovic, S., Knightly, E.W.: Routing primitives for wireless mesh networks: Design, analysis and experiments. In: IEEE INFOCOM, pp. 2793–2801 (2010), http://dx.doi.org/10.1109/INFCOM.2010.5462111
11. Musuvathi, M., Park, D.Y.W., Chou, A., Engler, D.R., Dill, D.L.: CMC: a pragmatic approach to model checking real code. In: Operating Systems Design and Implementation, OSDI 2002 (2002)
12. Perkins, C., Belding-Royer, E., Das, S.: Ad hoc on-demand distance vector (AODV) routing. RFC 3561 (2003), http://www.ietf.org/rfc/rfc3561.txt
13. de Renesse, R., Aghvami, A.H.: Formal verification of ad hoc routing protocols using SPIN model checker. In: Proceedings of IEEE MELECON 2004, pp. 1177–1182. IEEE (2004)
14. Wibling, O., Parrow, J., Pears, A.N.: Automatized Verification of Ad Hoc Routing Protocols. In: de Frutos-Escrig, D., Núñez, M. (eds.) FORTE 2004. LNCS, vol. 3235, pp. 343–358. Springer, Heidelberg (2004)

Modeling and Verification of a Dual Chamber Implantable Pacemaker*

Zhihao Jiang, Miroslav Pajic, Salar Moarref, Rajeev Alur,
and Rahul Mangharam

University of Pennsylvania, Philadelphia PA, USA

Abstract. The design and implementation of software for medical devices is challenging due to their rapidly increasing functionality and the tight coupling of computation, control, and communication. The safety-critical nature and the lack of existing industry standards for verification, make this an ideal domain for exploring applications of formal modeling and analysis. In this study, we use a dual chamber implantable pacemaker as a case study for modeling and verification of control algorithms for medical devices in UPPAAL. We begin with detailed models of the pacemaker, based on the specifications and algorithm descriptions from Boston Scientific. We then define the state space of the closed-loop system based on its heart rate and developed a heart model which can non-deterministically cover the whole state space. For verification, we first specify unsafe regions within the state space and verify the closed-loop system against corresponding safety requirements. As stronger assertions are attempted, the closed-loop unsafe state may result from healthy open-loop heart conditions. Such *unsafe transitions* are investigated with two clinical cases of Pacemaker Mediated Tachycardia and their corresponding correction algorithms in the pacemaker. Along with emerging tools for code generation from UPPAAL models, this effort enables model-driven design and certification of software for medical devices.

Keywords: Medical Devices, Implantable Pacemaker, Software Verification, Cyber-Physical Systems.

1 Introduction

Over the past four decades, cardiac rhythm management devices such as pacemakers have expanded their role from "keeping the patient alive" to "making the patient's life comfortable". The addition of more safety and efficacy features has resulted in increased complexity, inevitably leading to more safety violations. From 1996-2006, the percentage of software-related causes in medical device recalls have grown from 10% to 21% [1]. During the first half of 2010, the US Food and Drug Administration (FDA) issued 23 recalls of defective devices, all

---

* This research was partially supported by NSF research grants MRI 0923518, CNS 0931239, CNS 1035715 and CCF 0915777.

C. Flanagan and B. König (Eds.): TACAS 2012, LNCS 7214, pp. 188–203, 2012.

of which are categorized as *Class I*, meaning there is a "reasonable probability that use of these products will cause serious adverse health consequences or death." At least six of the recalls were caused by software defects [2]. Unlike other industries such as aviation and automotive, the safety concern in the medical device domain is focused on the physical plant, the patient in this case, rather than the controller. As a result, although in aviation and automotive industries, standards are enforced during software development, manufacturing, and post-market change [3,4], there are no well-established standards for development of software for medical devices. There is a pressing need for standards and tools to certify and verify the safety of software in medical devices. For device manufacturers, this has prompted recent interest in applying formal modeling and verification techniques in medical devices software development [5,6].

In this effort, we propose a Timed Automata representation of the heart and a dual chamber pacemaker. Our models and specifications are designed based on descriptions available from Boston Scientific [7,8], a leading manufacturer of pacemakers, and extensive medical literature on this topic. We then demonstrate how a model checker, like UPPAAL [9], can be used to find safety violations and prove the correctness of medical device algorithms. We define the state space of the closed-loop system based on its heart rate. Unsafe regions can then be specified and the closed-loop system is verified against corresponding safety requirements. We also define *unsafe transitions* as the controller drives the open-loop plant from a safe state into an unsafe closed-loop state. We focus on two cases of unsafe transitions which are referred to as "Pacemaker Mediated Tachycardia (PMT)". Modern pacemakers are equipped with correction algorithms to terminate these behaviors. We demonstrate how to identify known unsafe transitions and prove the correctness of corresponding correction algorithms using model checker. The UPPAAL model developed in this paper is freely available online [10]. These models can be used as a starting point for many purposes (e.g. to build models with costs and probabilities for quantitative analysis of the efficacy of pacemaker algorithms; development of patient-specific algorithms). In particular, the verified pacemaker model can be automatically translated into Stateflow charts in Simulink for test generation and code generation [11].

The paper is organized as follows: In Section 2, we introduce the physiological and timing basics of the heart and pacemaker. Section 3 presents UPPAAL models of the basic DDD pacemaker and the heart. In Section 4, we define unsafe regions and verify the basic pacemaker model against corresponding safety requirements. In Section 5, we proposed a procedure for identifying and verifying unsafe transitions and demonstrated using two cases of PMT.

2   Heart and Pacemaker Basics

The coordinated contraction of the heart is governed by its Electrical Conduction System (see Fig. 1). The Sinoatrial (SA) node, which is a collection of specialized tissue at the top of the right atrium, periodically spontaneously generates electrical pulses that can cause muscle contraction. The SA node is controlled by the nervous system and acts as the natural pacemaker of the heart. The electrical pulses first cause both atria to contract, forcing the blood into the ventricles. The electrical conduction is then delayed at the Atrioventricular (AV) node, allowing the ventricles to fill fully. Finally the fast-conducting His-Pukinje system spreads the electrical activation within both ventricles, causing simultaneous contraction of the ventricular muscles, and pumps the blood out of the heart.

Due to aging and/or diseases, the conduction properties of heart tissue may change. These changes may cause timing anomalies in heart rhythm, thus decrease the blood pumping efficiency of the heart. These timing anomalies are referred to as arrhythmias, and are categorized into *Tachycardia* and *Bradycardia*. Tachycardia features undesirable fast heart rate which impairs hemodynamics. Bradycardia features slow heart rate which results in insufficient blood supply. Bradycardia maybe due to failure of impulse generation with anomalies in the SA node, or failure of impulse propagation where the conduction from atria to the ventricles is delayed or blocked.

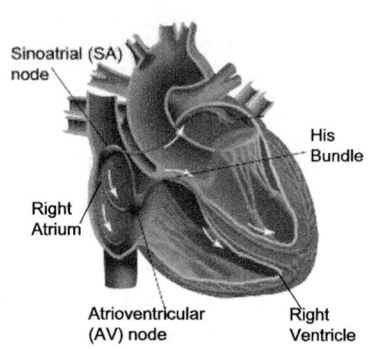

Fig. 1. Cardiac electrical system

Since the heart tissue can be activated by external electrical pulses, Bradycardia can be treated by providing electrical pulses when the heart rate is low. *Implantable Pacemakers* have been developed to deliver timely electrical pulses to the heart to maintain an appropriate heart rate and Atrial-Ventricular synchrony. Implantable pacemakers normally have two leads fixed on the wall of the right atrium and the right ventricle respectively. Activation of local tissue is sensed by the leads, triggering Atrial Sense (AS) and Ventricular Sense (VS) events. Atrial Pacing (AP) and Ventricular Pacing (VP) are delivered if no sensed events occur within deadlines.

In order to deal with different heart conditions, modern pacemakers are able to operate in different modes. The modes are labeled using a three character system. The first character describes the pacing locations, the second character describes the sensing locations, and the third character describes how the pacemaker software responds to sensing. In this work we describe the most commonly used mode of pacemaker, the dual-chamber DDD mode that paces both the atrium and the ventricle, senses both chambers, and sensing can both activate or inhibit further pacing. Similarly, the VDI mode paces only in the ventricle, senses both chambers, and inhibits pacing if event is sensed. [12]

3   System Modeling

3.1   Timed Automata and UPPAAL

Timed automaton [13] is an extension of a finite automaton with a finite set of real-valued clocks. It has been used for modeling and verifying systems which are triggered by events and have timing constraints between events. From the Boston Scientific pacemaker specification [7], the pacemaker can be modeled using this Extended Timed Automata notation, which is a subset of formal semantics in UPPAAL. UPPAAL ([9,14]) is a standard tool for modeling and verification of real-time systems, based on networks of timed automata. The graphical and text-based interface makes modeling more intuitive. Requirements can be specified using Computational Tree Logic (CTL) [15] and violations can be visualized in the simulation environment.

3.2 System Overview

The function of a pacemaker is to manage the timing relationship between the atrial and ventricular events. Thus Timed Automata is suitable for modeling both the deterministic behavior of a pacemaker and the non-deterministic behavior of the heart. The overview of the closed-loop system is showed in Fig. 2(a). The heart and the pacemaker communicate with each other using broadcast channels. The heart generates *Aget!* and *Vget!* actions, representing atrial and ventricular events that the pacemaker take as inputs. The pacemaker processes the signals and generates pacing actions *AP!* and *VP!* to the corresponding components in the heart.

3.3 Basic DDD Pacemaker Modeling

The DDD pacemaker has 5 basic timing cycles triggered by events, as shown in Fig. 2(b). We decomposed our pacemaker model into 5 components which correspond to the 5 counters. These components communicate with each other using broadcast channels and shared variables (as shown in Fig. 3).

Lower Rate Interval (LRI): This component keeps the heart rate above a minimum value. In DDD mode, the LRI component models the basic timing cycle which defines the longest interval between two ventricular events. The clock is reset when a ventricular event (VS, VP) is received. If no atrial event has been sensed (AS), the component will deliver atrial pacing (AP) after TLRI-TAVI. The UPPAAL design of LRI component is shown in Fig. 3(a).

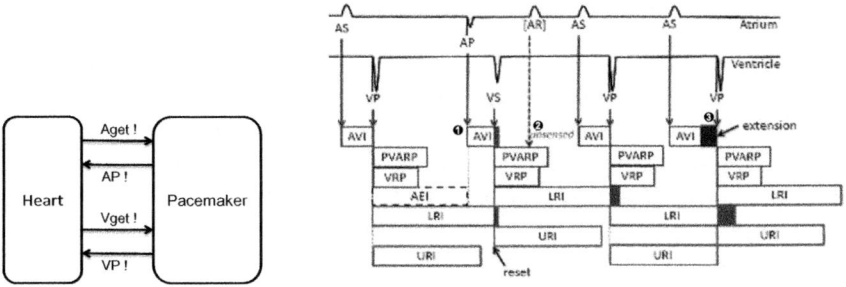

Fig. 2. (a) System Overview, (b) Basic 5 timing cycles of DDD pacemaker

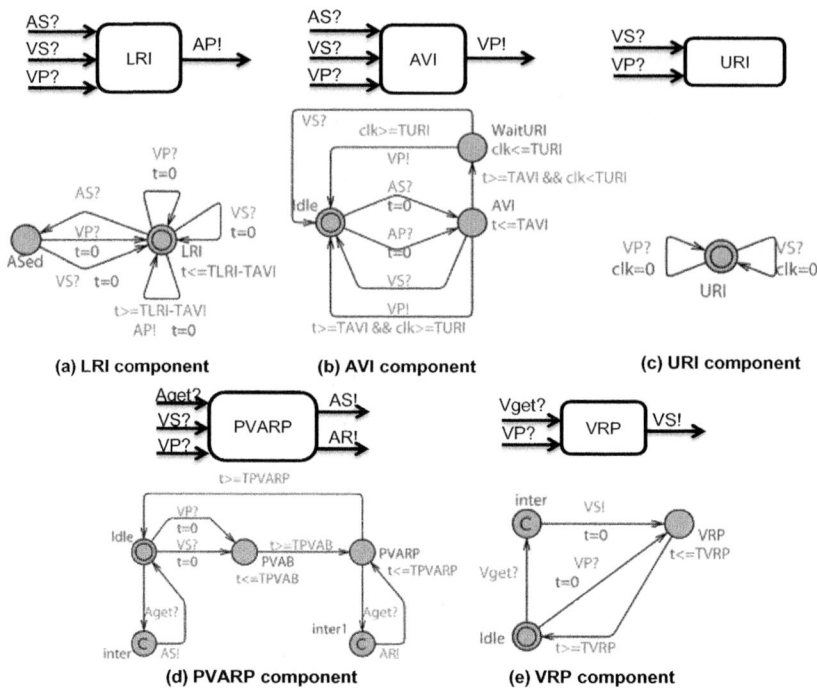

Fig. 3. Components of the pacemaker model in UPPAAL

Atrio-Ventricular Interval (AVI) and Upper Rate Interval (URI): The function of the AVI component is to maintain the appropriate delay between the atrial activation and the ventricular activation. It defines the longest interval between an atrial event and a ventricular event. If no ventricular event has been sensed (VS) within TAVI after an atrial event (AS, AP), the component will deliver ventricular pacing (VP). In order to prevent the pacemaker from pacing the ventricle too fast, a URI component uses a global clock *clk* to track the time after a ventricular event (VS, VP). The URI limits the ventricular pacing rate by enforcing a lower bound on the times between consecutive ventricle events. If the global clock value is less than TURI when the AVI component is about to deliver VP, AVI will hold VP and deliver it after the global clock reaches TURI. The UPPAAL design of AVI and URI component is shown in Fig. 3(b) and (c).

Post Ventricular Atrial Refractory Period (PVARP) and Post Ventricular Atrial Blanking (PVAB): Not all atrial events (*Aget!*) are recognized as Atrial Sense (*AS!*). After each ventricular event, there is a blanking period (PVAB) followed by a refractory period (PVARP) for the atrial events in order to filter noise. Atrial events during PVAB are ignored and atrial events during PVARP trigger *AR!* event which can be used in some advanced diagnostic algorithms. The UPPAAL design of PVARP component is shown in Fig. 3(d).

Ventricular Refractory Period (VRP): Correspondingly, the VRP follows each ventricular event (VP, VS) to filter noise and early events in the ventricular channel which could otherwise cause undesired pacemaker behavior. Fig. 3(e) shows the UPPAAL design of VRP component.

Parameter Selection: Each timing parameter of the pacemaker has a feasible range. However, after those parameters are programmed, they are fixed during pacemaker operation. Consider all possible combinations of feasible parameter values is infeasible. In this work, we only verify one instance of a DDD pacemaker with nominal values in clinical settings [8]. The values we choose are TAVI=150, TLRI=1000, TPVARP=100, TVRP=150, TURI=400, TPVAB=50.

3.4   Random Heart Model (RHM)

In order to verify pacemaker algorithm, we need to first define the state space for the closed-loop system. The state space definition should not only cover all pos-  sible pacemaker operations, but also be

Fig. 4. RHM for the Atrial Channel

physiologically intuitive for safety requirement specification. To this end, we define the state space of the closed-loop system by the atrial interval (interval between atrial events $\in \{AS, AP\}$) and ventricular interval (interval between ventricular events $\in \{VS, VP\}$). This heart rate representation enables us to define unsafe regions for bradycardia and tachycardia.

The Random Heart Model (RHM) is designed to cover open-loop heart behaviors. It non-deterministically generates an intrinsic heart event $Xget!$ within $[Xminwait, Xmaxwait]$ after each intrinsic heart event $Xget$ or pacing XP. Here we use two RHMs for the atrial and ventricular channel where X can be atrial (A) or ventricular (V). RHM covers all possible input to the pacemaker if the interval $[Xminwait, Xmaxwait]$ is set to $[0, \infty]$. It can also cover subset of possible heart conditions by assigning appropriate values to those two parameters. The UPPAAL model of the atrial RHM is shown in Fig. 4.

4   Verification Regarding Unsafe Regions

In this section, we define unsafe regions regarding bradycardia and tachycardia and specify two basic safety properties. These two basic safety properties are strict so that they must be satisfied by any pacemaker under all heart conditions. We then discuss refinement of the safe regions and make stronger assertions.

4.1   Lower Rate Limit

The most essential function for the pacemaker is to treat bradycardia by maintaining the ventricular rate above a certain threshold. We define the region where the ventricular rate is slow, as *unsafe*. The monitor Pvv is designed to measure interval between ventricular events and is shown in Fig. 5(a). The property $A[]$ ($Pvv.two\_a$ imply $Pvv.t \leq TLRI$) is satisfied by the basic DDD pacemaker.

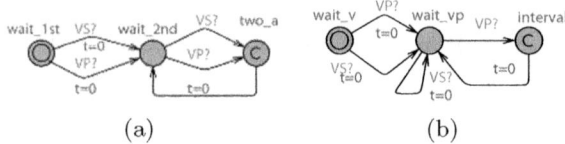

Fig. 5. (a) Monitor for LRL: Interval between two ventricular events should be less than TLRI, (b) Monitor for URL: Interval between a ventricular event and a VP should be longer than TURI

4.2   Upper Rate Limit

The pacemaker is not designed to treat tachycardia so it can only pace the heart to increase its rate and cannot slow it down. However, it is still important to guarantee it does not pace the ventricles beyond a maximum rate to ensure safe operation. To this effect, an upper rate limit is specified such that the pacemaker can increase the ventricular rate up to this limit.

We require that a ventricle pace (VP) can only occur at least *TURI* after a ventricle event (VS, VP). The monitor for the property is shown in Fig. 5(b) and the property *A[] (PURI_test.interval imply PURI_test.t≥ TURI)* is satisfied by the basic DDD pacemaker model.

5   Verification Regarding Unsafe Transitions

The two unsafe regions, introduced above, are intuitive but provide for loose safety properties. One may wonder if we can further reduce the safe region. When the closed-loop system is in some unsafe state, there are two possible scenarios. One is when, the open-loop plant without the controller, is also in unsafe state. In our case, if the heart is in tachycardia, the pacemaker is not supposed to react so that this case is of little value to us. The other scenario is that the open-loop plant is in a safe state and the controller is driving the closed-loop system into some unsafe states. We call this scenario *Unsafe Transition*. In our case, the pacemaker may increase the heart rate inappropriately, which is referred to as Pacemaker Mediate Tachycardia (PMT).

We now introduce two cases of PMT and their corresponding correction algorithms. Since one closed-loop state may correspond to multiple execution traces, these PMT scenarios will not be returned by the model checker as counterexamples of safety requirements. However, we can still identify known PMT by adding constraints to the heart model or developing more complex requirements.

5.1   Verification Procedure

The pacemaker manufacturers have developed anti-PMT algorithms to terminate different PMT scenarios. In this section, we propose a general procedure to identify PMT scenarios and verify the safety and correctness of anti-PMT algorithms. The general steps for the procedure include:

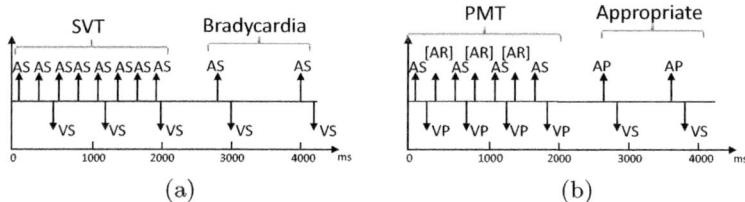

Fig. 6. (a) SVT with ODO pacemaker (b) SVT with DDD pacemaker

1. Show existence of PMT behaviors in the closed-loop system
2. Introduce anti-PMT algorithms and check whether the two basic safety requirements still hold
3. Prove correctness of anti-PMT algorithms by showing the non-existence of PMT scenarios

Here we use two well-identified PMT cases to demonstrate the methodology.

5.2   Verification of the Mode-Switch Algorithm

Supraventricular Tachycardia (SVT):   SVT is an arrhythmia which features an abnormally fast atrial rate. Typically the AV node, which has a long refractory period, can filter most of the fast atrial activations during SVT thus the ventricular rate remains relatively normal. Fig. 6(a) demonstrates a pacemaker event trace during SVT, with a ODO mode pacemaker which just sensing in both channels. In this particular case, every 3 atrial events (AS) correspond to 1 ventricular event (VS) during SVT.

As an arrhythmia, SVT is still considered as a safe heart condition since the ventricles operate under normal rate can still maintain adequate cardiac output. However, the AVI component of a dual chamber pacemaker is equivalent to a virtual pathway in addition to the intrinsic conduction pathway between the atria and the ventricles. The pacemaker tries to maintain 1:1 A-V conduction and thus increases the ventricular rate inappropriately. Fig. 6(b) shows the pacemaker trace of the same SVT case with DDD pacemaker. Although half of the fast atrial events are filtered by the PVARP period ([AR]s), the DDD pacemaker still drives the closed-loop system into 2:1 A-V conduction with faster ventricular rate, which is inappropriate. This problem can be resolved by switching pacemaker into single chamber mode to maintain appropriate ventricular rate.

Existence of PMT during SVT:
Since PMT during SVT is an unsafe transition, we need to first adjust the heart model so that the open-loop behaviors covers SVT and are in the safe region. To this end, the interval for the ventricular RHM is set to [500,800]. This rate is slow enough not to be considered as tachycardia, but faster than the Lower Rate Limit

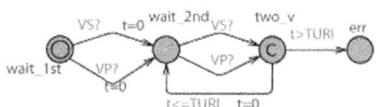

Fig. 7. Monitor for SVT: Check existence of an endless sequence where the ventricular event interval ≤TURI

of the pacemaker so that pacemaker should not intervene. The monitor Pv_v is designed to show existence of PMT during SVT. It goes to the error state if the ventricular rate drops below the Upper Rate Limit (Fig. 7).

The existence property $E[](notPv\_v.err)$ is specified, which verifies if there exists an execution in which the ventricular interval is always less or equal to TURI. The property is first verified on pacemaker without the mode-switch algorithm. The property is satisfied during verification.

Mode-Switch Algorithm: Intuitively, the mode-switch algorithm first detects SVT. After confirmed detection, it switches the pacemaker from a dual-chamber mode to a single-chamber mode. During the single-chamber mode, the A-V synchrony function of the pacemaker is deactivated thus the ventricular rate is decoupled from the fast atrial rate. After the algorithm determines the end of SVT, it will switch the pacemaker back to the dual chamber mode.

The mode-switch algorithm specification we use is the same as the one used in Boston Scientific pacemakers [8]. The algorithm first measures the interval between atrial events outside the blanking period (AS, AR). The interval is considered as *fast* if it is above a threshold (*Trigger Rate*) and *slow* otherwise (see Fig. 8 (1)). A counter increments for *fast* events and decrement for *slow* events (see Fig. 8 (2)). After the counter value reaches the *Entry Count*, the algorithm will start a *Duration* which is a time interval used to confirm the detection of SVT (see Fig. 8 (3)). In the *Duration*, the counter keeps counting. If the counter value is still positive after the *Duration*, the pacemaker will switch to the VDI mode (*Fallback mode*). In the VDI mode, the pacemaker only senses and paces the ventricle. At any time if the counter reaches zero, the *Duration* will terminate and the pacemaker is switched back to DDD mode.

In our UPPAAL model of the mode-switch algorithm, we use nominal parameter values from the clinical setting. We define *trigger rate* at 170bpm (350ms), *entry count* at 8, *duration* for 8 ventricular events and *fallback mode* as VDI.

In order to model both DDD and VDI modes and the switching between them, we made modifications to the AVI and LRI components. In each component

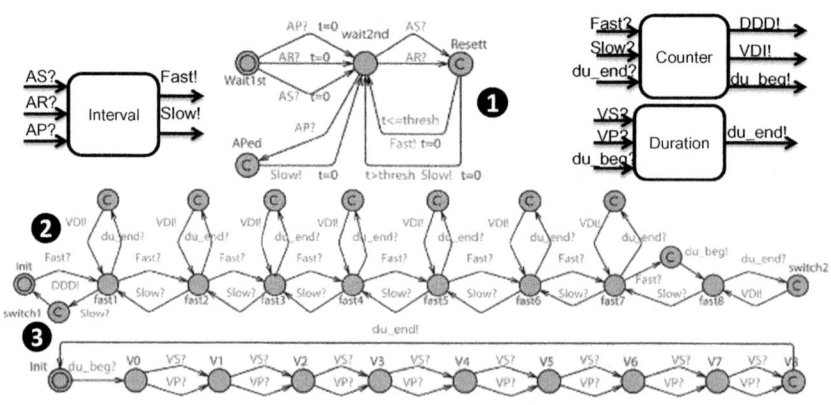

Fig. 8. Mode-Switch algorithm

two copies for both modes are modeled, and switch between each other when switching events (DDD, VDI) are received. During VDI mode, VP is delivered by the LRI component instead of the AVI component. The clock values are shared between both copies in order to preserve essential intervals even after switching. The modified AVI and LRI components are shown in Fig. 9.

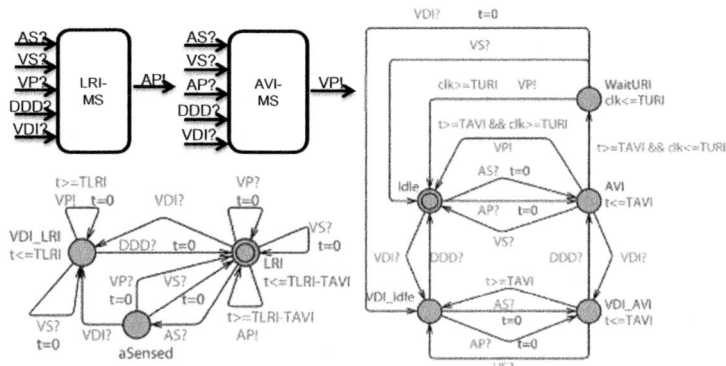

Fig. 9. New LRI & AVI components

Verification Against Basic Safety Requirements: We verify the same basic safety requirements on the pacemaker model with mode-switch algorithm. The Upper Rate Limit property still holds but the Lower Rate Limit property is violated. When the pacemaker is switching from VDI mode to DDD mode, the responsibility to deliver VP switched from LRI component to AVI component. Since the clock reference is different (Ventricular events in LRI and Atrial events in AVI), the clock value for delivering the next VP is not preserved. As a result, if an atrial event which triggered the mode-switch from VDI to DDD happens within [TLRI-TAVI, TLRI) after the last ventricular event, the next ventricular pacing will be delayed by at most TAVI time, which violates the Lower Rate Limit property (Fig. 11(a)).

Verification of the Algorithm: We now present the verification of the correctness of the mode-switch algorithm by checking the same existence property *E[] (not Pv_v.err)* on pacemaker with mode-switch algorithm. We expect the violation of this property, since during VDI mode the ventricular rate of the heart model is less than the Upper Rate Limit and will not trigger ventricular pacing. The counter example of the violation should show that mode-switch algorithm successfully switches the mode of the pacemaker to VDI mode. However, this property is still satisfied, indicating the mode-switch algorithm failed to eliminate the PMT scenario. Then we further restrict the atrial interval of RHM to [100, 200]. Since the atrial rate for the new heart model is always above the trigger rate, mode switch to VDI mode should always eventually happen. The monitor PMS for the new property is shown in Fig. 10.

   The property *A<> (PMS.err)* is not satisfied. The counter-example shows that some of the atrial events fall into the Post Ventricular Atrial Blanking period

Fig. 10. Monitor for Mode-Switch: Check if mode-switch to VDI mode will always eventually happen

(PVAB) and got ignored. As a result, two *fast* intervals may be considered as one *slow* interval (see Fig. 11(b)). If this happens more than one out of the *Entry Count*, mode-switch from DDD to VDI may never happen.

Discussion: We demonstrated that model checking techniques can be used to identify unknown violations which cannot be identified during open-loop testing, showing the necessity and usefulness of formal verification in medical device software development and certification. We also showed that adding new features to the verified system is a potential source for safety violations.

5.3   Verification of Endless Loop Tachycardia (ELT) algorithm

ELT overview: The AVI component of a dual-chamber pacemaker introduces a virtual A-V pathway which forms a loop with the intrinsic A-V conduction pathway (see Fig. 12(a)). In this scenario, a ventricular event (VS) triggers a V-A conduction through the intrinsic pathway (Marker 1 in Fig. 12(b)). The pacemaker registers this signal as an Atrial Sense (AS) (Marker 2 in Fig. 12(b)). This event triggers VP after TAVI, as if the signal conducts through the virtual A-V pathway (Marker 3 in Fig. 12(b)). The VP will trigger another V-A conduction and this VP-AS-VP-AS looping behavior will continue (see Fig. 12(b)). The interval between atrial events is TAVI plus the V-A conduction delay, which will drive the ventricular rate as high as the Upper Rate Limit.

From the pacemaker's point of view, the pacemaker paces the ventricles as specified for every AS. That is why open-loop testing is unable to detect this closed-loop behavior. Modern pacemakers are equipped with anti-ELT algorithms to identify and terminate potential ELT. One common algorithm identifies ELT by the ELT pattern and terminates ELT by increasing TPVARP time once to block the AS caused by the V-A conduction.

Existence of ELT: As another case of unsafe transition, we again constrain the open-loop heart model into healthy heart. We set both the atrial interval

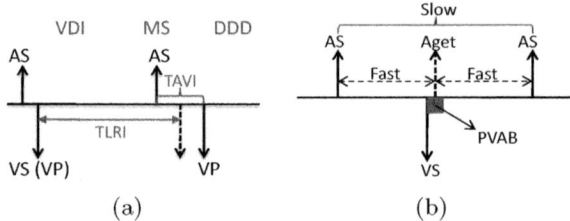

(a)   (b)

Fig. 11. (a) Safety Violation: VP is delayed due to the reset of timer during mode-switch, (b) Correctness Violation: The blocking period may block some atrial events, turning two *Fast* events to one *Slow* event

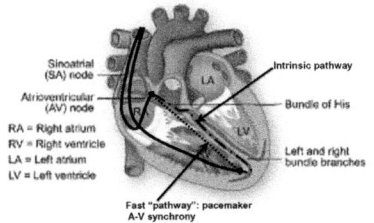
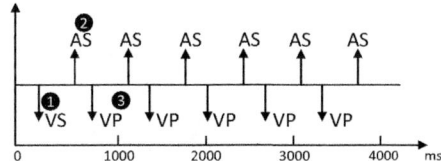

(a) Virtual circuit formed by the pacemaker and the heart

(b) Pacemaker trace for ELT initialized by a early ventricular signal

Fig. 12. Endless Loop Tachycardia case study demonstrating the situation when the pacemaker drives the heart into an unsafe state [16]

and the ventricular interval above TURI so that ELT behavior is not covered by the heart model. Two monitors were designed to show the existence of ELT. One monitor, *PELT_det*, shows the persistence of the VP-AS pattern and the other monitor, *Pvv*, shows that the ventricular rate is always no slower than the upper rate limit (Fig. 13). The existence property *E[] ((not PELT_det.err) && (not Pvv.err))* fails on pacemaker without an anti-ELT algorithm.

The reason for the failure is that in our closed-loop system, AS can only be triggered by *Aget* signal from the atrial heart model, where in ELT case the AS is triggered by backward V-A conduction, which is not covered by our heart model. In order to solve this problem, we model the A-V conduction of the heart in addition to the orignal RHM. The adjusted RHM and the conduction component is shown in Fig. 14. For each atrial event *Aget*, the conduction component generates *V_act* after certain delay and vice versa. The conduction is non-deterministic so that the old RHM is a special case for the new RHM. The PVARP and VRP components are also modified to accommodate new events *A_act* and *V_act*.

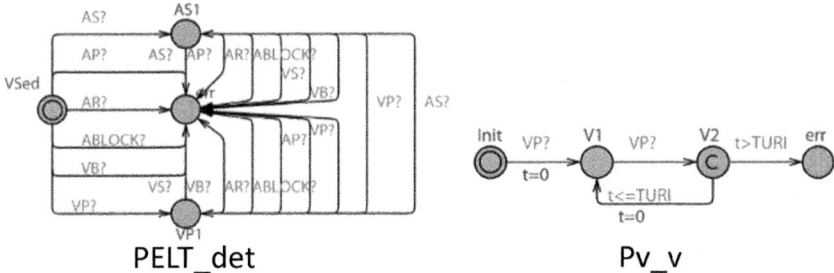

Fig. 13. Monitor for ELT: VP-AS pattern detection and Upper Rate detection

After introducing the conduction component, the existence property holds, indicating the closed-loop system with new heart model covers ELT.

The ELT-termination Algorithm: The ELT detection algorithm by Boston Scientific [7] utilizes these three features:

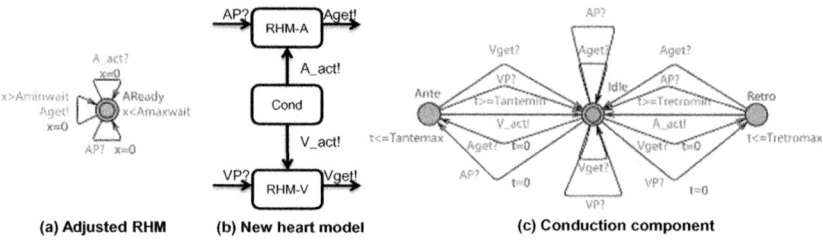

Fig. 14. Modified heart model and the conduction component

- Ventricular rate at Upper Rate Limit
- VP-AS pattern
- Fixed V-A conduction delay

The pacemaker first monitors VP-AS pattern with ventricular rate at upper rate limit. Then it compares the VP-AS interval with previous intervals. ELT is confirmed if the difference between the current VP-AS interval and the first VP-AS interval are within ±32ms for 16 consecutive times. Then the pacemaker increases the PVARP period to 500ms once so that the next AS will be blocked and will not trigger a VP. ELT will then be terminated.

As the V-A conduction delays are patient-specific, the algorithm compares VP-AS interval to a previously sensed value instead of an absolute value. Since we can not store past clock values in UPPAAL, we can not explicitly model this ELT detection algorithm. However, since the conduction delay in our heart model is within a known range, we can compare the VP-AS interval with this range. The VP-AS pattern detection module for our anti-ELT algorithm is shown in Fig. 15 (1). It detects the VP-AS pattern with ventricular rate at upper rate limit and sends out *VP_AS* event if the interval qualifies.

A counter counts the number of qualified VP-AS patterns. It increases the PVARP period to 500ms if eight consecutive VP-AS patterns are detected. (Fig. 15 (2)) The PVARP component is also modified so that the PVARP period can only be changed once by the anti-ELT algorithm. (Fig. 15 (3))

Verification Against Bottom-Line Safety Requirements: The two bottom-line safety requirements still hold when the anti-ELT algorithm is introduced.

Verification of the Algorithm: The existence property $E[]((not\ PELT\_det.err)$ && $(not\ Pvv.err))$ is not satisfied after the anti-ELT algorithm is introduced, indicating the algorithm successfully terminates ELT. We successfully reproduced the case when the algorithm works in the simulation environment of UPPAAL.

Discussion: In this case study, we showed that we may require the heart model to provide more physiological details when verifying more complex properties. We also observed some limitations of Timed Automata when modeling more complex algorithms.

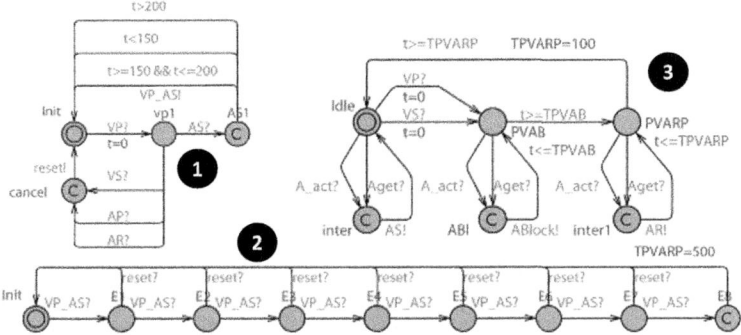

Fig. 15. Counter for VP-AS pattern

6 Related Work

Jee et. al present a safety assured development approach of real-time software using pacemaker as their case study in [17]. They formally model and verify a single chamber VVI pacemaker using UPPAAL and then implement it and check the preservation of properties transferred from model to implementation code. Tuan et. al propose an RTS formal model for pacemaker and its environment and verified it against number of safety properties and timed constraints using PAT model checker [18]. They have modeled the pacemaker for all 18 operating modes as described in Boston scientific, but their work lacks specification and analysis os complex behaviors of the pacemaker, such as mode-switch.

Wiggelinkhuizen uses mCRL2 and UPPAAL to formally model the pacemaker from the firmware design of Vitatron's DA+ pacemaker [19]. Two main approaches have been used to investigate the feasibility of applying formal model checking to the design of device firmware. The main approach consists of verifying the firmware model in context of a formal heart model and a formal model of a hardware module which fails for high heart rates because of the state explosion. Another approach is to verify a part of firmware design which was feasible and was able to detect a known deadlock rather soon.

Macedo et. al have developed a concurrent and distributed real-time model for a cardiac pacemaker through a pragmatic incremental approach. The models are expressed using the VDM and are validated primarily by scenario-based test, where test scenarios are defined to model interesting situations such as the absence of input pulses [20]. The models cover 8 modes of pacemaker operation.

Gomes et. al present a formal specification of pacemaker system using the Z notation in [21]. They have also tried to validate that the formal specification satisfies the informal requirements of Boston Scientific by using a theorem prover, ProofPower-Z. They have partially checked the consistency of their specification through reasoning. No validation experiment regarding safety conditions were performed yet. [21]

Mery et. al in [22], formally model all operational modes of a single electrode pacemaker system using event-B and prove them. They use an incremental proof-based approach to refine the basic abstract model of the system and add more functional and timing properties. They use the ProB tool to validate their models in different situations such as absence of input pulses.

7   Conclusion and Future Work

In this paper, we modeled a dual-chamber pacemaker with advanced features using Timed Automata. Timed automaton captures key features of the closed-loop system and enables the use of tools like UPPAAL in verification. We then verified one instance of a dual chamber pacemaker model with nominal parameter values since it is impossible to consider all possible combinations. We defined a heart rate representation of closed-loop state space and identified unsafe regions and unsafe transitions. We demonstrated that model checking techniques can be used to reveal safety violations which cannot be identified during open-loop testing. We also showed that adding features to previously verified system may result in safety violations. Furthermore, we showed that more complex heart model is need to provide more physiological insights during property specification. The UPPAAL model developed in this paper is freely available online [10]. We hope that these models can be used as a starting point for many purposes (e.g. to build models with costs and probabilities for quantitative analysis).

In this paper, we only verified the safety and correctness of pacemaker algorithms. However, the ultimate goal for a pacemaker is to maintain the efficiency of the heart. As future work, we would like to evaluate the efficiency of those algorithms by assigning costs for different heart conditions. The evaluation can be used to develop better treatment for general and specific patients. More complex heart models are therefore needed to provide physiological insights. However, rigorous heart model refinement should be considered to ensure model consistency. While Timed Automata is a good fit for the problem studied here, it also has some drawbacks as it can not capture certain behaviors of some advanced algorithms like memorizing difference of clocks, and is also not scalable enough. Our future work will also focus on improving the efficiency of verification toolchain for medical device certification.

References

[1] List of Device Recalls, U.S. Food and Drug Admin. (last visited July 19, 2010)
[2] Sandler, K., Ohrstrom, L., Moy, L., McVay, R.: Killed by Code: Software Transparency in Implantable Medical Devices. Software Freedom Law Center (2010)
[3] AUTOSAR website: http://www.autosar.org/
[4] AVSI website: http://www.avsi.aero
[5] Alur, R., Arney, D., Gunter, E.L., Lee, I., Lee, J., Nam, W., Pearce, F., Van Albert, S., Zhou, J.: Formal Specifications and Analysis of the Computer-Assisted Resuscitation Algorithm (CARA) Infusion Pump Control System. Intl. Journal on Software Tools for Technology Transfer (STTT) 5, 308–319 (2004)

[6] ten Teije, A., et al.: Improving medical protocols by formal methods. Artificial Intelligence in Medicine 36(3), 193–209 (2006)

[7] PACEMAKER System Specification. Boston Scientific (2007)

[8] The Compass - Technical Guide to Boston Scientific Cardiac Rhythm Management Products (2007)

[9] Larsen, K.G., Pettersson, P., Yi, W.: Uppaal in a Nutshell. International Journal on Software Tools for Technology Transfer (STTT), 134–152 (1997)

[10] Jiang, Z., Pajic, M., Moarref, S., Alur, R., Mangharam, R.: Pacemaker UPPAAL model download: http://www.seas.upenn.edu/~zhihaoj/VHM/PM_verify.zip

[11] Pajic, M., Jiang, Z., Sokolsky, O., Lee, I., Mangharam, R.: From Verification to Implementation: A Model Translation Tool and a Pacemaker Case Study. In: 18th IEEE Real-Time and Embedded Technology and Applications Symposium, IEEE RTAS (2012)

[12] Barold, S., Stroobandt, R., Sinnaeve, A.: Cardiac Pacemakers Step by Step. Blackwell Futura (2004)

[13] Alur, R., Dill, D.L.: A Theory of Timed Automata. Theoretical Computer Science 126, 183–235 (1994)

[14] Behrmann, G., David, A., Larsen, K.G.: A Tutorial on UPPAAL. In: Bernardo, M., Corradini, F. (eds.) SFM-RT 2004. LNCS, vol. 3185, pp. 200–236. Springer, Heidelberg (2004)

[15] Clarke, E.M., Allen Emerson, E.: Design and synthesis of synchronization skeletons using branching-time temporal logic. In: Logic of Programs, Workshop, pp. 52–71 (1982)

[16] Jiang, Z., Pajic, M., Mangharam, R.: Model-based Closed-loop Testing of Implantable Pacemakers. In: ICCPS 2011: ACM/IEEE 2nd Intl. Conf. on Cyber-Physical Systems (2011)

[17] Jee, E., Wang, S., Kim, J.K., Lee, J., Sokolsky, O., Lee, I.: A Safety-Assured Development Approach for Real-Time Software. In: The Proceedings of 16th IEEE International Conference on Embedded and Real-Time Computing Systems and Applications, pp. 133–142 (2010)

[18] Tuan, L.A., Zheng, M.C., Tho, Q.T.: Modeling and Verification of Safety Critical Systems: A Case Study on Pacemaker. In: Fourth International Conference on Secure Software Integration and Reliability Improvement, pp. 23–32 (2010)

[19] Wiggelinkhuizen, J.E.: Feasibility of Formal Model Checking in the Vitatron Environment. Master thesis, Eindhoven University of Technology (2007)

[20] Macedo, H.D., Larsen, P.G., Fitzgerald, J.S.: Incremental Development of a Distributed Real-Time Model of a Cardiac Pacing System Using VDM. In: Cuellar, J., Sere, K. (eds.) FM 2008. LNCS, vol. 5014, pp. 181–197. Springer, Heidelberg (2008)

[21] Gomes, A.O., Oliveira, M.V.M.: Formal Specification of a Cardiac Pacing System. In: Cavalcanti, A., Dams, D.R. (eds.) FM 2009. LNCS, vol. 5850, pp. 692–707. Springer, Heidelberg (2009)

[22] Mery, D., Singh, N.K.: Pacemaker's Functional Behaviors in Event-B. Research report, INRIA (2009)

Counter-Example Guided Fence Insertion under TSO

Parosh Aziz Abdulla[1], Mohamed Faouzi Atig[1], Yu-Fang Chen[2],
Carl Leonardsson[1], and Ahmed Rezine[3]

[1] Uppsala University, Sweden
[2] Academia Sinica, Taiwan
[3] Linköping University, Sweden

Abstract. We give a *sound* and *complete* fence insertion procedure for concurrent finite-state programs running under the classical TSO memory model. This model allows "write to read" relaxation corresponding to the addition of an unbounded store buffer between each processor and the main memory. We introduce a novel machine model, called the *Single-Buffer* (SB) semantics, and show that the reachability problem for a program under TSO can be reduced to the reachability problem under SB. We present a simple and effective backward reachability analysis algorithm for the latter, and propose a counter-example guided fence insertion procedure. The procedure is augmented by a *placement constraint* that allows the user to choose places inside the program where fences may be inserted. For a given placement constraint, we automatically infer all minimal sets of fences that ensure correctness. We have implemented a prototype and run it successfully on all standard benchmarks together with several challenging examples that are beyond the applicability of existing methods.

1 Introduction

Modern concurrent process architectures allow *weak (relaxed)* memory models, in which certain memory operations may overtake each other. The use of weak memory models makes reasoning about the behaviors of concurrent programs much more difficult and error-prone compared to the classical Sequentially Consistent (SC) memory model. In fact, several algorithms that are designed for the synchronization of concurrent processes, such as mutual exclusion and producer-consumer protocols, are not correct when run on weak memories [2]. One way to eliminate the non-desired behaviors resulting from the use of weak memory models is to insert memory *fence* instructions in the program code. In this work, a fence instruction forbids reordering between instructions issued by the same process. It does not allow any operation issued after the fence instruction to overtake an operation issued before it. Hence, a naive approach to correct a program running under a weak memory model is to insert a fence instruction after every operation. Adopting this approach results in significant performance degradation [13]. Therefore, it is important to optimize fence placement. A natural criterion is to provide *minimal* sets of fences whose insertion is sufficient for ensuring program correctness under the considered weak memory model (provided correctness under SC).

One of the most common relaxations corresponds to TSO (Total Store Ordering) that is adopted by Sun's SPARC multiprocessors. TSO is the kernel of many common

C. Flanagan and B. König (Eds.): TACAS 2012, LNCS 7214, pp. 204–219, 2012.

weak memory models [28,31], and is the latest formalization of the x86 memory model. In TSO, read operations are allowed to overtake write operations of the same process if they concern different variables. In this paper, we use the usual formal model of TSO, developed in e.g. [28,30], and assume it gives a faithful description of the actual hardware on which we run our programs. This model adds an unbounded FIFO buffer between each process and the main memory.

Our Approach. We present a sound and complete method for checking safety properties and for inserting fences in finite-state programs running on the TSO model. The procedure is parameterized by a fence placement constraint that allows to restrict the places inside the program where fences may be inserted. To cope with the unbounded store buffers in the case of TSO, we present a new semantics, called the *Single-Buffer (SB)* semantics, in which all the processes share one (unbounded) buffer. We show that the SB semantics is equivalent to the operational model of TSO (as defined in [30]). A crucial feature of the SB semantics is that it permits a natural ordering on the (infinite) set of configurations, and that the induced transition relation is monotonic wrt. this ordering. This allows to use general frameworks for *well quasi-ordered systems* [1] in order to derive verification algorithms for programs running on the SB model. In case the program fails to satisfy the specification with the current set of fences, our algorithm provides counter-examples (traces) that can be used to increase the set of fences in a systematic manner. Thus, we get a counter-example guided procedure for refining the sets of fences. We prove termination of the obtained procedure. Since each refinement step is performed based on an exact reachability analysis algorithm, the procedure will eventually return all minimal sets of fences (wrt. the given placement constraint) that ensure correctness of the program. Although we instantiate our framework to the case of TSO, the method can be extended to other memory models such as the PSO model.

Contribution. We present the first *sound* and *complete* procedure for fence insertion for programs under TSO. The main ingredients of the framework are the following: (i) A new semantical model, the so called SB model, that allows efficient infinite state model checking. (ii) A simple and effective backward analysis algorithm for solving the reachability problem under the SB semantics. (iii) The algorithm uses finite-state automata as a symbolic representation for infinite sets of configurations, and returns a symbolic counter-example in case the program violates its specification. (iv) A counter-example guided procedure that automatically infers all minimal sets of fences sufficient for correctness under a given fence placement policy. (v) Based on the algorithm, we have implemented a prototype, and run it successfully on several challenging concurrent programs, including some that cannot be handled by existing methods.

Proofs, implementation details and experimental results are in the appendix.

Related Work. To our knowledge, our approach is the first sound and complete automatic fence insertion method that discovers all minimal sets of fences for finite-state concurrent programs running under TSO. Since we are dealing with infinite-state verification, it is hard to provide methods that are both automatic and that return exact solutions. Existing approaches avoid solving the general problem by considering *under-approximations, over-approximations, restricted* classes of programs, *forbidding* sequential inconsistent behavior, or by proposing exact algorithms for which termination is *not* guaranteed. Under-approximations of the program behavior can be achieved through testing [9], bounded model checking [7,6], or by restricting the behavior of the program, e.g., through bounding the sizes of the buffers [18] or the number of switches [5]. Such techniques are useful in practice for finding errors. However, they are not able to check all possible traces and can therefore not tell whether the generated set of fences is sufficient for correctness. Recent techniques based on over-approximations [19] are valuable for showing correctness; however they are not complete and might not be able to prove correctness although the program satisfies its specification. Hence, the computed set of fences need not be minimal. Examples of restricted classes of programs include those that are free from different types of data races [27]. Considering only data-race free programs can be unrealistic since data races are very common in efficient implementations of concurrent algorithms. Another approach is to use monitors [3,8,10], compiler techniques [12], and explicit state model checking [16] to insert fences in order to remove all non-sequential consistent behaviors even if these will not violate the desired correctness properties. As a result, this approach cannot guarantee to generate minimal sets of fences to make programs correct because they also remove benign sequentially inconsistent behaviors. The method of [23] performs an exact search of the state space, combined with fixpoint acceleration techniques, to deal with the potentially infinite state space. However, in general, the approach does not guarantee termination. State reachability for TSO is shown to be non primitive recursive in [4] by reductions to/from lossy channel systems. The reductions involve nondeterministically guessing buffer contents, which introduces a serious state space explosion problem. The approach does not discuss fence insertion and cannot even verify the simplest examples. An important contribution of our work is the introduction of a single buffer semantics for avoiding the immediate state space explosion. In contrast to the above approaches, our method is efficient and performs *exact* analysis of the program on the given memory model. We show termination of the analysis. As a consequence, we are able to compute all *minimal* sets of fences required for correctness of the program.

2   Preliminaries

In this section we first introduce notations that we use throughout the paper, and then define a model for concurrent systems.

Notation. We use $\mathbb{N}$ to denote the set of natural numbers. For sets A and B, we use $[A \mapsto B]$ to denote the set of all total functions from A to B and $f : A \mapsto B$ to denote that f is a total function that maps A to B. For $a \in A$ and $b \in B$, we use $f[a \hookleftarrow b]$ to denote the function f' defined as follows: $f'(a) = b$ and $f'(a') = f(a')$ for all $a' \neq a$.

Let Σ be a finite alphabet. We denote by Σ^* (resp. Σ^+) the set of all *words* (resp. non-empty words) over Σ, and by ε the empty word. The length of a word $w \in \Sigma^*$ is denoted by $|w|$; we assume that $|\varepsilon| = 0$. For every $i : 1 \leq i \leq |w|$, let $w(i)$ be the symbol at position i in w. For $a \in \Sigma$, we write $a \in w$ if a appears in w, i.e., $a = w(i)$ for some $i : 1 \leq i \leq |w|$. For words w_1, w_2, we use $w_1 \cdot w_2$ to denote the concatenation of w_1 and w_2. For a word $w \neq \varepsilon$ and $i : 0 \leq i \leq |w|$, we define $w \odot i$ to be the suffix of w we get by deleting the prefix of length i, i.e., the unique w_2 such that $w = w_1 \cdot w_2$ and $|w_1| = i$.

A transition system $\mathcal{T}$ is a triple $(\mathsf{C}, \mathtt{Init}, \rightarrow)$ where C is a (potentially infinite) set of *configurations*, $\mathtt{Init} \subseteq \mathsf{C}$ is the set of *initial configurations*, and $\rightarrow \subseteq \mathsf{C} \times \mathsf{C}$ is the *transition relation*. We write $c \rightarrow c'$ to denote that $(c, c') \in \rightarrow$, and $\xrightarrow{*}$ to denote the reflexive transitive closure of $\rightarrow$. A configuration c is said to be *reachable* if $c_0 \xrightarrow{*} c$ for some $c_0 \in \mathtt{Init}$; and a set C of configurations is said to be *reachable* if some $c \in C$ is reachable. A *run* π of $\mathcal{T}$ is of the form $c_0 \rightarrow c_1 \rightarrow \cdots \rightarrow c_n$, where $c_i \rightarrow c_{i+1}$ for all $i : 0 \leq i < n$. Then, we write $c_0 \xrightarrow{\pi} c_n$. We use $target(\pi)$ to denote the configuration c_n. Notice that, for configurations c, c', we have that $c \xrightarrow{*} c'$ iff $c \xrightarrow{\pi} c'$ for some run π. The run π is said to be a *computation* if $c_0 \in \mathtt{Init}$. Two runs $\pi_1 = c_0 \rightarrow c_1 \rightarrow \cdots \rightarrow c_m$ and $\pi_2 = c_{m+1} \rightarrow c_{m+2} \rightarrow \cdots \rightarrow c_n$ are said to be *compatible* if $c_m = c_{m+1}$. Then, we write $\pi_1 \bullet \pi_2$ to denote the run $\pi = c_0 \rightarrow c_1 \rightarrow \cdots \rightarrow c_m \rightarrow c_{m+2} \rightarrow \cdots \rightarrow c_n$. Given an ordering $\sqsubseteq$ on C, we say that $\rightarrow$ is *monotonic* wrt. $\sqsubseteq$ if whenever $c_1 \rightarrow c_1'$ and $c_1 \sqsubseteq c_2$, there exists a c_2' s.t. $c_2 \xrightarrow{*} c_2'$ and $c_1' \sqsubseteq c_2'$. We say that $\rightarrow$ is *effectively monotonic* wrt. $\sqsubseteq$ if, given configurations c_1, c_1', c_2 as above, we can compute c_2' and a run π s.t. $c_2 \xrightarrow{\pi} c_2'$.

Concurrent Programs. We define *concurrent programs*, a model for representing shared-memory concurrent processes. A concurrent program $\mathcal{P}$ has a finite number of finite-state processes (threads), each with its own program code. Communication between processes is performed through a shared-memory that consists of a fixed number of shared variables (finite domains) to which all threads can read and write.

We assume a finite set X of *variables* ranging over a finite data domain V. A *concurrent program* is a pair $\mathcal{P} = (P, A)$ where P is a finite set of *processes* and $A = \{A_p | p \in P\}$ is a set of extended finite-state automata (one automaton A_p for each process $p \in P$). The automaton A_p is a triple $\left(Q_p, q_p^{init}, \Delta_p \right)$ where Q_p is a finite set of *local states*, $q_p^{init} \in Q_p$ is the *initial* local state, and Δ_p is a finite set of *transitions*. Each transition is a triple (q, op, q') where $q, q' \in Q_p$ and op is an *operation*. An operation is of one of the following five forms: (1) *"no operation"* nop, (2) *read operation* $\mathsf{r}(x, v)$, (3) *write operation* $\mathsf{w}(x, v)$, (4) *fence operation* fence, and (5) *atomic read-write operation* $\mathsf{arw}(x, v, v')$, where $x \in X$, and $v, v' \in V$. For a transition $t = (q, op, q')$, we use $source(t)$, $operation(t)$, and $target(t)$ to denote q, op, and q' respectively. We define $Q := \cup_{p \in P} Q_p$ and $\Delta := \cup_{p \in P} \Delta_p$. A *local state definition* $\underline{q}$ is a mapping $P \mapsto Q$ such that $\underline{q}(p) \in Q_p$ for each $p \in P$.

3   TSO Semantics

We describe the TSO model formalized in [28,30]. Conceptually, the model adds a FIFO buffer between each process and the main memory. The buffer is used to store the write operations performed by the process. Thus, a process executing a write instruction inserts it into its store buffer and immediately continues executing subsequent instructions. Memory updates are then performed by non-deterministically choosing a process and by executing the first write operation in its buffer (the left-most element in the buffer). A read operation by a process p on a variable x can overtake some write operations stored in its own buffer if all these operations concern variables that are different from x. Thus, if the buffer contains some write operations to x, then the read value must correspond to the value of the most recent write operation to x. Otherwise, the value is fetched from the memory. A fence means that the buffer of the process must be flushed before the program can continue beyond the fence. Notice that the store buffers of the processes are *unbounded* since there is *a priori* no limit on the number of write operations that can be issued by a process before a memory update occurs. Below we define the transition system induced by a program running under the TSO semantics. To do that, we define the set of configurations and transition relation. We fix a concurrent program $\mathcal{P} = (P, A)$.

Formal Semantics. A *TSO-configuration* c is a triple $(q, \underline{b}, mem)$ where q is a local state definition, $\underline{b} : P \mapsto (X \times V)^*$, and $mem : X \mapsto V$. Intuitively, $q(p)$ gives the local state of process p. The value of $\underline{b}(p)$ is the content of the buffer belonging to p. This buffer contains a sequence of write operations, where each write operation is defined by a pair, namely a variable x and a value v that is assigned to x. In our model, messages will be appended to the buffer from the right, and fetched from the left. Finally, mem defines the value of each variable in the memory. We use C_{TSO} to denote the set of TSO-configurations. We define the transition relation $\rightarrow_{TSO}$ on C_{TSO}. The relation is induced by (1) members of Δ; and (2) a set $\Delta' := \{ \text{update}_p | \ p \in P \}$ where update_p is an operation that updates the memory using the first message in the buffer of process p. For configurations $c = (q, \underline{b}, mem)$, $c' = (q', \underline{b}', mem')$, a process $p \in P$, and $t \in \Delta_p \cup \{ \text{update}_p \}$, we write $c \overset{t}{\rightarrow}_{TSO} c'$ to denote that one of the following conditions is satisfied:

- **Nop:** $t = (q, \text{nop}, q')$, $q(p) = q$, $q' = q[p \hookleftarrow q']$, $\underline{b}' = \underline{b}$, and $mem' = mem$. The process changes its local state while buffer and memory contents remain unchanged.
- **Write to store:** $t = (q, \text{w}(x, v), q')$, $q(p) = q$, $q' = q[p \hookleftarrow q']$, $\underline{b}' = \underline{b}[p \hookleftarrow \underline{b}(p) \cdot (x, v)]$, and $mem' = mem$. The write operation is appended to the tail of the buffer.
- **Update:** $t = \text{update}_p$, $q' = q$, $\underline{b} = \underline{b}'[p \hookleftarrow (x, v) \cdot \underline{b}'(p)]$, and $mem' = mem[x \hookleftarrow v]$. The write in the head of the buffer is removed and memory is updated accordingly.
- **Read:** $t = (q, \text{r}(x, v), q')$, $q(p) = q$, $q' = q[p \hookleftarrow q']$, $\underline{b}' = \underline{b}$, $mem' = mem$, and one of the following two conditions is satisfied:
  - **Read own write:** There is an $i : 1 \leq i \leq |\underline{b}(p)|$ such that $\underline{b}(p)(i) = (x, v)$, and $(x, v') \notin (\underline{b}(p) \odot i)$ for all $v' \in V$. If there is a write operation on x in the buffer of p then we consider the most recent of such a write operation (the right-most one in the buffer). This operation should assign v to x.
  - **Read memory:** $(x, v') \notin \underline{b}(p)$ for all $v' \in V$ and $mem(x) = v$. If there is no write operation on x in the buffer of p then the value v of x is fetched from memory.
- **Fence:** $t = (q, \text{fence}, q')$, $q(p) = q$, $q' = q[p \hookleftarrow q']$, $\underline{b}(p) = \varepsilon$, $\underline{b}' = \underline{b}$, and $mem' = mem$. A fence operation may be performed by a process only if its buffer is empty.
- **ARW:** $t = (q, \text{arw}(x, v, v'), q')$, $q(p) = q$, $q' = q[p \hookleftarrow q']$, $\underline{b}(p) = \varepsilon$, $\underline{b}' = \underline{b}$, $mem(x) = v$, and $mem' = mem[x \hookleftarrow v']$. The ARW operation corresponds to an atomic compare and swap (or test and set). It can be performed by a process only if its buffer is empty. The operation checks whether the value of x is v. In such a case, it changes its value to v'.

We use $c \rightarrow_{TSO} c'$ to denote that $c \overset{t}{\rightarrow}_{TSO} c'$ for some $t \in \Delta \cup \Delta'$. The set Init_{TSO} of *initial* TSO-configurations contains all configurations of the form $(q_{init}, \underline{b}_{init}, mem_{init})$ where, for all $p \in P$, we have that $q_{init}(p) = q_p^{init}$ and $b_{init}(p) = \varepsilon$. In other words, each process is in its initial local state and all the buffers are empty. On the other hand, the memory may have any initial value. The transition system induced by a concurrent system under the TSO semantics is then given by $(C_{TSO}, \text{Init}_{TSO}, \rightarrow_{TSO})$.

The TSO Reachability Problem. Given a set Target of local state definitions, we use *Reachable(TSO)* $(\mathcal{P})$ (Target) to be a predicate that indicates the reachability of the set $\{(q, \underline{b}, mem) \mid q \in \text{Target}\}$, i.e., whether a configuration c, where the local state definition of c belongs to Target, is reachable. The reachability problem for TSO is to check, for a given Target, whether *Reachable(TSO)* $(\mathcal{P})$ (Target) holds or not. Using standard techniques we can reduce checking safety properties to the reachability problem. More precisely, Target denotes "bad configurations" that we do not want to occur during the execution of the system. For instance, for mutual exclusion protocols, the bad configurations are those where the local states of two processes are both in the critical sections. We say that the "program is correct" to indicate that Target is not reachable.

4   Single-Buffer Semantics

The formal model of TSO [28,30] is quite powerful since it uses *unbounded perfect* buffers. However, the reachability problem remains decidable [4]. Our goal is to exploit this to design a practically efficient verification algorithm. To do that, we introduce a new semantics model, called the *Single-Buffer (SB)* model that weaves the buffers of all processes into one unified buffer. The SB model satisfies two important properties (1) it is equivalent to the TSO semantics wrt. reachability, i.e., Target is reachable in the TSO semantics iff it is reachable in the SB semantics; (2) the induced transition system is "monotonic" wrt. some pre-order (on configurations) so that the classical infinite state model checking framework of [1] can be applied. Fix a concurrent system $\mathcal{P} = (P, A)$.

Formal Semantics. A *SB-configuration* c is a triple $(q, b, \underline{z})$ where q is (as in the case of TSO-semantics) a local state definition, $b \in ([X \mapsto V] \times P \times X)^+$, and $\underline{z} : P \mapsto \mathbb{N}$. Intuitively, the (only) buffer contains triples of the form (mem, p, x) where mem defines variable values (encoding a memory snapshot), x is the latest variable that has been written into, and p is the process that performed the write operation. Furthermore, $\underline{z}$ represents a set of *pointers* (one per process) where, from the point of view of p, the word $b \odot \underline{z}(p)$ is the sequence of write operations that have not yet been used for memory updates and the first element of the triple $b(\underline{z}(p))$ represents the memory content. As we shall see below, the buffer will never be empty, since it is not empty in an initial configuration, and since no messages are ever removed from it during a run of the system (in the SB semantics, the update operation moves a pointer to the right instead of removing a message from the buffer). This implies (among other things) that the invariant $\underline{z}(p) > 0$ is always maintained. We use C_{SB} to denote the set of SB-configurations.

Let $c = (q, b, \underline{z})$ be an SB-configuration. For every $p \in P$ and $x \in X$, we use LastWrite(c, p, x) to denote the index of the most recent buffer message where p writes to x or the current memory of p if the aforementioned type of message does not exist in the buffer from the point of view of p. Formally, LastWrite(c, p, x) is the largest index i such that $i = \underline{z}(p)$ or $b(i) = (mem, p, x)$ for some mem.

We define the transition relation $\rightarrow_{SB}$ on the set of SB-configurations as follows. In a similar manner to the case of TSO, the relation is induced by members of $\Delta \cup \Delta'$. For configurations $c = (q, b, \underline{z})$, $c' = (q', b', \underline{z}')$, and $t \in \Delta_p \cup \{\text{update}_p\}$, we write $c \xrightarrow{t}_{SB} c'$ to denote that one of the following conditions is satisfied:

- Nop: $t = (q, \text{nop}, q')$, $\underline{q}(p) = q$, $\underline{q}' = \underline{q}[p \hookleftarrow q']$, $b' = b$ and $\underline{z}' = \underline{z}$. The operation changes only the local state of p.
- Write to store: $t = (q, \text{w}(x, v), q')$, $\underline{q}(p) = q$, $\underline{q}' = \underline{q}[p \hookleftarrow q']$, $b(|b|)$ is of the form (mem_1, p_1, x_1), $b' = b \cdot (mem_1 [x \hookleftarrow v], p, x)$, and $\underline{z}' = \underline{z}$. A new element is appended to the tail of the buffer. Values of variables in the new element are identical to those in the previous last element except that the value of x has been updated to v. Furthermore, we include the updating process p and the updated variable x.
- Update: $t = \text{update}_p$, $\underline{q}' = \underline{q}$, $b' = b$, $\underline{z}(p) < |b|$ and $\underline{z}' = \underline{z}[p \hookleftarrow \underline{z}(p) + 1]$. An update operation (as seen by p) is simulated by moving the pointer of p one step to the right. This means that we remove the oldest write operation that is yet to be used for a memory update. The removed element will now represent the memory contents from the point of view of p.
- Read: $t = (q, \text{r}(x, v), q')$, $\underline{q}(p) = q$, $\underline{q}' = \underline{q}[p \hookleftarrow q']$, $b' = b$, and $b(\text{LastWrite}(c, p, x)) = (mem_1, p_1, x_1)$ for some mem_1, p_1, x_1 with $mem_1(x) = v$.
- Fence: $t = (q, \text{fence}, q')$, $\underline{q}(p) = q$, $\underline{q}' = \underline{q}[p \hookleftarrow q']$, $\underline{z}(p) = |b|$, $b' = b$, and $\underline{z}' = \underline{z}$. The buffer should be empty from the point of view of p when the transition is performed. This is encoded by the equality $\underline{z}(p) = |b|$.
- ARW: $t = (q, \text{arw}(x, v, v'), q')$, $\underline{q}(p) = q$, $\underline{q}' = \underline{q}[p \hookleftarrow q']$, $\underline{z}(p) = |b|$, $b(|b|)$ is of the form (mem_1, p_1, x_1), $mem_1(x) = v$, $b' = b \cdot (mem_1 [x \hookleftarrow v'], p, x)$, and $\underline{z}' = \underline{z}[p \hookleftarrow \underline{z}(p) + 1]$. The fact that the buffer is empty from the point of view of p is encoded by the equality $\underline{z}(p) = |b|$. The content of the memory can then be fetched from the right-most element $b(|b|)$ in the buffer. To encode that the buffer is still empty after the operation (from the point of view of p) the pointer of p is moved one step to the right.

We use $c \rightarrow_{SB} c'$ to denote that $c \xrightarrow{t}_{SB} c'$ for some $t \in \Delta \cup \Delta'$. The set Init_{SB} of initial SB-configurations contains all configurations of the form $(\underline{q}_{init}, b_{init}, \underline{z}_{init})$ where $|b_{init}| = 1$, and for all $p \in P$, we have that $\underline{q}_{init}(p) = q_p^{init}$, and $\underline{z}_{init}(p) = 1$. In other words, each process is in its initial local state. The buffer contains a single message, say of the form $(mem_{init}, p_{init}, x_{init})$, where mem_{init} represents the initial value of the memory. The memory may have any initial value. Also, the values of p_{init} and x_{init} are not relevant since they will not be used in the computations of the system. The pointers of all the processes point to the first position in the buffer. According to our encoding, this indicates that their buffers are all empty. The transition system induced by a concurrent system under the SB semantics is then given by $(C_{SB}, \text{Init}_{SB}, \rightarrow_{SB})$.

The SB Reachability Problem. We define the predicate $Reachable(SB)(\mathcal{P})(\text{Target})$, and the reachability problem for the SB semantics, in a similar manner to TSO. The following theorem states equivalence of the reachability problems under TSO and SB semantics. Due to its technicality and lack of space, we leave the proof for the appendix.

Theorem 1. *For a concurrent program $\mathcal{P}$ and a local state definition* Target, *the reachability problems are equivalent under the TSO and SB semantics.*

5   The SB Reachability Algorithm

In this section, we present an algorithm for checking reachability of an (infinite) set of configurations characterized by a (finite) set Target of local state definitions. In addition to answering the reachability question, the algorithm also provides an "error trace" in case Target is reachable. First, we define an ordering $\sqsubseteq$ on the set of SB-configurations, and show that it satisfies two important properties, namely (i) it is a well quasi-ordering (wqo), i.e., for every infinite sequence $c_0, c_1, \ldots$ of SB-configurations, there are $i < j$ with $c_i \sqsubseteq c_j$; and (ii) the SB-transition relation $\rightarrow_{SB}$ is monotonic wrt. $\sqsubseteq$. The algorithm performs backward reachability analysis from the set of configurations with local state definitions that belong to Target. During each step of the search procedure, the algorithm takes the upward closure (wrt. $\sqsubseteq$) of the generated set of configurations. By monotonicity of $\sqsubseteq$ it follows that taking the upward closure preserves exactness of the analysis [1]. From the fact that we always work with upward closed sets and that $\sqsubseteq$ is a wqo it follows that the algorithm is guaranteed to terminate [1]. In the algorithm, we use a variant of finite-state automata, called *SB-automata*, to encode (potentially infinite) sets of SB-configurations.

Ordering.   For an SB-configuration $c = (q, b, \underline{z})$ we define $\texttt{ActiveIndex}(c) := \min\{\underline{z}(p) \mid p \in P\}$. In other words, the part of b to the right of (and including) $\texttt{ActiveIndex}(c)$ is "active", while the part to the left is "dead" in the sense that all its content has already been used for memory updates. The left part is therefore not relevant for computations starting from c.

Let $c = (q, b, \underline{z})$ and $c' = (q', b', \underline{z}')$ be two SB-configurations. Define $j := \texttt{ActiveIndex}(c)$ and $j' := \texttt{ActiveIndex}(c')$. We write $c \sqsubseteq c'$ to denote that (i) $q = q'$ and that (ii) there is an injection $g : \{j, j+1, \ldots, |b|\} \mapsto \{j', j'+1, \ldots, |b'|\}$ such that the following conditions are satisfied. For every $i, i_1, i_2 \in \{j, \ldots, |b|\}$, (1) $i_1 < i_2$ implies $g(i_1) < g(i_2)$, (2) $b(i) = b'(g(i))$, (3) $\texttt{LastWrite}(c', p, x) = g(\texttt{LastWrite}(c, p, x))$ for all $p \in P$ and $x \in X$, and (4) $\underline{z}'(p) = g(\underline{z}(p))$ for all $p \in P$. The first condition means that g is strictly monotonic. The second condition corresponds to that the *active* part of b is a *sub-word* of the *active* part of b'. The third condition ensures the last write indices wrt. all processes and variables are consistent. The last condition ensures each process points to identical elements in b and b'.

We get the following lemma from the fact that (i) the sub-word relation is a well-quasi ordering on finite words [15], and that (ii) the number of states and messages (associated with last write operations and pointers) that should be equal, is finite.

Lemma 1. *The relation $\sqsubseteq$ is a well-quasi ordering on SB-configurations.*

The following lemma shows effective monotonicity (cf. Section 2) of the SB-transition relation wrt. $\sqsubseteq$. As we shall see below, this allows the reachability algorithm to only work with upward closed sets. Monotonicity is used in the termination of the reachability algorithm. The effectiveness aspect is used in the fence insertion algorithm (cf. Section 6).

Lemma 2. $\rightarrow_{SB}$ *is effectively monotonic wrt.* $\sqsubseteq$.

The *upward closure* of a set C is defined as $C\!\uparrow := \{c' \mid \exists c \in C, c \sqsubseteq c'\}$. A set C is *upward closed* if $C = C\!\uparrow$.

SB-Automata. First we introduce an alphabet $\Sigma := ([X \mapsto V] \times P \times X) \times 2^P$. Each element $((mem, p, x), P') \in \Sigma$ represents a single position in the buffer of an SB-configuration. More precisely, the triple (mem, p, x) represents the message stored at that position and the set $P' \subseteq P$ gives the (possibly empty) set of processes whose pointers point to the given position. Consider a word $w = a_1 a_2 \cdots a_n \in \Sigma^*$, where a_i is of the form $((mem_i, p_i, x_i), P_i)$. We say that w is proper if, for each process $p \in P$, there is exactly one $i : 1 \leq i \leq n$ with $p \in P_i$. In other words, the pointer of each process is uniquely mapped to one position in w. A proper word w of the above form can be "decoded" into a (unique) pair $decoding(w) := (b, \underline{z})$, defined by (i) $|b| = n$, (ii) $b(i) = (mem_i, p_i, x_i)$ for all $i : 1 \leq i \leq n$, and (iii) $\underline{z}(p)$ is the unique integer $i : 1 \leq i \leq n$ such that $p \in P_i$ (the value of i is well-defined since w is proper). We extend the function to sets of words where $decoding(W) := \{decoding(w) \mid w \in W\}$.

An SB-automaton A is a tuple $(S, \Delta, S^{final}, h)$ where S is a finite set of *states*, $\Delta \subseteq S \times \Sigma \times S$ is a finite set of transitions, $S^{final} \subseteq S$ is the set of *final* states, and $h : (P \mapsto Q) \mapsto S$. The total function h defines a labeling of the states of A by the local state definitions of the concurrent program $\mathcal{P}$, such that each q is mapped to a state $h(q)$ in A. For a state $s \in S$, we define $L(A, s)$ to be the set of words of the form $w = a_1 a_2 \cdots a_n$ such that there are states $s_0, s_1, \ldots, s_n \in S$ satisfying the following conditions: (i) $s_0 = s$, (ii) $(s_i, a_{i+1}, s_{i+1}) \in \Delta$ for all $i : 0 \leq i < n$, (iii) $s_n \in S^{final}$, and (iv) w is proper. We define the *language* of A by $L(A) := \{(q, b, \underline{z}) \mid (b, \underline{z}) \in decoding(L(A, h(q)))\}$. Thus, the language $L(A)$ characterizes a set of SB-configurations. More precisely, the configuration $(q, b, \underline{z})$ belongs to $L(A)$ if $(b, \underline{z})$ is the decoding of a word that is accepted by A when A is started from the state $h(q)$ (the state labeled by q). A set C of SB-configurations is said to be *regular* if $C = L(A)$ for some SB-automaton A.

Operations on SB-Automata. We show that we can compute the operations (union, intersection, test emptiness, compute predecessor, etc.) needed for the reachability algorithm. First, observe that regular sets of SB-configurations are closed under union and intersection. For SB-automata A_1, A_2, we use $A_1 \cap A_2$ to denote an automaton A such that $L(A) = L(A_1) \cap L(A_2)$. We define $A_1 \cup A_2$ in a similar manner. We use A^0 to denote an (arbitrary) automaton whose language is empty. We can construct SB-automata for the set of initial SB-configurations, and for sets of SB-configurations characterized by local state definitions.

Lemma 3. *We can compute an SB-automaton A^{init} such that $L(A^{init}) = \text{Init}_{SB}$. For a set* Target *of local state definitions, we can compute an SB-automaton $A^{final}(\text{Target})$ such that $L(A^{final}(\text{Target})) := \{(q, b, \underline{z}) \mid q \in \text{Target}\}$.*

The following lemma tells us that regularity of a set is preserved by taking upward closure, and that we in fact can compute an automaton describing its upward closure.

Lemma 4. *For an SB-automaton A we can compute an SB-automaton $A\uparrow$ such that $L(A\uparrow) = L(A)\uparrow$.*

We define the *predecessor function* as follows. Let $t \in \Delta \cup \Delta'$ and let C be a set of SB-configurations. We define $\text{Pre}_t(C) := \{c \mid \exists c' \in C, c \xrightarrow{t}_{SB} c'\}$ to denote the set of immediate predecessor configurations of C w.r.t. the transition t. In other words, $\text{Pre}_t(C)$ is the set of configurations that can reach a configuration in C through a single execution of t. The following lemma shows that Pre preserves regularity, and that in fact we can compute the automaton of the predecessor set.

Lemma 5. *For a transition t and an SB-automaton A, we can compute an SB-automaton* $\mathrm{Pre}_t(A)$ *such that* $L(\mathrm{Pre}_t(A)) = \mathrm{Pre}_t(L(A))$.

Reachability Algorithm. The algorithm performs a symbolic backward reachability analysis [1], where we use SB-automata for representing infinite sets of SB-configurations. In fact, the algorithm also provides *traces* that we will use to find places inside the code where to insert fences (see Section 6). For a set Target of local state definitions, a *trace* δ to Target is a sequence of the form $A_0 t_1 A_1 t_2 \cdots t_n A_n$ where $A_0, A_1, \ldots, A_n$ are SB-automata, $t_1, \ldots, t_n$ are transitions, and (i) $L(A_0) \cap \mathrm{Init}_{SB} \neq \emptyset$; (ii) $A_i = \left(\mathrm{Pre}_{t_{i+1}}(A_{i+1})\right) \uparrow$ for all $i : 0 \leq i < n$ (even if $L(A_{i+1})$ is upward-closed, it is still possible that $L\left(\mathrm{Pre}_{t_{i+1}}(A_{i+1})\right)$ is not upward-closed; however due to

---

Algorithm 1: Reachability

input : A concurrent program $\mathcal{P}$ and a finite set Target of local state definitions.

output: *"unreachable"* if $\neg Reachable(SB)(\mathcal{P})(\text{Target})$ holds. A trace to Target otherwise.

1 $\mathcal{W} \leftarrow \left\{ A^{final}(\text{Target}) \right\}$;
2 $A^{\mathcal{V}} \leftarrow A^0$;
3 **while** $\mathcal{W} \neq \emptyset$ **do**
4   Pick and remove a trace δ from $\mathcal{W}$;
5     $A \leftarrow head(\delta)$;
6   **if** $L\left(A \cap A^{init}\right) \neq \emptyset$ **then return** δ;
7   **if** $L(A) \subseteq L\left(A^{\mathcal{V}}\right)$ **then discard** A;
8   **else**
9        $\mathcal{W} \leftarrow \left\{ \delta' \in \mathcal{W} \mid L(head(\delta')) \not\subseteq L(A) \right\} \cup \left\{ (\mathrm{Pre}_t(A)) \uparrow \cdot t \cdot \delta \mid t \in \Delta \cup \Delta' \right\}$;
10       $A^{\mathcal{V}} \leftarrow A^{\mathcal{V}} \cup A$
11 **return** *"unreachable"*;

---

monotonicity taking upward closure does not affect exactness of the analysis); and (iii) $A_n = A^{final}(\text{Target})$. In the following, we use $head(\delta)$ to denote the SB-automaton A_0. The algorithm inputs a finite set Target, and checks the predicate $Reachable(SB)(\mathcal{P})(\text{Target})$. If the predicate does not hold then Algorithm 1 simply answers *unreachable*; otherwise, it returns a *trace*. It maintains a *working* set $\mathcal{W}$ that contains a set of traces. Intuitively, in a trace $A_0 t_1 A_1 t_2 \cdots t_n A_n \in \mathcal{W}$, the automaton A_0 has been "detected" but not yet "analyzed", while the rest of the trace represents a sequence of transitions and SB-automata that has led to the generation of A_0. The algorithm also maintains an automaton $A^{\mathcal{V}}$ that encodes configurations that have already been analyzed.

Initially, $A^{\mathcal{V}}$ is an automaton recognizing the empty language, and $\mathcal{W}$ is the singleton $\left\{ A^{final}(\text{Target}) \right\}$. In other words, we start with a single trace containing the automaton representing configurations induced by Target (can be constructed by Lemma 3). At the beginning of each iteration, the algorithm picks and removes a trace δ (with head A) from the set $\mathcal{W}$. First it checks whether A intersects with A^{init} (can be constructed by Lemma 3). If yes, it returns the trace δ. If not, it checks whether A is covered by $A^{\mathcal{V}}$ (i.e., $L(A) \subseteq L\left(A^{\mathcal{V}}\right)$). If *yes* then A does not carry any new information and it (together with its trace) can be safely discarded. Otherwise, the algorithm performs the following operations: (i) it discards all elements of $\mathcal{W}$ that are covered by A; (ii) it adds

A to $A^{\mathcal{V}}$; and (iii) for each transition t it adds a trace $A_1 \cdot t \cdot \delta$ to $\mathcal{W}$, where we compute A_1 by taking the predecessor $\mathrm{Pre}_t(A)$ of A wrt. t, and then taking the upward closure (Lemmata 4 and 5). Notice that since we take the upward closure of the generated automata, and since $A^{final}(\mathtt{Target})$ accepts an upward closed set, then $A^{\mathcal{V}}$ and all the automata added to $\mathcal{W}$ accept upward closed sets. The algorithm terminates when $\mathcal{W}$ becomes empty.

Theorem 2. *The reachability algorithm always terminates with the correct answer.*

6   Fence Insertion

Our fence insertion algorithm is parameterized by a predefined *placement constraint* G where $G \subseteq Q$. The algorithm will place fences only after local states that belong to G. This gives the user the freedom to choose between the efficiency of the verification algorithm and the number of fences that are needed to ensure correctness of the program. The weakest placement constraint is defined by taking G to be the set of all local states of the processes, which means that a fence might be placed anywhere inside the program. On the other hand, one might want to place fences only after write operations, place them only before read operations, or avoid putting them within certain loops (e.g., loops that are known to be executed often during the runs of the program). For any given G, the algorithm finds the minimal sets of fences (if any) that are sufficient for correctness. First, we show how to use a trace δ to derive a *counter-example*: an SB-computation that reaches $\mathtt{Target}$. From the counter example, we explain how to derive a set of fences in G such that the insertion of at least one element of the set is necessary in order to eliminate the counter-example. Finally, we introduce the fence insertion algorithm.

Fences. We identify fences with local states. For a concurrent program $\mathcal{P} = (P, A)$ and a fence $f \in Q$, we use $\mathcal{P} \oplus f$ to denote the concurrent program we get by inserting a fence operation just after the local state f in $\mathcal{P}$. Formally, if $f \in Q_p$, for some $p \in P$, then $\mathcal{P} \oplus f := \left(P, \left\{ A'_{p'} \mid p' \in P \right\} \right)$ where $A'_{p'} = A_{p'}$ if $p \neq p'$. Furthermore, if $A_p = (Q_p, q_p^{init}, \Delta_p)$, then we define $A'_p = (Q_p \cup \{q'\}, q_p^{init}, \Delta'_p)$ with $q' \notin Q_p$, and $\Delta'_p = \Delta_p \cup \{(f, \mathtt{fence}, q')\} \cup \{(q', op, q'') \mid (f, op, q'') \in \Delta_p\} \setminus \{(f, op, q'') \mid (f, op, q'') \in \Delta_p\}$. We say F is *minimal* wrt. a set $\mathtt{Target}$ of local state definitions and a placement constraint G if $F \subseteq G$ and $Reachable(SB)(\mathcal{P} \oplus F \setminus \{f\})(\mathtt{Target})$ holds for all $f \in F$ but not $Reachable(SB)(\mathcal{P} \oplus F)(\mathtt{Target})$. We use $F_{min}^G(\mathcal{P})(\mathtt{Target})$ to denote the set of minimal sets of fences in $\mathcal{P}$ wrt. $\mathtt{Target}$ that respect the placement constraint G.

Counter-Example Generation. Consider a trace $\delta = A_0 t_1 A_1 t_2 \cdots t_n A_n$. We show how to derive a counter-example from δ. Formally, a counter-example is a run $c_0 \xrightarrow{t_1}_{SB} c_1 \xrightarrow{t_2}_{SB} \cdots \xrightarrow{t_m}_{SB} c_m$ of the transition system induced from $\mathcal{P}$ under the SB semantics, where $c_0 \in \mathtt{Init}_{SB}$ and $c_m \in \{(q, b, \bar{z}) \mid q \in \mathtt{Target}\}$. We assume a function *choose* that, for each automaton A, chooses a member of $L(A)$ (if $L(A) \neq \emptyset$), i.e., $choose(A) = w$ for some arbitrary but fixed $w \in L(A)$. We will define π using a sequence of configurations $c_0, \ldots, c_n$ where $c_i \in L(A_i)$ for $i : 0 \leq i \leq n$. Define

$c_0 := choose\left(A_0 \cap A^{init}\right)$. The first configuration c_0 in π is a member of the intersection of A_0 and A^{init} (this intersection is not empty by the definition of a trace). Suppose that we have computed c_i for some $i : 0 \leq i < n$. Since $A_i = \text{Pre}_{t_{i+1}}(A_{i+1})\uparrow$ and $c_i \in L(A_i)$, there exist $c_i' \in \text{Pre}_{t_{i+1}}(A_{i+1}) \subseteq L(A_i)$ and $d_{i+1} \in L(A_{i+1})$ such that $c_i' \sqsubseteq c_i$ and $c_i' \xrightarrow{t_{i+1}}_{SB} d_{i+1}$. Since there are only finitely many configurations that are smaller than c_i wrt. $\sqsubseteq$, we can indeed compute both c_i' and d_{i+1}. By Lemma 2, we know we can compute a configuration c_{i+1} and a run π_{i+1} such that $d_{i+1} \sqsubseteq c_{i+1}$ and $c_i \xrightarrow{\pi_{i+1}}_{SB} c_{i+1}$. Since $L(A_{i+1}\uparrow)$ is upward closed, we know that $c_{i+1} \in L(A_{i+1}\uparrow)$. We define $\pi := c_0 \bullet \pi_1 \bullet c_1 \bullet \pi_2 \bullet \cdots \bullet \pi_n \bullet c_n$. We use $\text{CounterEx}(\delta)$ to denote such a π.

Fence Inference. We will identify points along a counter-example $\pi = c_0 \xrightarrow{t_1}_{SB} c_1 \xrightarrow{t_2}_{SB} \cdots \xrightarrow{t_{n-1}}_{SB} c_{n-1} \xrightarrow{t_n}_{SB} c_n$ at which read operations overtake write operations and derive a set of fences such that any one of them forbids such an overtaking. We do this in several steps. Let c_i be of the form $\left(q_i, b_i, z_i\right)$. Define $n_i := |b_i|$. First, we define a sequence of functions $\alpha_0, \ldots, \alpha_n$ where α_i associates to each message in the buffer b_i the position in π of the write transition that gave rise to the message. Below we explain how to generate those α functions. The first message $b_i(1)$ in each buffer represents the initial state of memory. It has not been generated by any write transition, and therefore $\alpha_i(1)$ is undefined. Since b_0 contains exactly one message, $\alpha_0(j)$ is undefined for all j. If t_{i+1} is not a write transition then define $\alpha_{i+1} := \alpha_i$ (no new message is appended to the buffer, so all transitions associated to all messages have been defined). Otherwise, we define $\alpha_{i+1}(j) := \alpha_i(j)$ if $2 \leq j \leq n_i$ and define $\alpha_{i+1}(n_i + 1) := i+1$. In other words, a new message will be appended to the end of the buffer (placed at position $n_{i+1} = n_i + 1$); and to this message we associate $i+1$ (the position in π of the write transition that generated the message).

Next, we identify the write transitions that have been overtaken by read operations. Concretely, we define a function Overtaken such that, for each $i : 1 \leq i \leq n$, if t_i is a read transition then the value $\text{Overtaken}(\pi)(i)$ gives the positions of the write transitions in π that have been overtaken by the read operation. Formally, if t_i is not a read transition define $\text{Overtaken}(\pi)(i) := \emptyset$. Otherwise, assume that $t_i = (q, r(x,v), q') \in \Delta_p$ for some $p \in P$. We have $\text{Overtaken}(\pi)(i) := \left\{\alpha_i(j) \mid \text{LastWrite}(c_i, p, x) < j \leq n_i \wedge t_{\alpha_i(j)} \in \Delta_p\right\}$. In other words, we consider the process p that has performed the transition t_i and the variable x whose value is read by p in t_i. We search for pending write operations issued by p on variables different from x. These are given by transitions that (i) belong to p and (ii) are associated with messages inside the buffer that belong to p and that are yet to be used for updating the memory (they are in the postfix of the buffer to the right of $\text{LastWrite}(c_i, p, x)$).

Finally, we notice that, for each $i : 1 \leq i \leq n$ and each $j \in \text{Overtaken}(\pi)(i)$, the pair (j, i) represents the position j of a write operation and the position i of a read operation that overtakes the write operation. Therefore, it is necessary to insert a fence at least in one position between such a pair in order to ensure that we eliminate at least one of the overtakings that occur along π. Furthermore, we are only interested in local states that belong to the placement constraint G. To reflect this, we define $\text{Barrier}(G)(\pi) := \left\{q_k(p) \mid \exists i : 1 \leq i \leq n. \exists j \in \text{Overtaken}(\pi)(i). j \leq k < i\right\} \cap G$.

Algorithm. Our fence insertion algorithm (Algorithm 2) inputs a concurrent program $\mathcal{P}$, a placement constraint G, and a finite set Target of local state definitions, and returns all minimal sets of fences ($F_{min}^G(\mathcal{P})$(Target)). If this set is empty then we conclude that the program cannot be made correct by placing fences in G. In this case, and if $G = Q$ (or indeed, if G includes sources of all read operations or destinations of all write operations), the program is not correct even under SC-semantics (hence no set of fences can make it correct).

Theorem 3. *For a concurrent program $\mathcal{P}$, a placement constraint G, and a finite set Target, Algorithm 2 terminates and returns $F_{min}^G(\mathcal{P})$(Target).*

Algorithm 2: Fence Inference

---

input : concurrent program $\mathcal{P}$, placement
           constraint G, local state definitions
           Target.
output: $F_{min}^G(\mathcal{P})$(Target).

1   $\mathcal{W} \leftarrow \{\emptyset\}$;
2   $C \leftarrow \emptyset$;
3   **while** $\mathcal{W} \neq \emptyset$ **do**
4   Pick and remove a set F from $\mathcal{W}$;
5   **if** $Reachable(SB)(\mathcal{P} \oplus F)$(Target) $= \delta$ **then**
6       $F_B \leftarrow \text{Barrier}(G)(\text{CounterEx}(\delta))$;
7   **if** $F_B = \emptyset$ **then**
8   **return** $\emptyset$
9   **else foreach** $f \in F_B$ **do**
10         $F' \leftarrow F \cup \{f\}$;
11   **if** $\exists F'' \in C \cup \mathcal{W}. F'' \subseteq F'$ **then**
12   discard F'
13   **else** $\mathcal{W} \leftarrow \mathcal{W} \cup \{F'\}$
14   **else**
15       $C \leftarrow C \cup \{F\}$
16 **return** C;

Remark 1. If only a smallest minimal set is of interest, then it is sufficient to implement $\mathcal{W}$ as a queue and to return the first added element to C.

7   Experimental Results

We have evaluated our approach on several benchmark examples including some difficult problem sets that cannot be handled by *any previous approaches*. We have implemented Algorithm 2 in OCaml and run the experiments using a laptop computer with an Intel Core i3 2.26 GHz CPU and 4GB of memory. Table 1 summarizes our results. The placement constraint only allows fences immediately after write operations. The experiments were run in two modes: one until the first minimal set of fences is found, and one where all minimal sets of fences are found. For each concurrent program we give the program size (number of processes, number of states, variables and transitions), the total required time in seconds, the number of inserted fences in the smallest minimal fence set and the number of minimal fence sets.

    Our implementation is able to verify all above examples. This is beyond the capabilities of previous approaches. In particular, none of our examples is data-race free. Furthermore, some of our examples may generate an arbitrary number of messages inside the buffers and they may have sequential inconsistent behaviors. To the best of our knowledge, only the approaches in [19] and in [22] are potentially able to handle such general classes of problems. However, the approach of [22] does not guarantee termination. The work in [19] abstracts away the *order* between buffer messages, and hence it cannot handle examples where the order of messages sent to the buffer is crucial (such as the "Increasing Sequence" example in the table). See the appendix for further details.

Table 1. Analyzed concurrent programs

	Size Proc./States/Var./Trans	Total time seconds (one fence set)	Total time seconds (all fence sets)	Fences necessary (smallest set)	Number of minimal fence sets
1. Simple Dekker [31]	2/8/2/10	0.02	0.02	1 per process	1
2. Full Dekker [11]	2/14/3/18	0.28	0.28	1 per process	1
3. Peterson [29]	2/10/3/14	0.24	0.6	1 per process	1
4. Lamport Bakery [20]	2/22/4/32	52	5538	2 per process	4
5. Lamport Fast [21]	2/26/4/38	6.5	6.5	2 per process	1
6. CLH Queue Lock[25]	2/48/4/60	26	26	0	1
7. Sense Reversing Barrier [26]	2/16/2/24	1.1	1.1	0	1
8. Burns [24]	2/9/2/11	0.07	0.07	1 per process	1
9. Dijkstra [24]	2/14/3/24	9.5	10	1 per process	1
10. Tournament Barriers [14]	2/8/2/8	1.2	1.2	0	1
11. A Task Scheduling Algorithm	3/7/2/9	60	60	0	1
12. Increasing Sequence	2/26/1/44	25	27	0	1
13. Alternating Bit	2/8/2/12	0.2	0.2	0	1
14. Producer Consumer, v1, N=2	18/3/22	0.2	0.2	Erroneous	0
15. Producer Consumer, v1, N=3	22/4/28	4.5	4.5	Erroneous	0
16. Producer Consumer, v2, N=2	14/3/18	5.7	5.7	0	1
17. Producer Consumer, v2, N=3	16/4/22	580	583	0	1

8 Conclusion

We have presented a sound and complete method for automatic fence insertion in finite-state programs running under the TSO memory model, based on a new (so called) SB-semantics. We have automatically verified several challenging examples, including some that cannot be handled by existing approaches. The design of the new SB semantics is not a trivial task. For instance, "obvious" variants such as simply making the buffer in TSO "lossy", or removing the pointers or storing less information inside the messages of the SB-buffer would fail, since they yield either over- or under-approximations (even wrt. reachability properties). Also the ordering we define on SB configurations cannot be "translated back" to an ordering on TSO configuration (this would make it possible to apply our method directly on TSO rather than on the SB semantics). The reason is that standard proofs that show reductions between different semantics (models), where each configuration in one model is shown to be in (bi-)simulation with a configuration in the other model cannot be used here. Given an SB-configuration, it is not obvious how to define an "equivalent" TSO configuration, and vice versa. However (crucially, as shown in the proof of Theorem 1) we show that each computation in one semantics violating/satisfying a given safety property is simulated by a (whole) computation that violates/satisfies the same safety property in the other. Our method can be carried over to other memory models such as PSO in a straightforward manner. In the future, we plan to apply our techniques to more memory models and to combine with predicate abstraction for handling programs with unbounded data.

References

1. Abdulla, P.A., Cerans, K., Jonsson, B., Tsay, Y.-K.: General decidability theorems for infinite-state systems. In: LICS (1996)
2. Adve, S., Gharachorloo, K.: Shared memory consistency models: a tutorial. Computer 29(12) (1996)

3. Alglave, J., Maranget, L.: Stability in Weak Memory Models. In: Gopalakrishnan, G., Qadeer, S. (eds.) CAV 2011. LNCS, vol. 6806, pp. 50–66. Springer, Heidelberg (2011)
4. Atig, M.F., Bouajjani, A., Burckhardt, S., Musuvathi, M.: On the verification problem for weak memory models. In: POPL (2010)
5. Atig, M.F., Bouajjani, A., Parlato, G.: Getting Rid of Store-Buffers in TSO Analysis. In: Gopalakrishnan, G., Qadeer, S. (eds.) CAV 2011. LNCS, vol. 6806, pp. 99–115. Springer, Heidelberg (2011)
6. Burckhardt, S., Alur, R., Martin, M.: CheckFence: Checking consistency of concurrent data types on relaxed memory models. In: PLDI (2007)
7. Burckhardt, S., Alur, R., Martin, M.M.K.: Bounded Model Checking of Concurrent Data Types on Relaxed Memory Models: A Case Study. In: Ball, T., Jones, R.B. (eds.) CAV 2006. LNCS, vol. 4144, pp. 489–502. Springer, Heidelberg (2006)
8. Burckhardt, S., Musuvathi, M.: Effective Program Verification for Relaxed Memory Models. In: Gupta, A., Malik, S. (eds.) CAV 2008. LNCS, vol. 5123, pp. 107–120. Springer, Heidelberg (2008)
9. Burnim, J., Sen, K., Stergiou, C.: Testing concurrent programs on relaxed memory models. Technical Report UCB/EECS-2010-32, UCB (2010)
10. Burnim, J., Sen, K., Stergiou, C.: Sound and Complete Monitoring of Sequential Consistency for Relaxed Memory Models. In: Abdulla, P.A., Leino, K.R.M. (eds.) TACAS 2011. LNCS, vol. 6605, pp. 11–25. Springer, Heidelberg (2011)
11. Dijkstra, E.W.: Cooperating sequential processes. Springer-Verlag New York, Inc., New York (2002)
12. Fang, X., Lee, J., Midkiff, S.P.: Automatic fence insertion for shared memory multiprocessing. In: ICS. ACM (2003)
13. Fraser, K.: Practical lock-freedom. Technical Report UCAM-CL-TR-579, University of Cambridge, Computer Laboratory (2004)
14. Hensgen, D., Finkel, R., Manber, U.: Two algorithms for barrier synchronization. IJPP 17 (February 1988)
15. Higman, G.: Ordering by divisibility in abstract algebras. Proc. London Math. Soc. (3), 2(7) (1952)
16. Huynh, T.Q., Roychoudhury, A.: A Memory Model Sensitive Checker for C#. In: Misra, J., Nipkow, T., Karakostas, G. (eds.) FM 2006. LNCS, vol. 4085, pp. 476–491. Springer, Heidelberg (2006)
17. I. Inc. Intel^{TM}64 and IA-32 Architectures Software Developer's Manuals
18. Kuperstein, M., Vechev, M., Yahav, E.: Automatic inference of memory fences. In: FMCAD (2011)
19. Kuperstein, M., Vechev, M., Yahav, E.: Partial-coherence abstractions for relaxed memory models. In: PLDI (2011)
20. Lamport, L.: A new solution of dijkstra's concurrent programming problem. CACM 17 (August 1974)
21. Lamport, L.: A fast mutual exclusion algorithm (1986)
22. Linden, A., Wolper, P.: An Automata-Based Symbolic Approach for Verifying Programs on Relaxed Memory Models. In: van de Pol, J., Weber, M. (eds.) SPIN 2010. LNCS, vol. 6349, pp. 212–226. Springer, Heidelberg (2010)
23. Linden, A., Wolper, P.: A Verification-Based Approach to Memory Fence Insertion in Relaxed Memory Systems. In: Groce, A., Musuvathi, M. (eds.) SPIN 2011. LNCS, vol. 6823, pp. 144–160. Springer, Heidelberg (2011)
24. Lynch, N., Patt-Shamir, B.: Distributed Algorithms, Lecture Notes for 6.852 FALL 1992. Technical report, MIT, Cambridge, MA, USA (1993)
25. Magnusson, P., Landin, A., Hagersten, E.: Queue locks on cache coherent multiprocessors. In: IPPS. IEEE Computer Society (1994)

26. Mellor-Crummey, J.M., Scott, M.L.: Algorithms for scalable synchronization on shared-memory multiprocessors. ACM Trans. Comput. Syst. 9 (February 1991)

27. Owens, S.: Reasoning about the Implementation of Concurrency Abstractions on x86-TSO. In: D'Hondt, T. (ed.) ECOOP 2010. LNCS, vol. 6183, pp. 478–503. Springer, Heidelberg (2010)

28. Owens, S., Sarkar, S., Sewell, P.: A Better x86 Memory Model: x86-TSO. In: Berghofer, S., Nipkow, T., Urban, C., Wenzel, M. (eds.) TPHOLs 2009. LNCS, vol. 5674, pp. 391–407. Springer, Heidelberg (2009)

29. Peterson, G.L.: Myths About the Mutual Exclusion Problem. IPL 12(3) (1981)

30. Sewell, P., Sarkar, S., Owens, S., Nardelli, F.Z., Myreen, M.O.: x86-tso: A rigorous and usable programmer's model for x86 multiprocessors. CACM 53 (2010)

31. Weaver, D., Germond, T. (eds.): The SPARC Architecture Manual Version 9. PTR Prentice Hall (1994)

Java Memory Model-Aware Model Checking

Huafeng Jin, Tuba Yavuz-Kahveci, and Beverly A. Sanders

University of Florida

Abstract. The Java memory model guarantees sequentially consistent behavior only for programs that are data race free. Legal executions of programs with data races may be sequentially inconsistent but are subject to constraints that ensure weak safety properties. Occasionally, one allows programs to contain data races for performance reasons and these constraints make it possible, in principle, to reason about their correctness. Because most model checking tools, including Java Pathfinder, only generate sequentially consistent executions, they are not sound for programs with data races. We give an alternative semantics for the JMM that characterizes the legal executions as a least fixed point and show that this is an overapproximation of the JMM. We have extended Java Pathfinder to generate these executions, yielding a tool that can be soundly used to reason about programs with data races.

Keywords: model checking, relaxed memory model, benign data races.

1   Introduction

The memory model of a programming language defines which values a thread can see when reading a variable from shared memory. If the memory model is *sequential consistency* (SC), then the program behaves as if all of its reads and writes occur in some order consistent with the program order on individual threads, and each read of a variable sees the most recent write to that variable in the order.

Memory systems in most modern multi-core processors are not sequentially consistent and in addition, a variety of compiler optimizations that would be correct in a sequential program may introduce sequentially inconsistent behavior into a multi-threaded one. For example, consider the program in Fig. 1. In any sequentially consistent execution, depending on how the threads interleave, the x field of the single object involved would change from 0 to 3 at some point and then remain 3 thereafter. A common compiler optimization which causes no problems in a single threaded program might, however, replace the last read r1.x in Thread 1 with an assignment, r5 = r2. This admits executions where it appears that the value of r1.x changes from 0 to 3 and then back to 0. Such an execution is not sequentially consistent.

Sequential consistency is desirable because it corresponds with programmers' intuition. Also, it allows formal reasoning techniques and tools, most of which

C. Flanagan and B. König (Eds.): TACAS 2012, LNCS 7214, pp. 220–236, 2012.

assume sequential consistency, to be used. Most model checkers, for example, implicitly assume sequential consistency. If we used a model checker such as Java Pathfinder to check the scenario in Fig. 1, the legal, but the sequentially inconsistent execution described earlier would not be generated or checked.

Typical programming language memory models guarantee sequential consistency only for programs that are data race free. A *data race* is a pair of conflicting operations (i.e. the operations are performed by different threads, both access the same memory location and at least one is a write) that are not ordered by sufficient synchronization. Exactly what constitutes "sufficient synchronization" is defined by, and specific to, the memory model. The Java Memory Model (JMM), guarantees se-

Initially p == q, p.x == 0	
Thread 1	Thread 2
r1 = p;	r6 = p;
r2 = r1.x;	r6.x = 3;
r3 = q;	
r4 = r3.x;	
r5 = r1.x;	

Fig. 1. Execution trace with conflicting accesses to the same memory location. Variable names that begin with r are local variables of a thread. This example is from [11, §17.3].

quentially consistency only for programs that are data race free, but also constrains programs with data races in order to provide some weak security guarantees. If all of the legal executions, including the sequentially inconsistent ones, of a data racy program still satisfy the program's specification, then we can consider a data race to be benign. Occasionally, one may want to take advantage of this to improve performance. For example, intentional, benign data races can be found in the java.lang.String and java.util.ConcurrentHashMap classes.

The JMM is complicated and reasoning about programs with data races is difficult, thus tool support is desirable. We describe a JMM aware model checker, Java PathRelaxer (JPR) that is an extension of Java Pathfinder [22,15] and generates all of the legal executions of finite Java programs with data races so that their properties can be verified. The way the JMM defines legal executions in programs with data races does not lend itself to precise implementation with a model checker and has been shown [23] to be stricter than the designers intended. We use an alternate approach. Instead of defining a legal execution by the existence of a sequence of justifying executions as the JMM does, we compute a set of paths that is the least fixed point of a monotone function. We show that the set of paths generated by JPR is an overapproximation of the set of legal executions. Although the details of the formalization and implementation of JPR are specific for Java, the main ideas are applicable to other languages with a memory model based on the happens-before relation.

The main contributions of this paper are

- A new, fixed-point based, approach to the characterization of legal executions for relaxed memory models.
- A tool, JPR that generates all of the legal executions according to the fixed-point characterization.

- A proof that the fixed-point based approach is an overapproximation of the JMM, and thus JPR is sound for Java programs with data races.
- Insights into how the JMM maps (or does not map) into program constructs.

2   Background

Below, we give a brief overview of the formal definition of the Java Memory Model, including formal, JMM specific definitions of some concepts introduced previously. Our treatment follows that of [1], which is in turn based on the specification of the JMM given in [19,11].[1]

An *action* a is a memory-related operation with an arbitrary unique ID, *aid* that is performed by a thread *tid*, interacts with variable v or (monitor) lock m, and has a *kind*. The kind is one of the following: volatile read from v, volatile write to v, (non-volatile) read from v, (non-volatile) write to v, locking of lock m, unlocking of lock m, starting a thread, detecting termination of thread, and instantiating an object with a set of volatile fields *volatiles* and a set of non-volatile fields *fields* set to their default values. All of the action kinds, with the exception of read and write are *synchronization actions*.

Definition 1 (Execution).   *An   execution E is   described   by   a   tuple* $\langle A, P, \leq_{po}, \leq_{so}, W, V \rangle$ *where*

- *A is a finite set of actions*
- *P is a program*
- *$\leq_{po}$, the program order, is a partial order on A obtained by taking the union of total orders representing each thread's sequential semantics*
- *$\leq_{so}$, the synchronization order, is a total order over all of the synchronization actions in A*
- *V, the value written function, assigns a value to each write*
- *W, the write-seen function, assigns a write action to each read action so that the value obtained by a read action r is $V(W(r))$.*

A *sequentially consistent* (SC) execution is one where there exists a total order, $\leq_{sc}$, on the actions consistent with $\leq_{po}$ and $\leq_{so}$ and where a read r of variable v sees the results of the most recent preceding write w, i.e.

- $W(r) \leq_{sc} r$
- For all reads r of variable v: if $W(r) \leq_{sc} w \leq_{sc} r$ and w writes to v then $W(r) = w$.

---

[1] The most important differences between [19] and [1] are that the latter requires that the total order for SC executions be consistent with both the synchronization order and program order (as opposed to just the program order, correcting an apparent oversight in the JMM formulation), formulates the semantics in terms of finite executions, and ignores external actions.

The JMM relaxes SC because it is not required that W return the "most recent" write to the variable in question or that it is consistent for actions on different threads.

The synchronizes-with relation, $\leq_{sw}$, relates certain pairs of actions. For example, the action unlocking a monitor synchronizes-with any subsequent (according to $\leq_{so}$) unlock of the same monitor. Other pairs include writing a volatile variable and a subsequent read, the action of starting a thread and the first action of the newly started thread, etc. See [11, §17.4.4] for a complete list. We categorize the first action of a $\leq_{sw}$ pair as a *release* action, and the second as an *acquire* action. The *happens-before* order, $\leq_{hb}$, is a partial order on the actions in an execution obtained by taking the transitive closure of the union of $\leq_{sw}$ and $\leq_{po}$. A well-formed execution satisfies type safety and some unsurprising consistency requirements on the various partial and total orders. The two most important rules for our purposes are intra-thread consistency and happens-before consistency.

Definition 2 (Well-formed execution). *See [1, Definition 6] for the complete definition.*

7. *Program order is intra-thread consistent: for each thread t, the sequence of action kinds and values of actions performed by t in the program order $\leq_{po}$ is sequentially valid*[2] *with respect to P and t.*

9. *$\leq_{hb}$ is consistent with W: for all reads r of variable v, $r \not\leq_{hb} W(r)$ and there is no intervening write w to v, i.e. if $W(r) \leq_{hb} w \leq_{hb} r$ and w writes to v then $W(r) = w$.*

Two operations from different threads *conflict* if neither is a synchronization action, they access the same memory location and at least one is a write. A *data race* is defined to be a pair of conflicting operations *not* ordered by $\leq_{hb}$.[3]

A Java program is *correctly synchronized* if all of its SC executions are data race free. An important property of most programming language memory models, including the JMM [11,19] [1, Theorem 1], is that all legal executions of a well-formed correctly synchronized program behave as if they are sequentially consistent. This data race free guarantee (DRF) is important for programmers. Because "correctly synchronized" depends only on the sequentially consistent executions, detecting data races can be done with model checkers or other tools that assume sequential consistency. Java RaceFinder (JRF) [16,17], for example, extends Java Pathfinder to precisely detect data races in Java programs according to the memory model.

---

[2] Sequential validity essentially means that given the values obtained when a variable is read, each thread obeys the Java language semantics.

[3] Because reads and writes of volatile variables are synchronization actions, a volatile variable in Java can never be involved in a data race. Volatile variables can still be involved in non-deterministic behavior that is sometimes called a race condition. In this paper, we use the term race only in the context of data race as defined in the JMM.

While most Java programs should be data race free, the JMM attempts to define the semantics of programs with data races. The main goal was to provide a modicum of security guarantees even for incorrect programs with data races while still allowing as many optimizations as possible. Desirable properties include type safety and no *out-of-thin-air* values.

While the notion of out-of-thin-air value has not been precisely defined, the example in Fig. 2 [19] illustrates the idea and shows why well-formedness (Def. 2) and in particular happens-before consistency, does not suffice. In a sequentially consistent execution of the example in Fig. 2, the only values allowed are r1==r2==0. However, letting $W(A1) = B2$, $W(B1) = A2$, and $V(A2) =$val, and $V(B2) =$val, for *any* value val of the correct type, we have a well-formed execution where r1 == r2 == val, and in this situation, val is said to come out-of-thin-air.

To rule out such cases, the JMM requires *legal* executions to satisfy additional causality conditions intended to rule out so-called causal loops that could lead to self-justifying speculative executions. The idea is that a well-formed execution E is legal if there is (roughly speaking) a sequence of well-formed executions E_i with action sets A_i and a subset of actions C_i called the commit set where each committed read either sees a committed write or a write that happens-before it. It is required that $C_{i-1} \subseteq C_i$ and that the sequence eventually produces E with all of its actions committed.

Initially, x == y == 0

Thread 1	Thread 2
A1: r1 = x	B1: r2 = y
A2: y = r1	B2: x = r2

Fig. 2. The rules for a well-formed execution admit traces with r1 == r2 == val, for any arbitrary out-of-thin-air value val of the correct type

Definition 3 (Legal Execution). *[1, Definition 7] A well-formed execution* $E = \langle A, P, \leq_{po}, \leq_{so}, W, V \rangle$ *with happens-before order* $\leq_{hb}$ *is legal if there is a finite sequence of sets of actions* C_i *and well-formed executions* $E_i = \langle A_i, P, \leq_{po_i}, \leq_{so_i}, W_i, V_i \rangle$ *with happens-before order* $\leq_{hb_i}$ *such that* $C_0 = \phi$, $C_{i-1} \subseteq C_i$ *for all* $i > 0$, $\bigcup C_i = A$, *and for each* $i > 0$, *the following are satisfied:*

1. $C_i \subseteq A_i$
2. $\leq_{hb_i} |_{C_i} = \leq_{hb} |_{C_i}$
3. $\leq_{so_i} |_{C_i} = \leq_{so} |_{C_i}$
4. $V_i|_{C_i} = V|_{C_i}$
5. $W_i|_{C_{i-1}} = W|_{C_{i-1}}$
6. *For all reads* $r \in A_i - C_i$, $W_i(r) \leq_{hb_i} r$
7. *For all reads* $r \in C_i - C_{i-1}$, $W_i(r) \in C_{i-1}$ *and* $W(r) \in C_{i-1}$

For example, in Fig. 2, suppose that we want to commit the write action A2:y=r1;. Then $V(A2)$ is the value read in action A1:r1=x. The value of x must be obtained from a write that either happened-before $A1$ (the initialization action is the only option) or is already committed. In the former case, the value read is 0, in the latter case, it is the value written by $B2$. Similarly, the value written in $B2$ must be the value read in $B1$, which must be either committed or happen-before it. However, $A2$ was not committed, so the initialization action is the only option. Thus the only possible outcome is r1==r2==0. Clearly, understanding and using this definition is difficult for all but the most trivial programs.

3   Java PathRelaxer (JPR)

In order to check properties of a program with data races, we want to generate all the possible legal executions of the program under the JMM. To do this, we start with a set of legal executions, namely the sequentially consistent ones. Then, from those executions, we find which alternative writes could have been seen by a read, i.e. what are other possibilities for $W(r)$ that do not violate well-formedness, and use these to generate additional executions. The process is repeated until it converges. Completely out-of-thin-air values are avoided because each value seen by a read must have been written in some execution already generated. In the rest of this section, we describe how this process was implemented using model checking in JPR[4]. In Sect. 4, we formulate the process more formally as the computation of the least fixed-point of a monotone function and show that the set of executions generated is an overapproximation of the JMM.

JPR extends Java Pathfinder (JPF) [22,15]. JPF is an explicit-state model checker that analyzes Java bytecode. Its custom JVM provides an efficient representation of the explored state space and can potentially provide paths (or traces) corresponding to all possible interleaving of the threads. Assertions are checked at appropriate points during generation of paths. Generic properties such as deadlock freedom may also be checked. JPF has an extensible architecture via its Listener interface. While standard JPF explores paths corresponding only to sequentially consistent executions, JPR explores all paths allowed by the JMM.

The basic idea behind JPR is to maintain a map, *WriteSet*, that maps memory locations to sets of (write action, value written) pairs. For a read action of variable x, instead of the standard JPF behavior where the read sees the value of the most recent write to x on the current path (which also corresponds to sequentially consistent behavior), a value from an element of *WriteSet*(x) is chosen. By exploring all of the possible *WriteSet* entries at each point and discarding paths that do not correspond to a well-formed execution, an iteration of the JPR algorithm generates all of the well-formed paths consistent with a given *WriteSet*. It also returns a possibly expanded *WriteSet* containing all of the writes that occurred during its execution. By repeating the process until the *WriteSet* no longer changes, JPR generates a superset of the legal executions of the program.

The **JMMAwareJPF** algorithm given in Fig. 3 represents the overall structure of JPR. A JPR specific listener, **JMMListener** is registered with JPF, then JPF is invoked iteratively. **JMMListener** takes the *GlobalWriteSet* from

---

[4] A discussion of JPR focusing on more technical implementation issues related to extending JPF can be found in [13].

```
   JMMAwareJPF(Program)
 2    GlobalWriteSet_old ← GlobalWriteSet_new ← ∅
      converged ← false
 4    while ¬converged do
        Call JPF(JMMListener(GlobalWriteSet_old))
 6      GlobalWriteSet_new ← JMMListener.GlobalWriteSet_new
        if GlobalWriteSet_new == GlobalWriteSet_old then
 8        converged ← true
        else    //not converged
10          GlobalWriteSet_old ← GlobalWriteSet_new
      endwhile
```

Fig. 3. JMMAwareJPF, the top level algorithm in JPR

the previous iteration and returns a new, possibly extended *GlobalWriteSet*, terminating when *GlobalWriteSet* no longer changes. Initially, the *GlobalWriteSet* is empty.

JMMListener is described in Figs. 4 and 5. As various search related events in JPF occur (i.e. start search, advance state, backtrack, execute an instruction, as represented by the variable *searchEvent* in Fig. 4) occur, the corresponding code is executed. Σ is a representation of the current state and is pushed onto a stack when a search starts and when the state advances, and popped when the search backtracks. When the end of a path is reached, the path is tested to see if it is well-formed. If so, the *WriteSet* of the current path is unioned with the *GlobalWriteSet_new*, otherwise the current *WriteSet* and path are discarded.[5]

```
 1 JMMListener(GlobalWriteSet_old)
     GlobalWriteSet_new ← ∅                                      //New global WriteSet
 3   Σ : ⟨WriteSet, ActionSet, HBSet, ImposeSet, Read, Write, ThreadLast⟩
                                                                 //Current state metadata
 5   switch(searchEvent)
       case SEARCH STARTS:
 7       WriteSet ← GlobalWriteSet_old
         ActionSet ← HBSet ← ImposeSet ← ∅
 9       ∀loc : Read(loc) ← undef, Write(loc) ← undef
         ∀tid : ThreadLast(tid) ← undef
11       Stack.push(Σ)
       case STATE ADVANCES:
13       Stack.push(Σ)
       case STATE BACKTRACKS:
15       Σ ← Stack.pop()
         if END OF PATH then
17         if  path is well-formed  then
             GlobalWriteSet_new ← GlobalWriteSet_new ∪ WriteSet
19         else  ignore write set and discard path
       case INSTRUCTION EXECUTES:
21       See Fig. 5
```

Fig. 4. JMMListener algorithm

In JPR, JPF's state representation is extended with the additional information given below, where *Aid* is the domain of action IDs, *Val* is the domain of values, *Loc* is the domain of memory locations, etc. Action was defined in Sect. 2.

---

[5] Although not shown in the algorithm, because paths may be discarded, assertion violations are not reported until the end of the path is reached. This is a departure from standard JPF behavior, which reports assertion violations when they occur.

```
22  case EXECUTING ACTION where action = (aid, tid, kind, loc):
        ActionSet ← ActionSet ∪ {action} // add current action to action set
24      HBSet ← HBSet ∪ {(ThreadLast(tid), aid)}   //update ≤_hb due to ≤_po
        ThreadLast(tid) ← aid
26      if isRELEASE(kind) then
            if kind == VOLATILE WRITE writing val then
28              Write(aid) ← val
        else if isAQCUIRE(kind) then
30          // for each release action rel that syncs with action do
            for each rel = (raid, rtid, rkind, rloc) s.t. isRELEASE(rel)
32                    ∧ (raid, aid) ∈ HBSet do
                HBSet ← HBSet ∪ {(raid, aid)}   //update ≤_hb due to ≤_so
34          if kind == VOLATILE READ then
                //let latest denote the most recent volatile write that syncs with action
36              let latest = (lid, ltid, lkind, lloc) s.t. lkind == VOLATILE WRITE ∧
                        (lid, aid) ∈ HBSet ∧ (∄a_k : a_k ∈ Aid ∧ (a_k, aid) ∈ HBSet ∧ Path(a_k) > Path(lid))
38              //Save the write action and value in Read. This is always a past write.
                Read(aid) ← (lid, false, Write(lid))
40          else if kind == WRITE of value val then
                // if this write action is in the impose set, check for well-formedness
42              if for some val', (aid, val') ∈ ImposeSet then
                    if val' ≠ val then
44                      backtrack // value written is not the imposed value, abandon the path
                    else //check for ≤_hb consistency
46                      if ∃r ∈ ActionSet : Read(r.aid) == (aid, true, *) ∧ r.aid ≤_hb aid then
                            backtrack  //not ≤_hb consistent, abandon path
48              //else path is still well-formed, save values and continue
                Write(aid) ← val
50              WriteSet(loc) ← WriteSet(loc) ∪ {(aid, val)}
            else if kind == READ then
52              non−deterministically choose (w, val) ∈ WriteSet(loc) do
                    if w ∈ ActionSet|_aid then // this is a past read
54                      //check for ≤_hb consistency
                        if (∄wa : wa ∈ ActionSet ∧ wa.kind == WRITE ∧ wa.loc == loc
56                              ∧ w ≤_hb wa.aid ∧ wa.aid ≤_hb aid)  //≤_hb consistent past read
                        then
58                          Read(aid) ← (w, false, Write(w))
                        else   //≤_hb inconsistent past read
60                          continue with next write set entry
                    else   // potential candidate for a future read
62                      if (∄val' : val' ∈ Val ∧ (w, val') ∈ ImposeSet ∧ val' ≠ val) then
                            ImposeSet ← ImposeSet ∪ {(w, val)}
64                          Read(aid) ← (w, true, val) //true indicates future write
                        else //illegal future read, was in impose set with inconsistent value
66                          continue with next write set entry
```

Fig. 5. Continued from Fig. 4. The algorithm for enforcing JMM's semantics by keeping track of write sets and happens-before relation among the actions executed on this path.

- **Path**: Sequence of action ids that represent the current path of execution. For a given action id *aid*, *Path(aid)* represents the index of that action id, where *Path(aid)* is 1 for the id of the first executed action in *Path*.
- **WriteSet**: $Loc \rightarrow 2^{Aid \times Val}$ maps a memory location to a set of action ID, value pairs, where each action is a *WRITE*.
- **ActionSet**: 2^{Action} contains the actions that have been executed on the current path so far.
- **HBSet**: $2^{Aid \times Aid}$ is a set of pairs of action IDs where $\langle aid_1, aid_2 \rangle \in$ **HBSet*** if and only if both are in **ActionSet** and $aid_1 \leq_{hb} aid_2$ and where **HBSet*** is the transitive closure of the relation represented by **HBSet**.

- **ImposeSet**: $2^{Aid \times Val}$ is a set of action ID, value pairs, where each action is a $WRITE$. In a well-formed path, if a read action r obtains a value val from write action w which may be executed in the future, w must occur at some point in any well-formed path containing r, and it must actually write val. Thus the $ImposeSet$ maps write actions to values imposed on them by past reads.
- **Read**: $Aid \to Aid \times boolean \times Val$ maps $READ$ and $VOLATILE$ $READ$ action IDs to a triple containing the write action it sees, i.e. $W(rid)$ and the value it returns, $W(V(rid))$ for action id rid. The boolean value indicates whether the $W(rid)$ occured in the future on the current path.
- **Write**: $Aid \to Val$ maps $WRITE$ and $VOLATILE$ $WRITE$. action IDs to the value written by the corresponding action, i.e. $V(wid)$.
- **ThreadLast**: $Tid \to Aid$ maps a thread id to the latest action performed by the thread and is used to maintain the program order, $\leq_{po}$.

4   Properties of the JPR Algorithm

In this section, we discuss the properties of JPR and its basic algorithms. Most of the proofs and some lemmas are omitted for brevity but can be found in the companion technical report [12]. Executions are the abstraction used in the JMM and defined in Def. 1 while paths are the totally ordered sequences of actions generated by JPR. We say that path p corresponds to execution $E = \langle A, P, \leq_{po}, \leq_{so}, W, V \rangle$ where A is the set of actions that occur in p, P is prog, $\leq_{po}$ is the union over all threads of $\leq_{path}$ restricted to each thread, and $\leq_{so}$ is $\leq_{path}$ restricted to the synchronization actions in p. If a non-volatile read r uses WriteSet entry (w, val), then $W(r) = w$ and $V(w) = val$. $V(w)$ is well-defined since all reads of the same write action in a path must get the same value.

For a fixed program, $prog$, usually considered to be understood, and letting WS be the type of $WriteSet$, let $\mathrm{JPR}_{\mathrm{prog}} : WS \to WS * Paths$ be a function that takes a $ws \in WS$ and returns a new WS and a set of paths $paths$. $\mathrm{JPR}_{\mathrm{prog}}$ is a function represents an invocation of JPF seen in Fig. 3, where $Paths$ is the set of paths searched by JPF. For $ws \in WS$ and path p, we say that $ws \overset{\mathrm{JPR}}{\to} p$ if $p \in \mathrm{JPR}_{\mathrm{prog}}(ws).paths$. We say that $ws \overset{\mathrm{JPR}*}{\to} p$ if $\exists i \geq 0 : p \in (\mathrm{JPR}^i_{\mathrm{prog}}(ws)).paths^6$. For convenience, we overload $\overset{\mathrm{JPR}}{\to}$ and $\overset{\mathrm{JPR}*}{\to}$ and also say $ws \overset{\mathrm{JPR}}{\to} ws'$ or $ws \overset{\mathrm{JPR}*}{\to} ws'$ with the obvious meanings.

Lemma 1 (HBSet). *JPR accurately records $\leq_{hb}$ for any generated path p or prefix of a path. It is invariant that for $\forall a_i, a_j \in p : a_i \neq a_j : a_i \leq_{hb} a_j \equiv (a_i, a_j) \in HBSet \lor (\exists a_k : (a_i, a_k) \in HBSet \land (a_k, a_j) \in HBSet)$.*

Proposition 1 (Safety). *Let ws_{sc} be the set of (w, v) pairs seen in the sequentially consistent executions of prog. If $ws_{sc} \overset{\mathrm{JPR}*}{\to} p$, then p corresponds to a well-formed execution of prog.*

---

^{6} If $i = 0$, p must be empty.

Proposition 2 (Completeness). *$JPR_{prog}(ws)$ generates a path correspond-ing to every well-formed execution of prog satisfying ($\forall reads\ r \in A :$ $(W(r), V(W(r))) \in ws$).*

Lemma 2 (Monotonicity of JPR_prog). *JPR_{prog} is monotonic, i.e.*

- *$ws \subseteq ws'$ and $JPR_{prog}(ws) = (ws_1, paths)$ and $JPR_{prog}(ws') = (ws_1', paths')$ then $ws_1 \subseteq ws_1'$, and $paths \subseteq paths'$.*
- *$ws \subseteq ws_1$.*

Theorem 1 (Convergence). *For finite state, terminating program prog, Sup-pose that JPR_{prog} is applied iteratively starting with ws_0. The process will reach a fixed point $ws*$ in a finite number of steps and the resulting $ws*$ will be the least fixed point of JPR_{prog} at least ws_0.*

Proof. Noting that the (finite) set of $(ws, paths)$ pairs with subset inclusion form a complete lattice, the result from the Knaster-Tarski fixed point theorem and lemma 2. □

Theorem 2 (Overapproximation). *Let ws_{sc} be the smallest WriteSet con-taining all of the values seen in the set of sequentially consistent executions of finite state, terminating program prog and $ws_{sc}*$ be the least fixed point of JPR_{prog} at least ws_{sc}. Let $JPR_{prog}(ws_{sc}*).paths$ be the set of paths generated by ws_{sc}. Let $JmmLegal_{prog}$ be the set of legal paths. Then $JmmLegal_{prog} \subseteq JPR_{prog}(ws_{sc}*).paths$.*

The above results show that the set of paths generated by JPR_{prog} is an overap-proximation of the JMM. As a practical matter, this means that JPR is sound: if we show that a data race is benign by tesing with JPR then we can conclude that a precise tool (if one existed) would also find it benign. On the other hand, the overapproximation allows false alarms. Below, we discuss the source of the imprecision in JPR.

In the example shown in Fig. 6, JPR generates a path with result r1 $==$ r2 $==$ 1, and r3 $==$0. There is a valid path where action D2 writes 1, A1 reads D2, A2 writes 1, B1 reads A2, B2 writes 1. Then, on the next iteration, A1 reads B2 (and imposes 1), B1 reads A2, and then B2 successfully writes 1 as imposed by A1, while D1 reads the initialization action. However, this is not legal according to the JMM. In order for r1 $==$ r2 $==$ 1 to appear in a JMM-legal execution, D2 would need to be a committed action with $V(D2) == 1$. But then r3 must already be 1, so the execution is not legal. The value 1 is considered to come out-of-thin-air in any execution where r3 $== 0$. Note that this is the same program

Initially, x = y = z = 0			
Thread 1	Thread 2	Thread 3	Thread 4
A1: r1 = x	B1: r2 = y	C1: z = 1	D1: r3 = z
A2: y = r1	B2: x = r2		D2: x = r3

Fig. 6

Initially, x == y == 0

Thread 1	Thread 2
r1 = x;	r2 = y;
y = r1;	if(r2 < 2)
	x = 3;
	x = 2;

(a) r1 == r2 == 2 is allowed by approach **scope** but forbidden by approach **occurrence**.

Initially, x == y == 0

Thread 1	Thread 2
r1 = x;	r2 = y;
y = r1;	if(r2 == 2)
	x = 1;
	else
	x = 1;

(b) r1 == r2 == 1 is allowed by approach **occurrence** but forbidden by **scope**.

Fig. 7. ActionID examples

as Fig. 2 with the addition of Threads 3 and 4. In Fig. 2, JPR does not generate paths with out-of-thin-air values. Thus JPR may generate illegal paths with out-of-thin-air values only when the out-of-thin-air values actually do appear in some generated path. It does not generate completely arbitrary out-of-thin air values. JPR could be made more precise by tracking impose requirements across iterations and dependent actions at the cost of significantly increased time and space overhead.

5   Experience

One of the difficulties encountered when implementing JPR was the lack of a well-defined connection between the notion of executions used to define the JMM and actual Java programs. This manifested itself in the representation of the actionID. Within a single execution, the basic requirement of the actionIDs is uniqueness. However, both the JMM definition of legal executions (Def. 3) and JPR require that the identity of actions be compared across different executions and paths, i.e. we must be able to determine if, say, a read of x in one execution or path is the same action as a read of x in another by comparing their IDs. This becomes problematic for programs with branches.

We considered four approaches to identify actions. Let t be the thread, k be the kind, v be the variable, and val be the value read or written.

Occurrence. (k, t, v, n). n counts occurrences of k-actions by thread t on v.
Scope. (t, S, n). S refers to the lexical scope, repeated invocations of the same instruction, such as in a loop are differentiated by a sequence number n.
Value. (k, t, v, val). Actions with the same k,v, and t are distinguished by the value. This is the approach used in [7] is not adequate because actions are no longer uniquely identified if a thread writes the same value to a variable more than once.
Occurence-Val. (k, t, v, val, n). Adds an occurence count$n$ to **value**with the consequence that for a write w, $V(w)$ always maps to the same value, making legality rules 4 and 7 in Def. 3 redundant and inoperative, respectively.

	#thr	scope				occurrence				occurrence-val				JPF		
		iter	T	states	M	iter	T	states	M	iter	T	states	M	T	state	M
tc1	2	3	1.4	164	15	3	1.4	164	15	3	1.5	173	15	0.8	44	15
tc3	3	3	4.1	2315	25	3	4.1	2315	24	3	4.7	2582	25	0.9	349	15
tc5*	4	3	11.2	6326	26	3	12.3	6326	26	3	14.8	6877	26	1.2	1169	15
tc7	2	4	2.2	496	25	4	2.2	496	25	4	2.3	557	26	0.8	64	15
tc9	3	3	3.0	1737	15	3	3.0	1737	15	3	3.3	1929	15	1.0	279	15
tc9a	4	3	2.2	880	15	3	2.2	880	15	3	2.7	914	15	0.9	261	15
tc11	2	4	3.1	1147	26	4	3.2	1147	26	4	4.0	1452	25	0.9	95	15
tc13	2	3	1.2	32	15	3	1.2	32	15	3	1.2	32	15	0.8	24	15
tc17	2	3	1.9	565	15	3	1.9	565	15	3	1.9	641	15	0.8	72	15
tc19	3	3	5.2	2205	25	3	5.6	2205	25	3	5.5	2502	25	0.9	381	15
hash	2	3	1.5	237	15	3	1.5	237	15	3	1.5	237	15	0.7	60	15
hash	4	3	38.3	12442	33	3	38.2	12442	34	3	38.6	12442	34	1.7	3720	15
hash2	2	3	1.3	23	15	3	1.3	23	15	3	1.3	23	15	0.8	98	15
isprime	2	3	2.0	308	15	3	2.1	308	15	3	2.2	308	23	0.9	118	15
dcl	2	3	1.1	22	15	3	1.2	22	15	3	1.2	22	15	0.9	243	15
peterson	2	3	1.5	83	15	3	1.5	83	15	3	1.5	83	15	1.0	194	15
dekker	2	3	1.3	24	15	3	1.2	24	15	3	1.2	24	15	0.9	203	15

Fig. 8. Experimental results comparing the performance of JPR using ActionID approaches **scope**, **occurrence**, and **occurrence-val**, respectively. Column T represents the total time in seconds; column M represents the maximum memory consumption in megabytes. * means that JPR generates paths not allowed by JMM.

The different approaches yield different sets of legal executions. Consider Figs. 7b and 7a. Approach **occurrence** allows the outcome in Figs. 7b because both assignments to x are considered to be the same action; if committed, the assignments could be included in the justifying executions. However, it forbids the outcome in Fig. 7a since the assignment x = 2 in two different executions may have different actionIDs depending on whether or not the branch was taken. Approach **scope** allows the indicated outcome in Fig. 7a because regardless of the execution order, x = 2 is within the same lexical scope and can be committed and verified. It does not allow the outcome in Fig. 7b because the two x = 1 actions are within different scopes and if one is committed, it is impossible for the action to be included in subsequent verification executions. We have implemented **scope**, **occurrence**, and **occurrence-val** in JPR and compared these approaches for several examples. A thorough analysis of which ActionID approach would be more appropriate for JMM, however, is outside the scope of this paper; we limit our contribution to calling the issue to the research community's attention and implementing the three approaches in JPR.

We ran JPR on three groups of test programs. Representative results are listed in Fig. 8. The columns contain the number of threads, and for each action ID approach described above, the number of iterations of JPF required to converge, the total time, the number of states visited in the final iteration, and the maximum memory consumed, respectively. The final columns indicate the resource

```
   public final class String{
 2     private final char value[];    //final fields set in constructor
       private final int offset , count;
 4     private int hash;       //not final, default value is 0
       ...
 6     public int hashCode(){
         int h = hash;
 8       int len = count;
         if(h == 0&&len > 0){
10         int off = offset ;
           char val[] = value;
12         for(int i = 0; i < len; i++){h = 31*h + val[off++];}
           hash = h;
14       }
         h = hash;    //redundant read
16       return h;
       }
18 }
```

Fig. 9. The data races are benign line 15 is removed from the program. Otherwise, the races are not benign

usage for standard JPF for comparison purposes. All testing was performed on a 2.27 GHz Intel(R) Core(TM) i5 CPU, 4 GB main memory, with 64-bit Windows 7 operating system, JDK 1.6, and JPF version 6.

The first group, labeled $tc1$ through $tc20$ are the test cases derived from the JMM Causality Test Cases [14], which were designed to illustrate the properties of the JMM (even numbered test cases are omitted for brevity, correctly synchronized test cases are not interesting). For these, we output the paths generated by JPR and compared them with the legal executions according to JMM. All legal executions were generated with $tc5$ and $tc10$ generating forbidden executions. $tc5$ is the example in Fig. 6 and discussed in section 4. $tc10$ is similar.

The second group contains more realistic examples. In *hash*, the hashCode method (Fig. 9 with line 15 deleted) contains a racy lazy initialization of its hash field; the read of hash (Line 7) and the write of hash (Line 13) may form a data race. This race is benign because in all legal executions, even the sequentially inconsistent ones, a call to the hashCode method will always return the correct hash code value. The assertions applied in both the 2-thread version and 4-thread version of *hash* confirm this finding.

hash2 on the other hand, calls a slightly different version of hashCode (Fig. 9) where the returned value is reread from hash (Line 15). This is correct under sequential consistency, but under the JMM, the race is not benign; a thread calling hashCode could get the initial value 0 instead of the correct hash code. The assertions failed in this case.

In *isprime* [20, §2.6], data races occur when multiple threads read and write elements of a shared array without synchronization. Because accesses to array elements in Java do not have volatile semantics, these accesses are racy and reads may see stale values. In this program, reading a stale value affects performance but not overall correctness; it always correctly identifies the prime numbers. The assertion succeeded in this test case.

```
class Foo{
2    private Helper helper = null;
     public Helper getHelper() {
4      if (helper == null){
                //read helper without synchronization, if not null, return value imme-
   diately.
6        synchronized(this){ //if helper was null, acquire monitor and read it again
            if(helper == null){ //if it is still null
8              helper = new Helper();   //instantiate a Helper object
            }
10       } //release the monitor lock by leaving synchronized block
       }
12     return helper;
     }
14 }
```

Fig. 10. Double checked locking

The third group contains the well-known synchronization problems. *dcl* is the infamous *double-checked locking* (DCL) idiom [2] which attempts to reduce locking overhead by lazy initialization of an object, but fails to safely publish the object, allowing other threads to see a partially constructed object. In the test case, two threads call the getHelper() method of Foo shown in Fig. 10.

peterson and *dekker* are implementations of the classic mutual exclusion algorithms without using volatiles. They guarantee mutual exclusion under sequential consistency, but fail in relaxed memory models such as JMM. Assertions inserted to check non-interference in the critical sections in *peterson* and *dekker* failed as expected. The paths in which *dcl*, *peterson*, and *dekker* had assertion violations are legal according to JMM and therefore were detected by JPR but are not exhibited by sequentially consistent programs. Standard JPF cannot detect these problems.

6   Related Work

Ferrara [9] used a fixed point formulation to interpret the happens-before memory model. This work was done in the context of abstract interpretation, but was not implemented into a real tool. Botincan, et. al. [3] showed that the causality requirements of the JMM are undecidable.

Work has been done using various techniques to verify programs under relaxed hardware and programming language memory models. JUMBLE [10] is a dynamic analysis system that implements an adversarial memory by keeping track of a history of writes to racy variables. When a racy variable is read, the adversarial memory returns some past value that JMM allows and is likely to crash the program. Unlike JPR, this tool does not consider nonracy variables and cannot simulate reading from a future write, hence can only provide an under-approximation of JMM. RELAXER [6], a two-phase analysis tool, employs dynamic analysis in its first phase to detect races on SC executions and predicts potential happen-before cycles if run under one of TSO, PSO, or PSLO. In the second phase, it runs the tested program under the relaxed memory model with a controlled scheduler that realizes the one with happen-before cycle to check for program violations. JPR can be extended with a similar heuristic to prefer exploring paths that may end up with a happen-before cycles. We also mention that we have extended JPF to implement the TSO and PSO memory models. While not of significant practical interest, these could be implemented without requiring iteration, thus giving an illustration of the significant complexity of the JMM.

Burckhardt, Alur and Martin [4] applied a SAT-based bounded verification method to check concurrent data types under relaxed memory ordering models employed by multiprocessors while Burckhardt and Musuvathi [5] described a monitor algorithm that could be implemented by model checkers to verify relaxed memory models due to store buffers. The MemSAT system [21] system accepts a test program containing assertions and an axiomatic specification of a memory model and then uses a SAT solver to find a trace that satisfies the assertions and axioms, if there is one. Both the original JMM specification [11], and the modified version proposed by [1] were found to have surprising results when applied to the JMM Causality test cases. MemSAT is intended to be used with small "litmus test" programs to debug memory model specifications. In contrast, JPR is intended to reason about programs. It explores all possible paths according to the JMM and reports any assertion (program constrain violation) violations, which can help to decide whether the races are benign or not. JPR can be used with programs containing object instantiation, loops and other features that are not well supported in MemSAT. The authors of Java memory model developed a simple simulator for the JMM [18] which appears to be geared more towards understanding the memory model than serving as a tool for program analysis. De et al. [8] developed OpMM which uses a model checker similar to Java PathFinder for state exploration. In contrast to JPR, OpMM is an underapproximation of the JMM where read actions can see past writes that occur before it in a sequentially consistent execution. As an underapproximation, OpMM could be used for bug detection of racy programs, but not verification.

7   Conclusion

We have described JPR, an extension of JPF that generates an overapproximation of the JMM. With this extension, JPF can also be applied to the verification of Java programs with data races. Our approach runs the model checking algorithm in an iterative way to compute a least fixed point of a monotone function that can generate sequentially inconsistent executions.

Although, like any tool based on model checking, state-space explosion is a potential problem, we were able to successfully use the tool to show that data races in some examples are benign. We also demonstrated assertion violations in some programs, which are not detectable without awareness of the JMM.

Finally, we have shown that an operational semantics of JMM requires more precise definition of the action ID concept. We have proposed, implemented, and empirically compared three approaches. Although, drawing a conclusion on which of these approaches would be the most appropriate one is outside the scope of this paper, we hope to start a fruitful discussion on the topic.

References

1. Aspinall, D., Ševčík, J.: Formalising Java's Data Race Free Guarantee. In: Schneider, K., Brandt, J. (eds.) TPHOLs 2007. LNCS, vol. 4732, pp. 22–37. Springer, Heidelberg (2007)
2. Bacon, D., Bloch, J., Bogda, J., Click, C., Haahr, P., Lea, D., May, T., Maessen, J., Manson, J., Mitchell, J.D., Nilsen, K., Pugh, B., Sirer, E.G.: The "double-checked locking is broken" declaration,
   http://www.cs.umd.edu/~pugh/java/memoryModel/DoubleCheckedLocking.html
3. Botinčan, M., Glavan, P., Runje, D.: Verification of Causality Requirements in Java Memory Model Is Undecidable. In: Wyrzykowski, R., Dongarra, J., Karczewski, K., Wasniewski, J. (eds.) PPAM 2009. LNCS, vol. 6068, pp. 62–67. Springer, Heidelberg (2010)
4. Burckhardt, S., Alur, R., Martin, M.M.K.: Bounded Model Checking of Concurrent Data Types on Relaxed Memory Models: A Case Study. In: Ball, T., Jones, R.B. (eds.) CAV 2006. LNCS, vol. 4144, pp. 489–502. Springer, Heidelberg (2006)
5. Burckhardt, S., Musuvathi, M.: Effective Program Verification for Relaxed Memory Models. In: Gupta, A., Malik, S. (eds.) CAV 2008. LNCS, vol. 5123, pp. 107–120. Springer, Heidelberg (2008)
6. Burnim, J., Sen, K., Stergiou, C.: Testing concurrent programs on relaxed memory models. In: ISSTA (2011)
7. Cenciarelli, P., Knapp, A., Sibilio, E.: The Java Memory Model: Operationally, Denotationally, Axiomatically. In: De Nicola, R. (ed.) ESOP 2007. LNCS, vol. 4421, pp. 331–346. Springer, Heidelberg (2007)
8. De, A., Roychoudhury, A., D'Souza, D.: Java Memory Model aware software validation. In: PASTE (2008)
9. Ferrara, P.: Static analysis via abstract interpretation of the happens-before memory model. In: Proceedings of the 2nd International Conference on Tests and Proofs (2008)
10. Flanagan, C., Freund, S.N.: Adversarial memory for detecting destructive races. In: PLDI, pp. 244–254 (2010)
11. Gosling, J., Joy, B., Steele, G., Bracha, G.: Java Language Specification, 3rd edn. Addison-Wesley (2005)
12. Jin, H., Yavuz-Kahveci, T., Sanders, B.A.: Java memory model-aware model checking. Tech. Rep. REP-2011-516, Department of Computer and Information Science, University of Florida (2011), http://www.cise.ufl.edu/tr/REP-2011-516/
13. Jin, H., Yavuz-Kahveci, T., Sanders, B.A.: Java Path Relaxer: Extending JPF for JMM-aware model checking. In: JPF Workshop 2011 (2011)
14. JMM causality test cases,
    http://www.cs.umd.edu/ pugh/java/memoryModel/
    unifiedProposal/testcases.html
15. Java Pathfinder, http://babelfish.arc.nasa.gov/trac/jpf
16. Java Racefinder,
    http://babelfish.arc.nasa.gov/trac/jpf/wiki/projects/jpf-racefinder
17. Kim, K., Yavuz-Kahveci, T., Sanders, B.A.: JRF-E: Using model checking to give advice on eliminating memory model-related bugs. In: ASE (2010)
18. Manson, J., Pugh, W.: The Java Memory Model simulator. In: Workshop on Formal Techniques for Java-like Programs (2002)
19. Manson, J., Pugh, W., Adve, S.V.: The Java memory model. In: POPL 2005 (2005)

20. Oracle thread analyzer's user guide,
    http://download.oracle.com/docs/cd/E18659_01/html/821-2124/gecqt.html
21. Torlak, E., Vaziri, M., Dolby, J.: MemSAT: checking axiomatic specifications of
    memory models. In: PLDI (2010)
22. Visser, W., Havelund, K., Brat, G., Park, S., Lerda, F.: Model checking programs.
    Automated Software Engineering Journal 10(2) (April 2003)
23. Ševčík, J., Aspinall, D.: On Validity of Program Transformations in the Java Mem-
    ory Model. In: Ryan, M. (ed.) ECOOP 2008. LNCS, vol. 5142, pp. 27–51. Springer,
    Heidelberg (2008)

Compositional Termination Proofs
for Multi-threaded Programs

Corneliu Popeea and Andrey Rybalchenko

Technische Universität München

Abstract. Automated verification of multi-threaded programs is difficult. Direct treatment of all possible thread interleavings by reasoning about the program globally is a prohibitively expensive task, even for small programs. Rely-guarantee reasoning is a promising technique to address this challenge by reducing the verification problem to reasoning about each thread individually with the help of assertions about other threads. In this paper, we propose a proof rule that uses rely-guarantee reasoning for compositional verification of termination properties. The crux of our proof rule lies in its compositionality wrt. the thread structure of the program and wrt. the applied termination arguments – transition invariants. We present a method for automating the proof rule using an abstraction refinement procedure that is based on solving recursion-free Horn clauses. To deal with termination, we extend an existing Horn-clause solver with the capability to handle well-foundedness constraints. Finally, we present an experimental evaluation of our algorithm on a set of micro-benchmarks.

1 Introduction

Proving termination of various components of systems software is critical for ensuring the responsiveness of the entire system. Modern systems often contain multiple execution threads, yet most of the recent advances in automated termination proving for systems software focused on sequential programs, see e.g. [6,14,20]. Of course, in principle an existing termination prover for sequential programs can be applied to deal with non-cooperating threads by explicitly considering all possible thread interleavings, but such an approach is prohibitively expensive even for smallest programs.

Existing compositional methods for proving safety properties exploit thread structure to facilitate scalable reasoning and can deal with intricate thread interaction, see e.g. [2, 11, 12]. Unfortunately, these methods are not directly applicable for proving termination, since they rely on a finite-state abstraction for approximating the set of reachable program states [19].

Rely-guarantee reasoning [13] is a promising basis for the development of termination provers for multi-threaded programs. The method proposed in [7] relies on environment transitions that keep track of the interaction of a thread with its environment to prove termination properties of individual threads in a multi-threaded program. To ensure scalability, the underlying proof rule is

C. Flanagan and B. König (Eds.): TACAS 2012, LNCS 7214, pp. 237–251, 2012.

thread-modular and hence incomplete, i.e., it considers a restricted class of environment transitions that refer only to global variables. One practical consequence of such an incompleteness is that termination proofs of programs that use synchronization primitives like mutexes are out of scope. This limitation can be eliminated by gradually exposing additional local state when necessary [3]. Providing ranking functions for each thread that are preserved under environment transitions yields a complete method for proving termination that is compositional wrt. thread structure.

In this paper, we explore a combination of two dimensions of compositionality for proving termination of multi-threaded programs. We present a complete proof rule that is compositional wrt. the thread structure of the program and wrt. the termination argument that is used for each of the threads. Following the rely-guarantee reasoning approach, we use environment transitions to keep track of the effect of the threads from the environment of a given thread. As termination argument we rely on transition invariants and disjunctive well-foundedness [19], whose discovery can be efficiently automated in compositional fashion using abstract interpretation. The completeness of our proof rule is achieved by allowing the environment transitions to refer to both global and local variables of all threads. We also provide a specialized version of the proof rule for checking thread termination, i.e., termination of individual threads [7].

We demonstrate the potential for automation of our proof rule by transforming the recursive equations that represent the proof rule into a function whose least fixpoint characterizes the strongest proof of termination. We propose a transition predicate abstraction and refinement-based algorithm to obtain a sufficiently precise over-approximation of the least fixpoint and thus prove termination as follows. Since the fixpoint computation keeps track of both transition invariants and environment transitions, we obtain counterexamples that have a Horn-clause structure similar to a recent approach [11], yet generalizing from reachability to binary reachability. By analyzing the least solution to the counterexample we determine a well-founded over-approximation of discovered lasso-shaped thread interleavings. Given this over-approximation, we perform the actual transition predicate abstraction refinement by computing interpolating solutions to the Horn clauses. Technically, we extend the Horn clause solver presented in [11] with the treatment of well-foundedness constraints.

In summary, our paper makes the following contributions: i) a compositional proof rule that is complete for checking termination and thread termination of multi-threaded programs, ii) a method for automating the proof rule by using a corresponding predicate abstraction and refinement scheme, and iii) the implementation of our algorithm together with the evaluation of its feasibility on micro-benchmarks.

2   Preliminaries

Programs A *multi-threaded* program P consists of $N \geq 1$ threads. Let $1..N$ be the set $\{1, \ldots, N\}$. We assume that the *program* variables $V = (V_G, V_1, \ldots, V_N)$

are partitioned into *global* variables V_G, which are shared by all threads, and *local* variables $V_1, \ldots, V_N$, which are only accessible by the respective threads.

The set of *global states* G consists of the valuations of global variables, and the sets of *local states* $L_1, \ldots, L_N$ consist of the valuations of the local variables of respective threads. A *program state* is a valuation of the global variables and the local variables of all threads. We represent sets of program states using assertions over program variables. Binary relations between sets of program states are represented using assertions over unprimed and primed variables. Let $\rho_i^=$ stand for $V_i = V_i'$. Let $\models$ denote the satisfaction relation between (pairs) of states and assertions over program variables (and their primed versions). We use $\rightarrow$ as the logical implication operator as well as the logical consequence relation, and rely on the context for disambiguation.

The set of *initial* program states is denoted by φ_{init} . For each thread $i \in 1..N$ we have a finite set of *transitions* $\mathcal{R}_i$. Each transition is a binary relation between sets of program states. Furthermore, each $\rho \in \mathcal{R}_i$ can only change the values of the global variables and the local variables of the thread i (local variables of other threads do not change), i.e., we have $\rho \rightarrow \rho_i^=$. We write ρ_i for the union of the transitions of the thread i , i.e., $\rho_i = \bigvee \mathcal{R}_i$. The transition relation of the program is $\rho_P = \rho_1 \vee \cdots \vee \rho_N$.

Computations. A *computation* of P is a sequence of program states $s_1, s_2, \ldots$ such that s_1 is an initial state, i.e., $s_1 \models \varphi_{init}$, and each pair of consecutive states s_i and s_{i+1} in the sequence is connected by a transition ρ of some program thread, i.e., $(s_i, s_{i+1}) \models \rho$. A *path* is a sequence of transitions.

A program state is *reachable* if it appears in some computation. Let φ_{reach} denote the set of all reachable states. The program is *terminating* if it does not have any infinite computations. A *thread* i is *terminating*, as defined in [7], if there is no program computation that contains infinitely many transitions of i .

Auxiliary definitions. A binary relation φ is *well-founded* if it does not admit any infinite sequences. Let φ^+ denote the *transitive closure* of φ . A *transition invariant* T [19] is binary relation over program states that contains the transitive closure of the program transition relation restricted to reachable states, i.e., $\varphi_{reach} \wedge \rho_P^+ \rightarrow T$. Let $\varphi[z/w]$ denote a substitution that replaces w by z in φ . We assume that a sequence of substitutions is evaluated from left to right. Let $\circ$ be the *relational composition* function: $\varphi \circ \psi = \exists V'' : \varphi[V''/V'] \wedge \psi[V''/V]$. A *path relation* is a relational composition of transition relations along the path, i.e., for $\pi = \rho_1 \cdots \rho_n$ we have $\rho_\pi = \rho_1 \circ \ldots \circ \rho_n$. A path π is *feasible* if its path relation is not empty, i.e., $\exists V \exists V' : \rho_\pi$. Given a binary relation φ, we define an *image* function $Img(\varphi) = \exists V'' : \varphi[V''/V][V/V']$.

A *Horn clause* $b_1(w_1) \wedge \cdots \wedge b_n(w_n) \rightarrow b(w)$ consists of relation symbols $b_1, \ldots, b_n, b$, and vectors of variables $w_1, \ldots, w_n, w$. We say that a relation symbol b *depends* on the relation symbols $\{b_1, \ldots, b_n\}$. We distinguish interpreted theory symbols that we use to write assertions denoting sets (of pairs) of program states, e.g., the equality $=$ or the inequality $\leq$. A set of Horn clauses is *recursion-free* if it induces a well-founded dependency relation on non-interpreted relation symbols.

For assertions $T_1, \ldots, T_N$, $E_1, \ldots, E_N$ over V and V' ,
and well-founded relations $WF_1, \ldots, WF_m$

$$
\begin{array}{llll}
\text{C1:} & \varphi_{init} \wedge (\rho_i \vee E_i \wedge \rho_i^=) & \to T_i & \text{for } i \in 1..N \\
\text{C2:} & Img(T_i) \wedge (\rho_i \vee E_i \wedge \rho_i^=) & \to T_i & \text{for } i \in 1..N \\
\text{C3:} & T_i \circ (\rho_i \vee E_i \wedge \rho_i^=) & \to T_i & \text{for } i \in 1..N \\
\text{C4:} & (\bigvee_{i \in 1..N \setminus \{j\}} \varphi_{init} \wedge \rho_i) & \to E_j & \text{for } j \in 1..N \\
\text{C5:} & (\bigvee_{i \in 1..N \setminus \{j\}} Img(T_i) \wedge \rho_i) & \to E_j & \text{for } j \in 1..N \\
\text{C6:} & T_1 \wedge \cdots \wedge T_N & \to WF_1 \vee \cdots \vee WF_m &
\end{array}
$$

---

multi-threaded program P terminates

Fig. 1. Proof rule PROGTERM for compositional proving of program termination

3   Proof Rules

In this section we present compositional rules for proving termination of multi-threaded programs and their threads.

3.1   Compositional Termination of Multi-threaded Programs

In order to reason about termination properties of multi-threaded programs, we propose a proof rule with two auxiliary assertions per thread denoted as T_i and E_i. $T_1, \ldots, T_N$ stand for transition invariants for respective threads. $E_1, \ldots, E_N$ represent environment transitions considered during the computation of transition invariants.

See Figure 1 for the proof rule PROGTERM. The first five premises ensure that T_i is a transition invariant of a program P as follows. The premises C1 and C2 require that T_i over-approximates the transition relation of the thread i restricted to initial states and to arbitrary reachable states. The same two premises require that T_i also over-approximates environment transitions $E_i \wedge \rho_i^=$. The conjunction with $\rho_i^=$ ensures that local variables of thread i do not change when an environment transition is applied. In the premise C3, extending a relation from T_i with either a local or an environment transition results in a relation that is also present in T_i. Two premises are used to record environment transitions for thread j that are induced by transitions executed by thread i either when starting from initial states (premise C4) or from arbitrary reachable states (premise C5). While the first five premises ensure the soundness of the transition invariants, the last premise C6 requires the existence of a disjunctive well-founded relation, e.g., $WF_1 \vee \cdots \vee WF_m$, as a witness for program termination.

Example 1. See Figure 2 for Choice, a multi-threaded version of the example with the same name from [19]. We use a parallel assignment instruction $(x,y)=\ldots$ to simultaneously update both variables x and y . Our verification method represents the program using assertions. The tuple $V = (x, y, pc_1, pc_2)$

```
int x,y;

// Thread T1                        // Thread T2
a0: while (x>0 && y>0) {            b0: while (x>0 && y>0) {
a1:    (x,y) = (x-1,x);            b1:    (x,y) = (y-2,x+1);
a2: }                             b2: }
```

$\varphi_{init} = (\mathrm{pc}_1 = \mathrm{a0} \wedge \mathrm{pc}_2 = \mathrm{b0})$

$\rho_1 = (x > 0 \wedge y > 0 \wedge x' = x - 1 \wedge y' = x \wedge \mathrm{pc}_1 = \mathrm{a0} \wedge \mathrm{pc}_1' = \mathrm{a0} \wedge \rho_{\overline{2}}^{=})$

$\rho_2 = (x > 0 \wedge y > 0 \wedge x' = y - 2 \wedge y' = x + 1 \wedge \mathrm{pc}_2 = \mathrm{b0} \wedge \mathrm{pc}_2' = \mathrm{b0} \wedge \rho_{\overline{1}}^{=})$

Fig. 2. Example program T1 $\|$ T2 for which the value of either of the following expressions decreases: x, y or $x + y$. We use $\rho_{\overline{1}}^{=} = (\mathrm{pc}_1 = \mathrm{pc}_1')$ and $\rho_{\overline{2}}^{=} = (\mathrm{pc}_2 = \mathrm{pc}_2')$

includes pc_1 and pc_2, the program counter variables of the two threads. φ_{init} describes the initial states of the program, while ρ_1 and ρ_2 represent the transition relations of the two threads (simplified for clarity of illustration).

We show that Choice terminates by applying PROGTERM and considering the following auxiliary assertions.

$$T_1 = T_2 = (x > 0 \wedge y > 0 \wedge x' \le x - 1 \wedge y' \le x + 1) \vee$$
$$(x > 0 \wedge y > 0 \wedge x' \le y - 2 \wedge y' \le y - 1)$$
$$E_1 = (x > 0 \wedge y > 0 \wedge x' \le y - 2 \wedge y' \le x + 1)$$
$$E_2 = (x > 0 \wedge y > 0 \wedge x' \le x - 1 \wedge y' \le x)$$

The assertion E_1 approximates the effect of applying the transition from the second thread, while E_2 approximates the effect of applying the transition from the first thread. There exists a disjunctively well-founded relation, $(x > x' \wedge x \ge 0 \vee y > y' \wedge y \ge 0)$, that approximates the transition invariants T_1 and T_2.

This example also illustrates a limitation of the proof rule from [7], which requires that the environment transitions must be transitive. Our non-transitive environment transitions can be weakened to their transitive closure, but then the transitive closure E_1^+ is too weak to prove termination since the constraints $x' \le y - 2$ and $y' \le x + 1$ will be missing.

3.2   Compositional Thread Termination

For the cases where the program termination is a property too strong to hold (e.g., dispatch routines of event-based systems continuously accept incoming events), it is still the case that termination of critical threads is important [7]. See Figure 3 for a proof rule that relaxes the conditions from PROGTERM but still ensures the thread-termination property for some thread from a given program.

The rule THREADTERM relies on assertions T_k and E_k that satisfy the first five premises from the rule PROGTERM. For a thread k, an additional assertion $\hat{t}_k$ is used to keep track of a subset of the transition invariant relation T_k. Unlike

For assertions $T_1, \ldots, T_N$, $E_1, \ldots, E_N$, $\hat{t}_k$ over V and V'
that satisfy C1, C2, C3, C4, C5 and well-founded relations $WF_1, \ldots, WF_m$

$$
\begin{array}{lll}
\text{C1':} & \varphi_{init} \wedge \rho_k & \to \hat{t}_k \\
\text{C2':} & Img(T_k) \wedge \rho_k & \to \hat{t}_k \\
\text{C3':} & \hat{t}_k \circ (\rho_k \vee E_k \wedge \rho_k^=) \to \hat{t}_k \\
\text{C6':} & \hat{t}_k & \to WF_1 \vee \cdots \vee WF_m
\end{array}
$$

thread k terminates

Fig. 3. Proof rule THREADTERM for compositional proving of thread termination

the program-termination premises C1 and C2, the thread-termination premises C1' and C2' require that transitions captured by $\hat{t}_k$ always start with a local transition from thread k . With this restriction in place, the premise C3' is similar to C3: it extends $\hat{t}_k$ with either a local transition ρ_k or an environment transition. The assertion $\hat{t}_k$ keeps track of thread interleavings that may lead to computations of arbitrary length, and its disjunctive well-foundedness is required for thread-termination of thread k .

4   Proof Rule Automation

Next, we demonstrate the potential for automation of the proof rule PROGTERM. (Automation of the proof rule THREADTERM is similar.) We formulate an algorithm for proving program termination that consists of three steps. The first step uses abstraction functions to compute transition invariants for each thread. If this process discovers abstract transitions – components of transition invariants – whose intersection is not disjunctively well-founded, then the second step generates Horn clauses such that their satisfiability implies a disjunctively well-founded refined counterpart exists. The last step of the algorithm invokes a solving procedure for the Horn clauses and uses the obtained solution to refine the abstraction functions.

The entry point of our algorithm initializes the transition abstraction functions using empty sets of predicates. Then, it repeats a loop that invokes transition invariant computation that is followed by an abstraction refinement step.

Procedure ABSTTRANSENV. Figure 4 shows the reachability procedure that computes abstract transitions $\mathcal{T}_i$ and abstract environment transitions $\mathcal{E}_i$ following the conditions from the PROGTERM proof rule, where $\mathcal{T}_i$ and $\mathcal{E}_i$ correspond to auxiliary assertions T_i and E_i, respectively. These sets are initialized corresponding with the rules C1 and C4 in lines 1–4. The auxiliary procedure ADDIFNEW implements a fixpoint check. It takes as input a newly computed binary relation τ and a container set C . It checks if the binary relation contains pairs of states that have not been reached and recorded in the container. If this check succeeds, the new binary relation is added to the container and the

global variables
P - program with N threads
$\ddot{\mathcal{P}}_i, \ddot{\alpha}_i, \ddot{\mathcal{P}}_{i\triangleright j}, \ddot{\alpha}_{i\triangleright j}$ - transition predicates and transition abstraction functions
$\mathcal{T}_i, \mathcal{E}_i$ - abstract (environment) transitions of thread i
$Parent, ParentTId$ - parent relations
procedure ABSTTRANSENV
begin

1	**for each** $i \in 1..N$ and $\rho \in \mathcal{R}_i$ **do**	
2	$\quad$ ADDIFNEW($\ddot{\alpha}_i(\varphi_{init} \wedge \rho), \mathcal{T}_i, (\varphi_{init}, \rho), i$)	$(* \; C1 \; *)$
3	$\quad$ **for each** $j \in 1..N \setminus \{i\}$ **do**	
4	$\quad\quad$ ADDIFNEW($\ddot{\alpha}_{i\triangleright j}(\varphi_{init} \wedge \rho), \mathcal{E}_j, (\varphi_{init}, \rho), i$)	$(* \; C4 \; *)$
5	**repeat**	
6	$\quad$ $finished \; := \; true$	
7	$\quad$ **for each** $i \in 1..N$ and $\tau \in \mathcal{T}_i$ **do**	
8	$\quad\quad$ **for each** $\rho \in \mathcal{R}_i \cup \mathcal{E}_i$ **do**	
9	$\quad\quad\quad$ $\varphi \; := $ **if** $\rho \in \mathcal{R}_i$ **then** $true$ **else** $\rho_i^=$	
10	$\quad\quad\quad$ ADDIFNEW($\ddot{\alpha}_i(Img(\tau) \wedge \rho \wedge \varphi), \mathcal{T}_i, (Img(\tau), \rho), i$)	$(* \; C2 \; *)$
11	$\quad\quad\quad$ ADDIFNEW($\ddot{\alpha}_i(\tau \circ (\rho \wedge \varphi)), \mathcal{T}_i, (\tau, \rho), i$)	$(* \; C3 \; *)$
12	$\quad\quad$ **for each** $\rho \in \mathcal{R}_i$ and $j \in 1..N \setminus \{i\}$ **do**	
13	$\quad\quad\quad$ $\tau' \; := \ddot{\alpha}_{i\triangleright j}(Img(\tau) \wedge \rho)$	
14	$\quad\quad\quad$ ADDIFNEW($\tau', \mathcal{E}_j, (Img(\tau), \rho), i$)	$(* \; C5 \; *)$
15	$\quad\quad\quad$ ADDIFNEW($\ddot{\alpha}_j(\varphi_{init} \wedge \tau' \wedge \rho_j^=), \mathcal{T}_j, (\varphi_{init}, \tau'), j$)	$(* \; C1 \; *)$
16	**until** $finished$	

end

procedure ADDIFNEW
input
$\quad$ τ - binary relation
$\quad$ $\mathcal{C}$ - container set, either $\mathcal{T}_i$ or $\mathcal{E}_i$ for some $i \in 1..N$
$\quad$ P - parent pair
$\quad$ j - thread identifier from $1..N$
begin
$\quad$ **if** $\neg(\exists \rho \in \mathcal{C} : \tau \rightarrow \rho)$ **then**

17	$\quad\quad$ $\mathcal{C} \; := \{\tau\} \cup \mathcal{C}$
18	$\quad\quad$ $Parent(\tau) \; := P$
19	$\quad\quad$ $ParentTId(\tau) \; := j$
20	$\quad\quad$ $finished \; := false$
21	**end**

Fig. 4. Procedure ABSTTRANSENV computes abstract transitions $\mathcal{T}_i$ and $\mathcal{E}_i$. It uses a local auxiliary procedure ADDIFNEW

$Parent$ function is updated to keep track of the child-parent relation between transitions.

The second part of the procedure (see lines 5–16) ensures that the other conditions from the proof rule are satisfied. It computes an abstract reachable state using the $Img(\tau)$ operator and then applies the abstract transition ρ from this abstract state. The result is extended with a transition local to thread i or from its environment in line 10. Next, one-step environment transitions are generated and added both to $\mathcal{E}_j$ in line 14 and $\mathcal{T}_j$ in line 15. Whenever an

procedure TERMREFINE
input
   $\tau_1, \ldots, \tau_N$ - abstract error tuple
begin
1   $HC := \text{MKHC}(\tau_1) \cup \cdots \cup \text{MKHC}(\tau_N)$
2   $UnkRels := \{ \text{``}\rho\text{''}(V, V') \mid i \in 1..N \wedge \rho \in \mathcal{T}_i \cup \mathcal{E}_i\})$
3   $UnkRelsWF := \{ \text{``}\tau_1\text{''}(V, V'), \ldots, \text{``}\tau_N\text{''}(V, V')\}$
4   SOL $:= \text{SOLVEHC}^{WF}(HC, UnkRelsWF, UnkRels))$
5   **for** each $i \in 1..N$ and $\rho \in \mathcal{T}_i$ **do**
6      $\ddot{\mathcal{P}}_i := PredsOf(\text{SOL}(\text{``}\rho\text{''}(V, V'))) \cup \ddot{\mathcal{P}}_i$
7   **for** each $j \in 1..N$ and $\rho \in \mathcal{E}_j$ **do**
8      $i := ParentTId(\rho)$
9      $\ddot{\mathcal{P}}_{i \triangleright j} := PredsOf(\text{SOL}(\text{``}\rho\text{''}(V, V'))) \cup \ddot{\mathcal{P}}_{i \triangleright j}$
end

procedure SOLVEHC^{WF}
input
   HC - recursion-free Horn clauses
   $UnkRelsWF = \{ \text{``}\tau_1\text{''}(V, V'), \ldots, \text{``}\tau_N\text{''}(V, V')\}$,
   $UnkRels$ - unknown relations
output
   SOL - solution for HC such that $UnkRelsWF$ are well-founded relations
begin
10   SOL$^\mu := \text{SOLVEHC}^\mu(HC, UnkRelsWF)$
11   $\rho := \text{SOL}^\mu(\text{``}\tau_1\text{''}(V, V')) \wedge \cdots \wedge \text{SOL}^\mu(\text{``}\tau_N\text{''}(V, V'))$
12   **if** $\exists WF : well\text{-}founded(WF) \wedge (\rho \rightarrow WF)$ **then**
13      $HC^{WF} := \{ \text{``}\tau_1\text{''}(V, V') \wedge \cdots \wedge \text{``}\tau_N\text{''}(V, V') \rightarrow WF\} \cup HC$
14   SOL $:= \text{SOLVEHC}(HC^{WF}, UnkRels)$
15   **else**
16   **throw** UNSATISFIABLE
end

Fig. 5. Procedure TERMREFINE takes as argument an abstract error tuple. The quotation function " $\cdot$ " creates a relation symbol from a given abstract (environment) transition. The function *PredsOf* extracts atomic predicates from the solutions to the set of Horn clauses HC , while MKHC generates Horn clauses

iteration of the **repeat** loop finishes with the *finished* flag set to *true*, the proof rules are saturated and the procedure returns a fixpoint for $\mathcal{T}_i$ and $\mathcal{E}_i$.

Procedure TERMREFINE. See Figure 5 for the procedure that takes as input a tuple of abstract transitions and computes, if possible, predicates that witness the well-foundedness of these transitions. The procedure generates in line 1 a set of Horn clauses corresponding to the abstract transitions. It collects in the set *UnkRels* unknown relations corresponding to each abstract (environment) transition. The set *UnkRelsWF* is a subset of *UnkRels* that contains unknown relations with an additional requirement: the conjunction of their solutions must be a well-founded relation.

After invoking the solving procedure at line 4, the solution SOL is used to update the transition abstraction functions as follows. Atomic predicates from SOL("ρ"(V, V')) are added to the set of predicates $\ddot{\mathcal{P}}_i$ corresponding to the parent thread of ρ . Similarly, SOL("ρ"(V, V')) is used to update $\ddot{\mathcal{P}}_{i\triangleright j}$ if ρ is an environment transition that was constructed in thread i and applied in thread j . These new predicates guarantee that the same counterexample will not be encountered during the next iterations of reachability analysis.

Procedure SOLVEHC^{WF}. Figure 5 shows a procedure for solving recursion-free Horn clauses such that the solution for some unknown relations is well-founded.

Due to the recursion-free nature of the Horn clauses, computing the least (most precise) solution for the clauses is trivial. Note that the set of Horn clauses *HC* is guaranteed to have at least the solution mapping every unknown to a *true* predicate. At line 10, the least solution SOL$^\mu$ that is returned is defined for the unknown relations from *UnkRelsWF* . If the solution SOL$^\mu$ does not have an upper-bound that is well-founded (see line 12), then the abstraction refinement fails with an UNSATISFIABLE exception. The test at line 12 can be implemented using a rank synthesis tool, e.g. [18].

If a well-founded relation *WF* is detected, then an additional Horn clause ensures that the solution for the abstract transitions over-approximates it: $\{$"τ_1"$(V, V') \wedge \cdots \wedge$ "τ_N"$(V, V') \rightarrow WF\}$. Given the set of Horn clauses HC^{WF} , our algorithm invokes a solving procedure in line 9, e.g., SOLVEHC from [11].

To ensure the completeness of the refinement procedure (relative to the domain of linear inequalities), solutions to unknown relations "ρ"(V, V') are expressed in terms of all program variables V and their primed version V'. However, to facilitate thread-modular reasoning, it is desirable to return solutions for transitions in $\mathcal{T}_i$ that are expressed over V_G, V_i and solutions for environment transitions in $\mathcal{E}_i$ over global program variables V_G . The Horn clause solving procedure SOLVEHC ensures that, whenever it exists, a modular solution is returned. Both reasoning about safety [11] and reasoning about termination properties benefit from such modular solutions.

Example 2. The `LockDecrement` example was used to illustrate the thread-modular verification method from [7]. Our algorithm is able to verify a termination property for this example even if we drop two of the assumptions required by the verification method from [7].

The code for this example is shown in Figure 6 and consists of two threads T1 and T2. The thread T1 acquires a lock on line a0 and decrements the value of the variable x if it is positive. The thread T2 assigns a non-deterministically chosen value to x after it acquires the lock lck on line b1 . The program assumes that initially the lock is not taken, i.e., lck = 0 . The locking statement lock(lck) waits until the lock is released and then assigns the value 1 to lck , thus taking the lock. We are interested in proving thread-termination of the first thread T1, i.e., no computation of the program contains infinitely many steps from T1. Note that the non-termination of the thread T2 (and of the entire program T1 || T2) does not affect the thread-termination property of T1.

$$\varphi_{init} = (\text{lck} = 0 \wedge \text{pc}_1 = \text{a0} \wedge \text{pc}_2 = \text{b0})$$

```
// Thread T1
a0: lock(lck);
a1: while (x>0) {
a2:    t = x-1;
a3:    x = t;
a4: }
a5: unlock(lck);
a6:
```

$$\rho_{11} = (\text{lck} = 0 \wedge \text{lck}' = 1 \wedge skip(\text{x}, \text{t}) \wedge move_1(\text{a0}, \text{a1}) \wedge \rho_2^=)$$
$$\rho_{12} = (\text{x} > 0 \wedge skip(\text{x}, \text{lck}, \text{t}) \wedge move_1(\text{a1}, \text{a2}) \wedge \rho_2^=)$$
$$\rho_{13} = (\text{t}' = \text{x} - 1 \wedge skip(\text{x}, \text{lck}) \wedge move_1(\text{a2}, \text{a3}) \wedge \rho_2^=)$$
$$\rho_{14} = (\text{x}' = \text{t} \wedge skip(\text{lck}, \text{t}) \wedge move_1(\text{a3}, \text{a1}) \wedge \rho_2^=)$$
$$\rho_{15} = (\text{x} \leq 0 \wedge skip(\text{x}, \text{lck}, \text{t}) \wedge move_1(\text{a1}, \text{a5}) \wedge \rho_2^=)$$
$$\rho_{16} = (\text{lck}' = 0 \wedge skip(\text{x}, \text{t}) \wedge move_1(\text{a5}, \text{a6}) \wedge \rho_2^=)$$
$$\rho_2^= = (\text{pc}_2' = \text{pc}_2)$$

```
// Thread T2
b0: while (nondet()) {
b1:    lock(lck);
b2:    x = nondet();
b3:    unlock(lck);
b4: }
```

$$\rho_{21} = (\text{lck} = 0 \wedge \text{lck}' = 1 \wedge skip(\text{x}) \wedge move_2(\text{b0}, \text{b2}) \wedge \rho_1^=)$$
$$\rho_{22} = (skip(\text{lck}) \wedge move_2(\text{b2}, \text{b3}) \wedge \rho_1^=)$$
$$\rho_{23} = (\text{lck}' = 0 \wedge skip(\text{x}) \wedge move_2(\text{b3}, \text{b0}) \wedge \rho_1^=)$$
$$\rho_{24} = (skip(\text{x}, \text{lck}) \wedge move_2(\text{b0}, \text{b5}) \wedge \rho_1^=)$$
$$\rho_1^= = (\text{t}' = \text{t} \wedge \text{pc}_1' = \text{pc}_1)$$

Fig. 6. Example with global variables x, t, lck , with lck initially set to 0 (unlocked)

See Figure 6 for the program representation, where $\rho_{11}, \ldots, \rho_{16}$ give the transition relation of T1, while $\rho_{21}, \ldots, \rho_{24}$ represent the transition relation of T2. The notation $skip(\text{x}, \text{t})$ is a shorthand for a constraint indicating that the values of x and t are not changed, i.e., $(\text{x} = \text{x}' \wedge \text{t} = \text{t}')$, while $move_1(\text{a0}, \text{a1})$ denotes that the location of thread T1 is changed from a0 to a1, i.e., $(\text{pc}_1 = \text{a0} \wedge \text{pc}_1' = \text{a1})$.

Next, we illustrate how a termination proof for the LockDecrement program can be constructed automatically using a sequence of abstract reachability and abstraction refinement steps.

First abstract computation. The abstract computation approximates the transition relation of the program using two abstraction functions $\ddot{\alpha}_1$ and $\ddot{\alpha}_2$ that correspond to threads T1 and respectively T2. For this example, we assume that the abstraction functions initially track the value of the program counters of the two threads using the following sets of predicates: $\ddot{\mathcal{P}}_1 = \{\text{pc}_1 = \text{a0}, \ldots, \text{pc}_1' = \text{a0}, \ldots\}$ and $\ddot{\mathcal{P}}_2 = \{\text{pc}_2 = \text{b0}, \ldots, \text{pc}_2' = \text{b0}, \ldots\}$. The abstraction of a transition relation is computed as $\ddot{\alpha}_i(T) = \bigwedge \{\ddot{p} \in \ddot{\mathcal{P}}_i \mid T \rightarrow \ddot{p}\}$. Our algorithm starts by computing one-step relations from initial states (see premise C1). We obtain two abstract transitions: $m_1 = \ddot{\alpha}_1(\varphi_{init} \wedge \rho_{11}) = (\text{pc}_1 = \text{a0} \wedge \text{pc}_1' = \text{a1})$ and $n_1 = \ddot{\alpha}_2(\varphi_{init} \wedge \rho_{21}) = (\text{pc}_2 = \text{b0} \wedge \text{pc}_2' = \text{b2})$. The abstract transition m_1 groups sequences of concrete transitions that start with $\text{pc}_1 = \text{a0}$ and finish with $\text{pc}_1 = \text{a1}$, e.g., the sequence of four transitions $\rho_{11}\rho_{12}\rho_{13}\rho_{14}$.

Next, our algorithm computes successors for the abstract transitions that were already discovered (see premise C2). For m_1 , we compute the image of the relation using the operator $Img(m_1) = (\text{pc}_1 = \text{a1})$. Then, we compute a successor with regards to a transition ρ_{12} as follows: $m_2 = \ddot{\alpha}_1(Img(m_1) \wedge \rho_{12}) = (\text{pc}_1 = \text{a1} \wedge \text{pc}_1' = \text{a2})$. A second method to derive successors for an abstract transition is to extend it with another transition from the thread transition relation (see premise C3). For m_2 , we compute its extension using the relational composition operator, $m_2 \circ \rho_{13} = (pc_1 = a1 \wedge pc_1' = a2) \circ (x' = x \wedge lck' = lck \wedge t' = x - 1 \wedge pc_1 = a2 \wedge pc_1' = a3 \wedge \rho_2^=) = (pc_1 = a1 \wedge pc_1' = a3)$. From m_2, two new abstract transitions are generated: $m_3 = \ddot{\alpha}_1(m_2 \circ \rho_{13}) = (pc_1 = a1 \wedge pc_1' = a3)$ and $m_4 = \ddot{\alpha}_1(m_3 \circ \rho_{14}) = (pc_1 = a1 \wedge pc_1' = a1)$.

The abstract computation continues until no new abstract transitions are found or there is an indication of non terminating executions. For our example, the abstract transition m_4 is not well-founded, i.e., an infinite sequence of such transitions appears to be feasible.

First abstraction refinement. Since m_4 was obtained using abstraction, we need to check whether the evidence for non-termination is spurious. This is realized by computing a constraint that is satisfiable if and only if the abstraction can be refined to remove the spurious transition. This constraint portrays the way the abstract transition m_4 was computed using unknown predicates to denote abstract transitions. Since m_1 was computed as an abstraction from the initial state, $\varphi_{init} \wedge \rho_{11} \rightarrow m_1(V, V')$, the constraint maintains this requirement by replacing the abstract transition with an unknown predicate that denotes the abstract transition to be refined: $\varphi_{init} \wedge \rho_{11} \rightarrow \text{"}m_1\text{"}(V, V')$. The constraint is represented by a set of such implications, each one in the form of a Horn clause (with implicit universal quantifiers). We obtain HC_1 as follows.

$$HC_1 = \{\varphi_{init} \wedge \rho_{11} \rightarrow \text{"}m_1\text{"}(V, V'),$$
$$\text{"}m_1\text{"}(V'', V) \wedge \rho_{12} \rightarrow \text{"}m_2\text{"}(V, V'),$$
$$\text{"}m_2\text{"}(V'', V) \wedge \rho_{13} \rightarrow \text{"}m_3\text{"}(V'', V'),$$
$$\text{"}m_3\text{"}(V'', V) \wedge \rho_{14} \rightarrow \text{"}m_4\text{"}(V'', V') \}$$

The constraint HC_1 may have multiple solutions for the unknown predicates. Our algorithm uses two types of solutions for HC_1 . First, we need to ensure that there exists a well-founded transition that over-approximates $\text{"}m_4\text{"}(V, V')$. This is done by computing the least (most precise) solution for $\text{"}m_4\text{"}(V, V')$ over some domain of solutions. For our example, we obtain SOL^μ .

$$\text{SOL}^\mu(\text{"}m_1\text{"}(V, V')) = (\ x' = x \wedge lck = 0 \wedge lck' = 1 \wedge t' = t)$$
$$\text{SOL}^\mu(\text{"}m_2\text{"}(V, V')) = (\ x > 0 \wedge x' = x \wedge lck = lck' = 1 \wedge t' = t)$$
$$\text{SOL}^\mu(\text{"}m_3\text{"}(V, V')) = (\ x > 0 \wedge x' = x \wedge lck = lck' = 1 \wedge t' = x - 1)$$
$$\text{SOL}^\mu(\text{"}m_4\text{"}(V, V')) = (\ x > 0 \wedge x' = x - 1 \wedge lck = lck' = 1 \wedge t' = x - 1)$$

The solution $\text{SOL}^\mu(\text{"}m_4\text{"}(V, V'))$ is well-founded. While refining the abstraction functions using the solution SOL^μ would remove the spurious transition that is not well-founded, such a solution hinders the convergence of the abstraction refinement procedure. For the current example, collecting all predicates that appear in the least solutions of $\text{"}m_1\text{"}(V, V')$, $\text{"}m_2\text{"}(V, V')$, $\text{"}m_3\text{"}(V, V')$ and $\text{"}m_4\text{"}(V, V')$ would compromise the benefit of abstraction. Instead, our algorithm uses the method from [18] to generate a well-founded relation that approximates

the transitive closure of $\text{SOL}^\mu(\text{"}m_4\text{"}(V, V')) : WF_1(V, V') = (x' \geq 0 \wedge x' < x)$.
The next step is to add a well-foundedness condition to the set of Horn clauses
$HC_1^{WF} = HC_1 \cup \{\text{"}m_4\text{"}(V, V') \to WF_1\}$, and look for a solution of the extended
set of clauses. We obtain the solution SOL by invoking the solving algorithm
from [11] with input HC_1^{WF} .

$$\text{SOL}(\text{"}m_1\text{"}(V, V')) = (true) \qquad \text{SOL}(\text{"}m_2\text{"}(V, V')) = (x' \geq 1)$$
$$\text{SOL}(\text{"}m_3\text{"}(V, V')) = (t' \geq 0 \wedge t' < x) \quad \text{SOL}(\text{"}m_4\text{"}(V, V')) = (x' \geq 0 \wedge x' < x)$$

We collect the predicates that appear in the solutions of the abstract transitions
from the first thread and add them to the set of predicates $\ddot{\mathcal{P}}_1$ as follows: $\ddot{\mathcal{P}}_1 = \ddot{\mathcal{P}}_1 \cup \{x' \geq 1, t' \geq 0, t' < x, x' \geq 0, x' < x\}$.

Termination property of thread T2. We now illustrate the capability of our algorithm to identify termination bugs. The abstract computation of the second thread proceeds as follows: $n_1 = \ddot{\alpha}_2(\varphi_{init} \wedge \rho_{21}) = (pc_2 = b0 \wedge pc_2' = b2)$, $n_2 = \ddot{\alpha}_2(n_1 \circ \rho_{22}) = (pc_2 = b0 \wedge pc_2' = b3)$ and $n_3 = \ddot{\alpha}_2(n_2 \wedge \rho_{23}) = (pc_2 = b0 \wedge pc_2' = b0)$. The abstract transition n_3 is not well-founded. Our algorithm generates the following set of three Horn clauses:

$$HC_4 = \{\varphi_{init} \wedge \rho_{21} \to \text{"}n_1\text{"}(V, V'), \text{"}n_1\text{"}(V, V') \wedge \rho_{22}(V', V'') \to \text{"}n_2\text{"}(V, V''),$$
$$\text{"}n_2\text{"}(V, V') \wedge \rho_{23}(V', V'') \to \text{"}n_3\text{"}(V, V'') \} .$$

The least solution $\text{SOL}^\mu(\text{"}n_3\text{"}(V, V')) = (lck = 0 \wedge lck' = 0 \wedge pc_1 = a0 \wedge pc_1' = a0 \wedge pc_2 = b0 \wedge pc_2' = b0)$ is not well-founded. The algorithm returns a counterexample to termination of thread T2, the sequence of transitions $\rho_{21}\rho_{22}\rho_{23}$.

Termination property of thread T1. After a few more iterations, our algorithm proves the termination of the first thread. It constructs all abstract transitions applying exhaustively the steps above and it finds that all of them represent well-founded abstract transitions. The counterexample π is not encountered during the proof of thread-termination of T1, since it contains only transitions from the second thread. We conclude the presentation of this example by showing the final environment transitions that are part of the proof for termination of the first thread. They ensure two conditions: 1) environment transitions constructed from ρ_{22} are guarded by the relation $pc_2 = b2$, 2) environment transitions generated from other transitions of the second thread satisfy the constraint $x' \leq x$.

$$\ddot{\mathcal{P}}_{2\triangleright 1} = \{x' \leq x, lck = 0, lck' = 1, pc_2 = b2, pc_2' \neq b2\}$$
$$E_1 = (pc_2 = b2 \wedge pc_2' \neq b2) \vee (pc_2' \neq b2 \wedge x' \leq x) \vee$$
$$(lck = 0 \wedge lck' = 1 \wedge x' \leq x)$$

The termination property of T1 cannot be established using the method proposed in [7], which is restricted to thread-modular proofs. The program LockDecrement lifts two restrictions from the original example [7]: the decrement of the value of x (not an atomic operation) is split in two statements at labels a2 and a3 ; secondly, instead of an integer, the value of the lock variable is represented using a single bit, as our method supports mutexes.

Table 1. Experimental evaluation

Benchmark programs		Proving Termination		
Name	LOC	Thread 1	Thread 1+2	Comments
Figure 2 [7]	19	✓-Modular 0.1s	×	T2 is non-term.
Figure 6	21	✓-NonModular 0.4s	×	T2 is non-term.
Figure 2	9	✓-Modular 5.9s	✓-Modular 9.4s	T1+T2 proven term.
Figure 2-fixed [15]	38	✓-Modular 0.3s	×	T2 is non-term.
Figure 4-fixed [15]	168	✓-Modular 13.2s	×	T2 is non-term.

5   Experiments

We implemented the algorithms for proving termination properties in a tool for verification of multi-threaded C programs. Our tool uses a frontend based on CIL [16] to translate C programs to multi-threaded transition systems as formalized in Section 2. As explained before, our implementation makes use of a procedure for finding ranking functions [18] and a Horn clause solving algorithm over the linear inequality domain [11]. As an optimization, we use the frontend capability of generating a cutpoint for every iterating construct in the input program. The implementation only checks for well-foundedness of transitions with start and end locations compatible with a cutpoint.

We evaluated our implementation using a set of benchmark programs listed in Table 1. The first example was presented as Figure 2 in [7]. Relying on an atomicity assumption encoded in the program, our tool is able to find a modular proof for the thread-termination of Thread 1. The second thread contains a non-terminating loop and our tool returns a counterexample to its termination.

The second program (see Figure 6) removes the atomicity assumption from the previous program and represents the lock variable using a single bit (i.e., a mutex). Under these assumptions, the termination of thread 1 has a non-modular proof that our tool is able to discover automatically.

For the example shown in Figure 2, our tool is able to derive a modular proof for the termination of both threads. This is facilitated by the abstraction refinement queries formulated in terms of Horn clauses with preference for modular solutions. More elaborate non-modular proofs of termination also exists for this program and our tool is able to avoid them.

We include two more examples that illustrate fixes from the MOZILLA CVS repository for two vulnerabilities initially presented in a study of concurrency bugs [15]. The termination of the second thread from Figure 2-fixed depends on a fairness assumption and our tool reports a counterexample that represents an unfair path. In practice, fair termination is a desirable property, since it allows one to exclude from consideration pathological, unfair schedulers. Our approach can be extended to deal with fair termination via program transformation [17]. Fairness assumptions can be explicated in the program using additional counters and hence termination proving methods apply out of the box.

6   Related Work

Our paper builds on two lines of research. The first one established abstraction refinement [1] as a successful method for proving termination of sequential programs [5, 6]. Our second inspiration is the rely-guarantee method that was proposed in [13] for reasoning about multi-threaded programs. This reasoning method was automated and implemented for the first time in the Calvin model checker [8] for Java programs. Calvin's implementation makes use of environment assumptions specified by the programmer for each thread. More recently, discovery of environment assumptions for proving safety properties was automated using abstraction refinement [11]. Their abstraction refinement is based on a Horn-clause solving procedure, which we use in this paper for a different class of refinement queries (related to termination properties).

For proving termination of multi-threaded programs, Cook et al. proposed an automated method based on rely-guarantee reasoning [7]. This verification method was the first to compute environment assumptions for termination, but is restricted to properties with modular proofs and does not support synchronization primitives like mutexes. Cohen and Namjoshi present an algorithm that gradually exposes additional local state that is needed in non thread-modular safety proofs [2] and termination proofs [3]. Although their proof rules and general algorithms apply to non-finite analysis domains, they have so far been implemented and tested only for finite-state systems. Unlike our proposal that does not directly support fairness assumptions, a recent extension of the SPLIT model checker [4] incorporates fairness in a compositional algorithm without relying on the trivial solution that turns all local variables from the specification in global variables.

Recent work on compositional transition invariants deals with verification of sequential programs [14]. In this work, compositionality refers to the transitivity property of transition invariants and allows one to check the validity of a transition invariant candidate without going through the transitive closure computation. This notion of compositionality was successfully used to speed up the verification of sequential programs but is not applicable for structuring the verification of multi-threaded programs. In contrast, in our work and in the literature on the verification of multi-threaded programs it is common to use compositionality to describe verification methods that consider only one thread at a time (instead of taking the entire program, which is prohibitively expensive even for very small programs).

Thread-modular shape analysis [9] has been proposed for verification of heap-manipulating multi-threaded programs. The original formulation handled safety properties, while a follow-up work [10] proposed an automatic method for proving liveness properties of concurrent data-structure implementations. While this line of work handles only properties that have thread-modular proofs, we believe that the proof rule that we propose in this paper is a step towards automatic proving of properties that require non-modular reasoning about heap.

Acknowledgement. We thank Ruslán Ledesma-Garza and Kedar Namjoshi for comments and suggestions.

References

1. Clarke, E.M., Grumberg, O., Jha, S., Lu, Y., Veith, H.: Counterexample-Guided Abstraction Refinement. In: Emerson, E.A., Sistla, A.P. (eds.) CAV 2000. LNCS, vol. 1855, pp. 154–169. Springer, Heidelberg (2000)
2. Cohen, A., Namjoshi, K.S.: Local Proofs for Global Safety Properties. In: Damm, W., Hermanns, H. (eds.) CAV 2007. LNCS, vol. 4590, pp. 55–67. Springer, Heidelberg (2007)
3. Cohen, A., Namjoshi, K.S.: Local Proofs for Linear-Time Properties of Concurrent Programs. In: Gupta, A., Malik, S. (eds.) CAV 2008. LNCS, vol. 5123, pp. 149–161. Springer, Heidelberg (2008)
4. Cohen, A., Namjoshi, K.S., Sa'ar, Y.: A Dash of Fairness for Compositional Reasoning. In: Touili, T., Cook, B., Jackson, P. (eds.) CAV 2010. LNCS, vol. 6174, pp. 543–557. Springer, Heidelberg (2010)
5. Cook, B., Podelski, A., Rybalchenko, A.: Abstraction Refinement for Termination. In: Hankin, C., Siveroni, I. (eds.) SAS 2005. LNCS, vol. 3672, pp. 87–101. Springer, Heidelberg (2005)
6. Cook, B., Podelski, A., Rybalchenko, A.: Termination proofs for systems code. In: PLDI (2006)
7. Cook, B., Podelski, A., Rybalchenko, A.: Proving thread termination. In: PLDI (2007)
8. Flanagan, C., Freund, S.N., Qadeer, S.: Thread-Modular Verification for Shared-Memory Programs. In: Le Métayer, D. (ed.) ESOP 2002. LNCS, vol. 2305, pp. 262–277. Springer, Heidelberg (2002)
9. Gotsman, A., Berdine, J., Cook, B., Sagiv, M.: Thread-modular shape analysis. In: PLDI (2007)
10. Gotsman, A., Cook, B., Parkinson, M.J., Vafeiadis, V.: Proving that non-blocking algorithms don't block. In: POPL (2009)
11. Gupta, A., Popeea, C., Rybalchenko, A.: Predicate abstraction and refinement for verifying multi-threaded programs. In: POPL (2011)
12. Henzinger, T.A., Jhala, R., Majumdar, R.: Race checking by context inference. In: PLDI (2004)
13. Jones, C.B.: Tentative steps toward a development method for interfering programs. ACM Trans. Program. Lang. Syst. 5(4), 596–619 (1983)
14. Kroening, D., Sharygina, N., Tsitovich, A., Wintersteiger, C.M.: Termination Analysis with Compositional Transition Invariants. In: Touili, T., Cook, B., Jackson, P. (eds.) CAV 2010. LNCS, vol. 6174, pp. 89–103. Springer, Heidelberg (2010)
15. Lu, S., Park, S., Seo, E., Zhou, Y.: Learning from mistakes: a comprehensive study on real world concurrency bug characteristics. In: ASPLOS (2008)
16. Necula, G.C., McPeak, S., Rahul, S.P., Weimer, W.: CIL: Intermediate Language and Tools for Analysis and Transformation of C Programs. In: Horspool, R.N. (ed.) CC 2002. LNCS, vol. 2304, pp. 213–228. Springer, Heidelberg (2002)
17. Olderog, E.-R., Apt, K.R.: Fairness in parallel programs: The transformational approach. ACM Trans. Program. Lang. Syst. 10(3), 420–455 (1988)
18. Podelski, A., Rybalchenko, A.: A Complete Method for the Synthesis of Linear Ranking Functions. In: Steffen, B., Levi, G. (eds.) VMCAI 2004. LNCS, vol. 2937, pp. 239–251. Springer, Heidelberg (2004)
19. Podelski, A., Rybalchenko, A.: Transition invariants. In: LICS (2004)
20. Spoto, F., Mesnard, F., Payet, É.: A termination analyzer for Java bytecode based on path-length. ACM Trans. Program. Lang. Syst. (2010)

Deciding Conditional Termination*

Marius Bozga[1], Radu Iosif[1], and Filip Konečný[1,2]

[1] VERIMAG, CNRS, 2 av. de Vignate, 38610 Gières, France
{bozga,iosif}@imag.fr
[2] FIT BUT, IT4Innovations Centre of Excellence, Czech Republic
ikonecny@fit.vutbr.cz

Abstract. This paper addresses the problem of conditional termination, which is that of defining the set of initial configurations from which a given program terminates. First we define the dual set, of initial configurations, from which a non-terminating execution exists, as the greatest fixpoint of the pre-image of the transition relation. This definition enables the representation of this set, whenever the closed form of the relation of the loop is definable in a logic that has quantifier elimination. This entails the decidability of the termination problem for such loops. Second, we present effective ways to compute the weakest precondition for non-termination for difference bounds and octagonal (non-deterministic) relations, by avoiding complex quantifier eliminations. We also investigate the existence of linear ranking functions for such loops. Finally, we study the class of linear affine relations and give a method of under-approximating the termination precondition for a non-trivial subclass of affine relations. We have performed preliminary experiments on transition systems modeling real-life systems, and have obtained encouraging results.

1 Introduction

The termination problem asks whether every computation of a given program ends in a halting state. The universal termination asks whether a given program stops for every possible input configuration. Both problems are among the first ever to be shown undecidable, by A. Turing [24]. In many cases however, programs will terminate when started in certain configurations, and may[1] run forever, when started in other configurations. The problem of determining the set of configurations from which a program terminates on all paths is called *conditional termination*.

In program analysis, the presence of non-terminating runs has been traditionally considered faulty. However, more recently, with the advent of *reactive systems*, accidental termination can be an equally serious error. For instance, when designing a web server,

---

* This work was supported by the French National Project ANR-09-SEGI-016 VERIDYC, by the Czech Science Foundation (projects P103/10/0306 and 102/09/H042), the Czech Ministry of Education (projects COST OC10009 and MSM 0021630528), the Barrande project MEB021023, and the EU/Czech IT4Innovations Centre of Excellence CZ.1.05/1.1.00/02.0070.

[1] If the program is non-deterministic, the existence of a single infinite run, among other finite runs, suffices to consider an initial configuration non-terminating.

C. Flanagan and B. König (Eds.): TACAS 2012, LNCS 7214, pp. 252–266, 2012.

a developer would like to make sure that the main program loop will not exit unless a stopping request has been issued. These facts lead us to considering the *conditional non-termination* problem, which is determining the set of initial configurations which guarantee that the program will not exit.

In this paper we focus on programs that handle integer variables, performing linear arithmetic tests and (possibly non-deterministic) updates. A first observation is that the set of configurations guaranteeing non-termination is the greatest fixpoint of the pre-image of the program's transition relation[2] R. This set, called the *weakest recurrent set*, and denoted $wrs(R)$ in our paper, can be defined in first-order arithmetic, provided that the closed form of the infinite sequence of relations $\{R^i\}_{i \geq 0}$, obtained by composing the transition relation with itself $0, 1, 2, \ldots$ times, can also be defined using first-order arithmetic. Moreover, if the fragment of arithmetic we use has quantifier elimination, the weakest recurrent set can be expressed in a quantifier-free decidable fragment of arithmetic. This also means that the problem $wrs(R) \stackrel{?}{=} \emptyset$ is decidable, yielding universal termination decidability proofs for free.

Contributions of this Paper. The main novelty in this paper is of rather theoretical nature: we show that the non-termination preconditions for integer transition relations defined as either *octagons* or *linear affine loops with finite monoid property* are definable in quantifier-free Presburger arithmetic. Thus, the universal termination problem for such program loops is decidable. However, since quantifier elimination in Presburger arithmetic is a complex procedure, we have developed alternative ways of deriving the preconditions for non-termination, and in particular:

- for *difference bounds*, we reduce the problem of finding the weakest recurrent set to finding the maximal solution of a system of inequalities in the complete lattice of integers extended with $\pm\infty$, where the right-hand sides use addition and min operators. Efficient algorithms for finding such maximal solutions are based on policy iteration [14]. This encoding gives us a worst-case time complexity of $\mathcal{O}(n^2 \cdot 2^n)$ in the number of variables n, for the computation of the weakest recurrent set for difference bounds relations.
- for *octagonal relations*, we use a result from [5], namely that the sequence $\{R^i\}_{i \geq 0}$ is, in some sense, periodic. We give here a simple quantifier elimination method, targeted for the class of formulae defining weakest recursive sets. Moreover, we investigate the existence of linear ranking functions, and prove that, for each well-founded octagonal relations, there exists an effectively computable witness relation i.e., a well-founded relation that has a linear ranking function.
- for *linear affine relations*, weakest recurrent sets can be defined in Presburger arithmetic if we consider several restrictions concerning the transformation matrix. If the matrix A defining R has eigenvalues which are either zeros or roots of unity, all non-zero eigenvalues being of multiplicity one (these conditions are equivalent to the finite monoid property of [2,12]), then $wrs(R)$ is Presburger definable. Otherwise, if all non-zero eigenvalues of A are roots of unity, of multiplicities greater

---

[2] This definition is the dual of the *reachability set*, needed for checking safety properties: the reachability set is the least fixpoint of the post-image of the transition relation.

or equal to one, $wrs(R)$ can be expressed using polynomial terms. In this case, we can systematically issue termination preconditions, which are of significant practical importance, as noted in [11].

For space reasons, all proofs are deferred to [7].

Practical Applications. Unfortunately, in practice, the cases in which the closed form of the sequence $\{R^i\}_{i \geq 0}$ is definable in a logic that has quantifier elimination, are fairly rare. All relations considered so far are conjunctive, meaning that they can represent only simple program loops of the form while(condition){body}, where the loop body contains no further conditional constructs. In order to deal with more complicated program loops, one can use the results from this paper in several ways:

- use the decision procedures as a back-end of a termination analyzer, in order to detect spurious non-termination counterexamples consisting of a finite prefix (stem) and a conjunctive loop body (lasso). The spurious counterexamples can be discarded by intersecting the program model with the complement of the weak deterministic Büchi automaton representing the counterexample, as in [17].
- abstract a disjunctive loop body $R_1 \vee \ldots \vee R_n$ by a non-deterministic difference bounds or octagonal[3] relation $R^{\#} \supseteq R_{1,\ldots,n}$ and compute the weakest recurrent set of the latter. The complement of this set is a set of configurations from which the original loop terminates.
- attempt to compute a *transition invariant* i.e., an overapproximations of the transitive closure of the disjunctive loop body $(R_1 \vee \ldots \vee R_n)^+$ (using e.g., the semi-algorithmic unfolding technique described in [6]) and overapproximate it by a disjunction $R_1^{\#} \vee \ldots \vee R_m^{\#}$ of difference bounds or octagonal relations. Then compute the weakest recurrent set of each relation in the latter disjunction. If $wrs(R_1^{\#}) = \ldots = wrs(R_m^{\#}) = \emptyset$, the original loop terminates on any input, following the principle of transition invariants [20].

1.1 Related Work

The literature on program termination is vast. Most work focuses however on universal termination, such as the techniques for synthesizing linear ranking functions of Sohn and Van Gelder [22] or Podelski and Rybalchenko [19], and the more sophisticated method of Bradley, Manna and Sipma [9], which synthesizes lexicographic polynomial ranking functions, suitable when dealing with disjunctive loops. However, not every terminating program (loop) has a linear (polynomial) ranking function. In this paper we show that, for an entire class of non-deterministic linear relations, defined using octagons, termination is always witnessed by a computable octagonal relation that has a linear ranking function.

Another line of work considers the decidability of termination for simple (conjunctive) linear loops. Initially Tiwari [23] shows decidability of termination for affine linear loops interpreted over *reals*, while Braverman [10] refines this result by showing

---

[3] The linear affine relations considered in this paper are deterministic, which makes them unsuitable for abstraction.

decidability over *rationals* and over *integers*, for homogeneous relations of the form $C_1\mathbf{x} > 0 \wedge C_2\mathbf{x} \geq 0 \wedge \mathbf{x}' = A\mathbf{x}$. The non-homogeneous integer case seems to be much more difficult as it is closely related to the open *Skolem's Problem* [16]: given a linear recurrence $\{u_i\}_{i \geq 0}$, determine whether $u_i = 0$ for some $i \geq 0$.

To our knowledge, the first work on proving non-termination of simple loops is reported in [15]. The notion of *recurrent sets* occurs in this work, however without the connection with fixpoint theory, which is introduced in the present work. Finding recurrent sets in [15] is complete with respect to a predefined set of templates, typically linear systems of rational inequalities.

The work which is closest to ours is probably that of Cook et al. [11]. In this paper, the authors develop an algorithm for deriving termination preconditions, by first guessing a ranking function candidate (typically the linear term from the loop condition) and then inferring a supporting assertion, which guarantees that the candidate function decreases with each iteration. The step of finding a supporting assertion requires a fixpoint iteration, in order to find an invariant condition. Unlike our work, the authors of [11] do not address issues related to completeness: the method is not guaranteed to find the weakest precondition for termination, even in cases when this set can be computed. On the other hand, it is applicable to a large range of programs, extracted from real-life software. To compare our method with theirs, we tried all examples available in [11]. Since most of them are linear affine relations, we used our under-approximation method and have computed termination preconditions, which turn out to be slightly more general than the ones reported in [11].

2   Preconditions for Non-termination

In the rest of this paper we denote by $\mathbf{x} = \{x_1, \ldots, x_n\}$ the set of working variables, ranging over a domain of values denoted as $\mathcal{D}$. A *state* is a valuation $s : \mathbf{x} \to \mathcal{D}$, or equivalently, an n-tuple of values from $\mathcal{D}$. An *execution step* is a relation $R \subseteq \mathcal{D}^n \times \mathcal{D}^n$ defined by an *arithmetic formula* $\mathcal{R}(\mathbf{x}, \mathbf{x}')$, where the set $\mathbf{x}' = \{x_1', \ldots, x_n'\}$ denotes the values of the variables after executing R once. If s and s' are valuations of the sets $\mathbf{x}$ and $\mathbf{x}'$, we denote by $\mathcal{R}(s, s')$ the fact that $(s, s') \in R$. A relation R is said to be *consistent* if there exist states s, s' such that $\mathcal{R}(s, s')$.

Relational composition is defined as $R_1 \circ R_2 = \{(s, s') \in \mathcal{D}^n \times \mathcal{D}^n \mid \exists s'' \in \mathcal{D}^n . R_1(s, s'') \wedge R_2(s'', s')\}$. For any relation $R \in \mathcal{D}^n \times \mathcal{D}^n$, we consider R^0 to be the identity relation, and we define $R^{i+1} = R^i \circ R$, for all $i \geq 0$. The pre-image of a set $S \subseteq \mathcal{D}^n$ via R is the set $pre_R(S) = \{s \in \mathcal{D}^n \mid \exists s' \in S . \mathcal{R}(s, s')\}$. It is easy to check that $pre_R^i(S) = pre_{R^i}(S)$, for any $S \subseteq \mathcal{D}^n$ and for all $i \geq 0$. For any $i \geq 0$, we write $\mathcal{R}^i$ for the formula defining the relation R^i and $\mathcal{R}^{-i}(\top)$ for the formula defining the set $pre_{R^i}(\mathcal{D}^n)$.

Definition 1. *A relation R is said to be $*$-consistent if and only if, for any $k > 0$, there exists a sequence of states $s_1, \ldots, s_k$, such that $\mathcal{R}(s_i, s_{i+1})$, for all $i = 1, \ldots, k - 1$. R is said to be* well-founded *if and only if there is no infinite sequence of states $\{s_i\}_{i > 0}$, such that $\mathcal{R}(s_i, s_{i+1})$, for all $i > 0$.*

Notice that if a relation is not $*$-consistent, then it is also well-founded. However the dual is not true. For instance, the relation $R = \{(n, n-1) \mid n > 0\}$ is both $*$-consistent and well-founded.

Definition 2. *A set $S \subseteq \mathcal{D}^n$ is said to be a* non-termination precondition *for R if, for each state $s \in S$ there exists an infinite sequence of states $s_0, s_1, s_2, \ldots$ such that $s = s_0$ and $\mathcal{R}(s_i, s_{i+1})$, for all $i \geq 0$.*

If $S_0, S_1, \ldots$ are all non-termination preconditions for R, then the (possibly infinite) union $\bigcup_{i=0,1,\ldots} S_i$ is a non-termination precondition for R as well. The set $wnt(R) = \bigcup \{S \in \mathcal{D}^n \mid S$ is a non-termination precondition for $R\}$ is called the *weakest non-termination precondition* for R. A relation R is well-founded if and only if $wnt(R) = \emptyset$. A set S such that $S \cap wnt(R) = \emptyset$ is called a *termination precondition*.

Definition 3. *A set $S \subseteq \mathcal{D}^n$ is said to be* recurrent *for a relation $R \in \mathcal{D}^n \times \mathcal{D}^n$ if and only if $S \subseteq pre_R(S)$.*

Proposition 1. *Let $S_0, S_1, \ldots \in \mathcal{D}^n$ be a (possibly infinite) sequence of sets, all of which are recurrent for a relation $R \in \mathcal{D}^n \times \mathcal{D}^n$. Then their union $\bigcup_{i=0,1,\ldots} S_i$ is recurrent for R as well.*

The set $wrs(R) = \bigcup \{S \in \mathcal{D}^n \mid S$ is a recurrent set for $R\}$ is called the *weakest recurrent set* for R. By Proposition 1, $wrs(R)$ is recurrent for R. Next we define the weakest recurrent set as the greatest fixpoint of the transition relation's pre-image.

Lemma 1. *Given a relation $R \in \mathcal{D}^n \times \mathcal{D}^n$, the weakest recurrent set for R is the greatest fixpoint of the function $X \mapsto pre_R(X)$.*

As a consequence, we obtain $wrs(R) = \bigcap_{i>0} pre_R^i(\mathcal{D}^n)$, by the Kleene Fixpoint Theorem. Since $pre_R^i = pre_{R^i}$, we have $wrs(R) = \bigcap_{i>0} pre_{R^i}(\mathcal{D}^n)$. In other words, from any state in the weakest recurrent set for a relation, an iteration of any finite length of the given relation is possible. The following lemma shows that in fact, this is exactly the set of states from which an infinite iteration is also possible.

Lemma 2. *Given a relation $R \in \mathcal{D}^n \times \mathcal{D}^n$, the weakest recurrent set for R equals its weakest non-termination precondition.*

The characterization of weakest recurrent sets as greatest fixpoints of the pre-image function suggests a method for computing such sets. In this section we show that, for certain classes of relations, these sets are definable in Presburger arithmetic, which gives a decision procedure for the well-foundedness problem for certain classes of relations, and consequently, for the termination problem for several classes of program loops.

Definition 4. *Given a relation $R \in \mathcal{D}^n \times \mathcal{D}^n$ defined by an arithmetic formula $\mathcal{R}(\mathbf{x}, \mathbf{x}')$, the* closed form *of R is a formula $\mathcal{R}^{(k)}(\mathbf{x}, \mathbf{x}')$, with free variables $\mathbf{x} \cup \mathbf{x}' \cup \{k\}$, such that for every integer valuation $i > 0$ of k, $\mathcal{R}^{(i)}(\mathbf{x}, \mathbf{x}')$ defines the relation R^i.*

Since, by Lemma 1, we have $wrs(R) = \mathrm{gfp}(pre_R) = \bigcap_{i>0} pre_{R^i}(\mathcal{D}^n)$, using the closed form of R, one can now define:

$$wrs(R) \equiv \forall k > 0 \; \exists \mathbf{x}' . \; \mathcal{R}^{(k)}(\mathbf{x}, \mathbf{x}') \tag{1}$$

Because Presburger arithmetic has quantifier elimination, $wrs(R)$ can be defined in Presburger arithmetic[4] whenever $\mathcal{R}^{(k)}$ can. In [5] we show three classes of relations for which $\mathcal{R}^{(k)}$ is Presburger definable: difference bounds, octagonal and finite-monoid affine relations (the formal definitions of these classes are given in the next section). For each of these classes of loops termination is decidable, by the above argument.

3   Difference Bounds Relations

In this and the following sections, we assume that the variables $\mathbf{x} = \{x_1, \ldots, x_n\}$ range over integers i.e., that $\mathcal{D} = \mathbb{Z}$.

Definition 5. *A formula $\phi(\mathbf{x})$ is a* difference bounds constraint *if it is equivalent to a finite conjunction of atomic propositions of the form $x_i - x_j \leq a_{ij}$, for $1 \leq i, j \leq n, i \neq j$, where $a_{ij} \in \mathbb{Z}$.*

Given a difference bounds constraint ϕ, a *difference bounds matrix* (DBM) representing ϕ is a matrix $m_\phi \in \mathbb{Z}_\infty^{n \times n}$ such that $(m_\phi)_{ij} = a_{ij}$, if $x_i - x_j \leq a_{ij}$ is an atomic proposition in ϕ, and ∞, otherwise.

If ϕ is inconsistent (logically equivalent to false) we also say that m_ϕ is inconsistent. The next definition gives a canonical form for consistent DBMs.

Definition 6. *A consistent DBM $m \in \mathbb{Z}_\infty^{n \times n}$ is said to be* closed *if and only if $m_{ii} = 0$ and $m_{ij} \leq m_{ik} + m_{kj}$, for all $1 \leq i, j, k \leq n$.*

Given a consistent DBM m, we denote by m^* the (unique) closed DBM equivalent with it. It is well-known that, if m is consistent, then m^* is unique, and can be computed from m in time $\mathcal{O}(n^3)$, by the classical Floyd-Warshall algorithm. The closure of DBM provides an efficient means to compare difference bounds constraints.

Proposition 2 ([18]). *Given two consistent difference bounds constraints $\varphi(\mathbf{x})$ and $\psi(\mathbf{x})$, the following conditions are equivalent:*

- $\forall \mathbf{x} . \varphi(\mathbf{x}) \rightarrow \psi(\mathbf{x})$
- $(m_\varphi^*)_{ij} \leq (m_\psi^*)_{ij}$, *for all $1 \leq i, j \leq n$*

In the following, let R be a relation defined by a difference bounds constraint. It is easy to show that, for any $i \geq 0$, the relation R^i is a difference bounds relation as well – in other words, difference bounds relations are closed under composition. Moreover, if S is a set defined by a difference bounds constraint, then the set $pre_{R^i}(S)$ is defined by a difference bounds constraint as well. But since $wrs(R) = \bigcap_{i>0} pre_{R^i}(\mathbb{Z}^n)$, it turns out that $wrs(R)$ can be defined by a difference bounds constraint, since the class of difference bounds constraints is closed under (possibly infinite) intersections.

We are now ready to describe the procedure computing the weakest recurrent set for a difference bounds relation R. Since $wrs(R)$ is a (possibly inconsistent) difference bounds constraint, we use the template $\mu(\mathbf{x}, \mathbf{p}) \equiv \bigwedge_{1 \leq i \neq j \leq n} x_i - x_j \leq p_{ij}$, where p_{ij} are parameters ranging over $\mathbb{Z}_{\pm \infty}$ (we clearly do not need to track the constraints of the

---

[4] Or, for that matter, in any theory that has quantifier elimination.

form $x_i - x_i \leq p_{ii}$). Moreover, we assume that the template is closed (Definition 6), which can be encoded as a system of inequalities of the form:

$$p_{ij} \leq \min \{p_{ik} + p_{kj} \mid k \neq i, k \neq j\} \qquad (2)$$

Next, we compute the (symbolic) difference bounds constraint corresponding to the set $pre_R(\mu) \equiv \exists \mathbf{x}' . \mathcal{R}(\mathbf{x}, \mathbf{x}') \wedge \mu(\mathbf{x}', \mathbf{p})$. This step requires computing the closure of the DBM corresponding to $\mathcal{R} \wedge \mu$, and elimination of the $\mathbf{x}'$ variables. The result is a closed symbolic DBM π, whose entries are min-terms consisting of sums of p_{ij} and integer constants. Further, we encode the recurrence condition $\mu \subseteq pre_R(\mu)$, again as a system of inequalities (Proposition 2) of the form:

$$p_{ij} \leq \pi_{ij}, i \neq j \qquad (3)$$

By conjoining the inequalities (2) and (3), we obtain a system of inequalities with variables p_{ij}, whose right-hand sides are linear combinations of p_{ij} with addition and min. We are interested in the maximal solution of this system, which can be obtained using an efficient policy iteration algorithm [14] in the complete lattice of $\mathbb{Z}_{\pm\infty}$ with addition, min and max operators. This solution defines the weakest recurrent set for R, and consequently, the weakest precondition for non-termination of the R loop. Since $wrs(R)$ is a difference bounds constraint, for any relation R definable by a difference bounds constraint, the maximal solution of the system is unique. It is to be noted that, if for some $1 \leq i \neq j \leq n$ we obtain $p_{ij} = -\infty$, then the weakest recurrent set is empty i.e., the relation R is well-founded.

Lemma 3. *Computing the weakest recurrent set of a difference bounds relation can be done in time $\mathcal{O}(n^2 \cdot 2^n)$, where n is the number of variables.*

4   Octagonal Relations

Octagonal relations are a generalization of difference bounds relations.

Definition 7. *A formula $\phi(\mathbf{x})$ is an octagonal constraint if it is equivalent to a finite conjunction of terms of the form $\pm x_i \pm x_j \leq a_{ij}$, where $a_{ij} \in \mathbb{Z}$ and $1 \leq i, j \leq n$.*

We represent octagons as difference bounds constraints over the *dual* set of variables $\mathbf{y} = \{y_1, y_2, \ldots, y_{2n}\}$, with the convention that y_{2i-1} stands for x_i and y_{2i} for $-x_i$, respectively. For example, the octagonal constraint $x_1 + x_2 = 3$ is represented as $y_1 - y_4 \leq 3 \wedge y_2 - y_3 \leq -3$. To handle the dual variables in the following, we define $\bar{\imath} = i - 1$, if i is even, and $\bar{\imath} = i + 1$ if i is odd. We say that a DBM $m \in \mathbb{Z}_\infty^{2n \times 2n}$ is *coherent* iff $m_{ij} = m_{\bar{\jmath}\bar{\imath}}$ for all $1 \leq i, j \leq 2n$. The coherence property is needed because any atomic proposition $x_i - x_j \leq a$, in ϕ can be represented as both $y_{2i-1} - y_{2j-1} \leq a$ and $y_{2j} - y_{2i} \leq a$, $1 \leq i, j \leq n$. We denote by $\overline{\phi}$ the difference bounds formula $\phi[y_1/x_1, y_2/ - x_1, \ldots, y_{2n-1}/x_n, y_{2n}/ - x_n]$ with free variables $\mathbf{y}$. The following equivalence relates ϕ and $\overline{\phi}$:

$$\phi(\mathbf{x}) \Leftrightarrow (\exists y_2, y_4, \ldots, y_{2n} . \overline{\phi} \wedge \bigwedge_{i=1}^{n} y_{2i-1} + y_{2i} = 0)[x_1/y_1, \ldots, x_n/y_{2n-1}] \qquad (4)$$

Given a coherent DBM m representing $\overline{\phi}$, we say that m is *octagonal-consistent* if and only if ϕ is consistent. The following definition gives the canonical form of a DBM representing an octagonal-consistent constraint.

Definition 8. *An octagonal-consistent coherent DBM* $m \in \mathbb{Z}^{2n \times 2n}$ *is said to be* tightly closed *if and only if the following hold:*

1. $m_{ii} = 0, \forall 1 \leq i \leq 2n$
2. $m_{i\bar{i}}$ *is even,* $\forall 1 \leq i \leq 2n$
3. $m_{ij} \leq m_{ik} + m_{kj}, \forall 1 \leq i, j, k \leq 2n$
4. $m_{ij} \leq \lfloor \frac{m_{i\bar{i}}}{2} \rfloor + \lfloor \frac{m_{\bar{j}j}}{2} \rfloor, \forall 1 \leq i, j \leq 2n$

Given an octagonal-consistent DBM m, we denote by m^t the equivalent tightly closed DBM. The tight closure of an octagonal-consistent DBM m is unique and can be computed in time $\mathcal{O}(n^3)$ as $m^t_{i,j} = min \left\{ m^*_{i,j}, \left\lfloor \frac{m^*_{i,\bar{i}}}{2} \right\rfloor + \left\lfloor \frac{m^*_{\bar{j},j}}{2} \right\rfloor \right\}$ [1]. This generalizes to unbounded finite compositions of octagonal relations [4]:

$$\forall k \geq 0 \,.\, (m^t_{R^k})_{i,j} = min \left\{ (m^*_{\overline{R}^k})_{i,j}, \left\lfloor \frac{(m^*_{\overline{R}^k})_{i,\bar{i}}}{2} \right\rfloor + \left\lfloor \frac{(m^*_{\overline{R}^k})_{\bar{j},j}}{2} \right\rfloor \right\} \tag{5}$$

Notice that the above relates the entries of the tightly closed DBM representation of R^k with the entries of the closed DBM representation of the relation defined by $\overline{R}^k$.

We are now ready to introduce a result [5] that defines the "shape" of the closed form $\mathcal{R}^{(k)}$ for an octagonal relation R. Intuitively, for each $i \geq 0$, R^i is an octagon, whose bounds evolve in a periodic way. The following definition gives the precise meaning of periodicity for relations that have a matrix representation.

Definition 9. *An infinite sequence of matrices* $\{M_k\}_{k=1}^{\infty} \in \mathbb{Z}_{\infty}^{m \times m}$ *is said to be* ultimately periodic *if and only if:*

$$\exists b > 0 \,\exists c > 0 \,\exists \Lambda_0, \Lambda_1, \ldots, \Lambda_{c-1} \in \mathbb{Z}_{\infty}^{m \times m} \,.\, M_{b+(k+1)c+i} = \Lambda_i + M_{b+kc+i}$$

for all $k \geq 0$ *and* $i = 0, 1, \ldots, c - 1$. *The smallest* b, c *for which the above holds are called* prefix *and* period *of the* $\{M_k\}_{k=1}^{\infty}$ *sequence, respectively.*

A result reported in [5] is that the sequence $\{m^t_{R^i}\}_{i \geq 0}$ (5) of tightly closed matrices representing the sequence $\{R^i\}_{i \geq 0}$ of powers of a $*$-consistent octagonal relation R is ultimately periodic, in the sense of the above definition. The constants b and c from Definition 9 will also be called the *prefix and period of the octagonal relation R*, throughout this section.

For a set $\mathbf{v}$ of variables, let $U(\mathbf{v}) = \{\pm v_1 \pm v_2 \mid v_1, v_2 \in \mathbf{v}\}$ denote the set of octagonal terms over $\mathbf{v}$. As a first remark, by the periodicity of the sequence $\{m^t_{R^i}\}_{i \geq 0}$, the closed form of the subsequence $\{R^{b+c\ell}\}_{\ell \geq 0}$ (of $\{R^i\}_{i \geq 0}$) can be defined as:

$$\mathcal{R}_{b,c}^{(\ell)} \equiv \bigwedge_{u \in U(\mathbf{x} \cup \mathbf{x}')} u \leq a_u \ell + d_u \tag{6}$$

where $a_u = (\Lambda_0)_{ij}$, $d_u = (m^t_{R^b})_{ij}$ for all octagonal terms $u = y_i - y_j$. This is indeed the case, since the matrix sequence $\{m^t_{R^{b+c\ell}}\}_{\ell \geq 0}$ is ultimately periodic i.e., $m^t_{R^{b+c\ell}} = m^t_{R^b} + \ell \Lambda_0$, for all $\ell \geq 0$.

Lemma 4. *Let R be a $*$-consistent octagonal relation with prefix b, period c and let $\mathcal{R}_{b,c}^{(\ell)}$ be the closed form of $\{R^{b+c\ell}\}_{\ell \geq 0}$ as defined in (6). Then, R is well-founded iff there exists $u \in U(\mathbf{x})$ s.t. $a_u < 0$. Moreover, if R is not well-founded, $wrs(R) \equiv \mathcal{R}^{-b}(\top)$.*

The above lemma can be used to compute $wrs(R)$ for an octagonal relation R. First we need to check the $*$-consistency of R, using the method reported in [6]. Second, we compute the closed form (6) and check for the existence of a term $u \in U(\mathbf{x})$ such that $a_u < 0$, in which case R is well-founded. Finally, if this is not the case, then we compute $wrs(R) = \mathcal{R}^{-b}(\top)$.

4.1 On the Existence of Linear Ranking Functions

A ranking function for a given relation R constitutes a proof of the fact that R is well-founded. We distinguish here two cases. If R is not $*$-consistent, then the well-foundedness of R is witnessed simply by an integer constant $i > 0$ such that $R^i = \emptyset$. Otherwise, if R is $*$-consistent, we need a better argument for well-foundedness. In this section we show that, for any $*$-consistent well-founded octagonal relation R with prefix b, the (strenghtened) relation defined by $\mathcal{R}^{-b}(\top) \wedge \mathcal{R}$ is well-founded and has a linear ranking function, even when R alone does not have one. For space reasons, we do not give here all the details of the construction of such a function. However, the existence proof suffices, as one can use *complete* ranking function extraction tools (such as e.g. RankFinder [19]) in order to find them.

Definition 10. *Given a relation $R \subseteq \mathbb{Z}^n \times \mathbb{Z}^n$, a linear ranking function for R is a term $f(\mathbf{x}) = \sum_{i=1}^{n} a_i x_i$ such that, for all states $s, s' : \mathbf{x} \to \mathbb{Z}$:*

1. *f is decreasing: $\mathcal{R}(s, s') \to f(s) > f(s')$*
2. *f is bounded: $\mathcal{R}(s, s') \to (f(s) > h \wedge f(s') > h)$, for some $h \in \mathbb{Z}$.*

The main result of this section is the following:

Theorem 1. *Let $R \subseteq \mathbb{Z}^n \times \mathbb{Z}^n$ be a $*$-consistent and well-founded octagonal relation, with prefix $b \geq 0$. Then, the relation defined by $\mathcal{R}^{-b}(\top) \wedge \mathcal{R}$ is well founded and has a linear ranking function.*

The first part of the theorem is proved by the following lemma:

Lemma 5. *Let $R \subseteq \mathbb{Z}^n \times \mathbb{Z}^n$ be a relation, and $m > 0$ be an integer. Then $wrs(R) = \emptyset$ if and only if $wrs(R_m) = \emptyset$, where R_m is the relation defined by $\mathcal{R}^{-m}(\top) \wedge \mathcal{R}$.*

It remains to prove that the witness relation defined by $\mathcal{R}^{-b}(\top) \wedge \mathcal{R}$ has a linear ranking function, provided that it is well-founded. The proof is organized as follows. First we show that well-foundedness of an octagonal relation R is equivalent to the well-foundedness of its difference bounds representation $\overline{\mathcal{R}}$ (Lemma 6). Second, we use a result from [8], that the constraints in the sequence of iterated difference bounds relations $\{\overline{\mathcal{R}}^i\}_{i \geq 0}$ can be represented by a finite-state weighted automaton, called the *zigzag automaton* in the sequel. If the relation defined by $\overline{\mathcal{R}}$ is well-founded, then this weighted automaton must have a cycle of negative weight. The structure of this cycle, representing several of the constraints in $\overline{\mathcal{R}}$, is used to show the existence of the linear ranking function for the witness relation $\mathcal{R}^{-b}(\top) \wedge \mathcal{R}$.

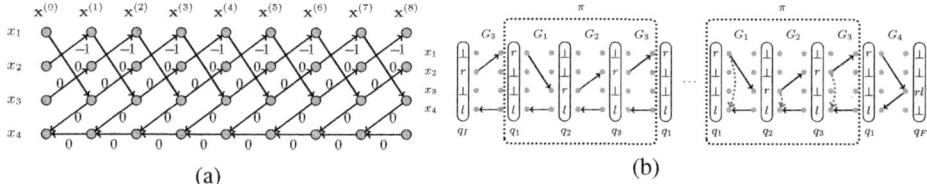

Fig. 1. (a) Unfolding of the constraint graph of the difference bounds relation $\mathcal{R} \equiv x_2 - x_1' \leq -1 \wedge x_3 - x_2' \leq 0 \wedge x_1 - x_3' \leq 0 \wedge x_4' - x_4 \leq 0 \wedge x_3' - x_4 \leq 0$. (b) A run of the zigzag automaton $\mathcal{A}_R$ over a path in the unfolded constraint graph of R.

Lemma 6. *Let $R \subseteq \mathbb{Z}^n \times \mathbb{Z}^n$ be an octagonal relation and R_{db} be the difference bounds relation defined by $\overline{\mathcal{R}}$. Then R is well-founded if and only if R_{db} is well-founded.*

The above lemma reduces the problem of showing existence of a ranking function for an octagonal relation $\mathcal{R}(\mathbf{x}, \mathbf{x}')$ to showing existence of a ranking function for its difference bounds encoding $\overline{\mathcal{R}}(\mathbf{y}, \mathbf{y}')$. Assume that $f(\mathbf{y})$ is a ranking function for $\overline{\mathcal{R}}$. Then $f[x_i / y_{2i-1}, -x_i / y_{2i}]_{i=1}^n$ is a linear ranking function for R. Hence, in the rest of this section, we consider without loss of generality that R is a difference bounds relation.

Zigzag Automata. For the later developments, we need to introduce the *zigzag automaton* corresponding to a difference bounds relation R. Intuitivelly, for any $i > 0$, the relation R^i can be represented by a constraint graph which is the i-times repetition of the constraint graph of R. The constraints induced by R^i can be represented as shortest paths in this graph, and can be recognized (in the classical automata-theoretic sense) by a weighted automaton $\mathcal{A}_R$ (see Fig. 1 for an example). The structure of this automaton is needed to show the existence of a linear ranking function.

The following lemma proves the existence of a negative weight cycle in the zigzag automata corresponding to well-founded difference bounds relation. The intuition behind this fact is that the rates of the DBM sequence $\{m_{R^i}\}_{i>0}$ are weights of optimal ratio (weight per length) cycles in the zigzag automaton. According to the previous section, if R is well-founded, there exists a negative rate for $\{m_{R^i}\}_{i>0}$, which implies the existence of a negative cycle in the zigzag automaton.

Lemma 7. *If R is a $*$-consistent well-founded difference bounds relation of prefix $b \geq 0$, and $\mathcal{A}_R$ is its corresponding zigzag automaton, then there exists a cycle π from a state q to itself, such that $w(\pi) < 0$ and there exists paths π_i from an initial state to q, and π_f from q to a final state, such that $|\pi_i| + |\pi_f| = b$.*

Next we prove the existence of a linear decreasing function, based on the existence of a negative weight cycle in the zigzag automaton.

Lemma 8. *If R is a $*$-consistent well-founded difference bounds relation of prefix $b \geq 0$, then there exists a linear function $f(\mathbf{x})$ such that, for all states $s, s' : \mathbf{x} \to \mathbb{Z}$ we have $\mathcal{R}^{-b}(\top)(s) \wedge \mathcal{R}(s, s') \to f(s) > f(s')$.*

Last, we prove that the function f of Lemma 8 is bounded, concluding that it is indeed a ranking function. Since each run in the zigzag automaton recognizes a path from some x_i to some x_j, a run that repeats a cycle can be decomposed into a prefix, the cycle itself and a suffix. The recognized path may traverse the cycle several times, however each exit point from the cycle must match a subsequent entry point. These paths from the exit to the corresponding entries give us the necessary lower bound. In fact, these paths appear already on constraint graphs that represent unfoldings of R^i, for any $i \geq b$. Hence the need for a strenghtened witness $\mathcal{R}^{-b}(\top) \wedge \mathcal{R}$, as $\mathcal{R}$ alone is not enough for proving boundedness of f.

Lemma 9. *If R is a $*$-consistent well-founded difference bounds relation of prefix b, and $f(\mathbf{x})$ is the linear decreasing function from Lemma 8, there exists an integer h such that, for all states $s, s' : \mathbf{x} \to \mathbb{Z}$, $(\mathcal{R}^b(\top)(s) \wedge \mathcal{R}(s, s')) \to (f(s) \geq h \wedge f(s') \geq h)$.*

As an experiment, we have tried the RANKFINDER [19] tool (complete for linear ranking functions), which failed to discover a ranking function on the relation $\mathcal{R}$ from Fig. 1. This comes with no surprise, since no linear decreasing function that is bounded after the first iteration exists. However, RANKFINDER finds a ranking function for the witness relation $\mathcal{R}^{-b}(\top) \wedge \mathcal{R}$ instead.

5 Linear Affine Relations

Let $\mathbf{x} = \langle x_1, \ldots, x_n \rangle^\top$ be a column vector of variables ranging over integers. A linear affine relation is a relation of the form $\mathcal{R}(\mathbf{x}, \mathbf{x}') \equiv C\mathbf{x} \geq \mathbf{d} \wedge \mathbf{x}' = A\mathbf{x} + \mathbf{b}$, where $A \in \mathbb{Z}^{n \times n}$, $C \in \mathbb{Z}^{p \times n}$ are matrices, and $\mathbf{b} \in \mathbb{Z}^n$, $\mathbf{d} \in \mathbb{Z}^p$ are column vectors of integer constants. Notice that we consider linear affine relations to be deterministic, unlike the octagonal relations considered in the previous. In the following, it is convenient to work with the equivalent homogeneous form:

$$\mathcal{R}(\mathbf{x}, \mathbf{x}') \equiv C_h\mathbf{x}_h \geq \mathbf{0} \wedge \mathbf{x}'_h = A_h\mathbf{x}_h$$

$$A_h = \begin{pmatrix} A & \mathbf{b} \\ 0 & 1 \end{pmatrix} \quad C_h = \begin{pmatrix} C & -\mathbf{d} \end{pmatrix} \quad \mathbf{x}_h = \begin{pmatrix} \mathbf{x} \\ x_{n+1} \end{pmatrix} \tag{7}$$

The closed form of a linear affine relation is defined by the following formula:

$$\mathcal{R}^{(k)}(\mathbf{x}, \mathbf{x}') \equiv \exists x_{n+1}, x'_{n+1}.\mathbf{x}'_h = A_h^k\mathbf{x}_h \wedge \forall 0 \leq \ell < k.CA_h^\ell\mathbf{x} \geq \mathbf{0} \wedge x_{n+1} = 1 \tag{8}$$

Intuitively, the first conjunct defines the (unique) outcome of iterating the relation $\mathbf{x}' = A\mathbf{x} + \mathbf{b}$ for k steps, while the second (universally quantified) conjunct ensures that the condition ($C\mathbf{x} \geq \mathbf{d}$) has been always satisfied all along the way. The definition of the weakest recursive set of a linear affine relation is (after the elimination of the trailing existential quantifier):

$$wrs(R)(\mathbf{x}) \equiv \exists x_{n+1} \forall k > 0 \,.\, C_h A_h^k\mathbf{x} \geq \mathbf{0} \wedge x_{n+1} = 1 \tag{9}$$

The main difficulty with the form (9) comes from the fact that the powers of a matrix A cannot usually be defined in a known decidable theory of arithmetic. In the following, we discuss the case of A having the finite monoid property [2,25], which leads to

$wrs(R)$ being Presburger definable. Further, we relax the finite monoid condition and describe a method for generating sufficient termination conditions, i.e. sets $S \in \mathbb{Z}^n$ such that $S \cap wrs(R) = \emptyset$.

Some basic notions of linear algebra are needed in the following. If $A \in \mathbb{Z}^{n \times n}$ is a square matrix, and $\mathbf{v} \in \mathbb{Z}^n$ is a column vector of integer constants, then any complex number $\lambda \in \mathbb{C}$ such that $A\mathbf{v} = \lambda\mathbf{v}$, for some complex vector $\mathbf{v} \in \mathbb{C}^n$, is called an *eigenvalue* of A. The vector $\mathbf{v}$ in this case is called an *eigenvector* of A. It is known that the eigenvalues of A are the roots of the *characteristic polynomial* $\det(A - \lambda I_n) = 0$, which is an effectively computable univariate polynomial in λ. A complex number r is said to be a *root of the unity* if $r^d = 1$ for some integer $d > 0$.

In the previous work of Weber and Seidl [25], Boigelot [2], and Finkel and Leroux [12], a restriction of linear affine relations has been introduced, with the goal of defining the closed form of relations in Presburger arithmetic. A matrix $A \in \mathbb{Z}^{n \times n}$ is said to have the *finite monoid property* if and only if its set of powers $\{A^i \mid i \geq 0\}$ is finite. A linear affine relation has the finite monoid property if and only if the matrix A defining the update has the finite monoid property.

Lemma 10 ([12,2]). *A matrix $A \in \mathbb{Z}^{n \times n}$ has the finite monoid property iff:*

1. *all eigenvalues of A are either zero or roots of the unity, and*
2. *all non-zero eigenvalues are of multiplicity one.*

Both conditions are decidable.

In the following, we drop the second requirement of Lemma 10, and consider only linear relations, such that all non-zero eigenvalues of A are roots of the unity. In this case, $\mathcal{R}^{(k)}$ cannot be defined in Presburger arithmetic any longer, thus we renounce defining $wrs(R)$ precisely, and content ourselves with the discovery of *sufficient conditions for termination*. Basically given a linear affine relation R, we aim at finding a disjunction $\phi(\mathbf{x})$ of linear constraints on $\mathbf{x}$, such that $\phi \wedge wrs(R)$ is inconsistent, without explicitly computing $wrs(R)$.

Lemma 11. *Given a square matrix $A \in \mathbb{Z}^{n \times n}$, whose non-zero eigenvalues are all roots of the unity. Then $(A^m)_{i,j} \in \mathbb{Q}[m]$, for all $1 \leq i, j \leq n$, are effectively computable polynomials with rational coefficients.*

We turn now back to the problem of defining $wrs(R)$ for linear affine relations R of the form (9). First notice that, if all non-zero eigenvalues of A are roots of the unity, then the same holds for A_h (7). By Lemma 11, one can find rational polynomials $p_{i,j}(k)$ defining $(A_h^k)_{i,j}$, for all $1 \leq i, j \leq n$. The condition (9) resumes to a conjunction of the form:

$$wrs(R)(\mathbf{x}) \equiv \bigwedge_{i=1}^{n} \forall k > 0 \cdot P_i(k, \mathbf{x}) \geq 0 \tag{10}$$

where each $P_i = a_{i,d}(\mathbf{x}) \cdot k^d + \ldots + a_{i,1}(\mathbf{x}) \cdot k + a_{i,0}(\mathbf{x})$ is a polynomial in k whose coefficients are the linear combinations $a_{i,d} \in \mathbb{Q}[\mathbf{x}]$. We are looking after a sufficient condition for termination, which is, in this case, any set of valuations of $\mathbf{x}$ that would invalidate (10). The following proposition gives sufficient invalidating clauses for each conjunct above. By taking the disjunction of all these clauses we obtain a sufficient termination condition for R.

Lemma 12. *Given a polynomial* $P(k, \mathbf{x}) = a_d(\mathbf{x}) \cdot k^d + \ldots + a_1(\mathbf{x}) \cdot k + a_0(\mathbf{x})$, *there exists* $n > 0$ *such that* $P(n, \mathbf{x}) < 0$ *if, for some* $i = 0, 1, \ldots, d$, *we have* $a_{d-i}(\mathbf{x}) < 0$ *and* $a_d(\mathbf{x}) = a_{d-1}(\mathbf{x}) = \ldots = a_{d-i+1}(\mathbf{x}) = 0$.

Example Consider the following program [11], and its linear transformation matrix A.

$$
\begin{aligned}
\text{while } & (x \geq 0) \\
& x' = x + y \\
& y' = y + z
\end{aligned}
\qquad
A = \begin{pmatrix} 1 & 1 & 0 \\ 0 & 1 & 1 \\ 0 & 0 & 1 \end{pmatrix}
\qquad
A^k = \begin{pmatrix} 1 & k & \frac{k(k-1)}{2} \\ 0 & 1 & k \\ 0 & 0 & 1 \end{pmatrix}
$$

The characteristic polynomial of A is $\det(A - \lambda I_3) = (1 - \lambda)^3$, hence the only eigenvalue is 1, with multiplicity 3. Then we compute A^k (see above), and $x' = x + k \cdot y + \frac{k(k-1)}{2} z$ gives the value of x after k iterations of the loop. Hence the (precise) non-termination condition is: $\forall k > 0 \, . \, \frac{z}{2} \cdot k^2 + (y - \frac{z}{2}) \cdot k + x \geq 0$. A sufficient condition for termination is: $(z < 0) \vee (z = 0 \wedge y < 0) \vee (z = 0 \wedge y = 0 \wedge x < 0)$ $\qquad\square$

We can generalize this method further to the case where all eigenvalues of A are of the form $q \cdot r$, with $q \in \mathbb{R}$ and $r \in \mathbb{C}$ being a root of the unity. The main reason for not using this condition from the beginning is that we are, to this point, unaware of its decidability status. With this condition instead, it is sufficient to consider only the eigenvalues with the maximal absolute value, and the polynomials obtained as sums of the polynomial coefficients of these eigenvalues. The result of Lemma 11 and the sufficient condition of Lemma 12 carry over when using these polynomials instead.

6   Experimental Evaluation

We have validated the methods described in this paper by automatically verifying termination of all the octagonal running examples, and of several integer programs synthesized from (i) programs with lists [3] and (ii) VHDL models [21]. We have first computed automatically their strongest summary relation $\mathcal{T}$, by adapting the method for reachability analysis for integer programs, described in [6], and implemented in the FLATA tool [13]. Then we automatically proved that $\mathcal{T}$ is contained in a disjunction of octagonal relations, which are found to be well-founded by the procedure described in Section 4.

We have first verified the termination of the LISTCOUNTER and LISTREVERSAL programs, which were obtained using the translation scheme from [3], which generates an integer program from a program manipulating dynamically allocated single-selector linked lists. Using the same technique, we also verified the COUNTER and SYNLIFO programs, obtained by translating VHDL designs of hardware counter and synchronous LIFO [21]. These models have infinite runs for any input values, which is to be expected, as they encode the behavior of synchronous reactive circuits.

Second, we have compared (Table 1) our method for termination of linear affine loops with the examples given in [11], and found the same termination preconditions

Table 1. Termination preconditions for several program fragments from [11]

PROGRAM	COOK ET. AL [11]	LINEAR AFFINE LOOPS

if (lvar $\geq$ 0)
while (lvar $< 2^{30}$)
    lvar = lvar $\ll$ 1;
	$lvar > 0 \lor lvar < 0 \lor lvar \geq 2^{30}$	$\neg(lvar=0) \lor lvar \geq 2^{30}$

while (x $\leq$ N)
    if (*) { x=2*x+y;
        y=y+1; }
    else x ++;
	$x > N \lor x + y \geq 0$	$x > N \lor x+y \geq 0$

while (x $\geq$ N) x = -2*x + 10;	$x > 5 \lor x + y \geq 0$	$x \neq \frac{10}{3} \iff$ true

//@ requires $n > 200$
x = 0;
while (1)
    if (x < n) { x=x+y;
        if (x $\geq$ 200) break; }
	$y > 0$	$y > 0$

as they do, with one exception, in which we can prove universal termination in integer input values (row 3 of Table 1). The last example from [11] is the Euclidean Greatest Common Divisor algorithm, for which we infer automatically the correct termination preconditions using a disjunctively well-founded octagonal abstraction of the transition invariant.

7   Conclusions

We have presented several methods for deciding conditional termination of several classes of program loops manipulating integer variables. The universal termination problem has been found to be decidable for octagonal relations and linear affine loops with the finite monoid property. In other cases of linear affine loops, we give sufficient termination conditions. We have implemented our method in the FLATA tool [13] and performed a number of preliminary experiments.

Acknowledgements. The authors would like to thank Eugene Asarin, Alexandre Donze, Hang Zhou, Mihai Moraru and Barbara Jobstmann for the insightful discussions on the matters of this paper, as well as the anonymous reviewers, for their interesting suggestions that led to improvements of our work.

References

1. Bagnara, R., Hill, P.M., Zaffanella, E.: An Improved Tight Closure Algorithm for Integer Octagonal Constraints. In: Logozzo, F., Peled, D.A., Zuck, L.D. (eds.) VMCAI 2008. LNCS, vol. 4905, pp. 8–21. Springer, Heidelberg (2008)
2. Boigelot, B.: Symbolic Methods for Exploring Infinite State Spaces, PhD Thesis, vol. 189. Collection des Publications de l'Université de Liège (1999)
3. Bouajjani, A., Bozga, M., Habermehl, P., Iosif, R., Moro, P., Vojnar, T.: Programs with Lists Are Counter Automata. In: Ball, T., Jones, R.B. (eds.) CAV 2006. LNCS, vol. 4144, pp. 517–531. Springer, Heidelberg (2006)
4. Bozga, M., Gîrlea, C., Iosif, R.: Iterating Octagons. In: Kowalewski, S., Philippou, A. (eds.) TACAS 2009. LNCS, vol. 5505, pp. 337–351. Springer, Heidelberg (2009)

5. Bozga, M., Iosif, R., Konečný, F.: Fast Acceleration of Ultimately Periodic Relations. In: Touili, T., Cook, B., Jackson, P. (eds.) CAV 2010. LNCS, vol. 6174, pp. 227–242. Springer, Heidelberg (2010)

6. Bozga, M., Iosif, R., Konečný, F.: Relational Analysis of Integer Programs. Technical Report TR-2011-14, Verimag, Grenoble, France (2011)

7. Bozga, M., Iosif, R., Konečný, F.: Deciding Conditional Termination. Technical Report TR-2012-1, Verimag, Grenoble, France (2012)

8. Bozga, M., Iosif, R., Lakhnech, Y.: Flat parametric counter automata. Fundamenta Informaticae 91, 275–303 (2009)

9. Bradley, A.R., Manna, Z., Sipma, H.B.: Linear Ranking with Reachability. In: Etessami, K., Rajamani, S.K. (eds.) CAV 2005. LNCS, vol. 3576, pp. 491–504. Springer, Heidelberg (2005)

10. Braverman, M.: Termination of Integer Linear Programs. In: Ball, T., Jones, R.B. (eds.) CAV 2006. LNCS, vol. 4144, pp. 372–385. Springer, Heidelberg (2006)

11. Cook, B., Gulwani, S., Lev-Ami, T., Rybalchenko, A., Sagiv, M.: Proving Conditional Termination. In: Gupta, A., Malik, S. (eds.) CAV 2008. LNCS, vol. 5123, pp. 328–340. Springer, Heidelberg (2008)

12. Finkel, A., Leroux, J.: How to Compose Presburger-Accelerations: Applications to Broadcast Protocols. In: Agrawal, M., Seth, A.K. (eds.) FSTTCS 2002. LNCS, vol. 2556, pp. 145–156. Springer, Heidelberg (2002)

13. http://www-verimag.imag.fr/FLATA.html

14. Gawlitza, T., Seidl, H.: Precise Fixpoint Computation Through Strategy Iteration. In: De Nicola, R. (ed.) ESOP 2007. LNCS, vol. 4421, pp. 300–315. Springer, Heidelberg (2007)

15. Gupta, A., Henzinger, T.A., Majumdar, R., Rybalchenko, A., Xu, R.-G.: Proving non-termination. SIGPLAN Not. 43, 147–158 (2008)

16. Halava, V., Harju, T., Hirvensalo, M., Karhumaki, J.: Skolem's problem – on the border between decidability and undecidability (2005)

17. Iosif, R., Rogalewicz, A.: Automata-Based Termination Proofs. In: Maneth, S. (ed.) CIAA 2009. LNCS, vol. 5642, pp. 165–177. Springer, Heidelberg (2009)

18. Miné, A.: The octagon abstract domain. Higher-Order and Symbolic Computation 19(1), 31–100 (2006)

19. Podelski, A., Rybalchenko, A.: A Complete Method for the Synthesis of Linear Ranking Functions. In: Steffen, B., Levi, G. (eds.) VMCAI 2004. LNCS, vol. 2937, pp. 239–251. Springer, Heidelberg (2004)

20. Podelski, A., Rybalchenko, A.: Transition invariants. In: LICS 2004, pp. 32–41 (2004)

21. Smrcka, A., Vojnar, T.: Verifying Parametrised Hardware Designs Via Counter Automata. In: Yorav, K. (ed.) HVC 2007. LNCS, vol. 4899, pp. 51–68. Springer, Heidelberg (2008)

22. Sohn, K., van Gelder, A.: Termination detection in logic programs using argument sizes. In: PODS 1991 (1991)

23. Tiwari, A.: Termination of Linear Programs. In: Alur, R., Peled, D.A. (eds.) CAV 2004. LNCS, vol. 3114, pp. 70–82. Springer, Heidelberg (2004)

24. Turing, A.M.: On computable numbers, with an application to the entscheidungsproblem. Proceedings of the London Mathematical Society 42, 230–265 (1936)

25. Weber, A., Seidl, H.: On finitely generated monoids of matrices with entries in n. In: ITA 1991, 19–38 (1991)

The AVANTSSAR Platform for the Automated Validation of Trust and Security of Service-Oriented Architectures

Alessandro Armando[1], Wihem Arsac[2], Tigran Avanesov[3], Michele Barletta[4],
Alberto Calvi[4], Alessandro Cappai[1], Roberto Carbone[1], Yannick Chevalier[5],
Luca Compagna[2], Jorge Cuéllar[6], Gabriel Erzse[7], Simone Frau[8],
Marius Minea[7], Sebastian Mödersheim[9], David von Oheimb[6],
Giancarlo Pellegrino[2], Serena Elisa Ponta[1,2], Marco Rocchetto[4],
Michael Rusinowitch[3], Mohammad Torabi Dashti[8],
Mathieu Turuani[3], and Luca Viganò[4]

[1] AI-Lab, DIST, Università di Genova, Italy
[2] SAP Research, Mougins, France
[3] LORIA & INRIA Nancy Grand Est, France
[4] Department of Computer Science, University of Verona, Italy
[5] IRIT, Université Paul Sabatier, France
[6] Siemens AG, Corporate Technology, Munich, Germany
[7] Institute e-Austria and Politehnica University, Timişoara, Romania
[8] Institute of Information Security, ETH Zurich, Switzerland
[9] IBM Zurich Research Laboratory, Switzerland and DTU, Lyngby, Denmark
www.avantssar.eu

Abstract. The AVANTSSAR Platform is an integrated toolset for the formal specification and automated validation of trust and security of service-oriented architectures and other applications in the Internet of Services. The platform supports application-level specification languages (such as BPMN and our custom languages) and features three validation backends (CL-AtSe, OFMC, and SATMC), which provide a range of complementary automated reasoning techniques (including service orchestration, compositional reasoning, model checking, and abstract interpretation). We have applied the platform to a large number of industrial case studies, collected into the AVANTSSAR Library of validated problem cases. In doing so, we unveiled a number of problems and vulnerabilities in deployed services. These include, most notably, a serious flaw in the SAML-based Single Sign-On for Google Apps (now corrected by Google as a result of our findings). We also report on the migration of the platform to industry.

1 Introduction

Driven by rapidly changing requirements and business needs, IT systems and applications are undergoing a paradigm shift: components are replaced by services distributed over the network, and composed and reconfigured dynamically in a demand-driven way into *Service-Oriented Architectures (SOAs)*.

C. Flanagan and B. König (Eds.): TACAS 2012, LNCS 7214, pp. 267–282, 2012.
© Springer-Verlag Berlin Heidelberg 2012

Deploying services in future network infrastructures such as SOAs or, even more generally, the Internet of Services (IoS), obviously entails a wide range of trust and security issues. Modeling and reasoning about these trust and security issues is complex due to three main characteristics of service orientation. First, SOAs are *heterogeneous*: their components are built using different technology and run in different environments, yet interact and may interfere with each other. Second, SOAs are also *distributed* systems, with functionality and resources distributed over several machines or processes. The resulting exponential state-space complexity makes their design and efficient validation difficult, even more so in hostile situations perhaps unforeseen at design time. Third, SOAs and their security requirements are *continuously evolving*: services may be composed at runtime, agents may join or leave, and client credentials are affected by dynamic changes in security policies (e.g., for incidents or emergencies). Hence, security policies must be regarded as part of the service specification and as first-class objects exchanged and processed by services. The *trust and security properties* that SOAs should provide to the users are, moreover, very diverse in type and scope, ranging from basic properties like confidentiality and authentication to complex dynamic and domain-specific requirements (e.g., non-repudiation or separation and binding of duty).

In this paper, we present the AVANTSSAR Platform, an integrated toolset for the formal specification and automated validation of trust and security of SOAs and, in general, of applications in the IoS. It has been developed in the context of the FP7 project "AVANTSSAR: Automated Validation of Trust and Security in Service-Oriented Architectures".

To handle the complexity of trust and security in service orientation, the platform integrates different technologies into a single tool, so they can interact and benefit from each other. More specifically, the platform comprises three *back-ends* (CL-AtSe [6,36], OFMC [15,22,33], and SATMC [3,5]), which operate on the same input specification (written in the *AVANTSSAR Specification Language ASLan*) and provide complementary automated reasoning techniques (including service orchestration and compositional reasoning, model checking, and abstraction-based validation). A *connectors layer* provides an interface to application-level specification languages (such as the standard BPMN, and our custom languages ASLan++, AnB and HLPSL++), which can be translated into the core ASLan language and vice versa.

We have applied the platform to a large number of exemplary industrial case studies, which we have collected into the *AVANTSSAR Library* of validated problem cases. In doing so, we have been able to uncover a number of problems and vulnerabilities in deployed services including, most notably, the detection of a serious flaw in the SAML-based SSO solution for Google Apps. Finally, we also report on our successful activities in migrating the platform to industry. As we describe in more detail in the following, to the best of our knowledge, no other tool exhibits the same scope and expressiveness while achieving the same performance and scalability.

We have implemented the AVANTSSAR Platform as a SOA itself, where each component service is offered as a web service. The platform also has a web-based graphical interface that allows the user to choose between three interaction modes of increasing level of sophistication and to execute, monitor and inspect the results of the platform in a user-friendly way. The web services and the associated documentation (a tutorial, guidelines, the Library and other examples, scientific papers and deliverables, and a users mailing list) are available at www.avantssar.eu, where one can also download the binaries and/or source code of the validation back-ends and play online with the platform through a prototype, web-based graphical user interface.

The platform is a successor to the AVISPA Tool [2], a push-button tool for the formal analysis of security protocols. The AVANTSSAR Platform significantly extends its predecessor's scope, effectiveness, and performance by scaling up to the trust and security of SOAs and the IoS. We thus expect that the AVANTSSAR Platform will inherit and considerably widen the user basis of AVISPA, which already comprises not only the members of the AVANTSSAR consortium but also several dozens of other academic and industrial practitioners, who have published a large number of works in which AVISPA is used. Our first, and positive, experience with the integration of the AVANTSSAR Platform within industrial practice indicates a strong potential for its wide take up.

It is important to note that this is the first comprehensive description of the platform, including the results of the experiments that we carried out. Descriptions of some of the different platform components have already been given and we will often refer to the corresponding documents for additional information.

2   The AVANTSSAR Platform

2.1   Description and Architecture

Fig. 1 shows the main components of the AVANTSSAR Platform, where the arrows represent the most general information flow, from input specification to validated output. In this flow, the platform takes as input specifications of the available services (including their security-relevant behavior and possibly the local policies they satisfy) together with a policy stating the functional and security requirements of the target service. In the orchestration phase, the platform applies automated reasoning techniques to build a validated orchestration of the available services that meets the security requirements. More specifically, the *Orchestrator* (short for Trust and Security Orchestrator) looks for a composition of the available services in a way that is expected but not yet guaranteed to satisfy the input policy (it may additionally receive as input a counterexample found by the Validator, if any) and outputs a specification of the target service that is guaranteed to satisfy the functional goals. Then, the *Validator* (short for Trust and Security Validator), which comprises the three back-ends CL-AtSe, OFMC and SATMC, checks whether the orchestration satisfies the security goals. If so, the orchestrated service is returned as output, otherwise, a counterexample is returned to the Orchestrator to provide a different orchestration, until it succeeds,

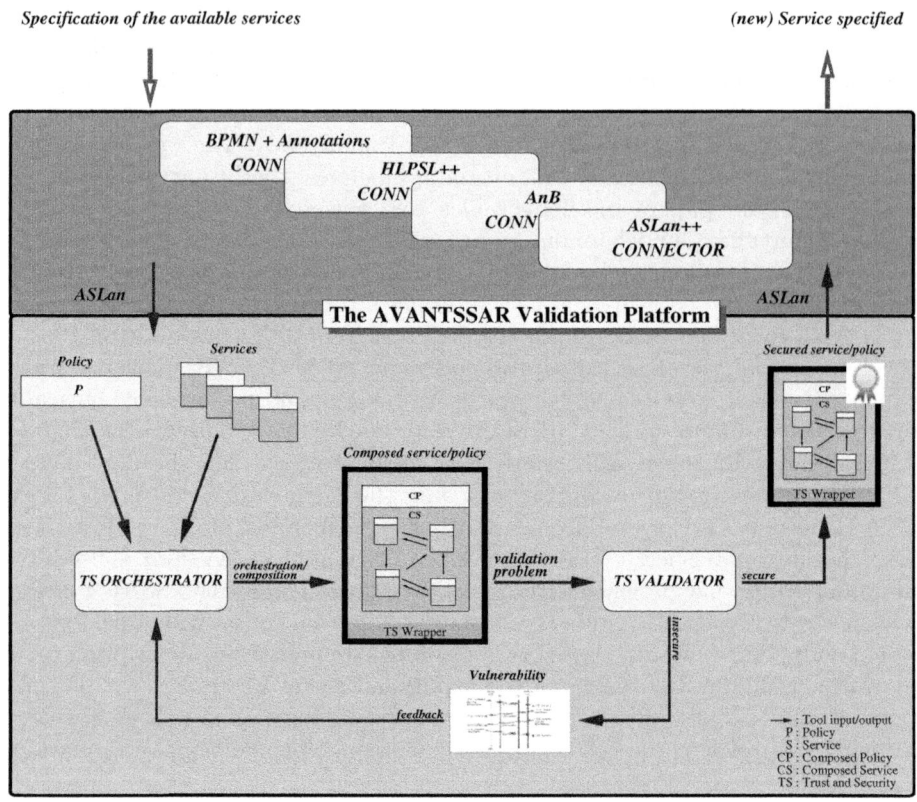

Fig. 1. The AVANTSSAR Validation Platform

or no suitable orchestration can be found. Instead of using the Orchestrator, a user may manually generate the target service and simply invoke the Validator, providing as input the service and its security goals. In this case, the platform outputs either validation success or the counterexample found.

To ease its usage and pave the way for its adoption by industry, the *connectors layer* of the platform provides a set of software modules that carry out both

(C1) the translation from application-level (e.g., our own ASLan++, AnB and HLPSL++) and industrially-suited specification languages (e.g., BPMN) into the low-level *AVANTSSAR Specification Language (ASLan)* [9], the common input language of formal analysis by the validator back-ends, and

(C2) the reverse translation from the common output format of the validator back-ends into a higher-level MSC-like output format to ease the interpretation of the results for the user.

Moreover, the connectors layer is open to the integration of other translations.

In the following subsections, we describe the different platform components in more detail, starting with the specification languages and the connectors layer, and then considering the Orchestrator and the Validator.

2.2   The Specification Languages ASLan and ASLan++

As observed in the introduction, modeling and reasoning about trust and security
of SOAs is complex due to the fact that SOAs are heterogeneous, distributed
and continuously evolving, and should guarantee security properties that are,
typically, very diverse. Besides the classical data security requirements including
confidentiality and authentication/integrity, more elaborate goals are authorization (with respect to a policy), separation or binding of duty, and accountability
or non-repudiation. Some applications may also have domain-specific goals (e.g.,
correct processing of orders). Finally, one may consider liveness properties under
certain fairness conditions) e.g., one may require that a web service for online
shopping eventually processes every order if the intruder cannot block the communication indefinitely. This diversity of goals cannot be formulated with a fixed
repertoire of generic properties (like authentication); instead, it suggests the need
for specification of properties in an expressive logic.

Various languages have been proposed to model trust and security of SOAs,
e.g., BPEL [34], π calculus [28], F# [17], to name a few. Each of them, however,
focuses only on some aspects of SOAs, and cannot cover all previously described
features, except perhaps in an artificial way. One needs a language fully dedicated to specifying trust and security aspects of services, their composition, the
properties that they should satisfy and the policies they manipulate and abide
by. Moreover, the language must go beyond static service structure: a key challenge is to integrate policies that are dynamic (e.g., changing with the workflow
context) with services that can be added and composed dynamically themselves.

We have designed ASLan so as to satisfy all these desiderata. At its core,
ASLan describes a *transition system*, where states are sets of typed ground terms
(facts), and transitions are specified as rewriting rules over sets of terms. A fact
iknows, true of any message (term) known to the intruder, is used to model communication as we consider a general *Dolev-Yao intruder* [26] that is in complete
control of the network and can compose, send, and intercept messages at will,
yet cannot break cryptography (following the perfect cryptography assumption).
A key feature of ASLan is the integration of this transition system that expresses
the dynamics of the model with *Horn clauses*, which are used to describe policies
in a clear, logical way. The execution model alternates transition steps with a
transitive closure of Horn clause applications. This allows us to model the effects of policies in different states: for instance, agents can become members of
a group or leave it, with immediate consequences for their access rights.

Moreover, to carry out the formal analysis of services, we need to model the
security goals. While this can be done by using different languages, in ASLan
we have chosen to employ a variant of linear temporal logic (LTL, e.g. [27]),
with backwards operators and ASLan facts as propositions. This logic gives us
the desired flexibility for the specification of complex goals, as illustrated by the
problem cases that are part of the AVANTSSAR Library.

ASLan is a low-level formal language and is thus easily usable only by experts,
so we have developed the higher-level language ASLan++ to achieve three main
design goals:

- the language should be expressive enough to model a wide range of SOAs while allowing for succinct specifications;
- it should facilitate the specification of services at a high level of abstraction in order to reduce model complexity as much as possible; and
- it should be close to specification languages for security protocols and web services, but also to procedural and object-oriented programming languages, so that it can be employed by users who are not formal specification experts.

For reasons of space, we refer to [13,37] for details on ASLan and ASLan++ including a tutorial with many modeling examples.

2.3   The Connectors Layer

As remarked above, writing formal specifications of complex systems at the low conceptual level of ASLan is not practically feasible and reasonable. The same applies to the activity of interpreting and understanding the raw output format returned by the validator back-ends. Industry, in particular, is used to higher-level modeling languages typically targeting very specific domain areas. That is why we have devised an open connectors layer, which currently comprises four connectors carrying out automatic translations.

The *ASLan++ connector* provides translations from ASLan++ specifications to ASLan and in the reverse direction for attack traces. Security protocol/service practitioners who are used to the more accessible but less expressive Alice-and-Bob notation or message sequence charts (MSCs) may prefer to use the *AnB connector*, which is based on an extended Alice-and-Bob notation [30,32,33], or the *HLPSL++ connector*, which is based on an extension of the High-Level Protocol Specification Language HLPSL [23], developed in the context of the AVISPA project [2,14].

Business process (BP) practitioners are used to standard languages such as the Business Process Modeling Notation (*BPMN*), the Business Process Execution Language (*BPEL*), etc. For them, even the usage of ASLan++ (or AnB or HLPSL++) may not be so easy, or they might already have specifications written in their favorite BP language that they do not wish to put aside to then repeat the modeling activity with another language. We have thus developed two connectors for BPMN (see [12]): a public connector that can be used in open-source environments such as Oryx to evaluate control flow properties of a BP modeled in BPMN, and a proprietary SAP NetWeaver BPM connector that is a plug-in of the SAP NetWeaver Development Studio that allows BP analysts to take advantage of the AVANTSSAR Platform via a security validation service. The business analyst provides the security requirements that are critical for the compliance of the BP (e.g., need-to-know in executing a task, data confidentiality with respect to certain users or roles) through easy-to-access UIs of the security validator that returns answers in a nice graphical BPMN-like format.

Connectors for other BP languages may be developed similarly. In fact, thanks to the openness of the connectors layer, new connectors for other application level and/or industrially-suited specification languages can be added by creating proper software modules implementing (C1) and (C2). To alleviate this task, we have devised, for both the common input language ASLan and the common output format of the validator back-ends, XML representations and software modules generating these XML representations [10].

2.4   The Orchestrator

Composability, one of the basic principles and design objectives of SOAs, expresses the need for providing simple scenarios where already available services can be reused to derive new added-value services. In their SOAP incarnation, based on XML messaging and relying on a rich stack of related standards, SOAs provide a flexible yet highly inter-operable solution to describe and implement a variety of e-business scenarios possibly bound to complex security policies.

It can be very complex to discover or even to adequately describe composition scenarios respecting overall security constraints. This motivates introducing automated solutions to scalable services composition. Two key approaches for composing web services have been considered, which differ by their architecture: *orchestration* is centralized and all traffic is routed through a *mediator*, whereas *choreography* is distributed and all web services can communicate directly.

Several "orchestration" notions have been advocated (see, e.g., [29]). However, in inter-organizational BPs it is crucial to protect sensitive data of each organization; and our main motivation is to take into account the security policies while computing an orchestration. The AVANTSSAR Platform implements an idea presented in [24] to automatically generate a mediator. We specify a web service profile from its *XML Schema* and *WS-SecurityPolicy* using first-order terms (including cryptographic functions). The mediator can use cryptography to produce new messages, and is constructed with respect to security goals using the techniques we developed for the verification of security protocols.

We highlight here the most important distinguishing features of our approach. First, several tools have addressed the WS orchestration problem but, to our knowledge, previous works abstract away the security policies attached to the services, while we consider them as an additional constraint. Second, most automatic orchestration approaches work by computing products of (communicating) finite-state automata, where messages are restricted to a finite alphabet. However, by specifying web services in ASLan, we can express a richer set of messages using first-order terms (including symbols for cryptographic functions). Third, we have applied the AVANTSSAR Orchestrator to several industrial case studies (cf. Table 1) that cannot be handled by other tools because the messages exchanged by services are too complex (e.g., they are non-atomic and built with cryptographic primitives) and require some automatic adaptation. For example, the Orchestrator has automatically generated a Security Server in the Digital Contract Signing case study (which originated from a commercial product), while in the Car Registration Process case study, the Orchestrator has been able to cope with additional constraints imposed by the authorization policies of the available services, specified as a set of Horn clauses.

Finally, and most importantly, the orchestration output can be automatically checked for security by the Validator as described below. If the specification meets the validation goals, i.e., no attack is found, the orchestration solution is considered as the final, validated, result of orchestration. Otherwise the Validator returns a goal violation report including an attack trace, which may be fed back to the Orchestrator, requesting it to backtrack and try an alternative solution.

2.5   The Validator

A specification in ASLan may be the result of an orchestration or of the translation of a specification given in some higher-level language such as ASLan++. The Validator takes any ASLan model of a system and its security goals and automatically checks whether the system meets its goals under the assumption that the network is controlled by a Dolev-Yao intruder.

Currently, the functionality of the Validator is supported by the three different back-ends CL-AtSe, OFMC and SATMC, but, again, the platform is open to the integration of additional validation back-ends.

The user can select which back-end is used for the validation process. By default, all three are invoked in parallel on the same input specification, so that the user can compare the results of the validation carried out by the complementary automated reasoning techniques that the back-ends provide (including compositional reasoning, model checking, and abstract interpretation).

CL-AtSe. The *Constraint-Logic-based Attack Searcher* for security protocols and services takes as input a service specified as a set of rewriting rules, and applies rewriting and constraint solving techniques to model all states that are reachable by the participants and decides if an attack exists with respect to the Dolev-Yao intruder. The main idea in CL-AtSe consists in running the services in all possible ways by representing families of traces with positive or negative constraints on the intruder knowledge, variable values or sets, etc. Each service step execution adds new constraints on the current intruder and environment state. Constraints are kept reduced to a normal form for which satisfiability is easily checked. This allows one to decide whether some security property has been violated up to this point. CL-AtSe requires a bound on the number of service calls in case the specification allows for loops in system execution. It implements several preprocessing modules to simplify and optimize input specifications before starting a verification. If a security property is violated then CL-AtSe outputs a trace that gives a detailed account of the attack scenario.

OFMC. The *Open-source Fixedpoint Model Checker* (which extends the *On-the-fly model checker*, the previous OFMC) consists of two modules. The *classical module* performs verification for a bounded number of transitions of honest agents using a constraint-based representation of the intruder behavior. The *fixedpoint module* allows verification without restricting the number of steps by working on an over-approximation of the search space that is specified by a set of

Horn clauses using abstract interpretation techniques and counterexample-based refinement of abstractions. Running both modules in parallel, OFMC stops as soon as the classic module has found an attack or the fixedpoint module has verified the specification, so as soon as there is a definitive result. Otherwise, OFMC can just report the bounded verification results and the potential attacks that the fixedpoint module has found. In case of a positive result, we can use the computed fixedpoint to automatically generate a proof certificate for the Isabelle interactive theorem prover. The idea behind the automatic proof generator OFMC/Isabelle [22] is to gain a high reliability, since after this step the correctness of the verification result no longer depends on the correctness of OFMC and the correct use of abstractions. Rather, it only relies on: (i) the correctness of the small Isabelle core that checks the proof generated by OFMC/Isabelle, and (ii) that the original ASLan specification (without over-approximations) indeed faithfully models the system and properties that are to be verified.

SATMC. The *SAT-based Model Checker* is an open, flexible platform for SAT-based bounded model checking of security services. Under the standard assumption of strong typing, SATMC performs a bounded analysis of the problem by considering scenarios with a finite number of sessions. At the core of SATMC lies a procedure that, given a security problem, automatically generates a propositional formula whose satisfying assignments (if any) correspond to counterexamples on the security problem of length bounded by some integer k. Intuitively, the formula represents all the possible evolutions, up to depth k, of the transition system described by the security problem. Finding attacks (of length k) on the service therefore reduces to solving propositional satisfiability problems. For this task, SATMC relies on state-of-the-art SAT solvers, which can handle propositional satisfiability problems with hundreds of thousands of variables and clauses or more. SATMC can also be instructed to perform an iterative deepening on the number k of steps. As soon as a satisfiable formula is found, the corresponding model is translated back into a *partial-order plan* (i.e., a partially ordered set of rules whose applications lead the system from the initial state to a state witnessing the violation of the expected security property).

As we remarked above, to the best of our knowledge, no other tool exhibits the same scope and expressiveness while achieving the same performance and scalability of the AVANTSSAR Platform. We have already discussed the expressiveness of the AVANTSSAR languages and the possibility of carrying out automated orchestration under security constraints, so now we briefly describe related work on automated analysis (and then discuss industrial case studies and industry migration in the following sections).

Service analysis methods based on abstract interpretation have become increasingly popular, e.g., [16,18,19,20,25,38]. For instance, TulaFale [16], a tool by Microsoft Research based on ProVerif [18], exploits abstract interpretation for verification of web services that use SOAP messaging, using logical predicates to relate the concrete SOAP messages to a less technical representation that is easier to reason about. ProVerif implements a form of static analysis based on abstract interpretation that supports unbounded verification but does not support the modeling of many aspects that occur in problems of real-world complexity such as revocation of keys at a key-server. In contrast, the AVANTSSAR Platform supports the formal modeling and automatic analysis of a large class of systems and properties, albeit for a bounded number of sessions. Two recent tools, namely the AIF framework [31] and StatVerif [1], have overcome some of the limitations of ProVerif, but they do not (yet) cover the full scope of what is specifiable and analyzable with the AVANTSSAR Platform.

2.6   The AVANTSSAR Platform: Web Services and Web Interface

We have implemented the AVANTSSAR Platform as a SOA itself, where each component service is offered as a web service (the URLs where each service, and its WSDL interface, can be accessed are given at www.avantssar.eu; binaries of each platform component are also available there, together with the source codes of OFMC and SATMC). The platform service is implemented in PHP5, by using the WSO2 Web Services Framework for PHP (WSO2 WSF/PHP) [39], an open source, enterprise grade, PHP extension for providing and consuming Web Services in PHP. The framework provides base communication functionality in SOAP, XML, and other message formats carried over various transports including HTTP, SMTP, XMPP and TCP. SOAP and HTTP are the standards used for the current Web Services implementation.

The platform also comes with a web-based graphical user interface that allows the user to execute, monitor and inspect the results of the platform in a user-friendly way. Scalable vector graphics and AJAX are suitably coupled to provide the user with an enhanced user experience. Fig. 2 shows a screenshot of the interface. Since the number of functionalities offered by the platform can discourage newcomers, the web interface supports three interaction modes with increasing level of sophistication: demo mode, basic mode, and expert mode.

3   AVANTSSAR Library and Experimental Results

As proof of concept, we have applied the AVANTSSAR Platform to the case studies that are now part of the so-called *AVANTSSAR Library*. In this way, we have been able to detect a considerable number of goal violations in the considered services and provide the required corrections. Moreover, the formal modeling of case studies has allowed us to consolidate our specification languages and has driven the evolution of the platform, both in terms of support for the new language and modeling features, as well as in efficiency improvements needed for the validation of the significantly more complex models. We expect that the library will provide a useful test suite for similar validation technologies.

As terminology, we say that an *application scenario* is composed of one or more *scenes* that focus on different use cases of the considered system, service, protocol, or the like. Each scene contains at least one goal formalizing a desired security property or security aspect, which we call a *problem case*.

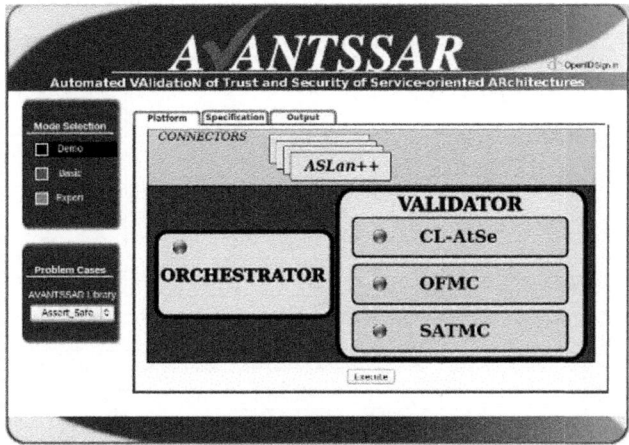

Fig. 2. The web interface of the AVANTSSAR Platform

The AVANTSSAR Library contains the formalization of 10 application scenarios of SOAs from the e-Business, e-Government and e-Health application areas. For these application scenarios we have written 26 specifications (in one of the application-level languages ASLan++, HLPSL++, annotated BPMN, or in the more low level specification language ASLan). Each of these specifications may address different security aspects, for a total of 94 problem cases. Among the 26 specifications, 4 involve orchestration, resulting in 13 problem cases that have to be orchestrated prior to validation.

Table 1 provides an overview of the problem cases formalized and validated by the AVANTSSAR Platform. It contains, for each application scenario, information about the connector used to translate high-level specifications into ASLan (for NW BPM see Section 4) and, if applicable, about the orchestration carried out (column "Orch."). For what concerns the families of problem cases, "f" indicates that a formalization of the problem case is present in the specification but was not validated, whereas "v" indicates its validation. Table 2 describes CPU times spent by each back-end on each application scenario. $S/NS/TOUT$ are abbreviations for *Supported/Not Supported/Timeout*; times are totals (in seconds) for successful runs. Moreover, for each scenario, the total number of Horn Clauses (HC) and transitions (i.e., ASLan steps) contained in the specifications are shown.

Since we lack space to describe all the application scenarios, problem cases and corresponding trust and security requirements in detail, we point the reader to [11] and here focus only on the SAML Single Sign-On scenario. It is representative for the effectiveness of the AVANTSSAR methods and tools, since we have succeeded in detecting vulnerabilities both in deployed SAML-based SSO solutions and in the use case described in the SAML Technical Overview [35]. Though well specified and thoroughly documented, the OASIS SAML security standard is written in natural language that is often subject to interpretation.

Table 1. The AVANTSSAR Library: formalization and validation status

Areas	Scenarios	Scene	Specification	Connector	Orch.	Federation	Authorization Policies	Accountability	Trust Management	Workflow Security	Privacy	Application Data Protection	Communication Security
E-Business													
Banking	Loan Origination	1	lop-scene1.aslan	No	No							v	
Services		2	lop-scene2.aslan	NW BPM	No		v					v	v
Electronic	Anonymous Shopping	1	IDMXScene1_Safe.aslan++	ASLan++	No				v			v	
Commerce		2	IDMXScene2_Safe.aslan++	ASLan++	No	v						v	
		3	IDMXScene3_Safe.aslan++	ASLan++	No		f					f	
E-Government													
Citizen and	Visa Application	1	PTD_VisaBank.aslan++	ASLan++	No		v	v	v	v	v	v	v
Service Portals	Car Registration	1	CRP.dyn.aslan++	ASLan++	Yes		v	v	v	v	v	v	v
Document	Public Bidding	1	pb_scene1.aslan++	ASLan++	No	f	f					f	f
Exchange		2	pb_scene2.aslan++	ASLan++	No	v	v					v	v
Procedures		3	pb-elig.aslan++	ASLan++	No	v							
		4	PB_alt.aslan	No	Yes	v	v					v	v
	Digital Contract Signing	1	dcs-scene1.aslan++	ASLan++	No	f	f			f		f	f
		2	dcs-scene2.aslan++	ASLan++	No	v	v			v		v	v
		3	dcs-scene3.aslan++	ASLan++	No	v	v			v		v	v
		4	DCS.ORCH.aslan	No	Yes	f							
		5	DCS-GoalStyleInput.ORCH.aslan	No	Yes	f							
E-Health													
Personal Health	Electronic Health Records	1	ECR.aslan++	ASLan++	No			v	v	v	v	v	v
Information	Process Task Delegation	1	PTD.aslan++	ASLan++	No	v	v			v	v	v	v
		2	PTD_PC.aslan++	ASLan++	No	v	v	v		v	v	v	v
	Access Control Management	1	eHRMS.txt	No	No		f					f	f
	SAML Single Sign-On	1	SP_init-FC-one_channel.hlpsl++	HLPSL++	No	v						v	v
		2	SP_init-BC-two_channels.hlpsl++	HLPSL++	No	v						v	v
		3	IdP_init-FC.hlpsl++	HLPSL++	No	v						v	v
		4	IdP_init-BC.hlpsl++	HLPSL++	No	v						v	v
		5	SAML-based_SSO_for_GoogleApp.hlpsl++	HLPSL++	No	v						v	v
		6	SAML-based_SSO_for_GoogleApp.aslan++	ASLan++	No	v						v	v

Table 2. CPU analysis times for each back-end on the application scenarios

Application Scenario	Dimensions		SATMC		OFMC		CL-AtSe	
	HC	Steps	Time	S/NS/TOUT	Time	S/NS/TOUT	Time	S/NS/TOUT
Anonymous Shopping	180	94	0	0/6/0	57.83	2/0/4	5.91	4/0/2
Car Registration	349	258	60.54	7/1/4	1001.31	2/1/9	69.35	10/0/2
Digital Contract Signing	238	52	10504.87	9/5/1	0	0/13/2	906.77	9/0/6
Electronic Health Records	89	48	19.33	1/1/0	5.08	1/0/1	125.37	1/1/0
Loan Origination	303	418	767.80	9/0/0	0	0/9/0	7175.26	6/0/3
Process Task Delegation	90	39	1.68	0/0/2	0	0/2/0	1092.43	2/0/0
Public Bidding	117	631	6747.37	12/0/3	9781.38	8/2/5	9298.7	14/1/0
SAML Single Sign-On	21	215	1989.49	15/0/1	22.77	1/15/0	1.85	1/15/0
Visa Application	38	19	44.83	1/0/0	3.12	1/0/0	9.86	1/0/0
Total	1425	1774	20135.84	52/15/16	10871.49	11/42/30	18685.50	51/17/15

Since the many configuration options, profiles, protocols, bindings, exceptions, and recommendations are laid out in different, interconnected documents, it is not always easy to establish which message fields are mandatory in a given profile and which are not. Moreover, SAML-based solution providers may have internal requirements that may result in small deviations from the standard. For instance, internal requirements (or DoS considerations) may lead the service provider to avoid checking the match between the ID field in the AuthResp and in the previously sent AuthReq. The consequences of such a choice must be examined in detail.

The SAML-based SSO for Google Apps in operation until June 2008 deviated from the standard in a few, seemingly minor ways. By using the AVANTSSAR Platform, we discovered a serious authentication flaw in the service, which a dishonest service provider could use to impersonate the victim user on Google Apps, granting unauthorized access to private data and services (email, docs, etc.) [5]. The vulnerability was detected by SATMC and the attack was reproduced in an actual deployment of SAML-based SSO for Google Apps. We readily informed Google and the US Computer Emergency Readiness Team (US-CERT) of the problem. Google developed a new version of the authentication service and asked their customers to update their applications accordingly. The vulnerability report released by US-CERT is available at http://www.kb.cert.org/vuls/id/612636. The severity of the vulnerability has been rated High by the National Institute of Standard and Technology (http://web.nvd.nist.gov/view/vuln/detail?vulnId=CVE-2008-3891).

By using the AVANTSSAR Platform we also discovered an authentication flaw in the prototypical SAML SSO use case (as described in the SAML Technical Overview) [4]. This flaw allows a malicious service provider to hijack a client authentication attempt and force the latter to access a resource without its consent. It also allows an attacker to launch Cross-Site Scripting (XSS) and Cross-Site Request Forgery (XSRF) attacks. This last type of attack is even more pernicious than classic XSRF, because XSRF requires the client to have an active session with the service provider, whereas in this case the session is created automatically, hijacking the client's authentication attempt. This may have serious consequences, as witnessed by the new XSS attack that we identified in the SAML-based SSO for Google Apps and that could have allowed a malicious web server to impersonate a user on any Google application. The problem has been reported to OASIS, and a proposal for an *errata* to the SAML standard is currently being discussed within OASIS (http://tools.oasis-open.org/issues/browse/SECURITY-12).

4   Technology Migration

Formal validation of trust and security will become a reality in SOAs and the IoS only if and when the available technologies will have migrated to industry and to standardization bodies (which are mostly driven by industry and influence the future of industrial development). Such a migration has to face the gap between advanced formal methods and their real exploitation within industry and standardization bodies.

To ease the adoption of formal methods, several obstacles have to be overcome, in particular: (i) the lack of automated technology supporting formal methods, (ii) the gap between the problem case that needs to be solved in industry and the abstract specification provided by formal methods, and (iii) the differences between formal languages and models and the languages used in industrial design and development environments (e.g., BPMN, Java, ABAP).

AVANTSSAR has addressed these issues by devising industrially-suited specification languages (model-driven languages), equipped with easy-to-use GUIs and translators to and from the core formal models, and migrating them to the selected development environments. This enables designers and developers from industry and standardization bodies to check more rapidly the correctness of the proposed solutions without having a strong mathematical background.

A concrete example is the industry migration of the AVANTSSAR Platform to the SAP environment. Two valuable migration activities have been carried out by building contacts with core business units. First, in the trail of the successful analysis of Google's SAML-based SSO, the AVANTSSAR Platform has been exploited to formally validate relevant scenarios where the SAP NetWeaver SAML Next Generation Single Sign On services (NW NG SSO) are employed. More than 50 formal specifications capturing these scenarios, the variety of configuration options, and SAP internal design and implementation choices have been formalized. Unsafe service compositions and configurations have been detected, and safe compositions and configurations have been put forward for use by SAP in setting up the NW NG SSO services on customer production systems.

The AVANTSSAR technology has been also integrated via a plug-in into the SAP NetWeaver BPM (NW BPM) product [7,8] to formally validate if a business process together with its access control policy complies with security-critical requirements, e.g., separation and binding of duty, need-to-know principle, etc. The plug-in provides a push-button technology with accessible user interfaces, bridging the gap between business process modeling languages and formal specifications. Thus, a BP modeler can easily specify the security goals to validate against the business process and access control policy; any violation of the security properties is depicted in a graphical way, enabling the modeler to take countermeasures.

5   Concluding Remarks

As exemplified by the case studies and success stories mentioned above, formal validation technologies can have a decisive impact for the trust and security of SOAs and the IoS. The research innovation put forth by the AVANTSSAR Platform aims at ensuring global security of dynamically composed services and their integration into complex SOAs by developing an integrated platform of automated reasoning techniques and tools. Similar technologies are being developed by other research teams (although none has yet the scale and depth of our platform, which is the reason why we could not compare scope and efficiency). Brought together, these research efforts will result in a new generation of tools for automated security validation at design time, which is a stepping stone for the development of similar tools for validation at service provision and consumption time. For instance, part of the AVANTSSAR consortium is developing a security testing toolset in the context of the FP7 project "SPaCIoS: Secure Provision and Consumption in the Internet of Services" (www.spacios.eu). These advances will significantly improve the all-round security of SOAs and the IoS, and thus boost their trustworthy development and public acceptance.

References

1. Arapinis, M., Ritter, E., Ryan, M.D.: StatVerif: Verification of Stateful Processes. In: Proc. CSF 2011, pp. 33–47. IEEE CS Press (2011)
2. Armando, A., Basin, D.A., Boichut, Y., Chevalier, Y., Compagna, L., Cuéllar, J., Drielsma, P.H., Héam, P.-C., Kouchnarenko, O., Mantovani, J., Mödersheim, S., von Oheimb, D., Rusinowitch, M., Santiago, J., Turuani, M., Viganò, L., Vigneron, L.: The AVISPA Tool for the Automated Validation of Internet Security Protocols and Applications. In: Etessami, K., Rajamani, S.K. (eds.) CAV 2005. LNCS, vol. 3576, pp. 281–285. Springer, Heidelberg (2005)
3. Armando, A., Carbone, R., Compagna, L.: LTL Model Checking for Security Protocols. Journal of Applied Non-Classical Logics 19(4), 403–429 (2009)
4. Armando, A., Carbone, R., Compagna, L., Cuéllar, J., Pellegrino, G., Sorniotti, A.: From Multiple Credentials to Browser-Based Single Sign-On: Are We More Secure? In: Camenisch, J., Fischer-Hübner, S., Murayama, Y., Portmann, A., Rieder, C. (eds.) SEC 2011. IFIP Advances in Information and Communication Technology, vol. 354, pp. 68–79. Springer, Heidelberg (2011)
5. Armando, A., Carbone, R., Compagna, L., Cuéllar, J., Tobarra Abad, L.: Formal Analysis of SAML 2.0 Web Browser Single Sign-On: Breaking the SAML-based Single Sign-On for Google Apps. In: Proc. FMSE 2008. ACM Press (2008)
6. Arora, C., Turuani, M.: Validating Integrity for the Ephemerizer's Protocol with CL-Atse. In: Cortier, V., Kirchner, C., Okada, M., Sakurada, H. (eds.) Formal to Practical Security. LNCS, vol. 5458, pp. 21–32. Springer, Heidelberg (2009)
7. Arsac, W., Compagna, L., Kaluvuri, S., Ponta, S.E.: Security Validation Tool for Business Processes. In: Proc. SACMAT 2011, pp. 143–144. ACM (2011)
8. Arsac, W., Compagna, L., Pellegrino, G., Ponta, S.E.: Security Validation of Business Processes via Model-Checking. In: Erlingsson, Ú., Wieringa, R., Zannone, N. (eds.) ESSoS 2011. LNCS, vol. 6542, pp. 29–42. Springer, Heidelberg (2011)
9. AVANTSSAR. Deliverable 2.1: Requirements for modelling and ASLan v.1 (2008)
10. AVANTSSAR. Deliverable 4.2: AVANTSSAR Validation Platform v.2 (2010)
11. AVANTSSAR. Deliverable 5.4: Assessment of the AVANTSSAR Validation Platform (2010)
12. AVANTSSAR. Deliverable 6.2.3: Migration to industrial development environments: lessons learned and best practices (2010)
13. AVANTSSAR. Deliverable 2.3: ASLan++ specification and tutorial (2011)
14. AVISPA: Automated Validation of Internet Security Protocols and Applications, http://www.avispa-project.org
15. Basin, D., Mödersheim, S., Viganò, L.: OFMC: A symbolic model checker for security protocols. IJIS 4(3), 181–208 (2005)
16. Bhargavan, K., Fournet, C., Gordon, A.D., Pucella, R.: TulaFale: A Security Tool for Web Services. In: de Boer, F.S., Bonsangue, M.M., Graf, S., de Roever, W.-P. (eds.) FMCO 2003. LNCS, vol. 3188, pp. 197–222. Springer, Heidelberg (2004)
17. Bhargavan, K., Fournet, C., Gordon, A.: Verified Reference Implementations of WS-Security Protocols. In: Bravetti, M., Núñez, M., Zavattaro, G. (eds.) WS-FM 2006. LNCS, vol. 4184, pp. 88–106. Springer, Heidelberg (2006)
18. Blanchet, B.: An efficient cryptographic protocol verifier based on Prolog rules. In: Proc. CSFW 2001, pp. 82–96. IEEE CS Press (2001)
19. Bodei, C., Buchholtz, M., Degano, P., Nielson, F., Riis Nielson, H.: Automatic validation of protocol narration. In: Proc. CSFW 2003, pp. 126–140. IEEE CS Press (2003)

20. Boichut, Y., Heam, P.-C., Kouchnarenko, O.: TA4SP: Tree Automata based on Automatic Approximations for the Analysis of Security Protocols (2004)
21. Boichut, Y., Heam, P.-C., Kouchnarenko, O., Oehl, F.: Improvements on the Genet and Klay Technique to Automatically Verify Security Protocols. In: Proc. AVIS 2004. ENTCS (2004)
22. Brucker, A., Mödersheim, S.: Integrating Automated and Interactive Protocol Verification. In: Degano, P., Guttman, J.D. (eds.) FAST 2009. LNCS, vol. 5983, pp. 248–262. Springer, Heidelberg (2010)
23. Chevalier, Y., Compagna, L., Cuéllar, J., Hankes Drielsma, P., Mantovani, J., Mödersheim, S., Vigneron, L.: A High Level Protocol Specification Language for Industrial Security-Sensitive Protocols. In: Proc. SAPS 2004, pp. 193–205 (2004)
24. Chevalier, Y., Mekki, M.A., Rusinowitch, M.: Automatic Composition of Services with Security Policies. In: Proc. WSCA, pp. 529–537. IEEE CS Press (2008)
25. Comon-Lundh, H., Cortier, V.: New Decidability Results for Fragments of First-order Logic and Application to Cryptographic protocols. TR LSV-03-3, Laboratoire Specification and Verification, ENS de Cachan, France (2003)
26. Dolev, D., Yao, A.: On the Security of Public-Key Protocols. IEEE Transactions on Information Theory 2(29) (1983)
27. Hodkinson, I., Reynolds, M.: Temporal Logic. In: Blackburn, P., van Benthem, J., Wolter, F. (eds.) Handbook of Modal Logic, pp. 655–720. Elsevier (2006)
28. Lucchi, R., Mazzara, M.: A pi-calculus based semantics for WS-BPEL. J. Log. Algebr. Program. 70(1), 96–118 (2007)
29. Marconi, A., Pistore, M.: Synthesis and Composition of Web Services. In: Bernardo, M., Padovani, L., Zavattaro, G. (eds.) SFM 2009. LNCS, vol. 5569, pp. 89–157. Springer, Heidelberg (2009)
30. Mödersheim, S.: Algebraic Properties in Alice and Bob Notation. In: Proc. Ares 2009, pp. 433–440. IEEE CS Press (2009)
31. Mödersheim, S.: Abstraction by Set-Membership: Verifying Security Protocols and Web Services with Databases. In: Proc. CCS 17, pp. 351–360. ACM Press (2010)
32. Mödersheim, S., Viganò, L.: Secure Pseudonymous Channels. In: Backes, M., Ning, P. (eds.) ESORICS 2009. LNCS, vol. 5789, pp. 337–354. Springer, Heidelberg (2009)
33. Mödersheim, S., Viganò, L.: The Open-source Fixed-point Model Checker for Symbolic Analysis of Security Protocols. In: Aldini, A., Barth, G., Gorrieri, R. (eds.) FOSAD 2007. LNCS, vol. 5705, pp. 166–194. Springer, Heidelberg (2009)
34. OASIS. Web Services Business Process Execution Language Version 2.0. (April 11, 2007), http://docs.asis-open.org/wsbpel/2.0/OS/wsbpel-v2.0-OS.pdf
35. OASIS. SAML v2.0 – Technical Overview (March 2007), http://www.oasis-open.org/committees/tc_home.php?wg_abbrev=security
36. Turuani, M.: The CL-Atse Protocol Analyser. In: Pfenning, F. (ed.) RTA 2006. LNCS, vol. 4098, pp. 277–286. Springer, Heidelberg (2006)
37. von Oheimb, D., Mödersheim, S.: ASLan++ — A Formal Security Specification Language for Distributed Systems. In: Aichernig, B.K., de Boer, F.S., Bonsangue, M.M. (eds.) FMCO 2010. LNCS, vol. 6957, pp. 1–22. Springer, Heidelberg (2011)
38. Weidenbach, C.: Towards an Automatic Analysis of Security Protocols in First-Order Logic. In: Ganzinger, H. (ed.) CADE 1999. LNCS (LNAI), vol. 1632, pp. 314–328. Springer, Heidelberg (1999)
39. WSO2. Web Services Framework for PHP (2006), http://wso2.org/projects/wsf/php

Reduction-Based Formal Analysis of BGP Instances

Anduo Wang[1], Carolyn Talcott[2], Alexander J.T. Gurney[1],
Boon Thau Loo[1], and Andre Scedrov[1]

University of Pennsylvania, SRI International
{anduo,boonloo}@cis.upenn.edu, clt@csl.sri.com,
agurney@seas.upenn.edu, scedrov@math.upenn.edu

Abstract. Today's Internet interdomain routing protocol, the Border Gateway
Protocol (BGP), is increasingly complicated and fragile due to policy misconfigurations by individual autonomous systems (ASes). These misconfigurations are
often difficult to manually diagnose beyond a small number of nodes due to the
state explosion problem. To aid the diagnosis of potential anomalies, researchers
have developed various formal models and analysis tools. However, these techniques do not scale well or do not cover the full set of anomalies. Current techniques use oversimplified BGP models that capture either anomalies within or
across ASes, but not the interactions between the two. To address these limitations, we propose a novel approach that reduces network size prior to analysis,
while preserving crucial BGP correctness properties. Using Maude, we have developed a toolkit that takes as input a network instance consisting of ASes and
their policy configurations, and then performs formal analysis on the reduced
instance for safety (protocol convergence). Our results show that our reduction-
based analysis allows us to analyze significantly larger network instances at low
reduction overhead.

1   Introduction

The Internet today runs on a complex routing protocol called the *Border Gateway Protocol* or *BGP* for short. BGP enables Internet Service Providers (ISPs) worldwide to
exchange reachability information to destinations over the Internet, and simultaneously,
each ISP acts as an autonomous system that imposes its own import and export policies
on route advertisements exchanged with its neighbors.

Over the past few years, there has been a growing consensus on the complexity and
fragility of BGP routing. Even when the basic routing protocol converges, conflicting
policy decisions among different ISPs have led to route oscillation and slow convergence. Several empirical studies (e.g. [12]) have shown that there are prolonged periods
in which the Internet cannot reliably route data packets to specific destinations, due to
routing errors induced by BGP.

Since protocol oscillations cause serious performance disruptions and router overhead, researchers devote significant attention to BGP stability (or "safety"). A BGP
system converges and is said to be safe, if it produces stable routing tables, given any

C. Flanagan and B. König (Eds.): TACAS 2012, LNCS 7214, pp. 283–298, 2012.

sequence of routing message exchanges. We broadly refer to any route misconfigurations that result in instability as *BGP anomalies* in this paper. To study potential configuration issues with BGP, the network community has studied small network instances. Sometimes, these come from a single network (the "internal BGP" or "iBGP" case), or they may relate to interaction between different networks ("external BGP" or "eBGP"). These small topology configurations (or "gadgets") serve as examples of safe systems, or counterexamples showing a safety problem such as lack of convergence.

Today, analyzing these gadgets is a manual and tedious process, let alone analyzing actual network instances that are orders of magnitude larger. Researchers check these gadgets by manually constructing "activation sequences" where the nodes make successive routing decisions that form an oscillation. To automate the process, in our prior work [18], we have developed an analysis toolkit using Maude [1] that automates the process of analyzing BGP instances using a *rewriting logic* [13] approach. While automated, this approach can only work for small network instances, since the approach is susceptible to the state explosion problem as the number of nodes increases. To address these challenges, this paper makes the following contributions.

First, we identify the key contributing attributes in BGP routing that lead to eBGP and iBGP anomalies resulting in route oscillations.

Second, we propose an efficient algorithm for reducing BGP instances, so that the network size can be reduced by merging nodes in such a way that the overall convergence properties remain the same. Our reduction uses the well-known *Stable Paths Problem* (SPP) [9] formalism for safety analysis of BGP configurations, where the entire instance is modeled in terms of the router-level topology and each router's policy-induced route preferences. We show how the reduction works for both inter-AS and intra-AS policy configurations using well-known gadgets as examples, and provide formal proofs that the reduction correctly preserves convergence properties in general for any arbitrary BGP instances. The reduction process not only reduces the state size for subsequent analysis, but also provides us the capability to reduce an existing BGP network instance into a known anomaly (i.e. misbehaving gadget), or determine equivalence between two configuration instances.

Finally, using Maude, we develop a tool that (1) takes as input router configurations, (2) extracts the SPP representation of a protocol by generating and comparing all possible routes against each AS policy, (3) applies the reduction step, and (4) performs an exhaustive state exploration on the reduced BGP instance to check for possible configuration anomalies that result in divergence.

Our results show that the reduction-based analysis is much more effective than the prior approach of doing exhaustive search on unreduced instances [18]. The data demonstrate that we have not only gained speed, but also the ability to analyze network instances that were previously infeasible to study.

For example, an instance which would naively take 221193ms to analyze, can now be reduced in 22ms to one which takes only 8ms to analyze, which makes the new method over 7000 times faster (See our technical report [17] for more details). There are also many instances whose analysis was infeasible with the previous approach, but which can now be tackled by reduction. The naive technique was limited to networks of no more than about 20 nodes (at a push, 25) whereas we now have no difficulty in scaling to instances with over a hundred nodes, and with a greater arc density, which are more characteristic of real networks.

2   Analyzing BGP Anomalies

BGP assumes a network model in which routers are grouped into various Autonomous Systems (ASes), each assumed to be under separate administrative control. An individual AS exchanges route advertisements with neighboring ASes using a *path-vector* protocol. Upon receiving a route advertisement, a BGP router may choose to accept or ignore the advertisement based on its *import policy*. If the route is accepted, the node stores the route as a possible candidate. Each node selects among all candidate routes the best route to each destination, based on its local *route preference policy*. Once a best route is selected, the node advertises it to its neighbors. A BGP node may choose to export only selected routes to its neighboring ASes based on its *export policy*. The determination of these three kinds of policy is up to the network operator: BGP allows considerable flexibility. Conflicting policies, within or between ASes, are the cause of protocol oscillation, as the protocol struggles and fails to satisfy all policies at once.

Router-to-router BGP sessions come in two flavors: external BGP (eBGP), which establishes routes between ASes; and internal BGP (iBGP), which distributes routes within an AS. All routers maintain internal state, including their roster of known paths for all destinations, and the list of other routers to whom they are connected. They communicate with one another by exchanging route advertisements. Not all routers communicate directly with routers outside their own AS. If they do, they are *border routers*; if they do not, they are *internal routers*. Additionally, some routers may have a special role as *route reflectors*, collating and distributing route advertisements on behalf of their *clients*, to avoid having to establish pairwise connections between all routers in an AS.

Table 1. Key attributes in BGP route selection

Stage	BGP route selection step
eBGP	**1. Highest LOC_PREF**
	2. Lowest AS path length
	3. Lowest origin type
iBGP	**4. Lowest MED (with same NEXT-HOP AS)**
	5. Closest exit point (lowest IGP cost)
	6. Lowest router ID (break tie)

Every BGP route is endowed with *attributes* that describe it. These are summarized in Table 1. We also characterize these by whether they are primarily associated with eBGP- or iBGP-level routing decisions. Whenever an AS receives a new route, it will compare the attributes of its current available routes (for a given destination) with the new route, and then decide whether the new route is selected as best route. The attributes are listed in the order in which they are compared during route selection: if the routes are tied at any stage, then BGP proceeds to consider the next attribute on the list.

The most important attribute in eBGP route selection is local preference (LOC_PREF). This is a value set by each router on routes it receives, according to (arbitrary) rules established by the network operator. If two routes have the same local preference, then the next tiebreaking attribute is the AS path length—the number of ASes through which this route passes—followed by the 'origin' code. The next step is to use the multi-exit discriminator (MED) attribute, the most important attribute in iBGP route selection, which says which individual link is preferred, out of the many links between this AS and its neighbor. If that was not enough to determine a single best route, BGP breaks ties by examining the shortest-path distance to the relevant border router. Finally, if all else fails, it uses the value of each router's unique identifier. This final step is meant to ensure that all possible routes can be placed in a total order, with no two routes being equivalent in preference.

Oscillation anomalies in BGP can be localized to the definition and use of particular attributes. This paper looks at three families of problems.

- In **eBGP anomalies**, routing policy conflicts occur at an inter-AS level. The typical causing attribute is LOC_PREF, because it is set arbitrarily at each AS, independently of any other.
- **iBGP anomalies** are limited to a single AS, and associated with MED. Due to a quirk in the decision procedure, it is possible for there to be three routes p, q, and r such that p is preferred to q, q to r, and r to p. The router will be unable to settle on a single choice, if there is feedback where its actions cause the visibility of those three routes to change.
- **iBGP-IGP anomalies** result from inconsistency between the semantics of route reflectors, and particular IGP distance values.

We will revisit these anomalies and give formal definitions in Section 4. We will also examine the correctness of network reduction with respect to these anomalies.

3 Network Reduction

Existing network analysis techniques do not scale well: Static analysis [6,5,8,2,7,14] by checking combinatorial structure that reflects routing oscillations is normally NP-complete; and dynamic analysis [3,11] by systematic exploration of the protocol state space will likewise suffer an exponential blow-up as problem size increases. As a result, analysis techniques normally assume an over-simplified BGP model that only covers a portion of the routing anomalies in Section 2.

To address these limitations, we propose network reduction that preservers correctness properties - a process that simplifies network instances. Network reduction can be viewed either as a pre-step prior to formal analysis in order to reduce analysis space; or a model construction step that extracts a simplified model from the real BGP instance.

In network reduction, the basic idea is to incrementally merge two network nodes into one while preserving network properties. To formally define reduction, we need to first represent a BGP instance in an abstract form that also captures each node's routing policy. We choose the extended stable paths problem (SPP) as the formal representation to include both eBGP and iBGP instances.

SPP is a well-established combinatorial model of BGP configurations that captures the outcomes of routing policy—which paths are preferred over which other paths, at each router—while avoiding the need for detailed modeling of the BGP decision process in all its complexity.

We extend SPP to define path preference in a more general way. The extended SPP is then used as the representation to implement reduction. In addition, we provide automatic generation of extended SPP for a BGP instance given its network topology and high-level routing policy (e.g. how the path attributes are configured/transformed). We will revisit this in Section 5.

3.1 Hierarchical Reduction

The SPP formalism captures the route preferences that exist for all routers, over their routes to a single fixed destination. [1] An SPP instance consists of a graph, together with each router's preferences over paths in the graph. We define this in a more general way than in previous work.

Definition 1. *An extended SPP instance is given by $G = (V, E, d, P, \prec)$, where V is the set of nodes, E is the set of directed arcs, $d \in V$ is the destination node, P is the set of all permitted paths to d, and the binary relation $\prec$ over P indicates when one route is preferred over another. Every path in P must be a simple path (that is, no node appears more than once).*

For a given extended SPP instance G as above, and a node i in V, write P^i for the subset of P consisting of paths from i to d. The SPP definition requires that $\prec$ be a transitive total order on each P^i, but our definition does not enforce that, and supports more routing policies. Routes from different source nodes are incomparable. Conventionally, '$p \prec q$' means that path p is preferred to path q, where both p and q are paths to d from the same source.

In this paper, we will use the symbol '$\circ$' for concatenation of arcs and paths. If (i, j) is an arc in E, and p is a path from j to d, then their concatenation $(i, j) \circ p$ is a path from i to d. Similarly, if p is a path from i to j, and q is a path from k to l, and (j, k) is an arc in E, then the concatenation is $p \circ (j, k) \circ q$ or just $p \circ q$.

This combinatorial definition washes away the some important features about how BGP operates and how paths are chosen: in particular, the distinction between external and internal BGP. The eBGP/iBGP distinction is critical for our reduction technique, because it is based on the observation that certain kinds of anomaly can be 'localized' to one or the other mode. Our reductions will operate on the iBGP level first, for each AS. After iBGP simplification, we simplify eBGP by reducing the extended SPP instance for the remaining network. This ordering also allows certain kinds of inconsistency, that can only occur in iBGP, to be detected and handled; we do not need to contaminate our other reductions with knowledge of these special cases. Since our reduction method includes steps that are specific to one or the other mode of operation— we assume that, in reduction, all we are faced with are extended SPP instances, derived from BGP configurations.

---

[1] Route preferences are configurable separately for each destination, so this assumption focuses the analysis rather than limits it.

3.2   Network Reduction

This subsection proposes sufficient conditions for two BGP nodes to be 'unifiable', meaning that they can be merged into one node. The reduction proceeds by repeatedly (1) locating two unifiable nodes, and rewriting their local configuration, and (2) rewriting the remainder of the BGP instance to reflect that local change. In the following, assume we are working with a given extended SPP instance $G = (V, E, d, P, \prec)$.

Locate unifiable nodes
We identify two special cases of unifiable nodes, which we call *duplicate* and *supplementary*. We first define an auxiliary notion: "node rewrite", based on which unifiable node conditions are defined.

Definition 2. *For two nodes i, k in V, rewrite i to k by rewriting P^i and $\prec$ as follows:*
1. *Check for any path p in P on which i and k both occur, if they occur only in an adjacent position, then proceed to the next step, otherwise abort the rewrite.*
2. *For every path p in P^i, if i or ik occurs, replace it by k.*
3. *For every two distinct paths p and q in P^i, that rewrite to p' and q' respectively, check whether p' and q' are equal. If they are, then abort the rewrite; otherwise, proceed to the next step.*
4. *Every preference $p \prec q$, where p and q are in P^i and rewrite to p' and q' respectively, becomes $p' \prec q'$.*

Step 1 ensures that if there is *any* permitted path on which i and k both occur, with some intervening nodes between them, then they are not considered for rewriting (unification). Therefore, after the first step of rewriting, the paths in P^i will still be simple (k will not occur twice). Step 3 ensures that after rewriting, no two paths in P^i can collapse into one. Based on this rewriting notion, the two unifiable node conditions are as follows.

Definition 3. *Two nodes i, j in V are* unifiable *if i is supplementary for j, or i and j are duplicate, where:*

1. A node i is *supplementary* for j if:
    1. i can be rewritten to j as defined in Definition 2.
    2. For every path p in P^i, there is some path q in P^j such that p and q are equal after rewriting.
    3. Whenever $p_1 \prec p_2$ in P^i, there are paths q_1 and q_2 in P^j such that $q_1 \prec q_2$; p_1 and q_1 are equal after rewriting; and p_2 and q_2 are equal after rewriting.
4. Two nodes i and j in V are *duplicate* if each is supplementary for the other.

Reduce BGP Instance
After locating two unifiable nodes i and j, we rewrite the entire extended SPP to reflect this unification. This completes one network reduction step.

First define a function θ_{ij} from V to $V \setminus \{i\}$ by $\theta_{ij}(i) = \theta_{ij}(j) = j$, and $\theta_{ij}(x) = x$ for all x not equal to either i or j. This function induces corresponding maps on E and P, as follows.

Definition 4. *If i and j are unifiable nodes in V, then G may be* reduced *to $G' = (V', E', d, P', \prec')$, where*

- $V' = V \setminus \{i\}$
- $E' = \{(\theta_{ij}(u), \theta_{ij}(v)) \mid (u, v) \in E \setminus \{(i, j), (j, i)\}\}$
- P' *consists of all paths in P after rewriting each node according to θ_{ij}, and eliding any (j, j) arc.*
- $p' \prec' q'$ *if and only if $p' \neq q'$ and there exist paths p and q in P such that p rewrites to p', q rewrites to q', and $p \prec q$.*

3.3   Examples: Reducing eBGP and iBGP Instances

We now illustrate the intuition of network reduction by applying reduction to various eBGP and iBGP instances.

Example 1. *Reducing eBGP instances* Two eBGP instances called *Bad gadget* and *Good gadget* are shown on the left of Figure 1. The topology of each eBGP instance is given by the network graph, whereas the routing policies are shown by the path preferences indicated alongside each network node. In each list, the more preferred paths are at the top, and paths that do not appear are not permitted. For example, in the good gadget, the policy for node 1 says it has two permitted paths, 1 3 0 and 1 0, where 1 3 0 is preferred to 1 0.

In both gadgets, nodes 3,4 are unifiable nodes according to Definition 3. After reduction, these nodes are merged into one, shown on the right hand side of Figure 1.

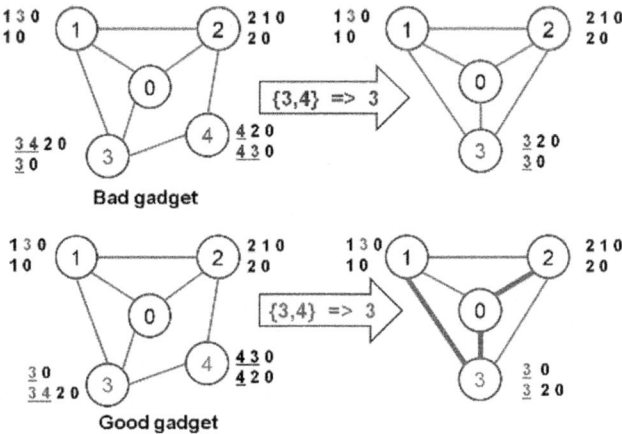

Fig. 1. Reducing bad/good gadget makes it easier to detect divergence/prove safety

The reason why the bad gadget is called 'bad' is that it suffers from permanent route oscillation: the preferences are incompatible, there is no stable solution, and the iterative attempt to find one does not terminate. The 'dispute wheel' pattern alluded to above is what causes the badness, and after reduction this pattern becomes clearer. In the reduced bad gadget instance, we can see that each of the three outer nodes prefers an indirect path (around the cycle and then in) over a direct one (straight to the destination). This is an order-three dispute wheel. The pattern was present in the original instance, but obscured by the presence of node 4. On the other hand, the 'good' gadget has a unique stable solution, which is found by iteration. We can identify the solution on the reduced instance (shown here in green), and the original instance also converges.

In addition to good and bad gadget, our technical report [17] shows an eBGP instance that is not reducible, and the reduction of an iBGP instance.

4   Correctness of Network Reduction

We have identified three types of routing anomalies in Section 2, and associated each of them with particular BGP attributes. In this section, we examine sufficient conditions by which each of these three can be avoided. These are *safety*, the standard property for convergence of a path-vector routing system; *acyclic preference*, for ensuring that iBGP configurations express a consistent choice function; and *IGP-iBGP consistency*, for avoiding intra-AS oscillation. We then show that our reduction is sound with respect to preservation of the first two properties, but it does not always preserve the third. Therefore, the third condition needs to be checked separately.

4.1   eBGP Correctness

The eBGP correctness property we consider is safety [15,10]. The progress of the BGP algorithm towards a solution depends on the timing of messages and other non-deterministic factors: we want to ensure that every execution schedule will result in a routing solution being found, regardless of the asynchronous nature of the protocol. The final state is characterized by *stability*, meaning that no future messages will affect which best paths are selected by each router.

Definition 5. *A BGP instance is **safe**, if under all possible executions, it converges to a stable state, where the best routes selected by all the routers form a policy-compliant routing tree.*

We show that our reduction preserves safety, using a structure called the *path digraph* [15]. This is derived from an SPP instance $(V, E, d, P, \prec)$. Compared with the extended SPP which is used to define reduction, SPP requires an additional constraint: $\prec$ totally orders each P^i where i is a node in V. This holds for instances which are restricted to the 'eBGP' attributes, plus the router identifier, in Table 1.

Definition 6. *Let $G = (V, E, d, P, \prec)$ be an SPP instance. The path digraph is a graph whose nodes are the elements of P, and where there is an arc (p, q) from p to q if either of these two cases holds:*

1. *If $q = r \circ p$ for some path r, there is a 'transmission arc'.*
2. *If p and q are two paths in P^i and $p \prec q$, there is a 'preference arc'.*

If the digraph is acyclic then the SPP has a unique stable solution, which can be found by iteration from any starting state. We will call an SPP instance *cyclic* (or *acyclic*) if its path digraph is cyclic (or acyclic).

The following proposition 1, proved by Sobrinho [15], relates cyclicity of the digraph to safety of the SPP, and therefore of the BGP configuration it represents.

Proposition 1. *If a SPP instance is acyclic, then it is safe. If an SPP instance is cyclic, then we can construct an execution trace that exhibits route oscillation.*

Our main result (Lemma 1) is that our reduction technique transforms cyclic SPPs into cyclic SPPs, and acyclic SPPs into acyclic SPPs. This means that we never have false positives or false negatives, with respect to this safety property, after applying the reduction.

Lemma 1. *Let $G = (V, E, d, P, \prec)$ be an SPP instance, containing unifiable nodes u and v, and let $G' = (V', E', d, P', \prec')$ be the result of applying the procedure of Definition 4 to unify those two nodes. Then G is cyclic if and only if G' is cyclic.*

Proof. See technical report [17] for more details.   □

Finally, the following theorem proves that network reduction is sound: to analyze G for safety, it is sufficient to analyze its reduction G'.

Theorem 1. *If G' is acyclic then G is safe; If G' is cyclic then in running G, there exists at least one execution trace that exhibits route oscillation.*

Proof. Obvious from Lemma 1 and Proposition 1.   □

4.2   iBGP correctness: Cyclic iBGP Route Preference

As previously noted, use of the MED attribute means that routes might not be totally ordered, and therefore Proposition 1 is inapplicable. We handle this case by employing a more general notion of route selection in our analysis, and can show that our reduction *does* preserve these kinds of preference cycle. The details are in technical report [17].

4.3   iBGP Correctness: IGP-iBGP Consistency Property

While BGP can choose the correct egress point in an AS, for each destination, establishment of the intra-AS path to that border router is the responsibility of another protocol (an interior gateway protocol or IGP). Problems can occur if the iBGP configuration does not match the distance values used in the IGP. Our network reduction is designed for analysis BGP routing policies, and is unaware of IGP-iBGP inconsistency (see technical report [17]). Therefore, to ensure the soundness of analysis, one should check IGP-iBGP consistency before applying network reduction, using pre-existing methods from the literature [16,4].

5   Network Reduction in Maude

To validate our reduction method, we have extended our library for analysis of BGP configurations [18] to support automatic abstraction from dynamic (BGP) configurations to static (extended SPP) configurations, reduction based on SPP configurations, and integration with dynamic exhaustive search analysis. Using the original library BGP instances up to 25 nodes have been successfully analyzed in *minutes*. Using our reduction technique, we are able to reduce and analyze various 100 nodes BGP instances within *seconds*. Our extended library consists of the following three components:

- **Dynamic Network Representation.** For a BGP instance, we require users to input routing policies, i.e., the values of the BGP attributes that cause anomalies. We also require users to input the network topology. Based on the routing policy and topology, we automatically generate the dynamic representation of the BGP instance. The dynamic representation includes configurations (snapshots of an executing instance) and rewrite rules describing a router's actions during execution of the BGP protocol. The dynamic representation can be used to compute the complete set of permitted paths, and route selection information.
- **Static Network Representation.** While the dynamic representation is good for simulating the dynamic behavior of a BGP system, it is not the right representation for network reduction. Thus we introduce a static representation of BGP instances corresponding to the extended SPP instance (Definition 1). For each router, its static representation consists of its complete set of permitted paths, and route selection result given any sub-set of the permitted paths. Our library provides functions to compute the static representation from the dynamic initial network state.
- **Network Reduction on Static Representation.** Our library implements the network reduction process described in Definition 4 that applies to the static (extended SPP) representation.

Our library is implemented in Maude [1], a language and tool based on rewriting logic. Rewriting logic [13] is a logical formalism that is based on two simple ideas: states of a system can be represented as elements of an algebraic data type, and the behavior of a system can be given by transitions between states described by local rewrite rules. A rewrite rule has the form '$t \implies t'$ if c' where t and t' are patterns (terms possibly containing variables) and c is a condition (a boolean term). Such a rule applies to a system state s if t can be matched to a part of s by supplying the right values for the variables, and if the condition c holds when supplied with those values. In this case the rule can be applied by replacing the part of s matching t by t' using the matching values for variables in t'. Maude provides a high performance rewriting engine featuring matching modulo associativity, commutativity, and identity axioms. Given a specification S of a concurrent system, Maude can execute this specification, allowing one to observe some possible behaviors of the system. One can also use the search functionality of Maude to check if a state meeting a given condition can be reached during any system execution.

The dynamic representation is a small extension of [18] to account for the MED attribute. In this paper we only discuss generation of the static representation and the implementation of the reduction process.

5.1   Computing the Static BGP Representation

We recall that the dynamic representation of a BGP router has the form [rid : asid |Nb: nbrs,LR: routes ,BR: best] where rid : asid is called the NodeInfo with rid the router ID, and asid the AS ID. The remaining three arguments represent the routers state: nbrs is a list of neighbor router IDs, routes is a list of routes, and best is the best route.

Recall that in Definition 4, we apply the network reduction to the static representation of a BGP system $G = (V, E, d, P, \prec)$. In this representation we need the following information: (1) the complete set of permitted paths P that the routers could ever generate in protocol execution; and (2) the $\prec$ relation that determines how each router selects the best route, given an arbitrary subset of permitted paths. To capture P and $\prec$, we introduce the static representation of a BGP system using the Maude constructor declaration:

```
op [_|Nb:_,perPath:_,pref:_] : NodeInfo List{NodeInfo} List{route} List{sel-fun}
    -> absNode .
```

Similar to the dynamic representation, the first two arguments (indicated by underscores) specify the router's ID, AS and neighbor information. What is different is the second two attributes: rather than keeping the dynamic routing table and best route attributes, we have the static permitted paths attribute perPath:, and the route preference attribute pref:. The value of perPath: is the list of paths that can be computed during BGP execution, and the value of pref: represents the preference function as a list of pairs, each consisting of a route set and the selected route.

A BGP system's static representation is computed from the specification of the dynamic representation in two steps. First, the complete set of permitted paths is computed by simulating route exchanges and computation on the dynamic representation using the the rewrite rule compute-spp:

```
rl [compute-spp]:
[from (S1 : AS1) to S2 : (S3 : AS3),1f2,[asp1],med1,S4]
[S2 : AS2 |Nb: nodes2, LR: lr2, BR: nilRoute ]
=>
if ((occurs(import((S1 : AS1),(S2 : AS2),((S3 : AS3),1f2,[asp1],med1,S4)),lr2)) or
    import(...) == nilRoute)
 then [S2 : AS2 |Nb: nodes2, LR: lr2, BR: nilRoute ]
 else
  [S2:AS2|Nb: nodes2,
        LR: update(import(...),lr2),
        BR: nilRoute ]
  generateMsg((S2:AS2),nodes2,export(import(...))))
fi .
```

Here, the left-hand matches a router S2 and a route message sent from its neighbor S1. The right-hand side says that S2 computes a new route import(...), and if either of the two conditions occurs(import(...), lr2) or import(...)==nilRoute holds, that is, if either the new route import(...) is already in routing table lr2, or if the new route is filtered out according to S2's routing policy, S2 is unchanged, and the routing message on the left-hand is consumed. Otherwise, the new route is inserted into the routing table (update(import(...),lr2)), and S2 applies its export policy export(import(...)) and then (if allowed by export policy and export does not result in nilRoute) S2 re-advertises this new route to all of its neighbors nodes2. Compared with the normal BGP protocol execution, this rule is simpler in the sense that it does not perform best route selection: Note that BR: is kept blank. Normal BGP execution is non-deterministic—depending on the result of route selection, one of three different types of actions are taken [18], and the system may converge to different final states or not terminate (route oscillation may happen due to conflicting best route selection). However, the process defined by rule compute-spp always terminates with the same final state, when the complete sets of permitted paths of all nodes are generated.

Second, based on the permitted paths, the route selection function pref: is computed as follows:

```
eq compSPP ([S1 : AS1 |Nb: nodes1,LR: lr1, BR: nilRoute] Network) =
    [ S1 : AS1 |Nb: nodes1, perPath: lr1, pref: compSPPNode(lr1)] compSPP(Network).
```

compSPP converts each dynamic router representation [S1:AS1|Nb:_,LR:_, BR:_] in the network to its static form [S1:AS1 |Nb:_, perPath:_, pref:_]. The critical part is to compute route selection compSPPNode(lr1), given the complete set of subsets of the permitted paths lr1, by applying the best route selection function select to each subset. The function select is defined in terms of path attributes. As example, we show here the encoding of the two eBGP attributes LOCAL_PREF and AS_PATH as follows:

```
op select : List{route} -> List{route} .
eq select(lr1) =
    select-as(select-lf(lr1, best-lf(lr1)),
              best-as(select-lf(lr1, best-lf(lr1)))) .
```

Here select first invokes best-lf to compute the lowest (best) LOCAL_PREF value in the permitted paths lr1, then select-lf selects from lr1 the set of routes with this lowest LOCAL_PREF value. Next, from these remaining routes, select invokes best-as to compute the best AS value and select-as to select the set of routes with such best AS value.

5.2   Reduction by Merging All Pairs of Unifiable Nodes

To reduce a BGP instance, we take its static representation - a set of routers of the form [S1:AS1 |Nb:_, perPath:_, pref:_] as input, and repeatedly merge pairs of unifiable nodes. For each router S1 in the Network, we look for its unifiable nodes, if such nodes exist, we unify S1 with the first unifiable node S2, and transform the rest of the network according to Definition 4 (e.g. the neighbors of S1, S2 now become neighbors of S1). This reduction process is implemented by the function mergeDupEach.[2]

First, mergeDupEach implements the process of unifying node S1 and its first-found unifiable node as follows:

```
eq mergeDupEach([[(S1:AS1) |Nb:nodes1, perPath:lr1, pref:lsel-fun1],
                ([[(S2:AS2)|Nb:nodes2, perPath:lr2, pref:lsel-fun2] C))
  =
  if (size(([[(S1 : AS1) |Nb: nodes1, perPath: lr1, pref: lsel-fun1] unify
            [(S2 : AS2) |Nb: nodes2, perPath: lr2, pref: lsel-fun2])) == 1)
    then
      (([[(S1:... ] unify [(S2:...]) C)
    else
      ([[(S2:...] mergeDupEach([[(S1:...], C))
  fi .
```

---

[2] Obviously, reduction always terminates. However, how the order of merging nodes affects the reduction process—whether reduction always converges to the same reduced network—is the subject of ongoing work, but does not affect correctness.

Here, the `if` condition tests if `S1,S2` are unifiable, and `mergeDupEach` tests nodes in the network `C` until a unifiable node is found. Then `mergeDupEachEachRW` is invoked.

```
eq mergeDupEachRW(abn, C) =
    replaceNode(mergeDupEach(abn,C), get-NodeInfo(abn), findNodeInfo(abn, C))
```

Here, network `C` is transformed by replacing information relating to `abn` by that of abn's first unifiable node `findNodeInfo(abn, C)`. The specific transformation is as follows:

```
eq replaceNode ([S0:AS]|...] C, (S1:AS1) , (S2:AS2)) =
    [(S0 : AS) |Nb: removeRepeatedNB (...,(S1:AS1),(S2:AS2)),
               perPath: (replacePerPath(...,(S1:AS1),(S2:AS2))),
               pref: replacePref(..., (S1:AS1), (S2:AS2))]
    replaceNode (C, (S1:AS1), (S2:AS2)) .
```

Here, each node `S0` in the network is transformed by rewriting its neighboring table (`NB:`), permitted path (`perPath:`), and route selection function (`pref:`).

Finally, putting it all together, `mergeDup` specifies the reduction process on the entire network `C` as follows:

```
eq mergeDup(S1 Oid1, C) =
  mergeDupEachRW(get-Node(S1, mergeDup(Oid1, C)),
                 mergeDup(Oid1, C) - get-Node(S1, mergeDup(Oid1, C))) .
```

Here `mergeDup` takes two inputs. The first argument `S Oid1` is the list of router IDs, and the second argument is the list of routers `[ S1:AS1 |Nb:_, perPath:_, pref:_]`. `mergeDup(Oid1, C)` denotes the set of remaining routers after reducing all nodes other than `S1`; if `S1` is in these remaining routers, then `get-Node(S1, mergeDup(Oid1, C))` denotes `S1` itself, otherwise (that is, if `S1` is removed in the reduction) the value is set to nil. In either case, `mergeDup(Oid1, C) - get-Node(S1, mergeDup(Oid1, C))` denotes the remaining routers other than `S1` after reducing all routers except `S1`. Based on these notions, the recursive definition of `mergeDup` says that, to merge all unifiable nodes, we only need to merge node `S1` into the routers that (1) are already reduced among themselves; and (2) do not contain `S1` itself.

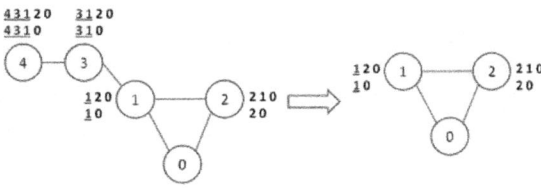

Fig. 2. Reduction example

As an example, to perform network reduction to the network on the left of Figure 2, we execute in Maude as follows:

```
red mergeDup(N1 N2 N3 N4, Network)
```

Where `Nodes` = `N1 N2 N3 N4` and `Network` = `[N0:0 |Nb:_,LR:_,BR:_] [N1 :` `1 |...] ... [N5 : 5 |...]`. The result is as follows:

```
[N0 : 0 |Nb: (N1 : 1) (N2 : 2),perPath: nil,pref: nil]
[N1 : 1 |Nb: N2 : 2,
         perPath:(N0 : 0,200,[2 0],[1,2],N2)
                 (N0 : 0,100,[0],[1,0],N0),
         pref: ...]
[N2 : 2 |Nb: N1 : 1, ...]
```

As expected, nodes N3, N4 are merged into N2.

6   Evaluation

In this section, we provide an empirical study to quantify the benefits of network reduction, by comparing the computation time required in safety analysis with and without network reduction.

Our safety analysis was performed via an exhaustive search strategy using Maude, as described in [18]. Oscillation is detected if the same best route is selected multiple times during protocol execution. To detect such recurring best routes, we use Maude to run the actual path vector protocol used in BGP, and simulate all possible sequences in which ASes receive routes. At each node, we use a monitor object to track the best routes that have been previously selected. We also attempt to apply our reduction technique and perform such analysis on the reduced version.

For the BGP instance shown in Figure 2, we note that in the reduced network (right), our analysis tool detects the same route oscillation pattern found in the original network (left), while requiring significantly less state space (reduction from 956 to 35) and analysis time (320ms to 8ms). In addition, we evaluate three common scenarios [10]: Bad gadget that exhibits permanent oscillation, Disagree transient oscillation, and Good gadget that is safe and no oscillation. The data table in this section shows the analysis results for Bad gadget scenario, indicating the performance requirements for the ordinary exhaustive search and for the reduction alternative, as well as the final safety outcome. For more details on the other scenarios, please see our technical report [17].

Table 2. Network Instances that Reduces to Bad Gadget

	Bad (reduced)	Bad-10	Bad-20	Bad-53	Bad-83	Bad-102
Search (Time)	30510ms	Unknown	Unknown	Unknown	Unknown	Unknown
SPP generation (Time)	0ms	3ms	44ms	134ms	246ms	273ms
Reduction (Time)	0ms	8ms	49ms	146ms	541ms	595ms
Search(State)	11118	Unknown	Unknown	Unknown	Unknown	Unknown
Oscillation?	Yes	Yes	Yes	Yes	Yes	Yes

Our analysis was carried out on a Intel 2.40GHz dual-core machine with 1.9GB memory, running Maude v2.4 on the Debian 5.0.6. operating system. Table 2 shows the analysis results for eBGP instances where the network size ranges from 10 (Bad-10) to 102 (Bad-102). For each network size, we embed in a bad gadget. After applying the reduction process, all BGP instances are reduced to a single bad gadget Bad (reduced). For each entry, Unknown means that the analysis cannot be completed within reasonable time (after running Maude for several hours).

We make the following observations from our results. First, reduction requires minimal time. Even for a large network of 102 nodes, reduction can be completed within one second. As input to the reduction process, the SPP formalism for a BGP instance is extracted as described in Section 5 where a static representation (corresponding to the SPP) is computed by simulating on the instance's dynamic representation (corresponding to the snapshot state). This is also an efficient process, requiring less than 300ms for the largest network. Overall, network reduction results in significant savings in both state and execution time during safety analysis. For example, while it was previously infeasible to complete the analysis of any network beyond 10 nodes due to the state explosion problem (depicted by Unknown), the reduced BGP instance can be analyzed in around 300 seconds (and 11118 states).

In our technique report [17], we present a similar comparison of analysis overhead for network instances that have the disagree and good gadget embedded. We similarly observe significant state and execution time savings via the use of reduction.

7   Conclusion

In this paper, we present a technique to reduce BGP instances, such that safety analysis can be performed efficiently on large networks. We prove correct our reduction technique, develop a reduction and BGP analysis tool using Maude, and demonstrate its effectiveness at reducing the state space and execution time required for analyzing BGP instances. As future work, we are (1) exploring the use of our tool on larger case studies drawn from real network configurations, (2) making the tool available with documentation, (3) optimizing the formal representation for more efficient analysis, and (4) possibly extending the library to detect iBGP cyclic preference, and IGP-iBGP inconsistency.

Acknowledgment. This research is funded in part by NSF grants (CCF-0820208, CNS-0830949, CNS-0845552, CNS-1040672, TC-0905607 and CPS-0932397), AFOSR grant FA9550-08-1-0352, and ONR grant N00014-11-1-0555.

References

1. Clavel, M., Durán, F., Eker, S., Lincoln, P., Martí-Oliet, N., Meseguer, J., Talcott, C.: All About Maude - A High-Performance Logical Framework. LNCS, vol. 4350. Springer, Heidelberg (2007)
2. Feamster, N., Johari, R., Balakrishnan, H.: Implications of autonomy for the expressiveness of policy routing. In: ACM SIGCOMM (2005)
3. Feldmann, A., Maennel, O., Mao, Z.M., Berger, A., Maggs, B.: Locating Internet routing instabilities. In: ACM SIGCOMM (2004)
4. Flavel, A., Roughan, M., Bean, N., Shaikh, A.: Where's Waldo? Practical Searches for Stability in iBGP. In: Proc. International Conference on Network Protocols, ICNP (October 2008)
5. Gao, L., Griffin, T.G., Rexford, J.: Inherently safe backup routing with BGP. In: IEEE INFOCOM (2001)

6. Gao, L., Rexford, J.: Stable Internet routing without global coordination. In: ACM SIGMETRICS (2000)
7. Griffin, T.G.: The stratified shortest-paths problem. In: COMSNETS (2010)
8. Griffin, T.G., Jaggard, A., Ramachandran, V.: Design principles of policy languages for path vector protocols. In: ACM SIGCOMM (2003)
9. Griffin, T.G., Shepherd, F.B., Wilfong, G.: The stable paths problem and interdomain routing. IEEE Trans. on Networking 10, 232–243 (2002)
10. Griffin, T.G., Wilfong, G.: An analysis of BGP convergence properties. In: SIGCOMM (1999)
11. Haeberlen, A., Avramopoulos, I., Rexford, J., Druschel, P.: NetReview: Detecting when interdomain routing goes wrong. In: NSDI (2009)
12. Labovitz, C., Malan, G.R., Jahanian, F.: Internet Routing Instability. TON (1998)
13. Meseguer, J.: Conditional Rewriting Logic as a Unified Model of Concurrency. Theoretical Computer Science 96(1), 73–155 (1992)
14. Schapira, M., Zhu, Y., Rexford, J.: Putting BGP on the right path: A case for next-hop routing. In: ACM SIGCOMM HotNets (October 2010)
15. Sobrinho, J.: Network routing with path vector protocols: theory and applications. In: SIGCOMM (2003)
16. Vutukuru, M., Valiant, P., Kopparty, S., Balakrishnan, H.: How to Construct a Correct and Scalable iBGP Configuration. In: IEEE INFOCOM, Barcelona, Spain (April 2006)
17. Wang, A., Talcott, C., Gurney, A.J.T., Loo, B.T., Scedrov, A.: Reduction-based formal analysis of BGP instances. University of Pennsylvania Department of Computer and Information Science Technical Report (2012),
http://netdb.cis.upenn.edu/papers/tacas12-TR.pdf
18. Wang, A., Talcott, C., Jia, L., Loo, B.T., Scedrov, A.: Analyzing BGP Instances in Maude. In: Bruni, R., Dingel, J. (eds.) FORTE 2011 and FMOODS 2011. LNCS, vol. 6722, pp. 334–348. Springer, Heidelberg (2011)

Minimal Critical Subsystems
for Discrete-Time Markov Models[*]

Ralf Wimmer[1], Nils Jansen[2], Erika Ábrahám[2],
Bernd Becker[1], and Joost-Pieter Katoen[2]

[1] Albert-Ludwigs-University Freiburg, Germany
{wimmer,becker}@informatik.uni-freiburg.de
[2] RWTH Aachen University, Germany
{nils.jansen,abraham,katoen}@informatik.rwth-aachen.de

Abstract. We propose a new approach to compute *counterexamples* for violated ω-regular properties of discrete-time Markov chains and Markov decision processes. Whereas most approaches compute a set of system paths as a counterexample, we determine a *critical subsystem* that already violates the given property. In earlier work we introduced methods to compute such subsystems based on a search for shortest paths. In this paper we use *SMT solvers* and *mixed integer linear programming* to determine *minimal* critical subsystems.

1 Introduction

Systems with uncertainties often act in safety-critical environments. In order to use the advantages of formal verification, formal models are needed. Popular modeling formalisms for such systems are *discrete-time Markov chains (DTMCs)* and—in the presence of non-determinism—*Markov decision processes (MDPs)*.

State-of-the-art model checking algorithms verify probabilistic safety properties like "The probability to reach a safety-critical state is at most 10^{-3}" or, more generally, ω-regular properties [1], efficiently by solving linear equation systems [2]. Thereby, if the property is violated, they do not provide any information about the reasons why this is the case. However, this is not only strongly needed for debugging purposes, but it is also exploited for abstraction refinement in CEGAR frameworks [3,4]. Therefore, in recent years much research effort has been made to develop algorithms for *counterexample generation* for DTMCs and MDPs (see, e.g., [5,6,7,8,9,10,11,12,13]). Most of these algorithms [6,7,8,9] yield *path-based* counterexamples, i.e., counterexamples in the form of a set of finite paths that all lead from the initial state to a safety-critical state and whose joint probability mass exceeds the allowed limit.

---

[*] This work was partly supported by the German Research Council (DFG) as part of the Transregional Collaborative Research Center "Automatic Verification and Analysis of Complex Systems" (SFB/TR 14 AVACS) and the DFG project "CE-Bug – Counterexample Generation for Stochastic Systems using Bounded Model Checking".

C. Flanagan and B. König (Eds.): TACAS 2012, LNCS 7214, pp. 299–314, 2012.
© Springer-Verlag Berlin Heidelberg 2012

Unfortunately, the number of paths needed for a counterexample is often very large or even infinite, in particular if the gap between the allowed probability and its actual value is small. The size of the counterexample may be several orders of magnitude larger than the number of system states, rendering the counterexample practically unusable for debugging purposes. Different proposals have been made to alleviate this problem: [6] represents the path set as a regular expression, [7] detects loops on paths, and [8] shrinks paths through strongly connected components into single transitions.

As an alternative to path-based counterexamples, the usage of small *critical subsystems* has been proposed in [5,10]. A critical subsystem is a part of the Markov chain such that the probability to reach a safety-critical state (or, more generally, to satisfy an ω-regular property) inside this part exceeds the bound. This induces a path-based counterexample by considering all paths leading through this subsystem. Contrary to the path-based representation, the size of a critical subsystem is bounded by the size of the model under consideration. Different heuristic methods have been proposed for the computation of small critical subsystems: The authors of [5] apply best first search to identify a critical subsystem, while in [10] a novel technique is presented that is based on a hierarchical abstraction of DTMCs in combination with heuristics for the selection of the states to be contained in the subsystem.

Both approaches use heuristic methods to select the states of a critical subsystem. However, we are not aware of any algorithm that is suited to compute a *minimal* critical subsystem, neither in terms of the number of states nor of the number of transitions. In this paper we fill this gap. We provide formulations as a SAT-modulo theories (SMT) problem and as a mixed integer linear program (MILP) which yield state-minimal critical subsystems of DTMCs and MDPs, respectively. We will present a number of optimizations which significantly speed up the computation times in many cases. Experimental results on some case studies are provided, which show the effectiveness of our approach. We show that our MILP approach yields significantly more compact counterexamples than the heuristic methods even if the MILPs cannot be solved to optimality due to time restrictions. We present our algorithms for probabilistic safety properties, but they can be extended to the more general case of arbitrary ω-regular properties.[1]

Structure of the Paper. In Section 2 we introduce the foundations of DTMCs, MDPs, and critical subsystems. In Sections 3 and 4 we present different approaches for the computation of state-minimal subsystems for DTMCs and MDPs. We discuss experimental results in Section 5 and finally draw a conclusion in Section 6.

2   Foundations

We first introduce discrete-time Markov chains and discrete-time Markov decision processes as well as critical subsystems for both models.

---

[1] They can be reduced to reachability after a product construction of a DTMC or MDP, resp., with a deterministic Rabin automaton, followed by a graph analysis [2]. For more details see [14].

Discrete-Time Markov Chains.

Definition 1. *A* discrete-time Markov chain (DTMC) *is a tuple $M = (S, s_I, P)$ with S being a finite set of states, $s_I \in S$ the initial state and $P : S \times S \to [0,1]$ the matrix of transition probabilities such that $\sum_{s' \in S} P(s, s') \leq 1$ for all $s \in S$.*[2]

Let in the following $M = (S, s_I, P)$ be a DTMC, $T \subseteq S$ a set of target states, and $\lambda \in [0,1]$ an upper bound on the allowed probability to reach a target state[3] in T from the initial state s_I. This property can be formulated by the PCTL formula $\mathcal{P}_{\leq \lambda}(\Diamond T)$. We assume this property to be violated, i.e., the actual probability of reaching T exceeds λ.

The probability to eventually reach a target state from a state s is the unique solution of a linear equation system [2, p. 760] containing an equation for each state $s \in S$: $p_s = 1$ if $s \in T$, $p_s = 0$ if there is no path from s to any state in T, and $p_s = \sum_{s' \in S} P(s, s') \cdot p_{s'}$ in all other cases.

Definition 2. *A* subsystem *of M is a DTMC $M' = (S', s'_I, P')$ such that $S' \subseteq S$, $s'_I \in S'$, and $P'(s, s') > 0$ implies $P'(s, s') = P(s, s')$ for all $s, s' \in S'$.*

We call a subsystem $M' = (S', s'_I, P')$ of M critical *if $s'_I = s_I$, $S' \cap T \neq \emptyset$, and the probability to reach a state in $S' \cap T$ from s'_I in M' is larger than λ.*

We want to identify a *minimal* critical subsystem (MCS) of M, which induces a counterexample for $\mathcal{P}_{\leq \lambda}(\Diamond T)$. Minimality can thereby be defined in terms of the number of *states* or the number of *transitions*. In this paper we restrict ourselves to state-minimal subsystems. However, our approaches can easily be adapted to transition minimality. In [4] it is shown that computing MCSs for arbitrarily nested PCTL formulae is NP-complete. It is unclear if this also holds for reachability properties.

We denote the *set of transitions* of M by $E_M = \{(s, s') \in S \times S \mid P(s, s') > 0\}$, the *set of successors* of state $s \in S$ by $\mathrm{succ}_M(s) = \{s' \in S \mid (s, s') \in E_M\}$, and its *predecessors* by $\mathrm{pred}_M(s) = \{s' \in S \mid (s', s) \in E_M\}$. A *finite path* π in M is a finite sequence $\pi = s_0 s_1 \ldots s_n$ such that $(s_i, s_{i+1}) \in E_M$ for all $0 \leq i < n$.

Definition 3. *Let $M = (S, s_I, P)$ be a DTMC with target states $T \subseteq S$. A state $s \in S$ is called* relevant *if there is a path $\pi = s_0 s_1 s_2 \ldots s_n$ with $s_0 = s_I$, $s_i \notin T$ for $0 \leq i < n$, $s_n \in T$ and $s = s_j$ for some $j \in \{0, \ldots, n\}$. A transition $(s, s') \in E_M$ is* relevant *if both s and s' are relevant and $s \notin T$.*

We denote the *set of relevant states* of M by S_M^{rel} and the *set of relevant transitions* by E_M^{rel}. States and transitions that are not relevant can be removed from all critical subsystems without changing the probability to reach a target

---

[2] Please note that we allow sub-stochastic distributions. Usually, the sum of probabilities is required to be exactly 1. This can be obtained by defining $M' = (S \cup \{s_\perp\}, s_I, P')$ with $s_\perp$ a fresh sink state, $P'(s, s') = P(s, s')$ for all $s, s' \in S$, $P'(s_\perp, s_\perp) = 1$, and finally $P(s, s_\perp) = 1 - P(s, S)$ and $P'(s_\perp, s) = 0$ for all $s \in S$.

[3] Model checking PCTL properties can be lead back to the problem of computing reachability probabilities.

state. Since we are interested in MCSs, we only have to take relevant states and transitions into account.

Let $E_M^- = \{(s, s') \in S \times S \mid (s', s) \in E_M\}$ be the *set of reversed transitions* of M. We consider the *directed graphs* $G = (S, E_M)$ and $G^- = (S, E_M^-)$.

Lemma 1. *A state $s \in S$ is* relevant *iff s is reachable from the initial state s_I in G and s is reachable from a target state in G^-. A transition $(s, s') \in E_M$ is* relevant *iff s is reachable from the initial state s_I in G and s' is reachable from a target state in G^-, and $s \notin T$.*

This lemma shows that the set S_M^{rel} of relevant states and the set E_M^{rel} of relevant transitions can be determined in linear time in the size of the DTMC by two simple graph analyses.

Markov Decision Processes. Extending DTMCs with non-determinism yields the class of Markov decision processes:

Definition 4. *A discrete-time Markov decision process (MDP) M is a tuple $M = (S, s_I, A, P)$ such that S is a finite set of states, $s_I \in S$ the initial state, A a finite set of actions, and $P : S \times A \times S \to [0, 1]$ a function such that $\sum_{s' \in S} P(s, a, s') \leq 1$ for all $a \in A$ and all $s \in S$.*

If $s \in S$ is the current state of an MDP M, its successor state is determined as follows: First a *non-deterministic* choice between the actions in A is made; say $a \in A$ is chosen. Then the successor state of s is determined *probabilistically* according to the distribution $P(s, a, \cdot)$.

We set $\text{succ}_M(s, a) = \{s' \in S \mid P(s, a, s') > 0\}$, $\text{pred}_M(s, a) = \{s' \in S \mid P(s', a, s) > 0\}$, and $E_M = \{(s, s') \in S \times S \mid \exists a \in A : P(s, a, s') > 0\}$. Relevant states S_M^{rel} and transitions E_M^{rel} are defined in the same way as for DTMCs.

Before probability measures can be defined for MDPs, the non-determinism has to be resolved. This is done by an entity called *scheduler*. For our purposes we do not need schedulers in their full generality, which are allowed to return a probability distribution over the actions A, depending on the path that led from the initial state to the current state. Instead, for reachability properties the following subclass suffices [2, Lemma 10.102]:

Definition 5. *Let $M = (S, s_I, A, P)$ be an MDP. A (deterministic memoryless)* scheduler *for M is a function $\sigma : S \to A$.*

Such a scheduler σ induces a DTMC $M^\sigma = (S, s_I, P^\sigma)$ with $P^\sigma(s, s') = P(s, \sigma(s), s')$. The probability of reaching a target state is now computed in this induced DTMC. The property $\mathcal{P}_{\leq \lambda}(\Diamond T)$ is satisfied in an MDP $M = (S, s_I, A, P)$ if it is satisfied in M^σ for all schedulers σ. Since all schedulers guarantee a reachability probability of at most λ, this implies that the maximal reachability probability is at most λ. If the property is violated, there is a non-empty set of schedulers for which the probability exceeds λ. We call them *critical schedulers*.

We want to compute a critical scheduler and a critical subsystem in the corresponding induced DTMC that is state- (transition-) minimal among all critical subsystems for all critical schedulers. Computing state-minimal critical subsystems for reachability properties of MDPs is NP-complete [4].

SAT-Modulo-Theories. SAT-modulo-theories (SMT) [15] refers to a generalization of the classical propositional satisfiability problem (SAT). Compared to SAT problems, in an SMT formula atomic propositions may be replaced by atoms of a given theory. For the computation of MCSs this theory is linear real arithmetic.

SMT problems are solved by the combination of a DPLL-procedure (as used for deciding SAT problems) with a theory solver that is able to decide the satisfiability of conjunctions of theory atoms. For a description of such a combined algorithm see [16].

Mixed Integer Linear Programming. In contrast to SMT, mixed integer linear programs consist only of a *conjunction* of linear inequalities. A subset of the variables occurring in the inequalities are restricted to take only integer values, which makes solving MILPs NP-hard.

Definition 6. *Let $A \in \mathbb{Q}^{n \times m}$, $B \in \mathbb{Q}^{k \times m}$, $b \in \mathbb{Q}^m$, $c \in \mathbb{Q}^n$, and $d \in \mathbb{Q}^k$. A mixed integer linear program (MILP) consists in computing $\min c^T x + d^T y$ such that $Ax + By \leq b$ and $x \in \mathbb{R}^n$, $y \in \mathbb{Z}^k$.*

MILPs are typically solved by a combination of a branch-and-bound algorithm with the generation of so-called cutting planes. These algorithms heavily rely on the fact, that relaxations of MILPs which result from removing the integrality constraints, can be solved efficiently. MILPs are widely used in operations research, hardware-software codesign and numerous other applications. Efficient open source as well as commercial implementations are available like SCIP or CPLEX. We refer the reader to, e. g., [17] for more information on solving MILPs.

3   Computing Minimal Critical Subsystems for DTMCs

The problem to find state-minimal critical subsystems for DTMCs can be specified as an SMT problem over linear real arithmetic, which we present in this section. As the experimental results were not satisfactory, we additionally elaborated MILP formulations of this problem. We also report on further optimizations that lead to a noticeable speed-up in many cases.

3.1   Formulation as an SMT Problem

We first specify an SMT formula over linear real arithmetic whose satisfying variable assignments correspond to the critical subsystems of M. The SMT formula is shown in Fig. 1. We use $\oplus$ for the binary XOR operator.

We introduce a variable $x_s \in [0,1] \subseteq \mathbb{R}$ for each relevant state $s \in S_M^{rel}$. We require in the formula that $x_s = 1$ or $x_s = 0$ holds. A state $s \in S_M^{rel}$ is contained in the subsystem iff $x_s = 1$ for the computed optimal satisfying assignment. In order to obtain a state-minimal critical subsystem, we have to minimize the number of x_s-variables to which the value 1 is assigned, or equivalently, the sum over all x_s-variables (line 1a). Besides the x_s variables we need one real-valued

$$\text{minimize} \quad \sum_{s \in S_M^{rel}} x_s \tag{1a}$$

$$\text{such that} \quad p_{s_I} > \lambda \tag{1b}$$

$$\forall s \in S_M^{rel} \cap T : \left((x_s = 0 \wedge p_s = 0) \oplus (x_s = 1 \wedge p_s = 1) \right) \tag{1c}$$

$$\forall s \in S_M^{rel} \setminus T : \left((x_s = 0 \wedge p_s = 0) \oplus \left(x_s = 1 \wedge p_s = \sum_{s' \in \text{succ}_M(s) \cap S_M^{rel}} P(s, s') \cdot p_{s'} \right) \right) . \tag{1d}$$

Fig. 1. SMT formulation for state-minimal critical subsystems of DTMCs

variable $p_s \in [0, 1] \subseteq \mathbb{R}$ for each state $s \in S_M^{rel}$ to which the probability of reaching a target state from s inside the subsystem is assigned.

If x_s is zero, the corresponding state s does not belong to the subsystem. Then its probability contribution is also zero. Target states that are contained in the subsystem have probability one (line 1c). Note that the MCS does not need to contain all target states. The probability of all non-target states in the subsystem is the weighted sum over the probabilities of the relevant successor states (line 1d). In order to obtain a critical subsystem we additionally have to require that $p_{s_I} > \lambda$ (line 1b).

The size of this formula is linear in the size of M. Since most of the state-of-the-art SMT solvers for linear real arithmetic cannot cope with the minimization of objective functions, we apply binary search in the range $\{1, \ldots, |S_M^{rel}|\}$ for the optimal value of the objective function. Starting with $k_l = 1$ and $k_u = |S_M^{rel}|$, we iteratively search for critical subsystems whose number of states is between k_l and $k_m := k_l + (k_u - k_l)/2$. If we find such a subsystem with k states, then we set k_u to $k - 1$. Otherwise we set k_l to $k_m + 1$. We repeat the search until $k_u < k_l$.

3.2   Formulation as a Mixed Integer Linear Program

The formulation as an SMT problem gives a good intuition how an MCS can be computed using solver technologies. However, as the experiments will show, the solution process is very time-consuming. This might be due to the fact that SMT solvers distinguish many cases while searching for a solution because of the involved disjunctions. We therefore reformulate the problem as an MILP that does not contain any disjunctions. The MILP is shown in Fig. 2.

In order to avoid the disjunctions of the SMT formulation, we need to explicitly require the variables x_s to be integer in contrast to the SMT formulation with $x_s \in [0, 1] \subseteq \mathbb{R}$. Hence, the MILP contains the variables $x_s \in [0, 1] \subseteq \mathbb{Z}$ and $p_s \in [0, 1] \subseteq \mathbb{R}$ for all states $s \in S_M^{rel}$.

The constraints can be translated as follows: For target states $s \in S_M^{rel} \cap T$, the condition (1c) of the SMT formulation corresponds to $p_s = x_s$ (line 2c). For the remaining states $s \in S_M^{rel} \setminus T$, we ensure by $p_s \leq x_s$ that the probability contribution of not selected states is zero (line 2d). For all non-target states s,

$$\text{minimize} \quad \left(-\frac{1}{2}p_{s_I} + \sum_{s \in S_M^{rel}} x_s\right) \tag{2a}$$

$$\text{such that} \quad p_{s_I} > \lambda \tag{2b}$$

$$\forall s \in S_M^{rel} \cap T : p_s = x_s \tag{2c}$$

$$\forall s \in S_M^{rel} \setminus T : p_s \le x_s \tag{2d}$$

$$p_s \le \sum_{s' \in \text{succ}_M(s) \cap S_M^{rel}} P(s,s') \cdot p_{s'}. \tag{2e}$$

Fig. 2. MILP formulation for state-minimal critical subsystems of DTMCs

an upper bound on the probability contribution p_s is given by the sum of the probabilities $p_{s'}$ of the relevant successor states s', weighted by the according transition probabilities $P(s, s')$ (line 2e). Together with the requirement that the probability of the initial state has to be larger than λ (line 2b) this describes the critical subsystems of the DTMC under consideration.

Using this formulation and the same objective function as in the SMT formula, the exact probability of reaching target states in the resulting MCS is not computed as a by-product. We would only compute a lower bound, because line (2e) is an inequality. However, we can achieve this by forcing the solver to maximize p_{s_I}. We change the objective function to $\min\left(-\frac{1}{2}p_{s_I} + \sum_{s \in S_M^{rel}} x_s\right)$. Then the solver computes not only an arbitrary MCS, but among all MCSs one with maximal probability, and assigns to the variable p_{s_I} its actual reachability probability. A factor $0 < c < 1$ is needed because if we only subtract the probability of the initial state, the solver may add an additional state if this results in $p_{s_I} = 1$. We chose $c = \frac{1}{2}$.

3.3 Optimizations

In the following we describe optimizations both of the SMT and the MILP formulation. They add redundant constraints to the problem. These constraints may help the solver to detect unsatisfiable branches in the search space earlier.

Successor and Predecessor Constraints. In order to guide the solver to choose states that form complete paths leading from the initial state to the set of target states, we firstly add the following optional constraints to the MILP formulation in Fig. 2:

$$\forall s \in S_M^{rel} \setminus T : -x_s + \sum_{s' \in (\text{succ}_M(s) \cap S_M^{rel}) \setminus \{s\}} x_{s'} \ge 0 \tag{3a}$$

$$\forall s \in S_M^{rel} \setminus \{s_I\} : -x_s + \sum_{s' \in (\text{pred}_M(s) \cap S_M^{rel}) \setminus \{s\}} x_{s'} \ge 0. \tag{3b}$$

The first set of constraints (3a), which we call *forward cuts*, states that each non-target state in the MCS must have a proper successor state which is also contained in the MCS. Proper in this case means that self-loops are ignored. The second set of constraints (3b), called *backward cuts*, requires that each non-initial state in the MCS has a proper predecessor in the MCS.

For MILP, forward and backward cuts do not modify the feasible solutions but add cutting planes which tighten the LP-relaxation of the MILP and may lead to better lower bounds on the optimal value.

For the SMT formulation similar constraints can be constructed. We omit their description here because in our experimental results they did not improve the performance. A reason for this phenomenon could be that these constraints come with an additional effort in propagation and theory solving that is not compensated by their positive effect of restricting the solution set.

SCC Constraints. The forward respectively backward cuts do not encode that all states of the MCS are forwards respectively backwards reachable: A satisfying assignment could define a loop to belong to the subsystem even if in the solution the states of the loop are connected neither to the initial nor to any target state.

To strengthen the effect of forward and backward cuts, we make use of strongly connected components. Formally, a *strongly-connected component (SCC)* of a DTMC $M = (S, s_I, P)$ is a maximal subset $C \subseteq S$ such that each state $s \in C$ is reachable from each state $s' \in C$ visiting only states from C. The *input states* $\text{In}(C)$ of an SCC C are those states which have an in-coming transition from outside the SCC, i.e., $\text{In}(C) = \{s \in C \mid \exists s' \in S \setminus C : P(s', s) > 0\}$. The *output states* of C, denoted $\text{Out}(C)$, are those states outside C which can be reached from C via a single transition. Hence, $\text{Out}(C) = \{s \in S \setminus C \mid \exists s' \in C : P(s', s) > 0\}$.

A state of an SCC can be reached from the initial state only through one of the SCC's input states. Therefore we define an *SCC input cut* for each SCC assuring that, if none of its input states is included in the MCS, then the MCS does not contain any states of the SCC. Line 4a shows the SMT variant of this constraint, whereas line 4b gives the corresponding MILP formulation:

$$\bigwedge_{s \in \text{In}(C)} x_s = 0 \quad \Rightarrow \quad \bigwedge_{s \in C \setminus \text{In}(C)} x_s = 0 \tag{4a}$$

$$\sum_{s \in C \setminus \text{In}(C)} x_s \leq |C \setminus \text{In}(C)| \cdot \sum_{s \in \text{In}(C)} x_s \,. \tag{4b}$$

Analogously, starting from a state inside an SCC, all paths to a target state lead through one of the SCC's output states. Therefore, if no output state of an SCC C is selected, we do not want to select any state of the SCC. Line 5a contains the SMT and line 5b the MILP formulation of this *SCC output cut*:

$$\bigwedge_{s \in \text{Out}(C)} x_s = 0 \quad \Rightarrow \quad \bigwedge_{s \in C} x_s = 0 \tag{5a}$$

$$\sum_{s \in C} x_s \leq |C| \cdot \sum_{s \in \text{Out}(C)} x_s \,. \tag{5b}$$

Complete Reachability Encoding. Although the SCC cuts further restrict the selection of unreachable states, they still do not encode reachability exactly: We could, for example, select a path from an input to an output state of an SCC and additionally select an unreachable loop inside the SCC.

For a complete encoding of *forward* reachability, we introduce a variable $r_s^{\rightarrow} \in [0,1] \subseteq \mathbb{R}$ for each state $s \in S_M^{rel}$. The values of these variables define a partial order on the states. We make use of this partial order to express forward reachability in critical subsystems: We encode that for each selected state s there is a path $s_0 \ldots s_n$ from the initial state $s_0 = s_I$ to $s_n = s$ such that $r_i^{\rightarrow} < r_{i+1}^{\rightarrow}$ for all $0 \le i < n$ and all states on the path are selected, i.e., $x_{s_i} = 1$ for all $0 \le i \le n$. Note that we can assign a proper value to $r_s^{\rightarrow}$ for each reachable state s, for example the value $n_s / |S_M^{rel}|$ with n_s being the size of the longest loop-free path leading from the initial state to s.

An SMT encoding of forward reachability can be defined as follows:

$$\forall s \in S_M^{rel} \setminus \{s_I\} : \left(\neg x_s \vee \bigvee_{s' \in \mathrm{pred}_M(s) \cap S_M^{rel}} (x_{s'} \wedge r_{s'}^{\rightarrow} < r_s^{\rightarrow}) \right). \qquad (6a)$$

The SMT encoding of *backward* reachability is analogous, using a variable $r_s^{\leftarrow} \in [0,1] \subseteq \mathbb{R}$ for each state $s \in S_M^{rel}$:

$$\forall s \in S_M^{rel} \setminus T : \left(\neg x_s \vee \bigvee_{s' \in \mathrm{succ}_M(s) \cap S_M^{rel}} (x_{s'} \wedge r_s^{\leftarrow} < r_{s'}^{\leftarrow}) \right). \qquad (7a)$$

For the MILP encoding of *forward* reachability, for each transition from a state s to s' we additionally use an integer variable $t_{s,s'}^{\rightarrow} \in [0,1] \subseteq \mathbb{Z}$. These variables correspond to the choice of the predecessor states in the disjunctions of the SMT encoding. Again, we encode that for each selected state s there is a path in the selected subsystem leading from the initial state to s. The variable $t_{s',s}^{\rightarrow}$ encodes if the transition from s' to s appears in that path.

$$\forall s \in S_M^{rel} \ \forall s' \in (\mathrm{succ}_M(s) \cap S_M^{rel}) : 2t_{s,s'}^{\rightarrow} \le x_s + x_{s'} \qquad (8a)$$

$$r_s^{\rightarrow} < r_{s'}^{\rightarrow} + (1 - t_{s,s'}^{\rightarrow}) \qquad (8b)$$

$$\forall s \in S_M^{rel} \setminus \{s_I\} : (1 - x_s) + \sum_{s' \in \mathrm{pred}_M(s) \cap S_M^{rel}} t_{s',s}^{\rightarrow} \ge 1. \qquad (8c)$$

Lines 8a and 8b encode that each transition from s to s' with $t_{s,s'}^{\rightarrow} = 1$ connects selected states with $r_s^{\rightarrow} < r_{s'}^{\rightarrow}$. Under this assumption, the constraints defined in line 8c imply by induction that for each selected state there is a reachable selected predecessor state.

Backward reachability is analogous using a variable $t_{s,s'}^{\leftarrow} \in [0,1] \subseteq \mathbb{Z}$ for each transition:

$$\forall s \in S_M^{rel} \ \forall s' \in (\mathrm{succ}_M(s) \cap S_M^{rel}) : 2t_{s,s'}^{\leftarrow} \le x_s + x_{s'} \qquad (9a)$$

$$r_s^{\leftarrow} < r_{s'}^{\leftarrow} + (1 - t_{s,s'}^{\leftarrow}) \qquad (9b)$$

$$\forall s \in S_M^{rel} \setminus T : (1 - x_s) + \sum_{s' \in \mathrm{succ}_M(s) \cap S_M^{rel}} t_{s,s'}^{\leftarrow} \ge 1. \qquad (9c)$$

$$\text{minimize} \quad \sum_{s \in S_M^{rel}} x_s \tag{10a}$$

$$\text{such that} \quad p_{s_I} > \lambda \tag{10b}$$

$$\forall s \in S_M^{rel} \cap T : \left((x_s = 0 \wedge p_s = 0) \oplus (x_s = 1 \wedge p_s = 1) \right) \tag{10c}$$

$$\forall s \in S_M^{rel} \setminus T : \left(\left(x_s = 0 \wedge p_s = 0 \right) \oplus \left(x_s = 1 \wedge \right. \right.$$
$$\left. \left. \bigvee_{a \in A} \left(a_s = a \wedge p_s = \sum_{s' \in \text{succ}_M(s,a) \cap S_M^{rel}} P(s,a,s') \cdot p_{s'} \right) \right) \right) . \tag{10d}$$

Fig. 3. SMT formulation for state-minimal critical subsystems of MDPs

These encodings come at the cost of new variables, but they cut all subsystems with unreachable states, especially unreachable loops which were not covered by the previous two encodings.

4   Computing Minimal Critical Subsystems for MDPs

In this section we describe how to extend our SMT- and MILP-based formulations to Markov decision processes. Using these formulations, we not only provide a state-minimal critical subsystem but also the corresponding critical scheduler.

4.1   SMT Formulation

The SMT formulation for MCSs for MDPs straightly follows the ideas for DTMCs. We additionally introduce a variable $a_s \in [0, |A| - 1] \subseteq \mathbb{R}$ for all states $s \in S_M^{rel} \setminus T$ which stores the action selected by a critical scheduler. If each action is assigned a unique number in the range $0, \dots, |A| - 1$, this again results in an SMT problem over linear real arithmetic, which is shown in Fig. 3.

4.2   MILP Formulation

The corresponding MILP for computing state-minimal critical subsystems for MDPs is shown in Fig. 4. We again have the decision variables $x_s \in [0,1] \subseteq \mathbb{Z}$ for $s \in S_M^{rel}$ and the probability variables $p_s \in [0,1] \subseteq \mathbb{R}$ for $s \in S_M^{rel}$. In contrast to the SMT formulation, for the MILP constraints we need a variable $a_s \in [0,1] \subseteq \mathbb{Z}$ for each state $s \in S_M^{rel}$ and each action $a \in A$. The variable a_s will carry the value 1 if the critical scheduler selects action a in state s, and 0 otherwise.

The main difference to the MILP of DTMCs is line (11f). If the current action is not selected, i.e., $a_s = 0$, the constraint is not a restriction for p_s. Otherwise, if $a_s = 1$, the constraint is equivalent to $p_s \leq \sum_{s' \in \text{succ}(s,a) \cap S_M^{rel}} P(s,a,s') \cdot p_{s'}$, which is the analogous constraint to the formulation for DTMCs.

$$\text{minimize} \quad \left(-\frac{1}{2}p_{s_I} + \sum_{s \in S_M^{rel}} x_s\right) \tag{11a}$$

$$\text{such that} \quad p_{s_I} > \lambda \tag{11b}$$

$$\forall s \in S_M^{rel} \cap T : x_s = p_s \tag{11c}$$

$$\forall s \in S_M^{rel} \setminus T : p_s \leq x_s \tag{11d}$$

$$x_s = \sum_{a \in A} a_s \tag{11e}$$

$$\forall s \in S_M^{rel} \setminus T \ \forall a \in A : p_s \leq \left(\sum_{s' \in \text{succ}_M(s,a) \cap S_M^{rel}} P(s,a,s') \cdot p_{s'}\right) + (1 - a_s) . \tag{11f}$$

Fig. 4. MILP formulation for state-minimal critical subsystems of MDPs

The redundant constraints that we have added to the SMT and MILP formulations for DTMCs in order to make the solution process more efficient can easily be transferred to MDPs. We omit them here due to space restrictions.

5 Experimental Evaluation

In order to evaluate the performance of our SMT and MILP formulations for state-minimal critical subsystems of DTMCs, we implemented a tool called SUB-SYS in C++ and applied it to two series of test cases. For all benchmarks, we used PRISM [18] models, which are available at http://prismmodelchecker.org.

(1) The *crowds protocol* [19] provides a mechanism for anonymous web browsing by routing messages through a network of N nodes. If a node wants to send a message, it has a probabilistic choice whether to deliver the message directly to its destination or to forward it to a randomly selected successor node. This procedure preserves anonymous sending of messages, as the original sender of a message cannot be determined. One instance consists of R rounds of message deliveries. In the following tables we denote the different instances by crowds$N$-R. The set T of target states contains all those states where a bad group member could identify the sender of a message.

(2) The synchronous *leader election protocol* [20] models the selection of a distinguished leader node in a ring of N identical network nodes. In each round, every node randomly selects an integer number in the range $\{0 \ldots K\}$. The node with the highest number becomes the leader, if this number is unique. Otherwise a new round starts. In the tables below, we denote the instances for different values of N and K by leader$N$-K.

All experiments were performed on a computer with four 2.3 GHz AMD Opteron Quad-Core CPUs and 64 GB memory, running Ubuntu 10.04 Linux in 64-bit mode. We aborted any experiment which did not finish within 7200 s or needed more than 4 GB of memory. A table entry "– TL –" means that the time limit was exceeded; the exceeding of the memory limit is denoted by "– ML –".

Table 1. Sizes of the benchmark models and comparison with the heuristic local search method of [10]

Model	$|S|$	$|E_M|$	$|T|$	λ	$|S_{\text{MCS}}|$	$|E_{\text{MCS}}|$	$|S_{\text{heur}}|$	$|E_{\text{heur}}|$
crowds2-3	183	243	26	0.09	22	27	23	27
crowds2-4	356	476	85	0.09	22	27	23	27
crowds2-5	612	822	196	0.09	22	27	23	27
crowds3-3	396	576	37	0.09	37	51	40	56
crowds3-4	901	1321	153	0.09	37	51	40	56
crowds3-5	1772	2612	425	0.09	37	51	40	56
crowds5-4	3515	6035	346	0.09	72	123	94	156
crowds5-6	18817	32677	3710	0.09	72	123	145	253
crowds5-8	68740	120220	19488	0.09	72	123	198	356
leader3-2	22	29	1	0.5	15	18	17	20
leader3-3	61	87	1	0.5	33	45	40	54
leader3-4	135	198	1	0.5	70	101	76	108
leader4-2	55	70	1	0.5	34	41	44	54
leader4-3	256	336	1	0.5	132	171	170	220
leader4-4	782	1037	1	0.5	395	522	459	605
leader4-5	1889	2513	1	0.5	946	1257	1050	1393
leader4-6	3902	5197	1	0.5	1953	2600	2103	2797

For solving the SMT formulae and the MILPs we used a number of state-of-the-art solvers, from which we selected, after a series of preliminary experiments, the most efficient ones, namely Z3 3.1 [21] (http://research.microsoft.com/en-us/um/redmond/projects/z3) as an SMT solver for linear real arithmetic, SCIP 2.0.2 (http://scip.zib.de) as a publicly available MILP solver, and CPLEX 12.3 (http://www-01.ibm.com/software/integration/optimization/cplex-optimizer) as a commercial MILP solver.

Table 1 contains statistics on our benchmarks. The columns contain (from left to right) the model name, the number of states, the number of transitions with non-zero probability, the number of target states, the probability bound, the number of states in the MCS, and finally the number of transitions in the MCS. The last two columns contain the sizes of heuristically computed critical subsystems. We used the local search approach of [10] to determine these subsystems. For all instances the heuristic tool terminated within 6 min.

We give the running times of Z3, SCIP, and CPLEX in Table 2. CPLEX supports a parallel mode, in which we started 16 threads in parallel. Therefore we give for CPLEX the accumulated times of all threads and, in parentheses, the actual time from start to termination of the tool. All times are given in seconds.

The block of columns entitled "w/o redundant constraints" contains the running times of the solvers without any optimizations. The block "optimal conf." lists the optimal times, i.e., the times achieved by adding the combination of optional constraints that leads to the smallest computation time. These running times can be obtained in general by using a portfolio approach which runs the different combinations of redundant constraints in parallel. Using Z3 did not

Table 2. Running times of Z3, Scip, and Cplex for computing MCSs

Model	w/o redundant constraints			optimal conf.		
	Z3	Scip	Cplex	Z3	Scip	Cplex
crowds2-3	6.80	0.16	1.33 (0.28)	4.82	0.12	0.06 (0.11)
crowds2-4	123.34	0.47	0.30 (0.24)	23.72	0.30	0.30 (0.24)
crowds2-5	293.94	0.90	0.56 (0.45)	152.28	0.60	0.56 (0.24)
crowds3-3	5616.39	0.64	0.49 (0.33)	640.61	0.35	0.38 (0.30)
crowds3-4	– TL –	4.29	5.53 (2.07)	– TL –	1.45	0.89 (0.58)
crowds3-5	– TL –	23.49	6.66 (2.77)	– TL –	5.58	1.51 (0.87)
crowds5-4	– TL –	743.84	14.23 (5.07)	– TL –	13.28	12.51 (4.89)
crowds5-6	– TL –	– TL –	302.03 (38.39)	– TL –	1947.46	100.26 (23.52)
crowds5-8	– TL –	– TL –	– TL –	– TL –	– TL –	1000.79 (145.84)
leader3-2	0.07	0.07	0.62 (0.22)	0.05	0.01	0.21 (0.13)
leader3-3	– TL –	91.89	0.43 (0.22)	– TL –	0.06	0.02 (0.06)
leader3-4	– TL –	2346.59	0.70 (0.36)	– TL –	0.40	0.07 (0.09)
leader4-2	3.57	0.23	0.45 (0.21)	1.38	0.07	0.24 (0.17)
leader4-3	– TL –	1390.79	22.33 (3.38)	– TL –	0.21	0.49 (0.37)
leader4-4	– TL –	– TL –	– TL –	– TL –	1.49	1.88 (1.21)
leader4-5	– TL –	– TL –	– TL –	– TL –	1.15	4.06 (2.80)
leader4-6	– TL –	– TL –	– ML –	– TL –	– TL –	8.70 (5.92)

lead to satisfying results although the optimizations, especially the reachability cuts, decreased the running times clearly. A significant speed-up is recognizable using the MILP solvers. The optimal running times were in some cases smaller by orders of magnitude considering in particular the large benchmarks for which the standard formulation could not be solved within the time limit.

To gain more insight into the effects of the different kinds of redundant constraints, we list more detailed results for crowds5-6 and leader4-5 in Table 3. The left block of columns contains the running times without SCC cuts. The column "–" contains the values without forward and backward cuts, the column "→" the values with forward cuts, "←" the values with backward cuts and "↔" with both forward and backward cuts. The values for the four different combinations of reachability cuts (none, only forward, only backward, and both) are listed in the according rows of the table.

Comparing the values for the reachability cuts, we can observe that they have a negative effect on the running times for crowds5-6 with MILP solvers. However, they speed up the solution of leader4-5 by a factor of more than 10^3, decreasing the solution times from more than 7200 s to less than 5 s. The same tendency can be observed for all crowds and leader instances, respectively.

The addition of backward cuts to crowds5-6 reduces the running time to about one third, and they typically decrease the times for most of the instances. Since the SCC cuts are even less effective, we only give the minimal value of the three cases (with SCC input, SCC output, and with both kinds of SCC cuts).

Table 3. Runtimes for crowds5-6 and leader4-5 with and without redundant constraints using CPLEX as MILP solver

	Reach	no SCC cuts				with SCC cuts			
		−	→	←	↔	−	→	←	↔
crowds5-6	none	302.03 (38.39)	367.49 (44.70)	**103.20** (23.52)	149.73 (26.07)	301.87 (38.64)	342.07 (42.76)	**100.26** (23.52)	138.36 (25.20)
	fwd	656.13 (120.04)	1292.59 (148.82)	651.47 (112.47)	966.57 (127.94)	634.40 (118.22)	833.37 (108.13)	646.64 (111.18)	925.95 (125.52)
	bwd	4043.93 (384.74)	3613.96 (358.49)	770.90 (121.02)	1070.50 (130.45)	3911.81 (375.48)	3603.49 (358.25)	756.28 (119.90)	1074.36 (130.72)
	both	2107.84 (251.41)	1403.44 (185.34)	5972.98 (546.83)	2191.83 (281.07)	1986.37 (238.58)	1379.78 (183.38)	5925.31 (542.18)	2210.68 (284.51)
leader4-5	none	− TL −	− TL −	− TL −	− TL −	− TL −	− TL −	− TL −	− TL −
	fwd	284.04 (40.02)	254.56 (35.97)	286.83 (40.85)	261.53 (36.46)	294.71 (41.15)	259.98 (36.21)	285.33 (40.57)	251.69 (35.80)
	bwd	6.30 (3.73)	6.29 (3.69)	6.27 (3.72)	6.10 (3.71)	5.89 (3.65)	5.73 (3.66)	5.78 (3.67)	5.95 (3.69)
	both	4.46 (2.77)	**4.06** (2.80)	4.34 (2.83)	4.56 (2.91)	**4.17** (2.77)	4.41 (2.83)	4.10 (2.84)	4.39 (2.90)

Fig. 5 shows the size of the MCS of crowds5-6 for different values of λ (red solid lines), comparing it with the size of heuristically computed critical subsystems using the local search of [10] (blue dotted lines). For $\lambda \geq 0.23$, we could only compute an upper bound (within 8 % from the optimal value) on the size of the MCS using our MILP formulation due to the timeout of 2 hours. Also the local search tool ran into a timeout for $\lambda \geq 0.35$, however, without yielding a critical subsystem.

6   Conclusion

In this paper we have shown how to compute state-minimal critical subsystems for DTMCs and MDPs using SMT- and MILP-based formulations. By adding redundant constraints, which trigger implications for SMT and tighten the LP-relaxation of the MILP, the solution process can be speeded up clearly. Thereby the MILP formulation is more efficient to solve by orders of magnitude compared to the SMT formulation. A topic for future research is to analyze the theoretical complexity of computing MCSs for DTMCs. We conjecture that this

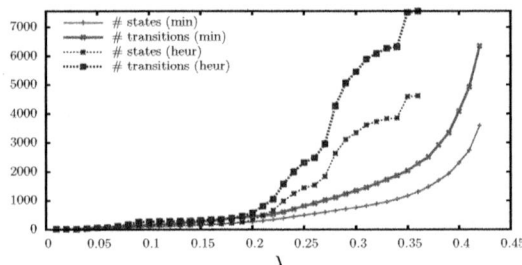

Fig. 5. Size of the MCS and heuristically determined critical subsystems for crowds5-6 and different values of λ

problem is NP-complete. Furthermore we plan to integrate the MILP approach into the hierarchical counterexample generation tool described in [10].

Acknowledgments. The authors thank the reviewers for pointing out the relevance of [4].

References

1. Bustan, D., Rubin, S., Vardi, M.Y.: Verifying ω-Regular Properties of Markov Chains. In: Alur, R., Peled, D.A. (eds.) CAV 2004. LNCS, vol. 3114, pp. 189–201. Springer, Heidelberg (2004)
2. Baier, C., Katoen, J.-P.: Principles of Model Checking. The MIT Press (2008)
3. Hermanns, H., Wachter, B., Zhang, L.: Probabilistic CEGAR. In: Gupta, A., Malik, S. (eds.) CAV 2008. LNCS, vol. 5123, pp. 162–175. Springer, Heidelberg (2008)
4. Chadha, R., Viswanathan, M.: A counterexample-guided abstraction-refinement framework for Markov decision processes. ACM TOCL 12(1), 1–45 (2010)
5. Aljazzar, H., Leue, S.: Directed explicit state-space search in the generation of counterexamples for stochastic model checking. IEEE Trans. on Software Engineering 36(1), 37–60 (2010)
6. Han, T., Katoen, J.-P., Damman, B.: Counterexample generation in probabilistic model checking. IEEE Trans. on Software Engineering 35(2), 241–257 (2009)
7. Wimmer, R., Braitling, B., Becker, B.: Counterexample Generation for Discrete-Time Markov Chains Using Bounded Model Checking. In: Jones, N.D., Müller-Olm, M. (eds.) VMCAI 2009. LNCS, vol. 5403, pp. 366–380. Springer, Heidelberg (2009)
8. Andrés, M.E., D'Argenio, P., van Rossum, P.: Significant Diagnostic Counterexamples in Probabilistic Model Checking. In: Chockler, H., Hu, A.J. (eds.) HVC 2008. LNCS, vol. 5394, pp. 129–148. Springer, Heidelberg (2009)
9. Günther, M., Schuster, J., Siegle, M.: Symbolic calculation of k-shortest paths and related measures with the stochastic process algebra tool CASPA. In: Proc. of DYADEM-FTS, pp. 13–18. ACM Press (2010)
10. Jansen, N., Ábrahám, E., Katelaan, J., Wimmer, R., Katoen, J.-P., Becker, B.: Hierarchical Counterexamples for Discrete-Time Markov Chains. In: Bultan, T., Hsiung, P.-A. (eds.) ATVA 2011. LNCS, vol. 6996, pp. 443–452. Springer, Heidelberg (2011)
11. Kattenbelt, M., Huth, M.: Verification and refutation of probabilistic specifications via games. In: Proc. of FSTTCS. LIPIcs, vol. 4, pp. 251–262. Schloss Dagstuhl – Leibniz-Zentrum für Informatik (2009)
12. Schmalz, M., Varacca, D., Völzer, H.: Counterexamples in Probabilistic LTL Model Checking for Markov Chains. In: Bravetti, M., Zavattaro, G. (eds.) CONCUR 2009. LNCS, vol. 5710, pp. 587–602. Springer, Heidelberg (2009)
13. Fecher, H., Huth, M., Piterman, N., Wagner, D.: PCTL model checking of Markov chains: Truth and falsity as winning strategies in games. Performance Evaluation 67(9), 858–872 (2010)
14. Wimmer, R., Becker, B., Jansen, N., Ábrahám, E., Katoen, J.-P.: Minimal critical subsystems as counterexamples for ω-regular DTMC properties. In: Brandt, J., Schneider, K. (eds.) Proc. of MBMV. Kovač-Verlag (2012)
15. de Moura, L.M., Bjørner, N.: Satisfiability modulo theories: introduction and applications. Communication of the ACM 54(9), 69–77 (2011)

16. Dutertre, B., de Moura, L.M.: A Fast Linear-Arithmetic Solver for DPLL(T). In: Ball, T., Jones, R.B. (eds.) CAV 2006. LNCS, vol. 4144, pp. 81–94. Springer, Heidelberg (2006)
17. Schrijver, A.: Theory of Linear and Integer Programming. Wiley (1986)
18. Kwiatkowska, M.Z., Norman, G., Parker, D.: PRISM 4.0: Verification of Probabilistic Real-Time Systems. In: Gopalakrishnan, G., Qadeer, S. (eds.) CAV 2011. LNCS, vol. 6806, pp. 585–591. Springer, Heidelberg (2011)
19. Reiter, M.K., Rubin, A.D.: Crowds: Anonymity for web transactions. ACM Trans. on Information and System Security 1(1), 66–92 (1998)
20. Itai, A., Rodeh, M.: Symmetry breaking in distributed networks. Information and Computation 88(1), 60–87 (1990)
21. de Moura, L.M., Bjørner, N.: Z3: An Efficient SMT Solver. In: Ramakrishnan, C.R., Rehof, J. (eds.) TACAS 2008. LNCS, vol. 4963, pp. 337–340. Springer, Heidelberg (2008)

Automatic Verification
of Competitive Stochastic Systems

Taolue Chen, Vojtěch Forejt, Marta Kwiatkowska,
David Parker, and Aistis Simaitis

Department of Computer Science, University of Oxford, Oxford, UK

Abstract. We present automatic verification techniques for the modelling and analysis of probabilistic systems that incorporate competitive behaviour. These systems are modelled as turn-based stochastic multi-player games, in which the players can either collaborate or compete in order to achieve a particular goal. We define a temporal logic called rPATL for expressing quantitative properties of stochastic multi-player games. This logic allows us to reason about the collective ability of a set of players to achieve a goal relating to the probability of an event's occurrence or the expected amount of cost/reward accumulated. We give a model checking algorithm for verifying properties expressed in this logic and implement the techniques in a probabilistic model checker, based on the PRISM tool. We demonstrate the applicability and efficiency of our methods by deploying them to analyse and detect potential weaknesses in a variety of large case studies, including algorithms for energy management and collective decision making for autonomous systems.

1 Introduction

Automatic verification techniques for probabilistic systems have been successfully applied in a variety of fields, from wireless communication protocols to dynamic power management schemes to quantum cryptography. These systems are inherently stochastic, e.g. due to unreliable communication media, faulty components or the use of randomisation. Automatic techniques such as *probabilistic model checking* provide a means to model and analyse these systems against a range of quantitative properties. In particular, when systems also exhibit *non-deterministic* behaviour, e.g. due to concurrency, underspecification or control, the subtle interplay between the probabilistic and nondeterministic aspects of the system often makes a manual analysis difficult and error-prone.

When modelling *open* systems, the designer also has to account for the behaviour of components it does not control, and which could have differing or opposing goals, giving rise to *competitive* behaviour. This occurs in many cases, such as security protocols and algorithms for distributed consensus, energy management or sensor network co-ordination. In such situations, it is natural to adopt a *game-theoretic* view, modelling a system as a game between different players. Automatic verification has been successfully deployed in this context, e.g. in the analysis of security [21] or communication protocols [20].

C. Flanagan and B. König (Eds.): TACAS 2012, LNCS 7214, pp. 315–330, 2012.

In this paper, we present an extensive framework for modelling and automatic verification of systems with both probabilistic *and* competitive behaviour, using *stochastic multi-player games* (SMGs). We introduce a *temporal logic rPATL* for expressing quantitative properties of this model and develop *model checking algorithms* for it. We then build a probabilistic model checker, based on the PRISM tool [22], which provides a high-level language for modelling SMGs and implements rPATL model checking for their analysis. Finally, to illustrate the applicability of our framework, we develop several large case studies in which we identify potential weaknesses and unexpected behaviour that would have been difficult to find with existing probabilistic verification techniques.

We model competitive stochastic systems as *turn-based* SMGs, where, in each state of the model, one player chooses between several actions, the outcome of which can be probabilistic. Turn-based games are a natural way to model many real-life applications. One example is when modelling several components executing concurrently under the control of a particular (e.g. round-robin, randomised) scheduler; in this case, nondeterminism in the model arises due to the choices made by each individual component. Another example is when we choose to explicitly model the (possibly unknown) scheduling of components as one player and the choices of components as other players.

In order to specify properties of the systems modelled, we formulate a temporal logic, rPATL. This is an extension of the logic PATL [14], which is itself a probabilistic extension of ATL [5] – a widely used logic for reasoning about multi-player games and multi-agent systems. rPATL allows us to state that a *coalition* of players has a *strategy* which can ensure that either the *probability* of an event's occurrence or an *expected reward* measure meets some threshold, e.g. "can processes 1 and 2 collaborate so that the probability of the protocol terminating within 45 seconds is at least 0.95, whatever processes 3 and 4 do?"

We place particular emphasis on *reward* (or, equivalently, *cost*) related measures. This allows us to reason quantitatively about a system's use of resources, such as time spent or energy consumed; or, we can use rewards as an algorithm design mechanism to validate, benchmark or synthesise strategies for components by rewarding or penalising them for certain behaviour. rPATL can state, for example, "can sensor 1 ensure that the expected energy used, *if the algorithm terminates*, is less than $75mJ$, for any actions of sensors 2, 3, and 4?". To the best of our knowledge, this is the first logic able to express such properties.

We include in rPATL three different *cumulative expected reward* operators. Cumulative properties naturally capture many useful system properties, as has been demonstrated for verification of other types of probabilistic models [17], and as proves to be true for the systems we investigate. Indicative examples from our case studies are "the maximum expected execution cost of a task in a Microgrid" and "the minimum expected number of messages required to reach a consensus". Several other reward-based objectives exist that we do not consider, including discounted rewards (useful e.g. in economics, but less so for the kind of systems we target) and long-run average reward (also useful, but practical implementations become complex in stochastic games [16]).

We also devise model checking algorithms for rPATL. A practical advantage of the logic is that, like for ATL, model checking reduces to analysing zero-sum two-player games. rPATL properties referring to the probability of an event are checked by solving simple stochastic two-player games, for which efficient techniques exist [15,16]. For reward-based properties, we present new algorithms.

Lastly, we develop and analyse several large case studies. We study algorithms for smart energy management [19] and distributed consensus in a sensor network [26]. In the first case, we use our techniques to reveal a weakness in the algorithm: we show that users may have a high incentive to deviate from the original algorithm, and propose modifications to solve the problem. For the consensus algorithm, we identify unexpected trade-offs in the performance of the algorithm when using our techniques to evaluate possible strategies for sensors.

Contributions. In summary, the contributions of this paper are:

- A comprehensive *framework* for analysis of competitive stochastic systems;
- A *logic* rPATL for specifying quantitative properties of stochastic multi-player games including, in particular, novel operators for *costs and rewards*, and their *model checking algorithms*;
- Implementation of a *tool* for modelling and rPATL model checking of SMGs;
- Development and analysis of several large new *case studies*.

An extended version of this paper, with proofs, is available as [12].

Related Work. There exist theoretical results on probabilistic temporal logics for a game-theoretic setting but, to our knowledge, this is the first work to consider a practical implementation, modelling and automated verification of case studies. [14] introduces the logic PATL, showing its model checking complexity via probabilistic parity games. [28] studies simulation relations preserved by PATL and [1] uses it in a theoretical framework for security protocol analysis. [6] presents (un)decidability results for another richer logic, with emphasis on the subtleties of nested properties. We note that all of the above, except [6], use concurrent, rather than turn-based, games and none consider reward properties.

Probabilistic model checking for a multi-agent system (a negotiation protocol) is considered in [8], but this is done by fixing a particular probabilistic strategy and analysing a Markov chain rather than a stochastic game. [13] describes analysis of a team formation protocol, which involves simple properties on stochastic two-player games. There has been much research on algorithms to solve stochastic games, e.g. [16,11,27], but these do not consider a modelling framework, implementation or case studies. Moreover, the reward-based properties that we introduce in this paper have not been studied in depth. In [25], a quantitative generalisation of the μ-calculus is proposed, and shown to be able to encode stochastic parity games. We also mention the tools MCMAS [23] and MOCHA [4], powerful model checkers for *non-probabilistic* multi-agent systems.

Finally, stochastic games are useful for *synthesis*, as in e.g. [9], which synthesises concurrent programs for randomised schedulers. Also, the tool GIST [10] is a stochastic game solver, but is targeted at synthesis problems, not modelling and verification of competitive systems, and only supports *qualitative* properties.

2 Preliminaries

We begin with some background on stochastic multi-player games. For a finite set X, we denote by $\mathcal{D}(X)$ the set of discrete probability distributions over X.

Definition 1 (SMG). *A (turn-based) stochastic multi-player game (SMG) is a tuple $\mathcal{G} = \langle \Pi, S, A, (S_i)_{i \in \Pi}, \Delta, AP, \chi \rangle$, where: Π is a finite set of players; S is a finite, non-empty set of states; A is a finite, non-empty set of actions; $(S_i)_{i \in \Pi}$ is a partition of S; $\Delta : S \times A \to \mathcal{D}(S)$ is a (partial) transition function; AP is a finite set of atomic propositions; and $\chi : S \to 2^{AP}$ is a labelling function.*

In each state $s \in S$ of the SMG $\mathcal{G}$, the set of *available* actions is denoted by $A(s) \overset{\text{def}}{=} \{a \in A \mid \Delta(s,a) \neq \bot\}$. We assume that $A(s) \neq \emptyset$ for all s. The choice of action to take in s is under the control of exactly one player, namely the player $i \in \Pi$ for which $s \in S_i$. Once action $a \in A(s)$ is selected, the successor state is chosen according to the probability distribution $\Delta(s,a)$. A *path* of $\mathcal{G}$ is a possibly infinite sequence $\lambda = s_0 a_0 s_1 a_1 \ldots$ such that $a_j \in A(s_j)$ and $\Delta(s_j, a_j)(s_{j+1}) > 0$ for all j. We use st_λ to denote $s_0 s_1 \ldots$, and $st_\lambda(j)$ for s_j. The set of all infinite paths is $\Omega_\mathcal{G}$ and the set of infinite paths starting in state s is $\Omega_{\mathcal{G},s}$.

A *strategy* for player $i \in \Pi$ in $\mathcal{G}$ is a function $\sigma_i : (SA)^* S_i \to \mathcal{D}(A)$ which, for each path $\lambda \cdot s$ where $s \in S_i$, assigns a probability distribution $\sigma_i(\lambda \cdot s)$ over $A(s)$. The set of all strategies for player i is denoted Σ_i. A strategy σ_i is called *memoryless* if $\forall \lambda, \lambda' : \sigma_i(\lambda \cdot s) = \sigma_i(\lambda' \cdot s)$, and *deterministic* if $\forall \lambda : \sigma_i(\lambda \cdot s)$ is a Dirac distribution. A *strategy profile* $\sigma = \sigma_1, \ldots, \sigma_{|\Pi|}$ comprises a strategy for all players in the game. Under a strategy profile σ, the behaviour of $\mathcal{G}$ is fully probabilistic and we define a *probability measure* $\mathrm{Pr}^\sigma_{\mathcal{G},s}$ over the set of all paths $\Omega_{\mathcal{G},s}$ in standard fashion (see, e.g. [11]). Given a random variable $X : \Omega_{\mathcal{G},s} \to \mathbb{R}$, we define the *expected value* of X to be $\mathbb{E}^\sigma_{\mathcal{G},s}[X] \overset{\text{def}}{=} \int_{\Omega_{\mathcal{G},s}} X \, d\mathrm{Pr}^\sigma_{\mathcal{G},s}$.

We also augment games with *reward structures* $r : S \to \mathbb{Q}_{\geq 0}$, mapping each state to a non-negative rational reward. To simplify presentation, we only use state rewards, but note that transition/action rewards can easily be encoded by adding an auxiliary state per transition/action to the model.

3 Property Specification: The Logic rPATL

We now present a temporal logic called rPATL (Probabilistic Alternating-time Temporal Logic with Rewards) for expressing quantitative properties of SMGs. Throughout the section, we assume a fixed SMG $\mathcal{G} = \langle \Pi, S, A, (S_i)_{i \in \Pi}, \Delta, AP, \chi \rangle$.

Definition 2 (rPATL). *The syntax of rPATL is given by the grammar:*

$$\phi ::= \top \mid a \mid \neg \phi \mid \phi \wedge \phi \mid \langle\!\langle C \rangle\!\rangle \mathsf{P}_{\bowtie q}[\psi] \mid \langle\!\langle C \rangle\!\rangle \mathsf{R}^r_{\bowtie x}[\mathsf{F}^\star \phi]$$
$$\psi ::= \mathsf{X} \phi \mid \phi \mathsf{U}^{\leq k} \phi \mid \phi \mathsf{U} \phi$$

where $a \in AP$, $C \subseteq \Pi$, $\bowtie \in \{<, \leq, \geq, >\}$, $q \in \mathbb{Q} \cap [0,1]$, $x \in \mathbb{Q}_{\geq 0}$, $\star \in \{0, \infty, c\}$, r is a reward structure and $k \in \mathbb{N}$.

rPATL is a CTL-style branching-time temporal logic, where we distinguish state formulae (ϕ) and path formulae (ψ). We adopt the coalition operator $\langle\!\langle C \rangle\!\rangle$ of ATL [5], combining it with the probabilistic operator $\mathsf{P}_{\bowtie q}[\cdot]$ from PCTL [18] and a generalised variant of the reward operator $\mathsf{R}^r_{\bowtie x}[\cdot]$ from [17].

An example of typical usage of the coalition operator is $\langle\!\langle\{1,2\}\rangle\!\rangle\mathsf{P}_{\geq 0.5}[\psi]$, which means "players 1 and 2 have a strategy to ensure that the probability of path formula ψ being satisfied is at least 0.5, regardless of the strategies of other players". As path formulae, we allow the standard temporal operators X ("next"), bounded $\mathsf{U}^{\leq k}$ ("bounded until") and U ("until").

Rewards. Before presenting the semantics of rPATL, we discuss the reward operators in the logic. We focus on *expected cumulative reward*, i.e. the expected sum of rewards cumulated along a path until a state from a specified target set $T \subseteq S$ is reached. To cope with the variety of different properties encountered in practice, we introduce three variants, which differ in the way they handle the case where T is *not* reached. The three types are denoted by the parameter $\star$, one of 0, ∞ or c. These indicate that, when T is not reached, the reward is zero, infinite or equal to the cumulated reward along the whole path, respectively.

Each reward type is applicable in different situations. If our goal is, for example, to minimise the expected time for algorithm completion, then it is natural to assume a value of infinity upon non-completion ($\star=\infty$). Consider, on the other hand, the case where we try to optimise a distributed algorithm by designing a reward structure that incentivises certain kinds of behaviour and then maximising it over the lifetime of the algorithm's execution. In this case, we might opt for type $\star=0$ to avoid favouring situations where the algorithm does not terminate. In other cases, e.g. when modelling algorithm's resource consumption, we might prefer to use type $\star=c$, to compute resources used regardless of termination.

We formalise these notions of rewards by defining *reward functions* that map each possible path in the game $\mathcal{G}$ to a cumulative reward value.

Definition 3 (Reward Function). *For an SMG $\mathcal{G}$, a reward structure r, type $\star \in \{0, \infty, c\}$ and a set $T \subseteq S$ of target states, the* reward function $rew(r, \star, T) : \Omega_{\mathcal{G}} \to \mathbb{R}$ *is a random variable defined as follows.*

$$rew(r, \star, T)(\lambda) \stackrel{\text{def}}{=} \begin{cases} g(\star) & \text{if } \forall j \in \mathbb{N} : st_\lambda(j) \notin T, \\ \sum_{j=0}^{k-1} r(st_\lambda(j)) & \text{otherwise, where } k = \min\{j \mid st_\lambda(j) \in T\}, \end{cases}$$

and where $g(\star) = \star$ *if* $\star \in \{0, \infty\}$ *and* $g(\star) = \sum_{j \in \mathbb{N}} r(st_\lambda(j))$ *if* $\star = c$. *The expected reward from a state $s \in S$ of $\mathcal{G}$ under a strategy profile σ is the expected value of the reward function,* $\mathbb{E}^\sigma_{\mathcal{G},s}[rew(r, \star, T)]$.

Semantics. Now, we define the semantics of rPATL. Formulae are interpreted over states of a game $\mathcal{G}$; we write $s \models \phi$ to indicate that the state s of $\mathcal{G}$ satisfies the formula ϕ and define $Sat(\phi) \stackrel{\text{def}}{=} \{s \in S \mid s \models \phi\}$ as the states satisfying ϕ. The meaning of atomic propositions and logical connectives is standard. For the $\langle\!\langle C \rangle\!\rangle\mathsf{P}_{\bowtie q}$ and $\langle\!\langle C \rangle\!\rangle\mathsf{R}^r_{\bowtie x}$ operators, we give the semantics via a reduction to a two-player game called a *coalition game*.

Definition 4 (Coalition Game). *For a* coalition *of players* $C \subseteq \Pi$ *of SMG* $\mathcal{G}$, *we define the* coalition game *of* $\mathcal{G}$ *induced by* C *as the stochastic two-player game* $\mathcal{G}_C = \langle \{1,2\}, S, A, (S_1', S_2'), \Delta, AP, \chi \rangle$ *where* $S_1' = \cup_{i \in C} S_i$ *and* $S_2' = \cup_{i \in \Pi \setminus C} S_i$.

Definition 5 (rPATL Semantics). *The* satisfaction relation $\models$ *for rPATL is defined inductively for every state* s *of* $\mathcal{G}$. *The semantics of* $\top$, *atomic propositions and formulae of the form* $\neg \phi$ *and* $\phi_1 \wedge \phi_2$ *is defined in the usual way. For the temporal operators, we define:*

$$s \models \langle\!\langle C \rangle\!\rangle \mathsf{P}_{\bowtie q}[\psi] \quad \Leftrightarrow \quad \textit{In coalition game } \mathcal{G}_C, \exists \sigma_1 \in \Sigma_1 \textit{ such that } \forall \sigma_2 \in \Sigma_2 \\ \mathrm{Pr}^{\sigma_1,\sigma_2}_{\mathcal{G}_C,s}(\psi) \bowtie q$$

$$s \models \langle\!\langle C \rangle\!\rangle \mathsf{R}^r_{\bowtie x}[\mathsf{F}^\star \phi] \Leftrightarrow \textit{In coalition game } \mathcal{G}_C, \exists \sigma_1 \in \Sigma_1 \textit{ such that } \forall \sigma_2 \in \Sigma_2 \\ \mathbb{E}^{\sigma_1,\sigma_2}_{\mathcal{G}_C,s}[rew(r,\star,Sat(\phi))] \bowtie x$$

where $\mathrm{Pr}^{\sigma_1,\sigma_2}_{\mathcal{G}_C,s}(\psi) \stackrel{\text{def}}{=} \mathrm{Pr}^{\sigma_1,\sigma_2}_{\mathcal{G}_C,s}(\{\lambda \in \Omega_{\mathcal{G}_C,s} \mid \lambda \models \psi\})$ *and for any path* λ *in* $\mathcal{G}$:

$$\lambda \models \mathsf{X}\phi \qquad \Leftrightarrow st_\lambda(1) \models \phi$$
$$\lambda \models \phi_1 \mathsf{U}^{\leq k} \phi_2 \Leftrightarrow st_\lambda(i) \models \phi_2 \textit{ for some } i \leq k \textit{ and } st_\lambda(j) \models \phi_1 \textit{ for } 0 \leq j < i$$
$$\lambda \models \phi_1 \mathsf{U} \phi_2 \qquad \Leftrightarrow \lambda \models \phi_1 \mathsf{U}^{\leq k} \phi_2 \textit{ for some } k \in \mathbb{N}.$$

Equivalences and Extensions. We can handle "negated path formulae" in a $\langle\!\langle C \rangle\!\rangle \mathsf{P}_{\bowtie q}$ operator by inverting the probability threshold, e.g.:

$$\langle\!\langle C \rangle\!\rangle \mathsf{P}_{\geq q}[\neg \psi] \equiv \langle\!\langle C \rangle\!\rangle \mathsf{P}_{\leq 1-q}[\psi].$$

This allows us to derive, for example, the G ("globally") and R ("release") operators. Also, from the determinacy result of [24] for zero-sum stochastic two-player games with Borel measurable payoffs, it follows that, e.g.:

$$\langle\!\langle C \rangle\!\rangle \mathsf{P}_{\geq q}[\psi] \equiv \neg \langle\!\langle \Pi \setminus C \rangle\!\rangle \mathsf{P}_{< q}[\psi]. \tag{1}$$

Finally, it is useful to consider "quantitative" versions of the $\langle\!\langle C \rangle\!\rangle \mathsf{P}$ and $\langle\!\langle C \rangle\!\rangle \mathsf{R}$ operators, in the style of PRISM [22], which return numerical values:

$$\langle\!\langle C \rangle\!\rangle \mathsf{P}_{\max=?}[\psi] \stackrel{\text{def}}{=} \mathrm{Pr}^{\max,\min}_{\mathcal{G}_C,s}(\psi) \stackrel{\text{def}}{=} \sup_{\sigma_1 \in \Sigma_1} \inf_{\sigma_2 \in \Sigma_2} \mathrm{Pr}^{\sigma_1,\sigma_2}_{\mathcal{G}_C,s}(\psi)$$

$$\langle\!\langle C \rangle\!\rangle \mathsf{R}^r_{\max=?}[\mathsf{F}^\star \phi] \stackrel{\text{def}}{=} \mathbb{E}^{\max,\min}_{\mathcal{G}_C,s}[rew(r,\star,Sat(\phi))] \tag{2}$$
$$\stackrel{\text{def}}{=} \sup_{\sigma_1 \in \Sigma_1} \inf_{\sigma_2 \in \Sigma_2} \mathbb{E}^{\sigma_1,\sigma_2}_{\mathcal{G}_C,s}[rew(r,\star,Sat(\phi))].$$

Example 1. Consider the SMG in Fig. 1. with $\Pi = \{1,2,3\}$. The player i controlling a state s is shown as $i{:}s$ in the figure, e.g. $S_1 = \{s_0, s_3\}$. We have actions $A = \{a, b\}$ and e.g. $\Delta(s_0, a)(s_1) = 0.7$. State s_3 is labelled with atomic proposition t. Consider

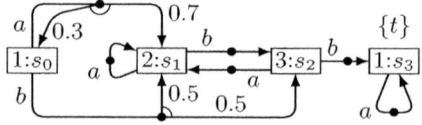

Fig. 1. Example SMG

the rPATL formulae $\langle\!\langle \{1,3\} \rangle\!\rangle \mathsf{P}_{\geq 0.5}[\mathsf{F}\, t]$ and $\langle\!\langle \{1,2\} \rangle\!\rangle \mathsf{P}_{\geq 0.5}[\mathsf{F}\, t]$. The first is satisfied in states $\{s_0, s_2, s_3\}$, the latter in s_3 only. Let r be a reward structure that assigns i to state s_i. rPATL formula $\langle\!\langle \{1,3\} \rangle\!\rangle \mathsf{R}^r_{\leq 2}[\mathsf{F}^\infty t]$ is true in states $\{s_2, s_3\}$. Formula $\langle\!\langle \{1\} \rangle\!\rangle \mathsf{R}^r_{\geq q}[\mathsf{F}^0 t]$ is false in all states for any $q > 0$ but $\langle\!\langle \{3\} \rangle\!\rangle \mathsf{R}^r_{\geq q}[\mathsf{F}^c t]$ is true in $\{s_0, s_1, s_2\}$ for any $q > 0$.

4   Model Checking for rPATL

We now discuss model checking for rPATL, the key part of which is computation of probabilities and expected rewards for stochastic two-player games. The complexity of the rPATL model checking problem can be stated as follows.

Theorem 1. (a) *Model checking an rPATL formula with no $\langle\!\langle C \rangle\!\rangle R^r_{\bowtie x}[F^0\phi]$ operator and where k for $U^{\le k}$ is given in unary is in* $\mathrm{NP} \cap \mathrm{coNP}$.
(b) *Model checking an unrestricted rPATL formula is in* $\mathrm{NEXP} \cap \mathrm{coNEXP}$.

Nevertheless, we present efficient and practically usable algorithms for model checking rPATL, in which computation of numerical values is done by evaluating fixpoints (up to a desired level of convergence[1]).

The basic algorithm for model checking an rPATL formula ϕ on an SMG $\mathcal{G}$ proceeds as for other branching-time logics, determining the set $Sat(\phi)$ recursively. Furthermore, as can be seen from the semantics, computing this set for atomic propositions or logical connectives is trivial. Thus, we only consider the $\langle\!\langle C \rangle\!\rangle P_{\bowtie q}$ and $\langle\!\langle C \rangle\!\rangle R^r_{\bowtie x}$ operators. Like for the logic PCTL, model checking of these reduces to computation of *optimal* probabilities or expected rewards, respectively, on the coalition game $\mathcal{G}_C$. For example, if $\triangleright \in \{\ge, >\}$, then:

$$s \models \langle\!\langle C \rangle\!\rangle P_{\triangleright q}[\psi] \;\Leftrightarrow\; \mathrm{Pr}^{\max,\min}_{\mathcal{G}_C,s}(\psi) \triangleright q$$

$$s \models \langle\!\langle C \rangle\!\rangle R^r_{\triangleright x}[F^\star \phi] \;\Leftrightarrow\; \mathbb{E}^{\max,\min}_{\mathcal{G}_C,s}[rew(r, \star, Sat(\phi))] \triangleright x.$$

Analogously, for operators $\le$ and $<$, we simply swap min and max in the above. The following sections describe how to compute these values.

4.1   Computing Probabilities

Below, we show how to compute the probabilities $\mathrm{Pr}^{\max,\min}_{\mathcal{G}_C,s}(\psi)$ where ψ is each of the temporal operators X, $U^{\le k}$ and U. We omit the dual case since, thanks to determinacy (see equation (1)), we have that $\mathrm{Pr}^{\min,\max}_{\mathcal{G}_C,s}(\psi) = \mathrm{Pr}^{\max,\min}_{\mathcal{G}_{\Pi \setminus C},s}(\psi)$. The following results follow in near identical fashion to the corresponding results for Markov decision processes [7]. We let opt^s denote max if $s \in S_1$ and min if $s \in S_2$. For the X operator and state $s \in S$:

$$\mathrm{Pr}^{\max,\min}_{\mathcal{G}_C,s}(X\,\phi) = opt^s_{a \in A(s)} \sum_{s' \in Sat(\phi)} \Delta(s, a)(s').$$

Probabilities for the $U^{\le k}$ operator can be computed recursively. We have that $\mathrm{Pr}^{\max,\min}_{\mathcal{G}_C,s}(\phi_1 U^{\le k} \phi_2)$ is equal to: 1 if $s \in Sat(\phi_2)$; 0 if $s \notin (Sat(\phi_1) \cup Sat(\phi_2))$; 0 if $k=0$ and $s \in Sat(\phi_1) \setminus Sat(\phi_2)$; and otherwise:

$$\mathrm{Pr}^{\max,\min}_{\mathcal{G}_C,s}(\phi_1 U^{\le k} \phi_2) = opt^s_{a \in A(s)} \sum_{s' \in S} \Delta(s, a)(s') \cdot \mathrm{Pr}^{\max,\min}_{\mathcal{G}_C,s'}(\phi_1 U^{\le k-1} \phi_2).$$

The unbounded case can be computed via *value iteration* [15], i.e. using:

$$\mathrm{Pr}^{\max,\min}_{\mathcal{G}_C,s}(\phi_1 U \phi_2) = \lim_{k \to \infty} \mathrm{Pr}^{\max,\min}_{\mathcal{G}_C,s}(\phi_1 U^{\le k} \phi_2).$$

---

[1] This is the usual approach taken in probabilistic verification tools.

In practice, this computation is terminated with a suitable convergence check (see, e.g. [17]). In addition, we mention that for the case $F\phi \equiv \top \cup \phi$, the computation can also be reduced to quadratic programming [16].

4.2 Computing Rewards

Now, we show how to compute the optimal values $\mathbb{E}_{\mathcal{G}_C,s}^{\max,\min}[rew(r, \star, Sat(\phi))]$ for different types of $\star$. As above, we omit the dual case where max and min are swapped. In this section, we fix a coalition game $\mathcal{G}_C$, a reward structure r, and a target set $T = Sat(\phi)$. We first make the following modifications to $\mathcal{G}_C$:

- labels are added to target and positive reward states: $AP := AP \cup \{t, a_{\text{rew}}\}$, $\forall s \in T : \chi(s) := \chi(s) \cup \{t\}$ and $\forall s \in S . r(s) > 0 : \chi(s) := \chi(s) \cup \{a_{\text{rew}}\}$;
- target states are made absorbing: $\forall s \in T : A(s) := \{a\}, \Delta(s,a)(s)=1, r(s)=0$.

Our algorithms, like the ones for similar properties on simpler models [7], rely on computing fixpoints of certain sets of equations. As in the previous section, we assume that this is done by *value iteration* with an appropriate convergence criterion. We again let opt^s denote max if $s \in S_1$ and min if $s \in S_2$.

An important observation here is that optimal expected rewards for $\star \in \{\infty, c\}$ can be achieved by memoryless, deterministic strategies. For $\star = 0$, however, finite-memory strategies are needed. See [12] for details.

The case $\star = c$. First, we use the results of [3] to identify the states from which the expected reward is infinite:

$$I := \{s \in S \mid \exists \sigma_1 \in \Sigma_1 \ \forall \sigma_2 \in \Sigma_2 \ \Pr_{\mathcal{G}_C,s}^{\sigma_1,\sigma_2}(inf(a_{rew})) > 0\}$$

where $inf(a_{rew})$ is the set of all paths that visit a state satisfying a_{rew} infinitely often. We remove the states of I from $\mathcal{G}_C$. For the other states, we compute the *least* fixpoint of the following equations:

$$f(s) = \begin{cases} 0 & \text{if } s \in T \\ r(s) + \text{opt}^s_{a \in A(s)} \sum_{s' \in S} \Delta(s,a)(s') \cdot f(s') & \text{otherwise} \end{cases} \tag{3}$$

The case $\star = \infty$. Again, we start by identifying and removing states with infinite expected reward; in this case: $I := \{s \in S \mid s \models \langle\!\langle\{1\}\rangle\!\rangle P_{<1}[F\,t]\}$. Then, for all other states s, we compute the *greatest* fixpoint, *over* $\mathbb{R}$, of equations (3). The need for the greatest fixpoint arises because, in the presence of zero-reward cycles, multiple fixpoints may exist. The computation is over $\mathbb{R}$ since, e.g. the function mapping all non-target states to ∞ may also be a fixpoint. To find the greatest fixpoint over $\mathbb{R}$, we first compute an over-approximation by changing all zero rewards to any $\varepsilon > 0$ and then evaluating the *least* fixpoint of (3) for the *modified* reward. Starting from the *new* initial values, value iteration now converges from above to the correct fixpoint [12]. For the simpler case of MDPs, an alternative approach based on removal of zero-reward end-components is possible [2], but this cannot be adapted efficiently to stochastic games.

The case $\star = 0$. As mentioned above, it does not suffice to consider memoryless strategies in this case. The optimal strategy may depend on the reward accumulated so far, $r(\lambda) \overset{\text{def}}{=} \sum_{s \in st_\lambda} r(s)$ for history λ. However, this is only needed until a certain reward bound B is reached, after which the optimal strategy picks actions that maximise the probability of reaching T (if multiple such actions exist, it picks the one with the highest expected reward). The bound B can be computed efficiently using algorithms for $\star = c$ and $\Pr^{\max,\min}_{\mathcal{G}_C,s}(\psi)$ and, in the worst case, can be exponential in the size of $\mathcal{G}$ (see [12]).

For clarity, we assume that rewards are integers. Let $R_{(s,k)}$ be the maximum expectation of $rew(r,0,T)$ in state s after history λ with $r(\lambda) = k$:

$$R_{(s,k)} \overset{\text{def}}{=} \max_{\sigma_1 \in \Sigma_1} \min_{\sigma_2 \in \Sigma_2} k \cdot \Pr^{\sigma_1,\sigma_2}_{\mathcal{G}_C,s}(\mathsf{F}\, t) + \mathbb{E}^{\sigma_1,\sigma_2}_{\mathcal{G}_C,s}[rew(r,0,T)],$$

and $r_{\max} = \max_{s \in S} r(s)$. The algorithm works as follows:

1. Using the results of [3], identify the states that have infinite reward:

$$I := \{s \in S \mid \exists \sigma_1 \in \Sigma_1 \; \forall \sigma_2 \in \Sigma_2 \; \Pr^{\sigma_1,\sigma_2}_{\mathcal{G}_C,s}(inf^t(a_{rew})) > 0\}\}$$

where $inf^t(a_{rew})$ is the set of all paths that visit a state satisfying $\mathsf{P}_{>0}[\mathsf{F}\, t]) \wedge a_{rew}$ infinitely often. Remove all states of I from the game.

2. For $B \leq k \leq B + r_{\max} - 1$ and for each state s:
   (a) Assign new reward $r'(s) = r(s) \cdot \Pr^{\max,\min}_{\mathcal{G}_C,s}(\mathsf{F}\, t)$;
   (b) Remove from $A(s)$ actions a that are sub-optimal for $\Pr^{\max,\min}_{\mathcal{G}_C,s}(\mathsf{F}\, t)$, i.e.:

$$\sum_{s' \in S} \Delta(s,a)(s') \cdot \Pr^{\max,\min}_{\mathcal{G}_C,s'}(\mathsf{F}\, t) < \Pr^{\max,\min}_{\mathcal{G}_C,s}(\mathsf{F}\, t)$$

   (c) Compute $R_{(s,k)}$ using the algorithm for $rew(r',c,T)$:

$$R_{(s,k)} = k \cdot \Pr^{\max,\min}_{\mathcal{G}_C,s}(\mathsf{F}\, t) + \mathbb{E}^{\max,\min}_{\mathcal{G}_C,s}[rew(r',c,T)].$$

3. Find, for all $0 \leq k < B$ and states s, the *least* fixpoint of the equations:

$$R_{(s,k)} = \begin{cases} k & \text{if } s \in T \\ \mathrm{opt}^s_{a \in A(s)} \sum_{s' \in S} \Delta(s,a)(s') \cdot R_{(s',k+r(s))} & \text{otherwise.} \end{cases}$$

4. The required values are then $\mathbb{E}^{\max,\min}_{\mathcal{G}_C,s}[rew(r,0,T)] = R_{(s,0)}$.

5   Implementation and Case Studies

Based on the techniques in this paper, we have built a probabilistic model checker for stochastic multi-player games as an extension of the PRISM tool [22]. For modelling of SMGs, we have extended the PRISM modelling language. This allows multiple parallel components (called modules) which can either operate asynchronously or by synchronising over common action labels. Now, a model

Table 1. Performance statistics for a representative set of models

Case study [parameters]		SMG statistics			Model checking		
		Players	States	Transitions	Prop. type	Constr. (s)	Model ch. (s)
mdsm [N]	5	5	743,904	2,145,120	$\langle\!\langle C\rangle\!\rangle R^r_{max=?}[F^0\phi]$	14.5	61.9
	6	6	2,384,369	7,260,756		55.0	221.7
	7	7	6,241,312	19,678,246		210.7	1,054.8
cdmsn [N]	3	3	1,240	1,240	$\langle\!\langle C\rangle\!\rangle P_{\bowtie q}[F^{\leq k}\phi]$	0.2	0.2
	4	4	11,645	83,252		0.8	0.8
	5	5	100,032	843,775		3.2	6.4
investor [vmax]	10	2	10,868	34,264	$\langle\!\langle C\rangle\!\rangle R^r_{min=?}[F^c\phi]$	1.4	0.7
	100	2	750,893	2,474,254		9.8	121.8
	200	2	2,931,643	9,688,354		45.9	820.8
team-form [N]	3	3	17,041	20,904	$\langle\!\langle C\rangle\!\rangle P_{max=?}[F\,\phi]$	0.3	0.5
	4	4	184,753	226,736		4.2	2.1
	5	5	2,366,305	2,893,536		36.9	12.9

also includes a set of *players*, each of which controls transitions for a disjoint subset of the modules and/or action labels. Essentially, we retain the existing PRISM language semantics (for Markov decision processes), but, in every state, each nondeterministic choice belongs to one player. For the current work, we detect and disallow the possibility of concurrent decisions between players.

Our tool constructs an SMG from a model description and then executes the algorithms from Sec. 4 to check rPATL formulae. Currently, we have developed an explicit-state model checking implementation, which we show to be efficient and scalable for various large models. It would also be relatively straightforward to adapt PRISM's symbolic model checking engines for our purpose, if required.

5.1　Experimental Results

We have applied our tool to the analysis of several large case studies: two developed solely for this work, and two others adapted from existing models. Experimental results from the new case studies are described in detail in the next sections. First, we show some statistics regarding the performance of our tool on a representative sample of models from the four case studies: Microgrid Demand-Side Management (*mdsm*) and Collective Decision Making for Sensor Networks (*cdmsn*), which will be discussed shortly; the Futures Market Investor (*investor*) example of [25]; and the team formation protocol (*team-form*) of [13]. Tab. 1 shows model statistics (number of players, states and transitions) and the time for model construction and checking a sample property on a 2.80GHz PC with 32GB RAM. All models and properties used are available online [29].

5.2　MDSM: Microgrid Demand-Side Management

Microgrid is an increasingly popular model for the future energy markets where neighbourhoods use electricity generation from local sources (e.g. wind/solar power) to satisfy local demand. The success of microgrids is highly dependent on *demand-side management*: active management of demand by users to avoid peaks. Thus, the infrastructure has to incentivise co-operation and discourage abuse. In this case study, we use rPATL model checking to analyse the MDSM infrastructure of [19] and identify an important incentive-related weakness.

The Algorithm. The system in [19] consists of N *households* (HHs) connected to a single *distribution manager* (DM). At every time-step, the DM randomly contacts a HH for submission of a load for execution. The probability of the HH generating a load is determined by a daily demand curve from [19]. The duration of a load is random between 1 and D time-steps. The cost of executing a load for a single step is the number of tasks currently running. Hence, the total cost increases quadratically with HHs executing more loads in a single step.

Each household follows a very simple algorithm, the essence of which is that, when it generates a load, if the cost is below an agreed limit c_{lim}, it executes it, otherwise it only does so with a pre-agreed probability P_{start}. In [19], the *value* for each household in a time-step is measured by $V = \frac{\text{loads executing}}{\text{cost of execution}}$ and it is shown (through simulations) that, *provided every household sticks to this algorithm*, the peak demand and the total cost of energy are reduced significantly while still providing a good (expected) value V for each household.

Modelling and Analysis. We modelled the system as an SMG with N players, one per household. We vary $N \in \{2, \dots, 7\}$ and fix $D=4$ and $c_{\text{lim}}=1.5$. We analyse a period of 3 days, each of 16 time-steps (using a piecewise approximation of the daily demand curve). First, as a benchmark, we assume that all households follow the algorithm of [19]. We define a reward structure r_i for the value V for household i at each step, and let $r_C = \sum_{i \in C} r_i$ be the total reward for coalition C. To compute the expected value per household, we use the rPATL query:

$$\tfrac{1}{|C|} \langle\!\langle C \rangle\!\rangle R^{r_C}_{\text{max}=?}[\, F^0 \ \text{time} = \text{max\_time}\,]$$

fixing, for now, C to be the set Π of all N players (households). We use this to determine the optimal value of P_{start} achievable by a memoryless strategy for each player, which we will then fix. These results are shown by the bold lines in Fig. 2. We also plot (as a dotted line) the values obtained if no demand-side management is applied.

Next, we consider the situation where the set of households C is allowed to deviate from the pre-agreed strategy, by choosing to ignore the limit c_{lim} if they wish. We check the same rPATL query as above, but now varying C to be coalitions of different sizes, $C \in \{\{1\}, \dots, \Pi\}$. The resulting values are also plotted in Fig. 2a, shown as horizontal dashes of width proportional to $|C|$: the shortest dash represents individual deviation, the longest is a collaboration of all HHs. The former shows the maximum value that can be achieved by following the optimal collaborative strategy, and in itself presents a benchmark for the performance of the original algorithm. The key result is that deviations by individuals or small coalitions guarantee a better expected value for the HHs than *any* larger collaboration: a highly undesired weakness for an MDSM system.

Fixing the Algorithm. We propose a simple punishment mechanism that addresses the problem: we allow the DM to cancel *one* job per step if the cost exceeds c_{lim}. The intuition is that, if a HH is constantly abusing the system,

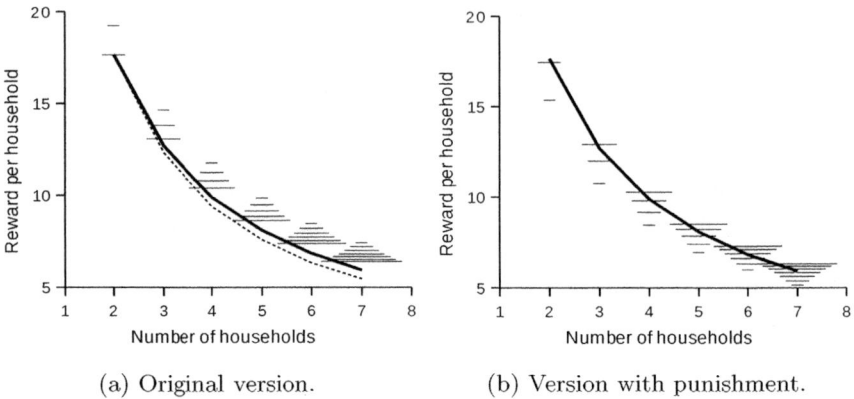

(a) Original version.   (b) Version with punishment.

Fig. 2. Expected value per household for MDSM. The bold line shows all households following the algorithm of [19]; the dotted line shows the case without DSM. Horizontal dashes show deviations by collaborations of increasing size (shortest dash: individual deviation; longest dash: deviation of all households).

its job could be cancelled. This modification inverts the incentives (see Fig. 2b). The best option now is full collaboration and small coalitions who deviate *cannot* guarantee better expected values any more.

5.3   CDMSN: Collective Decision Making for Sensor Networks

Sensor networks comprise a set of low-power, autonomous devices which must act collaboratively in order to achieve a particular goal. In this case study, we illustrate the use of rPATL model checking to aid the analysis and design of such systems by studying a distributed consensus algorithm for sensor networks [26].

The Algorithm. There are N sensors and a set of targets $K = \{k_1, k_2, \dots\}$, each with *quality* $Q_k \in [0, 1]$. The goal is for the sensors to agree on a target with maximum Q_k. Each sensor i stores a *preferred* target $p_i \in K$, its quality Q_{p_i} and an integer $l_i \in \{1, \dots, L\}$ to represent *confidence* in the preference. The algorithm has parameters η and λ, measuring the influence of target quality for the decision, and a parameter γ measuring the influence of the confidence level.

A sensor has three actions: *sleep*, *explore* and *communicate*. As proposed by [26], each sensor repeatedly *sleeps* for a random time t and then either *explores* (with probability $P_{\exp}$) or *communicates*. For the *explore* action, sensor i picks a target $k \in K$ uniformly at random and with probability $P_k = Q_k^\eta / (Q_k^\eta + Q_{p_i}^\eta)$ switches its preference (p_i) to k and resets confidence to 1. To *communicate*, it compares its preference with that of a random sensor j. If they agree, both confidences are increased. If not, with probability $P_s = Q_{p_j}^\lambda l_j^\gamma / (Q_{p_j}^\lambda l_j^\gamma + Q_{p_i}^\lambda l_i^\gamma)$, sensor i switches preference to p_j, resets confidence to 1 and increases sensor j's confidence; with probability $1 - P_s$, the roles of sensors i and j are swapped.

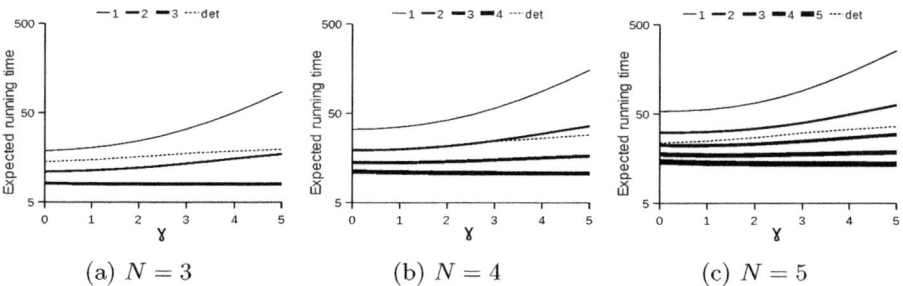

Fig. 3. Expected running time until the selection of the best quality target for different models and increasing sizes of coalition C. Dotted lines show optimal performance that can be achieved using the original algorithm from [26].

Modelling and Analysis. We have modelled the system as an SMG with N players, one per sensor. We consider models with $N=3, 4, 5$, three targets $K=\{k_1, k_2, k_3\}$ with qualities $Q_{k_1}=1$, $Q_{k_2}=0.5$, $Q_{k_3}=0.25$ and two confidence levels $l_i \in \{1,2\}$. As in [26], we assume a random scheduling and fix parameters $\eta=1$ and $\lambda=1$. In [26], two key properties of the algorithm are studied: *speed of convergence* and *robustness*. We consider the same two issues and evaluate alternative strategies for sensors (i.e. allowing sensors to execute any action when active). We also assume that only a subset C of the sensors are under our control, e.g., because the others are faulty. We use rPATL (with coalition C) to optimise performance, under the worst-case assumption about the other sensors.

First, we study the *speed of convergence* and the influence of parameter γ upon it. Fig. 3 shows the expected running time to reach the *best decision* (i.e. select k_1) for various values of γ and sizes of the coalition C. We use the reward structure: $r(s) = 1$ for all $s \in S$ and rPATL query:

$$\langle\!\langle C \rangle\!\rangle R^r_{\min=?}[\, F^\infty \, \textstyle\bigwedge_{i=1}^{|\Pi|} p_i = k_1 \,].$$

Fig. 3 also shows the performance of the original algorithm [26] (line 'det'). We make several important observations. First, if we lose control of a few sensors, we can still guarantee convergence time comparable to the original algorithm, indicating the fault tolerance potential of the system. On the other hand, the original version performs almost as well as the optimal case for large coalitions.

Secondly, we consider *robustness*: the ability to recover from a 'bad decision' (i.e., $\bigwedge_{i=1}^{|\Pi|} p_i = k_3$) to a 'good state' in n steps. We provide two interpretations of a 'good state' and show that the results for them are quite different.

(1) A 'good state': there exists a strategy for coalition C to make all sensors, *with probability > 0.9, select k_1 within 10 steps*. So robustness in rPATL is:

$$\langle\!\langle C \rangle\!\rangle P_{\max=?}[\, F^{\leq n} \, \langle\!\langle C \rangle\!\rangle P_{>0.9}[\, F^{\leq 10} \, \textstyle\bigwedge_{i=1}^{|\Pi|} p_i = k_1]\,].$$

(2) A 'good state': there exists a strategy for coalition C to make all sensors *select k_1 while using less than $0.5mJ$ of energy*. We use a reward structure

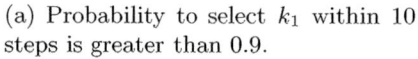

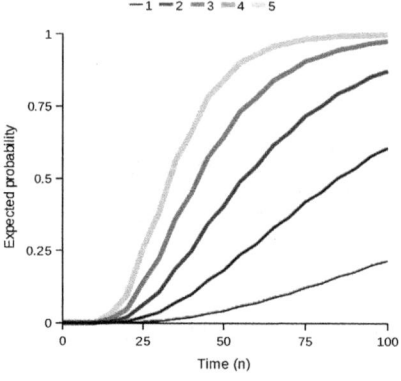

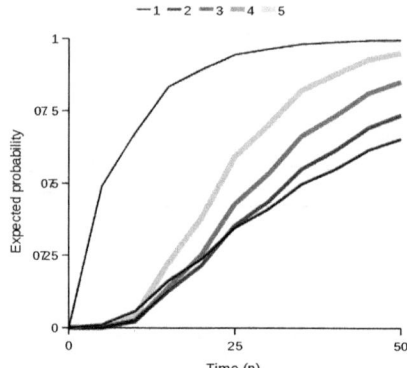

(a) Probability to select k_1 within 10 steps is greater than 0.9.

(b) Expected energy usage for coalition to select k_1 is less than $0.5mJ$.

Fig. 4. Minimum probability to *recover* from a state where all sensors prefer the lowest quality target, k_3, within n steps for different coalition sizes. Graphs (a) and (b) show results for two types of *recovery state* (see captions). $\gamma = 2$.

r_C representing energy usage by sensors in C: power consumption is $10mW$ for each communication and $1mW$ for each exploration, and each activity takes 0.1s. Then, robustness in rPATL is:

$$\langle\!\langle C \rangle\!\rangle P_{\max=?}[\, F^{\leq n}\, \langle\!\langle C \rangle\!\rangle R^{r_C}_{<50}[F^c \textstyle\bigwedge_{i=1}^{|\Pi|} p_i = k_1]]\,.$$

Fig. 4 shows, for each definition and for a range of values of n, the *worst-case* (minimum) value for the rPATL query from all possible 'bad states'. For (1), the results are intuitive: the larger the coalition, the faster it recovers. For (2), however, the one-sensor coalition outperforms all others. Also, we see that, in the early stages of recovery, 2-sensor coalitions outperform larger ones. This shows that small coalitions can be more resource efficient in achieving certain goals.

6   Conclusions

We have designed and implemented a framework for automatic verification of systems with both probabilistic and competitive behaviour, based on stochastic multi-player games. We proposed a new temporal logic rPATL, designed model checking algorithms, implemented them in a tool and then used our techniques to identify unexpected behaviour in several large case studies.

There are many interesting directions for future work, such as investigating extensions of our techniques to incorporate partial-information strategies or more complex solution concepts such as Nash, subgame-perfect or secure equilibria.

Acknowledgments. The authors are part supported by ERC Advanced Grant VERIWARE, the Institute for the Future of Computing at the Oxford Martin

School and EPSRC grant EP/F001096/1. Vojtěch Forejt is supported by a Royal Society Newton Fellowship. We also gratefully acknowledge the anonymous referees for their helpful comments.

References

1. Aizatulin, M., Schnoor, H., Wilke, T.: Computationally Sound Analysis of a Probabilistic Contract Signing Protocol. In: Backes, M., Ning, P. (eds.) ESORICS 2009. LNCS, vol. 5789, pp. 571–586. Springer, Heidelberg (2009)
2. de Alfaro, L.: Computing Minimum and Maximum Reachability Times in Probabilistic Systems. In: Baeten, J.C.M., Mauw, S. (eds.) CONCUR 1999. LNCS, vol. 1664, p. 66. Springer, Heidelberg (1999)
3. de Alfaro, L., Henzinger, T.A.: Concurrent omega-regular games. In: LICS (2000)
4. Alur, R., Henzinger, T.A., Mang, F., Qadeer, S., Rajamani, S., Tasiran, S.: MOCHA: Modularity in Model Checking. In: Vardi, M.Y. (ed.) CAV 1998. LNCS, vol. 1427, pp. 521–525. Springer, Heidelberg (1998)
5. Alur, R., Henzinger, T.A., Kupferman, O.: Alternating-time temporal logic. Journal of the ACM 49(5), 672–713 (2002)
6. Baier, C., Brázdil, T., Größer, M., Kucera, A.: Stochastic game logic. In: Proc. QEST 2007, pp. 227–236. IEEE (2007)
7. Baier, C., Katoen, J.P.: Principles of Model Checking. MIT Press (2008)
8. Ballarini, P., Fisher, M., Wooldridge, M.: Uncertain Agent Verification through Probabilistic Model-Checking. In: Barley, M., Mouratidis, H., Unruh, A., Spears, D., Scerri, P., Massacci, F. (eds.) SASEMAS 2004-2006. LNCS, vol. 4324, pp. 162–174. Springer, Heidelberg (2009)
9. Černý, P., Chatterjee, K., Henzinger, T.A., Radhakrishna, A., Singh, R.: Quantitative Synthesis for Concurrent Programs. In: Gopalakrishnan, G., Qadeer, S. (eds.) CAV 2011. LNCS, vol. 6806, pp. 243–259. Springer, Heidelberg (2011)
10. Chatterjee, K., Henzinger, T.A., Jobstmann, B., Radhakrishna, A.: GIST: A Solver for Probabilistic Games. In: Touili, T., Cook, B., Jackson, P. (eds.) CAV 2010. LNCS, vol. 6174, pp. 665–669. Springer, Heidelberg (2010)
11. Chatterjee, K.: Stochastic ω-Regular Games. Ph.D. thesis (2007)
12. Chen, T., Forejt, V., Kwiatkowska, M., Parker, D., Simaitis, A.: Automatic verification of competitive stochastic systems. Tech. Rep. RR-11-11, University of Oxford (2011)
13. Chen, T., Kwiatkowska, M., Parker, D., Simaitis, A.: Verifying Team Formation Protocols with Probabilistic Model Checking. In: Leite, J., Torroni, P., Ågotnes, T., Boella, G., van der Torre, L. (eds.) CLIMA XII 2011. LNCS, vol. 6814, pp. 190–207. Springer, Heidelberg (2011)
14. Chen, T., Lu, J.: Probabilistic alternating-time temporal logic and model checking algorithm. In: Proc. FSKD 2007, pp. 35–39. IEEE (2007)
15. Condon, A.: On algorithms for simple stochastic games. In: Advances in Computational Complexity Theory. DIMACS, vol. 13, pp. 51–73 (1993)
16. Filar, J., Vrieze, K.: Competitive Markov Decision Processes. Springer, Heidelberg (1997)
17. Forejt, V., Kwiatkowska, M., Norman, G., Parker, D.: Automated Verification Techniques for Probabilistic Systems. In: Bernardo, M., Issarny, V. (eds.) SFM 2011. LNCS, vol. 6659, pp. 53–113. Springer, Heidelberg (2011)

18. Hansson, H., Jonsson, B.: A logic for reasoning about time and reliability. Formal Aspects of Computing 6(5), 512–535 (1994)
19. Hildmann, H., Saffre, F.: Influence of variable supply and load flexibility on demand-side management. In: Proc. EEM 2011, pp. 63–68 (2011)
20. van der Hoek, W., Wooldridge, M.: Model checking cooperation, knowledge, and time - A case study. Research In Economics 57(3), 235–265 (2003)
21. Kremer, S., Raskin, J.-F.: A game-based verification of non-repudiation and fair exchange protocols. Journal of Computer Security 11(3), 399–430 (2003)
22. Kwiatkowska, M., Norman, G., Parker, D.: PRISM 4.0: Verification of Probabilistic Real-Time Systems. In: Gopalakrishnan, G., Qadeer, S. (eds.) CAV 2011. LNCS, vol. 6806, pp. 585–591. Springer, Heidelberg (2011)
23. Lomuscio, A., Qu, H., Raimondi, F.: MCMAS: A Model Checker for the Verification of Multi-Agent Systems. In: Bouajjani, A., Maler, O. (eds.) CAV 2009. LNCS, vol. 5643, pp. 682–688. Springer, Heidelberg (2009)
24. Martin, D.: The determinacy of Blackwell games. J. Symb. Log. 63(4) (1998)
25. McIver, A., Morgan, C.: Results on the quantitative mu-calculus qMu. ACM Transactions on Computational Logic 8(1) (2007)
26. Saffre, F., Simaitis, A.: Host selection through collective decision. ACM Transactions on Autonomous and Adaptive Systems, TAAS (to appear, 2012)
27. Ummels, M.: Stochastic Multiplayer Games: Theory and Algorithms. Ph.D. thesis, RWTH Aachen University (2010)
28. Zhang, C., Pang, J.: On Probabilistic Alternating Simulations. In: Calude, C.S., Sassone, V. (eds.) TCS 2010. IFIP AICT, vol. 323, pp. 71–85. Springer, Heidelberg (2010)
29. http://www.prismmodelchecker.org/files/tacas12smg/

Coupling and Importance Sampling
for Statistical Model Checking

Benoît Barbot, Serge Haddad, and Claudine Picaronny

LSV, ENS Cachan & CNRS & INRIA, Cachan, France
{barbot,haddad,picaronny}@lsv.ens-cachan.fr

Abstract. Statistical model-checking is an alternative verification technique applied on stochastic systems whose size is beyond numerical analysis ability. Given a model (most often a Markov chain) and a formula, it provides a confidence interval for the probability that the model satisfies the formula. One of the main limitations of the statistical approach is the computation time explosion triggered by the evaluation of very small probabilities. In order to solve this problem we develop a new approach based on importance sampling and coupling. The corresponding algorithms have been implemented in our tool COSMOS. We present experimentation on several relevant systems, with estimated time reductions reaching a factor of 10^{-120}.

Keywords: statistical model checking, rare events, importance sampling, coupling.

1 Introduction

Quantitative Model Checking. Model checking [13] is an efficient verification method to check that the behaviour of a system fulfills properties expressed by some temporal logic. It has been successfully implemented in a variety of tools, thanks to it algorithmic simplicity. Although a method initially dedicated to discrete event systems, it has been adapted to performance evaluation in order to check quantitative properties and in particular to estimate probabilities [18].

Statistical Model-Checking. Analysis of stochastic systems requires *numerical* or *statistical* techniques. Numerical methods give exact results (up to numerical approximations) but significantly restrict the class of analysable systems (manageable size, Markov properties, etc.). Otherwise, statistical method may be used. By simulating a big sample of trajectories of the system and computing the ratio of these trajectories that satisfy a given property, it produces a probabilistic framing of the expected value. To generate the sample we only need to have an operational stochastic semantic of the system. This usually requires a very small state space compared to the numerical method and allows to deal with huge models [20].

Rare Events. The main drawback of the statistical model-checking is its inefficiency in dealing with very small probabilities. The size of the sample of

C. Flanagan and B. König (Eds.): TACAS 2012, LNCS 7214, pp. 331–346, 2012.
© Springer-Verlag Berlin Heidelberg 2012

simulations required to estimate these small probabilities exceeds achievable capacities. This difficulty is known as the *rare event* problem. Several methods have been developed to cope with this problem whose main one is *importance sampling*. Importance sampling consists in modifying the model and in substituting to the indicator random variable related to the satisfaction of the formula, another variable with same mean and, in the favorable cases, reduced variance. Most of the techniques related to importance sampling are based on heuristics and cannot provide any confidence interval for the estimated probability.

Our Contribution. Here we propose a method based on importance sampling to estimate in a reliable way a very small probability[1].

We set up a theoretical framework using coupling theory [21], yielding an efficient importance sampling that guarantees a variance reduction and provides a confidence interval. This is done by performing numerical model checking on a small suitable reduction of the Markov chain associated with the system. The results are then used as parameters required for the importance sampling technique. Such a method deals with huge (possibly infinite) systems which are out of reach of numerical model checking and standard statistical model checking. It can be applied to a large variety of models compared to existing importance sampling methods which are usually put up in an ad-hoc way for particular families of models. Furthermore to the best of our knowledge, this is the first importance sampling method that provides a true (and not an approximate) confidence interval.

We implemented our method in the statistical model-checker COSMOS [4] using the tool PRISM for the numerical computation on the reduced model. We tested our tool on several models getting impressive time reductions.

Organisation. In section 2, we motivate this work and we give a state of the art related to rare event handling. Then we develop our method in section 3. Afterwards we present and discuss experimentation in section 4. Finally in section 5, we conclude and give some perspectives to this work. Due to lack of place, the proofs can be found in [6].

2   Motivation and State of the Art

The temporal logics for probabilistic systems include both the qualitative and quantitative aspects of the systems. For instance, such logics can express (1) boolean assertions like "the probability of failure of a fixed component is below some threshold" and (2) numerical indices like "the mean delivery time of a packet assuming three collisions". The semantics of such formula is based on the probability that a random path fulfills some property (in CSL [2]) or (in a more general setting) on the conditional expectation of a path random variable whose condition is the satisfaction of some property by the random path (in HASL [4]).

---

[1] We have presented in a previous paper [5] a preliminary approach of this method with stronger assumptions and without using the coupling theory.

Model checking of these logics can be performed in a numerical or in a statistical way. The former approach builds the underlying stochastic process of the model and then computes probabilities or expectations using direct or iterative methods. Such methods have been implemented efficiently in tools like PRISM [17], LiQuor [9] or MRMC [16].

However these methods have two drawbacks. On the one hand, they rely on strong assumptions about the stochastic process that must be a Markov chain (see for instance [2]) or at least a regenerative process (see for instance [1]). On the other hand they suffer from the combinatorial explosion of the size of the stochastic process w.r.t. the size of the model.

Models with huge stochastic process are handled by statistical model checking. The corresponding methods randomly generate a (large) set of execution paths and check whether the paths fulfill the formula. The result is a probabilistic estimation of the satisfaction given by a confidence interval [3]. In principle, it only requires to maintain a current state (and some numerical values in case of a non Markovian process). Furthermore no regenerative assumption is required and it is easier to parallelize the methods. Several tools include statistical model checking: COSMOS [4], GREATSPN [8], PRISM [17], UPPAAL [7], VESTA [23], YMER [25].

Model checking of probabilistic systems is particularly important for events which have disastrous consequences (loss of human life, financial ruin, etc.), but occur with very small probability. Unfortunately statistical model checking of *rare events* triggers a computation time explosion, forbidding its use. To illustrate this point, suppose one wants to estimate an unknown probability $p = 10^{-13}$ and one chooses to generate 10^{10} paths (which is already a large number) for such an estimation. With probability larger than 0.999 the result is 0, giving no information on the value of p. With probability smaller than 0.001 the result will be greater or equal than 10^{-10} which is a very crude estimation.

Thus *acceleration* techniques [22] have been introduced to cope with this problem. The two main families of methods are *splitting* and *importance sampling*.

Splitting methods [19] duplicate or eliminate paths during their generation depending on their intermediate behaviour. When generation is ended, the bias introduced by these operations is taken into account for the estimation of the probability. Splitting methods are by nature heuristics, model dependent and very few theoretical results are known.

Importance sampling methods [14] generate paths of a system whose probability distribution of transitions have been changed to increase the probability of the event to occur. A weight is then affected to each path to correct the introduced bias. The goal is to substitute to the Bernoulli random variable corresponding to the occurrence of the rare event, another one with same mean value (the probability of event occurrence) but smaller variance. In Markov chains, an optimal change of distribution exists leading to a zero variance but it requires more information than the searched value! However this optimal importance sampling allows to design efficient heuristics for some classes of models.

The modification of the distribution can be performed at the model level (called *static*) or at the Markov chain level (called *dynamic*). The static importance sampling requires no additional memory but in general provides a smaller reduction of variance than the dynamic importance sampling. More precisely, it is proved in [11] that asymptotic optimality (a weaker requirement than optimality) cannot be obtained even for very simple classes of models by static importance sampling. In full generality, the dynamic importance sampling [24] requires to maintain a memory whose size is proportional to the size of the Markov chain which is exactly what one wants to avoid. To deal with this problem, in [12] the authors develop the following method: (1) the possible distributions belong to the convex hull of a finite number of distributions, (2) the state space is partitioned and (3) a distribution is selected for each subset of this partition. They prove that for a simple class of models their method is asymptotically optimal. Other empirical approaches turn out to be efficient [15,10].

Summarizing, theoretical results (reduction of variance, asymptotical optimality, etc.) have been obtained for importance sampling. However none of these methods can produce a reliable confidence interval[2] for the mean value since the distribution of the modified random variable is unknown.

3   General Approach

3.1   Preliminaries

Definition 1. *A discrete time Markov chain (DTMC) C is defined as a set of states S, an initial state s_0, and a transition probability matrix $\mathbf{P}$ of size $S \times S$. The state of the chain at time n is a random variable X_n defined inductively by $\Pr(X_0 = s_0) = 1$ and $\Pr(X_{n+1} = s' \mid X_n = s, X_{n-1} = s_{n-1}, \ldots, X_0 = s_0) = \Pr(X_{n+1} = s' \mid X_n = s) = \mathbf{P}(s, s')$.*

Example 1.   *The figure 1(a) represents a Markov chain of a tandem queue system. This system contains two queues, the number of clients in the first queue is represented on the horizontal axis and the number of clients in the second one is represented on the vertical axis. In the initial state s_0, the two queues are empty. Given some state, a new client comes in the first queue with probability λ, a client leaves the first queue for the second one with probability ρ_1 and a client leaves the second queue and exits with probability ρ_2 $(\lambda + \rho_1 + \rho_2 = 1)$. An impossible event (due to the emptiness of some queue) corresponds to an event leaving unchanged the state. These loops are not represented in the figure.*

Usually the modeller does not specify its system with a Markov chain. He rather defines a higher level model $\mathcal{M}$ (a queueing network, a stochastic Petri net, etc.), whose operational semantic is a Markov chain $\mathcal{C}$.

In the context of model checking, the states of chain $\mathcal{C}$ are labelled with atomic propositions. The problem we address here is the computation of the probability that a random path starting from state s_0 satisfies a formula aUb where U is the *Until* operator and a, b are atomic propositions. Observe that in continuous time

---

[2] In contrast to the empirical confidence interval based on approximations by the normal distribution.

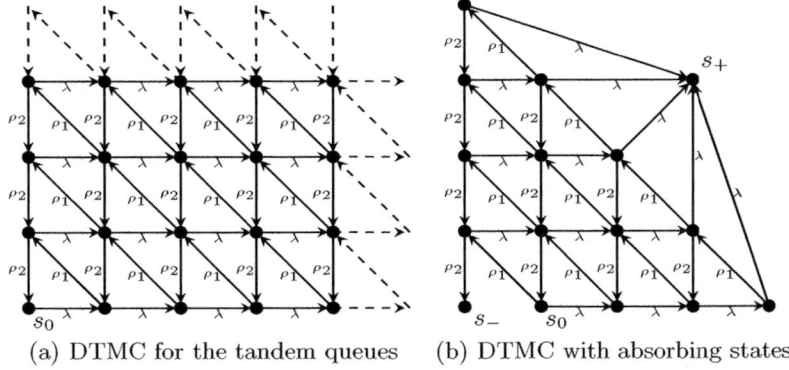

(a) DTMC for the tandem queues   (b) DTMC with absorbing states

Fig. 1. DTMC for tandem queues

Markov chains, this probability only depends on its embedded DTMC. Thus our results are also applicable in a continuous time setting. We (implicitly) transform $\mathcal{C}$ by lumping together all the states that satisfy b into an absorbing state s_+ (i.e. $\mathbf{P}(s_+, s_+) = 1$) and states that satisfy $\neg a \wedge \neg b$ into an absorbing state s_-. We assume that there is no terminal strongly connected component of $\mathcal{C}$ whose every state satisfies $a \wedge \neg b^3$. Hence in the modified chain, the probability to reach s_+ or s_- is equal to 1 and probability of satisfying the formula is the probability to reach s_+.

Example 1.   *The figure 1(b) shows the transformation of the tandem queues were the states have been lumped together w.r.t. the propositions a: There is at least one client in some queue and b: the sum of clients in both queues is equal to 5. The initial state s_0 is now the state with one client in the first queue (to avoid $s_0 = s_-$). We are looking for the probability to have simultaneously at least five clients between two idle periods.*

The statistical approach consists in generating K paths of the Markov chain which ends in an absorbing state. Let K_+ be the number of paths ending in the s_+ state. The random variable K_+ follows a binomial distribution with parameters p and K. Thus the random variable $\frac{K_+}{K}$ has a mean value p and a variance $\frac{p-p^2}{K}$. When K goes to infinity the variance goes to 0. In order to be more precise on the estimation, we introduce the notion of confidence interval.

Definition 2. *Let $X_1, \ldots, X_n$ be independent random variables following a common distribution including a parameter θ. Let $0 < \gamma < 1$ be a confidence level. Then a confidence interval for θ with level at least γ is given by two random variables $l(X_1, \ldots, X_n)$ and $u(X_1, \ldots, X_n)$ such that for all θ:*

$$\Pr\left(l(X_1, \ldots, X_n) \leq \theta \leq u(X_1, \ldots, X_n)\right) \geq \gamma$$

---

[3] There is currently no satisfactory solution for the statistical model checking of the unbounded until for chains that do not fulfill this assumption.

For standard parametrized distributions like the normal or the Bernoulli ones, it is possible to compute confidence intervals [3]. Thus, given a number of paths K and a confidence level $1 - \varepsilon$, the method produces a confidence interval. As discussed before when $p \ll 1$, the number of paths required for a small confidence interval is too large to be simulated.

The importance sampling method uses a modified transition matrix $\mathbf{P}'$ during the generation of paths. $\mathbf{P}'$ must satisfy:

$$\mathbf{P}(s, s') > 0 \Rightarrow \mathbf{P}'(s, s') > 0 \vee s' = s_- \tag{1}$$

which means that this modification cannot remove transitions that have not s_- as target, but can add new transitions. The method maintains a correction factor called L initialized to 1; this factor represents the *likelihood* of the path. When a path crosses a transition $s \to s'$ with $s' \neq s_-$, L is updated by $L \leftarrow L\frac{\mathbf{P}(s,s')}{\mathbf{P}'(s,s')}$. When a path reaches s_-, L is set to zero. If $\mathbf{P}' = \mathbf{P}$ (i.e. no modification of the chain), the value of L when the path reaches s^+ (resp. s^-) is 1 (resp. 0).

Let V_s (resp. W_s) be the random variable associated with the final value of L for a path starting in s in the original model (resp. in the modified one). By definition, $\mathbf{E}(V_{s_0}) = p$. The following proposition establishes the correctness of the method.

Proposition 1. $\mathbf{E}(W_{s_0}) = p$.

A good choice of $\mathbf{P}'$ should reduce the variance of W_{s_0} w.r.t. to variance of V_{s_0}. The following proposition shows that there exists a matrix $\mathbf{P}'$ which leads to a null variance. We denote the probability to reach s_+ starting from s by $\mu(s)$.

Proposition 2. *Let $\mathbf{P}'$ be defined by*

- $\forall s$ such that $\mu(s) \neq 0$, $\mathbf{P}'(s, s') = \frac{\mu(s')}{\mu(s)}\mathbf{P}(s, s')$
- $\forall s$ such that $\mu(s) = 0$, $\mathbf{P}'(s, s') = \mathbf{P}(s, s')$

Then for all s, we have $\mathbf{V}(W_s) = 0$.

This result has a priori no practical application since it requires the knowledge of μ for all states, whereas we only want to estimate $\mu(s_0)$!

The coupling method [21] is a classical method for comparing two stochastic processes, applied in different contexts (establishing ergodicity of a chain, stochastic ordering, bounds, etc.). In the sequel we will develop a new application for coupling. A coupling between two Markov chains is a chain whose space is a subset of the product of the two spaces which satisfies: (1) the projection of the product chain on any of its components behaves like the original corresponding chain, (2) an additional constraint which depends on the property to be proved (here related to the absorbing states).

Definition 3. *Let $\mathcal{C} = (S, \mathbf{P})$ and $\mathcal{C}' = (S', \mathbf{P}')$ be two Markov chains with s_+ and s_- two absorbing states of $\mathcal{C}$ and s'_+ and s'_- two absorbing states of $\mathcal{C}'$. A coupling between $\mathcal{C}$ and $\mathcal{C}'$ is a DTMC $\mathcal{C}^\otimes = (S^\otimes, \mathbf{P}^\otimes)$ such that :*

- $S^{\otimes} \subseteq S \times S'$
- $\forall s \neq s_1 \in S$, $\forall (s, s') \in S^{\otimes}$, $\mathbf{P}(s, s_1) = \sum_{s'_1 \in S'} \mathbf{P}^{\otimes}((s, s'), (s_1, s'_1))$ and
  $\forall s' \neq s'_1 \in S'$, $\forall (s, s') \in S^{\otimes}$, $\mathbf{P}'(s', s'_1) = \sum_{s_1 \in S} \mathbf{P}^{\otimes}((s, s'), (s_1, s'_1))$
- $\forall (s, s') \in S^{\otimes}$ $s' = s'_+ \Rightarrow s = s_+$

The set $S^{\otimes}$ defines a coupling relation between the two chains.

The following proposition allows to compare probabilities without any numerical computation. As before, $\mu(s)$ (resp. $\mu'(s')$) denotes the probability to reach the state s_+ (resp. s'_+) in $\mathcal{C}$ (resp. in $\mathcal{C}'$) starting from s (resp. from s').

Proposition 3. *Let $\mathcal{C}^{\otimes}$ be a coupling between $\mathcal{C}$ and $\mathcal{C}'$. Then, for all $(s, s') \in S^{\otimes}$, we have:*

$$\mu(s) \geq \mu'(s')$$

Example 1. *Let us illustrate coupling for the Markov chain represented in figure 2 and called $\mathcal{C}^{\bullet}$. This chain is obtained from the tandem queues by lumping together states which have the same number of clients and at least R clients in the second queue (in the figure $R = 2$). Its set of state is $S^{\bullet} = [0..N] \times [0..R]$. Here there is a coupling of this chain with itself defined by $S^{\otimes} = \{((n_1, n_2), (n'_1, n'_2)) \mid n_1 + n_2 \geq n'_1 + n'_2 \wedge n_1 \geq n'_1\}$.*

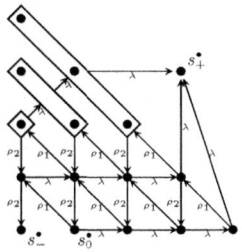

Fig. 2. Reduced DTMC

Lemma 1. *$S^{\otimes}$ is a coupling relation.*

Thus: $\forall ((n_1, n_2), (n'_1, n'_2)) \in S^{\otimes}$, $\mu^{\bullet}(n_1, n_2) \geq \mu^{\bullet}(n'_1, n'_2)$

3.2 An Importance Sampling Method with Variance Reduction and Confidence Interval

The proposed method combines statistical model checking on the original chain preceded by numerical model checking on a reduced chain whose formal definition is given below.

Definition 4. *Let $\mathcal{C}$ be a DTMC, a DTMC $\mathcal{C}^{\bullet}$ is called a reduction of $\mathcal{C}$ by a function f that maps S to $S^{\bullet}$, the state space of $\mathcal{C}^{\bullet}$, if, denoting $s^{\bullet}_- = f(s_-)$ and $s^{\bullet}_+ = f(s_+)$, the following assertions are satisfied:*

- $f^{-1}(s^{\bullet}_-) = \{s_-\}$ *and* $f^{-1}(s^{\bullet}_+) = \{s_+\}$.
- $s^{\bullet}_-$ *and* $s^{\bullet}_+$ *are absorbing states reached with probability 1.*
- *Let $s^{\bullet} \in S^{\bullet}$ and denote by $\mu^{\bullet}(s^{\bullet})$, the probability to reach $s^{\bullet}_+$ starting from $s^{\bullet}$. Then for all $s \in S$, we have $\mu^{\bullet}(f(s)) = 0 \Rightarrow \mu(s) = 0$.*

The two first assertions entail that the reduced chain has two absorbing states reached with probability 1 which are images of the absorbing states of the original chain. The last assertion requires that when from the image of some state s, one cannot reach $s^{\bullet}_+$, then one cannot reach s_+ from s. These (weak) assumptions ensure that the mapping f preserves the basic features of the original chain. Two states s and s' are *equivalent* if $f(s) = f(s')$, in other words f^{-1} define equivalence classes for this reduction.

Example 1. *In the example of tandem queues, the reduced chain* $C^\bullet$ *is obtained from the original chain by applying the following function to the state space.*

$$f(n_1, n_2) = \begin{cases} (n_1, n_2) & \text{if } n_2 \leq R \\ (n_1 + n_2 - R, R) & \text{otherwise} \end{cases}$$

The intuition behind this reduction is to block clients in the first queue when there are R clients in the second one, thus increasing the probability of a global overflow. Given some reduced chain $C^\bullet$, our goal is to replace the random variable (r.v.) V_{s_0} which takes value in $\{0, 1\}$ by a r.v. W_{s_0} which takes value in $\{0, \mu^\bullet(f(s_0))\}$. This requires that $\mu(s_0) \leq \mu^\bullet(f(s_0))$. By applying an homogeneity principle, we get the stronger requirement $\forall s \in S, \mu(s) \leq \mu^\bullet(f(s))$. In fact, the appropriate requirement which implies the previous one (see later proposition 4) is expressed by the next definition.

Definition 5. *Let C be a DTMC and $C^\bullet$ a reduction of C by f. $C^\bullet$ is a reduction with guaranteed variance if for all $s \in S$ such that $\mu^\bullet(f(s)) > 0$ we have :*

$$\sum_{s' \in S} \mu^\bullet(f(s')) \cdot \mathbf{P}(s, s') \leq \mu^\bullet(f(s)) \tag{2}$$

Given $s \in S$, let $h(s)$ be defined by $h(s) = \sum_{s' \in S} \frac{\mu^\bullet(f(s'))}{\mu^\bullet(f(s))} \mathbf{P}(s, s')$. We can now construct an efficient important sampling based on a reduced chain with guaranteed variance.

Definition 6. *Let C be a DTMC and $C^\bullet$ be a reduction of C by f with guaranteed variance. Then $\mathbf{P}'$ is transition matrix on S defined by:*
Let s be a state of S,

- *if $\mu^\bullet(f(s)) = 0$ then for all $s' \in S$, $\mathbf{P}'(s, s') = \mathbf{P}(s, s')$*
- *if $\mu^\bullet(f(s)) > 0$ then for all $s' \in S \setminus \{s_-\}$,*
  $\mathbf{P}'(s, s') = \frac{\mu^\bullet(f(s'))}{\mu^\bullet(f(s))} \mathbf{P}(s, s')$ and $\mathbf{P}'(s, s_-) = 1 - h(s)$.

The following proposition justifies the definition of $\mathbf{P}'$.

Proposition 4. *Let C be a DTMC and $C^\bullet$ be a reduction with guaranteed variance. The importance sampling based on matrix $\mathbf{P}'$ of definition 6 has the following properties:*

- *For all s such that $\mu(s) > 0$,*
  W_s is a random variable which has value in $\{0, \mu^\bullet(f(s))\}$.
- *$\mu(s) \leq \mu^\bullet(f(s))$ and $\mathbf{V}(W_s) = \mu(s)\mu^\bullet(f(s)) - \mu^2(s)$.*
- *One can compute a confidence interval for this importance sampling.*

Since $\mu(s_0) \ll 1$, $\mathbf{V}(V_{s_0}) \approx \mu(s_0)$. If $\mu(s_0) \ll \mu^\bullet(f(s_0))$, we obtain $\mathbf{V}(W_{s_0}) \approx \mu(s_0)\mu^\bullet(f(s_0))$, so the variance is reduced by a factor $\mu^\bullet(f(s_0))$. In the case where $\mu(s_0)$ and $\mu^\bullet(f(s_0))$ have same magnitude order, the reduction of variance is even bigger.

Unfortunately, Equation (2) requires to compute the function $\mu^\bullet$ in order to check that $C^\bullet$ is a reduction with guaranteed variance. We are looking for a structural requirement that does not involve the computation of $\mu^\bullet$.

Proposition 5. *Let C be a DTMC, $C^\bullet$ be a reduction of C by f. Assume there exists a family of functions $(g_s)_{s \in S}$, $g_s : \{t \mid \mathbf{P}(s,t) > 0\} \to S^\bullet$ such that:*

1. $\forall s \in S$, $\forall t^\bullet \in S^\bullet$, $\mathbf{P}^\bullet(f(s), t^\bullet) = \sum_{s' \mid g(s')=t^\bullet} \mathbf{P}(s, s')$
2. $\forall s, t \in S$ such that $\mathbf{P}(s,t) > 0$, $\mu^\bullet(f(t)) \leq \mu^\bullet(g_s(t))$

Then $C^\bullet$ is a reduction of C with guaranteed variance.

The family of functions (g_s) assigns to each transition of C starting from s a transition of $C^\bullet$ starting from $f(s)$. The first condition can be checked by straightforward examination of the probability transition matrices. The second condition still involves the mapping $\mu^\bullet$ but here there are only comparisons between its values. Thanks to proposition 3, it can be proved by exhibiting a coupling of C with itself.

We are now in position to describe the whole method for a model $\mathcal{M}$ with associated DTMC C.

1. Specify a model $\mathcal{M}^\bullet$ with associated DTMC $C^\bullet$, a function f and a family of functions $(g_s)_{s \in S}$. The specification of this family is done at the level of models $\mathcal{M}$ and $\mathcal{M}^\bullet$ as shown in the next example and in section 4.
2. Prove using a coupling on $C^\bullet$ that proposition 5 holds. Again the proof is performed at the level of models.
3. Compute function $\mu^\bullet$ with a numerical model checker applied on $\mathcal{M}^\bullet$.
4. Compute $\mu(s_0)$ with a statistical model checker applied on $\mathcal{M}$ using the importance sampling of definition 6.

The last two steps are done by tools. The second step is currently done by hand (see [6]) but could be handled by theorem provers. The only manual step is the specification of $\mathcal{M}^\bullet$ which requires to study $\mathcal{M}$ and the formula to be checked (see section 4).

Example 1.   *To apply the method on the example it remains to specify the family of functions $(g_s)_{s \in S}$.*

$$
\begin{aligned}
g_{(n_1,n_2)}(n_1, n_2) &= f(n_1, n_2) \\
g_{(n_1,n_2)}(n_1 + 1, n_2) &= f(n_1 + 1, n_2) \\
g_{(n_1,n_2)}(n_1 - 1, n_2 + 1) &= f(n_1 - 1, n_2 + 1) \\
g_{(n_1,n_2)}(n_1, n_2 - 1) &= \begin{cases} (n_1, n_2 - 1) & \text{if } n_2 \leq R \\ (n_1 + n_2 - R, R - 1) & \text{otherwise} \end{cases}
\end{aligned}
$$

The condition 2 always trivially holds except for the last case with $n_2 > R$. We have to check that $\mu^\bullet(n_1 + n_2 - 1 - R, R) \leq \mu^\bullet(n_1 + n_2 - R, R - 1)$. As $(n_1 + n_2 - R, R - 1), (n_1 + n_2 - 1 - R, R))$ belongs to the coupling relation the inequality holds.

3.3 Generalisation

We generalize the method but with no guarantee about the variance reduction.

Definition 7. *Let $\mathcal{C}$ be a DTMC and $\mathcal{C}^\bullet$ a reduction $\mathcal{C}$ of by f. We define a transition matrix $\mathbf{P}'$ on S by the following rules. Let $s \in S$:*

- *if $\mu^\bullet(f(s)) = 0$ then for all $s' \in S$, $\mathbf{P}'(s, s') = \mathbf{P}(s, s')$*
- *if $\mu^\bullet(f(s)) > 0$ and $h(s) \leq 1$ then for all $s' \in S \setminus \{s_-\}$,*
  $$\mathbf{P}'(s, s') = \frac{\mu^\bullet(f(s'))}{\mu^\bullet(f(s))}\mathbf{P}(s, s') \text{ and } \mathbf{P}'(s, s_-) = 1 - h(s)$$
- *if $h(s) > 1$, then for all $s' \in S$, $\mathbf{P}'(s, s') = \frac{\mu^\bullet(f(s'))}{h(s)\mu^\bullet(f(s))}\mathbf{P}(s, s')$*

When Equation 2 does not hold for state s, we have to "normalize" the matrix row $\mathbf{P}'(s, -)$. The next proposition characterises the range of the random variable W_s for this importance sampling. Thus the precision of the estimator highly depends on the shape of the (unknown) distribution of W_s beyond $\mu^\bullet(f(s))$.

Proposition 6. *Let $\mathcal{C}$ be a DTMC and $\mathcal{C}^\bullet$ his reduction. The importance sampling of the definition 7 has the following property: for all s such that $\mu(s) > 0$, W_s is a random variable which takes its values in $\{0\} \cup [\mu^\bullet(f(s)), \infty[$.*

4 Experimentation

4.1 Implementation

Tools. Our experiments [4] have been performed on a modified version of COSMOS (downloadable at http://www.lsv.ens-cachan.fr/Software/Cosmos).
COSMOS is a statistical model checker whose input model is a stochastic Petri net with general distributions and formulas are expressed by the Hybrid Automata Stochastic Logic [4]. The numerical model checking of the reduced model have been performed by PRISM whereas we have also used the statistical model checker PRISM 4.0 for comparisons with our method.

Adaptation of COSMOS. Since COSMOS takes as input a stochastic Petri net with a continuous time semantic, we have adapted our method to work with continuous time Markov chains. As discussed before, for formulas that we consider, this does not present serious difficulty.

The importance sampling increases the computation time of simulation. First we have to compute and store in an hash table the probability vector $\mu^\bullet$ of the reduced model in polynomial time w.r.t. the reduced Markov chain $\mathcal{C}^\bullet$. Then after a transition of the path we must compute $\mathbf{P}'(s, -)$ where s is the current state whose computation time is linear w.r.t. the number of events of $\mathcal{M}$.

---

[4] All the experiments have been performed on a computer with a 2.6Ghz processor and 48Go of memory without parallelism.

4.2   Example 1: Global Overflow in Tandem Queues

This example is a classical benchmark for importance sampling. We compare our results with those of [12] which provides an efficient solution (see section 2). We take the same parameters with $\lambda = 0.1$, $\rho_1 = \rho_2 = 0.45$, $N = 50$ and we also simulate 20000 paths. We set the parameter R to 4. The probability (computed by a numerical model checker) is 3.8×10^{-31}. The width of confidence interval produced by [12] is 6.4×10^{-32} whereas ours is six times smaller (9.63×10^{-33}).

We also compare our method to both numerical and standard statistical model checking done by PRISM. We change the parameter of the model to $\lambda = 0.32$, $\rho_1 = \rho_2 = 0.34$ in order to evaluate the methods for large values of N. Results are depicted in table 1. We set the value of R to the minimal value such that $\frac{\mu(s_0)}{\mu^{\bullet}(f(s_0))} < 1.5$. We found that R and N satisfy $R \approx 36.3 \log(N) - 126$. The narrowness of the obtained confidence interval confirms the validity of this choice. The reduced model has $\Theta(n \log(n))$ states whereas the initial one has $\Theta(n^2)$ states.

The standard statistical model checker fails to find a relevant confidence interval (i.e. the width of interval is half the value of the estimation) when $N \geq 100$ while the numerical model checker does not end when $N \geq 5000$. Our method can handle greater values of this parameter. We can approximate the number of paths required by standard statistical model checking and deduce the estimated corresponding computation time which is 10^{120} greater than ours!

Table 1. Experimental results for example 1

N	Size of $\mathcal{C}$	Prism num		Prism stat				Cosmos			
		T (s)	$\mu(s_0)$	T (s)	$\mu(s_0)$	Conf. Int. width	R	T $\mathcal{C}^{\bullet}$ (s)	T (s)	$\mu(s_0)$	Conf. Int. width
50	2601	0.3	0.0929	1.45	0.091	0.016	4	0.03	7	0.090	0.017
100	10 201	1.6	0.01177	2.7	0.015	0.007	30	1	36	0.01156	8.6E-4
500	251 001	126	2.06E-12	2.3	0	#	87	23	145	2.075E-12	1.72E-13
1000	1E6	860	2.87E-25	No path reaches the rare event			111	113	263	2.906E-25	2.52E-26
5000	25E6	>12h	#	#			150	3061	1099	7.10E-130	1.21E-130

4.3   Example 2 : Parallel Random Walk

The Petri net depicted in figure 3 models a parallel random walk of N walkers. A walk is done between position 1 and position L starting in position $L/2$ and ends up in the extremal positions. At every round, some random walker can randomly go in either direction. However when walkers i and $i + 1$ are in the same position, walker i can only go toward 1. We represent on this figure the walker i and his interactions with walker $i + 1$. Transition $A_{i,j}$ (resp. $R_{i,j-1}$) corresponds to a step toward L (resp. 1).

This model is a paradigm of failure tolerant systems in which each walker represents a process which finishes its job when it reaches position 1. Failures can occur and move the process away of its goal. When position L is reached the job is aborted. We want to evaluate the probability that a majority of players have reached position L.

This model has L^N states. In order to get a reduced model, we remove all synchronisation between walkers. Behaviours of all walkers are now independent and thus a state of the reduced system is now defined by the number of walkers in each position. The size of the reduced system is $\binom{N+L-1}{L-1}$.

Proposition 5 holds for this reduced model. Intuitively, removing synchronisation between walkers increases the probability to reach position L.

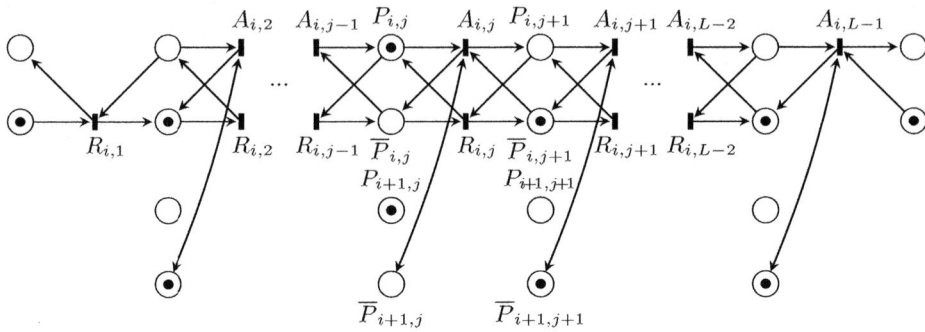

Fig. 3. The Petri net for example 2

Table 2 shows the experimental result with the following parameters $p = 0.3$, $q = 0.7$, $L = 15$. We stop the simulation when the confidence interval width reaches one tenth of the estimated value. Our method handles huge models (with size up to $8 \ 10^{12}$) with very small probabilities ($8 \ 10^{-18}$) whereas the standard statistical model checking and numerical model checkings fail due to either the low probability or the size of the system.

Table 2. Experimental results for example 2

N	size of $\mathcal{C}$	Prism num T (s)	$\mu(s_0)$	Prism stat T (s)	$\mu(s_0)$	Conf. Int.	Nb Traj.	Cosmos T $\mathcal{C}^\bullet$ (s)	T (s)	$\mu(s_0)$	Conf. Int.
1	15	≈ 0	0.00113	12	1.15E-3	1E-4	1	≈ 0	≈ 0	0.00113	0
5	7.5E5	6	1.88E-9	21	0	#	18000	0.5	13	1.94E-9	1.89E-10
6	1.1E7	127	1.14E-12	No path reaches the rare event			53000	1	57	1.17E-12	1.17E-13
7	1.7E8	2248	2.93E-12	#			50000	2.8	186	2.92E-12	2.89E-13
8	2.0E9	Out of memory		#			145000	7.9	1719	1.86E-15	1.86E-16
9	3.8E10	#		#			128000	24	3800	4.7E-15	4.75E-16
10	5.7E11	#		#			371000	71	26000	3.12E-18	3.11E-19
11	8.0E12	#		#			321000	228	67000	7.90E-18	7.89E-19

4.4 Example 3: Local Overflow in Tandem Queues

We consider the tandem queues system with a different property to check: The second queue contains N clients ($n_2 = N$) before the second queue is empty($n_2 = 0$). The state space is $S = \mathbb{N} \times [0..N]$ with initial state $(0, 1)$. Contrary to the first example $\mathcal{C}$ is now infinite but $\mathcal{C}^\bullet$ must be finite in order to apply the numerical model checker.

The reduced model behaves as the original one until the first queue contains R clients. Then the model behaves as if there is an infinite number of clients in the first queue. The corresponding Petri net is depicted in figure 4(a). The corresponding reduction function f (whose effect on the original chain is represented in figure 4(b)) is defined by:

$$f(n_1, n_2) = \begin{cases} (n_1, n_2) & \text{if } n_1 \leq R \\ (R, n_2) & \text{otherwise} \end{cases}$$

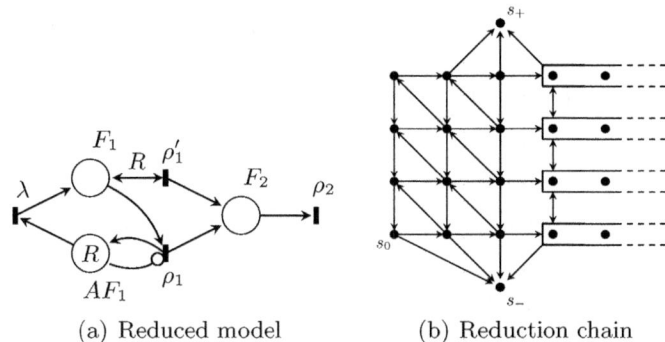

(a) Reduced model   (b) Reduction chain

Fig. 4. Petri net for example $3(R = 3, N = 5)$

Table 3. Experimental results for example 3

N	R	T (s)	Size of	$\mu^\bullet(f(s_0))$	Cosmos				Prism stat		
		$c^\bullet$	$c^\bullet$		$\mu(s_0)$	Conf. Int.	T (s)	Nb Traj.	T (s)	$\mu(s_0)$	Conf. Int.
25	12	≈ 0	338	1.16E-5	1.48E-6	2.83E-7	2	5000	33	1.1E-6	1.6E-6
50	29	≈ 0	1530	2.98E-10	3.81E-11	7.19E-12	13	5000	No path reaches the rare event		
100	66	1.44	6767	1.87E-19	4.22E-20	7.34E-21	17	3000	#		
500	370	1770	185871	1.03E-90	6.63E-91	8.05E-32	37	2000	#		
1000	740	24670	741741	3.24E-177	3.95E-179	4.00E-179	180	3000	#		

We found by running experiments on small values of N and R that for $R \geq 0.74 \times N$ we have $\mu(s_0) \geq \mu^\bullet(f(s_0))/10$. This example shows that we can apply our method on an infinite model subject to the specification of a finite reduced model. Observe that computation time reductions w.r.t. standard statistical model checking are still impressive.

4.5 Example 4: Bottleneck in Tandem Queues

We consider the tandem queues system with a different property to check: **The second queue is full** ($n_2 = N$) **before the first one** ($n_1 = N$). The reduced model is obtained by considering that the second queue is full when it contains $N - R$ clients or in an equivalent way that the second queue always contains at least R clients. The corresponding Petri net is depicted in figure 5(a).

The reduction function (whose effect on the original chain is represented in figure 5(b)) is defined by:

$$f(n_1, n_2) = \begin{cases} (n_1, R) & \text{if } n_2 \le R \\ (n_1, n_2) & \text{otherwise} \end{cases}$$

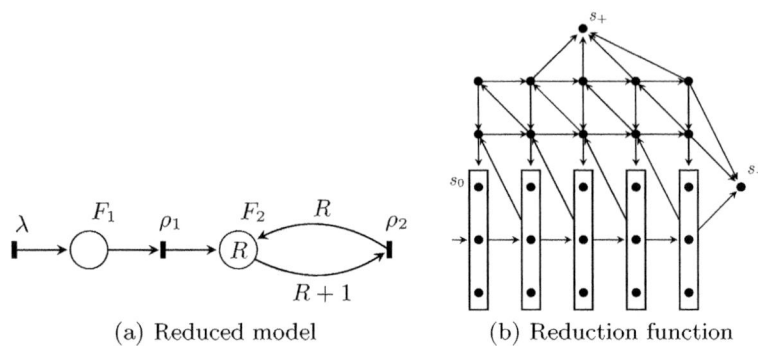

(a) Reduced model   (b) Reduction function

Fig. 5. Petri net for the tandem queues $(R = 2, N = 5)$

However, the experimental results are not satisfactory since $\mu(s_0) \ll \mu^\bullet(f(s_0))$ when R is small compared to N. This shows that designing a reduced model with relevant computation time reduction is sometimes tricky (and remains to be done for this example).

5   Conclusion

We proposed a method of statistical model checking which computes a reduced confidence interval for the probability of a rare event. Our method is based on importance sampling techniques. Other methods usually rely on heuristics and fail to provide a confidence interval. We have developed a theoretical framework ensuring the reduction of the variance and providing a confidence interval. This framework requires a structural analysis of the model but no numerical computation thanks to coupling theory. Our method is implemented in the statistical model checker COSMOS and we have done experiments with impressive results.

We plan to go further in four directions. First we want to deal with more complex infinite systems. Secondly we want to handle "bounded until" formulas requiring to deal with non Markovian systems. We also would mechanize the proofs of coupling since they consist to check parametrized inequalities. Finally we are looking for a class of models which structurally fulfill the required assumptions.

References

1. Amparore, E.G., Donatelli, S.: Model checking CSL^{TA} with deterministic and stochastic Petri nets. In: DSN, pp. 605–614 (2010)

2. Baier, C., Haverkort, B.R., Hermanns, H., Katoen, J.-P.: Model-checking algorithms for continuous-time Markov chains. IEEE Trans. Software Eng. 29(6), 524–541 (2003)
3. Bain, L.J., Engelhardt, M.: Introduction to Probability and Mathematical Statistics, 2nd edn. Duxbury Classic Series (1991)
4. Ballarini, P., Djafri, H., Duflot, M., Haddad, S., Pekergin, N.: HASL: An expressive language for statistical verification of stochastic models. In: VALUETOOLS 2011, Cachan, France (May 2011) (to appear)
5. Barbot, B., Haddad, S., Picaronny, C.: Échantillonnage préférentiel pour le model checking statistique. In: MSR 2011. Journal Européen des Systèmes Automatisés, vol. 45, pp. 237–252 (2011)
6. Barbot, B., Haddad, S., Picaronny, C.: Coupling and importance sampling for statistical model checking. Research Report LSV-12-01, Laboratoire Spécification et Vérification. ENS Cachan, France (January 2012)
7. Bengtsson, J., Larsen, K.G., Larsson, F., Pettersson, P., Yi, W.: UPPAAL - a tool suite for automatic verification of real-time systems. In: Hybrid Systems, pp. 232–243 (1995)
8. Chiola, G., Franceschinis, G., Gaeta, R., Ribaudo, M.: GreatSPN 1.7: Graphical editor and analyzer for timed and stochastic Petri nets. Perform. Eval. 24(1-2), 47–68 (1995)
9. Ciesinski, F., Baier, C.: LiQuor: A tool for qualitative and quantitative linear time analysis of reactive systems. In: QEST 2006, pp. 131–132 (2006)
10. Clarke, E.M., Zuliani, P.: Statistical Model Checking for Cyber-Physical Systems. In: Bultan, T., Hsiung, P.-A. (eds.) ATVA 2011. LNCS, vol. 6996, pp. 1–12. Springer, Heidelberg (2011)
11. de Boer, P.-T.: Analysis of state-independent importance-sampling measures for the two-node tandem queue. ACM Trans. Model. Comput. Simul. 16(3), 225–250 (2006)
12. Dupuis, P., Sezer, A.D., Wang, H.: Dynamic importance sampling for queueing networks. Annals of Applied Probability 17, 1306–1346 (2007)
13. Emerson, E.A., Clarke, E.M.: Characterizing Correctness Properties of Parallel Programs Using Fixpoints. In: de Bakker, J.W., van Leeuwen, J. (eds.) ICALP 1980. LNCS, vol. 85, pp. 169–181. Springer, Heidelberg (1980)
14. Glynn, P.W., Iglehart, D.L.: Importance sampling for stochastic simulations. Management Science (1989)
15. Heegaard, P.E., Sandmann, W.: Ant-based approach for determining the change of measure in importance sampling. In: Winter Simulation Conference, pp. 412–420 (2007)
16. Katoen, J.-P., Zapreev, I.S., Hahn, E.M., Hermanns, H., Jansen, D.N.: The ins and outs of the probabilistic model checker MRMC. In: International Conference on Quantitative Evaluation of Systems, pp. 167–176 (2009)
17. Kwiatkowska, M., Norman, G., Parker, D.: PRISM: Probabilistic Symbolic Model Checker. In: Field, T., Harrison, P.G., Bradley, J., Harder, U. (eds.) TOOLS 2002. LNCS, vol. 2324, pp. 113–140. Springer, Heidelberg (2002)
18. Kwiatkowska, M., Norman, G., Parker, D.: Stochastic Model Checking. In: Bernardo, M., Hillston, J. (eds.) SFM 2007. LNCS, vol. 4486, pp. 220–270. Springer, Heidelberg (2007)
19. L'Ecuyer, P., Demers, V., Tuffin, B.: Splitting for rare-event simulation. In: Winter Simulation Conference, pp. 137–148 (2006)

20. Legay, A., Delahaye, B., Bensalem, S.: Statistical Model Checking: An Overview. In: Barringer, H., Falcone, Y., Finkbeiner, B., Havelund, K., Lee, I., Pace, G., Roşu, G., Sokolsky, O., Tillmann, N. (eds.) RV 2010. LNCS, vol. 6418, pp. 122–135. Springer, Heidelberg (2010)
21. Lindvall, T.: Lectures on the coupling method. Dover (2002)
22. Rubino, G., Tuffin, B.: Rare Event Simulation using Monte Carlo Methods. Wiley (2009)
23. Sen, K., Viswanathan, M., Agha, G.: VESTA: A statistical model-checker and analyzer for probabilistic systems. In: QEST, pp. 251–252 (2005)
24. Srinivasan, R.: Importance sampling – Applications in communications and detection. Springer, Berlin (2002)
25. Younes, H.L.S.: Ymer: A Statistical Model Checker. In: Etessami, K., Rajamani, S.K. (eds.) CAV 2005. LNCS, vol. 3576, pp. 429–433. Springer, Heidelberg (2005)

Verifying pCTL Model Checking

Johannes Hölzl[*] and Tobias Nipkow

Institut für Informatik, Technische Universität München
www.in.tum.de/~hoelzl, www.in.tum.de/~nipkow

Abstract. Probabilistic model checkers like PRISM check the satisfiability of probabilistic CTL (pCTL) formulas against discrete-time Markov chains. We prove soundness and completeness of their underlying algorithm in Isabelle/HOL. We define Markov chains given by a transition matrix and formalize the corresponding probability measure on sets of paths. The formalization of pCTL formulas includes unbounded cumulated rewards.

1 Introduction

Modeling systems as discrete-time Markov chains is a popular technique to analyze probabilistic behavior of network protocols, algorithms, communication systems or biological systems. Probabilistic model checkers, like PRISM [13] or MRMC [10], interpret Markov chains and analyze quantitative properties, specified as probabilistic CTL (pCTL) formulas [6]. In this paper we formalize the background theory and the algorithm used by these probabilistic model checkers in the proof assistant Isabelle/HOL [20].

Our (almost) executable model checker is certainly not a rival to any of the existing model checkers. Instead, our work should be seen as a foundational contribution that paves the way towards a fruitful combination of automatic and interactive verification methods. Possible application scenarios include the following: interactive verification of parameterized systems, or a verified checker that checks individual runs of a hand coded model checker that produces a certificate. We discuss these in more detail in the conclusion. Quite apart from applications, we see our work as another building block in the larger undertaking of formalizing key areas of computer science (as is currently happening with compilers [15] and operating system kernels [11]).

This is the first time probabilistic model checking has been formalized in a proof assistant. So far, the necessary mathematical background theories were simply not available. We start from our recent formalization of measure theory in Isabelle/HOL, including the Lebesgue integral and Caratheodory's theorem [8]. Based on this, Section 3 formalizes the following material:

- infinite products of probability spaces,
- Markov chains defined by a transition matrix and the existence of their probability measure on paths,

---

[*] Supported by the DFG Graduiertenkolleg 1480 (PUMA).

C. Flanagan and B. König (Eds.): TACAS 2012, LNCS 7214, pp. 347–361, 2012.

– properties of paths reaching a set of states almost everywhere.

Now we have the necessary theory to define, in Section 4, syntax, semantics and a model checking algorithm for pCTL, and verify the algorithm wrt the semantics, following the standard literature [12].

2   Related Work

There are a number of formalizations and proofs of aspects of model checking: verification of a model checker for the modal μ-calculus [23], of partial order reduction [3], of two innermost loops of a model checker for real time CTL [21], of a CTL model checker [20], of a model checker for dynamic pushdown networks [14], and of the translation from LTL to Büchi automata [22]. But none involve probabilities.

The formalization of probability theory in HOL starts with Hurd's thesis [7]. He introduces measure theory, proves Caratheodory's theorem about the existence of measure spaces and uses it to introduce a probability space on infinite boolean sequences. He provides methods to generate discrete random variables with Bernoulli or uniform distribution. Based on this work Liu *et al.* [17] formalize the concept of Markov chains. Their theory does not provide everything we need: it lacks a probability measure on paths, and their measure space needs to be the type universe whereas we relax it to sets. Coble [4] and Mhamdi *et al.* [18] introduce generalized measure spaces on sets, the extended real numbers $\overline{\mathbb{R}}$ and the Lebesgue integral. However, there is no theorem to show the existence of a measure space.

3   Foundations

3.1   Isabelle/HOL

The formalizations presented in this paper are done in the Isabelle/HOL theorem prover. In this section we give an overview of our syntactic conventions.

The term syntax follows the λ-calculus, i.e. function application is juxtaposition as in $f\ t$. The notation $t :: \tau$ means that t has type τ. Types are built from the base types $\mathbb{B}$ (booleans), $\mathbb{N}$ (natural numbers), $\mathbb{R}$ (reals), $\overline{\mathbb{R}} = \mathbb{R} \cup \{\infty, -\infty\}$, and type variables ($\alpha$, β etc.) via the function type constructor $\Rightarrow$. In particular, (infinite) sequences have type $\mathbb{N} \Rightarrow \alpha$ and are usually denoted by ω.

Prepending an element x to a sequence ω is written as $x \cdot \omega$, i.e. $(x \cdot \omega)\ 0 = x$ and $(x \cdot \omega)\ (n+1) = \omega\ n$. The while-combinator $while :: (\alpha \Rightarrow \mathbb{B}) \Rightarrow (\alpha \Rightarrow \alpha) \Rightarrow \alpha \Rightarrow \alpha$ satisfies the standard recursion equation:

$$while\ P\ f\ x = if\ P\ x\ then\ while\ P\ f\ (f\ x)\ else\ x$$

We write $\bigtimes_{i \in I} A\ i := \{f \mid \forall i \in I.\ f\ i \in A\ i\}^1$ for the dependent function space (which is a set, not a type in HOL); if A is constant we write $I \rightarrow A$.

---

[1] We use $\bigtimes$ for products of sets, and $\prod$ for products of numbers.

To represent non-total functions in HOL we use the option data type

$$\alpha \ option = \text{\textbf{Some}} \ \alpha \ | \ \text{\textbf{None}}$$

whose values are **Some** x for $x :: \alpha$ and **None**. We introduce the option-monad to combine non-total functions to new non-total functions. The infix bind-operator $\gg\!\!=$ is defined by the equations $((\text{\textbf{Some}} \ x) \gg\!\!= f) = f \ x$ and $(\text{\textbf{None}} \gg\!\!= f) = \text{\textbf{None}}$. Notation *return* is equal to **Some**. Similar to Haskell's monad-syntax we use the *do*-syntax to represent chains of bind-operators, for example:

$$
\begin{array}{lll}
do & x \leftarrow f & \\
   & y \leftarrow g \ x & \\
   & let \ z = h \ x \ y & \\
   & return \ (x + y + z) &
\end{array}
\qquad \Longrightarrow \qquad
\begin{array}{l}
r \gg\!\!= (\lambda x. \\
\quad g \ x \gg\!\!= (\lambda y. \\
\qquad \text{\textbf{Some}} \ (x + y + h \ x \ y)))
\end{array}
$$

We use the option-monad not only to represent non-total functions, but also to write the algorithm in a more imperative style. The only non-total function in the pCTL model checking algorithm is Gauss-Jordan elimination.

3.2   Probability Space

Probabilistic CTL formulas are defined in terms of probabilities on sets of paths. To define a probability space on paths we use the measure theory formalized in [8]. This provides us with the concepts of extended real numbers, σ-algebras, measure spaces, the Lebesgue integral, the Lebesgue measure and, as a way to construct measures, Caratheodory's theorem. We write $\mathcal{B}$ for the Borel sets, $\sigma(\mathcal{A})$ for the σ-algebra generated by $\mathcal{A}$, $f \in \mathcal{A}_1 \rightarrow_M \mathcal{A}_2$ for a measurable function f mapping from $\mathcal{A}_1$ to $\mathcal{A}_2$, $\int_\omega f \ \omega d\mu$ for the Lebesgue integral, and $\text{AE}_\mu \omega. \ P \ \omega$ if the predicate P holds *almost everywhere*, i.e. the complement of P is a subset of a null set in μ. In the following two sections we will introduce the infinite product of probability spaces and based on this a probability measure on paths in Markov chains. This was not yet formalized in [8], and we are not aware of any formalization of these concepts in interactive theorem provers.

A probability space is a measure space which assigns 1 to the entire space:

Definition 1. *Probability space*

$$\text{\textit{prob-space}} \ (\Omega, \mathcal{A}, \mu) :\longleftrightarrow \text{\textit{measure-space}} \ (\Omega, \mathcal{A}, \mu) \wedge \mu \ \Omega = 1$$

Our first step in introducing a probability space for paths is to formalize the infinite product of probability spaces $(\Omega \ i, \mathcal{A} \ i, \mu \ i)$, for i in some index set I. We used the proof in [2] as the base of our formalization of infinite products. The space of infinite products of probability spaces is the function space $\Omega_P := \bigtimes_{i \in I} \Omega \ i$. The generating set of infinite products is the collection of all embeddings of finite products:

Definition 2. *Embedding of finite products and the product σ-algebra $\mathcal{A}_P$*

$$
\begin{array}{ll}
\text{\textit{emb}} \ J \ F := & \{\omega \in \Omega_P \mid \forall i \in J. \ \omega \ i \in F \ i\} \\
\mathcal{A}_P \quad := & \sigma(\{\text{\textit{emb}} \ J \ F \mid \text{finite} \ J \wedge J \subseteq I \wedge (\forall i \in J. \ F \ i \in \mathcal{A} \ i)\})
\end{array}
$$

With Caratheodory's theorem we show that a probability measure on $\mathcal{A}_P$ exists which maps *emb J F* to the product of the real numbers $\mu\ i\ (F\ i)$, the property we want to have from a product space.

Theorem 3. *Probability measure on* $\mathcal{A}_P$
  There exists a unique probability measure μ_P *on* $\mathcal{A}_P$

$$prob\text{-}space\ (\Omega_P, \mathcal{A}_P, \mu_P)$$

with: *If* $J \subseteq I$ *is finite and* $F \in \underset{i \in J}{\times} \mathcal{A}\ i$ *then*

$$\mu_P(\textit{emb } J\ F) = \prod_{i \in J} \mu\ i\ (F\ i)\ .$$

We choose such a probability measure μ_P with $I := \mathbb{N}$ and $\mathcal{A}\ i := \lambda_{[0;1[}$, the Borel-Lebesgue measure restricted to $[0;1[$. Hence μ_P is now a probability measure on sequences $\mathbb{N} \to [0;1[$. From the equation in Theorem 3 we have

$$\mu_P\left(\underset{i}{\times} F\ i\right) = \prod_i \lambda_{[0;1[}\ (F\ i)\ .$$

Hence the elements in the product space induce countably many, independent random variables with a continuous, uniform distribution. The formalization in [7] only provides a probability measure on sequences $\mathbb{N} \to \mathbb{B}$, which induces random variables with a discrete distribution.

3.3 Markov Chains

We introduce Markov chains as probabilistic automata, i.e. as discrete-time time-homogeneous finite-space Markov processes. A Markov chain is defined by its state space S and an associated transition matrix τ. We assume no initial distribution or starting state, however when measuring paths we always provide a starting state. A path on a Markov chain is a function $\mathbb{N} \to S$, i.e. an infinite sequence of states visited in the Markov chain.

Definition 4. *Markov chain*

$$\begin{aligned}
\textit{markov-chain } S\ \tau :&\longleftrightarrow \textit{finite } S \wedge S \neq \emptyset \\
&\wedge\ \forall s, s' \in S.\ 0 \leq \tau\ s\ s' \\
&\wedge\ \forall s \in S.\ \left(\textstyle\sum_{s' \in S} \tau\ s\ s'\right) = 1
\end{aligned}$$

For the rest of the paper we assume a Markov chain with state space S and transition matrix τ. We write $E(s)$ for the set of all successor states, i.e. all $s' \in S$ with $\tau\ s\ s' \neq 0$. Note that a path ω does not require that $\omega\ (i+1)$ is a successor of $\omega\ i$. Our first goal is to define a probability space $(\Omega, \mathcal{T}, \mu_s)$ on the space of all paths $\mathbb{N} \to S$. We call the set of all paths starting with a common prefix, namely ω', $\textit{Cy } \omega'\ n := \{\omega \in \mathbb{N} \to S \mid \forall i < n.\ \omega\ i = \omega'\ i\}$, a *cylinder*. The probability measure on paths assigns to cylinders the product of the transition probabilities:

Definition 5. *Path σ-algebra, and pre-measure μ'_P*

$$\Omega \qquad\qquad := \mathbb{N} \rightarrow S$$
$$\mathcal{T} \qquad\qquad := \sigma(\{Cy\ \omega\ n \mid \omega \in \Omega\})$$
$$\forall \omega \in \mathbb{N} \rightarrow S, s \in S, n.\ \ \mu'_s\ (Cy\ \omega\ n) := \prod_{i<n} \mathcal{T}\ ((s{\cdot}\omega)\ i)\ (\omega\ i)$$

Note that μ'_s explicitly carries the starting state, hence we assign to $Cy\ \omega\ n$ the transition probability for the steps $s \rightarrow_\mathcal{T} \omega\ 0 \rightarrow_\mathcal{T} \omega\ 1 \rightarrow_\mathcal{T} \cdots \rightarrow_\mathcal{T} \omega\ (n-1)$. Before we use this as a probability space we need to show that μ'_s can be extended to a probability measure. To this end, we provide a function *path* which constructs a path out of a sequence $\mathbb{N} \rightarrow [0; 1[$, and show that this function is measurable.

As S is finite and not empty we know that there exists a bijective function mapping from $\{0, \ldots, |S| - 1\}$ to S, we define *order* to be such a function. Using *order* we introduce *select* which splits $[0; 1[$ into disjoints intervals of size $\tau\ s\ s'$, see Fig. 1. The recursive function *path* now walks along a sequence X of values in the unit interval and selects the next state.

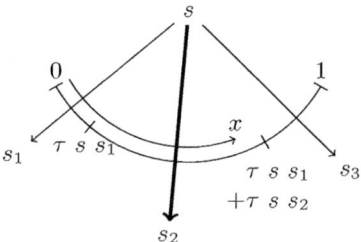

Fig. 1. The next state after s selected by x is $s_2 = $ *select* $s\ x$

Definition 6. *Path selection*

$$select\ s\ x \qquad := order\ \left(\min \left\{ i \mid x < \sum_{j \leq i} \tau\ s\ (order\ j) \right\} \right)$$
$$path\ s\ X\ 0 \qquad := select\ s\ (X\ 0)$$
$$path\ s\ X\ (n+1) := select\ (path\ s\ X\ n)\ (X\ (n+1))$$

The set $T\ s\ s' := \{x \in [0; 1[\mid select\ s\ x = s'\}$ is measurable and $\lambda_{[0;1[}(T\ s\ s') = \tau\ s\ s'$. We represent the inverse image of cylinders over *path* with *emb* and T:

Lemma 7. *For all states s, paths ω, and prefix length n:*

$$\{\omega \in \Omega_P \mid path\ s\ \omega \in Cy\ \omega\ n\} = emb\ \{0, 1, \ldots, n-1\}\ (\lambda i.\ T\ ((s{\cdot}\omega)\ i)\ (\omega\ i))$$

As $Cy\ \omega\ n$ are the generators of $\mathcal{T}$ and *emb* is measurable in $\mathcal{A}_P$, *path* s is in $\mathcal{A}_P \rightarrow_M \mathcal{T}$. With this we show that $\mu_s\ A := \mu_P\{\omega \in \Omega_P \mid path\ s\ \omega \in A\}$ defines a probability measure, and with Lemma 7 we show that μ_P extends μ'_P.

Theorem 8. μ_s *is the unique probability measure on paths which extends* μ'_s

$$\textsf{prob-space } (\Omega, \mathcal{T}, \mu_s)$$

$$\forall \omega \in \Omega, s \in S, n.\ \mu_s\ (\textsf{Cy}\ \omega\ n) = \prod_{i < n} \tau\ ((s \cdot \omega)\ i)\ (\omega\ i)$$

The Markov chain induces *iterative equations* on the measure μ_s, the Lebesgue integral and the AE-quantifier, relating properties about s to properties of $E(s)$, states that are not successors of s are ignored. These equations are often useful in inductive proofs, and already give a hint how to implement a probabilistic model checker. In the rest of the paper we write the AE-quantifier on the path measure μ_s as $\mathrm{AE}_s\,\omega.\ P\,\omega$ instead of $\mathrm{AE}_{\mu_s}\,\omega.\ P\,\omega$.

Theorem 9. *Iterative equations for* μ_s, *the Lebesgue integral and* AE_s
    If s, A, P, *and* f *are measurable, i.e.* s *is in* S, A *and* $\{\omega \in \Omega \mid P\,\omega\}$ *are in* $\mathcal{T}$, *and* f *is in* $\mathcal{T} \rightarrow_M \mathcal{B}$ *then the following equations hold:*

$$\mu_s\ A\ =\ \sum_{s' \in E(s)} \tau\ s\ s' \cdot \mu_{s'}\{\omega \in \Omega \mid s' \cdot \omega \in A\}$$

$$\int_\omega f\ \omega d\mu_s\ =\ \sum_{s' \in E(s)} \tau\ s\ s' \cdot \int_\omega f\ (s' \cdot \omega) d\mu_{s'}$$

$$AE_s\,\omega.\ P\,\omega \longleftrightarrow \forall s' \in E(s).\ AE_{s'}\,\omega.\ P\,(s' \cdot \omega)$$

We prove the iterative equation for μ_s by proving the equality when A is a cylinder, with the uniqueness of measures [8] follows that they are equal for all measurable sets A. Based on this the integral equation is shown for simple functions, and then for $\mathcal{B}$-measurable functions.
    A state s' is reachable in Φ starting in s iff there is a non-zero probability to reach s' by only going through the specific set of states Φ. The starting state s and the final state s' are not necessary in Φ.

Definition 10. *Reachability of states*

$$\textsf{reachable } \Phi\ s := \{s' \in S \mid \exists \omega \in \Omega, n.\ (\forall i \le n.\ \omega\ i \in E((s \cdot \omega)\ i)) \wedge \\ (\forall i < n.\ \omega\ i \in \Phi) \wedge \omega\ n = s'\}$$

Reachability is a purely qualitative property, as it is defined on the graph of non-zero transitions. Hence an upper bound R of *reachable* $\Phi\ s$ is given when all successor states of $R \cap \Phi$ are in R again.

Lemma 11. *Sets closed under* E *contain* reachable

$$s \in R \cap \Phi \wedge (\forall t \in R \cap \Phi.\ E(t) \subseteq R) \wedge R \subseteq S \wedge \Phi \subseteq S \\ \longrightarrow \textsf{reachable } \Phi\ s \subseteq R$$

The until-operator introduces a similar concept on paths. Its definition does not assume that a state is a successor state of the previous one, as this is already ensured by the probability measure μ_s.

Definition 12. *Until on paths*

$$until \; \Phi \; \Psi := \{\omega \in \Omega \mid \exists n. \; (\forall i < n. \; \omega \; i \in \Phi) \land \omega \; n \in \Psi\}$$

Can we compute the probability of $\mu_s(until \; \Phi \; \Psi)$ by only using *reachable*? It is easy to show that $\mu_s(until \; \Phi \; \Psi) = 0$ iff $(reachable \; \Phi \; s) \cap \Psi = \emptyset$. But is there also a method to characterize $\mu_s(until \; \Phi \; \Psi) = 1$ in terms of *reachable*? For this we need to introduce state fairness. A path ω is *state fair* w.r.t. s and t if t appears infinitely often as the successor of s in ω, provided that s appears infinitely often. The definition and proofs about state fairness are based on Baier [1].

Definition 13. *State fairness*

$$fair \; s \; t :=$$
$$\{\omega \in \Omega \mid (\exists n. \; \forall i \geq n. \; \omega \; i \neq s) \lor (\forall n. \; \exists i \geq n. \; \omega \; i = s \land \omega \; (i+1) = t)\}$$

Baier [1] defines state fairness and a more general version called p-fairness, but we only need state fairness. We show that almost every path is state fair for each state and its successors.

Lemma 14. *Almost every path is state fair*

$$\forall s \in S. \; AE_s \, \omega. \; \forall s' \in S. \; \forall t' \in E(s'). \; s \cdot \omega \in fair \; s' \; t'$$

Using this we prove that starting in a state s almost every path fulfills $until \; \Phi \; \Psi$ if (1) all states reachable by Φ are in Φ or Ψ and (2) each state reachable from s has again the possibility to reach Ψ. This theorem allows us to prove that $until \; \Phi \; \Psi$ holds almost everywhere by a reachability analysis on the graph, and hence $\mu_s(until \; \Phi \; \Psi) = 1$.

Theorem 15. *Reachability implies until*

$$s \in \Phi \land \Phi \subseteq S \land reachable \; (\Phi \setminus \Psi) \; s \subseteq \Phi \cup \Psi$$
$$\land \; \forall t \in (reachable \; (\Phi \setminus \Psi) \; s \cup \{s\}) \setminus \Psi. \; reachable \; (\Phi \setminus \Psi) \; t \cap \Psi \neq \emptyset$$
$$\longrightarrow AE_s \, \omega. \; s \cdot \omega \in until \; \Phi \; \Psi$$

The hitting time on a path ω is the first index at which a state from a set Φ occurs.

Definition 16. *hitting-time* $\Phi \; \omega = \min\{i \mid \omega \; i \in \Phi\}$

For the computation of rewards it is important to know if the expected hitting time is finite. Standard textbook proofs assume an irreducible chain. We took such a proof from [16], and adapted it to our setting. Instead of a irreducible chain we assume Φ is always reached from s. We show that the expected hitting time of Φ for paths starting in s is finite if almost every path starting in s reaches Φ.

Theorem 17. *Finite expected hitting time*
   If s is in S and $AE_s \, \omega. \; s \cdot \omega \in until \; S \; \Phi$ then

$$\int_\omega hitting\text{-}time \; \Phi \; (s \cdot \omega) d\mu_s \neq \infty$$

4 Verifying pCTL Model Checking

4.1 pCTL Formulas

We do not introduce a labeled Markov chain as [12] does, instead we define labels to be subsets of S. We introduce a Markov chain with rewards as a Markov chain with ρ, the rewards associated per state, and ι, the rewards associated per transitions. These rewards are non-negative, real numbers.

Definition 18. *Markov chain with rewards*

$$
\begin{aligned}
\textit{rewarded-markov-chain } S \ \tau \ \rho \ \iota := \ &\textit{markov-chain } S \ \tau \\
&\wedge \ \forall s \in S. \ 0 \leq \rho \ s \\
&\wedge \ \forall s, s' \in S. \ 0 \leq \iota \ s \ s'
\end{aligned}
$$

For the rest of the paper we assume a Markov chain with rewards, with the state space S, the transition matrix τ, and the reward functions ρ and ι.

The pCTL syntax is introduced as an inductive data type.

Definition 19. *pCTL syntax*

$$
\begin{aligned}
\textit{sform} :=& \ \textit{label } \mathcal{P}(S) \mid \neg \textit{sform} \mid \textit{sform} \wedge \textit{sform} \\
&\mid P^{\bowtie \mathbb{R}} \ \textit{pform} \mid E^{\bowtie \mathbb{R}} \ \textit{eform} \\
\textit{pform} :=& \ X \ \textit{sform} \mid \textit{sform } U^{\leq \mathbb{N}} \ \textit{sform} \mid \textit{sform } U^{\infty} \ \textit{sform} \\
\textit{eform} :=& \ C^{<\mathbb{N}} \mid I^{=\mathbb{N}} \mid F^{\infty} \ \textit{sform} \\
\bowtie :=& \ \leq \mid < \mid = \mid > \mid \geq
\end{aligned}
$$

Informally, a state s fulfills $P^{\bowtie r} \ \Phi$ (or $E^{\bowtie r} \ \Phi$) if the probability (expected reward) of the paths starting in s and fulfilling Φ is related with $\bowtie r$. A path fulfills $X \ \Phi$ if its second state fulfills Φ. A path fulfills $\Phi \ U^{\leq k} \ \Psi$ (or $\Phi \ U^{\infty} \ \Psi$, the unbounded until) if it stays in Φ, until it reaches Ψ in at least k steps (at some step). The reward $C^{<k}$ sums all state and transitions rewards for the first k steps, $I^{=k}$ is the state reward at step k, and the unbounded cumulated reward $F^{\infty} \ \Phi$ sums rewards until Φ is reached, if it is never reached it is infinity. We define now semantics to assign a formal meaning to the pCTL syntax, cf. [6,12].

Definition 20. *pCTL semantics*

$$
\begin{aligned}
[\![\textit{label } S']\!] \quad &:= \quad \{s \in S \mid s \in S'\} \\
[\![\neg \ \Phi]\!] \quad &:= \quad S \setminus [\![\Phi]\!] \\
[\![\Phi \ \wedge \ \Psi]\!] \quad &:= \quad [\![\Phi]\!] \cap [\![\Psi]\!] \\
[\![P^{\bowtie r} \ \Phi]\!] \quad &:= \quad \{s \in S. \ \mu_s \{\omega \in \Omega \mid [\![\Phi, s{\cdot}\omega]\!]_P\} \bowtie r\} \\
[\![E^{\bowtie r} \ \Phi]\!] \quad &:= \quad \{s \in S \mid \int_\omega [\![\Phi, s{\cdot}\omega]\!]_E d\mu_s \bowtie r\}
\end{aligned}
$$

$$
\begin{aligned}
[\![X \ \Phi, \omega]\!]_P \quad &:\longrightarrow \quad \omega \ 1 \in [\![\Phi]\!] \\
[\![\Phi \ U^{\leq k} \ \Psi, \omega]\!]_P \quad &:\longrightarrow \quad \exists n \leq k. \ \omega \ n \in [\![\Psi]\!] \wedge (\forall i < n. \ \omega \ i \in [\![\Phi]\!]) \\
[\![\Phi \ U^{\infty} \ \Psi, \omega]\!]_P \quad &:\longrightarrow \quad \exists n. \ \omega \ n \in [\![\Psi]\!] \wedge (\forall i < n. \ \omega \ i \in [\![\Phi]\!])
\end{aligned}
$$

$$
\begin{aligned}
[\![C^{<k}, \omega]\!]_E \quad &:= \quad \sum_{i<k} \rho \ (\omega \ i) + \iota \ (\omega \ i) \ (\omega \ (i+1)) \\
[\![I^{=k}, \omega]\!]_E \quad &:= \quad \rho \ (\omega \ k) \\
[\![F^{\infty} \ \Phi, \omega]\!]_E \quad &:= \quad
\begin{cases}
[\![C^{<\textit{hitting-time } [\![\Phi]\!]} \ \omega, \omega]\!]_E & \textit{if} \ \ \exists i. \ \omega \ i \in [\![\Phi]\!] \\
\infty & \textit{otherwise}
\end{cases}
\end{aligned}
$$

We see that $[\![\Phi]\!]$ is a subset of S and hence also finite. The set $\{\omega \in \Omega \mid [\![\Phi, \omega]\!]_P\}$ is measurable in $\mathcal{T}$, and $\lambda\omega.\ [\![\Phi, \omega]\!]_E \in \mathcal{T} \to_M \mathcal{B}$, i.e. is Borel-measurable on $\mathcal{T}$. So the probability for $[\![P^{\bowtie r}\ \Phi]\!]$, and the integral for $[\![E^{\bowtie r}\ \Phi]\!]$ are well-defined.

4.2   Verifying the Algorithm

The model checking algorithm *Sat* for pCTL formulas is based on three methods:

- Iterative methods to compute the probability of bounded until and the expectation of bounded rewards
- Reachability analysis on the graph of non-zero transitions to compute the sets $[\![P^{=0}(\Phi\ U^\infty\ \Psi)]\!]$ and $[\![P^{=1}(\Phi\ U^\infty\ \Psi)]\!]$.
- Solving systems of linear equations for the unbounded until operator and unbounded rewards. This requires the previous methods to construct a system of linear equations with a unique solution.

Solving systems of linear equations may (in general) fail. To cater for this possibility we use option values in our computation and formulate our algorithm with the help of the *do*-syntax (recall Section 3.1).

The definition and the correctness proof of the algorithm *Sat* is by induction over the syntax of pCTL formulas. For a better overview of the formalization we split the definition of *Sat* into multiple parts interleaved with the necessary auxiliary definitions. The final soundness theorem states that *Sat* Φ returns a result and computes the set of states s for which $s \in [\![\Phi]\!]$ holds, i.e. *Sat* Φ = *Some* $[\![\Phi]\!]$.

The definition of *Sat* on *label* S', $\neg\Phi$, $\Phi \wedge \Psi$, and $P^{\bowtie r}(X\Phi)$ is easy. The soundness proof of the first three is done automatically, the last one needs Theorem 9.

Definition 21. *Computing pCTL-satisfiability (1)*

$$
\begin{aligned}
&\textit{Sat (label } S') &&:= \textit{return } \{s \in S \mid s \in S'\} \\
&\textit{Sat } (\neg\ \Phi) &&:= \textit{do} \\
&&&\quad F \leftarrow \textit{Sat } \Phi \\
&&&\quad \textit{return } (S \setminus F) \\
&\textit{Sat } (\Phi \wedge \Psi) &&:= \textit{do} \\
&&&\quad F_1 \leftarrow \textit{Sat } \Phi \\
&&&\quad F_2 \leftarrow \textit{Sat } \Psi \\
&&&\quad \textit{return } (F_1 \cap F_2) \\
&\textit{Sat } (P^{\bowtie r}\ (X\ \Phi)) &&:= \textit{do} \\
&&&\quad F \leftarrow \textit{Sat } \Phi \\
&&&\quad \textit{return } \left\{s \in S \mid \left(\textstyle\sum_{s' \in F} \tau\ s\ s'\right) \bowtie r\right\}
\end{aligned}
$$

The iterative methods to compute bounded until (*ProbUb k s S_1 S_2*), cumulative expectation (*ExpC k s*) and state expectation (*ExpI k s*) are simply defined by recursion on the bounding value k. Soundness is proved by induction on the bounding value k and using the iterative equations given by Theorem 9.

Definition 22. *Computing pCTL-satisfiability (2)*

$ProbUb\ 0\ s\ S_1\ S_2 \qquad := if\ s \in S_2\ then\ 1\ else\ 0$
$ProbUb\ (k + 1)\ s\ S_1\ S_2 := if\ s \in S_1 \setminus S_2\ then\ \sum_{s' \in S} \tau\ s\ s' \cdot ProbUb\ k\ s'\ S_1\ S_2$
$\qquad\qquad\qquad\qquad\qquad else\ (if\ s \in S_2\ then\ 1\ else\ 0)$

$ExpC\ 0\ s \qquad\qquad := 0$
$ExpC\ (k + 1)\ s \qquad := \rho\ s + \sum_{s' \in S} \tau\ s\ s' \cdot (\iota\ s\ s' + ExpC\ k\ s')$

$Expl\ 0\ s \qquad\qquad := \rho\ s$
$Expl\ (k + 1)\ s \qquad := \sum_{s' \in S} \tau\ s\ s' \cdot Expl\ k\ s'$

$Sat\ (P^{\bowtie r}\ (\Phi\ U^{\leq k}\ \Psi)) \quad := do$
$\qquad\qquad\qquad\qquad\qquad F_1 \leftarrow Sat\ \Phi$
$\qquad\qquad\qquad\qquad\qquad F_2 \leftarrow Sat\ \Psi$
$\qquad\qquad\qquad\qquad\qquad return\ \{s \in S \mid ProbUb\ k\ s\ F_1\ F_2 \bowtie r\}$
$Sat\ (E^{\bowtie r}\ (C^{<k})) \qquad := return\ \{s \in S \mid ExpC\ k\ s \bowtie r\}$
$Sat\ (E^{\bowtie r}\ (I^{=k})) \qquad := return\ \{s \in S \mid Expl\ k\ s \bowtie r\}$

Our next step is to check the unbounded until operator. Here we compute the probability $P_{\Phi,\Psi}(s) := \mu_s\{\omega \in \Omega \mid [\![\Phi\ U^\infty\ \Psi, s\cdot\omega]\!]_P\}$ for each state s by setting up a system of linear equations. From Theorem 9 and the behavior of the unbounded until operator we derive a system of linear equations for $P_{\Phi,\Psi}(s)$.

$$P_{\Phi,\Psi}(s) = \begin{cases} \sum_{s' \in E(s)} \tau\ s\ s' \cdot P_{\Phi,\Psi}(s') & if \quad s \in \Phi \setminus \Psi \\ 1 & if \quad s \in \Psi \\ 0 & otherwise \end{cases}$$

We show that such a linear equation system has a unique solution, with two conditions: (1) the solutions are equal on Ψ and (2) the solutions are equal in all states which never reach Ψ, i.e. $P_{\Phi,\Psi}(s) = 0$. We proved this lemma following the uniqueness proof in [6].

Lemma 23. *Unique solution*

$$\Phi \subseteq S \wedge \Psi \subseteq N \subseteq S$$
$$\wedge \quad \forall s \in S.\ P_{\Phi,\Psi}(s) = 0 \longrightarrow s \in N$$
$$\wedge \quad \forall s \in S \setminus N.\ l_1\ s - c\ s = \sum_{s' \in S} \tau\ s\ s' \cdot l_1\ s'$$
$$\wedge \quad \forall s \in S \setminus N.\ l_2\ s - c\ s = \sum_{s' \in S} \tau\ s\ s' \cdot l_2\ s'$$
$$\wedge \quad \forall s \in N.\ l_1\ s = l_2\ s$$
$$\longrightarrow \forall s \in S.\ l_1\ s = l_2\ s$$

To find a solution of such a system of linear equations, we formalized Gauss-Jordan elimination on matrices represented as functions [19]. Then we adapted this to use states as indices instead of natural numbers. Correctness says that if *gauss-jordan M a* returns *Some x*, then x is a solution to the equation system $M \cdot x = a$.

Lemma 24. *Gauss-Jordan elimination*

$$\textit{gauss-jordan } M \ a = \textit{Some } x \longrightarrow \forall s \in S. \left(\sum_{s' \in S} M \ s \ s' \cdot x \ s'\right) = a \ s$$

Before we use the uniqueness of our system of linear equationss, Lemma 23 requires us to compute the states with $P_{\Phi,\Psi}(s) = 0$ before the algorithm builds the system of linear equations. *Prob0* computes the set of all states with $P_{\Phi,\Psi}(s) > 0$ and returns the complement. The set of all s with $P_{\Phi,\Psi}(s) > 0$ is computed by starting with $R = \Psi$ and adding states to R which are in Φ and are predecessors of a state in R. With Lemma 11 we know that R contains all reachable states, hence $P_{\Phi,\Psi}(s) > 0$ for all $s \in R$. The termination measure for the *while*-combinator is the difference $S \setminus R$, with each step either states are added, or the loop terminates.

Definition 25. *Compute* $[\![P^{=0}(\Phi \ U^\infty \ \Psi)]\!]$

$$\textit{pred } \Phi \ R \ := \{s \in \Phi \mid R \cap E(s) \neq \emptyset\}$$
$$\textit{Prob0 } \Phi \ \Psi := S \ \setminus \ \textit{while } (\lambda R. \ \neg \textit{pred } \Phi \ R \subseteq R) \ (\lambda R. \ R \cup \textit{pred } \Phi \ R) \ \Psi$$

The system of linear equations solved by *gauss-jordan M a* needs to be in the right form, i.e. the matrix M contains all variable coefficients and a all constants. We introduce *LES F* to define the matrix of the linear equation system $l \ s = (\sum_{s \in S} \tau \ s \ s' \cdot l \ s') + a \ s$ for $s \notin F$, and $l \ s = a \ s$ if $s \in F$.

Definition 26. *Linear Equation System to Compute Unbounded Until*

$$\textit{LES F r c} := \textit{if } r \in F \textit{ then } (\textit{if } c = r \textit{ then } 1 \textit{ else } 0)$$
$$\textit{else } (\textit{if } c = r \textit{ then } \tau \ r \ c - 1 \textit{ else } \tau \ r \ c)$$

Combining all this we can finally compute the probability of a unbounded until formula. We prove its soundness using Lemmas 24 and 23, and Theorem 9.

Definition 27. *Computing pCTL-satisfiability (3)*

$$\textit{Sat } (P^{\bowtie r} \ (\Phi \ U^\infty \ \Psi)) := \textit{do}$$
$$F_1 \leftarrow \textit{Sat } \Phi$$
$$F_2 \leftarrow \textit{Sat } \Psi$$
$$p \leftarrow \textit{gauss-jordan } (\textit{LES } (F_2 \cup \textit{Prob0 } F_1 \ F_2))$$
$$(\lambda s. \ \textit{if } s \in F_2 \textit{ then } 1 \textit{ else } 0)$$
$$\textit{return } \{s \in S \mid p \ s \bowtie r\}$$

The last equation of *Sat* computes the unbounded reward $E^{\bowtie r}(F^\infty \ \Phi)$. Similar to the unbounded until operator, we introduce a system of linear equations for $R_\Phi(s) := \int_\omega [\![F^\infty \ \Phi, s \cdot \omega]\!]_E d\mu_s$. With Theorem 17 we know that $R_\Phi(s)$ is finite if $P_{S,\Phi}(s) = 1$. If $P_{S,\Phi}(s) < 1$ there is a non-zero probability that Φ is never reached, and hence $R_\Phi(s) = \infty$.

$$R_\Phi(s) = \begin{cases} \sum_{s' \in E(s)} \tau \ s \ s' \cdot (\rho \ s + \iota \ s \ s' + R_\Phi(s')) & \text{if} \quad P_{S,\Phi}(s) = 1 \wedge s \notin \Phi \\ 0 & \text{if} \quad s \in \Phi \\ \infty & \text{otherwise} \end{cases}$$

To be usable with *LES*, we rewrite the first equation into:

$$R_\Phi(s) - \left(\rho\ s + \sum_{s' \in E(s)} \tau\ s\ s' \cdot \iota\ s\ s' \right) = \sum_{s' \in E(s)} \tau\ s\ s' \cdot R_\Phi(s') \ .$$

The Gauss-Jordan elimination we use works only on real numbers, luckily we can replace ∞ by 0 and replace it again after we solved the equation system. This is sound since for each s and $s' \in E(s)$ with $R_\Phi(s') = \infty$ either $s \in \Phi$ or $R_\Phi(s) = \infty$ hold. The states s with $P_{S,\Phi}(s) = 1$ are computed by *Prob1*, building on *Prob0*.

Definition 28. *Compute* $[\![P^{=1}(\Phi\ U^\infty\ \Psi)]\!]$

$$Prob1\ \Phi\ \Psi := Prob0\ (\Phi \setminus \Psi)\ (Prob0\ \Phi\ \Psi)$$

We know that the resulting states only reach states which again reach Ψ, hence the assumptions of Theorem 15 are fulfilled, and we know that *Prob1 S Φ* is the set of all states s with $P_{S,\Phi}(s) = 1$. With all this, we can formalize the last equation for *Sat*.

Definition 29. *Computing pCTL-satisfiability (4)*

$Sat\ (E^{\bowtie r}\ (F^\infty\ \Phi)) := do$
$\qquad F \leftarrow Sat\ \Phi$
$\qquad let\ Y = Prob1\ S\ F$
$\qquad l \leftarrow gauss\text{-}jordan\ (LES\ (S \setminus (Y \setminus F)))$
$\qquad\qquad (\lambda s.\ if\ i \in Y \setminus F\ then\ -(\rho\ s + (\sum_{s' \in S} \cdot\ \tau\ s\ s' \cdot \iota\ s\ s'))$
$\qquad\qquad\qquad else\ 0)$
$\qquad let\ e = (\lambda s.\ if\ s \in Y\ then\ l\ s\ else\ \infty)$
$\qquad return\ \{s \in S \mid e\ s \bowtie r\}$

Finally we show the soundness of *Sat* by induction on the structure of Φ. If we assume that *Sat* terminates with a result F, then F is the same set as defined by the semantic.

Theorem 30. *Soundness of Sat*

$$Sat\ \Phi = Some\ F \longrightarrow [\![\Phi]\!] = F$$

Now we turn to completeness. The only case in which *Sat* returns *None* is when the Gauss-Jordan elimination does not find a unique solution. Hence we need the property that if a unique solution exists, then *gauss-jordan* returns this solution.

Theorem 31. *Completeness of gauss-jordan*
  If there is a unique solution x for $M \cdot x = a$:

$$\forall s \in S.\ \sum_{s' \in S} M\ s\ s' \cdot x\ s' = a\ s$$

$$\forall y. \left(\forall s \in S. \sum_{s' \in S} M \ s \ s' \cdot y \ s' = a \ s \right) \longrightarrow \forall s \in S. \ x \ s = y \ s$$

then gauss-jordan returns a result:

$$\exists x'. \ \textit{gauss-jordan } M \ a = \textit{Some } x'$$

With this and Lemma 23 we prove that *Sat* always returns a result:

Theorem 32. *Completeness of Sat*

$$\exists F. \ \textit{Sat } \Phi = \textit{Some } F \ .$$

Using Theorem 30 we finally show

Corollary 33. *Soundness and completeness of Sat*

$$\textit{Sat } \Phi = \textit{Some } [\![\Phi]\!] \ .$$

5   Discussion

We used the tutorial [12] as a guideline to formalize the pCTL model checking algorithm. Most parts of the soundness proof are straightforward. Three parts, however, required a more substantial formalization of the background theory:

- The correctness of *Prob1* is based on Theorem 15, which required us to formalize state fairness as found in [1].
- For the unbounded until and the unbounded rewards we solve a linear equation system. We needed to show that the solution of this equation system is unique, for which we followed the original proof from [6].
- The unbounded reward for a state can only be characterized as a linear equation if the reward is finite. We needed Theorem 17 to show that the reward is finite, if the final states are almost always reached.

Technically, the largest difference between our work and Kwiatkowska *et. al.* [12] is the construction of the probability space of paths: we use infinite products of probability spaces, whereas they use Caratheodory on semi-rings of sets. We do not need to show that the probability of cylinders is countably additive, this is generically done for infinite products. We want to reuse the infinite products for continuous-time Markov chains and Markov decision processes. With Caratheodory on semi-rings of sets it would be necessary to show countably additivity for each of them. Nevertheless, we intend to formalize the latter construction, too, as it is a valuable addition to our library.

The equations we give for the algorithm are not directly executable by the code generator in Isabelle [5]. We use sets in our equations, and the adaption of Gauss-Jordan elimination uses an arbitrary mapping from $\{0, \ldots, |S| - 1\}$ to S. One method to obtain a executable version is to create a copy Sat_L of Sat operating on lists instead of subsets of S. We assume as input a list of states $xs := [s_0, s_1, \ldots s_n]$, and define the Markov chains on $S := \textit{set-of } xs$. It should be straightforward to show that $\textit{Sat } \Phi = \textit{Some } F$ implies $\textit{set-of } (Sat_L \ \Phi) = F$. The biggest hurdle is the while-combinator in *Prob0* and the adaption of Gauss-Jordan elimination.

6   Conclusion

The formalization of pCTL model checking in a proof assistant opens up a number of possible application scenarios:

Model Checking as an Isabelle Proof Method. Once we have made our pCTL model checker executable as explained in Section 5, we can call it as an automated proof method for pCTL formulas within Isabelle. Of course this is only practical for small examples, for larger ones an external pCTL model checker would be used as an oracle that must be trusted.

Certified Model Checking. Result checking is an established technique where, rather than verifying an algorithm, each execution of the algorithm is checked. This requires the algorithm to return a checkable certificate. A particularly successful example of such a system architecture is CeTA [24], a checker for termination proofs which regularly finds bugs in termination proof tools. CeTA is verified in Isabelle and an efficient Haskell program is extracted that can check large proof certificates.

Verification of Parametrized Models. The Markov chain may depend on parameters like the number of parallel processes. Such parameterized models can be model checked only for fixed parameter values. Our theory allows one to formalize and verify such parameterized models for all possible parameter values interactively. As case studies we formalized IPv4 address allocation in the ZeroConf protocol and anonymity of the Crowds protocol [9]. The formalizations we describe in Section 3 where essential for these case studies.

The formalization is available in the AFP [9,19]. It has about 4480 lines: 3670 lines for the formalization of DTMCs, 270 lines for Gauss-Jordan elimination, and 1140 lines for pCTL model checking.

Our future goal is to formalize more probabilistic models with the corresponding model checking algorithms, like pCTL for Markov decision processes, continuous stochastic logic for continuous-time Markov chains and probabilistic timed CTL for probabilistic timed automata.

References

1. Baier, C.: On the Algorithmic Verification of Probabilistic Systems. Habilitation, Universität Mannheim (1998)
2. Bauer, H.: Probability Theory. de Gruyter (1995)
3. Chou, C.T., Peled, D.: Formal verification of a partial-order reduction technique for model checking. Journal of Automated Reasoning 23(3-4), 265–298 (1999)
4. Coble, A.R.: Anonymity, Information, and Machine-Assisted Proof. Ph.D. thesis, King's College, University of Cambridge (2009)
5. Haftmann, F., Nipkow, T.: Code Generation via Higher-Order Rewrite Systems. In: Blume, M., Kobayashi, N., Vidal, G. (eds.) FLOPS 2010. LNCS, vol. 6009, pp. 103–117. Springer, Heidelberg (2010)
6. Hansson, H., Jonsson, B.: A logic for reasoning about time and reliability. Tech. Rep. SICS/R90013, Swedish Institute of Computer Science (December 1994)

7. Hurd, J.: Formal Verification of Probabilistic Algorithms. Ph.D. thesis, University of Cambridge (2002)
8. Hölzl, J., Heller, A.: Three Chapters of Measure Theory in Isabelle/HOL. In: van Eekelen, M.C.J.D., Geuvers, H., Schmaltz, J., Wiedijk, F. (eds.) ITP 2011. LNCS, vol. 6898, pp. 135–151. Springer, Heidelberg (2011)
9. Hözl, J., Nipkow, T.: Markov models. In: Klein, G., Nipkow, T., Paulson, L. (eds.) The Archive of Formal Proofs, formal proof development (January 2012), http://afp.sf.net/entries/Markov_Models.shtml
10. Katoen, J.P., Zapreev, I.S., Hahn, E.M., Hermanns, H., Jansen, D.N.: The ins and outs of the probabilistic model checker MRMC. Performance Evaluation 68, 90–104 (2011)
11. Klein, G., Elphinstone, K., Heiser, G., Andronick, J., Cock, D., Derrin, P., Elkaduwe, D., Engelhardt, K., Kolanski, R., Norrish, M., Sewell, T., Tuch, H., Winwood, S.: seL4: Formal verification of an OS kernel. In: Proc. 22nd ACM Symposium on Operating Systems Principles 2009, pp. 207–220 (2009)
12. Kwiatkowska, M., Norman, G., Parker, D.: Stochastic Model Checking. In: Bernardo, M., Hillston, J. (eds.) SFM 2007. LNCS, vol. 4486, pp. 220–270. Springer, Heidelberg (2007)
13. Kwiatkowska, M., Norman, G., Parker, D.: PRISM 4.0: Verification of Probabilistic Real-Time Systems. In: Gopalakrishnan, G., Qadeer, S. (eds.) CAV 2011. LNCS, vol. 6806, pp. 585–591. Springer, Heidelberg (2011)
14. Lammich, P., Müller-Olm, M., Wenner, A.: Predecessor Sets of Dynamic Pushdown Networks with Tree-Regular Constraints. In: Bouajjani, A., Maler, O. (eds.) CAV 2009. LNCS, vol. 5643, pp. 525–539. Springer, Heidelberg (2009)
15. Leroy, X.: A formally verified compiler back-end. J. Automated Reasoning 43, 363–446 (2009)
16. Levin, D.A., Peres, Y., Wilmer, E.L.: Markov chains and mixing times. AMS (2006)
17. Liu, L., Hasan, O., Tahar, S.: Formalization of Finite-State Discrete-Time Markov Chains in HOL. In: Bultan, T., Hsiung, P.-A. (eds.) ATVA 2011. LNCS, vol. 6996, pp. 90–104. Springer, Heidelberg (2011)
18. Mhamdi, T., Hasan, O., Tahar, S.: Formalization of Entropy Measures in HOL. In: van Eekelen, M.C.J.D., Geuvers, H., Schmaltz, J., Wiedijk, F. (eds.) ITP 2011. LNCS, vol. 6898, pp. 233–248. Springer, Heidelberg (2011)
19. Nipkow, T.: Gauss-Jordan elimination for matrices represented as functions. In: Klein, G., Nipkow, T., Paulson, L. (eds.) The Archive of Formal Proofs, formal proof development (August 2011), http://afp.sf.net/entries/Gauss-Jordan-Elim-Fun.shtml
20. Nipkow, T., Paulson, L.C., Wenzel, M.: Isabelle/HOL. LNCS, vol. 2283. Springer, Heidelberg (2002)
21. Reif, W., Schellhorn, G., Vollmer, T., Ruf, J.: Correctness of efficient real-time model checking. J. UCS 7(2), 194–209 (2001)
22. Schimpf, A., Merz, S., Smaus, J.-G.: Construction of Büchi Automata for LTL Model Checking Verified in Isabelle/HOL. In: Berghofer, S., Nipkow, T., Urban, C., Wenzel, M. (eds.) TPHOLs 2009. LNCS, vol. 5674, pp. 424–439. Springer, Heidelberg (2009)
23. Sprenger, C.: A Verified Model Checker for the Modal μ-Calculus in Coq. In: Steffen, B. (ed.) TACAS 1998. LNCS, vol. 1384, pp. 167–183. Springer, Heidelberg (1998)
24. Thiemann, R., Sternagel, C.: Certification of Termination Proofs Using CeTA. In: Berghofer, S., Nipkow, T., Urban, C., Wenzel, M. (eds.) TPHOLs 2009. LNCS, vol. 5674, pp. 452–468. Springer, Heidelberg (2009)

Parameterized Synthesis[*]

Swen Jacobs[1] and Roderick Bloem[2]

[1] École Polytechnique Fédérale de Lausanne (EPFL), Switzerland
swen.jacobs@epfl.ch
[2] IAIK, Graz University of Technology, Austria
roderick.bloem@iaik.tugraz.at

Abstract. We study the synthesis problem for distributed architectures with a parametric number of finite-state components. Parameterized specifications arise naturally in a synthesis setting, but thus far it was unclear how to decide realizability and how to perform synthesis. Using a classical result from verification, we show that for specifications in LTL\X, parameterized synthesis of token ring networks is equivalent to distributed synthesis of a network consisting of a few copies of a single process. Adapting a result from distributed synthesis, we show that the latter problem is undecidable. We then describe a semi-decision procedure based on bounded synthesis and show applicability on a simple case study. Finally, we sketch a general framework for parameterized synthesis based on cut-off results for verification.

1 Introduction

Synthesis is the problem of turning a temporal logical specification into a reactive system [1,2]. In synthesis, parameterized specifications occur very naturally. For instance, Piterman, Pnueli, and Sa'ar illustrate their GR(1) approach with two parameterized examples of an arbiter and an elevator controller [3]. Similarly, the case studies given in [4,5] consist of a parameterized specification of the AMBA bus arbiter. A simple example of a parameterized specification may be

$$\forall i. \ \mathsf{G}(r_i \to \mathsf{F} \, g_i) \ \land \ \forall i \neq j. \ \mathsf{G}(\neg g_i \lor \neg g_j).$$

This specification describes an arbiter serving an arbitrary number of clients, say n. Client i receives an input r_i for requests and controls an output g_i for grants. The specification states that for each client i, a request r_i is eventually followed by a grant g_i, but grants never occur simultaneously.

Previous approaches have focused on the synthesis of such systems for a fixed n. The question whether such a specification is realizable for *any* n is natural: it occurs, for instance, in the work on synthesis of processes for the leader election problem by Katz and Peled [6]. Only an answer to this question can determine whether a parameterized specification is correct. A further natural question is

---

[*] This work was supported by the Austrian Science Fund (FWF) under the RiSE National Research Network (S11406) and by the Swiss NSF Grant #200021_132176.

C. Flanagan and B. König (Eds.): TACAS 2012, LNCS 7214, pp. 362–376, 2012.
© Springer-Verlag Berlin Heidelberg 2012

how to construct a parameterized system, i.e., a recipe for quickly constructing a system for an arbitrary n. Such a construction would avoid the steep increase of runtime and memory use with n that current tools incur [4,5,7].

Parameterized systems have been studied extensively in the context of verification. It is well known that the verification of such systems is undecidable [8,9], although it can be decided for some restricted cases. In particular, for restricted topologies, the problem of verifying a network of isomorphic processes of arbitrary size can be reduced to the verification of a small network [10,11]. As a corollary, synthesis of a network of an arbitrary number of processes can be reduced to synthesis of a small network, as long as the restricted topology is respected. In this paper, we focus on token ring topologies [10].

The question of synthesis of token rings is thus equivalent to the synthesis of a small network of isomorphic processes. This question is closely related to that of distributed synthesis [12,13,14]. Distributed synthesis is undecidable for all systems in which processes are incomparable with respect to their information about the environment. Our problem is slightly different in that we only consider specifications in LTL\X and that our synthesis problem is *isomorphic*, i.e., processes have to be identical. Unfortunately, this problem, and thus the original problem of parameterized synthesis, is also undecidable.

Having obtained a negative decidability result, we turn our attention to a semi-decision procedure, namely bounded synthesis [15,16], an approach that searches for systems with a bounded number of states. We modify this approach to deal with isomorphic token-passing systems. Bounded synthesis reduces the problem of realizability to an SMT formula, a model of which gives an implementation of the system. Using Z3 [17], we show that a simple parameterized arbiter can be synthesized in reasonable time. Finally, we sketch a framework that extends our approach to the more general topologies of [11], and other classes of systems and specifications, in particular those that allow a cut-off for the corresponding verification problem.

2 Preliminaries

We consider the synthesis problem for distributed systems, with specifications in (fragments of) LTL. Given a system architecture A and a specification φ, we want to find implementations of all system processes in A, such that their composition satisfies φ.

Architectures. An *architecture* A is a tuple (P, env, V, I, O), where P is a finite set of processes, containing the environment process env and system processes $P^- = P \setminus \{env\}$, V is a set of boolean system variables, $I = \{I_i \subseteq V \mid i \in P^-\}$ assigns a set I_i of boolean input variables to each system process, and $O = \{O_i \subseteq V \mid i \in P\}$ assigns a set O_i of boolean output variables to each process, such that $\bigcup_{i \in P} O_i = V$. In contrast to output variables, inputs may be shared between processes. Wlog., we use natural numbers to refer to system processes, and assume $P^- = \{1, \ldots, k\}$ for an architecture with k system processes.

Implementations. An *implementation* $\mathcal{T}_i$ of a system process i with inputs I_i and outputs O_i is a labeled transition system (LTS) $\mathcal{T}_i = (T_i, t_i, \rho_i, o_i)$, where T_i is a set of states including the initial state t_i, $\rho_i : T_i \times \mathcal{P}(I_i) \to T_i$ a transition function, and $o_i : T_i \to \mathcal{P}(O_i)$ a labeling function.

The *composition* of the set of system process implementations $\{\mathcal{T}_1, \ldots, \mathcal{T}_k\}$ is the LTS $\mathcal{T}_A = (T_A, t_0, \rho, o)$, where the states are $T_A = T_1 \times \cdots \times T_k$, the initial state $t_0 = (t_1, \ldots, t_k)$, the labeling function $o : T_A \to \mathcal{P}(\bigcup_{1 \leq i \leq k} O_i)$ with $o(t_1, \ldots, t_k) = o_1(t_1) \cup \cdots \cup o_k(t_k)$, and finally the transition function $\rho : T_A \times \mathcal{P}(O_{env}) \to T_A$ with

$$\rho((t_1, \ldots, t_k), e) = (\rho_1(t_1, (o(t_1, \ldots, t_k) \cup e) \cap I_1), \ldots, \rho_k(t_k, (o(t_1, \ldots, t_k) \cup e) \cap I_k)),$$

i.e., every process advances according to its own transition function and input variables, where inputs from other system processes are interpreted according to the labeling of the current state.

A *run* of an LTS (T, t_0, ρ, o) is an infinite sequence $(t^0, e^0), (t^1, e^1), \ldots$, where $t^0 = t_0$, $e^i \subseteq O_{env}$ and $t^{i+1} = \rho(t^i, e^i)$. An LTS *satisfies* a formula φ if for every run, the sequence $o(t^0) \cup e^0, o(t^1) \cup e^1, \ldots$ is a model of φ.

Asynchronous Systems. An *asynchronous system* is an LTS such that in every transition, only a subset of the system processes changes their state. This is decided by a *scheduler*, which can choose in every step which of the processes (including the environment) is allowed to make a step. In our setting, we will assume that the environment is always scheduled, and consider the scheduler as a part of the environment.

Formally, O_{env} contains additional scheduling variables $s_1, \ldots, s_k$, and $s_i \in I_i$ for every i. We require $\rho_i(t, I) = t$ for any i and set of inputs I with $s_i \notin I$.

Token Rings. We consider a class of architectures called *token rings*, where the only communication between system processes is a token. At any time only one process can possess the token, and a process i which has the token can pass it to process $i + 1$ by raising an output $\mathsf{send}_i \in O_i \cap I_{i+1}$. For processes in token rings of size k, addition and subtraction is done modulo k.

We assume that token rings are implemented as asynchronous systems, where in every step only one system process may change its state, except for token-passing steps, in which both of the involved processes change their state.

Distributed Synthesis. The *distributed synthesis problem* for a given architecture A and a specification φ, is to find implementations for the system processes of A, such that the composition of the implementations $\mathcal{T}_1, \ldots, \mathcal{T}_k$ satisfies φ, written $A, (\mathcal{T}_1, \ldots, \mathcal{T}_k) \models \varphi$. A specification φ is *realizable* with respect to an architecture A if such implementations exist. Synthesis and checking realizability of LTL specifications have been shown to be undecidable for architectures in which not all processes have the same information wrt. environment outputs in the synchronous case [13], and even for all architectures with more than one system process in the asynchronous case [14].

Bounded Synthesis. The *bounded synthesis problem* for given architecture A, specification φ and a family of bounds $\{b_i \in \mathbb{N} \mid i \in P^-\}$ on the size of system processes as well as a bound b_A for the composition $\mathcal{T}_A$, is to find implementations $\mathcal{T}_i$ for the system processes such that their composition $\mathcal{T}_A$ satisfies φ, with $|T_i| \leq b_i$ for all process implementations, and $|T_A| \leq b_A$.

3   Parameterized Synthesis

In this section, we introduce the parameterized synthesis problem. Using a classical result for the verification of token rings by Emerson and Namjoshi [10], we show that parameterized synthesis for token ring architectures and specifications in LTL\X can be reduced to distributed synthesis of isomorphic processes in a ring of fixed size. We then show that for this class of architectures and specifications, the isomorphic distributed synthesis problem is still undecidable.

3.1   Definition

Parameterized Architectures and Specifications. Let $\mathcal{A}$ be the set of all architectures. A *parameterized architecture* is a function $\Pi : \mathbb{N} \to \mathcal{A}$. A *parameterized token ring* is a parameterized architecture R with $R(n) = (P_n, env, V_n, I_n, O_n)$, where

- $P_n = \{env, 1, \ldots, n\}$,
- I_n is such that all system processes are assigned isomorphic sets of inputs, consisting of the token-passing input send_{i-1} from process $i-1$ and a set of inputs from the environment, distinguished by indexing each input with i.
- Similarly, O_n assigns isomorphic, indexed sets of outputs to all system processes, with $\mathsf{send}_i \in O_n(i)$, and every output of *env* is indexed with all values from 1 to n.

A *parameterized specification* φ is an LTL specification with indexed variables, and universal quantification over indices. We say that a parameterized architecture Π and a process implementation $\mathcal{T}$ *satisfy* a parameterized specification (written $\Pi, \mathcal{T} \models \varphi$) if for any n, $\Pi(n), (\mathcal{T}, \ldots, \mathcal{T}) \models \varphi$.

Example 1. Consider the parameterized token ring R_{arb} with $R_{arb}(n) = (P_n, env, V_n, I_n, O_n)$, where

$$P_n = \{env, 1, \ldots, n\} \tag{1}$$

$$V_n = \{r_1, \ldots, r_n, g_1 \ldots, g_n, \mathsf{send}_1, \ldots, \mathsf{send}_n\} \tag{2}$$

$$I_n(i) = \{r_i, \mathsf{send}_{i-1}\} \tag{3}$$

$$O_n(env) = \{r_1, \ldots, r_n\} \tag{4}$$

$$O_n(i) = \{g_i, \mathsf{send}_i\} \tag{5}$$

The architecture $R(n)$ defines a token ring with n system processes, with each process i receiving an input r_i from the environment and another input send_{i-1} from the previous process in the ring, and an output send_i to the next process, as well as an output g_i to the environment.

An instance of this parameterized architecture for $n = 4$ is depicted in Fig. 1. Together with the parameterized specification from Section 1, we will use it in Section 5 to synthesize process implementations for a parameterized arbiter.

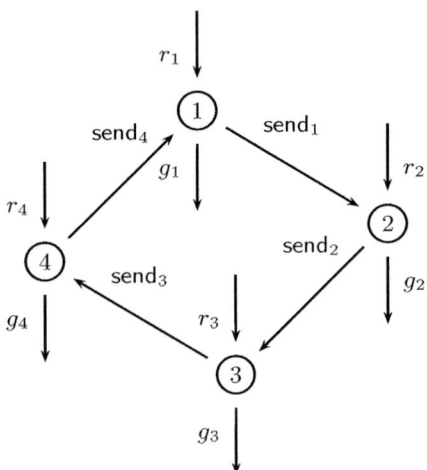

Fig. 1. Token ring architecture with 4 processes

Isomorphic and Parameterized Synthesis. The *isomorphic synthesis problem* for an architecture A and a specification φ is to find an implementation $\mathcal{T}$ for all system processes $(1, \ldots, k)$ such that $A, (\mathcal{T}, \ldots, \mathcal{T}) \models \varphi$. The *parameterized synthesis problem* for a parameterized architecture Π and a parameterized specification φ is to find an implementation $\mathcal{T}$ for all system processes such that $\Pi, \mathcal{T} \models \varphi$. The *parameterized (isomorphic) realizability problem* is the question whether such an implementation exists.

3.2   Reduction of Parameterized to Isomorphic Synthesis

Emerson and Namjoshi [10] have shown that verification of LTL\X properties for implementations of parameterized token rings can be reduced to verification of a small ring with up to five processes, depending on the form of the specification.

Theorem 1 ([10]). *Let R be a parameterized token ring, $\mathcal{T}$ an implementation of the isomorphic system processes that ensures fair token passing, and φ a parameterized specification. Then*

a) *If $\varphi = \forall i.\ f_i$, where f_i is a formula that only refers to variables indexed by i, then $R, \mathcal{T} \models \varphi \iff R(2), \mathcal{T} \models \varphi$*

b) If $\varphi = \forall i.\ f_{i,i+1}$, where $f_{i,i+1}$ is a formula that only refers to variables indexed by i and $i+1$, then $R, \mathcal{T} \models \varphi \iff R(3), \mathcal{T} \models \varphi$

c) If $\varphi = \forall i \neq j.\ f_{i,j}$, where $f_{i,j}$ is a formula that only refers to variables indexed by i and j, then $R, \mathcal{T} \models \varphi \iff R(4), \mathcal{T} \models \varphi$

d) If $\varphi = \forall i \neq j.\ f_{i,i+1,j}$, where $f_{i,i+1,j}$ is a formula that only refers to variables indexed by i, $i+1$, and j, then $R, \mathcal{T} \models \varphi \iff R(5), \mathcal{T} \models \varphi$

This theorem implies that verification of such structures is decidable. For synthesis, we obtain the following corollary:

Corollary 1. *For a given parameterized token ring R and parametric specification φ, parameterized synthesis can be reduced to isomorphic synthesis in rings of size 2 (3, 4, 5) for specifications of type a) (b, c, d, resp.).*

In the following, we will show that this reduction in general does not make the synthesis problem decidable.

3.3   Decidability

The parameterized synthesis problem is closely related to the distributed synthesis problem [12,13]. We will use a modification of the original undecidability proof for distributed systems to show undecidability of isomorphic synthesis in token rings, which in turn implies undecidability of parameterized synthesis.

Theorem 2. *The isomorphic realizability problem is undecidable for token rings with 2 or more processes and specifications in $LTL\backslash X$.*

Proof. The proof follows that of Pnueli and Rosner [12] (see also Finkbeiner and Schewe [13]). The original proof is for two synchronous processes, neither of which can observe the inputs or outputs of the other. The proof builds a specification that allows a single implementation, and forces the two processes to each simulate a Turing machine and halt. Thus, it is realizable iff the Turing machine halts, which shows undecidability. We will show that we can specify (in $LTL\backslash X$) an asynchronous system in a token ring that simulates the behavior of these two synchronous processes. The proof works for rings of arbitrary size, if we assume that the specification is the same for all processes.

In the original proof, each process has a start signal that triggers the processes to output the next configuration of the Turing machine. The specification assumes that the number of start signals for the two processes is never different by more than one and requires that the configurations that are output by the two processes are either equal (if the number of start signals is equal) or that they are successors (if the number of start signals is off by one). This is easily specified because the processes are synchronized by a global clock.

We need to modify the original proof such that it works for asynchronous systems, and the specification can be written without the X operator.

This can be achieved by forcing the asynchronous system to simulate a synchronous system by using the token for synchronization: We augment the specification to assume that the token starts at a designated process, say 1. A clock cycle consists of a full cycle of the token, and we require that each process changes its output only once in each cycle. Thus, the asynchronous system simulates a synchronized system, where the synchronous states consist of the state of the asynchronous system immediately after the token passes to 1. Using tok_1 to identify these states, it is now possible to correlate the states of the simulated system: for instance, $\mathsf{X}\, q_i$ for the synchronous system corresponds to $\neg\mathsf{tok}_1\,\mathsf{W}\,(\mathsf{tok}_1 \wedge q_i)$ for the asynchronous system, and $\mathsf{G}\, q_i$ corresponds to $\mathsf{G}(\neg\mathsf{tok}_1 \implies \neg\mathsf{tok}_1\,\mathsf{W}\,\mathsf{tok}_1 \wedge q_i)$. This allows us to translate the construction in [12] to our setting, and remove all occurrences of X in the specification.

Finally, the token cannot be used to pass any additional information (beyond the synchronization): the only freedom a process has is *when* to pass the token, and by lack of a global clock and visibility of the input and output signals of the other processes, a given process cannot measure this time or observe any changes of the system during this time.

Thus, our asynchronous system simulates the synchronous system from [12] and is realizable iff the Turing machine halts.   □

Combining Theorems 1 and 2, we obtain the following result.

Theorem 3. *The parametric realizability problem is undecidable for token rings and specifications of type (a), (b), (c), or (d).*

Proof. By Theorem 1, the isomorphic realizability problem for a specification of type (a) and two processes can be reduced to a parameterized realizability problem of type (a). Since the former problem is undecidable, so is the latter. The proof for cases (b)–(d) is analogous.   □

4   Bounded Isomorphic Synthesis

The reduction from Section 3 allows us to reduce parameterized synthesis to isomorphic synthesis with a fixed number of processes. Still, the problem does not fall into a class for which the distributed synthesis problem is decidable.

For distributed architectures that do not fall into decidable classes, Finkbeiner and Schewe have introduced the semi-decision procedure of *bounded synthesis* [15,16], which converts an undecidable distributed synthesis problem into a sequence of decidable synthesis problems, by bounding the size of the implementation. In the following, we will show how to adapt bounded synthesis for isomorphic synthesis in token rings, which by Corollary 1 amounts to parameterized synthesis in token rings.

4.1   Bounded Synthesis

The bounded synthesis procedure consists of three main steps:

Step 1: Automata translation. Following an approach by Kupferman and Vardi [18], the LTL specification φ (including fairness assumptions like fair scheduling) is translated into a universal co-Büchi-automaton $\mathcal{U}$ which accepts an LTS $\mathcal{T}$ iff $\mathcal{T}$ satisfies φ.

Step 2: SMT Encoding. Existence of an LTS which satisfies φ is encoded into a set of SMT constraints over the theory of integers and free function symbols. States of the LTS are represented by natural numbers, state labels as free functions of type $\mathbb{N} \to \mathbb{B}$, and the global transition function as a free function of type $\mathbb{N} \times \mathbb{B}^{|O_{env}|} \to \mathbb{N}$. Transition functions of individual processes are defined indirectly by introducing projections $d_i : \mathbb{N} \to \mathbb{N}$, mapping global to local states. To ensure that local transitions of process i only depend on inputs in I_i, we add a constraint

$$\forall i. \; \forall t, t'. \; \forall I, I'. \; d_i(t) = d_i(t') \wedge I \cap I_i = I' \cap I_i \; \to \; d_i(\tau(t, I)) = d_i(\tau(t', I')).$$

To obtain an interpretation of these symbols that satisfies the specification φ, additional annotations of states are introduced. This includes labels $\lambda_q^{\mathbb{B}} : \mathbb{N} \to \mathbb{B}$ and free functions $\lambda_q^{\#} : \mathbb{N} \to \mathbb{N}$, which are defined such that (i) $\lambda_q^{\mathbb{B}}(t)$ is true iff the product of $\mathcal{T}$ and $\mathcal{U}$ contains a path from an initial state to a state (t, q) with $q \in Q$, i.e., the product automaton can reach a state in which $\mathcal{U}$ is in q, among other states, and (ii) valuations of the $\lambda_q^{\#}$ must be increasing along paths of $\mathcal{U}$, and strictly increasing for transitions that enter a rejecting state of $\mathcal{U}$. Together, this ensures that an LTS satisfying these constraints cannot have runs which enter rejecting states infinitely often (and thus would be rejected by $\mathcal{U}$).

Step 3: Iteration for Increasing Bounds. To obtain a decidable problem, we restrict the number of states in the LTS that we are looking for, which allows us to instantiate all quantifiers over state variables t, t' explicitly with all values in the given range. If the constraints are unsatisfiable for a given bound, we increase it and try again. If they are satisfiable, we obtain a model, giving us an implementation for the system processes such that φ is satisfied.

4.2 Adaption to Token Rings

We adapt the bounded synthesis approach for synthesis in token rings, and introduce some optimizations we found vital for a good performance of the synthesis method.

Additional Constraints and Optimizations. We use some of the general modifications and optimizations mentioned in [16]:

– We use an additional constraint to ensure that the resulting system implementation is asynchronous. In general, we could directly add a constraint $\forall i. \; \forall I. \; s_i \notin I \to d_i(\tau(t, I)) = d_i(t)$ (where I is a set of inputs and s_i is the scheduling variable for process i). For the particular case of token rings we use a modified version, explained below.

- We use symmetry constraints to encode that all processes should be isomorphic. Particularly, we use the same function symbols for state labels of all system processes, and special constraints for the local transition functions, also explained below.
- We use the semantic variant where environment inputs are not stored in system states, but are directly used in the transition term that computes the following state. This results in an implementation which is a factor of $|O_{env}|$ smaller.[1]

Encoding Token Rings. For the particular case of token rings, we use the following modifications to the SMT encoding:

- We want to obtain an asynchronous system in which the environment is always scheduled, along with exactly one system process. Thus, we do not need $|P|$ scheduling variables, but can encode the index of the scheduled process into a binary representation with $log_2(|P^-|)$ inputs.
- We encode the special features of token rings: i) exactly one process should have the token at any time, ii) only a process which has the token can send it, iii) if process i is scheduled, currently has the token and wants to send it, then in the next state process $i+1$ has the token and process i does not, and iv) if process i has the token and does not send it (or is not scheduled), it also has the token in the next state. Properties ii) – iv) are encoded in the following constraints, where $\mathsf{tok}_i((d_i(t))$ is true in state t iff process i has the token, $\mathsf{send}(d_i(t))$ is true iff i is ready to send the token, and $\mathsf{sched}_i(I)$ is true iff the scheduling variables in I are such that process i is scheduled:

$$\forall i.\ \forall t.\ \forall I.\ \ \mathsf{tok}(d_i(t))\ \rightarrow\ (\mathsf{send}(d_i(t)) \wedge \mathsf{sched}_i(I)) \vee \mathsf{tok}(d_i(\tau(t,I)))$$
$$\forall i.\ \forall t.\ \qquad\ \neg\mathsf{tok}(d_i(t))\ \rightarrow\ \neg\mathsf{send}(d_i(t))$$
$$\forall i.\ \forall t.\ \forall I.\ \ \mathsf{send}(d_i(t)) \wedge \mathsf{sched}_i(I)\ \rightarrow\ \neg\mathsf{tok}(d_i(\tau(t,I)))$$
$$\forall i.\ \forall t.\ \forall I.\ \ \mathsf{send}(d_{i-1}(t)) \wedge \mathsf{sched}_i(I)\ \rightarrow\ \mathsf{tok}(d_i(\tau(t,I)))$$

We do not encode property i) directly, because it is implied by the remaining constraints whenever we start in a state where only one process has the token.
- Token passing is an exception to the rule that only the scheduled process changes its state: if process i is scheduled in state t, and both $\mathsf{tok}(d_i(t))$ and $\mathsf{send}(d_i(t))$ hold, then in the following transition both processes i and $i+1$ will change their state. The constraint which ensures that only scheduled processes may change their state is modified into

$$\forall i.\ \forall t.\ \forall I.\ \ \neg\mathsf{sched}_i(I) \wedge \neg(\mathsf{sched}_{i-1}(I) \wedge \mathsf{tok}(d_{i-1}(t)) \wedge \mathsf{send}(d_{i-1}(t)))$$
$$\rightarrow\ d_i(\tau(t,I)) = d_i(t)$$

- Finally, we need to restrict local transitions in order to obtain isomorphic processes. The general rule is that local transitions of process i should only

---

[1] The different semantics (compared to the input-preserving LTSs used in [15,16]) is already reflected in our definition of LTSs and satisfaction of LTL formulas.

depend on the local state and inputs in I_i. With our definition, token passing is an exception to this rule. The resulting constraints for local transitions are:

$$\forall i > 1.\ \forall t, t'.\ \forall I, I'.\ \ d_1(t) = d_i(t') \wedge \text{sched}_1(I) \wedge \text{sched}_i(I')$$
$$\rightarrow d_1(\tau(t, I)) = d_i(\tau(t', I'))$$

$$\forall i > 1.\ \forall t, t'.\ \forall I, I'.\ \ d_1(t) = d_i(t') \wedge \text{send}(d_n(t)) \wedge \text{send}(d_{i-1}(t'))$$
$$\wedge\ \text{sched}_n(I) \wedge \text{sched}_{i-1}(I') \wedge I \cap I_1 = I' \cap I_i$$
$$\rightarrow d_1(\tau(t, I)) = d_i(\tau(t', I'))$$

Fairness of Scheduling and Token Passing. A precondition of Thm. 1 is that the implementation needs to ensure fair token-passing. Thus, we always add

$$\forall i.\ \text{fair\_scheduling} \rightarrow (\mathsf{G}(\text{tok}_i \rightarrow \mathsf{F}\ \text{send}_i))$$

to φ, where fair_scheduling stands for $\forall j.\ \mathsf{G}\,\mathsf{F}\ \text{sched}_j$. Note that with this condition, the formula does not fall into any of the cases from Thm. 1. However, in the model of Emerson and Namjoshi, fairness of scheduling is an implicit assumption, since otherwise fairness of token passing will also be violated. Thus, by adding this formula, we are making explicit two of the assumptions of Emerson and Namjoshi, and this formula does not need to be taken into account when choosing which case of the theorem needs to be applied.

Similarly, the fair_scheduling assumption needs to be added to any liveness conditions of the specification, as without fair scheduling in general liveness conditions cannot be guaranteed. As before, this does not need to be taken into account considering Thm. 1.

Correctness and Completeness of Bounded Synthesis for Token Rings. Based on completeness of the original bounded synthesis approach (and correct modeling of the features of token rings), we obtain

Corollary 2. *If a given specification φ is satisfiable in a token ring of a given size n, then the bounded synthesis algorithm, adapted to token rings, will eventually find this implementation.*

Finally, based on the correctness of our adaption of bounded synthesis, and Corollary 1, we obtain

Theorem 4. *If a given specification φ falls into class a (b,c,d) of Thm. 1 and the adapted bounded synthesis algorithm finds an implementation that satisfies φ in a token ring of size 2 (3,4,5), then this implementation satisfies φ in token rings of arbitrary size.*

5   Synthesizing a Parameterized Arbiter

In this section, we show how parameterized synthesis can be used to obtain process implementations for token ring architectures. Our example is a parameterized arbiter in a token ring as depicted in Fig. 1, with the following specification:

$$\forall i \neq j. \ \ \mathsf{G} \neg (g_i \wedge g_j)$$
$$\forall i. \qquad (\mathsf{G}(r_i \to \mathsf{F}\, g_i))$$

Every process i has an input r_i for requests from the environment, which it can grant by activating an output g_i. We want grants of all processes to be mutually exclusive, and every request to be eventually followed by a grant. The specification satisfies case c) in Theorem 1, i.e., a ring of size 4 is sufficient to synthesize implementations that satisfy the specification for rings of any size.

According to the adapted bounded synthesis approach from Sect. 4.2, we need to add the token fairness requirement, and add the fair scheduling assumption to all liveness constraints. This results in the extended specification

$$\forall i \neq j. \ \ \mathsf{G} \neg (g_i \wedge g_j)$$
$$\forall i. \qquad \mathsf{fair\_scheduling} \to (\mathsf{G}(r_i \to \mathsf{F}\, g_i))$$
$$\forall i. \qquad \mathsf{fair\_scheduling} \to (\mathsf{G}(\mathsf{tok}_i \to \mathsf{F}\,\mathsf{send}_i)).$$

We translate the specification into a universal co-Büchi automaton, shown for 2 processes in Fig. 2. This automaton translates to a set of first-order constraints for the annotations of an LTS implementing φ, a part of which is shown in Fig. 3 (only constraints for states $0, 1, 3, 5$ of the automaton are shown). These constraints, together with general constraints for asynchronous systems, isomorphic processes, token rings, and size bounds, are handed to Z3 [17]. For correctly chosen bounds ($|T_A| \leq 4$ and $|T_p| \leq 2$), we obtain a model of the process implementation in ~10 seconds (on an Intel Core i7 CPU @ 2.67 GHz). The solution is very simple: every process needs only 2 states, with send_i and g_i signals high iff the process has the token. In the parallel composition of 4 such processes, only 4 global states are reachable. Theorem 4 guarantees that with this process implementation, φ will be satisfied for any instance of the architecture.

Note that synthesis is "easy" in this case because we can restrict it to a small ring of 4 processes, and have a rather simple specification. For 5 processes (and $|T_A| \leq 5$), Z3 already needs ~100 seconds to solve the resulting constraints. We expect similar increases in needed time for specifications with more system variables.

The translation of specifications into SMT constraints is currently not fully automated. We leave the development of an automatic tool and its application to more complex case studies for future work.

6   A Framework for Parameterized Synthesis

Our approach for reduction of parameterized synthesis to distributed/isomorphic synthesis is not limited to token rings. In the following, we sketch a framework which allows us to lift decision procedures for the verification of parameterized systems to semi-decision procedures for their synthesis.

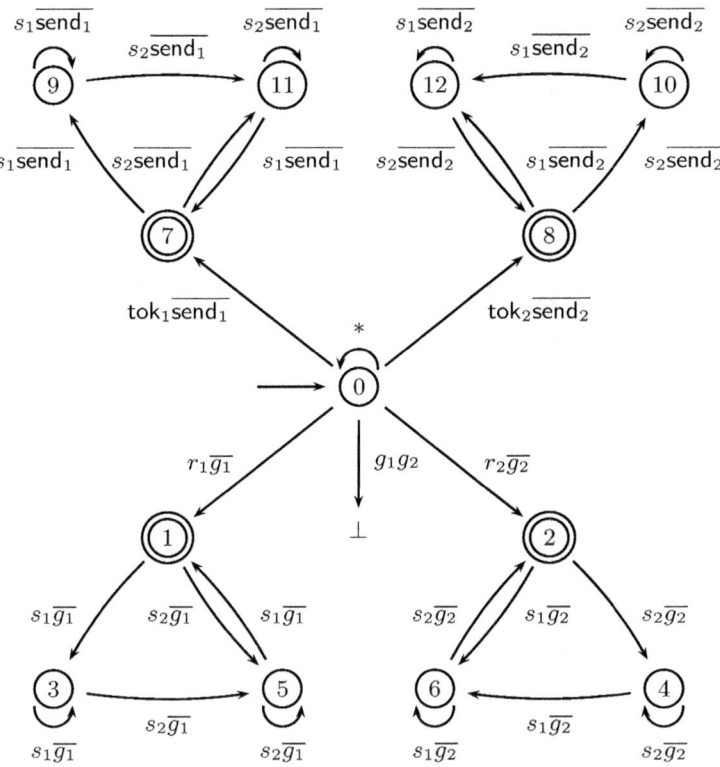

Fig. 2. Universal co-Büchi automaton for specification φ

6.1   General Token-Passing Systems

Clarke, Talupur, Touilli, and Veith [11] have extended the results of Emerson and Namjoshi to arbitrary token-passing networks. They reduce the parameterized verification problem to a finite set of model checking problems, where the number of problems and the size of systems to be checked depends on the architecture of the parameterized system and on the property to be proved.

To lift these results to the synthesis of parameterized token-passing systems in general, we need to adapt the bounded synthesis algorithm further, such that it searches for a process implementation which satisfies the required properties for all verification problems in the set determined by architecture and specification. This requirement can easily be encoded into corresponding constraints for the SMT solver, but may of course increase complexity of synthesis significantly.

Encoding of token-passing into SMT constraints must be adapted to the possibility that processes may be able to choose which other process will receive the token. Furthermore, Clarke et al. [11] have the assumption that the system satisfies fair token passing. For synthesis, we must strengthen the given specification of the system such that it will satisfy this property. For general

$$\lambda_0^{\mathbb{B}}(0)$$
$$\mathsf{tok}(d_1(0)) \wedge \forall i \neq 1. \ \neg\mathsf{tok}(d_i(0))$$

$\forall t. \ \forall I.$ $\lambda_0^{\mathbb{B}}(t) \ \rightarrow \ \lambda_0^{\mathbb{B}}(\tau(t,I)) \wedge \lambda_0^{\#}(\tau(t,I)) \geq \lambda_0^{\#}(t)$

$\forall i \neq j. \ \forall t.$ $\lambda_0^{\mathbb{B}}(t) \ \rightarrow \ \neg(g(d_i(t)) \wedge g(d_j(t)))$

$\forall i. \ \forall t. \ \forall I.$ $\lambda_0^{\mathbb{B}}(t) \wedge \mathsf{sched}_1(I) \wedge r_i \in I \ \rightarrow \ \lambda_1^{\mathbb{B}}(t) \wedge \lambda_1^{\#}(\tau(t,I)) > \lambda_0^{\#}(t)$

$\forall i \neq j. \ \forall t. \ \forall I.$ $\lambda_1^{\mathbb{B}}(t) \wedge \neg\mathsf{sched}_2(I) \wedge \neg g(d_1(t)) \ \rightarrow \ \lambda_3^{\mathbb{B}}(t) \wedge \lambda_3^{\#}(\tau(t,I)) \geq \lambda_1^{\#}(t)$

$\forall i \neq j. \ \forall t. \ \forall I.$ $\lambda_1^{\mathbb{B}}(t) \wedge \mathsf{sched}_2(I) \wedge \neg g(d_1(t)) \ \rightarrow \ \lambda_5^{\mathbb{B}}(t) \wedge \lambda_5^{\#}(\tau(t,I)) \geq \lambda_1^{\#}(t)$

$\forall i \neq j. \ \forall t. \ \forall I.$ $\lambda_3^{\mathbb{B}}(t) \wedge \neg\mathsf{sched}_2(I) \wedge \neg g(d_1(t)) \ \rightarrow \ \lambda_3^{\mathbb{B}}(t) \wedge \lambda_3^{\#}(\tau(t,I)) \geq \lambda_3^{\#}(t)$

$\forall i \neq j. \ \forall t. \ \forall I.$ $\lambda_5^{\mathbb{B}}(t) \wedge \neg\mathsf{sched}_1(I) \wedge \neg g(d_1(t)) \ \rightarrow \ \lambda_5^{\mathbb{B}}(t) \wedge \lambda_5^{\#}(\tau(t,I)) \geq \lambda_5^{\#}(t)$

$\forall i \neq j. \ \forall t. \ \forall I.$ $\lambda_5^{\mathbb{B}}(t) \wedge \mathsf{sched}_1(I) \wedge \neg g(d_1(t)) \ \rightarrow \ \lambda_1^{\mathbb{B}}(t) \wedge \lambda_1^{\#}(\tau(t,I)) > \lambda_5^{\#}(t)$

$\dots$ $\dots$

Fig. 3. Constraints that are equivalent to realizability of φ

token-passing networks, the assumption that every process that holds the token will always eventually send it may not be enough to ensure fair token passing. One possibility to ensure fair token-passing in general networks is to require

$$\forall i. \forall j. \ \mathsf{G}(\mathsf{tok}_i \ \rightarrow \ \mathsf{F} \ \mathsf{send}_{(i,j)}),$$

where i quantifies over all processes as usually, j over all processes which can receive the token from process i, and $\mathsf{send}_{(i,j)}$ means that i sends the token to j.

6.2 Other Results with Cutoffs

In the literature, there is a vast body of work on the verification of parameterized systems, much of it going beyond token-passing systems (e.g., [19,20]). In particular, many of these results prove a cutoff for the given class of systems and specifications [21,22,23], making the verification problem decidable.

In principle, any verification result that provides a cutoff, i.e., reduces the verification of LTL properties for parameterized systems to the verification of a finite set of fixed-size systems, can be used in a similar way to obtain a semi-decision procedure for the parameterized synthesis problem. Our limitation is the ability to encode the special features of the class of systems in decidable first-order constraints, and the specifications under consideration are omega regular.

Approaches that detect a cutoff for a given system implementation dynamically [24,25] (i.e., not determined by architecture and specification) are less suited for our framework: they could in principle be integrated with our approach, but cutoff detection would have to be interleaved with generation of candidate implementations, making it hard or impossible to devise a complete synthesis approach.

7 Conclusions

We have stated the problem of parameterized realizability and parameterized synthesis: whether and how a parameterized specification can be turned into a simple recipe for constructing a parameterized system. The realizability problem asks whether a parameterized specification can be implemented for any number of processes, i.e., whether the specification is correct. The answer to the synthesis question gives a recipe that can quickly be turned into a parameterized system, thus avoiding the steeply rising need for resources associated with synthesis for increasing n using classical, non-parameterized methods.

We have considered the problem in detail for token rings, and to some extent for general token-passing topologies. Using results from parameterized verification, we showed that the parameterized synthesis problem reduces to distributed synthesis of a small network of isomorphic processes with fairness constraints on token passing. Unfortunately, the synthesis problem remains undecidable.

Regardless of this negative result, we managed to synthesize an actual—albeit very small—example of a parameterized arbiter. To this end, we used Schewe and Finkbeiner's results on bounded synthesis. In theory, this approach will eventually find an implementation if it exists. In practice, this currently only works for small implementations. One line of future work will be on making synthesis feasible for larger systems, possibly as an extension of the lazy synthesis approach [7].

For unrealizable specifications, our approach will run forever. It is an interesting question whether it could be combined with incomplete methods to check unrealizability.

We note that the topologies we considered do limit communication between processes and therefore also the possible solutions. For our running example, processes give grants only when they hold the token. Obviously, this means that response time increases linearly with the number of processes, something that can be avoided in other topologies. The use of more general results on parameterized verification may widen the class of topologies that we can synthesize.

Acknowledgments. Many thanks to Leonardo de Moura for his help with little known features of Z3. We thank the members of ARiSE, particularly Helmut Veith, for stimulating discussions on parameterized synthesis, and Bernd Finkbeiner for discussions on distributed and bounded synthesis. Finally, thanks to Hossein Hojjat for useful comments on a draft of this paper.

References

1. Church, A.: Logic, arithmetic and automata. In: Proceedings International Mathematical Congress (1962)
2. Pnueli, A., Rosner, R.: On the synthesis of a reactive module. In: Proc. Symposium on Principles of Programming Languages (POPL 1989), pp. 179–190 (1989)
3. Piterman, N., Pnueli, A., Sa'ar, Y.: Synthesis of Reactive(1) Designs. In: Emerson, E.A., Namjoshi, K.S. (eds.) VMCAI 2006. LNCS, vol. 3855, pp. 364–380. Springer, Heidelberg (2005)
4. Bloem, R., Galler, S., Jobstmann, B., Piterman, N., Pnueli, A., Weiglhofer, M.: Specify, compile, run: Hardware form PSL. In: 6th International Workshop on Compiler Optimization Meets Compiler Verification. Electronic Notes in Theoretical Computer Science (2007)

5. Bloem, R., Galler, S., Jobstmann, B., Piterman, N., Pnueli, A., Weiglhofer, M.: Automatic hardware synthesis from specifications: A case study. In: Proceedings of the Design, Automation and Test in Europe, pp. 1188–1193 (2007)
6. Katz, G., Peled, D.: Synthesizing Solutions to the Leader Election Problem Using Model Checking and Genetic Programming. In: Namjoshi, K., Zeller, A., Ziv, A. (eds.) HVC 2009. LNCS, vol. 6405, pp. 117–132. Springer, Heidelberg (2011)
7. Finkbeiner, B., Jacobs, S.: Lazy Synthesis. In: Kuncak, V., Rybalchenko, A. (eds.) VMCAI 2012. LNCS, vol. 7148, pp. 219–234. Springer, Heidelberg (2012)
8. Apt, K., Kozen, D.: Limits for automatic verification of finite-state concurrent systems. Information Processing Letters 22, 307–309 (1986)
9. Suzuki, I.: Proving properties of a ring of finite state machines. Information processing Letters 28, 213–214 (1988)
10. Emerson, E.A., Namjoshi, K.S.: On reasoning about rings. International Journal of Foundations of Computer Science 14, 527–549 (2003)
11. Clarke, E., Talupur, M., Touili, T., Veith, H.: Verification by Network Decomposition. In: Gardner, P., Yoshida, N. (eds.) CONCUR 2004. LNCS, vol. 3170, pp. 276–291. Springer, Heidelberg (2004)
12. Pnueli, A., Rosner, R.: Distributed systems are hard to synthesize. In: Proc. Foundations of Computer Science (FOCS), pp. 746–757 (1990)
13. Finkbeiner, B., Schewe, S.: Uniform distributed synthesis. In: Logic in Computer Science (LICS), pp. 321–330. IEEE Computer Society Press (2005)
14. Schewe, S., Finkbeiner, B.: Synthesis of Asynchronous Systems. In: Puebla, G. (ed.) LOPSTR 2006. LNCS, vol. 4407, pp. 127–142. Springer, Heidelberg (2007)
15. Schewe, S., Finkbeiner, B.: Bounded Synthesis. In: Namjoshi, K.S., Yoneda, T., Higashino, T., Okamura, Y. (eds.) ATVA 2007. LNCS, vol. 4762, pp. 474–488. Springer, Heidelberg (2007)
16. Finkbeiner, B., Schewe, S.: Bounded synthesis. Software Tools for Technology Transfer (to appear)
17. de Moura, L., Bjørner, N.: Z3: An Efficient SMT Solver. In: Ramakrishnan, C.R., Rehof, J. (eds.) TACAS 2008. LNCS, vol. 4963, pp. 337–340. Springer, Heidelberg (2008)
18. Kupferman, O., Vardi, M.Y.: Safraless decision procedures. In: FOCS, pp. 531–542 (2005)
19. Clarke, E.M., Grumberg, O., Jha, S.: Verifying parameterized networks. ACM Trans. Program. Lang. Syst. 19(5), 726–750 (1997)
20. Zuck, L.D., Pnueli, A.: Model checking and abstraction to the aid of parameterized systems (a survey). Computer Languages, Systems & Structures 30(3-4), 139–169 (2004)
21. German, S.M., Sistla, A.P.: Reasoning about systems with many processes. J. ACM 39(3), 675–735 (1992)
22. Emerson, E.A., Kahlon, V.: Reducing Model Checking of the Many to the Few. In: McAllester, D. (ed.) CADE 2000. LNCS, vol. 1831, pp. 236–254. Springer, Heidelberg (2000)
23. Kahlon, V., Ivančić, F., Gupta, A.: Reasoning About Threads Communicating Via Locks. In: Etessami, K., Rajamani, S.K. (eds.) CAV 2005. LNCS, vol. 3576, pp. 505–518. Springer, Heidelberg (2005)
24. Hanna, Y., Basu, S., Rajan, H.: Behavioral automata composition for automatic topology independent verification of parameterized systems. In: ESEC/SIGSOFT FSE, pp. 325–334 (2009)
25. Kaiser, A., Kroening, D., Wahl, T.: Dynamic Cutoff Detection in Parameterized Concurrent Programs. In: Touili, T., Cook, B., Jackson, P. (eds.) CAV 2010. LNCS, vol. 6174, pp. 645–659. Springer, Heidelberg (2010)

QuteRTL: Towards an Open Source Framework for RTL Design Synthesis and Verification

Hu-Hsi Yeh[1], Cheng-Yin Wu[2], and Chung-Yang (Ric) Huang[1,2]

[1] Department of Electrical Engineering, National Taiwan University, Taipei, Taiwan
[2] Graduate Institute of Electronics Engineering, National Taiwan University, Taipei, Taiwan

Abstract. We build an open-source RTL framework, QuteRTL, which can serve as a front-end for research in RTL synthesis and verification. Users can use QuteRTL to read in RTL Verilog designs, obtain CDFGs, generate hierarchical or flattened gate-level netlist, and link to logic synthesis/ optimization tools (e.g. Berkeley ABC). We have tested QuteRTL on various RTL designs and applied formal equivalence checking with third party tool to verify the correctness of the generated netlist. In addition, we also define interfaces for the netlist creation and formal engines. Users can easily adopt other parsers into QuteRTL by the netlist creation interface, or call different formal engines for verification and debugging by the formal engine interface. Various research opportunities are made possible by this framework, such as RTL debugging, word-level formal engines, design abstraction, and a complete RTL-to-gate tool chain, etc. In this paper, we demonstrate the applications of QuteRTL on constrained random simulation and property checking.

Keywords: Synthesis, Verification, Open Source, Framework.

1 Introduction

In a typical EDA (Electronic Design Automation) software, a quality front-end is necessary for reading in complex design and extracting significant information for later executions. A quality front-end should be capable of reading in all the defined descriptions and translating them into efficient data structures. Traditional academic tools, such as SIS [1], VIS [2], and MVSIS [3], focus on the Boolean-level optimization algorithms that can improve the quality of circuits in various aspects. They are robust enough and, at the same time, scalable for practical use. In the past decades, people from industry and academia have adopted and developed their synthesis and verification tools from these tools. However, as the design paradigm moves to Register-Transfer-Level (RTL) and up, most of the new research have to deal with the high-level design constructs, syntax, and semantics. Without a robust front-end, the applicability of these tools will be limited.

Recently, Berkeley ABC [4], which is a software system for synthesis and verification, has become very popular in both academia and industry. It proposes: (1) fast and scalable logic optimization based on And-Inverter Graphs (AIGs), (2)

C. Flanagan and B. König (Eds.): TACAS 2012, LNCS 7214, pp. 377–391, 2012.

optimal-delay DAG-based technology mapping for standard cells and FPGAs, and (3) innovative algorithms for integrated sequential optimization and verification. However, it still has incomplete support on design formats; for example, it cannot read in most of the descriptions in RTL Verilog, hierarchical BLIF and BLIF-MV, and it mainly handles the specialized format — BLIF, which is bit-level. Therefore, we need to resort to other tools to translate the RTL design into the BLIF format. Consequently, we will then lose most of the high-level design intents such as FSM, counter, and control/data separation, etc., which can be useful in guiding the design verification.

On the other hand, there are also some open-source front-ends, including VIS and Icarus Verilog [5]. The front-end of VIS acts as an intermediate role to translate designs into BLIF format. It does not completely keep the high-level design intents and does not have complete support for HDL. On the other hand, Icarus Verilog aims at simulation and FPGA synthesis. It still has some known and unknown bugs and the author continues releasing patches.

We implement a quick and quality RTL front-end (QuteRTL) which supports most of the synthesizable RTL Verilog with different library formats and can synthesize the design to word-level circuit netlist. The key features of QuteRTL include: (1) complete Verilog support, (2) flexible design view: word-level or bit level; hierarchical or flatten, (3) formally verified by commercial equivalent checker, and (4) complete netlist creation interface for other parsers (e.g. VHDL/System Verilog parser) and engine interface for external solvers (e.g. BDD, MiniSAT [6], and Boolector [7]).

As an exemplar application of the QuteRTL framework, we publish an Automatic Target Constraint Generation (ATCG) technique in [8] to address the bottleneck in the constrained random simulation flow. Instead of focusing on the constraint solving techniques as other research [9, 10] do, we propose an alternative approach to alleviate the burden of the users by automatically generating high-quality constraints with the support of QuteRTL. In another application, we devise a property-specific sequential invariant extraction algorithm in [11] to improve the performance of the SAT-based unbounded model checking (UMC). We first utilize QuteRTL to extract the property-related predicates and their corresponding high-level design constructs such as FSMs and counters. Thus, we can quickly identify the sequential invariants and then utilize them to refine the inductive hypothesis [12] in induction-based UMC, and to improve the accuracy of reachable state approximation in interpolation-based UMC [13, 14].

The rest of the paper is organized as follows: in Section 2, we first give an introduction of the architecture and interfaces of QuteRTL. Section 3 presents the tool implementation and data structure, and Section 4 presents the applications of QuteRTL. In Section 5, we give a user guide and some demo examples for general users. Finally, we conclude the paper in Section 6.

2   Architecture of QuteRTL Framework

In this section, we will present our RTL synthesis and verification framework —
QuteRTL. Section 2.1 gives an overview of the framework while Section 2.2
describes the design and engine interfaces of QuteRTL. Finally, Section 2.3 provides
a comparison between QuteRTL and other open-source front-ends

2.1   Overview of QuteRTL Framework

Figure 1 shows the architecture of QuteRTL framework, which can be separated into
two parts, RTL synthesis and circuit verification/debugging. In the RTL synthesis
part, the RTL design is first translated into some intermediate representations, for
example, Control-Data Flow Graph (CDFG). Then, QuteRTL resolves such
temporary models by elaborating an equivalent circuit netlist and extracting plenty of
design intents, including hierarchy information, FSM, counter structures, etc. These
intents can help both test pattern generation and safety/liveness property checking in
the circuit verification/debugging part. For general users, we release the source code
of our parsers, netlist creation procedure and interface functions.

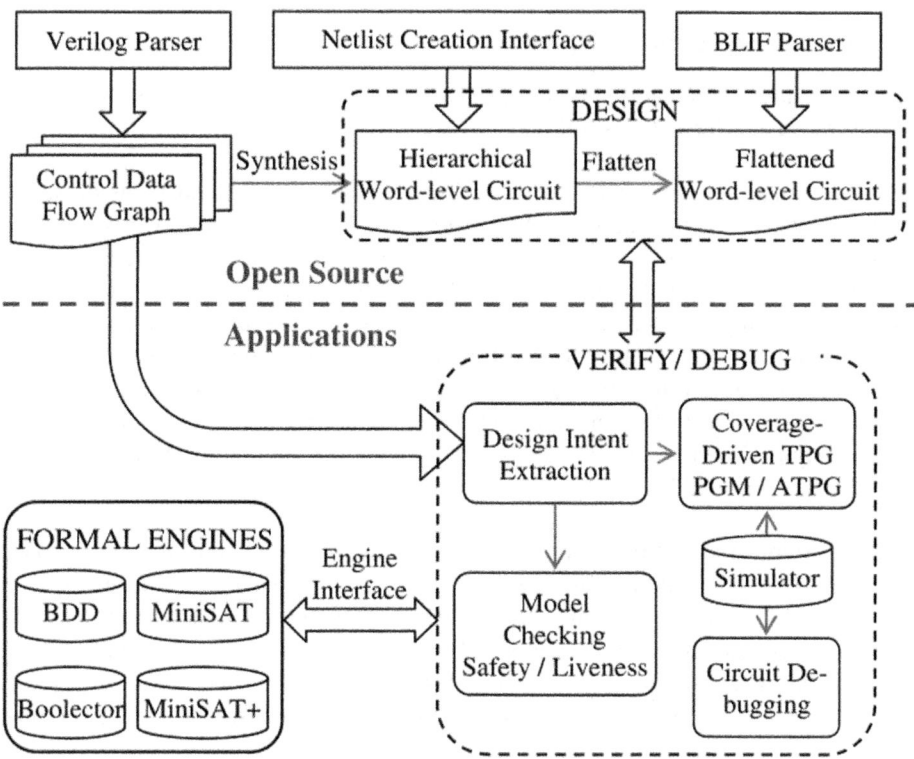

Fig. 1. Architecture of QuteRTL

In the view of design, we have both hierarchical and flattened word-level circuit structure in QuteRTL. Using the hierarchical structure, we can analyze designs more systematically and identify predicates easily in the original RTL. For example, QuteRTL can determine the independence between modules with the hierarchy, and then the information is utilized to alleviate design complexity. For formal engines, the search space can be pruned significantly; for simulators, the efficiency can be improved by the divide-and-conquer algorithm. On the other hand, for circuit redundancy elimination and global optimization, QuteRTL can flatten the design into a single circuit netlist. When flattening the hierarchical design, it will collect the necessary cells in depth first search from PO to PI, and remove redundant cells, which come from bad coding styles or function-less buffers.

In the view of circuit netlist, most logic optimization tools perform their algorithm on bit-level logic netlist, but they rarely handle the word-level netlist. The proposed tool in the paper can completely translate netlist into what logic optimization tools support. That is, QuteRTL can output both the word-level or the bit-level netlist, or even the mixed-level netlist. In addition, it can utilize some word-level circuit components to assist logic optimization tools. For example, QuteRTL can use high-speed adders, says carry look-ahead adders, to substitute carry ripple adders, or Booth's multipliers for high speed designs.

2.2   Supported Features of QuteRTL

Various features are supported by QuteRTL. To illustrate them more clearly and succinctly, we categorize them as follows:

Design Formats. As shown in Figure 1, QuteRTL supports several kinds of design input formats, which include not only Verilog but also other well-known word-level or Boolean network, for instance BLIF and BTOR. Moreover, we provide a complete set of interface functions for interactive netlist creation. The biggest advantage is that anyone can simply call our netlist creation functions to build up a hierarchical word-level network in QuteRTL despite what input formats of the designs are. Hence, any word-level or Boolean network can be intuitively constructed in QuteRTL with the help of these interface functions. On the contrary, QuteRTL also supports corresponding design output formats, including both hierarchical and flattened structural Verilog, and BLIF.

Design Intent Extraction. Design intents contain useful information to help optimization or verification tools improve design and dependability, but many tools and research abandon the information when they proceed. In QuteRTL, the synthesized circuit can be easily annotated to the original RTL structure before logic optimization. Thus, we can extract some design intents from the circuit netlist and CDFG. These design intents include local FSMs, counters, constraints, and invariants.

Interface for Verification and Debugging. After constructing the target design, we can adopt the following interface functions to verify or debug the properties. We split these functions into two parts:

1. Property Specification Interface: We support various types of assertion specifications. These assertions including simple CTL safety and liveness properties in either AG(p) or EG(p) format, where p can be specified as an auxiliary Boolean output signal formulated from the design; for instance $a + b < c$ or $x * y > 10$. Besides, part of System Verilog Assertions (SVA) semantics is also supported for common industrial instances.

2. Engine Interface: Formal engines are crucial to both verification and debugging, especially in formal approaches. However, every engine embraces its individual interface functions, so users need to use the respective interface functions when applying different solvers. It causes maintainability problems in the interfaces for the solvers. Therefore, we integrate those interfaces into a union set of functions that are conformable to different verification and debugging needs in QuteRTL. The integrated engine interface makes the usage of formal engines simple and unified. That is, users can specify which formal engine they expect to adopt in their applications.

2.3 Comparison with other Open Source RTL Front-End

In this subsection, we discuss the comparison between QuteRTL and other open-source front-ends, including VIS and Icarus Verilog. The VIS group releases a Verilog HDL front-end VL2MV, which compiles a subset of Verilog into an intermediate format BLIF-MV (a multi-valued extension of BLIF). With the support of VL2MV, VIS is able to synthesize finite state systems and verify properties of such system. Besides, VL2MV extracts a set of interacting FSMs which preserve the behavior of the source Verilog defined in terms of the simulated results. However, the front-end does not guarantee the extracted FSMs are optimal, and is not able to handle full set Verilog language due to its dynamic nature.

Another open source RTL front-end Icarus Verilog aims at simulation and FPGA synthesis. It can support richer syntax for simulation in RTL language, and generate the text or waveform output of the simulation results. Icarus Verilog is intended to work mainly as a simulator, although its synthesis capabilities are improving. However, the tool focuses on generating specific netlist format for FPGA synthesis, and it is hard to utilize novel formal techniques on the specific netlist.

To apply modern formal techniques to industrial RTL design, we implement a quick and quality RTL front-end QuteRTL. It can synthesize most of the synthesizable RTL with different library (Verilog and Liberty) formats into word-level circuit netlist. For design verification, QuteRTL also supports other design input formats, for example BLIF and BTOR, etc. Besides, users can easily implement novel formal techniques on the word-level circuit netlist, for example UMC, property directed reachability (PDR) [15], etc.

3   Tool Implementation

In this section, we describe the implementation of QuteRTL, which consists of a Verilog parser, an RTL synthesizer, and a circuit flattening procedure.

3.1   Parser and Preprocessor

Verilog Parser. We use Lex and Yacc to implement the Verilog parser. If the syntax of the design conforms to Verilog Backus-Naur Form (BNF), the parser will parse corresponding syntax trees for a start. It also checks the grammars of the syntaxes and lints for Verilog. Then we construct CDFG of the design from the syntax trees for each module. For the purpose of design synthesis and verification, we focus on the synthesizable Verilog subset, which includes synthesizable "for loop", "task" and "function" declarations, etc.

Preprocessor. The preprocessor mainly handles macro substitution, hierarchy construction, and parameter overriding. After generating the CDFGs, we first perform a simple substitution and expand the occurrence of each argument in macro using the replacement text. For the modules containing macros, we revise their CDFGs. Next, we construct a hierarchical tree to represent the relation of the module instances in the design, and then perform parameter overriding from top to down in the hierarchical tree to set up the overridden parameter for each module instance. After the steps, the CDFGs and hierarchical tree are ready for synthesis.

3.2   RTL Synthesis and Circuit Flattening

Data Structure of Circuit Netlist. Figure 2 shows the data structure of circuit netlist in QuteRTL. We use three components–Cell, InPin, and OutPin to describe a circuit. The Cell contains OutPin(s) to fan out to other cells and an InPin list to receive multiple fanins from other Cells to construct the circuit netlist. The pins can be multiple bits to describe word-level netlist.

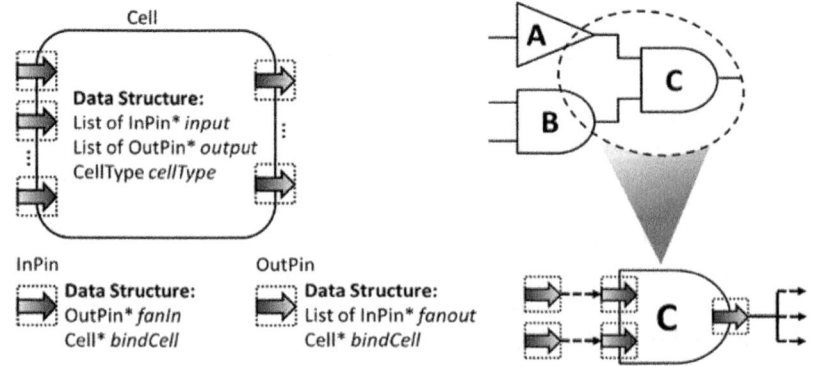

Fig. 2. The data structure of circuit netlist in QuteRTL

The types of cells can be classified as follows:

- Operator cell: arithmetic, relational, equality, logical, bit-wise, reduction, shift cell, and multiplexer
- IO cell: primary input, primary output, and primary inout
- Sequential cell: flip-flop and latch
- Module cell: module instantiation
- Modeling cell: bit-merging, bit-splitting, bus, memory, bufif, etc.

The operator cells are synthesized from the common operators in Verilog. For example, the multiplexers are synthesized from conditional operator (?:) or conditional block (if, case, etc.). For the instances used in a module, we model them as module cells in the hierarchical view of design. Besides, to support the specific elements in circuit, we create some modeling cells for net, bus, memory, and high impedance. Please note that the pins in word-level netlist are multiple bits, so we use bit-merging (bit-splitting) cells to concatenate (slice) pins to form specific fanins to other cells.

RTL Synthesis Procedure. The synthesis procedure translates CDFGs to the circuit netlist data structures we defined above. The synthesizer first traverses the CDFG of each module and flattens each variable to the data structure "SynVar". Figure 3 gives an example to show the relations between RTL and SynVar. In the data structure, each node contains the data and conditional fanins, which are respectively synthesized from data predicate list (DPL) and control predicate list (CPL) of the variable. The tree structure represents the priority of control predicates in nodes, and then we connect these pins with multiplexers. If the variable is in a sequential block or is not fully assigned in a combinational block in the original Verilog code, the output of the last multiplexer will be connected to a sequential cell (flip-flop/latch). Finally, the synthesized circuit netlist is illustrated in Figure 4. In order to back-annotate the netlist information to the original RTL code, we just synthesize the RTL design to an equivalent circuit netlist without optimizing the netlist during this procedure.

Circuit Flattening. The circuit flattening is to generate a flattened circuit netlist which is functionally equivalent to the hierarchical netlist. The implementation includes the concretion of instance models (i.e. module cells) and the removal of redundant cells (ex. buffers, non-fan-out cells). First, we traverse the hierarchical tree built in preprocessor, and duplicate the non-IO cells (except top level module) to a new flattened module. Simultaneously, we make the connections between cells within the same hierarchical module, and record the connections between different hierarchical modules. After duplicating all necessary cells, we connect the cells in different hierarchical and then traverse the flattened netlist to remove the redundant cells.

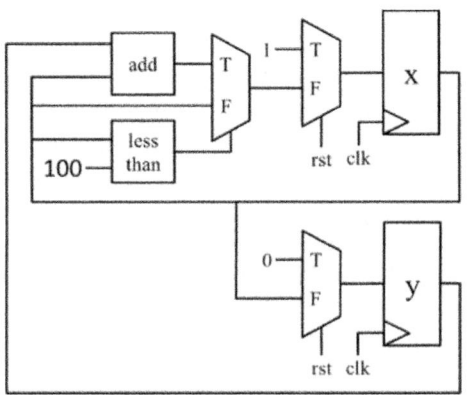

```
always@ (posedge clk) begin
    if (rst) begin
        x <= 1;
        y <= 0;
    end
    else begin
        y <= x;
        if (x < 100)
            x <= y+x;
    end
end
```

Fig. 3. RTL synthesis procedure

Fig. 4. The synthesized circuit

4   Applications of QuteRTL

In this section, we introduce two applications of QuteRTL, which include intent extraction in Section 4.1 and model checking in Section 4.2.

4.1   Intent Extraction

For FSM extraction, we categorize the types of FSM as either explicit FSMs or implicit FSMs according to the definition of state values. In an explicit FSM, its state values are explicit defined as parameters or constants, while there are no explicit state values defined in an implicit FSM, where the state values are implicitly embedded in conditions or expressions. In our implementation of the extractor, we extract both of them and identify counters. Note that we extract the explicit FSMs based on the coding styles [16] and implicit FSMs from the transition relations computed by BDDs [17]. The extraction algorithm is mainly performed in the three steps: candidate state variable extraction, state transition extraction, and state transition graph (STG) construction. We briefly express these steps as follows:

1. Candidate state variable identification: In sequential blocks of Verilog, we first treat the variables in left hand side of assignments as possible state variables. Then, we traverse the data dependency list of the possible state variable to find a loop of assign statements to identify the candidate state variables.
3. State transition extraction: In this step, we extract the state transition relation from each candidate state variable. For explicit FSM, we can extract a set of state pair (S_i, S_j), which represents the state transition from S_i to S_j. While for implicit FSM, we traverse the assignments of the candidate state variables to build the state transition relation in binary decision diagram (BDD).
4. State transition graph construction: For explicit FSM, we use the set of state pair to construct the STG. In order to extract the STG of implicit FSM, we will traverse the BDD to enumerate all transition conditions and relations.

Further, these extracted FSMs are utilized to identify the sequential invariants and then improve the property proving capabilities in [11]. On the other hand, for constrained random simulation, we proposed an ATCG technique [8] based on QuteRTL. In that work, we extract compact constraints for a set of coverage holes from the circuit netlist and CDFG. The experimental results show that the extracted constraints indeed help simulation achieves the highest coverage and smallest runtime when compared to both random and directed simulations.

4.2   Model Checking

The powerful characteristics of our QuteRTL that retain word-level information with high-level design intent provide us an adequate circuit abstraction level for researching on word-level verification and debugging problems. With the prosperous SMT solvers, it becomes practical and ideal to apply model checking on our word-level netlist with a word-level solver.

There are basically two approaches to implement a model checking algorithm on QuteRTL. First, we can adopt the provided engine interface functions to realize a new model checking algorithm. This is commonly used by almost all the verification algorithms we have implemented. Second, we can dump out word-level netlist from QuteRTL and then call the solvers by their supported interfaces. When transforming word-level functions into CNF for Boolean SAT engines, such as adder, multiplier, comparators, etc., we perform naïve bit-blasting technique with better encodings.

Traditional SAT-based model checking algorithms, including bounded model checking (BMC), k-induction, and their extensions such as simple-path and interpolation-based, can be simply implemented with circuit traversal and transforming individual gate function into corresponding solver input formula (e.g. CNF). Without loss of generosity, all the Boolean model checking algorithms can be implemented on QuteRTL. Moreover, our word-level framework provides even better capability in coping with more complex designs and realistic properties by abstraction and refinement techniques, for instance, predicate abstraction, interpolation, design intent extraction, and probabilistic inferences.

5   Availability for General Users

For general users, we release our RTL front-end source code and the compiled QuteRTL executable in the following website:

```
http://dvlab.ee.ntu.edu.tw/~publication/QuteRTL/
```

In this section, we first give a brief overview to the command-line interface of QuteRTL. Then we show some examples related to what QuteRTL can do for general users through our user-friendly command-line interface. Users can also download these examples in our website, which include a general RTL to gate synthesis flow, an example to construct hierarchical word-level netlist, and a property checking instance.

5.1   A Brief Description to QuteRTL Command-Line Interface

Similar to most tools from EDA vendors, QuteRTL supports friendly command-line interface for users. Our commands are usually composed by one or two mandatory key words followed by a set of required/optional parameters. For example, command to parse an input design from a single file or filelist is "REAd DEsign   [-Verilog | -Blif] <[-Filelist] (string filename)>". We can see the command is named by "REAd DEsign", where the upper case letters are mandatory for command-line parser. Parameters in square brackets indicate optional arguments, and those in angle brackets indicate required arguments. More detailed description to our command rules can be found in our website, and we will mention some of them in our examples later.

Besides, there is a command "HELp" for showing all available commands, or showing detailed usage of each command (for instance, "HELp REAd DEsign").

5.2   Example: RTL to Gate Synthesis Flow

In the first example, we are going to show the synthesis flow of QuteRTL. The adopted designs are "i2c" and "usb_phy" from OpenCore [18]. We present the commands of the flow in Figure 5. Note that users can run the series of commands from a batch file using "dofile" command or execute argument "-f" to specify the batch file.

In the first line of Figure 5, we record the commands we are going to execute throughout the program into a log file, which can be used as batch file in the future run. Then we parse the Verilog design from the file list. Note that users must write all related files in the file list for QuteRTL once. After the Verilog design is parsed, the command "syn" performs synthesis procedure to transform the design into a word-level circuit netlist, and the command "flat" flatten the design into a single flattened module. Note that internal signals in the flattened module will be renamed.

After our front-end processing, QuteRTL can output either hierarchical design by using the command "write design" or flattened circuit by using the command "write ckt". As shown in Figure 5, after performing circuit flattening, we output a hierarchical word-level netlist in Verilog format (i2c_design.v) in line 5, and a flattened one (i2c_ckt.v) in line 6. Besides, we can also output the BLIF format of the design (i2c_ckt.blif). The BLIF format is a suitable input for other research on Boolean network, and easily transformed into other related formats, for instance, AIG. Accompanied with the macro library file "lib2.v", users can run the Cadence Conformal LEC [19] equivalence checker script to check the equivalence among original design and all generated output designs. Figure 6 shows the partial synthesized word-level circuit of "i2c". In the figure, rectangles, trapezoids, and ellipses respectively represent flip-flops, multiplexers, and other operating gates.

```
// Example : flow_i2c.dofile
1.  set log -cmd flow_i2c.dofile
2.  read des -f filelist
3.  syn
4.  flat
5.  write des i2c_design.v
6.  write ckt i2c_ckt.v
7.  write ckt -blif i2c_ckt.blif
8.  q -f
```

```
// Example : flow_usb.dofile
1.  set logfile -cmd flow_usb.dofile
2.  read design -f filelist
3.  synthesis
4.  flatten
5.  write design usb_design.v
6.  write ckt usb_ckt.v
7.  write ckt usb_ckt.blif -blif
8.  quit -f
```

Fig. 5. Batch files for RTL to gate synthesis flow example

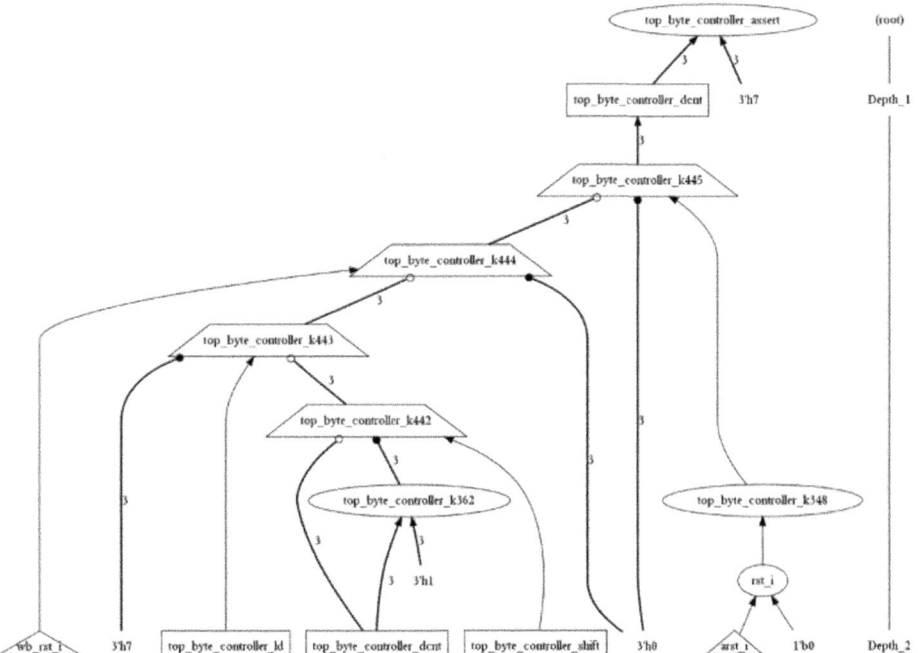

Fig. 6. The part synthesis word-level circuit of i2c

5.3   Example: Hierarchical Word-Level Netlist Creation

As shown in Fig 1, QuteRTL has a complete set of interface functions for netlist creation. Especially, we also support users to construct design through our command-line interface. It is especially convenient to build small designs for instant experiments.

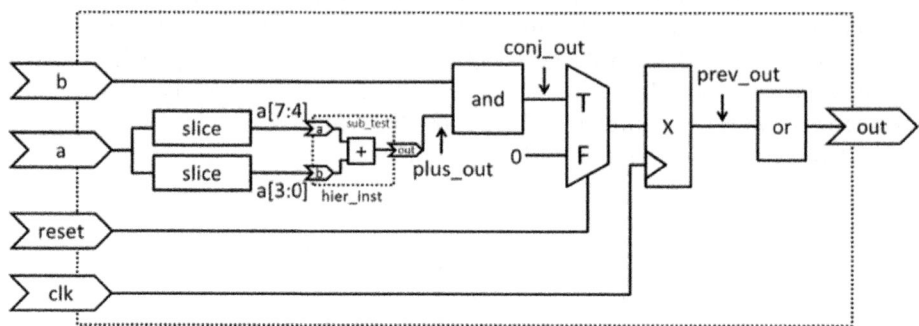

Fig. 7. An example of circuit netlist

Suppose we want to construct the circuit netlist in Figure 7. We present two scripts: "construct_flat.dofile" for constructing the design with only one module, and "construct_hier.dofile" for constructing a hierarchical design, which is functionality identical to the former. The batch files are shown in Figure 8 and 9, respectively. (A portion of commands in "construct_hier.dofile" is omitted in Figure 9 due to space concerns.)

```
// Example : construct_flat.dofile
1.  create design flat_design   10. define net a[3:0] 4
2.  define net -PI clk 1         11. define net a[7:4] 4
3.  define net -PI reset 1       12. define cell SLICE a[7:4] a 7 4
4.  define net -PI a 8           13. define cell SLICE a[3:0] a 3 0
5.  define net -PI b 4           14. define cell ADD plus_out a[7:4] a[3:0]
6.  define net -PO out 1         15. define cell AND conj_out plus_out b
7.  define net prev_out 4        16. define cell DFF prev_out conj_out clk reset
8.  define net plus_out 4        17. define cell OR out prev_out
9.  define net conj_out 4
```

Fig. 8. Batch file for design construction with single module

```
// Example : construct_hier.dofile
1.  create design hier_test 10. define net -PI clk 1
2.  define module sub_test  11. define net -PI reset 1
3.  define net a 3 0          ...
4.  define net a 4          25. define cell or out prev_out
5.  define net -PI a 4      26. define inst sub_test hier_inst a[7:4] a[3:0] plus_out
6.  define net -PI b 4      27. flat
7.  define net -PO out 4    28. write des
8.  define cell add out a b 29. write ckt
9.  change module          30. write ckt -blif
```

Fig. 9. Batch file for hierarchical design construction with multiple modules

At first, we create a new design named "flat_design" (line 1) in Figure 8. Then we use the command "DEFine NET" to create word-level nets with widths. Parameter "[-PI | -PO | -PIO]" is used if such the net is also an I/O port. Note that some illegal names to Verilog, e.g. "a[3:0]" in line 10, will be renamed by QuteRTL; hence it is convenient for general users. Then we construct cells from line 12 to the end, which include a register with synchronous reset in line 16 (we omit the reset value and use default value). In this example, although all nets are defined before cells, actually the only restriction is that all the I/O nets of the defined cell should be defined before. Hence, users can construct a netlist with great flexibility in QuteRTL. Note that commands for cell definition can be comparably complex, due to different type of word-level cells. Users can type "HELp DEFine CELL" to see the detailed usages in the command line mode.

Next, we construct a hierarchical design with the batch file "construct_hier.dofile" in Figure 9. After constructing design "hier_test" in line 1, we define a sub-module "sub_test" in line 2. Now, our current scope is transformed into module "sub_test". Hence all nets and cells defined in line 3-8 will be constructed in module "sub_test". After "sub_test" is constructed, a simple command "CHAnge MODule" will bring us back to the parent module, which is "hier_test" in the case. Note that it is impossible to enter into sub-module "sub_test" again for incremental construction further. Once a sub-module is defined, we expect that it will eventually be instantiated in other modules. The command for module instantiation is "DEFine INST", as shown in line 26, where an instance named "hier_inst" is constructed. In this command, I/O nets defined after the instance name, namely, "a[7:4], a[3:0], and plus_out", will be connected to I/O ports of module "sub_test" in the order identical to the I/O port defined in "sub_test" previously. Hence, I/O relation of "hier_inst" will be "plus_out" = "a[7:4]" + "a[3:0]".

Note that when building a hierarchical design through those commands, users can write out the hierarchical Verilog directly, or write out circuit after flatten, as introduced in Section 5.2.

5.4   Example: Property Checking

One of the important applications to QuteRTL is word-level verification and debugging. In the last example, we utilize QuteRTL to perform safety property checking on a simple traffic light controller. We simplify the design to only two primary inputs (clock and reset) and only one output (time_left), which shows clock cycles left before the light changes to the next. As light is changed, we reset "time_left" to the number of cycles, which is the time to keep the same light: 60 for RED, 40 for GREEN and 5 for YELLOW. Initially, light is RED and time_left is zero, so the light will turn GREEN in the next cycle. We illustrate the state transition graph of the design in Figure 10.

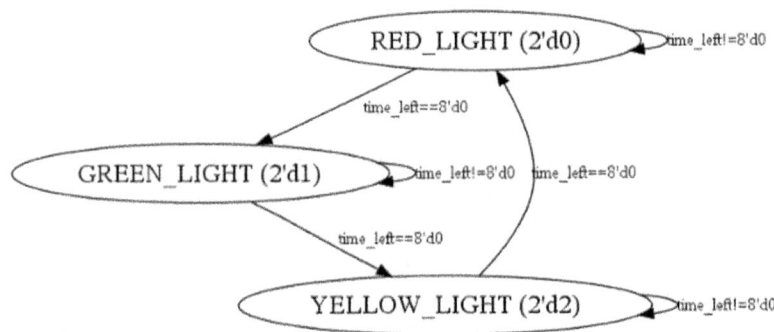

Fig. 10. State transition graph for traffic light design example

We adopt three safety properties for verifying the traffic light design: First, we assert that time_left should never exceed 60, as the longest time in the same light is 60. Second, we assert the light will never turn YELLOW, which should be proven false for this design. Finally, we assert light will never turn to an unknown state, which is encoded as "2'd3" in the design.

Figure 11 show the batch file for property checking. In line 4, 5, and 6, we set three formulas as the three safety properties, and then three "MODel CHecking" commands will call the formal engine to check whether these properties are true or not. Users can see the first and third properties are proved, and the second one is disproved with a 42-cycle trace from initial state.

```
// Example : traffic_light_check.dofile
1.  read des Traffic.v                    // Property Checking : AG(formula 1)
2.  syn                            7.  model check 1
3.  fla                                   // Property Checking : AG(formula 2)
4.  set formula 1 "time_left <= 8'd60"  8.  model check 2
5.  set formula 2 "Light_Sign != 2'd2"     // Property Checking : AG(formula 3)
6.  set formula 3 "Light_Sign != 2'd3"  9.  model check 3
```

Fig. 11. Batch file for property checking on traffic light design example

In this example, we do not specify anything but property to model checker; however, QuteRTL allows users to change solvers and model checking algorithms by setting parameters in the command, or even to specify a file for counter-example trace dump in Value Change Dump (VCD) file format. Figure 12 shows the waveform of a counterexample for the second property. It disproves the property with a 42-cycle trace from initial state.

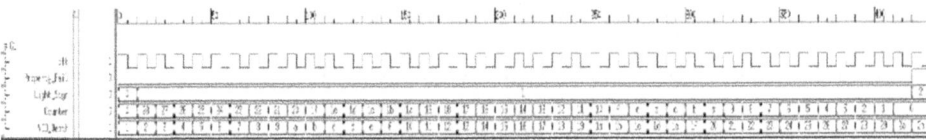

Fig. 12. The waveform of a counterexample of formula 2

6   Conclusion

We construct an open source framework for RTL design synthesis and verification, and verify the correctness and robustness of the framework with a third party tools — Cadence Conformal LEC and Berkeley ABC. With the framework, various research directions on RTL can be made possible. In the future, we will develop some techniques on RTL design debugging with the extracted design intents.

References

1. SIS,
   http://embedded.eecs.berkeley.edu/pubs/
   downloads/sis/index.html
2. VIS, http://vlsi.colorado.edu/~vis/
3. MVSIS,
   http://embedded.eecs.berkeley.edu/Respep/Research/mvsis/
4. Berkeley ABC, http://www.eecs.berkeley.edu/~alanmi/abc/
5. Icarus Verilog, http://iverilog.icarus.com/
6. MiniSAT, http://minisat.se/
7. Boolector, http://fmv.jku.at/boolector/
8. Yeh, H.-H., Huang, C.-Y.: Automatic Constraint Generation for Guided Random Simulation. In: Asia and South Pacific Design Automation Conference, pp. 613–618 (2010)
9. Kitchen, N., Kuehlmann, A.: Stimulus generation forconstrained random simulation. In: International Conference on Computer-Aided Design, pp. 258–265 (2007)
10. Wu, B.-H., Yang, C.-J., Tso, C.-C., Huang, C.-Y.: Toward an Extremely-High-Throughput and Even-Distribution Pattern Generator for the Constrained Random Simulation Techniques. In: International Conference on Computer-Aided Design, pp. 602–607 (2011)
11. Yeh, H.-H., Wu, C.-Y., Huang, C.-Y.: Property-Specific Sequential Invariant Extraction for SAT-based Unbounded Model Checking. In: International Conference on Computer-Aided Design, pp. 674–678 (2011)
12. Thalmaier, M., Nguyen, M.D., Wedler, M., Stoffel, D., Bormann, J., Kunz, W.: Analyzing k-step induction to compute invariants for SAT-based property checking. In: Design Automation Conference, pp. 176–181 (2010)
13. McMillan, K.L.: Interpolation and SAT-Based Model Checking. In: Hunt Jr., W.A., Somenzi, F. (eds.) CAV 2003. LNCS, vol. 2725, pp. 1–13. Springer, Heidelberg (2003)
14. Vizel, Y., Grumberg, O.: Interpolation-Sequence Based Model Checking. In: Formal Methods in Computer Aided Design, pp. 1–8 (2009)
15. Een, N., Mishchenko, A., Brayton, R.: Efficient Implementation of Property Directed Reachability. In: Formal Methods in Computer Aided Design, pp. 125–134 (2011)
16. Liu, C.-N., Jou, J.-Y.: A FSM Extractor from HDL Description at RTL Level. In: Asia Pacific Conference on Hardware Description Languages, pp. 33–38 (1998)
17. Touati, H., Savoj, H., Lin, B., Brayton, R.K., Sangiovanni-Vincentelli, A.: Implicit State Enumeration of Finite State Machines using BDDs. In: International Conference on Computer-Aided Design, pp. 130–133 (1990)
18. OpenCores, http://www.opencores.org
19. Cadence Conformal LEC, http://www.cadence.com/products/

Template-Based Controller Synthesis for Timed Systems

Bernd Finkbeiner and Hans-Jörg Peter

Reactive Systems Group
Universität des Saarlandes, Germany

Abstract. We present an effective controller synthesis method for real-time systems modeled as timed automata with safety requirements. Under the realistic assumption of partial observability, the problem is undecidable in general, and prohibitively expensive (2EXPTIME-complete) if a bound on the granularity of the controller is set in advance. We investigate the synthesis of controllers from templates, given as timed automata with parametric control structure. Template-based synthesis is significantly cheaper (PSPACE-complete) than standard synthesis and produces much simpler controllers. We present an efficient symbolic synthesis algorithm based on automatic abstraction refinement and report on encouraging experimental results from an implementation in the timed verification and synthesis tool SYNTHIA.

1 Introduction

In controller synthesis, we automatically transform a given model of a plant and a safety requirement into a finite-state controller that monitors and affects the ongoing behavior of the plant to ensure the safety of its operation. There has been a lot of recent progress [1,8,5,24,23] in synthesizing controllers for real-time systems, where the plant is given as a timed automaton. Notably, the UPPAAL-TIGA tool [5], which is based on the popular timed model checker UPPAAL, has extended the highly efficient state-space traversal based on symbolic zone representations from verification to synthesis.

Unfortunately, timed controller synthesis quickly turns into an intractable problem if one makes the realistic assumption that the controller does not have access to the full state of the plant, but rather only sees a subset of the events. Under *partial observability*, the controller synthesis problem is undecidable in general, and remains prohibitively expensive (2EXPTIME-complete) if the problem is made decidable by fixing a bound on the granularity of the controller in advance [6]. Furthermore, since the size of the controller may be doubly-exponential in the size of the plant, it is often infeasible to actually construct the controller.

In this paper, we propose a new synthesis approach, where the size and general shape of the controller is fixed in advance in the form of a *template*. A template is a timed or untimed automaton with parametric control structure. Figures 1

C. Flanagan and B. König (Eds.): TACAS 2012, LNCS 7214, pp. 392–406, 2012.

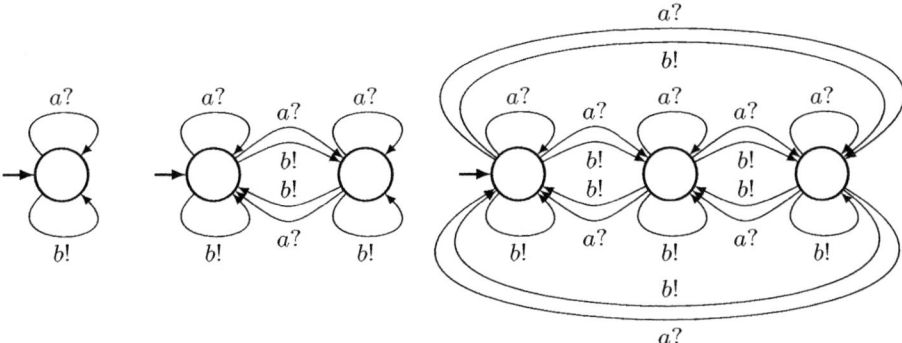

Fig. 1. Full template with one, two, and three locations

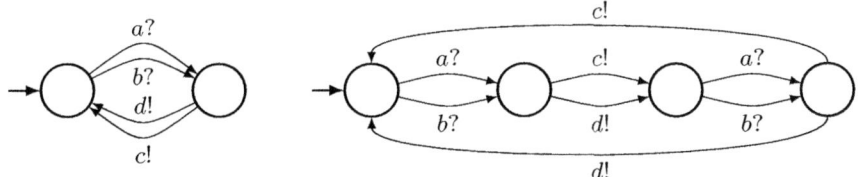

Fig. 2. Cyclic-executive template with two and four locations

and 2 show two example template families. In the *full template*, shown in Figure 1, every pair of locations is connected by an edge for every possible action. In the *cyclic-executive template*, shown in Figure 2, the controller implements some schedule according to which the actions are handled; the controller alternates between waiting for an uncontrollable action and responding with some controllable action. The template families are organized according to the number of locations. Typically, we start the synthesis process with small templates and then iteratively increase the size until an optimal controller is found. A controller *matches* a template if it can be obtained by removing a subset of the edges. We encode the presence of edges using Boolean parameter variables and combine the resulting parametric timed automaton with the plant automaton: any valuation of the Boolean parameters under which only safe states are reachable represents a correct controller.

Template-based synthesis has several attractive features: Since the observations of the controller are limited by the template, template-based synthesis naturally solves the controller synthesis problem with partial observability. The size of the controller is also limited by the size of the template. Because the templates model standard types of controllers, the synthesized controllers are well-structured, resembling a manually built controller.

In terms of complexity, it is not surprising that template-based synthesis is much simpler than standard synthesis. The problem is PSPACE-complete, matching the complexity of model checking. Template-based synthesis can in fact be understood as parametric model checking, where we verify a timed automaton that is parametrized with Boolean variables.

The technical challenge in developing fast algorithms for template-based synthesis thus lies in the efficient manipulation of the potential valuations of the parameters. In the paper, we present a solution to this challenge based on automatic abstraction refinement. Starting with an initial abstraction that considers all parameter valuations, a refinement loop incrementally focuses the search towards smaller and smaller sets of parameter valuations. The loop terminates as soon as, for the remaining parameter valuations, only safe states are reachable. New refinements are computed by identifying situations where an unsafe state is reachable for a subset of the parameter valuations.

Our experimental results indicate that template-based synthesis is not only an effective solution to the controller synthesis problem with partial observability, it is an attractive alternative to standard synthesis also in the simpler case of complete observability, outperforming tools like UPPAAL-TIGA on benchmarks where structurally simple controllers, corresponding to the available templates, exist.

Related Work. The basic timed controller synthesis problem for timed automata [2] with complete observability was defined by Maler et al. [21,4] in the setting of turn-based timed games. In their fundamental work, the decidability of the problem was shown by demonstrating that the standard discrete attractor construction [26] on the region graph suffices to obtain winning strategies. Henzinger and Kopke showed that this construction is theoretically optimal by proving that the problem is EXPTIME-complete [17]. D'Souza and Madhusudan investigated the complexity of timed controller synthesis against external specifications [12]. Bouyer et al. continued this line of research and also investigated the impact of partial information [6].

A first more practical approach to timed controller synthesis, implemented in the tool SYNTHKRO, was proposed by Altisen and Tripakis [1]. The approach requires, however, an expensive preprocessing step. Cassez et al. presented a symbolic algorithm, implemented in UPPAAL-TIGA, that avoids the upfront state explosion by combining the backward attractor construction with a forward zone graph exploration [8,5]. Our timed verification and synthesis tool SYNTHIA [23] is based on abstraction refinement techniques that combine symbolic representations for the discrete and the continuous state components [14] and exploit the compositional structure of the timed system [24]. We have implemented the approach presented in this paper as an extension of SYNTHIA.

Compared to the significant body of work on timed controller synthesis with complete observability, there has been comparatively little work on the more realistic setting of partial observability. Fundamental results on the undecidability of the general problem and the complexity for fixed granularity are due to Bouyer et al. [6]. Cassez et al. proposed a pragmatic approach to handle partial information, which restricts the choices and the observability of the controller so that a zone-based synthesis algorithm remains possible [9]. An extension of this work uses *alternating timed simulation relations* to efficiently control partially observable systems [10]. This approach has also been implemented in UPPAAL-TIGA.

Template-based synthesis is related to the *bounded synthesis* approach [25], where one fixes the size (but not the structure) of the controller. Bounded synthesis has so far been limited to purely discrete systems. There are efficient algorithms for bounded synthesis based on SMT-solving [16], antichains [15], and BDDs [13], which, however, unfortunately do not seem to have straightforward extensions to the timed case. Another interesting restriction on the type of controllers to be considered has been proposed by Lustig et al.: *synthesis from component libraries* [20] attempts to construct a controller by assembling routines from a given library. The difference to template-based synthesis is that the synthesized controller is a combination of predefined components rather than an instantiation of a parametric template. Currently, this approach is also limited to discrete systems.

2   Timed Systems

Timed Automata.   The components of a timed system are represented by *timed automata*. A timed automaton [2] is a tuple $\mathcal{A} = (L, l_0, \Sigma, \Delta, X, I)$, where L is a finite set of (control) locations, $l_0 \in L$ is the initial location, Σ is a finite set of actions, $\Delta \subseteq (L \times \Sigma \times \mathcal{C}(X) \times 2^X \times L)$ is an edge relation, X is a finite set of real valued clocks, $I : L \to \mathcal{C}(X)$ maps each location to an invariant, and $\mathcal{C}(X)$ is the set of clock constraints over X. A (rectangular) clock constraint $\varphi \in \mathcal{C}(X)$ is of the form

$$\varphi = \textbf{true} \mid x \leq c \mid c \leq x \mid x < c \mid c < x \mid \varphi_1 \wedge \varphi_2,$$

where x is a clock in X and c is a constant in $\mathbb{N}_0$. A *clock valuation* $t : X \to \mathbb{R}_{\geq 0}$ assigns a nonnegative value to each clock and can also be represented by a $|X|$-dimensional vector $t \in \mathcal{R}$, where $\mathcal{R} = \mathbb{R}^X_{\geq 0}$ denotes the set of all clock valuations.

The states of a timed automaton are pairs (l, t) of locations and clock valuations. Timed automata have two types of transitions: *timed transitions*, where only time passes and the location remains unchanged, and *discrete transitions*, where no time passes, the current location can be changed and some clocks can be reset to zero. In a timed transition, denoted by $(l, t) \xrightarrow{d} (l, t + d \cdot 1)$, the same nonnegative value $d \in \mathbb{R}_{\geq 0}$ is added to all clocks such that, for each $0 \leq d' \leq d$, $t + d'$ satisfies the location invariant $I(l)$. A discrete transition, denoted by $(l, t) \xrightarrow{a} (l', t')$ for some action $a \in \Sigma$, corresponds to an edge $\delta = \langle l, a, \varphi, \lambda, l' \rangle$ of Δ such that t satisfies the clock constraint φ of δ, and $t' = t[\lambda := 0]$ is obtained from t by setting the clocks in λ to 0 and satisfies the location invariant $I(l')$. For two states s and s', we write $s \xrightarrow{\delta} s'$ if there is a delay $d \in \mathbb{R}_{\geq 0}$ and an edge δ with action a such that there is an s'' with $s \xrightarrow{d} s''$ and $s'' \xrightarrow{a} s'$.

We say that a state s is *reachable* if there is a finite sequence of transitions of the form $s_0 \xrightarrow{\delta_0} s_1 \ldots s_{n-1} \xrightarrow{\delta_{n-1}} s$ such that $\delta_0, \ldots, \delta_{n-1} \in \Delta$ are edges in Δ,

$s_0 = (l_0, \mathbf{0})$ is the initial state (where $\mathbf{0}$ is the zero vector), and for all $1 \leq i \leq n$, the individual $s_i = (l_i, t_i)$ are states of the automaton. We define $\mathrm{Reach}(\mathcal{A})$ as the set of all forward reachable states of a timed automaton $\mathcal{A}$.

Composition. Timed automata can be composed to networks, in which the automata run in parallel and synchronize on shared actions. For two timed automata $\mathcal{A} = (L_1, l_0^1, \Sigma_1, \Delta_1, X_1, I_1)$ and $\mathcal{A}' = (L_2, l_0^2, \Sigma_2, \Delta_2, X_2, I_2)$ with disjoint clock sets $X_1 \cap X_2 = \emptyset$, the *parallel composition* $\mathcal{A}_1 \| \mathcal{A}_2$ is the timed automaton $(L_1 \times L_2, (l_0^1, l_0^2), \Sigma_1 \cup \Sigma_2, \Delta, X_1 \cup X_2, I)$, where $I(l_1, l_2) = I_1(l_1) \wedge I_2(l_2)$ for all $l_1 \in L_1$ and $l_2 \in L_2$, and Δ is the smallest set that contains

- for $a \in \Sigma_1 \cap \Sigma_2$, $\langle (l_1, l_2), a, \varphi_1 \wedge \varphi_2, \lambda_1 \cup \lambda_2, (l_1', l_2') \rangle$ if $\langle l_1, a, \varphi_1, \lambda_1, l_1' \rangle \in \Delta_1$ and $\langle l_2, a, \varphi_2, \lambda_2, l_2' \rangle \in \Delta_2$,
- for $a \in \Sigma_1 \setminus \Sigma_2$, $\langle (l_1, l_2), a, \varphi_1, \lambda_1, (l_1', l_2) \rangle$ if $\langle l_1, a, \varphi_1, \lambda_1, l_1' \rangle \in \Delta_1$, and
- for $a \in \Sigma_2 \setminus \Sigma_1$, $\langle (l_1, l_2), a, \varphi_2, \lambda_2, (l_1, l_2') \rangle$ if $\langle l_2, a, \varphi_2, \lambda_2, l_2' \rangle \in \Delta_2$.

Finite Semantics. The decidability of the reachability problem of timed automata relies on the existence of the *region equivalence relation* [2] on $\mathcal{R}$ which has a finite index.

For a timed automaton $\mathcal{A} = (L, l_0, \Sigma, \Delta, X, I)$, we call the value of a clock $x \in X$ *maximal* if it is strictly greater than the highest constant c_{max} any clock is compared to. We say that two clock valuations $t_1, t_2 \in \mathcal{R}$ are in the same *clock region*, denoted $t_1 \sim_R t_2$, if

- the set of clocks with maximal value is the same in t_1 and in t_2 ($\forall x \in X$: $t_1(x) > c_{max} \Leftrightarrow t_2(x) > c_{max}$), and
- t_1 and t_2 agree (1) on the integer parts of the clock values, (2) on the relative order of the noninteger parts of the clock values, and (3) on the equality of the noninteger parts of the clock values with 0. That is, for all clocks x and y with nonmaximal value, it holds that (1) $\lfloor t_1(x) \rfloor = \lfloor t_2(x) \rfloor$, (2) $\hat{t}_1(x) \leq \hat{t}_1(y) \Leftrightarrow \hat{t}_2(x) \leq \hat{t}_2(y)$, and (3) $\hat{t}_1(x) = 0$ if, and only if, $\hat{t}_2(x) = 0$, where $\hat{t}_i(x) = t_i(x) - \lfloor t_i(x) \rfloor$ for $i \in \{1, 2\}$.

We denote with $[t]_R = \{ t' \in \mathcal{R} \mid t \sim_R t' \}$ the clock region t belongs to. We say that two states $s_1 = (l_1, t_1)$ and $s_2 = (l_2, t_2)$ of $\mathcal{A}$ are *region-equivalent*, denoted by $s_1 \sim_R s_2$, if their locations are the same ($l_1 = l_2$) and the clock valuations are in the same clock region ($t_1 \sim_R t_2$), and denote with $[(l, t)]_R = \{ (l, t') \in L \times \mathcal{R} \mid t \sim_R t' \}$ the equivalence class of region-equivalent states that (l, t) belongs to.

Regions are a suitable semantics for the abstraction of timed automata because they essentially preserve the language: if there is a discrete transition $s \xrightarrow{a} s'$ from a state s to a state s' of a timed automaton, then there is, for all states r with $r \sim_R s$, a state r' with $r' \sim_R s'$ such that $r \xrightarrow{a} r'$ is a discrete transition with the same label. For timed transitions, a slightly weaker property holds: if there is a timed transition $s \xrightarrow{t} s'$ from a state s to a state s', then there is, for all states r with $r \sim_R s$, a state r' with $r' \sim_R s'$ such that there is a timed transition $r \xrightarrow{t'} r'$ (but possibly with $t' \neq t$).

The *finite semantics* of a timed automaton $\mathcal{A} = (L, l_0, \Sigma, \Delta, X, I)$ is the finite graph $\llbracket \mathcal{A} \rrbracket = (Q, q_0, T)$ where

- the symbolic state set $Q = \{[(l, t)]_R \mid (l, t) \in L \times \mathcal{R}\}$ of $[\![\mathcal{A}]\!]$ is the set of equivalence classes of region-equivalent states of $\mathcal{A}$, with
- the initial state $q_0 = [(l_0, t_0)]_R$, and
- the set $T = \{(q, q') \in Q \times Q \mid \exists r \in q, \; r' \in q', a \in \Sigma \cup \mathbb{R}_{\geq 0}. \, r \xrightarrow{a} r'\}$ of transitions.

The finite semantics is reachability-preserving:

Lemma 1. *[2] For a timed automaton $\mathcal{A} = (L, l_0, \Sigma, \Delta, X, I)$ there is a finite path from a state (l, t) to a state (l', t') if, and only if, there is a finite path from $\left[(l, t)\right]_R$ to $\left[(l', t')\right]_R$ in $[\![\mathcal{A}]\!]$.*

Assuming a binary encoding of the constants in the clock constraints, the number of states of the finite semantics is exponential in the number of clocks and in the magnitude of the constants:

Lemma 2. *[2] For a timed automaton $\mathcal{A} = (L, l_0, \Sigma, \Delta, X, I)$, with c_x as the maximal constant appearing in any constraint of $\mathcal{A}$, the number of states of $[\![\mathcal{A}]\!]$ is bounded by*

$$|L| \cdot |X|! \cdot 2^{|X|-1} \cdot \prod_{x \in X} O(c_x) = |L| \cdot |X|! \cdot O(c_x)^{|X|}.$$

As it turns out, the finite semantics is a theoretically optimal state space representation for deciding reachability:

Theorem 1. *[2] For a timed automaton $\mathcal{A}$ and a set of states B, testing whether $\mathsf{Reach}(\mathcal{A}) \cap B = \emptyset$ is PSPACE-complete.*

In practice, instead of deciding $\mathsf{Reach}(\mathcal{A}) \cap B = \emptyset$ based on an explicit construction of the finite semantics, tools like SYNTHIA or UPPAAL use the much coarser *clock zones* as the fundamental representation of clock values.

3   Template-Based Controller Synthesis

In this section, we formalize controller templates and the template instantiation problem. A *controller template* is a tuple $(\mathcal{T}, P, \Pi)$ consisting of a timed automaton $\mathcal{T} = (L, l_0, \Sigma, \Delta, X, I)$, a finite set of Boolean parameters P, and a total function $\Pi : \mathcal{P} \to 2^{\Delta}$ defining which edges are enabled for a given parameter valuation, where $\mathcal{P} = P \to \mathbb{B}$ is the set of all parameter valuations. In the following, we will assume that the timed automaton modeling the environment (or plant) is already integrated (by parallel composition) in $\mathcal{T}$. As usual, we assume that the controller does neither reset plant clocks, inhibit plant actions, nor introduce timelocks.

Definition 1. *For a controller template $(\mathcal{T}, P, \Pi)$ with $\mathcal{T} = (L, l_0, \Sigma, \Delta, X, I)$ and a set of bad states B, the instantiation problem asks for a parameter valuation $p \in \mathcal{P}$ such that $\mathcal{I} = (L, l_0, \Sigma, \Pi(p), X, I)$ and $\mathsf{Reach}(\mathcal{I}) \cap B = \emptyset$.*

or starts with the bad states and propagates, in a backward manner, those combinations of states and parameter valuations that have a path to the bad states.

To accommodate both directions, we define a successor and a predecessor propagation function $\mathsf{Succ}, \mathsf{Pred} : 2^{S \times \mathcal{P}} \to 2^{S \times \mathcal{P}}$ with

$$\mathsf{Succ}(Y) = \left\{ (s', \boldsymbol{p}) \in S \times \mathcal{P} \mid \exists \delta \in \Pi(\boldsymbol{p}) : \exists s \in S : (s, \boldsymbol{p}) \in Y \wedge s \xrightarrow{\delta} s' \right\} \text{ and}$$

$$\mathsf{Pred}(Y') = \left\{ (s, \boldsymbol{p}) \in S \times \mathcal{P} \mid \exists \delta \in \Pi(\boldsymbol{p}) : \exists s' \in S : (s', \boldsymbol{p}) \in Y' \wedge s \xrightarrow{\delta} s' \right\}.$$

The set FR of forward-reachable states and parameter valuations and the set BR of backward-reachable states and parameter valuations are obtained by the following fixed point computations (the index identifies the round of the fixpoint iteration):

$$
\begin{aligned}
\mathsf{FR}_0 &= \{(l_0, \mathbf{0})\} \times \mathcal{P} & \mathsf{BR}_0 &= B \times \mathcal{P} \\
\mathsf{FR}_{i+1} &= \mathsf{Succ}(\mathsf{FR}_i) \cup \mathsf{FR}_i & \mathsf{BR}_{i+1} &= \mathsf{Pred}(\mathsf{BR}_i) \cup \mathsf{BR}_i \\
\mathsf{FR} &= \lim_i \mathsf{FR}_i & \mathsf{BR} &= \lim_i \mathsf{BR}_i.
\end{aligned}
$$

Clearly, if there is some $(s, \boldsymbol{p}) \in \mathsf{FR}_i$ then this means that state s is reached after $i \in \mathbb{N}$ forward steps for parameter valuation $\boldsymbol{p}$, which corresponds to a path $s_0 \xrightarrow{\delta_1} s_1 \xrightarrow{\delta_2} \ldots \xrightarrow{\delta_i} s$, where each $\delta_1, \delta_2, \ldots, \delta_i$ is in $\Pi(\boldsymbol{p})$. Dually, if there is some state $(s, \boldsymbol{p}) \in \mathsf{BR}_i$ then this means that state s is reached after $i \in \mathbb{N}$ backward steps for parameter valuation $\boldsymbol{p}$, which corresponds to a path $s \xrightarrow{\delta_1} s_1 \xrightarrow{\delta_2} \ldots \xrightarrow{\delta_i} b$, where $b \in B$ and each $\delta_1, \delta_2, \ldots, \delta_i$ is in $\Pi(\boldsymbol{p})$.

We can obtain the feasible instantiations either by looking for parameter valuations in FR that are not paired up with bad states, or by looking for parameter valuations in BR that are not paired up with the initial state. Both constructions identify the same set of feasible instantiations.

Theorem 3. *The set*

$$G = \{\boldsymbol{p} \in \mathcal{P} \mid (B \times \{\boldsymbol{p}\}) \cap \mathsf{FR} = \emptyset\} = \{\boldsymbol{p} \in \mathcal{P} \mid ((l_0, \mathbf{0}), \boldsymbol{p}) \notin \mathsf{BR}\}$$

consists of exactly the feasible instantiations.

In practice, neither construction performs well. The problem is that it is difficult and expensive to maintain the correlation between parameter valuations and reachable states; typically, each parameter valuation results in a different set of states.

Instead of directly computing the precise set of parameter valuations, in the next subsection, we will present an abstraction technique that allows us to reason about approximations of parameter valuations.

4.2 The Focus Abstraction

We now consider an abstraction of the template based on a given set $P \subseteq \mathcal{P}$ of parameter valuations, which we call *focus*. We use the parameter valuations in P to obtain an over- or underapproximation of the sets FR and BR, by considering P as an equivalence class: we require that a transition must exist for some or all parameter valuations in P, respectively. In the following, we use an overapproximation for the forward construction and an underapproximation for the backward construction; obviously, all constructions can also be dualized. We obtain the following approximate successor and predecessor functions: $\overline{\mathsf{Succ}}^P, \underline{\mathsf{Pred}}^P : 2^S \to 2^S$ with

$$\overline{\mathsf{Succ}}^P(Y) = \left\{ s' \in S \mid \exists \boldsymbol{p} \in P : \exists \delta \in \Pi(\boldsymbol{p}) : \exists s \in Y : s \xrightarrow{\delta} s' \right\} \text{ and}$$

$$\underline{\mathsf{Pred}}^P(Y') = \left\{ s \in S \mid \forall \boldsymbol{p} \in P : \exists \delta \in \Pi(\boldsymbol{p}) : \exists s' \in Y' : s \xrightarrow{\delta} s' \right\}.$$

Replacing the precise Succ and Pred operators in the fixed point construction from Subsection 4.1, we obtain two new fixed point constructions for the approximations $\overline{\mathsf{FR}}^P$ and $\underline{\mathsf{BR}}^P$:

$$
\begin{aligned}
\overline{\mathsf{FR}}_0^P &= \{(l_0, \boldsymbol{0})\} & \underline{\mathsf{BR}}_0^P &= B \\
\overline{\mathsf{FR}}_{i+1}^P &= \overline{\mathsf{Succ}}^P(\overline{\mathsf{FR}}_i^P) \cup \overline{\mathsf{FR}}_i^P & \underline{\mathsf{BR}}_{i+1}^P &= \underline{\mathsf{Pred}}^P(\underline{\mathsf{BR}}_i^P) \cup \underline{\mathsf{BR}}_i^P \\
\overline{\mathsf{FR}}^P &= \lim_i \overline{\mathsf{FR}}_i^P & \underline{\mathsf{BR}}^P &= \lim_i \underline{\mathsf{BR}}_i^P.
\end{aligned}
$$

Clearly, if there is some state $s \in \overline{\mathsf{FR}}_i^P$ then this means that state s is reached after $i \in \mathbb{N}$ forward steps for a set of parameter valuations P, which corresponds to a path $s_0 \xrightarrow{\delta_1} s_1 \xrightarrow{\delta_2} \ldots \xrightarrow{\delta_i} s$, where, for each δ_i, there is a $\boldsymbol{p}_i \in P$ such that δ_i in $\Pi(\boldsymbol{p}_i)$. Dually, if there is some state $s \in \underline{\mathsf{BR}}_i^P$ then this means that state s is reached after $i \in \mathbb{N}$ backward steps for a set of parameter valuations P, which corresponds to a path $s \xrightarrow{\delta_1} s_1 \xrightarrow{\delta_2} \ldots \xrightarrow{\delta_i} b$, where $b \in B$ and, for each δ_i and each $\boldsymbol{p} \in P$, we have δ_i in $\Pi(\boldsymbol{p})$.

The following lemma clarifies the relationships between the approximate and precise versions of FR and BR: $\overline{\mathsf{FR}}^P$ overapproximates FR on P, $\underline{\mathsf{BR}}^P$ underapproximates BR on P.

Lemma 3. *For every set $P \subseteq \mathcal{P}$ of parameter valuations, it holds that*

$$\overline{\mathsf{FR}}^P \supseteq \{s \in S \mid \exists \boldsymbol{p} \in P : (s, \boldsymbol{p}) \in \mathsf{FR}\} \text{ and}$$

$$\underline{\mathsf{BR}}^P \subseteq \{s \in S \mid \exists \boldsymbol{p} \in P : (s, \boldsymbol{p}) \in \mathsf{BR}\}.$$

Combining Lemma 3 with Theorem 3, we obtain that the focus abstraction allows us to approximate the set of feasible instantiations: A set of parameter valuations P definitely represents feasible instantiations if no bad states appear in $\overline{\mathsf{FR}}^P$. Dually, the parameter valuations in P definitely represent infeasible instantiations if the initial state appears in $\underline{\mathsf{BR}}^P$. Hence, we obtain the following lower and upper bounds for the set of feasible instantiations.

Theorem 4. *Let G be the precise set of feasible instantiations. For every set $P \subseteq \mathcal{P}$, it holds that*

$$\left\{ \boldsymbol{p} \in P \mid B \cap \overline{\mathsf{FR}}^P = \emptyset \right\} \subseteq G \subseteq \left\{ \boldsymbol{p} \in \mathcal{P} \mid \boldsymbol{p} \in P \Rightarrow (l_0, \boldsymbol{0}) \notin \underline{\mathsf{BR}}^P \right\}.$$

In the next subsection, we will describe an automatic refinement algorithm for the Focus abstraction.

4.3   Abstraction Refinement

We now describe a refinement procedure that computes an increasingly precise approximation of the set of feasible instantiations. The procedure starts with the set $\mathcal{P}$ of all parameter valuations, and then splits the set into smaller and smaller subsets, until either a feasible instance is found, or it is established that no feasible instance exists.

Algorithm 1. Solve(P): The algorithm computes a safe subset of a given set P of parameter valuations, or returns **fail** if no safe subset exists.

```
 1: if P = ∅ then
 2:     return fail
 3: else if (l0, 0) ∈ BR^P then
 4:     return fail
 5: else if FR‾^P ∩ BR^P = ∅ then
 6:     return P
 7: else
 8:     P1 := Refine(P)
 9:     R1 := Solve(P1)
10:     if R1 ≠ fail then
11:         return R1
12:     else
13:         P2 := P \ P1
14:         return Solve(P2)
```

The procedure is shown as Algorithm 1. The input to the procedure is the current focus P, for which we initially use $\mathcal{P}$. Unless the (un)reachability of some bad state can be surely established, after each refinement step, Solve recurs on the refined focus. In each call of Solve, the set of bad states are augmented with the states in $\underline{\mathsf{BR}}^P$. This is justified by the following lemmas, which state that the old underapproximation $\underline{\mathsf{BR}}^P$ is a subset of the new underapproximation $\underline{\mathsf{BR}}^{P'}$ for a refinement $P' \subset P$, and that excluding $\underline{\mathsf{BR}}^P$ from $\overline{\mathsf{FR}}^P$ does not affect the resulting upper bound on the feasible instantiations.

Lemma 4. *For two sets $P, P' \subseteq \mathcal{P}$ of parameter valuations such that $P' \subset P$, it holds that $\underline{\mathsf{BR}}^P \subseteq \underline{\mathsf{BR}}^{P'}$.*

Lemma 5. *For every set $P \subseteq \mathcal{P}$ of parameter valuations, it holds that*

$$\left\{ \boldsymbol{p} \in P \mid B \cap \overline{\mathsf{FR}}^P = \emptyset \right\} = \left\{ \boldsymbol{p} \in P \mid \underline{\mathsf{BR}}^P \cap \overline{\mathsf{FR}}^P = \emptyset \right\}.$$

It remains to specify the function Refine, which is called in procedure Solve to find an appropriate subset of $\mathcal{P}$ to split on. Since $\mathcal{P}$ is finite, we could, in principle, choose any strict (and non-empty) subset of $\mathcal{P}$ during the refinement step. In the following we describe a heuristic choice that has proved useful in practice: we choose a set of parameter valuations that are guaranteed to increase $\underline{\mathsf{BR}}^P$ in the next iteration.

Suppose the termination conditions of procedure Solve are not true yet, i.e., the initial state is not in $\underline{\mathsf{BR}}^P$ and there are still states in $\overline{\mathsf{FR}}^P \cap \underline{\mathsf{BR}}^P$. We choose a state $s \in \overline{FR}^P \setminus \underline{\mathsf{BR}}^P$ and a state $s' \in \underline{BR}^P$, such that there exists a transition $\delta \in \Pi(\boldsymbol{p})$ that leads from s to s' for some $\boldsymbol{p} \in P$, but not for all $\boldsymbol{p} \in P$. The refinement proceeds with the parameter valuations that allow a transition from s to s':

$$\mathsf{Refine}(P) = \{\boldsymbol{p} \in P \mid \exists \delta \in \Pi(\boldsymbol{p}) : s \xrightarrow{\delta} s'\}$$

Since such a pair s, s' of states can be found until the termination conditions of procedure Solve become true, we obtain that Refine always ensures progress of our refinement algorithm.

Lemma 6. *For every set of parameter valuations $P \subseteq \mathcal{P}$, if $P \neq \emptyset$, $(l_0, \mathbf{0}) \notin \underline{\mathsf{BR}}^P$, and $\overline{\mathsf{FR}}^P \cap \underline{\mathsf{BR}}^P \neq \emptyset$, then there is a state $s \in \overline{\mathsf{FR}}^P \setminus \underline{\mathsf{BR}}^P$ and a state $s' \in \underline{\mathsf{BR}}^P$ such that*

$$\emptyset \subset \mathsf{Refine}(P) \subset P.$$

Putting everything together, we obtain the following correctness theorem for Solve($\mathcal{P}$), where Lemma 6 guarantees termination and Theorem 4 guarantees soundness of the result.

Theorem 5. *Called with the set $\mathcal{P}$ of parameter valuations, Procedure Solve($\mathcal{P}$) terminates after at most $|\mathcal{P}|$ refinement steps and either computes a feasible template instantiation or reports failure, in which case no feasible template instantiation exists.*

5   Experimental Results

In this section, we report on experimental results based on a prototype implementation of the symbolic instantiation algorithm from Section 4.

Implementation. We have implemented the symbolic instantiation algorithm from Section 4 in the SYNTHIA tool [23]. SYNTHIA is a verification and synthesis tool for timed automata extended with bounded integer variables. The tool provides facilities for automatic abstraction refinement and combines reduced ordered binary decision diagrams (ROBDDs) [7] with difference bound matrices (DBMs) [11] to obtain symbolic state representations for both discrete and continuous state components.

While standard SYNTHIA already includes a game-based synthesis algorithm for timed controllers with complete observability, we only use the verification functionality of SYNTHIA for template-based synthesis. For a template instantiation problem given by a controller template $(\mathcal{T}, P, \Pi)$ and a set of bad states, we encode the Boolean parameters P by global integer variables whose initial values are left undefined.

The algorithm from Section 4 is realized as a specialized refinement procedure for SYNTHIA's standard (location-based) abstraction refinement loop. Whenever an edge for refinement is found, we identify the parameter valuations associated with that edge and split the global abstraction with these valuations. In the subsequent refinement step, we focus (i.e., we restrict the forward exploration) on the identified parameters of the last refinement.

Benchmarks. In the *Chinese Juggler* benchmark [19], a performer needs to stabilize spinning plates to prevent them from falling. After a certain amount of time has passed since a plate was stabilized, it can nondeterministically become unstable. If no restabilization takes place, it ultimately falls down. The plates have different sizes, and hence, different times to become unstable. It takes the performer one time unit to stabilize a certain plate. During that time, he cannot stabilize another plate. The controller synthesis task consists in finding a safe strategy for the performer such that no plate will ever fall down. The benchmark size is parametrized in the number of plates n. For the template-based synthesis, we use a generic cyclic-executive template (cf. Section 1) with n locations. In each step of the cyclic execution, the controller decides which plate should be stabilized next.

In the *Dam* benchmark, a controller is to be synthesized that determines the speed of the inflow to a dam. The controller can either stop the inflow or choose between a slow or a fast inflow speed. The bounded reachability requirement is that the fill level should reach a certain value between a minimal and maximal bound. While a fast speed might reach the desired fill level more quickly, the variance of the actual inflow is larger so that the maximal level might be exceeded. On the other hand, being in slow mode, it takes longer to reach the desired fill level, but the variance is not so high so that it is always possible to exactly reach a desired fill level. Thus, being in one mode all the time is not feasible, since a feasible controller must alternate between fast and slow at least once to fulfill the requirement. The benchmark size is parametrized in the degree of precision in which the fill level and the inflow amount is digitized. For the template-based synthesis, we use a controller template that models a parametric two-phase program: in the first phase, a certain inflow speed is set until a threshold of the current fill level is passed. Then, the controller enters the second phase with a possibly different speed. The controller stops as soon as a desired fill level is reached. The first and the second speed, as well as the phase-switching threshold are parameters, for which feasible instantiations are to be found.

Results. We compare the performance of the template-based extension of SYN-THIA against standard SYNTHIA and UPPAAL-TIGA 0.16 [5]. In Table 1, the columns show, from left to right, the benchmark instance, the number of refinement steps and abstract locations in the final abstraction of the parameter synthesis algorithm, the running time and memory consumption of our template-based implementation, the performance of SYNTHIA's standard controller syn-

Table 1. Experimental evaluation of template-based synthesis. We compare the performance of the template-based extension of SYNTHIA against standard SYNTHIA and UPPAAL-TIGA.

Benchmark	Template-based SYNTHIA				Standard SYNTHIA				UPPAAL-TIGA		
	Steps	Abs	Time	Mem	Steps	Abs	Time	Mem	States	Time	Mem
Juggler 2	6	19	0	53	2	5	1	53	57	0	6
Juggler 3	38	136	0	61	6	9	1	61	477	0	6
Juggler 4	110	421	2	87	TIMEOUT				6755	6	57
Juggler 5	423	1899	59	247	TIMEOUT				81292	1095	79
Juggler 6	1445	8335	1932	1335	TIMEOUT				TIMEOUT		
Juggler 7	TIMEOUT				TIMEOUT				TIMEOUT		
Dam 5	58	100	1	80	230	149	4	80	88592	2	65
Dam 25	268	380	13	87	1115	718	1182	91	3114648	307	443
Dam 50	530	730	87	105	TIMEOUT				13545848	5018	2355
Dam 75	793	1080	329	111	TIMEOUT				TIMEOUT		
Dam 100	1055	1430	927	143	TIMEOUT				TIMEOUT		
Dam 125	1318	1780	1949	149	TIMEOUT				TIMEOUT		
Dam 150	1580	2130	3483	153	TIMEOUT				TIMEOUT		
Dam 175	1843	2480	5127	213	TIMEOUT				TIMEOUT		
Dam 200	TIMEOUT				TIMEOUT				TIMEOUT		

thesis algorithm, UPPAAL-TIGA's number of explored states, running time and memory consumption. For UPPAAL-TIGA, we measured the performance for various parameters and always selected the best results. Running times are given in seconds, memory consumption in MB, the time limit was set to 2 hours, and the memory limit was set to 4 GB. All experiments were conducted on a 2.6 GHz AMD Opteron computer running Ubuntu 10.04.

For both benchmarks, template-based SYNTHIA clearly outperforms the game-based synthesis techniques implemented in standard SYNTHIA and UPPAAL-TIGA. A closer look at the *Chinese Juggler* example reveals that a major source of complexity results from the subtraction operation that occurs in the backwards computation of the winning states. Subtraction is expensive because it does not preserve convexity, and therefore requires a split into multiple zones. The much better performance of template-based synthesis is due to the fact that template-based synthesis is based on model checking, rather than game solving, and model checking does not require such nonconvex operations. In the *Dam* example, we observe that the size of the abstraction, and, thus, the running time, of the template-based approach increases polynomially in the size of the benchmark, while both standard SYNTHIA and UPPAAL-TIGA suffer from an exponential blow-up.

6   Conclusion

Our results demonstrate that template-based synthesis is an attractive alternative to the standard game-based approach to timed synthesis. Template-based synthesis has the better worst-case complexity, is easier to implement with symbolic data structures such as DBMs, and produces nicely structured controllers with a small number of locations.

In future work, we plan to expand the class of templates considered by the synthesis algorithm. Particularly interesting is the introduction of parameters in the clock constraints. Results from parametric timed model checking [3] indicate that the analysis of such templates is in general undecidable. However, subclasses of parametric timed automata, such as L/U automata [18], for which the emptiness problem is decidable, are promising candidates for a more expressive and yet computationally feasible class of templates.

The long-term goal is to obtain a succinct but comprehensive library of standard templates that serves as a basis for a fully automatic template-based synthesis approach.

Acknowledgments. The authors would like to thank Alexandre David for pointing out the *Chinese Juggler* benchmark and Christoph Scholl for helpful comments on an early draft of the paper.

This work was supported by the German Research Foundation (DFG) as part of the Transregional Collaborative Research Center "Automatic Verification and Analysis of Complex Systems" (SFB/TR 14 AVACS).

References

1. Altisen, K., Tripakis, S.: Tools for controller synthesis of timed systems. In: 2nd Workshop on Real-Time Tools, RT-TOOLS (2002)
2. Alur, R., Dill, D.L.: A theory of timed automata. Theoretical Computer Science 126(2), 183–235 (1994)
3. Alur, R., Henzinger, T.A., Vardi, M.Y.: Parametric real-time reasoning. In: STOC, pp. 592–601 (1993)
4. Asarin, E., Maler, O., Pnueli, A., Sifakis, J.: Controller synthesis for timed automata. In: Lafay, J.-F. (ed.) Proc. 5th IFAC Conference on System Structure and Control, pp. 469–474. Elsevier (1998)
5. Behrmann, G., Cougnard, A., David, A., Fleury, E., Larsen, K.G., Lime, D.: UPPAAL-Tiga: Time for Playing Games! In: Damm, W., Hermanns, H. (eds.) CAV 2007. LNCS, vol. 4590, pp. 121–125. Springer, Heidelberg (2007)
6. Bouyer, P., D'Souza, D., Madhusudan, P., Petit, A.: Timed Control with Partial Observability. In: Hunt Jr., W.A., Somenzi, F. (eds.) CAV 2003. LNCS, vol. 2725, pp. 180–192. Springer, Heidelberg (2003)
7. Bryant, R.E.: Graph-based algorithms for boolean function manipulation. IEEE Trans. Computers 35(8), 677–691 (1986)
8. Cassez, F., David, A., Fleury, E., Larsen, K.G., Lime, D.: Efficient on-the-Fly Algorithms for the Analysis of Timed Games. In: Abadi, M., de Alfaro, L. (eds.) CONCUR 2005. LNCS, vol. 3653, pp. 66–80. Springer, Heidelberg (2005)
9. Cassez, F., David, A., Larsen, K.G., Lime, D., Raskin, J.-F.: Timed control with observation based and stuttering invariant strategies. In: [22], pp. 192–206
10. Chatain, T., David, A., Larsen, K.G.: Playing games with timed games. In: Giua, A., Silva, M., Zaytoon, J. (eds.) Proceedings of the 3rd IFAC Conference on Analysis and Design of Hybrid Systems (ADHS 2009), Zaragoza, Spain (September 2009)

11. Dill, D.L.: Timing Assumptions and Verification of Finite-State Concurrent Systems. In: Sifakis, J. (ed.) CAV 1989. LNCS, vol. 407, pp. 197–212. Springer, Heidelberg (1990)

12. D'Souza, D., Madhusudan, P.: Timed Control Synthesis for External Specifications. In: Alt, H., Ferreira, A. (eds.) STACS 2002. LNCS, vol. 2285, pp. 571–582. Springer, Heidelberg (2002)

13. Ehlers, R.: Symbolic Bounded Synthesis. In: Touili, T., Cook, B., Jackson, P. (eds.) CAV 2010. LNCS, vol. 6174, pp. 365–379. Springer, Heidelberg (2010)

14. Ehlers, R., Mattmüller, R., Peter, H.-J.: Combining Symbolic Representations for Solving Timed Games. In: Chatterjee, K., Henzinger, T.A. (eds.) FORMATS 2010. LNCS, vol. 6246, pp. 107–121. Springer, Heidelberg (2010)

15. Filiot, E., Jin, N., Raskin, J.-F.: An Antichain Algorithm for LTL Realizability. In: Bouajjani, A., Maler, O. (eds.) CAV 2009. LNCS, vol. 5643, pp. 263–277. Springer, Heidelberg (2009)

16. Finkbeiner, B., Schewe, S.: SMT-based synthesis of distributed systems. In: Proceedings of the 2nd Workshop on Automated Formal Methods (AFM 2007), November 6, pp. 69–76. ACM Press, Atlanta (2007)

17. Henzinger, T.A., Kopke, P.W.: Discrete-time control for rectangular hybrid automata. Theoretical Computer Science 221(1-2), 369–392 (1999)

18. Hune, T., Romijn, J., Stoelinga, M., Vaandrager, F.W.: Linear Parametric Model Checking of Timed Automata. In: Margaria, T., Yi, W. (eds.) TACAS 2001. LNCS, vol. 2031, pp. 189–203. Springer, Heidelberg (2001)

19. Larsen, K.G., Behrmann, G., Skou, A.: Exercises for Uppaal,
http://www.cs.aau.dk/~bnielsen/TOV08/ESV04/exercises

20. Lustig, Y., Vardi, M.Y.: Synthesis from Component Libraries. In: de Alfaro, L. (ed.) FOSSACS 2009. LNCS, vol. 5504, pp. 395–409. Springer, Heidelberg (2009)

21. Maler, O., Pnueli, A., Sifakis, J.: On the Synthesis of Discrete Controllers for Timed Systems (An Extended Abstract). In: Mayr, E.W., Puech, C. (eds.) STACS 1995. LNCS, vol. 900, pp. 229–242. Springer, Heidelberg (1995)

22. Namjoshi, K.S., Yoneda, T., Higashino, T., Okamura, Y. (eds.): ATVA 2007. LNCS, vol. 4762. Springer, Heidelberg (2007)

23. Peter, H.-J., Ehlers, R., Mattmüller, R.: Synthia: Verification and Synthesis for Timed Automata. In: Gopalakrishnan, G., Qadeer, S. (eds.) CAV 2011. LNCS, vol. 6806, pp. 649–655. Springer, Heidelberg (2011)

24. Peter, H.-J., Mattmüller, R.: Component-based abstraction refinement for timed controller synthesis. In: Baker, T.P. (ed.) IEEE Real-Time Systems Symposium, pp. 364–374. IEEE Computer Society (2009)

25. Schewe, S., Finkbeiner, B.: Bounded synthesis. In: [22], pp. 474–488

26. Thomas, W.: On the Synthesis of Strategies in Infinite Games. In: Mayr, E.W., Puech, C. (eds.) STACS 1995. LNCS, vol. 900, pp. 1–13. Springer, Heidelberg (1995)

Zeno: An Automated Prover for Properties of Recursive Data Structures

William Sonnex, Sophia Drossopoulou, and Susan Eisenbach

Imperial College London

Abstract. Zeno is a new tool for the automatic generation of proofs of simple properties of functions over recursively defined data structures. It takes a Haskell program and an assertion as its goal and tries to contruct a proof for that goal. If successful, it converts the proof into Isabelle code. Zeno searches for a proof tree by iteratively reducing the goal into a conjunction of sub-goals, terminating when all leaves are proven true.

This process requires the exploration of many alternatives. We have adapted known, and developed new, heuristics for the reduction of the search space. Our new heuristics aim to promote the application of function definitions, and avoid the repetition of similar proof steps.

We compare with the rippling based tool IsaPlanner and the industrial strength tool ACL2s on the basis of a test suite from the IsaPlanner website. We found that Zeno compared favourably with these tools both in terms of theorem proving power and speed.

1 Introduction

Proving algebraic properties of recursive functions usually requires inductive reasoning. SMT solvers[6], while successfully applied in imperative program verification[1,2], can only construct such proofs when supplied with induction schemata. Recent work[12] automatically sets up the base case and the induction step, and then passes the proof obligation to an SMT solver, and has been successful in proving several such properties. Nevertheless, such an approach runs into difficulties with proofs which require several inductive sub-proofs.

Such cases require proof systems which explicitly handle induction, such as ACL2s[3,7] or IsaPlanner[8]. ACL2 is an industrial strength proof system based on the Boyer-Moore technique, recently extended to ACL2s, the "Sedan Edition". IsaPlanner is a proof-planning framework for the Isabelle[13] proof system.

To address the huge search space ensuing from the fact that at each proof step several induction steps and case-splits are applicable, ACL2 uses *recursion-analysis*[3] while IsaPlanner enumerates every free variable or potential split. IsaPlanner features the rippling technique for applying function definitions, "preferring" steps which make it possible to apply the induction hypothesis. IsaPlanner can also discover auxiliary lemmas needed for a larger proof by appealing to *proof critics* when a proof search is unable to progress [9].

C. Flanagan and B. König (Eds.): TACAS 2012, LNCS 7214, pp. 407–421, 2012.
© Springer-Verlag Berlin Heidelberg 2012

We propose a novel approach, which differs from those above in the following aspects: First, in contrast to rippling, we "prefer" steps which make it possible to apply function definitions, and thus we "bring the proof forwards". Second, through intelligently chosen generalization or CUT steps, we introduce intermediary auxiliary lemmas which the tool tries to prove. Third, we avoid revisiting proof steps which have recently been tried out, and thus we reduce the search to finite space. Furthermore, we adopted some known techniques, e.g. a search for counterexamples before trying to prove new sub-goals.

To support our approach, we introduce a concept called a *critical term*, which is either a variable which appears in the original term (guiding the tool to apply induction on this variable), or a new term which was not a part of the original term (guiding the tool to apply a case-split on this new term), or a "non-minimal term" (guiding the tool to discover an auxiliary lemma). We also introduce *critical paths*, which reflect the cases already visited in a proof branch and avoid applying steps whose paths expand those of earlier steps.

Based on these ideas, we built Zeno, a fully automated verification tool which requires no extra lemmas to be supplied by the user, and often discovers the necessary auxiliary lemmas. Zeno supports **HC**, a minimal functional language with a small language of properties which allows for algebraic properties with entailment. From the constructed proof tree, Zeno creates a proof in Isabelle.

We evaluated Zeno against IsaPlanner and ACL2s using a test suite from the IsaPlanner website, and found that Zeno could prove strictly more properties than either, and with similar computation times.

This paper is organised as follows. **Section 2** defines the input language **HC**. **Section 3** describes the steps Zeno uses to construct its proofs. **Section 4** describes the heuristics which trim the search space. **Section 5** compares Zeno, IsaPlanner and ACL2s and discusses our Isabelle proof output. In **Section 6** we conclude and discuss future work.

Download files and instructions are at `haskell.org/haskellwiki/Zeno`, and try out Zeno online at `tryzeno.org`.

2   Zeno's Internal Functional Language HC

In this section we describe **HC**, Zeno's internal language. **HC** is annotated with labels, which are used by the heuristics for trimming the search space. These labels will not be of interest before **Section 4**, and are written in this colour.

Fig. 1 describes **HC**, a slightly simplified version of GHC Core, the internal language of the Glasgow Haskell Compiler. **HC** is created from GHC Core through an almost direct translation through the GHC API - Zeno uses GHC for parsing and type-checking. For simplicity, in this paper we do not present polymorphic typing, even though Zeno is able to handle it.

Fig. 2 contains an example Haskell program, while **Fig. 3** contains its representation in **HC** (GHC has inlined the definition of (`&&`) in `ord`). We use infix operator syntax in **HC** in the same way as Haskell, as well as the built-in Boolean data type and list type and syntax - `[]` for the empty list and (`:`) for cons.

Fig. 1. Zeno's internal language **HC**

x, y ∈ *Var*	**f, g** ∈ *Fun*	K ∈ *Con*	**T** ∈ *TypeVar*	i ∈ *Id*

$E \in Expr$	::=	Var<$Path^*$> | Fun | Con	Variable/Function/Constructor
	|	($Expr\ Expr$)	Application
	|	$\backslash Var$ -> $Expr$	Lambda abstraction
	|	**case**<Id> $Expr$ **of** { Alt^* }	Pattern
Alt	::=	$Con\ Var^*$ -> $Expr$	A pattern match
	|	_ -> $Expr$	_ is the default pattern
$Bind$	::=	**let** $Fun = Expr$	Non-recursive definition
	|	**letrec** $Fun = Expr$	(Mutually) recursive
		(**and** $Fun = Expr$)*	definitions
$TypeDef$	::=	**data** $TypeVar = Con\ Type^*$	
		(| $Con\ Type^*$)*	Data-type definition
$\tau \in Type$	::=	$TypeVar$	Simple type
	|	$Type$ -> $Type$	Function type
$Prog$	::=	$TypeDef^*\ Bind^*$	An **HC** program
$p \in Path$	::=	[] | $Id : Path$	A critical path
$P \in Prop$	::=	**all x*** . Cls | Cls	Properties
$\Phi \in Cls$	::=	$Prop^*$ ==> Eq | Eq	Clauses
$\varphi \in Eq$	::=	$Expr = Expr$	Equations

Fig. 2. Example program in Haskell

```
data Nat   = Zero | Succ Nat

(<=)  :: Nat -> Nat -> Bool
Zero <= _ = True; Succ x <= Zero = False
Succ x <= Succ y = x <= y

ord  :: [Nat] -> Bool
ord [] = True; ord [x] = True
ord (x:y:ys) = x <= y && ord (y:ys)

ins  :: Nat -> [Nat] -> [Nat]
ins n [] = [n]
ins n (x:xs) | n <= x = n:x:xs | otherwise = x:(ins n xs)

sort  :: [Nat] -> [Nat]
sort [] = []; sort (x:xs) = ins x (sort xs)
```

Fig. 3. The interpretation of **Fig. 2** in **HC**, all uses of a variable have implicitly empty paths

```
data Nat = Zero | Succ Nat

letrec (<=) = \x -> \y -> case<lq1> x of
  { Zero -> True; Succ x' -> case<lq2> y of
    { Zero -> False; Succ y' -> x' <= y' } }

letrec ord = \ns -> case<o1> ns of
  { [] -> True; x:xs -> case<o2> xs of
    { [] -> True; y:ys -> case<o3> (x <= y) of
      { True -> ord (y:ys); False -> False } } }

letrec ins = \n -> \ns -> case<i1> ns of
  { [] -> n:[]; x:xs -> case<i2> (n <= x) of
    {  True -> n:x:xs; False -> x:(ins n xs) } }

letrec sort = \ns -> case<s1> ns of
  { [] -> True; x:xs -> ins x (sort xs) }
```

Fig. 1 also defines the language in which we express properties P. These have the obvious meaning, where free variables are implicitly universally quantified. Thus, `ord (sort as) = True` asserts that `sort` returns an ordered list.

Fig. 4 defines reduction, $\overline{P} \vdash E \rightsquigarrow E'$, which means that E reduces to E' given the facts $\overline{P}$. The first rule uses call-by-value reduction ($\overset{bv}{\rightsquigarrow} \subseteq Expr \times Expr$). For example, as shown in **Fig. 5**, `ord (ins b (d:ds))` reduces to `ord (b:d:ds)`, using an intermediate step which applies `b <= d = False`. Even though our input syntax is Haskell, the evaluation is eager - thus our proofs talk about finite structures only. Some expressions, *e.g.* pattern matching, are not conducive to proofs. To distinguish those expressions that *are* conducive, we introduce in **Fig. 4** *terms*, $Term \subseteq Expr$, which are expressions with a name leftmost, and *normal terms*, $NormalTerm \subseteq Term$, which cannot be further reduced to other terms.

Notation. We use $\rightsquigarrow_+$ for the transitive, and $\rightsquigarrow_*$ for the reflexive transitive closure of $\rightsquigarrow$. For symbols s ranging over S, we use $\overline{s}$ to range over $\wp(S)$, e.g. $\overline{P} \in \wp(Prop)$. Functions are lifted to sets in the obvious way, e.g. for $\overline{s} \in \wp(S)$ and $f \in S \rightarrow X$, we have $\overline{f(s)} \in \wp(X)$. We use a syntactic notion of expression equality, where we ignore critical pairs, since, as we will see, these are annotations and do not change the semantics of an expression; for example, `f x`$<p_2, p_3>$ = `f x`$<p_1>$. Set membership operators between expressions ($E \in E'$) denote the reflexive sub-expression relationship (ignoring critical paths), e.g. `f x`$<p_1> \in$ `g (f x`$<p_2>$`) y`.

Fig. 4. Reduction modulo rewriting, $Terms$ and $NormalTerms$

$$\frac{E \overset{bv}{\leadsto} E'}{\overline{P} \vdash E \leadsto E'} \qquad \frac{(E' = E'') \in \overline{P} \ \vee \ (E'' = E') \in \overline{P}}{\overline{P} \vdash E \leadsto E[E' := E'']_c}$$

$$Term = Var \cup Fun \cup Con \cup \{ (E\ E') \mid E \in Term,\ E' \in Expr \}$$

$$NormalTerm = \{ E \in Term \mid \nexists E' \in Term\ .\ E \overset{bv}{\leadsto}_+ E' \}$$

Fig. 5. An example of reduction modulo rewriting

`(b <= d) = False ⊢ ord (ins b (d:ds))`	Starting expression
`⤳* ord (case<i2> b <= d of { False -> b:d:ds; ... })`	Unfold ins definition
`⤳ ord (case<i2> False of { False -> b:d:ds; ... })`	Apply fact as rewrite
`⤳* ord (b:d:ds)`	Reduce pattern match

---

3 Proof Steps

In this section we discuss the individual proof steps used in Zeno's proofs. We define these steps through the rules in **Fig. 6**.

Zeno constructs proof-trees by applying these rules "backwards". As usual, in a given situation, several different rules may be applicable, and a rule may be applicable in several different ways. Zeno searches for a proof in a depth-first manner. We reduced the search space considerably by prioritizing some rules over others, and by restricting the applicability of some of the rules by requiring further conditions. These further conditions are expressed through premises in the rules written in this colour. In this section we ignore the extra conditions, and will discuss them in **Section 4**.

Fig. 7 describes parts of Zeno's proof that `ord (sort as) = True`, i.e. that our insertion sort function produces ordered lists. For simplicity, we write E to mean E `= True` and `not` E to mean E `= False`, e.g. `b <= d` means `(b <= d) = True`. We use Greek letters between α and μ to denote particular steps in the proof.

Steps (EQL) and (CON) are the only two not to follow from a sub-proof and so can close a proof branch. (EQL) means that both sides of the property consequent are syntactically equal so the goal is true, e.g. in step $[\delta]$ and $[\theta]$ of our example we close these branches as we have `True = True` as our goal. (EXP) applies our previously defined reduction rule to a property, e.g. in step $[\lambda]$ we apply the rewrite shown in **Fig. 5**. (FAC) means that with expression application on both sides of our goal equation it suffices to prove equality between both functions and both arguments respectively - known as "factoring". (USE) converts an antecedent in $Prop$ to one in Eq, i.e. `all` $\overline{x}$. $\overline{P}' ==> \varphi'$ is converted to $\varphi'[x := E_x]$, by choosing a value for each quantified variable ($\overline{x}$) and proving all its antecedents ($\overline{P}'$); φ' can now be used in a later step like (EXP) or (CON).

Fig. 6. Zeno's proof steps

$$(\text{EQL}) \; \frac{E =_c E'}{\vdash \overline{P} \Longrightarrow (E = E')} \qquad (\text{CON}) \; \frac{\text{K} \neq \text{K}' \qquad (\text{K} \, \overline{E} = \text{K}' \, \overline{E}') \in \overline{P}}{\vdash \overline{P} \Longrightarrow \varphi}$$

$$(\text{EXP}) \; \frac{\overline{P} \vdash E \leadsto_* E' \qquad E' \in NormalTerm}{\vdash \overline{P} \Longrightarrow \varphi} \qquad (\text{FAC}) \; \frac{\vdash \overline{P} \Longrightarrow (E_f = E'_f)}{\vdash \overline{P} \Longrightarrow (E_a = E'_a)}{\vdash \overline{P} \Longrightarrow (E_f \, E_a = E'_f \, E'_a)}$$

$$(\text{USE}) \; \frac{\vdash \left\{ \varphi' \overline{[\text{x} := E\text{x}]} \right\} \cup \overline{P} \Longrightarrow \varphi \qquad \left(\text{all } \overline{\text{x}} \, . \, \overline{P}' \Longrightarrow \varphi' \right) \in \overline{P}}{foreach \; \left(\text{all } \overline{\text{y}} \, . \, \overline{P}'' \Longrightarrow \varphi'' \right) \in \overline{P}' \, . \, \begin{cases} \vdash (\overline{P} \cup \overline{P}''') \Longrightarrow \varphi''' \\ \text{where } \overline{P}''' = \overline{P}'' \overline{[\text{x} := E\text{x}]} \, \overline{[\text{y} := E\text{y}]} \\ \varphi''' = \varphi'' \overline{[\text{x} := E\text{x}]} \, \overline{[\text{y} := E\text{y}]} \end{cases}}{\vdash \overline{P} \Longrightarrow \varphi}$$

$$(\text{GEN}) \; \frac{\vdash \Phi[E := x]_c \qquad \text{fresh } \text{x} : \tau \qquad E : \tau \qquad E \in gens(\Phi)}{\vdash \Phi} \qquad (\text{CUT}) \; \frac{\vdash \overline{P} \Longrightarrow E = E' \qquad \vdash (\overline{P} \cup \{E = E'\}) \Longrightarrow \varphi}{\langle E, \_ \rangle \in cases(\overline{P} \Longrightarrow \varphi)}{\text{K} \in cons(\text{T}) \qquad (E', \_) = inst(\text{K})}{\vdash \overline{P} \Longrightarrow \varphi}$$

$$(\text{CASE}) \; \frac{foreach \atop \text{K} \in cons(\text{T})} \cdot \begin{Bmatrix} \vdash (\overline{P} \cup \{E = E_\text{K}'\}) \Longrightarrow \varphi \\ \text{where} \\ (E_\text{K}, \_) = inst(\text{K}) \\ E_\text{K}' = addHistory(E_\text{K}, \{p\}) \end{Bmatrix}}{E : \text{T} \qquad \langle E, p \rangle \in cases(\overline{P} \Longrightarrow \varphi)}{\vdash \overline{P} \Longrightarrow \varphi}$$

$$(\text{IND}) \; \frac{foreach \atop \text{K} \in cons(\text{T})} \cdot \begin{Bmatrix} \vdash \overline{P}_h \cup \overline{P}[\text{x} := E_\text{K}']_c \Longrightarrow \varphi[\text{x} := E_\text{K}']_c \\ \text{where} \\ (E_\text{K}, \overline{\text{r}}) = inst(\text{K}) \qquad E_\text{K}' = addHistory(E_\text{K}, p) \\ \overline{\text{y}} = (FV(\overline{P} \Longrightarrow \varphi)) \backslash \{\text{x}\} \quad \overline{\text{y}'} \text{ all fresh} \\ \overline{P}_h = \{ \text{ all } \overline{\text{y}'}.(\overline{P}' \Longrightarrow \varphi) \overline{[\text{y} := \text{y}']}[\text{x} := \text{r} < p >]_c \mid \text{r} \in \overline{\text{r}} \} \end{Bmatrix}}{E : \text{T} \qquad \langle \text{x}, p \rangle \in inds(\overline{P} \Longrightarrow \varphi)}{\vdash \overline{P} \Longrightarrow \varphi}$$

Fig. 7. Parts of Zeno's proof for `ord (sort as)`. Proof steps are annotated by the name of the rule applied, and by a different Greek letter.

$$[\kappa](\textsc{case})\ \dfrac{[\mu](\textsc{exp})\ \dfrac{\dfrac{\cdots}{\vdash \ldots,\ \texttt{not (b <= d) ==> ord (d:(ins b bs))}}}{\vdash \ldots,\ \texttt{not (b <= d) ==> ord (ins b (d:ds))}} \qquad [\lambda](\textsc{exp})\ \dfrac{\dfrac{\cdots}{\vdash \ldots,\ \texttt{b <= d ==> ord (b:d:ds)}}}{\vdash \ldots,\ \texttt{b <= d ==> ord (ins b (d:ds))}}}{\vdash \texttt{all b' . ord ds ==> ord (ins b' ds), ord (d:ds) ==> ord (ins b (d:ds))}}$$

$$[\beta](\textsc{exp})\ \dfrac{[\delta](\textsc{eql})\ \dfrac{}{\vdash \texttt{True}}}{\vdash \texttt{ord (sort [])}} \qquad [\eta](\textsc{exp})\ \dfrac{[\theta](\textsc{eql})\ \dfrac{}{\vdash \texttt{True ==> True}}}{\vdash \texttt{ord [] ==> ord (ins b [])}}$$

$$[\alpha](\textsc{ind})\ \dfrac{[\beta] \quad [\gamma](\textsc{exp})\ \dfrac{[\epsilon](\textsc{gen})\ \dfrac{[\zeta](\textsc{ind})\ \dfrac{[\eta] \qquad [\kappa]}{\vdash \texttt{ord cs ==> ord (ins b cs)}}}{\vdash \texttt{ord (sort bs) ==> ord (ins b (sort bs))}}}{\vdash \texttt{ord (sort bs) ==> ord (sort (b:bs))}}}{\vdash \texttt{ord (sort as)}}$$

(GEN) and (CUT) both discover necessary sub-lemmas of our goal. Generalisation replaces an expression with a fresh variable of the same type - it corresponds to $\forall$-elimination. E.g., step $[\epsilon]$ "discovers" the sub-lemma `ord cs ==> ord (ins b cs)` by generalising `sort bs`. (CUT) is cumulative transitivity; it adds a new antecedent by proving it from the existing ones - this proof is our discovered sub-lemma.

The partial function *inst* in **Fig. 8** takes an expression of function type and applies fresh argument variables until the expression is simply typed - returning the new simply typed expression and the set of every recursively typed fresh variable applied, i.e. those variables whose type is the simple type of the returned expression. For example $inst(\texttt{(:)}) = (\texttt{b:bs}, \{\texttt{bs}\})$; this follows from $inst(\texttt{b:}) = (\texttt{b:bs}, \{\texttt{bs}\})$; which, in its turn, follows from $inst(\texttt{b:bs}) = (\texttt{b:bs}, \emptyset)$.

(CASE)-splitting proves a goal by choosing a simply typed expression (E : `T`) and proving a branch for each value this expression could take, viz. each constructor of its type ($cons(\texttt{T})$). The value this expression has been assigned down each branch is added as an antecedent. In step $[\kappa]$ we case-split upon `b <= d` creating two branches - $[\mu]$ `(b <= d) = False`, and $[\lambda]$ `(b <= d) = True`.

(IND) applies structural induction on a variable `x`, proving branch for every constructor of its type where an inductive hypothesis is added for every recursive variable in that constructor - $\overline{P_h}$ is the set of all these new hypotheses. Every free variable that is not `x` becomes $\forall$-quantified in our new hypotheses. In step $[\alpha]$ we apply induction on `as`, creating two branches - $[\beta]$ `as = []` and $[\gamma]$ `as = (b:bs)` which gains the hypothesis `ord (sort bs)`. In step $[\zeta]$ we apply induction on

Fig. 8. Instantiation function

$$rtype : Type \to TypeVar \qquad rtype(\text{T}) = \text{T} \qquad rtype(\tau_1 \text{ -> } \tau_2) = rtype(\tau_2)$$

$$inst : Expr \rightharpoonup Expr \times \wp(Var)$$

$$inst(E) = \begin{cases} (E, \emptyset) & \text{if } E : \text{T} \\ (E', \{\text{x}\} \cup \bar{\text{x}}) & \text{if } E : (rtype(\tau) \text{ -> } \tau) \\ & \text{where } \text{x} : rtype(\tau) \text{ is fresh, } (E', \bar{\text{x}}) = inst(E \text{ x}) \\ (E', \bar{\text{x}}) & \text{if } E : (\tau_a \text{ -> } \tau_r), \ \tau_a \neq rtype(\tau_r) \\ & \text{where } \text{x} : \tau_a \text{ is fresh, } (E', \bar{\text{x}}) = inst(E \text{ x}) \end{cases}$$

cs, creating branches $[\eta]$ and $[\kappa]$, where the latter gains a hypothesis in which b has been replaced by a fresh b' which is $\forall$-quantified.

Soundness. We believe, but have not yet proven, that the proof-steps presented in this section are sound, *i.e.*, that any property provable using these steps is provable in first order logic enhanced with structural induction, provided that induction or case splits are only applied on terms guaranteed to terminate. Unsoundness through non-terminating functions does not arise in our case, because, as we will see in **Section 4**, the calculation of critical terms for expressions containing such functions would not terminate. Thus, when faced with non-terminating functions, Zeno might loop for ever but will not produce erroneous proofs. We want to adopt termination checkers in further work. Moreover, created proofs are checked by Isabelle; this gives a strong guarantee of soundness.

4   Heuristics

In **Section 3** we discussed the proof rules without discussing the highlighted, further conditions. In this section we describe the most important heuristics which trim the search space, and in particular the further conditions.

4.1   Prioritize (EQL) and (CON), and Counterexamples

The steps (EQL) and (CON) are applied whenever possible, as they immediately close their proof branch.

When generating a new proof goal, before attempting to prove it, Zeno searches for counterexamples, and abandons the proof search if it finds any. Our approach is similar to SmallCheck[14], in that both use execution to generate values, but differs in that SmallCheck uses depth of recursion to restrict to a finite set, whereas we use our critical pair technique, described later on. In contrast, ACL2s generates a constant number of random values, much more like QuickCheck[5].

4.2   Applying (CUT) Only When (CASE) is also Possible

In principle, (CUT) is applicable at *any* point during proof search, and for *any* intermediate goal $E = E'$, which follows from the current antecedents, and which implies the current goal. This makes (CUT) highly non-deterministic.

Fig. 9. A (CUT) step

$$[\nu](\text{CUT}) \quad \dfrac{[\sigma] \; \dfrac{\cdots}{\vdash \ldots,\; \texttt{not(b<=d) ==> d<=b}} \qquad [\xi] \; \dfrac{\cdots}{\vdash \ldots,\; \texttt{not(b<=d), d<=b ==> ord [d, b]}}}{\vdash \ldots,\; \texttt{not(b<=d) ==> ord [d, b]}}$$

Our heuristic resticts the applicability of (CUT), and the search for an appropriate intermediate goal, by requiring that this step should be chosen only when a (CASE) would have been applicable too (i.e. when $E.. \in cases(...)$), and only when the intermediate goal $E = E'$ can be inferred from the current antecedents. Thus, (CUT) discovers necessary sub-lemmas.

In **Fig. 9** we use a (CUT) in the proof of not (b <= d) ==> ord [d,b]; the latter is a subproof for $[\mu]$ from **Fig. 7**. At $[\nu]$, a case analysis on b<=d is possible, however, since b<=d follows from the antecedents, our heuristic prefers a (CUT) instead. In the process, it proves the sublemma that not (b <= d) ==> d <= b.

4.3 Critical Terms

Critical pairs[1], defined in **Fig. 11**, are pairs of terms and paths, and are used to select between induction, case analysis, generalization and cut steps. We will first discuss the first component of these pairs, i.e. the *critical terms*, and then, in 4.4 we will refine the picture and introduce the *critical paths*. As our running example we use **Fig. 10**, which revisits the example from **Fig. 7**, this time annotated with critical paths.

For a motivation for critical terms, consider the bottom of **Fig. 10**, annotated with $[\alpha]$, where – ignoring the empty critical path <> – the aim is to prove that ord (sort as). At this point, it would be possible to apply induction on as, or case analysis on as, or generalization on sort as. Nevertheless, Zeno only considers the (IND) step. It does this based on the *critical term* of the expression.

We focus in **Fig. 11** on the term E' such that $pair(\overline{P}, E) = \langle E', ...\rangle$. Then, E' is crucial for the evaluation of E, i.e. the evaluation of E can continue only if we have some more information about the value of E' than currently available in E' itself or in $\overline{P}$. For example, because sort as $\leadsto_*$ case as of { ... }, and ord (sort as) $\leadsto_*$ case as of { ... }, we have $pair(\emptyset, \texttt{sort as}) = \langle\texttt{as}, ...\rangle$ and $pair(\emptyset, \texttt{ord (sort as)}) = \langle\texttt{as}, ...\rangle$. Also, ord (ins b (d:ds)) $\leadsto_*$ case b <= d of { ... }, and thus $pair(..., \texttt{ord (ins b (d:ds))}) = \langle\texttt{b <= d}, ...\rangle$. With *pairs* (**Fig. 11**) critical pairs are lifted to equations, clauses and properties, allowing for more than one critical pair per goal.

Now we focus again on the proof steps in **Fig. 6**, and consider the extra conditions in (IND), (CASE), (GEN), which restrict the applicability of these steps. For example, induction is applicable only on $inds(...)$. In **Fig. 12** we define functions $inds$, $cases$ and $gens$. We will discuss paths in the next section, but we can see already that if a critical term is a variable, then Zeno uses it for

---

[1] Not to be confused with the *critical pairs* of term rewriting.

Fig. 10. First part of the example from **Fig. 7** revisited.

Here Φ_α, Φ_γ, Φ_ϵ, and Φ_ζ, stand for the proof goals at $[\alpha]$, $[\gamma]$, $[\epsilon]$, and $[\zeta]$. The goals are annotated with paths.

We use names for paths: $p_1 = \texttt{o1:s1:[]}$, $p_2 = \texttt{o1:i1:s1:[]}$, $p_3 = \texttt{o1:i1:[]}$, and $p_4 = \texttt{s1:[]}$.

Then we have $p_1 \sqsubseteq p_2$, and $p_1 \not\sqsubseteq p_3$, and $p_4 \sqsubseteq p_3$.

Also, $inds(\Phi_\alpha) = pairs(\Phi_\alpha) = \{ \langle \texttt{as}, p_1 \rangle \}$, and $cases(\Phi_\alpha) = gens(\Phi_\alpha) = \emptyset$.

Also, $gens(\Phi_\epsilon) = \{ \texttt{isort bs<}p_1\texttt{>} \}$, and $inds(\Phi_\epsilon) = cases(\Phi_\epsilon) = \emptyset$.

Also, $pairs(\Phi_\zeta) = \{ \langle \texttt{bs<}p_1\texttt{>}, p_2 \rangle, \langle \texttt{bs<}p_1\texttt{>}, p_1 \rangle \}$, and $pairs(\Phi_\zeta) \cap MinPairs = \emptyset$.

Thus, $pairs(\Phi_\zeta) \cap MinPairs = \{ \langle \texttt{cs<}p_1\texttt{>}, p_3 \rangle, \langle \texttt{cs<}p_1\texttt{>}, p_4 \rangle \}$.

Thus, $inds(\Phi_\zeta) = \{ \langle \texttt{cs<}p_1\texttt{>}, p_3 \rangle \}$, and $cases(\Phi_\zeta) = gens(\Phi_\zeta) = \emptyset$.

$$[\kappa](\text{CASE}) \frac{[\mu] \qquad [\lambda]}{\begin{array}{l} \vdash (\texttt{all b' . ord ds<}p_1,p_3\texttt{> ==> ord (ins b' ds<}p_1,p_3\texttt{>)}), \\ \quad \texttt{ord (d<}p_1,p_3\texttt{>:ds<}p_1,p_3\texttt{>)} \\ \quad \texttt{==> ord (ins b<}p_1\texttt{> (d<}p_1,p_3\texttt{>:ds<}p_1,p_3\texttt{>))} \end{array}}$$

$$[\zeta](\text{IND}) \frac{[\eta] \qquad [\kappa]}{\begin{array}{l} \vdash \texttt{ord cs<}p_1\texttt{> ==>} \\ \quad \texttt{ord (ins b<}p_1\texttt{> cs<}p_1\texttt{>)} \quad (\Phi_\zeta) \end{array}}$$

$$[\epsilon](\text{GEN}) \frac{}{\begin{array}{l} \vdash \texttt{ord (sort bs<}p_1\texttt{>) ==>} \\ \quad \texttt{ord (ins b<}p_1\texttt{> (sort bs<}p_1\texttt{>))} \quad (\Phi_\epsilon) \end{array}}$$

$$[\gamma](\text{EXP}) \frac{}{\begin{array}{l} \vdash \texttt{ord (sort bs<}p_1\texttt{>) ==>} \\ \quad \texttt{ord (sort (b<}p_1\texttt{>:bs<}p_1\texttt{>))} \quad (\Phi_\gamma) \end{array}}$$

$$[\alpha](\text{IND}) \frac{[\beta]}{\vdash \texttt{ord (sort as<>)} \quad (\Phi_\alpha)}$$

Fig. 11. Defining critical pairs

$$Pair = NormalTerm \times Path$$

$$pair : \wp(Prop) \times Expr \rightharpoonup Pair$$

$$pair(\overline{P}, E) = \begin{cases} \langle E, [] \rangle & \text{if } E \in NormalTerm, \ E : \texttt{T}, \ \nexists \texttt{K}.\exists \overline{E}.E = \texttt{K} \ \overline{E} \\ \langle E', [\texttt{i}] \rangle & \text{if } \overline{P} \vdash E \rightsquigarrow_* \texttt{case<i> } E' \texttt{ of \{ ... \}}, \ E' \notin E, \\ & \quad E' \in NormalTerm \\ \langle E'', \texttt{i} : p \rangle & \text{if } \overline{P} \vdash E \rightsquigarrow_* \texttt{case<i> } E' \texttt{ of \{ ... \}}, \ E' \in E \\ & \quad \text{where } \langle E'', p \rangle = pair(\overline{P}, E') \end{cases}$$

$$\begin{aligned} exprs &: Prop &&\rightarrow \wp(Expr) \\ exprs(E_1 &= E_2) &&= \{ E_1, \ E_2 \} \\ exprs(\overline{P} &\texttt{ ==> } \varphi) &&= \bigcup exprs(P) \cup exprs(\varphi) \\ exprs(\texttt{all } \overline{\texttt{x}} \ . \ \Phi) &&&= \{ E \in exprs(\Phi) \mid \nexists \texttt{x} \in \overline{\texttt{x}} \ . \ \texttt{x} \in E \} \end{aligned}$$

$$\begin{aligned} pairs &: Cls &&\rightarrow \wp(Pair) \\ pairs(\overline{P} &\texttt{ ==> } \varphi) &&= \{ pair(\overline{P}, E) \mid E \in exprs(\overline{P} \texttt{ ==> } \varphi) \} \end{aligned}$$

induction, if it is not a sub-term of the current goal, then Zeno uses it for case analysis. This is why, in step $[\alpha]$ of **Fig. 10** the only applicable step is induction on `as`. Similarly, in step $[\kappa]$ of **Fig. 10** a case-split on `b <= d` is applicable.

4.4 Critical Paths

Although critical terms are essential in pruning the search space, they do not prevent repeated application of what is essentially the same step. Consider, e.g., step $[\gamma]$ in **Fig. 10**; here $pair(..., \text{ord (sort bs)}) = \langle \text{bs}, ...\rangle$. Naïve application of critical terms as we have considered them so far would be applying induction again, and in fact, would be applying induction for ever!

For this, we built into Zeno a way of remembering which cases in a function definition it has tried so far, and then avoiding covering the same cases when selecting the next step. We use the notion of *path*, which consists of a sequence of labels; the labels indicate cases in the definition of functions. For this, the full syntax definition in **Fig. 1** prescribes distinct labels for each case in a function definition, c.f. the labels o1, o2, c1 etc., in **Fig. 3**. Furthermore, **Fig. 1** prescribes each variable in an expression to be decorated with a set of paths. We call these paths the *history* of an expression, as they record for each of these variables the reason why this this variable has been introduced. The variables in the function definitions and in the original property have an empty history (e.g. we start with `ord (sort as<>)` in $[\alpha]$ in **Fig. 10**). Then, as the proof progresses, new variables gain history.

We now read the full definition of critical pairs in **Fig. 11**. We call the second component of the critical pair the *intention* of the pair. It is a path, which describes the function cases that would be covered if that term were used in the next step of the proof. For example, $pair(\emptyset, \text{ord (sort as<>)}) = \langle \text{as}, p_1\rangle$ where $p_1 = \text{o1:s1: []}$, which means that selecting the variable `as` would progress the proof along the cases o1 and s1. Also, $pair(..., \text{ord (sort bs<}p_1\text{>)}) = \langle \text{bs<}p_1\text{>}, p_2\rangle$, where $p_2 = \text{o1:i1:s1: []}$. Finally, $pair(..., \Phi_\gamma) = \{\ \langle \text{bs<}p_1\text{>}, p_1\rangle, \langle \text{bs<}p_1\text{>}, p_2\rangle\ \}$.

When we use a critical pair in a (CASE) or (IND) step we store its intention in the corresponding variables in the new goal (see **Fig. 6**, *addHistory* is given in **Fig. 12**). For example, in step $[\gamma]$ in **Fig. 10**, we store the path p_1 in bs.

In **Fig. 12** we define the partial order $\sqsubseteq$ on paths, where $p \sqsubseteq p'$ if p' contains all cases from p extended at places with further cases. For example, $\text{o1:s1: []} \sqsubseteq \text{o1:i1:s1: []}$. We also define $MinimalPairs$, so as to remove any pairs where the history of the term "covers" (in the sense of $\sqsubseteq$) the intention of the pair. Thus, $MinPairs$ are critical pairs which do *not* represent a previously applied similar path. For example, the pair $\langle \text{bs<}p_1\text{>}, p_2\rangle$ is *not* minimal.

For induction and case splits, we only take the minimal pairs, c.f. **Fig. 12**, and thus we avoid chosing paths which have been essentially covered in earlier proof steps. Therefore, at $[\gamma]$ in **Fig. 10** induction and case analysis are not applicable, but generalization is. On the other hand, the purpose of generalisation, when discovering intermediary lemmas, w.r.t. to critical paths is to shorten the critical path of the critical pair of an expression. Thus, in **Fig. 10** in $[\gamma]$ we apply generalisation and turn Φ_ϵ into Φ_ζ which has the critical pair $\langle \text{cs<}p_1\text{>}, p_3\rangle$, in $MinPairs$ since $p_1 \not\sqsubseteq p_3$ - here generalisation has shortened p_2 to p_3, yielding a less complex step ($p_3 \sqsubseteq p_2$) and no longer covered by the previous p_1 step.

Fig. 12. Critical-pair-based heuristics: *inds*, *cases* and *gens*, and auxiliary definitions

$$p_1.p_2...p_n \sqsubseteq q_1.p_1.q_2.p_2...p_n.q_{n+1} \quad \text{where } p_i, q_i \in Path$$

$$\begin{aligned}
&history : Expr && \to \wp(Path)\\
&history(E) && = \bigcup\{\ \overline{p} \mid \texttt{x<}\overline{p}\texttt{>} \in FV(E)\ \}\\[1ex]
&addHistory : Expr \times \wp(Path) && \to Expr\\
&addHistory(E,\overline{p}) && = E\{\ [\texttt{x<}\overline{p}'\texttt{>} := \texttt{x<}\overline{p} \cup \overline{p}'\texttt{>}] \mid \texttt{x<}\overline{p}'\texttt{>} \in FV(E)\ \}\\[1ex]
&MinPairs && \subseteq \wp(Pair)\\
&MinPairs && = \{\ \langle E,p\rangle \in Pair \mid \nexists p' \in history(E)\ .\ p' \sqsubseteq p\ \}\\[1ex]
&maxPaths : \wp(Pair) && \to \wp(Pair)\\
&maxPaths(\overline{\pi}) && = \{\ \langle E,p\rangle \in \overline{\pi} \mid \nexists\langle E,p'\rangle \in \overline{\pi}\ .\ p \sqsubseteq p'\ \}\\[1ex]
&inds : Cls && \to \wp(Pair)\\
&inds(\Phi) && = maxPaths(\{\ \langle \texttt{x},p\rangle \in (pairs(\Phi) \cap MinPairs)\ \})\\[1ex]
&cases : Cls && \to \wp(Pair)\\
&cases(\Phi) && = maxPaths(\{\ \langle E,p\rangle \in (pairs(\Phi) \cap MinPairs)\\
& && \qquad \mid \nexists E_\Phi \in exprs(\Phi)\ .\ E \in E_\Phi\ \})\\[1ex]
&gens : Cls && \to \wp(Expr)\\
&gens(\Phi) && = \{\ E \mid E_c \in E \in E_\Phi \in exprs(\Phi),\quad E : \texttt{T}\\
& && \qquad \langle E_c,p\rangle \in (pairs(\Phi) \setminus MinPairs),\\
& && \qquad \nexists E' : \texttt{T}'\ .\ E_c \in E' \in E \wedge E' \neq E\ \}
\end{aligned}$$

Fig. 13. Partial definition of substitution preserving critical paths

$E =_c E'$ iff $[\![E]\!] = [\![E']\!]$ where $[\![]\!]$ removes all stored critical paths

$$E[E' := E'']_c = addHistory(E'', history(E) \cup history(E')) \quad \text{if } E =_c E'$$

$$\vdots$$

$$(\backslash\texttt{x -> } E)[E' := E'']_c = \begin{cases} \backslash\texttt{x -> } E[E' := E'']_c & \text{if } \texttt{x} \notin E' \wedge \texttt{x} \notin E''\\ \backslash\texttt{x -> } E & \text{otherwise}\end{cases}$$

$$\vdots$$

Notice, that although induction is not applicable on Φ_γ, it is applicable on Φ_ζ: Our critical pairs technique blocked induction until we had generalised to our intermediary lemma, and then re-enabled it.

Substitution and comparison in the presence of paths. In order to preserve the history of an expression after substitution - such as in (GEN)eralisation or (IND)uction - we have defined "capture avoiding substitution preserving critical paths" ($E[E' := E'']_c$) in **Fig. 13**. The first line is all we have changed from regular capture avoiding substitution; we have left out most of the definition since it is as you would expect and we give the rule for abstraction as an example. We define this, and some of our earlier rules in terms of "equality modulo critical paths" ($=_c$), which is syntactic equality ignoring critical paths stored in variables, as these are an annotation and do not affect execution.

5   Comparisons and the Output of Isabelle Proofs

We now compare Zeno, IsaPlanner[4,8,10], ACL2s[3,7], and Dafny's extension with induction [12], in terms of their respective performance on the 87 lemmas from a test-suite from the IsaPlanner website[2], which also appears in [10].

Of the 87 lemmas, 2 are false. Zeno can prove 82 lemmas, IsaPlanner can prove 47, while ACL2s can prove 74. All lemmas unprovable by Zeno are unprovable for the other tools too. ACL2s can prove 28 lemmas unprovable by Isaplanner, and Isaplanner can prove 1 lemma unprovable by ACL2s; the latter over a binary tree – something with a more natural representation in IsaPlanner. See below for percentages of proven, and lists of, unproven lemmas:

Tool	Percent. Proven	Id's of unproven lemmas
Dafny+indct.	53.5%	45 - 85
IsaPlanner	55%	48 - 85
ACL2s	86%	47, 50, 54, 56, 67, 72, 73, 74, 81, 83, 84, 85
Zeno	96%	72, 74, 85

Zeno's proofs take between 0.001s and 2.084s on an Intel Core i5-650 processor. ACL2s and Isaplanner produce proofs in similar times. The Haskell code to test Zeno and the LISP code to test ACL2s can be found at `tryzeno.org/comparison`. As functions in ACL2s are untyped, we supplied the type information ourselves through proven theorems. Without this information, ACL2s is unable to prove 6, 7, 8, 9, 15, 18, and 21.

To avoid ending up "training" Zeno towards the specific IsaPlanner test-suite, we developed a suite of 71 further lemmas. We tried to find lemmas to differentiate the tools, and show their respective strengths. Indeed, we found lemmas provable by ACL2s but not by Zeno, and lemmas provable by IsaPlanner but not by ACL2s, but none which IsaPlanner could prove and Zeno couldn't. Below we discuss five lemmas from our suite:

---

[2] `http://dream.inf.ed.ac.uk/projects/lemmadiscovery/results/`
   `case-analysis-rippling.txt`

Nr.	Lemma	Nr.	Lemma
P1	`sort (sort xs) = sort xs`	P2	`x * (S 0) = x`
P3	`x * (y + z) = (x * y) + (x * z)`	P4	`x ^ (y + z) = (x ^ y) * (x ^ z)`
P5	`even(x), even(y) ==> even(x+y)`		

Zeno can prove P1, while IsaPlanner and ACL2s can not. Zeno takes around 2s to find the proof, has a proof tree 14 steps deep and discovers the sub-lemmas `ins x (ins y xs) = ins y (ins x xs)` and `sort (ins x xs) = ins x (sort xs)`. Both Zeno and IsaPlanner can prove P2, P3, P4, but ACL2s can not. Because Zeno does not support strong induction, it cannot prove P5, while ACL2s can.

Our (CUT) makes Zeno robust to multiple names of the same function. For example, when we defined `ord` so that it uses `leq` for comparison, while `ins` uses `<=`, where `leq` and `<=` are semantically equivalent, then ACL2s goes into an infinite loop, whereas Zeno applies several (CUT) steps, discovers the sub-lemmas `leq x y ==> x <= y`, and `x <= y ==> leq x y`, and `not (x <= y) ==> leq y x`, and `leq x y, leq y z ==> x <= z`, and finally finds the proof.

Isabelle Output Zeno translates its internal proof tree into an Isar[16] proof - every sub-goal of its proof becomes a new line and each step has a natural counterpart in Isabelle. The **HC** functions and data-types are easily converted into Isabelle's ML. Zeno's internal proof is purely backwards reasoning; we kept high-level structure of the proof output this way but we mixed in forwards reasoning for small internal sections of non-branching proof steps. At certain points we restart the output in a new sub-lemma, to keep the proof tree from becoming too deep and to display important sub-lemmas to the user. All lemmas from this section have had their proof output checked by Isabelle.

6   Conclusions and Future Work

We have described Zeno's proof steps and its heuristics; in particular, how critical pairs guide the selection of proof rules and avoid revisiting earlier proof steps. Zeno requires no further lemmas to be suggested to it, and indeed often discovers interesting auxiliary lemmas. We found that Zeno compared favourably with other tools both in terms of theorem proving power and speed.

On the other hand, Zeno cannot use auxiliary lemmas, while IsaPlanner and ACL2s can, and therefore are better suited for larger, human guided proofs. In particular, ACL2s has been used in the verification of properties of real-world systems. Integration of Zeno into a proof system like Isabelle, perhaps as a tactic for IsaPlanner, would be a useful next step – we also plan to adapt Zeno so that it can use background lemmas. Furthermore, ACL2s can prove properties over full first-order logic whereas Zeno lacks negation and existentials - another potential extension. Currently, Zeno does not check for function termination before attempting a proof, leaving us with a potential infinite loop in the critical pair discovery step; we plan to integrate techniques for termination checking.

The three properties from our test suite which Zeno is unable to prove require auxiliary lemmas which are not generalisations of sub-goals. Therefore, we want to develop intelligent methods of finding such necessary lemmas, through techniques like IsaCoSy's[11] random perturbation of property terms.

Acknowledgements. We thank D. Ancona, SLURP research group, K. Broda, M. Johansson, R. Leino, and most particularly T. Allwood for useful feedback.

References

1. Barnett, M., Chang, B.-Y.E., DeLine, R., Jacobs, B., Leino, K.R.M.: Boogie: A Modular Reusable Verifier for Object-Oriented Programs. In: de Boer, F.S., Bonsangue, M.M., Graf, S., de Roever, W.-P. (eds.) FMCO 2005. LNCS, vol. 4111, pp. 364–387. Springer, Heidelberg (2006)
2. Barnett, M., Leino, K., Schulte, W.: The Spec# Programming System: An Overview. In: Barthe, G., Burdy, L., Huisman, M., Lanet, J.-L., Muntean, T. (eds.) CASSIS 2004. LNCS, vol. 3362, pp. 49–69. Springer, Heidelberg (2005)
3. Boyer, R.S., Moore, J.S.: A theorem prover for a computational logic. In: CADE (1990)
4. Bundy, A., Stevens, A., Harmelen, F.V., Ireland, A., Smaill, A.: Rippling: A Heuristic for Guiding Inductive Proofs. Art. Intell. (62) (1993)
5. Claessen, K., Hughes, J.: Quickcheck: a lightweight tool for random testing of Haskell programs. In: ICFP, pp. 268–279 (2000)
6. de Moura, L., Bjørner, N.: Z3: An Efficient SMT Solver. In: Ramakrishnan, C.R., Rehof, J. (eds.) TACAS 2008. LNCS, vol. 4963, pp. 337–340. Springer, Heidelberg (2008)
7. Dillinger, P.C., Manolios, P., Vroon, D., Moore, J.S.: ACL2s: "The ACL2 Sedan". In: ICSE, pp. 59–60 (2007)
8. Dixon, L., Fleuriot, J.: IsaPlanner: A Prototype Proof Planner in Isabelle. In: Baader, F. (ed.) CADE 2003. LNCS (LNAI), vol. 2741, pp. 279–283. Springer, Heidelberg (2003)
9. Ireland, A., Bundy, A.: Productive use of failure in inductive proof. Journal of Automated Reasoning 16, 16–1 (1995)
10. Johansson, M., Dixon, L., Bundy, A.: Case-Analysis for Rippling and Inductive Proof. In: Kaufmann, M., Paulson, L.C. (eds.) ITP 2010. LNCS, vol. 6172, pp. 291–306. Springer, Heidelberg (2010)
11. Johansson, M., Dixon, L., Bundy, A.: Conjecture Synthesis for Inductive Theories. Journal of Automated Reasoning 47, 251–289 (2011)
12. Leino, K.R.M.: Automating Induction with an SMT Solver. In: Kuncak, V., Rybalchenko, A. (eds.) VMCAI 2012. LNCS, vol. 7148, pp. 315–331. Springer, Heidelberg (2012)
13. Paulson, L.C.: The foundation of a generic theorem prover. Journal of Automated Reasoning 5 (1989)
14. Runciman, C., Naylor, M., Lindblad, F.: Smallcheck and lazy Smallcheck: automatic exhaustive testing for small values. In: First ACM SIGPLAN Symposium on Haskell, pp. 37–48 (2008)
15. Walther, C., Schweitzer, S.: About VeriFun. In: Baader, F. (ed.) CADE 2003. LNCS (LNAI), vol. 2741, pp. 322–327. Springer, Heidelberg (2003)
16. Wenzel, M.: Isar - A Generic Interpretative Approach to Readable Formal Proof Documents. In: Bertot, Y., Dowek, G., Hirschowitz, A., Paulin, C., Théry, L. (eds.) TPHOLs 1999. LNCS, vol. 1690, pp. 167–183. Springer, Heidelberg (1999)
17. Xu, D., Peyton-Jones, S., Claesen, K.: Static Contract Checking for Haskell. In: POPL (2009)

A Proof Assistant for Alloy Specifications

Mattias Ulbrich, Ulrich Geilmann, Aboubakr Achraf El Ghazi,
and Mana Taghdiri

Karlsruhe Institute of Technology, Germany
{mulbrich,geilmann,elghazi,taghdiri}@ira.uka.de

Abstract. Alloy is a specification language based on a relational first-order logic with built-in operators for transitive closure, set cardinality, and integer arithmetic. The Alloy Analyzer checks Alloy specifications automatically with respect to bounded domains. Thus, while suitable for finding counterexamples, it cannot, in general, provide correctness proofs. This paper presents Kelloy, a tool for verifying Alloy specifications with respect to potentially infinite domains. It describes an automatic translation of the full Alloy language to the first-order logic of the KeY theorem prover, and an Alloy-specific extension to KeY's calculus. It discusses correctness and completeness conditions of the translation, and reports on our automatic and interactive experiments.

1   Introduction

Due to their expressive logics, theorem provers have been successfully used to prove detailed properties of complex system specifications. However, they are often considered too expensive to use frequently during software development. The cost is twofold: (1) the proof process is often interactive and requires the user to be an expert in both the problem domain being analyzed and the theorem prover being used, and (2) the input language is often a low-level logic that makes specifications unintuitive and error-prone.

Lightweight formal methods [17], on the other hand, promote checking software partially, yet frequently, during design and implementation. Alloy [16], for example, has been successfully used for checking several software systems against their requirements (see [7] for some examples). The main reasons for Alloy's popularity are its concise language, simple semantics, and fully automatic analyzer. Alloy provides a first-order logic based on relations that is augmented with built-in transitive closure (reachability), set cardinality, and basic integer arithmetic operators, which makes it suitable for specifying structure-rich systems. Alloy specifications are automatically analyzed using a SAT solver by requiring a bound on the number of elements of each relation. Consequently, while the counterexamples are non-spurious, lack of a counterexample, in general, does not constitute a proof of correctness. Thus, for safety-critical systems, the user must perform a second round of analysis in which he specifies the problem again, in the input language of a theorem prover for full verification.

This paper introduces Kelloy, an engine for verifying Alloy specifications, with the goal of bridging the gap between lightweight formal methods and theorem

C. Flanagan and B. König (Eds.): TACAS 2012, LNCS 7214, pp. 422–436, 2012.

provers. To reduce the cost of using the underlying theorem prover, namely KeY [4], it (1) provides a fully automatic translation of Alloy to KFOL—the first-order logic of KeY, (2) defines an Alloy-specific extension to KeY's calculus and a reasoning strategy that improves KeY's capability in finding proofs automatically, and (3) simplifies user interaction by generating intermediate proof obligations that are easy to understand.

KeY is attractive because it combines an interactive proof assistant with an automatic engine, and its calculus strategy is extensible: one can easily add new calculus rules and assign them to different proof search heuristic strategies. Kelloy translates Alloy's operators to function symbols, thus generating formulas that conform to the high-level structure of the analyzed Alloy specification. This makes the tool easy to interact with.

Our target logic is first-order because it is semi-decidable which indicates a higher automation potential in practice than higher order logics. KeY has built-in integer support and provides a set of rules implementing Peano arithmetic [5] extended to integers. Therefore, unlike previous approaches [2, 13, 14], we are able to translate the entire Alloy language including integer expressions, cardinality, and the ordering module (see Section 4.5). Our calculus cannot be complete because integers are not FOL-axiomatizable, but it suffices in almost all practical cases [10, p.153]. Although we target KFOL, our resulting formulas (with minor modifications) can be verified by any prover for first-order logic, which supports integers.

While some Alloy specifications can be verified automatically, specifications that extensively use quantifiers or transitive closure result in proof obligations that are too difficult to discharge fully automatically. In a user-guided, semi-automatic proof system such as KeY, however, it often suffices if the user provides only a few inputs (e.g. in the form of quantifier instantiations or induction hypotheses) to help the system find a proof. Since trying to prove an invalid assertion is particularly costly, we suggest Kelloy be used after the assertion has been checked by the Alloy Analyzer. The user can increase the analysis bounds to gain more confidence in the correctness of the assertion before using Kelloy for full verification.

In this paper, we describe a fully automatic translation of Alloy to KFOL, and establish that the translation is correct (for finite and infinite domains) and complete (for finite domains). We also describe an Alloy-specific extension to KeY's calculus and reasoning strategy. We evaluate the approach by conducting both automatic and interactive proofs. Out of a total of 22 proved Alloy assertions in 10 specifications, 12 were proved without any user interaction. We also evaluate the impact of the user's experience on the interactive proof process.

2   Related Work

Several approaches address the verification of Alloy specifications. Prioni [2] is the closest to ours: it translates Alloy to a first-order logic in which function symbols represent Alloy operators. Prioni's theorem prover, Athena [3], has a polymorphic type system that allows a more succinct representation of operators, but cannot handle infinite sets. Therefore, unlike Kelloy that verifies assertions in both finite and infinite domains, Prioni only analyzes finite domains.

Another approach [12] to verifying Alloy specifications is via a translation to omega closure fork algebras [11]. Since the target system is an equational calculus, the translation eliminates all Alloy quantifiers, leading to intermediate expressions that are extremely hard to understand [14]. To reduce the cost of user interaction, Dynamite [14] has been developed, which targets a calculus in fork algebra that supports quantifiers. Unlike Kelloy that uses a first-order logic, Dynamite uses the higher-order logic of the PVS theorem prover [20].

To our knowledge, unlike Kelloy, Prioni and Dynamite do not support the complete Alloy language; they cannot handle integers and set cardinality. However, they integrate the Alloy Analyzer in their interactive proof processes by checking user-provided intermediate hypotheses using the Analyzer first. A similar feature can be added to Kelloy as well.

In [8,9], SMT solvers are used to verify Alloy specifications fully automatically. However, since Alloy is undecidable, in many cases, SMT solvers fail to verify valid specifications. As described in [7], these approaches are complementary to a full verification, but semi-automatic engine like Kelloy.

In [18], proof obligations for Event-B [1]—a set-theoretical language supporting integer expressions, cardinalities, and binary relations—are translated to KFOL. Similar to Kelloy, this approach targets a first-order theory that resembles the constructs of the source language. This work, however, targets an untyped logic, provides no calculus rules, no tool support, and no discussion of soundness and completeness of the translation. Furthermore, relations of higher arities and the transitive closure operator are not supported by Event-B and thus not covered by this work.

3   Background

3.1   Alloy and the Alloy Analyzer

Alloy [16] is a first-order relational logic with built-in transitive closure, set cardinality, and integer arithmetic operators. Our analysis introduces function symbols for the Alloy constructs of Fig. 1. In addition to the core Alloy logic, this subset of Alloy—called Alloy0—contains commonly-used Alloy constructs. It therefore enables us to generate formulas that closely conform to the structure of the Alloy specification being analyzed, and thus simplifies user interactions. Alloy constructs not present in Alloy0 are desugared by Kelloy.

As shown in Fig. 1, an Alloy0 problem consists of declarations and an assertion to check[1]. The Alloy Analyzer checks assertions with respect to a user-provided, *finite scope*—an upper bound on the number of elements of each type—fully automatically. While the reported counterexamples are guaranteed to be non-spurious, absence of a counterexample does not constitute proof.

---

[1]   An Alloy problem with facts $f_1, \ldots, f_n$ and an assertion a is an Alloy0 problem with an assertion $(f_1$ **and** ... **and** $f_n)$ **implies** a.

$$problem ::= dcl^* \; assertion$$
$$dcl ::= \textbf{sig} \; id \; [(\textbf{in} \mid \textbf{extends}) \; type]$$
$$\mid rel : type \; [\rightarrow type]^+$$
$$assertion ::= formula$$
$$exp ::= type \mid var \mid rel \mid \textbf{none} \mid exp \; \textbf{+} \; exp$$
$$\mid exp \; \textbf{-} \; exp \mid exp \; \textbf{\&} \; exp \mid exp.exp$$
$$\mid exp \rightarrow exp \mid \text{\textasciitilde} exp \mid \; \hat{} \; exp \mid \textbf{Int} \; intExp$$
$$intExp ::= number \mid \textbf{\#} exp \mid \textbf{int} \; var$$
$$\mid intExp \; (\textbf{+} \mid \textbf{-} \mid \textbf{*}) \; intExp$$
$$\mid (\textbf{sum} \; var : exp \mid intExp)$$

$$formula ::= intExp \; intComp \; intExp$$
$$\mid exp \; (\textbf{in} \mid \textbf{=}) \; exp \mid \textbf{not} \; formula$$
$$\mid formula \; (\textbf{and} \mid \textbf{or} \mid \textbf{implies}) \; formula$$
$$\mid (\textbf{lone} \mid \textbf{some} \mid \textbf{one}) \; exp$$
$$\mid \textbf{all} \; var : \textbf{set} \; exp \mid formula$$
$$\mid \textbf{some} \; var : \textbf{set} \; exp \mid formula$$
$$intComp ::= \; \textbf{<} \mid \textbf{>} \mid \textbf{=}$$
$$type ::= id \mid \textbf{Int}$$
$$rel ::= id$$
$$var ::= id$$

Fig. 1. Abstract syntax for Alloy0

Declarations. Alloy0 types represent sets of uninterpreted atoms. The signature `sig A` declares A as a top-level type. `sig B in A` declares B as a subtype (subset) of A. The `extends` keyword has the same effect with the additional constraint that extensions of a type are mutually disjoint. Relations can have arbitrary arities and are declared as $f : A_1 \rightarrow \ldots \rightarrow A_n$.

Expressions. Alloy0 expressions evaluate to relations. Sets are unary relations and scalars are singleton unary relations. The built-in relation `none` denotes the empty set. Operators `+`, `-`, and `&` denote union, difference, and intersection, respectively. For relations `r` and `s`, relational join (composition), Cartesian product, and transpose are denoted by `r.s`, `r -> s`, and `~r`, respectively. Transitive closure `^r` denotes the smallest transitive relation that contains `r`.

Integer expressions evaluate to integer values ($\mathbb{Z}$) and are constructed from numbers, arithmetic operators `+`, `-`, and `*`, set cardinality `#`, and `sum`. The built-in signature `Int` denotes the set of integer atoms. The cast operators `int` and `Int` give the integer value corresponding to an integer atom, and vice versa. The expression (`sum x: A | ie`) gives the sum of the integer expression `ie` for all distinct bindings of the variable `x` in the unary relation `A`.

Formulas. Basic Alloy0 formulas are constructed using the subset (`in`), equality, and integer comparison operators, and combined using the usual logical operators. The formulas `lone e`, `some e`, and `one e` constrain the cardinality of a relational expression `e` to be at most one, at least one, and exactly one, respectively. The quantifiers `all` and `some` denote the universal and existential quantifiers, and are supported by the Alloy Analyzer if the quantified variable is either a scalar or can be skolemized [16]. In our analysis, however, relations are first-order constructs, and thus quantifiers in the general form of $Q \; x : \textbf{set} \; e \mid F$ are allowed, where e is an expression of an arbitrary arity, x ranges over all subsets of e, and Q is either the universal or existential quantifier.

3.2   The KeY Proof System

KeY [4] is a deductive theorem prover based on a sequent calculus for JavaDL— a first-order dynamic logic for the Java programming language—which allows both automatic and interactive proofs. Kelloy uses the many-sorted, first-order subset of JavaDL, called KFOL.

Declarations. KFOL declarations consist of a set of types, a set of function symbols, and a set of predicate symbols. We write $f : T_1 \times \ldots T_n \to T$ to declare an n-ary function symbol f that takes arguments of types $T_1, \ldots, T_n$, and returns a value of type T. A constant symbol c of type T is a function with no arguments, and is denoted by $c : T$. A predicate symbol p that takes arguments of types $T_1, \ldots, T_n$ is denoted by $p: T_1 \times \ldots \times T_n \to \mathtt{Prop}$.

Expressions and Formulas. The set of all expressions for a declaration is denoted by $\mathtt{Expr}$, and the set of all formulas by $\mathtt{Frml}$. Expressions are constructed from function applications, and formulas from predicate applications. The equality predicate $=$ is built-in. In addition, KFOL provides boolean constants **true** and **false**, and the propositional connectives $\wedge$, $\vee$, $\Rightarrow$, $\Leftrightarrow$, and the negation $\neg$ to combine formulas. Universal and existential quantification are written, respectively, as $\forall x : T \mid \phi$ and $\exists x : T \mid \phi$ for a variable symbol x of type T and a formula ϕ. KFOL has a built-in type int along with the binary function symbols $+, -, * : int \times int \to int$, and the predicate symbol $< : int \times int \to \mathtt{Prop}$ which are all written in infix notation.

For a KFOL declaration C, a set of formulas A and a formula f we write $A \models_C f$ to denote that f is a logical consequence of A in C.

4   Axiomatization of Alloy0

As shown in Fig. 2, Alloy0 specifications are translated to KFOL using the translation function T which takes two arguments: a well-typed Alloy0 problem P, and a set fin of signatures that are marked by the user to be considered as finite. This is because some specifications require a finite setting (see Section 4.3). The translation $T[<p, \ fin>]$ returns a ternary tuple $<C, \ k_d, \ k_a>$ where C denotes a set of KFOL constant declarations that represents the signatures and relations of P, k_d is a KFOL formula that encodes the declaration constraints and the finite types (using the $finite_1$ predicate described in Section 4.3), and k_a denotes a KFOL formula encoding the assertion. Instead of reducing Alloy0 constructs to their definitions, our translation uses function symbols. This increases the automation level, makes the formulas easy-to-understand, and clarifies their correspondence to the original Alloy0 formulas. The semantics of these KFOL functions are defined by a set of KFOL axioms Ax. To prove the intended assertion in P, we invoke KeY on the proof obligation for $Ax \models_C k_d \Rightarrow k_a$.

4.1   Declarations

Let max denote the maximum relation arity used in the analyzed Alloy0 problem. For every $1 \leq n \leq max$, we introduce a KFOL type Rel_n to denote the set of n-ary relations. Atoms of the universe are denoted by the KFOL type $Atom$.

For every arity n, the membership predicate $in_n: Atom^n \times Rel_n \to \mathtt{Prop}$ allows the construction of the KFOL formula $in_n(a_1, \ldots, a_n, r)$ which denotes the

$T : problem \times \mathcal{P}(type) \to \mathcal{P}(\text{KFOL-decl}) \times \texttt{Frml} \times \texttt{Frml}$
$S : dcl \to \text{KFOL-decl}$
$F : dcl \cup formula \to \texttt{Frml}$
$N : type \cup rel \cup var \to \text{KFOL-const} \cup \text{KFOL-var}$
$Ax : \mathcal{P}(\texttt{Frml})$

$T[<d_1 \ldots d_n, a, \ fin>] = <\bigcup_{i=1}^{i=n}\{S[d_i]\}, \ \bigwedge_{i=1}^{i=n} F[d_i] \wedge \bigwedge_{S \in fin} finite_1(N[S]), \ F[a]>$
$S[\texttt{sig } A] = N[A] : Rel_1$
$S[\texttt{sig } A \ (\texttt{in}|\texttt{extends}) \ B] = N[A] : Rel_1$
$S[r\!:\!A_1 \to \ldots \to A_n] = N[r] : Rel_n$
$F[\texttt{sig } A] = \bigwedge_S disj_1(N[A], N[S])$ for any top-level signature $S \neq A$
$F[\texttt{sig } A \ \texttt{in } B] = subset_1(N[A], N[B])$
$F[\texttt{sig } A \ \texttt{extends } B] = F[\texttt{sig } A \ \texttt{in } B] \wedge$
                     $\bigwedge_S disj_1(N[A], N[S])$ for any extension S of B where $S \neq A$
$F[r\!:\!A_1 \to .. \to A_n] = subset_n(N[r], prod_{1 \times (n-1)}(N[A_1], (.., prod_{1 \times 1}(N[A_{n-1}], N[A_n]))))$

KFOL Axioms:
$\forall r, s\!:\! Rel_n \mid subset_n(r, s) \Leftrightarrow \forall a_{1:n}\!:\! Atom \mid in_n(a_{1:n}, r) \Rightarrow in_n(a_{1:n}, s)$
$\forall r, s\!:\! Rel_n \mid disj_n(r, s) \Leftrightarrow \forall a_{1:n}\!:\! Atom \mid \neg(in_n(a_{1:n}, r) \wedge in_n(a_{1:n}, s))$

Fig. 2. Translation rules for Alloy0 declarations. $\mathcal{P}(S)$ denotes the powerset of a set S.

membership of an n-ary tuple $<a_1, \ldots, a_n>$ in an n-ary relation r. We write $a_{i:j}$ as a shorthand for $a_i, \ldots, a_j$. The uninterpreted function $sin_1 : Atom \to Rel_1$ relates atoms and singleton sets.

In Fig. 2, the auxiliary translation function S translates any Alloy0 signature or relation r of arity n to a constant of type Rel_n with a unique name $N[r]$. The auxiliary translation function F constrains subtypes to be subsets of their parents using the uninterpreted predicate $subset_n : Rel_n \times Rel_n \to \texttt{Prop}$. Top-level signatures as well as extensions of a common signature are constrained to be mutually disjoint using the uninterpreted predicate $disj_n : Rel_n \times Rel_n \to \texttt{Prop}$. The semantics of these predicates are given by the axioms of Fig. 2. The type information of a relation r is encoded by constraining $N[r]$ to be a subset of the Cartesian product of its column types. Function $prod_{n \times m} : Rel_n \times Rel_m \to Rel_{n+m}$ denotes the Cartesian product of two relations of arities n and m, and is defined in Fig. 4.

Fig. 3 provides an example of translating Alloy0 declarations to KFOL. Fig. 3(a) gives a simple representation of a rooted, weighted, directed graph in Alloy0 and Fig. 3(b) gives its KFOL translation. The line numbers denote which Alloy0 statement produces each KFOL formula.

It should be noted that our type system is less precise than that of Alloy0; we encode some type-related properties as additional formulas that are incorporated as assumptions. Using the same type system would require a vast number of operators to be defined for all types, and a completely untyped system would not be compatible with our distinction of atoms and relations. Our calculus represents a useful compromise where arity information is captured syntactically by types, but the signature hierarchy is enforced semantically by formulas.

		1	$WT : Rel_1$
		2	$Node : Rel_1$
1	**sig** WT	1,2	$disj_1(WT, Node)$
2	**sig** Node	3	$edges : Rel_3$
3	**edges**: Node→Node→WT	3	$subset_3(edges, prod_{1\times2}(Node, prod_{1\times1}(Node, WT)))$
4	**sig** Root **in** Node	4	$Root : Rel_1$
		4	$subset_1(Root, Node)$
	(a)		(b)

Fig. 3. An example of translating declarations: (a) Alloy0, (b) KFOL

4.2   Relational Expressions

We use the auxiliary translation function $E : exp \cup intExp \rightarrow$ **Expr** to translate Alloy expressions. A basic expression, namely a type, relation, or variable t is translated to its KFOL counterpart $N[t]$. The translation of other relational expressions is given in Fig. 4. Integer expressions are discussed in Section 4.3.

The Alloy0 relation **none** is translated to a KFOL constant $none_1 : Rel_1$ and is axiomatized to be empty. Relational operators are translated to KFOL functions whose names are subscripted by the arity information of their arguments. The semantics of these functions are defined by axioms over the predicates in_n.

Most axioms of Fig. 4 directly define the Alloy0 semantics of the corresponding operators. Due to the compactness of FOL, however, transitive closure cannot be characterized by a recursively enumerable set of first-order axioms [19]. Such an axiomatization is possible for finite interpretations [6], but because we are interested in infinite systems as well, those results are not applicable to our approach.

We define transitive closure using a primitive recursive function $itrJoin_2$ that uses the built-in integer type of KeY. This translation is comprehensible for users and allows us to define canonical induction calculus rules. As shown in Fig. 4, for a binary relation r and any integer $i \geq 0$, the KFOL expression $itrJoin_2(r, i)$ evaluates to a relation that contains the pairs (a, b) where b is reachable from a by following 0 to i steps in r.

4.3   Integer Expressions and Cardinality

The Alloy Analyzer calculates arithmetic expressions with respect to a fixed bitwidth, and thus calculations are subject to overflow. When verifying specifications, however, overflow is often not intended and integers are assumed to represent the infinite set of mathematical integers. Therefore, we translate Alloy0 integer expressions using KFOL's *int* type that models the semantics of mathematical integers, thus deliberately deviating from the Alloy semantics. Integer numbers, arithmetic expressions, and comparisons in Alloy0 are translated to their counterparts in KFOL.

The Alloy Analyzer requires all relations to be finite, and thus the cardinality operator is defined for all expressions. In our translation, however, relations

$E : exp \cup intExp \rightarrow \mathbf{Expr}$

$E[\mathbf{none}] = none_1$

$E[x_n \mathbf{+} y_n] = union_n(E[x_n], E[y_n])$

$E[x_n \mathbf{-} y_n] = diff_n(E[x_n], E[y_n])$

$E[x_n \mathbf{\&} y_n] = intersect_n(E[x_n], E[y_n])$

$E[x_n . y_m] = join_{n \times m}(E[x_n], E[y_m])$

$E[x_n \mathbf{->} y_m] = prod_{n \times m}(E[x_n], E[y_m])$

$E[(\mathbf{sum}\ v : x_1 \mid ie)] = \Sigma_{i=1}^{i=card_1(E[x_1])} E[ie][sin_1(ordInv_1(E[x_1], i))/N[v]]$

$E[\tilde{\ } x_2] = transpose_2(E[x_2])$

$E[\hat{\ } x_2] = tc_2(E[x_2])$

$E[\mathbf{Int}\ ie] = sin_1(i2a(E[ie]))$

$E[\mathbf{int}\ v] = a2i(ordInv_1(E[v], 1))$

$E[i_1 \overset{*}{\underline{\ }} i_2] = E[i_1] \overset{*}{\underline{\ }} E[i_2]$

$E[\mathbf{\#} x_n] = card_n(E[x_n])$

KFOL Axioms:

$\forall a{:}\ Atom \mid in_1(a, none_1) \Leftrightarrow \mathbf{false}$

$\forall a, b{:}\ Atom \mid in_1(b, sin_1(a)) \Leftrightarrow a = b$

$\forall r, s{:}\ Rel_n, a_{1:n}{:}\ Atom \mid in_n(a_{1:n}, union_n(r, s)) \Leftrightarrow in_n(a_{1:n}, r) \lor in_n(a_{1:n}, s)$

$\forall r, s{:}\ Rel_n, a_{1:n}{:}\ Atom \mid in_n(a_{1:n}, diff_n(r, s)) \Leftrightarrow in_n(a_{1:n}, r) \land \neg in_n(a_{1:n}, s)$

$\forall r, s{:}\ Rel_n, a_{1:n}{:}\ Atom \mid in_n(a_{1:n}, intersect_n(r, s)) \Leftrightarrow in_n(a_{1:n}, r) \land in_n(a_{1:n}, s)$

$\forall r{:}\ Rel_n, s{:}\ Rel_m, a_{1:n+m-2}{:}\ Atom \mid in_{n+m-2}(a_{1:n+m-2}, join_{n \times m}(r, s))$
$$\Leftrightarrow (\exists b{:}\ Atom \mid in_n(a_{1:n-1}, b, r) \land in_m(b, a_{n:n+m-2}, s))$$

$\forall r{:}\ Rel_n, s{:}\ Rel_m, a_{1:n+m}{:}\ Atom \mid$
$$in_{n+m}(a_{1:n+m}, prod_{n \times m}(r, s)) \Leftrightarrow in_n(a_{1:n}, r) \land in_m(a_{n+1:n+m}, s)$$

$\forall r{:}\ Rel_2, a_1, a_2{:}\ Atom \mid in_2(a_1, a_2, transpose_2(r)) \Leftrightarrow in_2(a_2, a_1, r)$

$\forall r{:}\ Rel_2, a_{1:2}{:}\ Atom \mid in_2(a_{1:2}, tc_2(r)) \Leftrightarrow \exists i{:}\ int \mid i \geq 0 \land in_2(a_{1:2}, itrJoin_2(r, i))$

$\forall r{:}\ Rel_2, i{:}\ int^{\geq 0} \mid itrJoin_2(r, 0) = r\ \land$
$$itrJoin_2(r, i+1) = union_2(itrJoin_2(r, i), join_{2 \times 2}(r, itrJoin_2(r, i))))$$

$\forall r{:}\ Rel_n, a_{1:n}{:}\ Atom \mid (finite_n(r) \land in_n(a_{1:n}, r)) \Rightarrow 1 \leq ord_n(r, a_{1:n}) \leq card_n(r)$

$\forall r{:}\ Rel_n, i{:}\ int \mid (finite_n(r) \land 1 \leq i \leq card_n(r))$
$$\Rightarrow \exists a_{1:n}{:}\ Atom \mid in_n(a_{1:n}, r) \land ord_n(r, a_{1:n}) = i$$

$\forall r{:}\ Rel_n, a_{1:n}, b_{1:n}{:}\ Atom \mid (finite_n(r) \land in_n(a_{1:n}, r) \land in_n(b_{1:n}, r)$
$$\land\ ord_n(r, a_{1:n}) = ord_n(r, b_{1:n})) \Rightarrow (a_1 = b_1 \land \ldots \land a_n = b_n)$$

$\forall r{:}\ Rel_1, a{:}\ Atom \mid in_1(a, r) \Rightarrow ordInv_1(r, ord_1(r, a)) = a$

$\forall a{:}\ Atom \mid in_1(a, N[Int]) \Rightarrow i2a(a2i(a)) = a$

$\forall i{:}\ int \mid in_1(i2a(i), N[Int]) \land a2i(i2a(i)) = i$

Fig. 4. Translation rules for Alloy0 expressions. x_i and y_i represent Alloy0 expressions of arity i. The expression $e[e_1/e_2]$ substitutes e_1 for all occurrences of e_2 in e.

are potentially infinite, and thus cardinality is defined only for those that are explicitly known to be finite. For this purpose, we introduce a family of finiteness predicates $finite_n{:}\ Rel_n \rightarrow \mathbf{Prop}$ that hold if the user marks a signature as finite, or if finiteness can be inferred[2]. Unlike the Alloy Analyzer that finitizes relations by user-provided, specific upper bounds, Kelloy considers *all* finite domains for those relations that are flagged as finite.

As shown in Fig. 4, we translate Alloy0's cardinality operator to a KFOL function $card_n : Rel_n \rightarrow int^{\geq 0}$ which yields the cardinality of an n-ary relation r if it is finite, and is unspecified otherwise. $card_n$ is computed using an ordering function $ord_n : Rel_n \times Atom^n \rightarrow int^{>0}$—a bijection from the elements of a finite

---

[2] Kelloy includes a set of axioms to infer finiteness. For example, the singleton $sin_1(a)$ is always finite, and the union of two finite relations is finite.

$$F : dcl \cup formula \to \texttt{Frml}$$
$$F[x_n \ \texttt{in} \ y_n] = subset_n(E[x_n], E[y_n])$$
$$F[x_n = y_n] = (E[x_n] = E[y_n])$$
$$F[\texttt{all} \ a : \ \texttt{set} \ x_n \mid g] = (\forall \, N[a]: Rel_n \mid subset_n(N[a], E[x_n]) \Rightarrow F[g])$$
$$F[\texttt{some} \ a : \ \texttt{set} \ x_n \mid g] = (\exists \, N[a]: Rel_n \mid subset_n(N[a], E[x_n]) \wedge F[g])$$

$$F[\texttt{one} \ x_n] = one_n(E[x_n])$$
$$F[\texttt{lone} \ x_n] = lone_n(E[x_n])$$
$$F[\texttt{some} \ x_n] = some_n(E[x_n])$$

KFOL Axioms:
$$\forall r, s: Rel_n \mid r = s \Leftrightarrow \forall a_{1:n}: Atom \mid in_n(a_{1:n}, r) \Leftrightarrow in_n(a_{1:n}, s)$$
$$\forall r: Rel_n \mid one_n(r) \Leftrightarrow some_n(r) \wedge lone_n(r)$$
$$\forall r: Rel_n \mid lone_n(r) \Leftrightarrow$$
$$\qquad \forall a_{1:n}, b_{1:n}: Atom \mid in_n(a_{1:n}, r) \wedge in_n(b_{1:n}, r) \Rightarrow (a_1 = b_1) \wedge \ldots \wedge (a_n = b_n)$$
$$\forall r: Rel_n \mid some_n(r) \Leftrightarrow \exists a_{1:n}: Atom \mid in_n(a_{1:n}, r)$$

Fig. 5. Translation rules for Alloy0 formulas. x_n denotes an Alloy0 expression of arity n.

relation r to the inclusive integer interval $[1, \ldots, card_n(r)]$. It is easy to show that if the axioms for ord_n, as shown in Fig. 4 hold, $card_n(r)$ gives the cardinality of r. We also define the function $ordInv_1 : Rel_1 \times int \to Atom$ to denote the inverse of $ord_1(r)$ for any unary relation r.

The Alloy0 signature Int is translated like other top-level signatures to a constant function symbol $N[Int] : Rel_1$. The Alloy0 cast operators Int and int are translated using the bijections $i2a : int \to Atom$ and $a2i : Atom \to int$ that give the atom corresponding to an integer value and vice versa. Since in Alloy0, the int operator is only applicable to scalar variables, the atom corresponding to v in the expression (int v) can be retrieved by $ordInv_1(E[v], 1)$.

The sum construct is translated using the cardinality function and KFOL's bounded sum operator. Note that, due to the underspecification of $card_1$ and $ordInv_1$, the result of the sum operation is unspecified if $E[S]$ is not finite.

4.4 Formulas

Alloy0 formulas are translated using the auxiliary translation function F given in Fig. 5. Subset and equality formulas are translated using the $subset_i$ predicates and KFOL's built-in (polymorphic) equality. Negation, conjunction, disjunction, and implication operators in Alloy0 are translated to their counterparts in KFOL, and skipped in Fig. 5 in the interest of space. For an Alloy0 expression x of arity n, a multiplicity formula (mult x) is translated to a predicate $mult_n(E[x])$ in KFOL where $mult$ stands for the multiplicities $some$, $lone$, and one. Further axioms give the semantics of these predicates. Universal and existential quantifiers in Alloy0 are translated to those in KFOL where the bounding expression is incorporated into the body of the quantifier.

4.5 The Ordering Module

The Alloy Analyzer provides some library modules that can be used in Alloy problems. Most library functions are inlined and treated like other expressions. The ordering module, however, triggers special optimizations in the Analyzer.

Since this module is widely used, we also treat it specially. The declaration ord[S] defines a total order[3] on a signature S, which is represented by Alloy0 relations next:S->S, first:S, and last:S which, respectively, denote the successor of an element, and the smallest and the largest elements of the order. These relations are translated to KFOL constants $nextS : Rel_2$, $firstS : Rel_1$, and $lastS : Rel_1$, respectively.

If $finite_1(N[S])$ holds, the previously defined ord_1 function induces an ordering. When $N[S]$ is not finite, $nextS$ relates each element to its immediate successor and thus makes $N[S]$ countable (i.e. isomorphic to the natural numbers). In this case, we extend the axioms for ord_1 to define a bijection from $N[S]$ to $int^{>0}$. The semantics of $nextS$ is then given by:

$$\forall a, b \colon Atom \ |(in_1(a, N[S]) \land in_1(b, N[S]))$$
$$\Rightarrow (in_2(a, b, nextS) \Leftrightarrow ord_1(N[S], b) = ord_1(N[S], a) + 1)$$

Ordered signatures in Alloy0 cannot be empty. This is encoded as $\neg(N[S] = none_1)$. The constant $firstS$ yields the element associated with 1, and $lastS$ yields the one associated with $card_1(N[S])$ if $N[S]$ is finite, and the empty set otherwise.

$$firstS = sin_1(ordInv_1(N[S], 1)) \qquad \neg finite_1(N[S]) \Rightarrow lastS = none_1$$
$$finite_1(N[S]) \Rightarrow (lastS = sin_1(ordInv_1(N[S], card_1(N[S]))))$$

Properties about the elements of a countable set are often proved using induction. KeY provides an induction scheme for its integer type which can be used for this purpose.

4.6 Theoretical Properties

This section discusses the correctness and completeness of our translation. In KFOL, the semantics of the built-in integers is set to $\mathbb{Z}$. KeY's calculus contains a set of inference rules to deal with arithmetic expressions. The calculus, however, cannot be complete because according to Gödel's incompleteness theorem, there is no sound and complete calculus for integer arithmetic [4, §2.7].

The first two theorems state the properties for Alloy0 problems without integers and the third one handles the integer case. The proof sketches for the theorems can be found elsewhere [21].

In the following, we use $sigs(P)$ to denote the set of all signatures (not including the signature Int) defined in an Alloy0 problem P.

Theorem 1 (Correctness). *Let P be an Alloy0 problem that does not contain any integer expression (neither of type int nor Int) and $fin \subseteq sigs(P)$ a set of signatures. Let $T[<P, fin>] = <C, k_d, k_a>$ be the translation of P. If $Ax \models_C k_d \Rightarrow k_a$, then P is valid in all structures that interpret the signatures in fin as finite.*

---

[3] The Alloy Analyzer treats signatures as finite, so the last element of the order does not have a next element.

This theorem implies that the Alloy Analyzer will not produce a counterexample for an Alloy0 problem (not containing integers) that has been proved by Kelloy. If $fin = \emptyset$, Thm. 1 implies that our translation is correct with respect to all structures, i.e. both finite and infinite ones. The Alloy Analyzer, on the other hand, interprets Alloy0 problems only with respect to finite structures.

Completeness, however, holds only for finite structures. In first-order logic, it is not possible to formalize that one type is the powerset of another type. Consequently, our axioms cannot guarantee that the KFOL type Rel_n represents the set of all n-ary relations. This limitation did not appear problematic in practice, but restricts the completeness theorem to the case of finite structures.

Theorem 2 (Relative completeness). *Let P be an Alloy0 problem that does not contain any integer expression (neither of type* int *nor* Int*).*
Let $T[<P, \ sigs(P)>] = <C, \ k_d, \ k_a>$ be the translation of P with all the signatures finitized. If P has no counterexample which interprets all signatures as finite, then $Ax \models_C k_d \Rightarrow k_a$.

The next theorem considers Alloy0 problems that contain integer expressions. The Alloy Analyzer—due to its methodology—finitizes all domains. Hence, it cannot check problems for validity with respect to $\mathbb{Z}$, but only with respect to a fixed bitwidth. In contrast, integers in KFOL are never interpreted bounded. Therefore, we cannot establish a relationship between arbitrary Alloy0 and KFOL counterexamples. For example, an Alloy0 formula that specifies that there is a maximum integer value is not valid with respect to $\mathbb{Z}$ and thus cannot be proved by Kelloy. However, the Alloy Analyzer will not produce a counterexample either since the formula is valid in all structures with a finite integer domain. Consequently, we generalize the correctness and completeness results by fixing the semantics of the Alloy signature Int to $\mathbb{Z}$:

Theorem 3 (Correctness and Completeness modulo integers). *Let P be an Alloy0 problem (which may contain integer expressions) and $fin \subseteq sigs(P)$ a set of signatures. Let $T[<P, \ fin>] = <C, \ k_d, \ k_a>$ be the translation of P.*
If $Ax \models_C k_d \Rightarrow k_a$, then P is valid in all structures that interpret the signatures in fin as finite and interpret the signature Int as $\mathbb{Z}$.
If $fin = sigs(P)$ and P has no counterexample which interprets all signatures in $sig(P)$ as finite and the signature Int as $\mathbb{Z}$, then $Ax \models_C k_d \Rightarrow k_a$.

5   Reasoning Strategy

The KeY system uses a sequent calculus [15] for proving. A proof-tree is constructed by applying the calculus rules to a proof sequent. This can either be done manually through the GUI, or automatically by KeY's proof search strategy that assigns priorities to all applicable rules and instantiates quantifiers heuristically. We extend the existing strategy by incorporating new (Alloy-specific) calculus rules. All axioms from the previous sections are implemented as rules for the calculus. For example, axioms that follow the form $\forall x \colon T \mid P(x) \Rightarrow (F(x) \Leftrightarrow G(x))$

become *conditional rewrite rules* that replace $F(x)$ with $G(x)$ when the guard $P(x)$ is known to hold. Since all axioms become rules, the set of Axioms Ax is no longer included in the proof obligation.

The axiom rules rewrite all invocations of the predicates $subset_n$, $disj_n$, $lone_n$, one_n, $some_n$, and relational equality to their quantified definitions. We consider this undesirable for two reasons: (1) formulas grow considerably in size and are thus hard to understand, and (2) when quantifiers cannot be eliminated by skolemization, they require the strategy to provide a suitable instantiation which is a heuristic task.

Our strategy addresses this by only expanding predicates to their definitions when skolemization is applicable. Otherwise, *lemma rules* are used to exploit the semantics of the predicates without rewriting them. For example, a lemma lets us conclude $in_n(a_{1:n}, s_n)$ from $in_n(a_{1:n}, r_n)$ and $subset_n(r_n, s_n)$. This maintains the structural correspondence between formulas and the Alloy0 specification, and allows reasoning on the abstraction level of relations. Overall, the strategy features around 500 lemma rules that have been proved using KeY to follow from the axioms.

The recursive nature of the definition of transitive closure (tc_2) poses a special challenge during proving. In order to simplify proofs and to increase the automation level, we use additional lemmas to capture several simple properties about tc_2, such as its transitivity. Such lemmas are useful for proving some assertions that involve transitive closure. For some cases, however, an induction scheme is required. We can use induction on the integers in the definition of $itrJoin_2$. However, formulas generated this way get cumbersome quickly. We therefore define special induction schemes that are more intuitive, and thus easier to use. Further information can be found elsewhere [21].

6   Evaluation

In this section, we summarize our experimental results of proving Alloy assertions with Kelloy. We proved a total of 22 assertions in 10 Alloy problems of varying sizes and complexity[4]. The chosen collection of Alloy problems is included in the Alloy Analyzer 4.1 distribution and involves all relevant aspects of the language, including transitive closure, integer arithmetic, and the ordering module. In the following, we first elaborate on the automatically proved assertions and show the impact of our reasoning strategy. We then report on the interactive proofs.

6.1   Automation

Out of the 22 assertions, 12 have been proved completely automatically by Kelloy. The remaining 10 assertions required manual guidance as discussed in the next section. Table 1 shows runtime measured on an Intel Core2Quad, 2.8GHz, 8GB memory, and the number of proof steps (i.e. the number of single rule applications) required for each automatic proof.

---

[4] All Alloy problems and proofs can be found at http://asa.iti.kit.edu/306.php

Table 1. Automatically-proved assertions (time in seconds, time-out after 2h)

		KELLOY STRATEGY		BASIC STRATEGY	
PROBLEM	ASSERTION	TIME (STEPS)	RESULT	TIME (STEPS)	RESULT
address book	delUndoesAdd	9.3 (2476)	proved	27.1 (5475)	proved
	addIdempotent	0.1 (113)	proved	5.0 (1176)	proved
abstract memory	writeRead	0.8 (567)	proved	1.0 (597)	proved
	writeIdempotent	14.0 (4482)	proved	6.5 (1009)	proved
media assets	hidePreservesInv	0.0 (39)	proved	0.1 (70)	proved
	pasteNotAffectsHidden	15.9 (2619)	proved	time-out (–)	failed
mark sweep	soundness1	3.0 (1195)	proved	time-out (–)	failed
grandpa	noSelfFather	0.0 (77)	proved	0.0 (77)	proved
	noSelfGrandpa	26.5 (3144)	proved	39.8 (3276)	proved
filesystem	FileInDir	0.5 (160)	proved	time-out (–)	failed
	SomeDir	0.2 (205)	proved	time-out (–)	failed
birthday	addWorks	0.1 (129)	proved	1.2 (506)	proved

Kelloy's strategy uses numerous lemmas to maintain the structure of formulas and to allow reasoning on the abstraction level of relations. To evaluate the impact of these lemmas on the automation level, we compared Kelloy's strategy to a basic strategy that applies all the axiom rules, but none of the lemmas. Table 1 also shows these numbers.

Out of the 12 assertions automatically proved by Kelloy, 4 could not be proved automatically by the basic strategy. Furthermore, although for the remaining assertions, the basic strategy suffices, Kelloy performs most of the proofs faster and requires fewer proof steps. Exceptions include *writeIdempotent* for which the basic strategy is superior, and *noSelfFather* and *writeRead* for which the two strategies perform equally well. These assertions involve simple formulas for which rewriting function symbols as their quantified definitions and using the default quantifier instantiation is sufficient.

6.2   Interactive Proofs

10 of the verified Alloy assertions required user interaction to guide the prover. During interactive proving, the user manually applies rules to the sequent (KeY's GUI makes this quite convenient). The automatic proof search strategy can be invoked anytime on the subgoals of a proof. The strategy then either proves the subgoal, or stops when the maximum number of steps (set by the user) is reached. It is a common practice to frequently invoke the strategy and only focus on those cases that the prover cannot solve on its own.

The manual rule applications can be categorized into three groups of descending complexity: (1) Hypothesis introduction: for example as an induction hypothesis or for a case distinction. (2) Prover guidance: rule applications that allow the strategy to solve a subgoal (more quickly). These include, for example, quantifier instantiations and case distinctions on formulas from the sequent. (3) Non-essential steps: simple steps that the strategy would find automatically but the user prefers to do manually to keep track of the proof.

The complexity of the proofs for the 10 interactively-proved assertions differ considerably. 7 assertions required only very few (max. 10) interactive steps. One example of such a proof is the *completeness* assertion for the mark and sweep

Alloy problem: only one, yet quite a complex step to handle transitive closure had to be done interactively. The remaining 3 assertions required between 36 and 291 interactive steps. The assertion stating the correctness of Dijkstra's deadlock prevention algorithm had the most complex proof: we introduced seven intermediate hypotheses that were proved using induction. Overall, the proof took 18875 steps out of which 7/219/65 were manual steps of the categories 1/2/3. A proof of this complexity can be conducted by an experienced user in roughly one work day.

In order to evaluate the impact of the user's expertise on the interactive proof process, we asked an Alloy user with no previous experience in KeY to prove a soundness assertion for the mark and sweep Alloy problem. Out of 1389 steps, 207 (2/57/148) have been performed manually. The proof, including a proof-sketch on paper, was conducted within two work days.

In comparison to that, an experienced user in both Alloy and KeY, proved the same assertion in 4 hours with only 10 (5/1/4) interactive steps out of a total of 9372 steps. The higher number of total steps, but drastically smaller number of interactive steps show that it requires some experience to effectively leverage the automation strategy. However, the experiment indicates that the proof process is intuitive enough for a user with no prior experience in Kelloy.

7   Conclusion

We presented Kelloy, a tool for full verification of Alloy problems. We formally defined a translation of the Alloy language to the first-order logic of the theorem prover KeY and discussed its correctness and completeness. To our knowledge, Kelloy is the only system that provides proof capability for the whole Alloy language (including integers, cardinality, and the ordering module).

We used Kelloy to prove some challenging Alloy assertions semi-automatically. Our experiments showed that usually only structurally complex systems or systems that involve inductive properties require user interaction. Moreover, considerable parts of a proof can be automated while the user only performs central steps interactively. In many cases, however, conducting a proof using Kelloy requires in-depth knowledge of the analyzed Alloy problem; the required time and effort depend on the user's experience in the tool. Kelloy is thus intended to be used in conjunction with automatic approaches as described in [7].

The presentation of proof obligations in Kelloy resembles the original Alloy structure such that even a non-expert in KeY can conduct interactive proofs. The effort of interaction might be further lowered in the future, for example by pretty-printing expressions in the Alloy syntax or integration of the Alloy Analyzer for counterexample generation and visualization.

Several program analysis tools (see [7] for some examples) use Alloy as their specification languages. Incorporating Alloy into KeY raises the opportunity of full verification of programs that contain Alloy specifications, leveraging both the expressiveness of Alloy and the dynamic logic of KeY. Pursuing this idea is left for future work.

Acknowledgement. We thank Peter H. Schmitt and the anonymous reviewers for their helpful comments. This work was funded in part by the MWK-BW grant 655.042/taghdiri/1.

References

1. Abrial, J.-R., Hallerstede, S.: Refinement, decomposition, and instantiation of discrete models: Application to Event-B. Fundamenta Informaticae (2007)
2. Arkoudas, K., Khurshid, S., Marinov, D., Rinard, M.: Integrating model checking and theorem proving for relational reasoning. In: RMICS (2003)
3. Athena, http://people.csail.mit.edu/kostas/dpls/athena/
4. Beckert, B., Hähnle, R., Schmitt, P.H. (eds.): Verification of Object-Oriented Software. LNCS (LNAI), vol. 4334. Springer, Heidelberg (2007)
5. Buss, S.R.: First-order proof theory of arithmetic. In: Handbook of Proof Theory, pp. 79–147. Elsevier (1998)
6. van Eijck, J.: Defining (reflexive) transitive closure on finite models, http://homepages.cwi.nl/~jve/papers/08/pdfs/FinTransClosRev.pdf
7. El Ghazi, A.A., Geilmann, U., Ulbrich, M., Taghdiri, M.: A Dual-Engine for Early Analysis of Critical Systems. In: DSCI (2011)
8. El Ghazi, A.A., Taghdiri, M.: Analyzing Alloy Constraints using an SMT Solver: A Case Study. In: AFM (2010)
9. El Ghazi, A.A., Taghdiri, M.: Relational Reasoning via SMT Solving. In: Butler, M., Schulte, W. (eds.) FM 2011. LNCS, vol. 6664, pp. 133–148. Springer, Heidelberg (2011)
10. Fortune, S., Leivant, D., O'Donnell, M.: The expressiveness of simple and second-order type structures. J. ACM (1983)
11. Frias, M., Pombo, C.G.L.: Interpretability of first-order linear temporal logics in fork algebras. In: Journal of logic and algebraic programming (2006)
12. Frias, M.F., Pombo, C.G.L., Aguirre, N.M.: An Equational Calculus for Alloy. In: Davies, J., Schulte, W., Barnett, M. (eds.) ICFEM 2004. LNCS, vol. 3308, pp. 162–175. Springer, Heidelberg (2004)
13. Frias, M., Pombo, C.G.L., Baum, G., Aguirre, N.M., Maibaum, T.: Taking Alloy to the movies. In: Araki, K., Gnesi, S., Mandrioli, D. (eds.) FME 2003. LNCS, vol. 2805, pp. 678–697. Springer, Heidelberg (2003)
14. Frias, M.F., Pombo, C.G.L., Moscato, M.M.: Alloy Analyzer+PVS in the Analysis and Verification of Alloy Specifications. In: Grumberg, O., Huth, M. (eds.) TACAS 2007. LNCS, vol. 4424, pp. 587–601. Springer, Heidelberg (2007)
15. Gentzen, G.: Untersuchungen über das logische Schließen. Mathematische Zeitschrift (1935)
16. Jackson, D.: Software Abstractions: Logic, Language and Analysis. MIT Press (2006)
17. Jackson, D., Wing, J.: Lightweight formal methods. IEEE Computer (1996)
18. Köker, C.: Discharging Event-B proof obligations. Studienarbeit, Universität Karlsruhe, TH (2008)
19. Lev-Ami, T., Immerman, N., Reps, T.W., Sagiv, M., Srivastava, S., Yorsh, G.: Simulating reachability using first-order logic with applications to verification of linked data structures. Logical Methods in Computer Science 5(2) (2009)
20. Shankar, N., Owre, S., Rushby, J., Stringer-Calvert, D.: PVS Prover Guide. Computer Science Laboratory, SRI International (1999)
21. Ulbrich, M., Geilmann, U., Ghazi, A.A.E., Taghdiri, M.: On proving alloy specifications using KeY. Tech. Rep. 2011-37, Karlsruhe Institute of Technology (2011)

Reachability under Contextual Locking

Rohit Chadha[1], P. Madhusudan[2], and Mahesh Viswanathan[2]

[1] LSV, ENS Cachan & CNRS & INRIA
[2] University of Illinois, Urbana-Champaign

Abstract. The pairwise reachability problem for a multi-threaded program asks, given control locations in two threads, whether they can be simultaneously reached in an execution of the program. The problem is important for static analysis and is used to detect statements that are concurrently enabled. This problem is in general undecidable even when data is abstracted and when the threads (with recursion) synchronize only using a finite set of locks. Popular programming paradigms that limit the lock usage patterns have been identified under which the pairwise reachability problem becomes decidable. In this paper, we consider a new natural programming paradigm, called contextual locking, which ties the lock usage to calling patterns in each thread: we assume that locks are released in the same context that they were acquired and that every lock acquired by a thread in a procedure call is released before the procedure returns. Our main result is that the pairwise reachability problem is polynomial-time decidable for this new programming paradigm as well.

1 Introduction

In static analysis of sequential programs [7], such as control-flow analysis, data-flow analysis, points-to analysis, etc., the semantics of the program and the data that it manipulates is *abstracted*, and the analysis concentrates on computing fixed-points over a lattice using the control-flow in the program. For instance, in flow-sensitive context-sensitive points-to analysis, a finite partition of the heap locations is identified, and the analysis keeps track of the set of possibilities of which variables point may point to each heap-location partition, propagating this information using the control-flow graph of the program. In fact, several static analysis questions can be formulated as reachability in a pushdown system that captures the control-flow of the program (where the stack is required to model recursion) [10].

In concurrent programs, abstracting control-flow is less obvious, due to the various synchronization mechanisms used by threads to communicate and orchestrate their computations. One of the most basic questions is *pairwise reachability*: given two control locations pc_1 and pc_2 in two threads of a concurrent program, are these two locations simultaneously reachable? This problem is very basic to static analysis, as many analysis techniques would, when processing pc_1, take into account the *interference* of concurrent statements, and hence would

C. Flanagan and B. König (Eds.): TACAS 2012, LNCS 7214, pp. 437–450, 2012.
© Springer-Verlag Berlin Heidelberg 2012

like to know if a location like pc_2 is concurrently reachable. Data-races can also be formulated using pairwise reachability, as it amounts to asking whether a read/write to a location (or an abstract heap region) is concurrently reachable with a write to the same location (or region). More sophisticated verification techniques like deductive verification can also utilize such an analysis. For instance, in an Owicki-Gries style proof [8] of a concurrent program, the invariant at pc_1 must be proved to be stable with respect to concurrent moves by the environment, and hence knowing whether pc_2 is concurrently reachable will help determine whether the statement at pc_2 need be considered for stability.

Pairwise reachability of control locations is hence an important problem. Given that individual threads may employ recursion, this problem can be *modeled* as reachability of *multiple* pushdown systems that synchronize using the synchronization constructs in the concurrent program, such as locks, barriers, etc. However, it turns out that even when synchronization is limited to using just locks, pairwise reachability is *undecidable* [9]. Consequently, recently, many natural restrictions have been identified under which pairwise reachability is decidable.

One restriction that yields a decidable pairwise reachability problem is *nested locking* [5,4]: if each thread performs only nested locking (i.e. locks are released strictly in the reverse order in which they are acquired), then pairwise reachability is known to be decidable [5]. The motivation for nested locking is that many high-level locking constructs in programming languages naturally impose nested locking. For instance the synchronize(o) { ...} statement in Java acquires the lock associated with o, executes the body, and releases the lock, and hence nested synchronized blocks naturally model nested locking behaviors. The usefulness of the pairwise reachability problem was demonstrated in [5] where the above decision procedure for nested locking was used to find bugs in the Daisy file system. Nested locking has been generalized to the paradigm of *bounded lock chaining* for which pairwise reachability has also been proved to be decidable [2,3].

In this paper, we study a different restriction on locking, called *contextual locking*. A program satisfies contextual locking if each thread, in every context, acquires new locks and releases all these locks before returning from the context. Within the context, there is *no requirement* of how locks are acquired and released; in particular, the program can acquire and release locks in a non-nested fashion or have unbounded lock chains.

The motivation for contextual locking comes from the fact that this is a very natural restriction. First, note that it's very natural for programmers to release locks in the same context they were acquired; this makes the acquire and release occur in the same syntactic code block, which is a very simple way of managing lock acquisitions.

Secondly, contextual locking is very much encouraged by higher-level locking constructs in programming languages. For example, consider the code fragment of a method, in Java [6] shown in Figure 1. The above code takes the lock associated with *done* followed later by a lock associated with object r. In order

```
public void m() {
  synchronized(done) {
    ...
    synchronized(r) {
      ...
      while (done=0)
      try {
        done.wait();
      }
  ...
}
```

Fig. 1. Synchronized blocks in Java

to proceed, it wants *done* to be equal to 1 (a signal from a concurrent thread, say, that it has finished some activity), and hence the thread waits on *done*, which releases the lock for *done*, allowing other threads to proceed. When some other thread issues a *notify*, this thread wakes up, reacquires the lock for *done*, and proceeds.

Notice that despite having synchronized blocks, the wait() statement causes releases of locks in a non-nested fashion (as it exhibits the sequence *acq lock_done; acq lock_r; rel lock_done; acq lock_done; rel lock_r; rel lock_done;*). However, note that the code above does satisfy *contextual locking*; the locks *m* acquires are all released before the exit, because of the synchronized-statements. Thus, we believe that contextual locking is a natural restriction that is adhered to in many programs.

The main result of this paper is that pairwise reachability is decidable under the restriction of contextual locking. It is worth pointing out that this result does not follow from the decidability results for nested locking or bounded lock chains [5,2]. Unlike nested locking and bounded lock chains, contextual locking imposes no restrictions on the locking patterns in the absence of recursive function calls; thus, programs with contextual locking may not adhere to the nested locking or bounded lock chains restrictions. Second, the decidability of nested locking and bounded lock chains relies on a non-trivial observation that the number of context switches needed to reach a pair of states is bounded by a value that is *independent* of the size of the programs. However, such a result of a bounded number of context switches does not hold for programs with contextual locking. Thus, the proof techniques used to establish decidability are different as well.

We conclude this introduction with a brief outline of the proof ideas behind our decidability result. We observe that if a pair of states is simultaneously reachable by some execution, then they are also simultaneously reachable by what we call a *well bracketed computation*. A concurrent computation of two threads is not well bracketed, if in the computation one process, say $\mathcal{P}_0$, makes a call which is followed by the other process ($\mathcal{P}_1$) making a call, but then $\mathcal{P}_0$ returns from its call before $\mathcal{P}_1$ does (but after $\mathcal{P}_1$ makes the call). We then observe that every well bracketed computation of a pair of recursive programs can simulated by a single recursive program. Thus, decidability in polynomial time follows from observations about reachability in pushdown systems [1].

The rest of the paper is organized as follows. Section 2 introduces the model of concurrent pushdown systems communicating using locks and presents its semantics. Our main decidability result is presented in Section 3. Conclusions are presented in Section 4.

2   Model

Pushdown Systems. For static analysis, recursive programs are usually modeled as pushdown systems. Since we are interested in modeling threads in concurrent programs we will also need to model locks for communication between threads. Formally,

Definition 1. *Given a finite set* Lcks, *a pushdown system (PDS)* $\mathcal{P}$ *using* Lcks *is a tuple* (Q, Γ, qs, δ) *where*

- Q *is a finite set of control states.*
- Γ *is a finite stack alphabet.*
- qs *is the initial state.*
- $\delta = \delta_{\mathsf{int}} \cup \delta_{\mathsf{cll}} \cup \delta_{\mathsf{rtn}} \cup \delta_{\mathsf{acq}} \cup \delta_{\mathsf{rel}}$ *is a finite set of transitions where*
    - $\delta_{\mathsf{int}} \subseteq Q \times Q.$
    - $\delta_{\mathsf{cll}} \subseteq Q \times (Q \times \Gamma).$
    - $\delta_{\mathsf{rtn}} \subseteq (Q \times \Gamma) \times Q.$
    - $\delta_{\mathsf{acq}} \subseteq Q \times (Q \times \mathsf{Lcks}).$
    - $\delta_{\mathsf{rel}} \subseteq (Q \times \mathsf{Lcks}) \times Q.$

For a PDS $\mathcal{P}$, the semantics is defined as a transition system. The configuration of a PDS $\mathcal{P}$ is the product of the set of control states Q and the stack which is modeled as word over the stack alphabet Γ. For a thread $\mathcal{P}$ using Lcks, we have to keep track of the locks being held by $\mathcal{P}$. Thus the set of configurations of $\mathcal{P}$ using Lcks is $\mathsf{Conf}_{\mathcal{P}} = Q \times \Gamma^* \times 2^{\mathsf{Lcks}}$ where 2^{Lcks} is the powerset of Lcks.

Furthermore, the transition relation is no longer just a relation between configurations but a binary relation on $2^{\mathsf{Lcks}} \times \mathsf{Conf}_{\mathcal{P}}$ since the thread now *executes* in an *environment*, namely, the free locks (i.e., locks not being held by any other thread). Formally,

Definition 2. *A PDS* $\mathcal{P} = (Q, \Gamma, qs, \delta)$ *using* Lcks *gives a labeled transition relation* $\longrightarrow_{\mathcal{P}} \subseteq (2^{\mathsf{Lcks}} \times (Q \times \Gamma^* \times 2^{\mathsf{Lcks}})) \times \mathsf{Labels} \times (2^{\mathsf{Lcks}} \times (Q \times \Gamma^* \times 2^{\mathsf{Lcks}}))$ *where* $\mathsf{Labels} = \{\mathsf{int}, \mathsf{cll}, \mathsf{rtn}\} \cup \{\mathsf{acq}(l), \mathsf{rel}(l) \mid l \in \mathsf{Lcks}\}$ *and* $\longrightarrow_{\mathcal{P}}$ *is defined as follows.*

- $\mathsf{fr} : (q, w, \mathsf{hld}) \xrightarrow{\mathsf{int}}_{\mathcal{P}} \mathsf{fr} : (q', w, \mathsf{hld})$ *if* $(q, q') \in \delta_{\mathsf{int}}.$
- $\mathsf{fr} : (q, w, \mathsf{hld}) \xrightarrow{\mathsf{cll}}_{\mathcal{P}} \mathsf{fr} : (q', wa, \mathsf{hld})$ *if* $(q, (q', a)) \in \delta_{\mathsf{cll}}.$
- $\mathsf{fr} : (q, wa, \mathsf{hld}) \xrightarrow{\mathsf{rtn}}_{\mathcal{P}} \mathsf{fr} : (q', w, \mathsf{hld})$ *if* $((q, a), q') \in \delta_{\mathsf{rtn}}.$
- $\mathsf{fr} : (q, w, \mathsf{hld}) \xrightarrow{\mathsf{acq}(l)}_{\mathcal{P}} \mathsf{fr} \setminus \{l\} : (q', w, \mathsf{hld} \cup \{l\})$ *if* $(q, (q', l)) \in \delta_{\mathsf{acq}}$ *and* $l \in \mathsf{fr}.$
- $\mathsf{fr} : (q, w, \mathsf{hld}) \xrightarrow{\mathsf{rel}(l)}_{\mathcal{P}} \mathsf{fr} \cup \{l\} : (q', w, \mathsf{hld} \setminus \{l\})$ *if* $((q, l), q') \in \delta_{\mathsf{rel}}$ *and* $l \in \mathsf{hld}.$

2.1   Multi-pushdown Systems

Concurrent programs are modeled as multi-pushdown systems. For our paper, we assume that threads in a concurrent program communicate only through locks which leads us to the following definition.

Definition 3. *Given a finite set* Lcks, *a n-pushdown system (n-PDS)* CP *communicating via* Lcks *is a tuple* $(\mathcal{P}_1, \ldots, \mathcal{P}_n)$ *where each* $\mathcal{P}_i$ *is a PDS using* Lcks.

Given a n-PDS CP, we will assume that the set of control states and the stack symbols of the threads are mutually disjoint.

Definition 4. *The semantics of a n-PDS* $CP = (\mathcal{P}_1, \ldots, \mathcal{P}_n)$ *communicating via* Lcks *is given as a labeled transition system* $T = (S, s_0, \longrightarrow)$ *where*

- *S is said to be the set of configurations of* CP *and is the set* $(Q_1 \times \Gamma_1^* \times 2^{\mathsf{Lcks}}) \times \cdots \times (Q_n \times \Gamma_n^* \times 2^{\mathsf{Lcks}})$ *where Q_i is the set of states of $\mathcal{P}_i$ and Γ_i is the stack alphabet of $\mathcal{P}_i$.*
- *s_0 is the initial configuration and is* $((qs_1, \epsilon, \emptyset), \cdots, (qs_m, \epsilon, \emptyset))$ *where qs_i is the initial state of $\mathcal{P}_i$.*
- *The set of labels on the transitions is* Labels $\times \{1, \ldots, n\}$ *where* Labels $=$ $\{\mathsf{int}, \mathsf{cll}, \mathsf{rtn}\} \cup \{\mathsf{acq}(l), \mathsf{rel}(l) \mid l \in \mathsf{Lcks}\}$. *The labeled transition relation* $\xrightarrow{(\lambda, i)}$ *is defined as follows*

$$((q_1, w_1, \mathsf{hld}_1), \cdots (q_n, w_n, \mathsf{hld}_n)) \xrightarrow{(\lambda, i)} ((q_1', w_1', \mathsf{hld}_1'), \cdots (q_n', w_n', \mathsf{hld}_n'))$$

iff

$$\mathsf{Lcks} \setminus \cup_{1 \leq r \leq n} \mathsf{hld}_r : (q_i, w_i, \mathsf{hld}_i) \xrightarrow{\lambda}_{\mathcal{P}_i} \mathsf{Lcks} \setminus \cup_{1 \leq r \leq n} \mathsf{hld}_r' : (q_i', w_i', \mathsf{hld}_i')$$

and for all $j \neq i$, $q_j = q_j'$, $w_j = w_j'$ and $\mathsf{hld}_j = \mathsf{hld}_j'$.

Notation: Given a configuration $s = ((q_1, w_1, \mathsf{hld}_1), \cdots, (q_n, w_n, \mathsf{hld}_n))$ of a n-PDS CP, we say that $\mathsf{Conf}_i(s) = (q_i, w_i, \mathsf{hld}_i), \mathsf{CntrlSt}_i(s) = q_i, \mathsf{Stck}_i(s) = w_i, \mathsf{LckSt}_i(s) = \mathsf{hld}_i$ and $\mathsf{StHt}_i(s) = |w_i|$, the length of w_i.

Computations. A *computation* of the n-PDS CP, Comp, is a sequence $s_0 \xrightarrow{(\lambda_1, i_1)} \cdots \xrightarrow{(\lambda_m, i_m)} s_m$ such that s_0 is the initial configuration of CP. The *label of the computation* Comp, denoted Label(Comp), is said to be the word $(\lambda_1, i_1) \cdots (\lambda_m, i_m)$. The transition $s_j \xrightarrow{(\mathsf{cll}, i)} s_{j+1}$ is said to be a *procedure call by thread i*. Similarly, we can define *procedure return, internal action, acquisition of lock l* and *release of lock l by thread i*. A procedure return $s_j \xrightarrow{(\mathsf{rtn}, i)} s_{j+1}$ is said to *match* a procedure call $s_\ell \xrightarrow{(\mathsf{cll}, i)} s_{\ell+1}$ iff $\ell < j$, $\mathsf{StHt}_i(s_\ell) = \mathsf{StHt}_i(s_{j+1})$ and for all $\ell + 1 \leq p \leq j$, $\mathsf{StHt}_i(s_{\ell+1}) \leq \mathsf{StHt}_i(s_p)$.

Example 1. Consider the two-threaded program showed in Figure 2. For sake of convenience, we only show the relevant actions of the programs. Figure 3 shows computations whose labels are as follows:

$$\mathsf{Label}(\mathsf{Comp1}) = (\mathsf{cll},0)(\mathsf{acq}(l1),0)(\mathsf{cll},1)(\mathsf{acq}(l2),0)(\mathsf{rel}(l1),0)(\mathsf{acq}(l1),1)$$
$$(\mathsf{rel}(l2),0)(\mathsf{rtn},0)(\mathsf{rel}(l1),1)(\mathsf{rtn},1)$$

and

$$\mathsf{Label}(\mathsf{Comp2}) = (\mathsf{cll},0)(\mathsf{acq}(l1),0)(\mathsf{cll},1)(\mathsf{acq}(l2),0)(\mathsf{rel}(l1),0)(\mathsf{acq}(l1),1)$$
$$(\mathsf{rel}(l1),1)(\mathsf{rtn},1)(\mathsf{rel}(l2),0)(\mathsf{rtn},0).$$

respectively.

```
int a(){
    acq l1;
    acq l2;
    if (..) then{
       ....
       rel l2;
       ....                    int b(){
       rel l1;                     acq l1;
       };                          rel l1;
    else{                          return j;
       .....                   };
       rel l1
       .....                   public void P1() {
       rel l2                      l=a();
       };                      }
    return i;
};

public void P0() {
  n=a();
}
```

Fig. 2. A 2-threaded programs with threads P0 and P1

2.2   Contextual Locking

In this paper, we are considering the pairwise reachability problem when the threads follow the discipline of *contextual locking*. Informally, this means that –

- every lock acquired by a thread in a procedure call must be released before the corresponding return is executed, and
- the locks held by a thread just before a procedure call is executed are not released during the execution of the procedure.

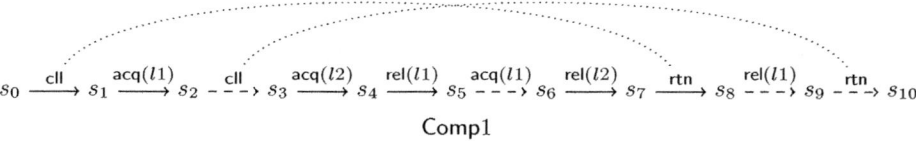

Fig. 3. Computations Comp1 and Comp2. Transitions of $P0$ are shown as solid edges while transition of $P1$ are shown as dashed edges; hence the process ids are dropped from the label of transitions. Matching calls and returns are shown with dotted edges.

Formally,

Definition 5. *A thread i in a n-PDS $CP = (\mathcal{P}_1, \ldots, \mathcal{P}_n)$ is said to follow contextual locking if whenever $s_\ell \xrightarrow{(\text{cll},i)} s_{\ell+1}$ and $s_j \xrightarrow{(\text{rtn},i)} s_{j+1}$ are matching procedure call and return along a computation $s_0 \xrightarrow{(\lambda_1,i)} s_1 \cdots \xrightarrow{(\lambda_m,i)} s_m$, we have that*

$$\text{LckSt}_i(s_\ell) = \text{LckSt}_i(s_j) \ \text{ and } \ \text{for all } \ell \le r \le j. \ \text{LckSt}_i(s_\ell) \subseteq \text{LckSt}_i(s_r).$$

Example 2. Consider the 2-threaded program shown in Figure 2. Both the threads P0 and P1 follow contextual locking. The program P2 in Figure 4 does not follow contextual locking.

```
int a(){
    acq l1;
    rel l2;
    return i;
};
public void P2(){
acq l2;
n=a();
rel l1;
}
```

Fig. 4. A program that does not follow contextual locking

Example 3. Consider the 2-threaded program in Figure 5. The two threads P3 and P4 follow contextual locking as there is no recursion! However, the two

```
public void P3(){                    public void P4(){
  acq 11;                              acq 13;
  while (true){                        while (true){
    acq 12;                              acq 11;
    rel 11;                              rel 13;
    acq 13;                              acq 12;
    rel 12;                              rel 11;
    acq 11;                              acq 13;
    rel 13;                              rel 12;
  }                                    }
}                                    }
```

Fig. 5. A 2-threaded program with unbounded lock chains

threads do not follow either the discipline of nested locking [5] or of bounded lock chaining [2]. Hence, algorithms of [5,2] cannot be used to decide the pairwise reachability question for this program. Notice that the computations of this pair of threads require an unbounded number of context switches as the two threads proceed in lock-step fashion. The locking pattern exhibited by these threads can present in any program with contextual locking as long as this pattern is within a single calling context (and not across calling contexts). Such locking patterns when used in a non-contextual fashion form the crux of undecidability proofs for multi-threaded programs synchronizing with locks [5].

3   Pairwise Reachability

The pairwise reachability problem for a multi-threaded program asks whether two given states in two threads can be simultaneously reached in an execution of the program. Formally,

> Given a n-PDS $\mathcal{CP} = (\mathcal{P}_1, \ldots, \mathcal{P}_n)$ communicating via Lcks, indices $1 \leq i, j \leq n$ with $i \neq j$, and control states q_i and q_j of threads $\mathcal{P}_i$ and $\mathcal{P}_j$ respectively, let $Reach(\mathcal{CP}, q_i, q_j)$ denote the predicate that there is a computation $s_0 \longrightarrow \cdots \longrightarrow s_m$ of $\mathcal{CP}$ such that $\mathsf{CntrlSt}_i(s_m) = q_i$ and $\mathsf{CntrlSt}_j(s_m) = q_j$. The pairwise control state reachability problem asks if $Reach(\mathcal{CP}, q_i, q_j)$ is true.

The pairwise reachability problem for multi-threaded programs communicating via locks was first studied in [9], where it was shown to be undecidable. Later, Kahlon et. al. [5] showed that when the locking pattern is restricted the pairwise reachability problem is decidable. In this paper, we will show that the problem is decidable for multi-threaded programs in which each thread follows contextual locking. Before we show this result, note that it suffices to consider programs with only two threads [5].

Proposition 1. *Given a n-PDS $CP = (\mathcal{P}_1, \ldots, \mathcal{P}_n)$ communicating via Lcks, indices $1 \le i, j \le n$ with $i \ne j$ and control states q_i and q_j of $\mathcal{P}_i$ and $\mathcal{P}_j$ respectively, let $CP_{i,j}$ be the 2-PDS $(\mathcal{P}_i, \mathcal{P}_j)$ communicating via Lcks. Then $Reach(CP, q_i, q_j)$ iff $Reach(CP_{i,j}, q_i, q_j)$.*

Thus, for the rest of the section, we will only consider 2-PDS.

3.1 Well-Bracketed Computations

The key concept in the proof of decidability is the concept of well-bracketed computations, defined below.

Definition 6. *Let $CP = (\mathcal{P}_0, \mathcal{P}_1)$ be a 2-PDS via Lcks and let $\mathsf{Comp} = s_0 \xrightarrow{(\lambda_1, i_1)} \cdots \xrightarrow{(\lambda_m, i_m)} s_m$ be a computation of CP. Comp is said to be non-well-bracketed if there exist $0 \le \ell_1 < \ell_2 < \ell_3 < m$ and $i \in \{0, 1\}$ such that*

- *$s_{\ell_1} \xrightarrow{(\mathsf{cll}, i)} s_{\ell_1 + 1}$ and $s_{\ell_3} \xrightarrow{(\mathsf{retn}, i)} s_{\ell_3 + 1}$ are matching call and returns of $\mathcal{P}_i$, and*
- *$s_{\ell_2} \xrightarrow{(\mathsf{cll}, i)} s_{\ell_2 + 1}$ is a procedure call of thread $\mathcal{P}_{1-i}$ whose matching return either occurs after $\ell_3 + 1$ or does not occur at all.*

Furthermore, the triple (ℓ_1, ℓ_2, ℓ_3) is said to be a witness of non-well-bracketing of Comp.

   Comp is said to be well-bracketed if it is not non-well-bracketed.

Example 4. Recall the 2-threaded program from Example 1 shown in Figure 2. The computation Comp1 (see Figure 3) is non-well-bracketed, while the computation Comp2 (see Figure 3) is well-bracketed. On the other hand, all the computations of the 2-threaded program in Example 3 (see Figure 5) are well-bracketed as the two threads are non-recursive.

The importance of well-bracketing for contextual locking is that if there is a computation that simultaneously reaches control states $p \in \mathcal{P}_0$ and $q \in \mathcal{P}_1$ then there is a well-bracketed computation that simultaneously reaches p and q.

Lemma 1. *Let $CP = (\mathcal{P}_0, \mathcal{P}_1)$ be a 2-PDS communicating via Lcks such that each thread follows contextual locking. Given control states $p \in \mathcal{P}_0$ and $q \in \mathcal{P}_1$, we have that $Reach(CP, p, q)$ iff there is a well-bracketed computation $s_0^{wb} \longrightarrow \cdots \longrightarrow s_r^{wb}$ of CP such that $\mathsf{CntrlSt}_0(s_r^{wb}) = p$ and $\mathsf{CntrlSt}_1(s_r^{wb}) = q$.*

Proof. Let $\mathsf{Comp}_{nwb} = s_0 \xrightarrow{(\lambda_1, i_1)} \cdots \xrightarrow{(\lambda_m, i_m)} s_m$ be a non-well-bracketed computation that simultaneously reaches p and q. Let ℓ_{mn} be smallest ℓ_1 such that there is a witness (ℓ_1, ℓ_2, ℓ_3) of non-well-bracketing of Comp_{nwb}. Observe now that it suffices to show that there is another computation Comp_{mod} of the same length as Comp_{nwb} that simultaneously reaches p and q and

- either Comp_{mod} is well-bracketed,
- or if Comp_{mod} is non-well-bracketed, then for each witness $(\ell'_1, \ell'_2, \ell'_3)$ of non-well-bracketing of Comp_{mod}, it must be the case $\ell'_1 > \ell_{\mathsf{mn}}$.

We show how to construct Comp_{mod}. Observe first that any witness $(\ell_{\mathsf{mn}}, \ell_2, \ell_3)$ of non-well-bracketing of Comp_{nwb} must necessarily agree in the third component ℓ_3. Let ℓ_{rt} denote this component. Let ℓ_{sm} be the smallest ℓ_2 such that $(\ell_{\mathsf{mn}}, \ell_2, \ell_{\mathsf{rt}})$ is a witness of non-well-bracketing of Comp_{mod}. Thus, the transition $s_{\ell_{\mathsf{mn}}} \longrightarrow s_{\ell_{\mathsf{mn}}+1}$ and $s_{\ell_{\mathsf{rt}}} \longrightarrow s_{\ell_{\mathsf{rt}}+1}$ are matching procedure call and return of some thread $\mathcal{P}_r$ while the transition $s_{\ell_{\mathsf{sm}}} \longrightarrow s_{\ell_{\mathsf{sm}}+1}$ is a procedure call by thread $\mathcal{P}_{1-r}$ whose return happens only after ℓ_{rt}. Without loss of generality, we can assume that $r = 0$.

Let $u, (\mathsf{cll}, 0), v_1, (\mathsf{cll}, 1), v_2, (\mathsf{rtn}, 0)$ and w be such that $\mathsf{Label}(\mathsf{Comp}_{nwb}) = u(\mathsf{cll}, 0)v_1(\mathsf{cll}, 1)v_2(\mathsf{rtn}, 0)w$. and length of u is $\ell_{\mathsf{mn}} + 1$, of $u(\mathsf{cll}, 0)v_1$ is $\ell_{\mathsf{sm}} + 1$. and of $u(\mathsf{cll}, 0)v_1(\mathsf{cll}, 1)v_2$ is $\ell_{\mathsf{rt}} + 1$. Thus, $(\mathsf{cll}, 0)$ and $(\mathsf{rtn}, 0)$ are matching call and return of thread $\mathcal{P}_0$ and $(\mathsf{cll}, 1)$ is a call of the thread $\mathcal{P}_1$ whose return does not happen in v_2.

We construct Comp_{mod} as follows. Intuitively, Comp_{mod} is obtained by "rearranging" the sequence $\mathsf{Label}(\mathsf{Comp}_{nwb}) = u(\mathsf{cll}, 0)v_1(\mathsf{cll}, 1)v_2(\mathsf{rtn}, 0)w$ as follows. Let $v_2|0$ and $v_2|1$ denote all the "actions" of thread $\mathcal{P}_0$ and $\mathcal{P}_1$ respectively in v_2. Then Comp_{mod} is obtained by rearranging $u(\mathsf{cll}, 0)v_1(\mathsf{cll}, 1)v_2(\mathsf{rtn}, 0)w$ to $u(\mathsf{cll}, 0)v_1(v_2|0)(\mathsf{rtn}, 0)(\mathsf{cll}, 1)(v_2|1)w$. This is shown in Figure 6.

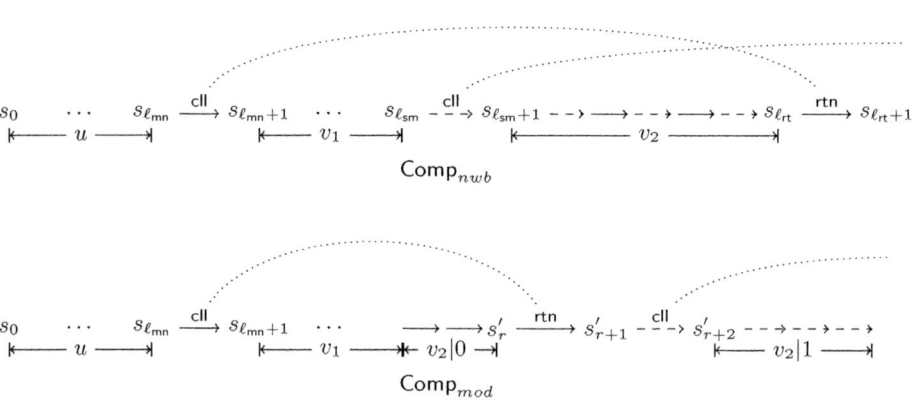

Fig. 6. Computations Comp_{nwb} and Comp_{mod}. Transitions of $\mathcal{P}_0$ are shown as solid edges and transitions of $\mathcal{P}_1$ are shown as dashed edges; hence process ids are dropped from the label of transitions. Matching calls and returns are shown with dotted edges. Observe that all calls of $\mathcal{P}_1$ in v_1 have matching returns within v_1.

The fact that if Comp_{mod} is non-well-bracketed, then there is no witness $(\ell'_1, \ell'_2, \ell'_3)$ of non-well-bracketing with $\ell'_1 \leq \ell_{\mathsf{mn}}$ will follow from the following observations on $\mathsf{Label}(\mathsf{Comp}_{nwb})$.

† v_1 cannot contain any returns of $\mathcal{P}_1$ which have a matching call that occurs in u (by construction of ℓ_{mn}).

†† All calls of $\mathcal{P}_1$ in v_1 must return either in v_1 or after c' is returned. But the latter is not possible (by construction of ℓ_{sm}). Thus, all calls of $\mathcal{P}_1$ in v_1 must return in v_1.

Formally, Comp_{mod} is constructed as follows. We fix some notation. For each $0 \leq j \leq m$, let $\mathsf{Conf}_0^j = \mathsf{Conf}_0(s_j)$ and $\mathsf{Conf}_1^j = \mathsf{Conf}_1(s_j)$. Thus $s_j = (\mathsf{Conf}_0^j, \mathsf{Conf}_1^j)$.

1. The first $\ell_{sm} + 1$ transitions of Comp_{mod} are the same as Comp_{nwb}, i.e., initially $\mathsf{Comp}_{mod} = s_0 \longrightarrow \cdots \longrightarrow s_{\ell_{sm}}$.

2. Consider the sequence of transitions $s_{\ell_{sm}} \xrightarrow{(\lambda_{sm+1}, i_{sm+1})} \cdots \xrightarrow{(\lambda_{rt+1}, i_{rt+1})} s_{\ell_{rt+1}}$ in Comp. Let k be the number of transitions of $\mathcal{P}_0$ in this sequence and let $\ell_{sm} \leq j_1 < \cdots < j_k \leq \ell_{rt}$ be the indices such that $s_{j_n} \xrightarrow{(\lambda_{j_n+1}, 0)} s_{j_n+1}$. Note that it must be the case that for each $1 \leq n \leq k$

$$\mathsf{Conf}_0^{\ell_{sm}} = \mathsf{Conf}_0^{j_1}, \quad \mathsf{Conf}_0^{j_n+1} = \mathsf{Conf}_0^{j_{n+1}} \quad \text{and} \quad \mathsf{Conf}_0^{j_k+1} = \mathsf{Conf}_0^{rt+1}.$$

For $1 \leq n \leq k$, let

$$s'_{\ell_{sm}+n} = (\mathsf{Conf}_0^{j_n+1}, \mathsf{Conf}_1^{\ell_{sm}}).$$

Observe now that, thanks to contextual locking, the set of locks held by $\mathcal{P}_1$ in $\mathsf{Conf}_1^{\ell_{sm}}$ is a subset of the locks held by $\mathcal{P}_1$ in $\mathsf{Conf}_1^{\ell_{j_n+1}}$ for each $1 \leq n \leq k$. Thus we can extend Comp_{mod} by applying the k transitions of $\mathcal{P}_0$ used to obtain $s_{j_n} \longrightarrow s_{j_n+1}$ in Comp_{nwb}. In other words, Comp_{mod} is now

$$s_0 \longrightarrow \cdots \longrightarrow s_{\ell_{sm}} \xrightarrow{(\lambda_{j_1+1}, 0)} s'_{\ell_{sm}+1} \cdots \xrightarrow{(\lambda_{j_k+1}, 0)} s'_{\ell_{sm}+k}.$$

Note that $s'_{\ell_{sm}+k} = (\mathsf{Conf}_0^{rt+1}, \mathsf{Conf}_1^{\ell_{sm}})$. Thus the set of locks held by $\mathcal{P}_0$ in $s'_{\ell_{sm}+k}$ is exactly the set of locks held by $\mathcal{P}_0$ at $\mathsf{Conf}_0^{\ell_{mn}}$.

3. Consider the sequence of transitions $s_{\ell_{sm}} \xrightarrow{(\lambda_{sm+1}, i_{sm+1})} \cdots \xrightarrow{(\lambda_{rt+1}, i_{rt+1})} s_{\ell_{rt+1}}$ in Comp. Let t be the number of transitions of $\mathcal{P}_1$ in this sequence and let $\ell_{sm} \leq j_1 < \cdots < j_t \leq \ell_{rt}$ be the indices such that $s_{j_n} \xrightarrow{(\lambda_{j_n+1}, 1)} s_{j_n+1}$. Note that it must be the case that for each $1 \leq n \leq t$,

$$\mathsf{Conf}_1^{j_1} = \mathsf{Conf}_1^{\ell_{sm}}, \quad \mathsf{Conf}_1^{j_n+1} = \mathsf{Conf}_1^{j_{n+1}} \quad \text{and} \quad \mathsf{Conf}_1^{j_t+1} = \mathsf{Conf}_1^{rt+1}.$$

For $1 \leq n \leq t$, let

$$s'_{\ell_{sm}+k+n} = (\mathsf{Conf}_0^{rt+1}, \mathsf{Conf}_1^{j_n+1}).$$

Observe now that, thanks to contextual locking, the set of locks held by $\mathcal{P}_0$ in $\mathsf{Conf}_0^{\ell_{rt}+1}$ is exactly the set of locks held by $\mathcal{P}_0$ at $\mathsf{Conf}_0^{\ell_{mn}}$ and the latter is a subset of the locks held by $\mathcal{P}_0$ in $\mathsf{Conf}_1^{\ell_{j_n+1}}$ for each $1 \leq n \leq t$. Thus we can extend Comp_{mod} by applying the t transitions of $\mathcal{P}_1$ used to obtain $s_{j_n} \longrightarrow s_{j_n+1}$ in Comp_{nwb}. In other words, Comp_{mod} is now

$$s_0 \longrightarrow \cdots \longrightarrow s'_{\ell_{sm}+k} \xrightarrow{(\lambda_{j_1+1}, 1)} s'_{\ell_{sm}+k+1} \cdots \xrightarrow{(\lambda_{j_t+1}, 1)} s'_{\ell_{sm}+k+t}.$$

Observe now that the extended Comp_{mod} is a sequence of $\mathsf{rt}+1$ transitions and that the final configuration of Comp_{mod}, $s'_{\ell_{sm}+k} \overset{(\lambda_{j_1+1},1)}{=} (\mathsf{Conf}_0^{\mathsf{rt}+1}, \mathsf{Conf}_1^{\mathsf{rt}+1})$ is exactly the configuration $s_{\mathsf{rt}+1}$.

4. Thus, now we can extend Comp_{mod} as

$$s_0 \longrightarrow \cdots \longrightarrow s'_{\ell_{sm}+k+t} = s_{\mathsf{rt}+1} \overset{(\lambda_{\mathsf{rt}+2},i_{\mathsf{rt}+2})}{\longrightarrow} \cdots \overset{(\lambda_m,i_m)}{\longrightarrow} s_m.$$

Clearly Comp_{mod} has the same length as Comp_{nwb} and simultaneously reaches p and q.

The lemma follows.   □

3.2   Algorithm for Deciding the Pairwise Reachability

We are ready to show that the problem of checking pairwise reachability is decidable.

Theorem 1. *There is an algorithm that given a 2-threaded program $\mathcal{CP} = (\mathcal{P}_0, \mathcal{P}_1)$ communicating via Lcks and control states p and q of $\mathcal{P}_0$ and $\mathcal{P}_1$ respectively decides if $\mathsf{Reach}(\mathcal{P}, p, q)$ is true or not. Furthermore, if m and n are the sizes of the programs $\mathcal{P}_0$ and $\mathcal{P}_1$ and ℓ the number of elements of Lcks, then this algorithm has a running time of $2^{O(\ell)}O((mn)^3)$.*

Proof. The main idea behind the algorithm is to construct a single PDS $\mathcal{P}_{comb} = (Q, \Gamma, qs, \delta)$ which simulates all the well-bracketed computations of $\mathcal{CP}$. $\mathcal{P}_{comb}$ simulates a well-bracketed computation as follows. The set of control states of $\mathcal{P}_{comb}$ is the product of control states of $\mathcal{P}_0$ and $\mathcal{P}_1$. The single stack of $\mathcal{P}_{comb}$ keeps track of the stacks of $\mathcal{P}_0$ and $\mathcal{P}_1$: it is the sequence of those calls of the well-bracketed computation which have not been returned. Furthermore, if the stack of $\mathcal{P}_{comb}$ is w then the stack of $\mathcal{P}_0$ is the projection of w onto the stack symbols of $\mathcal{P}_0$ and the stack of $\mathcal{P}_1$ is the projection of w onto the stack symbols of $\mathcal{P}_1$. Thus, the top of the stack is the most recent unreturned call and if it belongs to $\mathcal{P}_i$, well-bracketing ensures that no previous unreturned call is returned without returning this call.

Formally, $\mathcal{P}_{comb} = (Q, \Gamma, qs, \delta)$ is defined as follows. Let $\mathcal{P}_0 = (Q_0, \Gamma_0, qs_0, \delta_0)$ and $\mathcal{P}_1 = (Q_1, \Gamma_1, qs_1, \delta_1)$. Without loss of generality, assume that $Q_0 \cap Q_1 = \emptyset$ and $\Gamma_0 \cap \Gamma_1 = \emptyset$.

- The set of states Q is $(Q_0 \times 2^{\mathsf{Lcks}}) \times (Q_1 \times 2^{\mathsf{Lcks}})$.
- $\Gamma = \Gamma_0 \cup \Gamma_1$.
- $qs = ((qs_0, \emptyset), (qs_1, \emptyset))$.
- δ consists of three sets $\delta_{\mathsf{int}}, \delta_{\mathsf{cll}}$ and δ_{rtn} which simulate the internal actions, procedure calls, and returns and lock acquisitions and releases of the threads as follows. We explain here only the simulation of actions of $\mathcal{P}_0$ (the simulation of actions of $\mathcal{P}_1$ is similar).

- *Internal actions.* If (q_0, q_0') is an internal action of $\mathcal{P}_0$, then for each $\mathsf{hld}_0, \mathsf{hld}_1 \in 2^{\mathsf{Lcks}}$ and $q_1 \in Q_1$

$$(((q_0, \mathsf{hld}_0), (q_1, \mathsf{hld}_1)), ((q_0', \mathsf{hld}_0), (q_1, \mathsf{hld}_1))) \in \delta_{\mathsf{int}}.$$

- *Lock acquisitions.* Lock acquisitions are also modeled by δ_{int}. If $(q_0, (q_0', l))$ is a lock acquisition action of thread $\mathcal{P}_0$, then for each $\mathsf{hld}_0, \mathsf{hld}_1 \in 2^{\mathsf{Lcks}}$ and $q_1 \in Q_1$,

$$(((q_0, \mathsf{hld}_0), (q_1, \mathsf{hld}_1)), ((q_0', \mathsf{hld}_0 \cup \{l\}), (q_1, \mathsf{hld}_1))) \in \delta_{\mathsf{int}} \text{ if } l \notin \mathsf{hld}_0 \cup \mathsf{hld}_1.$$

- *Lock releases.* Lock releases are also modeled by δ_{int}. If $((q_0, l), q_0')$ is a lock release action of thread $\mathcal{P}_0$, then for each $\mathsf{hld}_0, \mathsf{hld}_1 \in 2^{\mathsf{Lcks}}$ and $q_1 \in Q_1$,

$$(((q_0, \mathsf{hld}_0), (q_1, \mathsf{hld}_1)), ((q_0', \mathsf{hld}_0 \setminus \{l\}), (q_1, \mathsf{hld}_1))) \in \delta_{\mathsf{int}} \text{ if } l \in \mathsf{hld}_0.$$

- *Procedure Calls.* Procedure calls are modeled by δ_{cll}. If $(q_0, (q_0', a))$ is a procedure call of thread $\mathcal{P}_0$ then $\mathsf{hld}_0, \mathsf{hld}_1 \in 2^{\mathsf{Lcks}}$ and $q_1 \in Q_1$,

$$(((q_0, \mathsf{hld}_0), (q_1, \mathsf{hld}_1)), (((q_0', \mathsf{hld}_0), (q_1, \mathsf{hld}_1)), a)) \in \delta_{\mathsf{cll}}.$$

- *Procedure Returns.* Procedure returns are modeled by δ_{rtn}. If $(q_0, (q_0', a))$ is a procedure call of thread $\mathcal{P}_0$ then $\mathsf{hld}_0, \mathsf{hld}_1 \in 2^{\mathsf{Lcks}}$ and $q_1 \in Q_1$,

$$((((q_0, \mathsf{hld}_0), (q_1, \mathsf{hld}_1)), a), ((q_0', \mathsf{hld}_0), (q_1, \mathsf{hld}_1))) \in \delta_{\mathsf{rtn}}.$$

It is easy to see that (p, q) is reachable in $\mathcal{CP}$ by a well-bracketed computation iff there is a computation of $\mathcal{P}_{comb}$ which reaches $((p, \mathsf{hld}_0), (q, \mathsf{hld}_1))$ for some $\mathsf{hld}_0, \mathsf{hld}_1 \in 2^{\mathsf{Lcks}}$. The complexity of the results follows from the observations in [1] and the size of $\mathcal{P}_{comb}$. $\square$

4   Conclusions

The paper investigates the problem of pairwise reachability of multi-threaded programs communicating using only locks. We identified a new restriction on locking patterns, called contextual locking, which requires threads to release locks in the same calling context in which they were acquired. Contextual locking appears to be a natural restriction adhered to by many programs in practice. The main result of the paper is that the problem of pairwise reachability is decidable in polynomial time for programs in which the locking scheme is contextual. Therefore, in addition to being a natural restriction to follow, contextual locking may also be more amenable to practical analysis. We observe that these results do not follow from results in [5,4,2,3] as there are programs with contextual locking that do not adhere to the nested locking principle or the bounded lock chaining principle. The proof principles underlying the decidability results are also different. Our results can also be mildly extended to handling programs that release locks a bounded stack-depth away from when they were acquired (for example, to handle procedures that call a function that acquires a lock, and calls another to release it before it returns).

There are a few open problems immediately motivated by the results in this paper. First, decidability of model checking with respect to fragments of LTL under the contextual locking restriction remains open. Next, while our paper establishes the decidability of pairwise reachability, it is open if the problem of checking if 3 (or more) threads simultaneously reach given local states is decidable for programs with contextual locking. Finally, from a practical standpoint, one would like to develop analysis algorithms that avoid to construct the cross-product of the two programs to check pairwise reachability.

For a more complete account for multi-threaded programs, other synchronization primitives such as thread creation and barriers should be taken into account. Combining lock-based approaches such as ours with techniques for other primitives is left to future investigation.

Acknowledgements. P. Madhusudan was supported in part by NSF Career Award 0747041. Mahesh Viswanathan was supported in part by NSF CNS 1016791 and NSF CCF 1016989.

References

1. Bouajjani, A., Esparza, J., Maler, O.: Reachability Analysis of Pushdown Automata: Application to Model-Checking. In: Mazurkiewicz, A., Winkowski, J. (eds.) CONCUR 1997. LNCS, vol. 1243, pp. 135–150. Springer, Heidelberg (1997)
2. Kahlon, V.: Boundedness vs. unboundedness of lock chains: Characterizing decidability of pairwise CFL-Reachability for threads communicating via locks. In: Proceedings of the IEEE Symposium on Logic in Computer Science, pp. 27–36 (2009)
3. Kahlon, V.: Reasoning about Threads with Bounded Lock Chains. In: Katoen, J.-P., König, B. (eds.) CONCUR 2011 – Concurrency Theory. LNCS, vol. 6901, pp. 450–465. Springer, Heidelberg (2011)
4. Kahlon, V., Gupta, A.: An automata-theoretic approach for model checking threads for LTL properties. In: Proceedings of the IEEE Symposium on Logic in Computer Science, pp. 101–110 (2006)
5. Kahlon, V., Ivančić, F., Gupta, A.: Reasoning About Threads Communicating via Locks. In: Etessami, K., Rajamani, S.K. (eds.) CAV 2005. LNCS, vol. 3576, pp. 505–518. Springer, Heidelberg (2005)
6. Lea, D.: Concurrent Programming in Java: Design Principles and Patterns. Addison-Wesley (1999)
7. Muchnick, S.S.: Advanced compiler design and implementation. Morgan Kaufmann Publishers Inc. (1997)
8. Owicki, S.S., Gries, D.: An axiomatic proof technique for parallel programs i. Acta Informatica 6, 319–340 (1976)
9. Ramalingam, G.: Context-sensitive synchronization-sensitive analysis is undecidable. ACM Transactions on Programming Languages and Systems 22(2), 416–430 (2000)
10. Reps, T.W., Horwitz, S., Sagiv, S.: Precise interprocedural dataflow analysis via graph reachability. In: Proceedings of the ACM Symposium on the Principles of Programming Languages, pp. 49–61 (1995)

Bounded Phase Analysis of Message-Passing Programs*,**

Ahmed Bouajjani and Michael Emmi***

LIAFA, Université Paris Diderot, France
{abou,mje}@liafa.jussieu.fr

Abstract. We describe a novel technique for bounded analysis of asynchronous message-passing programs with ordered message queues. Our bounding parameter does not limit the number of pending messages, nor the number of "contexts-switches" between processes. Instead, we limit the number of process communication cycles, in which an unbounded number of messages are sent to an unbounded number of processes across an unbounded number of contexts. We show that remarkably, despite the potential for such vast exploration, our bounding scheme gives rise to a simple and efficient program analysis by reduction to sequential programs. As our reduction avoids explicitly representing message queues, our analysis scales irrespectively of queue content and variation.

1 Introduction

Software is becoming increasingly concurrent: reactivity (e.g., in user interfaces, web servers), parallelization (e.g., in scientific computations), and decentralization (e.g., in web applications) necessitate asynchronous computation. Although shared-memory implementations are often possible, the burden of preventing unwanted thread interleavings without crippling performance is onerous. Many have instead adopted asynchronous programming models in which processes communicate by posting messages/tasks to others' message/task queues— [19] discuss why such models provide good programming abstractions. Single-process systems such as the JavaScript page-loading engine of modern web browsers [1], and the highly-scalable Node.js asynchronous web server [11], execute a series of short-lived tasks one-by-one, each task potentially queueing additional tasks to be executed later. This programming style ensures that the overall system responds quickly to incoming events (e.g., user input, connection requests). In the multi-process setting, languages such as Erlang and Scala have adopted message-passing as a fundamental construct with which highly-scalable and highly-reliable distributed systems are built.

Despite the increasing popularity of such programming models, little is known about precise algorithmic reasoning. This is perhaps not without good reason:

---

* Partially supported by the project ANR-09-SEGI-016 Veridyc.
** An extended version of this paper is online [6]
*** Supported by a Fondation Sciences Mathématiques de Paris post-doctoral fellowship.

C. Flanagan and B. König (Eds.): TACAS 2012, LNCS 7214, pp. 451–465, 2012.
© Springer-Verlag Berlin Heidelberg 2012

decision problems such as state-reachability for programs communicating with unbounded reliable queues are undecidable [10], even when there is only a single finite-state process (posting messages to itself). Furthermore, the known decidable under-approximations (e.g., bounding the size of queues) represent queues explicitly, are thus doomed to combinatorial explosion as the size and variability of queue content increases.

Some have proposed analyses which abstract message arrival order [23, 14, 13], or assume messages can be arbitrarily lost [2, 3]. Such analyses do not suffice when correctness arguments rely on reliable messaging—several systems specifically do ensure the ordered delivery of messages, including Scala, and recent web-browser specifications [1]. Others have proposed analyses which compute finite symbolic representations of queue contents [5, 8]. Known bounded analyses which model queues precisely either bound the maximum capacity of message-queues, ignoring executions which exceed the bound, or bound the total number of process "contexts" [21, 16], where each context involves a single process sending and receiving messages. For each of these bounding schemes there are trivial systems which cannot be adequately explored, e.g., by sending more messages than the allowed queue-capacity, having more processes than contexts, or by alternating message-sends to two processes—we discuss such examples in Section 3. All of the above techniques represent queues explicitly, though perhaps symbolically, and face combinatorial explosion as queue content and variation increase.

In this work we propose a novel technique for bounded analysis of asynchronous message-passing programs with reliable, ordered message queues. Our bounding parameter, introduced in Section 3, is not sensitive to the capacity nor content of message queues, nor the number of process contexts. Instead, we bound the number of process communication cycles by labeling each message with a monotonically-increasing phase number. Each time a message chain visits the same process, the phase number must increase. For a given parameter k, we only explore behaviors of up to k phases—though k phases can go a long way. In the leader election distributed protocol [24] for example, each election round occurs in 2 phases: in the first phase each process sends *capture* messages to the others; in the second phase some processes receive *accept* messages, and those that find themselves majority-winners broadcast *elected* messages. In these two phases an unbounded number of messages are sent to an unbounded number of processes across an unbounded number of process contexts!

We demonstrate the true strength of phase-bounding by showing in Sections 4 and 5 that the bounded phase executions of a message-passing program can be concisely encoded as a non-deterministic sequential program, in which message-queues are not explicitly represented. Our so-called "sequentialization" sheds hope for scalable analyses of message-passing programs. In a small set of simple experiments (Section 4), we demonstrate that our phase-bounded encoding scales far beyond known explicit-queue encodings as queue-content increases, and even remains competitive as queue-content is fixed while the number of phases grows. By reducing to sequential programs, we leverage highly-developed sequential program analysis tools for message-passing programs.

2   Asynchronous Message-Passing Programs

We consider a simple multi-processor programming model in which each processor is equipped with a procedure stack and a queue of pending tasks. Initially all processors are idle. When an idle processor's queue is non-empty, the oldest task in its queue is removed and executed to completion. Each task executes essentially a recursive sequential program, which besides accessing its own processor's global storage, can *post* tasks to the queues of any processor, including its own. When a task does complete, its processor again becomes idle, chooses the next pending task to execute to completion, and so on. The distinction between queues containing messages and queues containing tasks is mostly aesthetic, but in our task-based treatment queues are only read by idle processors; reading additional messages during a task's execution is prohibited. While in principle many message-passing systems, e.g., in Erlang and Scala, allow reading additional messages at any program point, we have observed that common practice is to read messages only upon completing a task [25].

Though similar to [23]'s model of asynchronous programs, the model we consider has two important distinctions. First, tasks execute across potentially several processors, rather than only one, each processor having its own global state and pending tasks. Second, the tasks of each processor are executed in exactly the order they are posted. For the case of single-processor programs, [23]'s model can be seen as an abstraction of the model we consider, since there the task chosen to execute next when a processor is idle is chosen non-deterministically among all pending tasks.

2.1   Program Syntax

Let Procs be a set of procedure names, Vals a set of values, Exprs a set of expressions, Pids a set of processor identifiers, and let T be a type. Figure 1 gives the grammar of *asynchronous message-passing programs*. We intentionally leave the syntax of expressions e unspecified, though we do insist Vals contains `true` and `false`, and Exprs contains Vals and the *(nullary) choice operator* $\star$.

Each program P declares a single global variable g and a procedure sequence, each $p \in$ Procs having a single parameter 1 and top-level statement denoted s_p; as statements are built inductively by composition with control-flow statements, s_p describes the entire body of p. The set of program statements s is denoted Stmts. Intuitively, a **post** ρ p e statement is an asynchronous call to a procedure p with argument e to be executed on the processor identified by ρ; a *self-post* to one's own processor is made by setting ρ to _. A program in which all **post** statements are self-posts is called a *single-processor program*, and a program without **post** statements is called a *sequential program*.

The programming language we consider is simple, yet very expressive, since the syntax of types and expressions is left free, and we lose no generality by considering only single global and local variables. Our extended report [6] lists several syntactic extensions which we use in the source-to-source translations of the subsequent sections, and which easily reduce to the syntax of our grammar.

$$P ::= \textbf{var } g\!:\!T \ (\textbf{proc } p \ (\textbf{var } 1\!:\!T) \ s)^*$$
$$s ::= s;\ s \ | \ \textbf{skip} \ | \ x := e$$
$$| \ \textbf{assume } e$$
$$| \ \textbf{if } e \textbf{ then } s \textbf{ else } s$$
$$| \ \textbf{while } e \textbf{ do } s$$
$$| \ \textbf{call } x := p \ e$$
$$| \ \textbf{return } e$$
$$| \ \textbf{post } \rho \ p \ e$$
$$x ::= g \ | \ 1$$

DISPATCH

$$\langle g, \varepsilon, f \cdot q \rangle \xrightarrow{\text{S}} \langle g, f, q \rangle$$

COMPLETE
$$f = \langle \ell, \textbf{return } e;\ s \rangle$$
$$\langle g, f, q \rangle \xrightarrow{\text{S}} \langle g, \varepsilon, q \rangle$$

SELF-POST
$$s_1 = \textbf{post } \_ \ p \ e;\ s_2$$
$$\ell_2 \in e(g, \ell_1) \qquad f = \langle \ell_2, s_p \rangle$$
$$\langle g, \langle \ell_1, s_1 \rangle \, w, q \rangle \xrightarrow{\text{S}} \langle g, \langle \ell_1, s_2 \rangle \, w, q \cdot f \rangle$$

Fig. 1. The grammar of asynchronous message-passing programs P. Here T is an unspecified type, and e, p, and ρ range, resp., over expressions, procedure names, and processor identifiers.

Fig. 2. The single-processor transition rules $\to^{\text{S}}$; see our extended report [6] for the standard sequential statements.

2.2   Single-Processor Semantics

A *(procedure) frame* $f = \langle \ell, s \rangle$ is a current valuation $\ell \in \text{Vals}$ to the procedure-local variable 1, along with a statement $s \in \text{Stmts}$ to be executed. (Here s describes the entire body of a procedure p that remains to be executed, and is initially set to p's top-level statement s_p; we refer to initial procedure frames $t = \langle \ell, s_p \rangle$ as *tasks*, to distinguish the frames that populate processor queues.) The set of all frames is denoted Frames.

A *processor configuration* $\kappa = \langle g, w, q \rangle$ is a current valuation $g \in \text{Vals}$ to the processor-global variable g, along with a procedure-frame stack $w \in \text{Frames}^*$ and a pending-tasks queue $q \in \text{Frames}^*$. A processor is idle when $w = \varepsilon$. The set of all processor configurations is denoted Pconfigs. A processor configuration map $\xi : \text{Pids} \to \text{Pconfigs}$ maps each processor $\rho \in \text{Pids}$ to a processor configuration $\xi(\rho)$. We write $\xi\,(\rho \mapsto \kappa)$ to denote the configuration ξ updated with the mapping $(\rho \mapsto \kappa)$, i.e., the configuration ξ' such that $\xi'(\rho) = \kappa$, and $\xi'(\rho') = \xi(\rho')$ for all $\rho' \in \text{Pids} \setminus \{\rho\}$.

For expressions without program variables, we assume the existence of an evaluation function $\llbracket \cdot \rrbracket_{\text{e}} : \text{Exprs} \to \wp(\text{Vals})$ such that $\llbracket \star \rrbracket_{\text{e}} = \text{Vals}$. For convenience we define $e(g, \ell) \stackrel{\text{def}}{=} \llbracket e[g/\text{g}, \ell/1] \rrbracket_{\text{e}}$ to evaluate the expression e in a global valuation g by substituting the current values for variables g and 1. As these are the only program variables, the substituted expression $e[g/\text{g}, \ell/1]$ has no free variables.

Figure 2 defines the transition relation $\to^{\text{S}}$ for the asynchronous behavior of each processor; the standard transitions for the sequential statements are listed in our extended report [6]. The SELF-POST rule creates a new frame to execute the given procedure, and places the new frame in the current processor's pending-tasks queue. The COMPLETE rule returns from the final frame of a task, rendering the processor idle, and the DISPATCH rule schedules the least-recently posted task on a idle processor.

SWITCH

$$\frac{\rho_2 \in \mathsf{enabled}(m, \xi)}{\langle \rho_1, \xi, m \rangle \xrightarrow{}_{M} \langle \rho_2, \xi, m \rangle}$$

STEP

$$\frac{\xi_1(\rho) \xrightarrow{S} \kappa \qquad \xi_2 = \xi_1 \, (\rho \mapsto \kappa)}{\rho \in \mathsf{enabled}(m_1, \xi_1) \qquad m_2 = \mathsf{step}(m_1, \xi_1, \xi_2)}}{\langle \rho, \xi_1, m_1 \rangle \xrightarrow{}_{M} \langle \rho, \xi_2, m_2 \rangle}$$

POST

$$\frac{\begin{array}{c} \xi_1(\rho_1) = \langle g_1, \langle \ell_1, \mathbf{post}\ \rho_2\ p\ e;\ s \rangle\ w_1, q_1 \rangle \\ \xi_1(\rho_2) = \langle g_2, w_2, q_2 \rangle \\ \rho_1 \neq \rho_2 \qquad \ell_2 \in e(g_1, \ell_1) \qquad f = \langle \ell_2, s_p \rangle \\ \rho_1 \in \mathsf{enabled}(m_1, \xi_1) \qquad m_2 = \mathsf{step}(m_1, \xi_1, \xi_3) \\ \xi_2 = \xi_1 \, (\rho_1 \mapsto \langle g_1, \langle \ell_1, s \rangle\ w_1, q_1 \rangle) \\ \xi_3 = \xi_2 \, (\rho_2 \mapsto \langle g_2, w_2, q_2 \cdot f \rangle) \end{array}}{\langle \rho_1, \xi_1, m_1 \rangle \xrightarrow{}_{M} \langle \rho_1, \xi_3, m_2 \rangle}$$

Fig. 3. The multi-processor transition relation $\rightarrow_M$ parameterized by a scheduler $M = \langle D, \mathsf{empty}, \mathsf{enabled}, \mathsf{step} \rangle$

```
// translation of var g: T
var G[k]: T

// translation of
// proc p (var l: T) s
proc p (var l: T, phase: k) s

// translation of g
G[phase]

// translation of call x := p e
call x := p (e,phase)

// translation of post _ p e
if phase+1 < k then
    call p (e,phase+1)
```

Fig. 4. The k-phase sequential translation $(\!(P)\!)_k$ of a single-processor asynchronous message-passing program P

2.3 Multi-processor Semantics

In reality the processors of multi-processor systems execute independently in parallel. However, as long as they either do not share memory, or access a sequentially consistent shared memory, it is equivalent, w.r.t. the observations of any single processor, to consider an *interleaving semantics*: at any moment only one processor executes. In order to later restrict processor interleaving, we make explicit the *scheduler* which arbitrates the possible interleavings. Formally, a *scheduler* $M = \langle D, \mathsf{empty}, \mathsf{enabled}, \mathsf{step} \rangle$ consists of a data type D of scheduler objects $m \in D$, a scheduler constructor $\mathsf{empty} \in D$, a scheduler decision function $\mathsf{enabled} : (D \times (\mathsf{Pids} \to \mathsf{Pconfigs})) \to \wp(\mathsf{Pids})$, and a scheduler update function $\mathsf{step} : (D \times (\mathsf{Pids} \to \mathsf{Pconfigs}) \times (\mathsf{Pids} \to \mathsf{Pconfigs})) \to D$. The arguments to $\mathsf{enabled}$ allow a scheduler to decide which processors are enabled depending on the execution history. A scheduler is *deterministic* when $|\mathsf{enabled}(m, \xi)| \leq 1$ for all $m \in D$ and $\xi : \mathsf{Pids} \to \mathsf{Pconfigs}$, and is *non-blocking* when for all m and ξ, if there is some $\rho \in \mathsf{Pids}$ such that $\xi(\rho)$ is either non-idle or has pending tasks, then there exists $\rho' \in \mathsf{Pids}$ such that $\rho' \in \mathsf{enabled}(m, \xi)$ and $\xi(\rho')$ is either non-idle or has pending tasks. A *configuration* $c = \langle \rho, \xi, m \rangle$ is a currently executing processor $\rho \in \mathsf{Pids}$, along with a processor configuration map ξ, and a scheduler object m.

Figure 3 defines the multi-processor transition relation $\rightarrow_M$, parameterized by a scheduler M. The SWITCH rule non-deterministically schedules any enabled processor, while the STEP rule executes one single-processor program step on the currently scheduled processor, and updates the scheduler object. Finally, the POST rule creates a new frame to execute the given procedure, and places the the new frame on the target processor's pending-tasks queue.

Until further notice, we assume M is a completely non-deterministic scheduler; i.e., all processors are always enabled. In Section 5 we discuss alternatives.

An *M-execution* of a program P *(from c_0 to c_j)* is a configuration sequence $c_0 c_1 \ldots c_j$ such that $c_i \to_M c_{i+1}$ for $0 \le i < j$. An *initial condition* $\iota = \langle \rho_0, g_0, \ell_0, p_0 \rangle$ is a processor identifier ρ_0, along with a global-variable valuation $g_0 \in \mathsf{Vals}$, a local-variable valuation $\ell_0 \in \mathsf{Vals}$, and a procedure $p_0 \in \mathsf{Procs}$. A configuration $c = \langle \rho_0, \xi, \mathsf{empty} \rangle$ of a program P is $\langle \rho_0, g_0, \ell_0, p_0 \rangle$-*initial* when $\xi(\rho_0) = \langle g_0, \varepsilon, \langle \ell_0, s_{p_0} \rangle \rangle$ and $\xi(\rho) = \langle g_0, \varepsilon, \varepsilon \rangle$ for all $\rho \ne \rho_0$. A configuration $\langle \rho, \xi, m \rangle$ is g_f-*final* when $\xi(\rho') = \langle g_f, w, q \rangle$ for some $\rho' \in \mathsf{Pids}$, and $w, q \in \mathsf{Frames}^*$. We say a global valuation g is M-*reachable* in P from ι when there exists an M-execution of P from some c_0 to some c such that c_0 is ι-initial and c is g-final[1].

Definition 1. *The* state-reachability problem *is to determine for an initial condition ι, valuation g, and program P, whether g is reachable in P from ι.*

3 Phase-Bounded Execution

Because processors execute tasks precisely in the order which they are posted to their unbounded task-queues, our state-reachability problem is undecidable, even with only a single processor accessing finite-state data [10]. Since it is not algorithmically possible to consider every execution precisely, in what follows we present an incremental under-approximation. For a given bounding parameter k, we consider a subset of execution (prefixes) precisely; as k increases, the set of considered executions increases, and in the limit as k approaches infinity, every execution of any program is considered—though for many programs, every execution is considered with a finite value of k.

In a given execution, a *task-chain* $t_1 t_2 \ldots t_i$ from t_1 to t_i is a sequence of tasks[2] such that the execution of each t_j posts t_{j+1}, for $0 < j < i$, and we say that t_1 is an *ancestor* of t_i. We characterize execution prefixes by labeling each task t posted in an execution with a *phase number* $\varphi(t) \in \mathbb{N}$:

$$\varphi(t) = \begin{cases} 0 & \text{if } t \text{ is initially pending.} \\ \varphi(t') & \begin{array}{l} \text{if } t \text{ is posted to processor } \rho \text{ by } t', \\ \text{and } t \text{ has no phase-}\varphi(t') \text{ ancestor on processor } \rho. \end{array} \\ \varphi(t') + 1 & \text{if } t \text{ is posted by } t', \text{ otherwise.} \end{cases}$$

For instance, considering Figure 5a, supposing all on a single processor, an initial task A_1 posts A_2, A_3, and A_4, then A_2 posts A_5 and A_6, and then A_3 posts A_7, which in turn posts A_8 and A_9. Task A_1 has phase 0. Since each post is made to the same processor, the phase number is incremented for each posted task. Thus the phase 1 tasks are $\{A_2, A_3, A_4\}$, the phase 2 tasks are $\{A_5, A_6, A_7\}$, and the phase 3 tasks are $\{A_8, A_9\}$. Notice that tasks of a given phase only execute

---

[1] In the presence of the **assume** statement, only the values reached in completed executions are guaranteed to be valid.

[2] We assume each task in a given execution has implicitly a unique task-identifier.

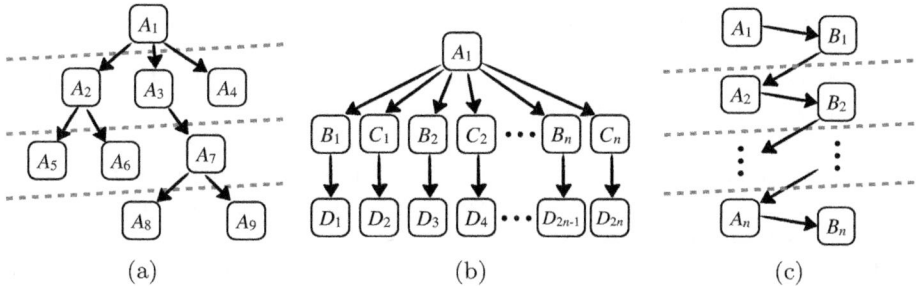

Fig. 5. Phase-bounded executions with processors A, B, C, and D; each task's label (e.g., A_i) indicates the processor it executes on (e.g., A). Arrows indicate the posting relation, indices indicate execution order on a given processor, and dotted lines indicate phase boundaries.

after all tasks of the previous phase have completed, i.e., execution order is in phase order; only executing tasks up to a given phase does correspond to a valid execution prefix.

Definition 2. *An execution is k-phase when $\varphi(t) < k$ for each executed task t.*

The execution in Figure 5a is a 4-phase execution, since all tasks have phase less than 4. Despite there being an arbitrary number $4n + 1$ of posted tasks, the execution in Figure 5b is 1-phase, since there are no task-chains between same-processor tasks. Contrarily, the execution in Figure 5c requires n phases to execute all $2n$ tasks, since every other occurrence of an A_i task creates a task-chain between A-tasks.

Note that bounding the number of execution phases does not necessarily bound the total number of tasks executed, nor the maximum size of task queues, nor the amount of switching between processors. Instead, a bound k restricts the maximum length of task chains to $k \cdot |\text{Pids}|$. In fact, phase-bounding is incomparable to bounding the maximum size of task queues. On the one hand, every execution of a program in which one root task posts an arbitrary, unbounded number of tasks to other processors (e.g., in Figure 5b) are explored with 1 phase, though no bound on the size of queues will capture all executions. On the other hand, all executions with a single arbitrarily-long chain of tasks (e.g., in Figure 5c) are explored with size 1 task queues, though no limited number of phases captures all executions. In the limit as the bounding parameter increases, either scheme does capture all executions.

Theorem 1 (Completeness). *For every execution h of a program P, there exists $k \in \mathbb{N}$ such that h is a k-phase execution.*

4   Phase-Bounding for Single-Processor Programs

Characterizing executions by their phase-bound reveals a simple and efficient technique for bounded exploration. This seems remarkable, given that phase-bounding explores executions in which arbitrarily many tasks execute, making the task queue arbitrarily large. The first key ingredient is that once the number of phases is bounded, each phase can be executed in isolation. For instance, consider again the execution of Figure 5a. In phase 1, the tasks A_2, A_3, and A_4 pick up execution from the global valuation g_1 which A_1 left off at, and leave behind a global valuation g_2 for the phase 2 tasks. In fact, given the sequence of tasks in each phase, the only other "communication" between phases is a single passed global valuation; executing that sequence of tasks on that global valuation is a faithful simulation of that phase.

The second key ingredient is that the ordered sequence of tasks executed in a given phase is exactly the ordered sequence of tasks posted in the previous phase. This is obvious, since tasks are executed in the order they are posted. However, combined with the first ingredient we have quite a powerful recipe. Supposing the global state g_i at the beginning of each phase i is known initially, we can simulate a k-phase execution by executing each task posted to phase i as soon as it is posted, with an independent virtual copy of the global state, initially set to g_i. That is, our simulation will store a vector of k global valuations, one for each phase. Initially, the ith global valuation is set to the state g_i in which phase i begins; tasks of phase i then read from and write to the ith global valuation. It then only remains to ensure that the global valuations g_i used at the beginning of each phase $0 < i < k$ match the valuations reached at the end of phase $i - 1$.

This simulation is easily encoded into a non-deterministic sequential program with k copies of global storage. The program begins by non-deterministically setting each copy to an arbitrary value. Each task maintains their current phase number i, and accesses the ith copy of global storage. Each posted task is simply called instead of posted, its phase number set to one greater than its parent—posts to tasks with phase number k are ignored. At the end of execution, the program ensures that the ith global valuation matches the initially-used valuation for phase $i+1$, for $0 \leq i < k-1$. When this condition holds, any global valuation observed along the execution is reachable within k phases in the original program. Figure 4 lists a code-to-code translation which implements this simulation.

Theorem 2. *A global-valuation g is reachable in a k-phase execution of a single-processor program P if and only if g is reachable in $(\!(P)\!)_k$—the k-phase sequential translation of P.*

When the underlying sequential program model has a decidable state-reachability problem, Theorem 2 gives a decision procedure for the phase-bounded state-reachability problem, by applying the decision procedure for the underlying model to the translated program. This allows us for instance to derive a decidability result for programs with finite data domains.

Corollary 1. *The k-phase state-reachability problem is decidable for single-processor programs with finite data domains.*

More generally, given any underlying sequential program model, our translation makes applicable any analysis tool for said model to message-passing programs, since the values of the additional variables are either from the finite domain $\{0, \ldots, k-1\}$, or in the domain of the original program variables.

Note that our simulation of a k-phase execution does not explicitly store the unbounded task queue. Instead of storing a multitude of possible unbounded task sequences, our simulation stores exactly k global state valuations. Accordingly, our simulation is not doomed to the unavoidable combinatorial explosion encountered by storing (even bounded-size) task queues explicitly. To demonstrate the capability of our advantage, we measure the time to verify two fabricated yet illustrative examples (listed in full in our extended report [6], comparing our bounded-phase encoding with a bounded task-queue encoding. In the bounded task-queue encoding, we represent the task-queue explicitly by an array of integers, which stores the identifiers of posted procedures[3]. When control of the initial task completes, the program enters a loop which takes a procedure identifier from the head of the queue, and calls the associated procedure. When the queue reaches a given bound, any further posted tasks are ignored.

The first program $P_1(i)$, parameterized by $i \in \mathbb{N}$, has a single Boolean global variable b, i procedures named $p_1, \ldots, p_i$, which assert b to be false and set b to true, and i procedures named $q_1, \ldots, q_i$ which set b to false. Initially, $P_1(i)$ sets b to false, and enters a loop in which each iteration posts some p_j followed by some q_j. Since a q_j task must be executed between each p_j task, each of the assertions are guaranteed to hold. Figure 6a compares the time required to verify $P_1(i)$ (using the BOOGIE verification engine [4]) for various values of i, and various bounds n on loop unrolling. Note that although every execution of $P_1(i)$ has only 2 phases, to explore all n loop iterations in any given execution, the size of queues must be at least $2n$, since two tasks are posted per iteration. Even for this very simple program, representing (even bounded) task-queues explicitly does not scale, since the number of possible task-queues grows astronomically as the size of task-queues grow. This ultimately prohibits the bounded tasks-queue encodings from exploring executions in which more than a mere few simple tasks execute. On the contrary, our bounded-phase simulation easily explores every execution up to the loop-unrolling bound in a few seconds.

To be fair, our second program P_2 is biased to support the bounded task-queue encoding. Following the example of Figure 5c, P_2 again has a single Boolean global variable b, and two procedures: p_1 asserts b to be false, sets b to true, and posts p_2, while p_2 sets b to false and posts p_1. Initially, the program P_2 sets b to false and posts a single p_1 task. Again here, since a p_2 task must execute between each p_1 task, each of the assertions are guaranteed to hold. Figure 6b compares the time required to verify P_2 for various bounds n on the number of tasks explored[4]. Note that although every execution of P_2 uses only size 1 task-queues, to explore all n tasks in any given execution, the number of phases must be at least n, since each task must execute in its own phase. Although

---

[3] For simplicity our examples do not pass arguments to tasks; in general, one should also store in the task-queue array the values of arguments passed to each posted procedure.

[4] The number n of explored tasks is controlled by limiting the number of loop unrollings in the bounded task-queue encoding, and limiting the recursion depth, and phase-bound, in the bounded-phase encoding.

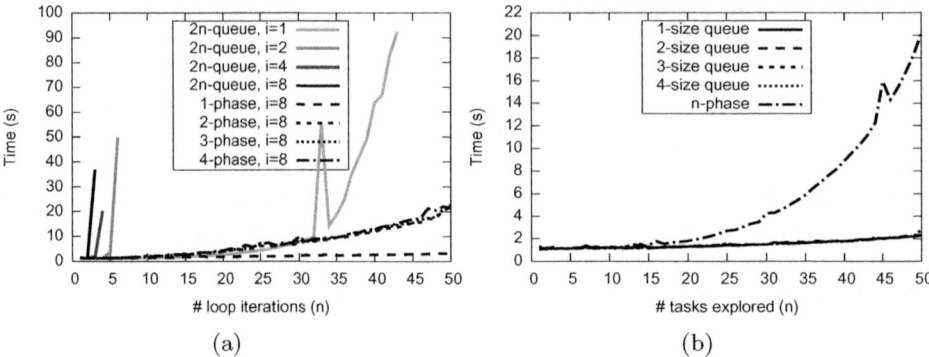

Fig. 6. Time required to verify (a) the program $P_1(i)$, and (b) the program P_2 with the BOOGIE verification engine using various encodings (bounded queues, bounded phase), and various loop unrolling bounds. Time-out is set to 100s.

verification time for the bounded-phase encoding does increase with n faster than the bounded task-queue encoding—as expected—due to additional copies of the global valuation, and more deeply in-lined procedures, the verification time remains manageable. In particular, the time does not explode uncontrollably: even 50 tasks are explored in under 20s.

5   Phase-Bounding for Multi-processor Programs

Though state-reachability under a phase bound is immediately and succinctly reducible to sequential program analysis for single-processor programs, the multi-processor case is more complicated. The added complexity arises due to the many orders in which tasks on separate processors can contribute to others' task-queues. As a simple example, consider the possible bounded-phase executions of Figure 5b with four processors, A, B, C, and D. Though B's tasks $B_1, \ldots, B_n$ must be executed in order, and C's tasks $C_1, \ldots, C_n$ must also be executed in order, the order of D's tasks are not pre-determined: the arrival order of D's tasks depends on how B's and C's tasks *interleave*. Suppose for instance B_1 executes to completion before C_1, which executes to completion before B_2, and so on. In this case D's tasks arrive to D's queue, and ultimately execute, in the index order $D_1, D_2, \ldots$ as depicted. However, there exist executions for every possible order of D's tasks respecting $D_1 < D_3 < \ldots$ and $D_2 < D_4 < \ldots$ (where $<$ denotes an ordering constraint)—many possible orders indeed! In fact, due to the capability of such unbounded interleaving, the problem of state-reachability under a phase-bound is undecidable for multi-processor programs, even for programs with finite data domains.

Theorem 3. *The k-phase bounded state-reachability problem is undecidable for multi-processor programs with finite data domains.*

Note that Theorem 3 holds independently of whether memory is shared between processors: the fact that a task-queue can store any possible (unbounded) shuffling of tasks posted by two processors lends the power to simulate Post's correspondence problem [20].

Theorem 3 insists that phase-bounding alone will not lead to the elegant encoding to sequential programs which was possible for single-processor programs. If that were possible, then the translation from a finite-data program would lead to a finite-data sequential program, and thus a decidable state-reachability problem. Since a precise algorithmic solution to bounded-phase state-reachability is impossible for multi-processor programs, we resort to a further incremental yet orthogonal under-approximation, which limits the number of considered processor interleavings. The following development is based on delay-bounded scheduling [12].

We define a *delaying scheduler* $M = \langle D, \text{empty}, \text{enabled}, \text{step}, \text{delay} \rangle$, as a scheduler $\langle D, \text{empty}, \text{enabled}, \text{step} \rangle$, along with a function delay : $(D \times \text{Pids} \times (\text{Pids} \rightarrow \text{Pconfigs})) \rightarrow D$. Furthermore, we extend the transition relation of Figure 3 with a postponing rule of Figure 7 which we henceforth refer to as a *delay (operation)*, saying that processor ρ is delayed. Note that a delay operation may or may not change the set of enabled processors in any given step, depending on the scheduler. A delaying scheduler is *delay-accessible* when for every configuration c_1 and non-idle or task-pending processor ρ, there exists a sequence $c_1 \rightarrow_M \cdots \rightarrow_M c_j$ of DELAY-steps such that ρ is enabled in c_j. Given executions h_1 and h_2 of (delaying) schedulers M_1 and M_2 resp., we write $h_1 \sim h_2$ when h_1 and h_2 are identical after projecting away delay operations.

Definition 3. *An execution with at most k delay operators is called k-delay.*

Consider again the possible executions of Figure 5b, but suppose we fix a deterministic scheduler M which without delaying would execute D's tasks in index order: $D_1, D_2, \ldots$; furthermore suppose that delaying a processor ρ in phase i causes M to execute the remaining phase i tasks of ρ in phase $i + 1$, while keeping the tasks of other processors in their current phase. Without using any delays, the execution of Figure 5b is unique, since M is deterministic. However, as Figure 8 illustrates, using a single delay, it is possible to also derive the order $D_1, D_3, \ldots, D_{2n-1}, D_2, D_4, \ldots, D_{2n}$ (among others): simply delay processor C once before C_1 posts D_2. Since this forces the D_{2i} tasks posted by each C_i to occur in the second phase, it follows they must all happen after the D_{2i-1} tasks posted by each B_i.

Theorem 4 (Completeness). *Let M be any delay-accessible scheduler. For every execution h of a program P, there exists an M-execution h' and $k \in \mathbb{N}$ such that h' is a k-delay execution and $h' \sim h$.*

Note that Theorem 4 holds for *any* delay-accessible scheduler M—even deterministic schedulers. As it turns out there is one particular scheduler M_{bfs} for which we know a convenient sequential encoding, and this scheduler is described in our extended report [6]. For the moment, the important points to note are

462   A. Bouajjani and M. Emmi

DELAY
$$\frac{m_2 = \mathsf{delay}(m_1, \rho, \xi)}{\langle \rho, \xi, m_1 \rangle \xrightarrow[M]{} \langle \rho, \xi, m_2 \rangle}$$

Fig. 7. The delay operation

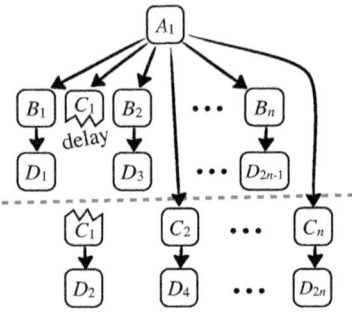

Fig. 8. A 2-phase delaying execution varying the 1-phase execution of Figure 5b

```
// translation of var g: T
var G[k+d]: T
var shift[Pids][k], delay: d
var ancestors[Pids][k+d]: B

// translation of proc p (var l: T) s
proc p (var l: T, pid: Pids, phase: k)

// translation of g
G[ phase + shift[pid][phase] ]

// code to be sprinkled throughout
while * and delay < d do
    shift[pid][phase]++; delay++

// translation of call x := p e
call x := p (e,pid,phase)

// translation of post ρ p e
let p = phase + shift[pid][phase] in
let p' = p + (if ancestors[ρ][p] then 1 else 0) in
if p' < k then
    ancestors[ρ][p' + shift[ρ][p']] := true;
    call p (e, ρ, p')
    ancestors[ρ][p' + shift[ρ][p']] := false
```

Fig. 9. The k-phase d-delay sequential translation $((P))_{k,d}^{\mathrm{bfs}}$ of a multi-processor message-passing asynchronous program P

that M_{bfs} is deterministic, non-blocking, and delay-accessible. Essentially, determinism allows us to encode the scheduler succinctly in a sequential program; the non-blocking property ensures this scheduler does explore some execution, rather than needlessly ceasing to continue; delay-accessibility combined with Theorem 4 ensure the scheduler is complete in the limit. Figure 9 lists a code-to-code translation which encodes bounded-phase and bounded-delay exploration of a given program according to the M_{bfs} scheduler as a sequential program.

Our translation closely follows the single-processor translation of Section 4, the key differences being:

- the phase of a posted task is not necessarily incremented, since posted tasks may not have same-processor ancestors in the current phase, and
- at any point, the currently executing task may increment a delay counter, causing all following tasks on the same processor to shift forward one additional phase.

As the global values reached by each processor at the end of each phase $i - 1$ must be ensured to match the initial values of phase i, for $0 < i < k + d$, so must the values for the shift counter: an execution is only valid when for each processor $\rho \in \mathsf{Pids}$ and each phase $0 < i < k$, shift$[\rho][i - 1]$ matches the initial value of shift$[\rho][i]$.

Theorem 5. *A global valuation g is reachable in a k-phase d-delay M_{bfs}-execution of a multi-processor program P if and only if g is reachable in $((P))_{k,d}^{bfs}$.*

As is the case for our single-processor translation, our simulation does not explicitly store the unbounded tasks queue, and is not doomed to combinatorial explosion faced by storing tasks-queues explicitly.

6   Related Work

Our work follows the line of research on compositional reductions from concurrent to sequential programs. The initial so-called "sequentialization" [22] explored multi-threaded programs up to one context-switch between threads, and was later expanded to handle a parameterized amount of context-switches between a statically-determined set of threads executing in round-robin order [21, 18]. [17] later extended the approach to handle programs parameterized by an unbounded number of statically-determined threads, and shortly after, [12] further extended these results to handle an unbounded amount of dynamically-created tasks, which besides applying to multi-threaded programs, naturally handles asynchronous event-driven programs [23]. [9] pushed these results even further to a sequentialization which attempts to explore as many behaviors as possible within a given analysis budget. Each of these sequentializations necessarily do provide a bounding parameter which limits the amount of interleaving between threads or tasks, but none are capable of precisely exploring tasks in creation order, which is abstracted away from their program models [23]. [15]'s sequentialization is sensitive to task priorities, their reduction assumes a finite number of statically-determined tasks.

In a closely-related work, [16] propose a "context-bounded" analysis of shared-memory multi-pushdown systems communicating with message-queues. According to this approach, one "context" involves a single process reading from its queue, and posting to the queues of other processes, and the number of contexts per execution is bounded. Our work can be seen as an extension in a few ways. First, and most trivially, in their setting a process cannot post to its own message queue; this implies that at least $2k$ contexts must be used to simulate k phases of a single-processor program. Second, there are families of 1-phase executions which require an unbounded number of task-contexts to capture; the execution order $D_1 D_2 D_3 \dots D_{2n}$ of Figure 5b is such an example. We conjecture that bounded phase and delay captures context-bounding—i.e., there exists a polynomial function $f : \mathbb{N} \to \mathbb{N}$ such that every k-context bounded execution of any program P is also a $f(k)$-phase and delay bounded execution. Finally, though phase-bounding leads to a convenient sequential encoding, we are unaware whether a similar encoding is possible for context-bounding.

[5] and   [7] have proposed analyses of message-passing programs by computing explicit finite symbolic representations of message-queues. As our sequentialization does not represent queues explicitly, we do not restrict the content

of queues to conveniently-representable descriptions. Furthermore, reduction to sequential program analyses is easily implementable, and allows us to leverage highly-developed and optimized program analysis tools.

7 Conclusion

By introducing a novel phase-based characterization of message-passing program executions, we enable bounded program exploration which is not limited by message-queue capacity nor the number of processors. We show that the resulting phase-bounded analysis problems can be solved by concise reduction to sequential program analysis. Preliminary evidence suggests our approach is at worst competitive with known task-order respecting bounded analysis techniques, and can easily scale where those techniques quickly explode.

Acknowledgments. We thank Constantin Enea, Cezara Dragoi, Pierre Ganty, and the anonymous reviewers for helpful feedback.

References

[1] HTML5: A vocabulary and associated APIs for HTML and XHTML, http://dev.w3.org/html5/spec/Overview.html

[2] Abdulla, P.A., Jonsson, B.: Verifying programs with unreliable channels. In: LICS 1993: Proc. 8th Annual IEEE Symposium on Logic in Computer Science, pp. 160–170. IEEE Computer Society (1993)

[3] Abdulla, P.A., Bouajjani, A., Jonsson, B.: On-the-Fly Analysis of Systems with Unbounded, Lossy FIFO Channels. In: Vardi, M.Y. (ed.) CAV 1998. LNCS, vol. 1427, pp. 305–318. Springer, Heidelberg (1998)

[4] Barnett, M., Leino, K.R.M., Moskal, M., Schulte, W.: Boogie: An intermediate verification language, http://research.microsoft.com/en-us/projects/boogie/

[5] Boigelot, B., Godefroid, P.: Symbolic verification of communication protocols with infinite state spaces using QDDs. Formal Methods in System Design 14(3), 237–255 (1999)

[6] Bouajjani, A., Emmi, M.: Bounded phase analysis of message-passing programs (2011), http://hal.archives-ouvertes.fr/hal-00653085/en

[7] Bouajjani, A., Habermehl, P.: Symbolic reachability analysis of fifo-channel systems with nonregular sets of configurations. Theor. Comput. Sci. 221(1-2), 211–250 (1999)

[8] Bouajjani, A., Habermehl, P., Vojnar, T.: Verification of parametric concurrent systems with prioritised FIFO resource management. Formal Methods in System Design 32(2), 129–172 (2008)

[9] Bouajjani, A., Emmi, M., Parlato, G.: On Sequentializing Concurrent Programs. In: Yahav, E. (ed.) SAS 2011. LNCS, vol. 6887, pp. 129–145. Springer, Heidelberg (2011)

[10] Brand, D., Zafiropulo, P.: On communicating finite-state machines. J. ACM 30(2), 323–342 (1983)

[11] Dahl, R.: Node.js: Evented I/O for V8 JavaScript, http://nodejs.org/

[12] Emmi, M., Qadeer, S., Rakamarić, Z.: Delay-bounded scheduling. In: POPL 2011: Proc. 38th ACM SIGPLAN-SIGACT Symposium on Principles of Programming Languages, pp. 411–422. ACM (2011)

[13] Ganty, P., Majumdar, R.: Algorithmic verification of asynchronous programs. CoRR, abs/1011.0551 (2010), http://arxiv.org/abs/1011.0551

[14] Jhala, R., Majumdar, R.: Interprocedural analysis of asynchronous programs. In: POPL 2007: Proc. 34th ACM SIGPLAN-SIGACT Symposium on Principles of Programming Languages, pp. 339–350. ACM (2007)

[15] Kidd, N., Jagannathan, S., Vitek, J.: One Stack to Run Them All: Reducing Concurrent Analysis to Sequential Analysis Under Priority Scheduling. In: van de Pol, J., Weber, M. (eds.) SPIN 2010. LNCS, vol. 6349, pp. 245–261. Springer, Heidelberg (2010)

[16] La Torre, S., Madhusudan, P., Parlato, G.: Context-Bounded Analysis of Concurrent Queue Systems. In: Ramakrishnan, C.R., Rehof, J. (eds.) TACAS 2008. LNCS, vol. 4963, pp. 299–314. Springer, Heidelberg (2008)

[17] La Torre, S., Madhusudan, P., Parlato, G.: Model-Checking Parameterized Concurrent Programs Using Linear Interfaces. In: Touili, T., Cook, B., Jackson, P. (eds.) CAV 2010. LNCS, vol. 6174, pp. 629–644. Springer, Heidelberg (2010)

[18] Lal, A., Reps, T.W.: Reducing concurrent analysis under a context bound to sequential analysis. Formal Methods in System Design 35(1), 73–97 (2009)

[19] Miller, M.S., Tribble, E.D., Shapiro, J.S.: Concurrency Among Strangers. In: De Nicola, R., Sangiorgi, D. (eds.) TGC 2005. LNCS, vol. 3705, pp. 195–229. Springer, Heidelberg (2005)

[20] Post, E.L.: A variant of a recursively unsolvable problem. Bull. Amer. Math. Soc. 52(4), 264–268 (1946)

[21] Qadeer, S., Rehof, J.: Context-Bounded Model Checking of Concurrent Software. In: Halbwachs, N., Zuck, L.D. (eds.) TACAS 2005. LNCS, vol. 3440, pp. 93–107. Springer, Heidelberg (2005)

[22] Qadeer, S., Wu, D.: KISS: Keep it simple and sequential. In: PLDI 2004: Proc. ACM SIGPLAN Conference on Programming Language Design and Implementation, pp. 14–24. ACM (2004)

[23] Sen, K., Viswanathan, M.: Model Checking Multithreaded Programs with Asynchronous Atomic Methods. In: Ball, T., Jones, R.B. (eds.) CAV 2006. LNCS, vol. 4144, pp. 300–314. Springer, Heidelberg (2006)

[24] Svensson, H., Arts, T.: A new leader election implementation. In: Erlang 2005: Proc. ACM SIGPLAN Workshop on Erlang, pp. 35–39. ACM (2005)

[25] Trottier-Hebert, F.: Learn you some Erlang for great good!, http://learnyousomeerlang.com/

Demonstrating Learning of Register Automata[*]

Maik Merten[1], Falk Howar[1], Bernhard Steffen[1], Sofia Cassel[2],
and Bengt Jonsson[2]

[1] Technical University Dortmund, Chair for Programming Systems, Dortmund,
D-44227, Germany
{maik.merten,falk.howar,steffen}@cs.tu-dortmund.de
[2] Dept. of Information Technology, Uppsala University, Sweden
{sofia.cassel,bengt.jonsson}@it.uu.se

Abstract. We will demonstrate the impact of the integration of our most recently developed learning technology for inferring Register Automata into the LearnLib, our framework for active automata learning. This will not only illustrate the unique power of Register Automata, which allows one to faithfully model data independent systems, but also the ease of enhancing the LearnLib with new functionality.

1 Introduction

Active automata learning (aka regular extrapolation) has been proposed to semi-automatically infer formal behavioral models of underspecified systems. The resulting formal behavioral models can be used, e.g., for documentation or regression testing and thus can be an enabler for continued system evolution. Automated mediation between networked systems by automatically synthesizing connectors from behavioral models is a current research interest. This approach is currently in development in the CONNECT project [6].

To cater the various use-cases of automata-learning, LearnLib has been created to offer a versatile library of learning algorithms and related tools. Result of an extensive reengineering effort, the *Next Generation LearnLib* [8] implements a flexible component-based approach that supports quick iteration and refinement of learning setups (in the following the Next Generation LearnLib will simply be referred to as "LearnLib").

The reengineered LearnLib has seen continued evolution, for which a complete account will not be provided in this paper. We will rather focus on two main innovations:

- The modeling paradigm in LearnLib Studio underwent significant changes. For example, in the old modeling paradigm a dedicated setup phase would precede the actual learning process. This has been replaced by on-the-fly configuration during the learning phase itself, with only minimal static configuration needed beforehand.

---

[*] This work is supported by the European FP 7 project CONNECT (IST 231167).

C. Flanagan and B. König (Eds.): TACAS 2012, LNCS 7214, pp. 466–471, 2012.

- LearnLib has been outfitted with support for the Register Automata automaton model [4], which is a simple extension of finite automata with data from infinite domains. It can model data-independent systems [7], i.e., systems that do not compute or manipulate data but manage their adequate distribution, such as protocols and mediators, in an intuitive way.

The demonstration will not only cover these two points in isolation, but will also highlight the ease of integration of new functionality into the overall learning framework.

In the remainder of the paper, we will recall the basics of active automata learning in Section 2, followed by an introduction to Register Automata in Section 3. A learning solution for Register Automata will be presented in Section 4, demonstrating the innovations outlined above. In Section 5 a conclusion is provided, with references showcasing the broad application scope of the presented tool environment.

2 Active Automata Learning

Angluin's seminal algorithm L^* [2] defines two query types to gather information about the System Under Learning (SUL):

- Membership Queries (MQs) are traces of symbols from a predefined alphabet of inputs of the SUL. The learning algorithm will construct such input traces, execute these on the SUL and capture system output. From the gathered information a hypothesis model is generated.
- Equivalence Queries (EQs) compare the produced hypotheses with the target system. If the model is not accurate, a counterexample will be provided revealing a difference between the current hypothesis and the SUL. Evaluating counterexamples the learning algorithm will produce refined hypothesis models using additional MQs. Once no counterexample can be produced the learning procedure has produced an accurate model and can be stopped.

With those two query types, L^* is guaranteed to produce a minimal and correct model. In current practice, however, EQs can only be implemented approximately for a large class of systems, e.g., with additional invocations of the target system.

The original L^* algorithm has originally been presented for DFAs, but has since been adapted to Mealy Machines, which are a better fit for learning actual reactive systems as they can encode system output in a natural way. A major and recent increase in expressiveness is achieved with Register Automata [5], which are described in the following section.

3 Register Automata

Register Automata are an extension of finite automata with data from infinite domains and are, e.g., well-suited for describing communication protocols. Register Automata are defined as follows:

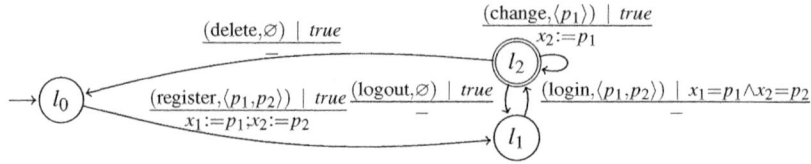

Fig. 1. Partial RA model for a fragment of XMPP

Definition 1. *Let a symbolic input be a pair $(a, \bar{p})$, of a parameterized input a of arity k and a sequence of symbolic parameters $\bar{p} = \langle p_1, \ldots, p_k \rangle$ Let further $X = \langle x_1, \ldots, x_m \rangle$ be a finite set of registers. A guard is a conjunction of equalities and negated equalities, e.g., $p_i \neq x_j$, over formal parameters and registers. An assignment is a partial mapping $\rho : X \to X \cup P$ for a set P of formal parameters.*

Definition 2. *A Register Automaton (RA) is a tuple $\mathcal{A} = (A, L, l_0, X, \Gamma, \lambda)$, where*

- A *is a finite set of actions.*
- L *is a finite set of locations.*
- $l_0 \in L$ *is the initial location.*
- X *is a finite set of registers.*
- Γ *is a finite set of transitions, each of which is of form $\langle l, (a, \bar{p}), g, \rho, l' \rangle$, where l is the source location, l' is the target location, $(a, \bar{p})$ is a parameterized action, g is a guard, and ρ is an assignment.*
- $\lambda : L \mapsto \{+, -\}$ *maps each location to either $+$ (accept) or $-$ (reject).* □

Let us define the semantics of an RA $\mathcal{A} = (A, L, l_0, X, \Gamma, \lambda)$. A X-valuation, denoted by v, is a (partial) mapping from X to D. A *state* of $\mathcal{A}$ is a pair $\langle l, v \rangle$ where $l \in L$ and v is a X-valuation. The *initial state* is $\langle l_0, v_0 \rangle$, i.e., the pair of initial location and empty valuation.

A *step* of $\mathcal{A}$, denoted by $\langle l, v \rangle \xrightarrow{(a, \bar{d})} \langle l', v' \rangle$, transfers $\mathcal{A}$ from $\langle l, v \rangle$ to $\langle l', v' \rangle$ on input $(a, \bar{d})$ if there is a transition $\langle l, (a, \bar{p}), g, \rho, l' \rangle \in \Gamma$ such that (1) g is modeled by $\bar{d}$ and v, i.e., if it becomes true when replacing all p_i by d_i and all x_i by $v(x_i)$, and such that (2) v' is the updated X-valuation, where $v'(x_i) = v(x_j)$ wherever $\rho(x_i) = x_j$, and $v'(x_i) = d_j$ wherever $\rho(x_i) = p_j$.

An example instance of a Register Automaton is provided in Figure 1, which models a subset of the XMPP instant messaging protocol focused on aspects of user authentication. In location l_0 no user account exists. With the parameterized action *register* a new account can be created, with the parameters p_1 and p_2 denoting a username and password. When executing the *register* action, the parameter values are copied into the registers x_1 and x_2 respectively. This action is unconditionally invocable in l_0, meaning that its guard is *true*. In contrast, the *login* in action of l_1 has a guard that specifies that the parameters p_1 and p_2 provided with the login action have to match the register contents of x_1 and x_2, i.e., the credentials provided during login have to match the ones stored in the registers. The other system actions represent logging out, changing the account password, and deleting the user account.

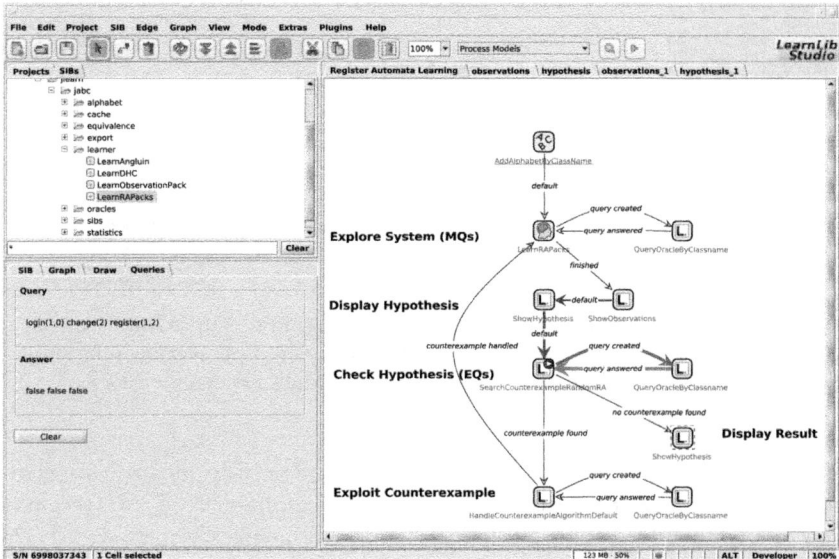

Fig. 2. A modeled learning setup created in LearnLib Studio. The model is currently executed, with bold edges denoting the path of execution. The current query and its answer are made visible in the panel on the lower left side.

4   The Tool Demo

We will present the LearnLib framework and highlight recently integrated innovations. LearnLib Studio allows the model-based composition and execution of learning setups, where LearnLib components are made available as reusable building blocks. In Figure 2 a learning setup tooled for learning register automata is shown, configured to learn the system shown in Figure 1. The result of executing this learning setup is presented in Figure 3.

The reworked modeling approach: In the learning setup presented in Figure 2 the distinct pattern specific to active automata learning can be witnessed: a learning algorithm is invoked, resulting in MQs an "oracle" has to answer. In this context, oracles are components that execute queries on a target system and gather the invocation results, meaning they produce an answer for a given query. Once the learning algorithm has formed a hypothesis it can be displayed and the data structure gathering the observations can be visualized, e.g., for debugging purposes. In the displayed setup, EQs are approximated by a random walk conformance test, which generates additional queries that are answered by the system oracle. If no counterexample can be found, the learning procedure will terminate, displaying the final result. Otherwise the counterexample will be consumed and analyzed, which can result in the production of additional queries. Once these have been answered and the counterexample is exploited so that a refined hypothesis can be produced, the learning process will restart the learning algorithm.

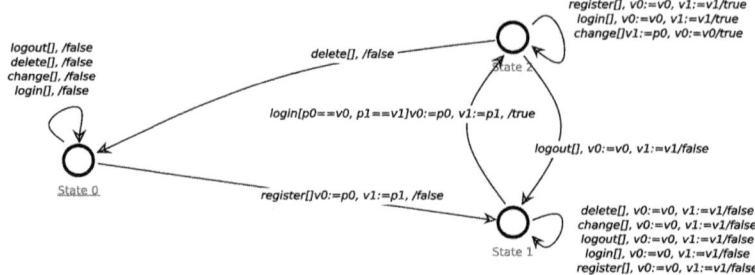

Fig. 3. The resulting model from executing the learning setup. Guards are in square brackets, the next register contents are explicitly denoted on every transition as is the acceptance status of the following state. Register values are denoted as v_i.

The general workflow is reusable in nature. The only application-specific parts that have to be provided are the definition of the alphabet and the system oracle, which has to interface the SUL. These application-specific parts can be loaded in a standardized way with parameterized building blocks.

Compared to the previous modeling style supported by LearnLib Studio, the new modeling style is much improved in terms of usability and flexibility: in the old modeling paradigm, a dedicated setup phase would create a fixed configuration, connecting, e.g., the learning algorithm to the SUL oracle. This setup would then be instantiated and started, with limited ways to influence the behavior of the setup afterwards. In the new modeling paradigm, only an alphabet has to be specified beforehand, after which the learning algorithm can directly begin operation. Learning queries are delegated on-the-fly according to the setup execution flow. This greatly increases flexibility, as, for instance, setups can decide at runtime what oracle instances are used to answer queries.

Integration of Register Automata learning: Thanks to the flexible component-based approach of LearnLib, components for learning this new model type could be integrated into the general framework without changes to the LearnLib architecture.

In the learning setup presented above, the only components that are specific for the RA machine model are the building blocks encapsulating the learning algorithm and the equivalence approximation. The overall infrastructure provided by the LearnLib is reused to a high degree when using RAs despite adapting a much richer automata model. The overall flavor of how learning setups can be created is completely unchanged compared to how setups are specified for other machine models, abstracting from the details of the underlying data structures and algorithms and thus shielding the user from additional complexity.

Due to being integrated into the component framework, learning setups for Register Automata can use all facilities LearnLib offers for debugging, visualization and statistics. Thus our extension provides a powerful and unique framework for learning data independent systems [7].

5 Conclusion

The LearnLib framework and accompanying tools provide a rich environment for experimentation with the new Register Automata formalism. Embedded into a flexible component model, much functionality is shared between learning setups for different machine models, which enables a high degree of reuse. Users already versed in the operation of the LearnLib Studio will not have to learn a new style of modeling when adapting Register Automata, while users new to LearnLib Studio are only exposed to a limited set of generic concepts that are easy to understand.

The LearnLib is mature and used by several independent research groups. It has, e.g., been used to infer the behavior of a electronic passports [1], in security research [3], and it is a central enabler within the CONNECT framework.

LearnLib is available for download at http://www.learnlib.de and free for all academic purposes.

References

1. Aarts, F., Schmaltz, J., Vaandrager, F.: Inference and Abstraction of the Biometric Passport. In: Margaria, T., Steffen, B. (eds.) ISoLA 2010, Part I. LNCS, vol. 6415, pp. 673–686. Springer, Heidelberg (2010)
2. Angluin, D.: Learning Regular Sets from Queries and Counterexamples. Information and Computation 75(2), 87–106 (1987)
3. Bossert, G., Hiet, G., Henin, T.: Modelling to Simulate Botnet Command and Control Protocols for the Evaluation of Network Intrusion Detection Systems. In: 2011 Conference on Network and Information Systems Security (SAR-SSI), pp. 1–8 (May 2011)
4. Cassel, S., Howar, F., Jonsson, B., Merten, M., Steffen, B.: A Succinct Canonical Register Automaton Model. In: Bultan, T., Hsiung, P.-A. (eds.) ATVA 2011. LNCS, vol. 6996, pp. 366–380. Springer, Heidelberg (2011)
5. Howar, F., Steffen, B., Cassel, S., Jonsson, B.: Inferring Canonical Register Automata. In: Kuncak, V., Rybalchenko, A. (eds.) VMCAI 2012. LNCS, vol. 7148, pp. 251–266. Springer, Heidelberg (2012)
6. Issarny, V., Steffen, B., Jonsson, B., Blair, G.S., Grace, P., Kwiatkowska, M.Z., Calinescu, R., Inverardi, P., Tivoli, M., Bertolino, A., Sabetta, A.: CONNECT Challenges: Towards Emergent Connectors for Eternal Networked Systems. In: ICECCS, pp. 154–161 (2009)
7. Lazić, R., Nowak, D.: A Unifying Approach to Data-Independence. In: Palamidessi, C. (ed.) CONCUR 2000. LNCS, vol. 1877, pp. 581–595. Springer, Heidelberg (2000)
8. Merten, M., Steffen, B., Howar, F., Margaria, T.: Next Generation LearnLib. In: Abdulla, P.A., Leino, K.R.M. (eds.) TACAS 2011. LNCS, vol. 6605, pp. 220–223. Springer, Heidelberg (2011)

Symbolic Automata: The Toolkit

Margus Veanes and Nikolaj Bjørner

Microsoft Research, Redmond, WA

Abstract. The symbolic automata toolkit lifts classical automata analysis to work modulo rich alphabet theories. It uses the power of state-of-the-art constraint solvers for automata analysis that is both expressive and efficient, even for automata over large finite alphabets. The toolkit supports analysis of finite symbolic automata and transducers over strings. It also handles transducers with registers. Constraint solving is used when composing and minimizing automata, and a much deeper and powerful integration is also obtained by internalizing automata as theories. The toolkit, freely available from Microsoft Research[1], has recently been used in the context of web security for analysis of potentially malicious data over Unicode characters.

Introduction. The distinguishing feature of the toolkit is the use and operations with symbolic labels. This is unlike classical automata algorithms that mostly work assuming a finite alphabet. Adtantages of a symbolic representation are examined in [4], where it is shown that the symbolic algorithms consistently outperform classical algorithms (often by orders of magnitude) when alphabets are large. Moreover, symbolic automata can also work with infinite alphabets. Typical alphabet theories can be *arithmetic* (over integers, rationals, bit-vectors), *algebraic data-types* (for tuples, lists, trees, finite enumerations), and *arrays*. Tuples are used for handling alphabets that are cross-products of multiple sorts. In the following we describe the core components and functionality of the tool. The main components are $\mathsf{Automaton}\langle T\rangle$, basic automata operations modulo a Boolean algebra T; $\mathsf{SFA}\langle T\rangle$, symbolic finite automata as theories modulo T; and $\mathsf{SFT}\langle T\rangle$, symbolic finite transducers as theories modulo T. We illustrate the tool's API using code samples from the distribution.

Automaton$\langle T\rangle$. The main building block of the toolkit, that is also defined as a corresponding generic class, is a (symbolic) automaton over T: $\mathsf{Automaton}\langle T\rangle$.

The type T is assumed to be equipped with effective Boolean operations over T: $\wedge$, $\vee$, $\neg$, $\bot$, $is_\bot$ that satisfy the standard axioms of Boolean algebras, where $is_\bot(\varphi)$ checks if a term φ is false (thus, to check if φ is true, check $is_\bot(\neg\varphi)$). The main operations over $\mathsf{Automaton}\langle T\rangle$ are $\cap$ (intersection), $\cup$ (union) $\complement$ (complementation), $A \equiv \emptyset$ (emptiness check). As an example of a simple symbolic operation consider products: when A, B are of type $\mathsf{Automaton}\langle T\rangle$, then $A \cap B$ has the transitions $\langle(p,q), \varphi \wedge \psi, (p',q')\rangle$ for each transition $\langle p, \varphi, p'\rangle \in A$, and

---

[1] The binary release is available from `http://research.microsoft.com/automata`

C. Flanagan and B. König (Eds.): TACAS 2012, LNCS 7214, pp. 472–477, 2012.

$\langle q, \psi, q' \rangle \in B$. Infeasible and unreachable transitions are pruned by using the $is_{\perp}$ tester. Note that Automaton$\langle T \rangle$ is also a Boolean algebra (using the operations $\cap, \cup, \complement, \equiv \emptyset$). Consequently, the tool supports building and analyzing nested automata Automaton$\langle$Automaton$\langle T \rangle \rangle$.

The tool provides a Boolean algebra solver CharSetSolver that uses specialized BDDs (see [4]) of type CharSet. This solver is used to efficiently analyze .Net regexes with Unicode character encoding. The following code snippet illustrates its use, as well as some other features like visualization.

```
CharSetSolver solver = new CharSetSolver(CharacterEncoding.Unicode);      // charset solver
string a = @"^[A-Za-z0-9]+@(([A-Za-z0-9\-])+\.)+([A-Za-z\-])+$";          // .Net regex
string b = @"^\d.*$";                                                      // .Net regex
Automaton<CharSet> A = solver.Convert(a);                // create the equivalent automata
Automaton<CharSet> B = solver.Convert(b);
Automaton<CharSet> C = A.Minus(B, solver);                    // construct the difference
var M = C.Determinize(solver).Minimize(solver);      // determinize then minimize the automaton
solver.ShowGraph(M, "M.dgml");                               // save and visualize
string s = solver.GenerateMember(M); //generate some member, e.g.  "HV7@9.2.8.-d2bVu0YH.z1f.R"
```

The resulting graph from line 8 is shown below.

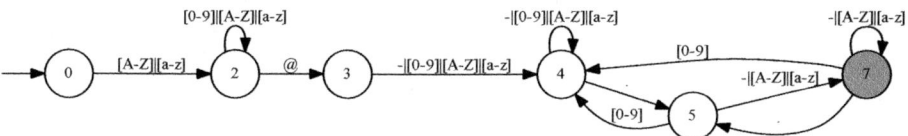

SFA$\langle T \rangle$. A symbolic finite automaton SFA$\langle T \rangle$ is an extension of Automaton$\langle T \rangle$ with a logical evaluation context of an SMT (Satisfiability Modulo Theories) solver that supports operations that go beyond mere Boolean algebraic operations. The main additional solver operations are: *assert* (to assert a logical formula), *push/pop* (to manage scopes of assertions), *get_model*: $T \rightarrow \mathcal{M}$ to obtain a model for a satisfiable formula. A model $\mathcal{M}$ is a dictionary from the free constants in the asserted formulas to values. The method *assert_theory* takes an SFA$\langle T \rangle$ A and adds the *theory of* A to the solver. It relies on a built-in theory of *lists* and uses it to define a *symbolic language acceptor* for A that is a unary relation symbol acc_A such that $acc_A(s)$ holds iff s is accepted by A.

The following code snippet illustrates the use of SFAs together with Z3 as the constraint solver. The class Z3Provider is a conservative extension of Z3 that extends its functionality for use in the automata toolkit. The sample is similar (in functionality) to the one above, but uses the Z3 Term type rather than CharSet for representing predicates over characters.

```
Z3Provider Z = new Z3Provider();
string a = @"^[A-Za-z0-9]+@(([A-Za-z0-9\-])+\.)+([A-Za-z\-])+$";           // .Net regex
string b = @"^\d.*$";                                                       // .Net regex
var A = new SFAz3(Z, Z.CharacterSort, Z.RegexConverter.Convert(a));         // SFA for a
var B = new SFAz3(Z, Z.CharacterSort, Z.RegexConverter.Convert(b));         // SFA for b
A.AssertTheory(); B.AssertTheory();                    // assert both SFA theories to Z3
Term inputConst = Z.MkFreshConst("input", A.InputListSort);   // declare List<char> constant
var assertion   = Z.MkAnd(A.MkAccept(inputConst),             // get solution for inputConst
                          Z.MkNot(B.MkAccept(inputConst)));   // accepted by A but not by B
var model       = Z.GetModel(assertion, inputConst);          //   retrieve satisfying model
string input    = model[inputConst].StringValue;              //   the witness in L(A)-L(B)
```

SFA acceptors can be combined with arbitrary other constraints. This feature is used in Pex[2] for path analysis of string manipulating .Net programs that use regex matching in branch conditions.

SFT$\langle T \rangle$. A symbolic finite transducer over labels in T (SFT) is a finite state symbolic input/output* automaton. A transition of an SFT$\langle T \rangle$ has the form (p, φ, out^*, q) where φ is an input character predicate and $out*$ is a sequence of output terms that may depend on the input character. For example, a transition $(p, x > 10, [x + 1, x + 2], q)$ means that, in state p, if the input symbol is greater than 10, then output $x + 1$ followed by $x + 2$ and continue from state q.

An SFT is a generalization of a classical finite state transducer to operate modulo a given label theory. The core operations overs SFTs are the following: *union* $T \cup T$, (relational) *composition* $T \circ T$, *domain restriction* $T \upharpoonright A$, *subsumption* $T \preceq T$ and *equivalence* $T \equiv T$. These operation form (under some conditions, such as *single-valuedness* of SFTs) a decidable algebra over SFTs. The theory and the algorithms of SFTs are studied in [7].

Bek is a domain specific language for string-manipulating functions. It is to SFTs as regular expressions are to SFAs. The toolkit includes a parser for Bek as well as customized visualization support using the graph viewer of Visual Studio. Key scenarios and applications of Bek for security analysis of sanitation routines are discussed in [3]. The following snippet illustrates using the library for checking idempotence of a decoder P (it decodes any consecutive digits d_1 and d_2 between '5' and '9' to their ascii letter $dec(d_1, d_2)$, e.g. $dec('7', '7') = 'M'$, thus $P("7777") = "MM"$. The Bek program decoder is first converted to an equivalent SFT (where the variable b is eliminated).

```
string bek = @"program P(input) {                          // The Bek program P
    return iter(c in input) [b := 0;] {                // P decodes certain digit pairs
        case (b == 0): if ((c>='5')&&(c<='9')) { b:=c; } else { yield(c); }
        case (true): if ((c>='5')&&(c<='9')) { yield(dec(b,c));b:=0; } else { yield(c); }
    } end { case (b != 0): yield (b);};}";
Z3Provider Z = new Z3Provider();                           // analysis uses the Z3 provider
var f = BekConverter.BekToSTb(Z, bek).ToST().Explore();    //         convert P to an SFT f
var fof = f + f;                                           //         self-compostion of f
if (!f.Eq1(fof)) {                                         //         check idempotence of f
    var w = f.Diff(fof);                   // find a witness where f and fof differ
    string input = w.Input.StringValue;    //                              e.g. "5555"
    string output1 = w.Output1.StringValue;  //                     e.g. f("5555") == "77"
    string output2 = w.Output2.StringValue; }  //                 e.g. f(f("5555")) == "M"
```

Users and tool availability. This is the first public release of the toolkit. It has so far been used at Microsoft, and part of the tool (Rex) is also an integrated part of the parameterized unit testing tool Pex. Applications that illustrate some key usage scenarios, are also used from the web services http://www.rise4fun.com/rex and http://www.rise4fun.com/bek. The tool has been used in numerous experiments, some of which are described in [4,3], that show scalability and applicability to concrete real-life scenarios.

---

[2] http://research.microsoft.com/pex/

Tool Overview. An overview of the toolkit is illustrated in the diagram below. The core components are in bold. Arrows indicate dependencies between the components. They are labeled by the main relevant functionality.

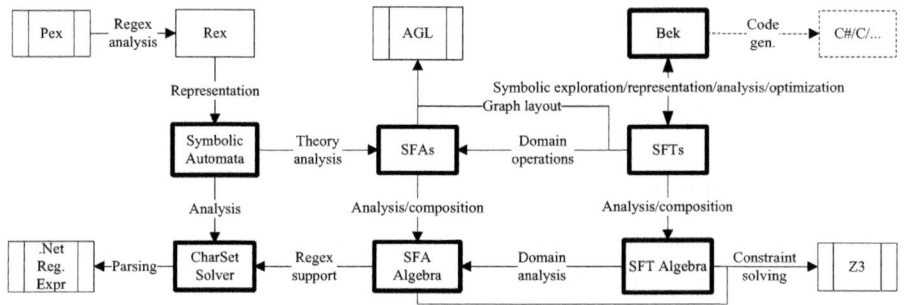

Related Tools. String analysis has recently received increased attention, with several automata-based analysis tools. We make a systematic comparison of related techniques in [4]. Tools include the Java String Analyzer [1], with the dk.brics.automaton library as a constraint solver for finite alphabets. It compresses contiguous character ranges. Hampi [5] solves bounded length string constraints over finite alphabets using a reduction to bit-vectors. Kaluza extends Hampi to systems of constraints with multiple variables and concatenation [6]. MONA [2] uses MTBDDs for encoding transitions. BDDs are used in the PHP string analysis tool in [8].

References

1. Christensen, A.S., Møller, A., Schwartzbach, M.I.: Precise Analysis of String Expressions. In: Cousot, R. (ed.) SAS 2003. LNCS, vol. 2694, pp. 1–18. Springer, Heidelberg (2003)
2. Henriksen, J.G., Jensen, J., Jørgensen, M., Klarlund, N., Paige, B., Rauhe, T., Sandholm, A.: Mona: Monadic Second-Order Logic in Practice. In: Brinksma, E., Steffen, B., Cleaveland, W.R., Larsen, K.G., Margaria, T. (eds.) TACAS 1995. LNCS, vol. 1019, pp. 89–110. Springer, Heidelberg (1995)
3. Hooimeijer, P., Livshits, B., Molnar, D., Saxena, P., Veanes, M.: Fast and precise sanitizer analysis with bek. In: USENIX Security Symposium (August 2011)
4. Hooimeijer, P., Veanes, M.: An Evaluation of Automata Algorithms for String Analysis. In: Jhala, R., Schmidt, D. (eds.) VMCAI 2011. LNCS, vol. 6538, pp. 248–262. Springer, Heidelberg (2011)
5. Kiezun, A., Ganesh, V., Guo, P.J., Hooimeijer, P., Ernst, M.D.: HAMPI: a solver for string constraints. In: ISSTA (2009)
6. Saxena, P., Akhawe, D., Hanna, S., Mao, F., McCamant, S., Song, D.: A Symbolic Execution Framework for JavaScript (March 2010)
7. Veanes, M., Hooimeijer, P., Livshits, B., Molnar, D., Bjørner, N.: Symbolic finite state transducers: Algorithms and applications. In: POPL 2012 (January 2012)
8. Yu, F., Alkhalaf, M., Bultan, T.: STRANGER: An Automata-Based String Analysis Tool for PHP. In: Esparza, J., Majumdar, R. (eds.) TACAS 2010. LNCS, vol. 6015, pp. 154–157. Springer, Heidelberg (2010)

A   Bek

Bek is a domain specific language for writing common string functions. With Bek, you can answer questions like: Do these two programs output the same strings? Given a target string, is there an input string such that the program produces the target string? Does the composition of two programs produce a desired result? Does the order of composition matter? Bek has been specifically tailored to capture common idioms in string manipulating functions.

A.1   UTF8Encode Example

Bek includes a fairly complete set of arithmetic operations that are, by default, over 16-bit bit-vectors, since the most common case of analysis is over strings that use UTF-16 encoding. The example shows a concrete representation of a UTF8 encoding routine written in Bek. It takes a UTF-16 encoded string and transforms it into the corresponding UTF8 encoded string. The encoder "raises an exception" when invalid surrogate pairs are detected. These exception cases define, in terms of the generated SFTs, partial behavior, i.e., that the input is not accepted by the SFT.

```
// UTF8 encoding from UTF16 strings, hs is the lower two bits of the previous high surrogate
               // this encoder raises an exception when an invalid surrogate is detected
program UTF8Encode(input){
return iter(c in input)[HS:=false; hs:=0;]
{
case (HS):                              // the previous character was a high surrogate
  if (!IsLowSurrogate(c)) { raise InvalidSurrogatePairException; }
  else {
    yield ((0x80|(hs << 4))|((c>>6)&0xF), 0x80|(c&0x3F));
    HS:=false; hs:=0;
  }
case (!HS):                             // the previous character was not a high surrogate
  if (c <= 0x7F) { yield(c); }                          // one byte: ASCII case
  else if (c <= 0x7FF) {                                // two bytes
    yield(0xC0 | ((c>>6) & 0x1F), 0x80 | (c & 0x3F)); }
  else if (!IsHighSurrogate(c)) {
    if (IsLowSurrogate(c)) { raise InvalidSurrogatePairException; }
    else { //three bytes
      yield(0xE0| ((c>>12) & 0xF), 0x80 | ((c>>6) & 0x3F), 0x80 | (c&0x3F));} }
  else {
    yield (0xF0|((((1+((c>>6)&0xF))>>2)&7), (0x80|((((1+((c>>6)&0xF))&3)<<4))|((c>>2) & 0xF));
    HS:=true; hs:=c&3; }
} end {
case (HS): raise InvalidSurrogatePairException;
case (true): yield();
};
}
```

The following code is a unit test from the automata toolkit. Assume that the above Bek program is in the file "UTF8Encode.bek". The code does the following. First, it converts the Bek program into a symbolic transducer *stb* (that allows branching conditions in rules). It then eliminates the registers *hs* and *HS* by fully exploring *stb*. Then the domain of the resulting sft is restricted with the regular expression that excludes the empty input string. The theory of the resulting sft is asserted as a background theory extension of the solver. New uninterpreted constants are defined for input and output lists of the sft. Then the Z3 provider is used to generate (50) solutions. Old solutions are pruned from iterated calls to the solver.

```
public void TestUTF8Encode() {
    Z3Provider solver = new Z3Provider();
    var stb = BekConverter.BekFileToSTb(solver, "UTF8Encode.bek");
    var sft = stb.Explore();
    //sft.ShowGraph(); //saves the sft in DGML format and opens it in Visual Studio.

    var restr = sft.ToST().RestrictDomain(".+");
    restr.AssertTheory();

    Term inputConst = solver.MkFreshConst("input", restr.InputListSort);
    Term outputConst = solver.MkFreshConst("output", restr.OutputListSort);

    solver.AssertCnstr(restr.MkAccept(inputConst, outputConst));

    //validate correctness for some values against the actual UTF8Encode
    int K = 50;
    for (int i = 0; i < K; i++) {
      var model = solver.GetModel(solver.True, inputConst, outputConst);
      string input = model[inputConst].StringValue;
      string output = model[outputConst].StringValue;

      Assert.IsFalse(string.IsNullOrEmpty(input));
      Assert.IsFalse(string.IsNullOrEmpty(output));

      byte[] encoding = Encoding.UTF8.GetBytes(input);
      char[] chars = Array.ConvertAll(encoding, b => (char)b);
      string output_expected = new String(chars);

      Assert.AreEqual<string>(output_expected, output);

      // exclude this solution, before picking the next one
      solver.AssertCnstr(solver.MkNeq(inputConst, model[inputConst].Value));
    }
}
```

The whole unit test takes a few seconds to complete. In this case the unit test simply tests on 50 random samples that the encoder does not differ from the built-in implementation.

A.2   Bek on Rise4Fun.com

The web-site http://rise4fun.com/bek illustrates several examples of Bek programs. It runs the Symbolic Automata Toolkit with the Bek extensions and converts Bek programs into ECMA script (Java script) and also shows a graphical representation of an graph representation of the transducer.

B   Rex on Rise4Fun.com

The web-site http://rise4fun.com/rex illustrates several examples of Rex as a game. The game is to guess a secret regular expression. The user enters a candidate expression, and the Symbolic Automata Toolkit is used to find strings that are (1) accepted by both languages (if any), (2) accepted by one and rejected by the other (if any), and (3) rejected by both languages.

McScM: A General Framework for the Verification of Communicating Machines[*]

Alexander Heußner[1], Tristan Le Gall[2], and Grégoire Sutre[3]

[1] Université Libre de Bruxelles, Brussels, Belgium
[2] CEA, LIST, DILS/LMeASI, Gif-sur-Yvette, France
[3] Univ. Bordeaux & CNRS, LaBRI, UMR 5800, Talence, France

Abstract. We present McScM, a platform for implementing and comparing verification algorithms for the class of finite-state processes exchanging messages over reliable, unbounded FIFO channels. McScM provides tools for the safety verification and controller synthesis of these infinite-state models. Our verification tool implements several model-checking techniques: CEGAR with different abstraction-refinement methods, abstract interpretation, abstract regular model checking, and lazy abstraction. Seen as a general framework for the class of transition systems with finite control/infinite data, McScM delivers the basic infrastructure for implementing verification algorithms, and privileges to conveniently implement new ideas on a high level of abstraction. It also allows us to compare and benchmark different algorithmic approaches with the same backend.

1 Introduction

The automatic verification of distributed algorithms and communication protocols is one of the most crucial tasks in software/hardware development and maintenance. It is also one of the hardest, e.g., as one cannot directly infer the global behaviour of a distributed system from its local components due to asynchronous communication. This renders already simple analysis, verification, and synthesis questions hard problems in theory. However, in practice, this leads to a growing demand for versatile tools that also apply semi-algorithmic solutions, approximations, abstractions, and heuristics.

We focus on the safety verification of *communicating finite-state machines* (CM), an infinite-state formalism that consists of a set of local, finite state machines that communicate via global, asynchronous, reliable and unbounded FIFO channels. The latter are demanded in practice by, e.g., distributed applications based on TCP, the Sockets API, or MPI. Note that CM do not demand the channels to be a priori point-to-point. The safety verification question demands, given a CM and a set of "bad" states, whether no execution of this CM reaches the bad states. This is known to be undecidable [3].

---

[*] This work was partially supported by the ANR project Vacsim (ANR-11-INSE-004).

C. Flanagan and B. König (Eds.): TACAS 2012, LNCS 7214, pp. 478–484, 2012.

As for other classes of Turing-complete infinite-state models, there are two ways to tackle this problem. The first one is to restrict CM to a decidable fragment, e.g., by imposing a bound on the size of the FIFO channels, or assuming the channels to lose messages [1]. The second one is to provide only "partial" results, either by semi-algorithmic methods that may not terminate, or over-approximative approaches that may be inconclusive. Despite the rich theoretical work concerning CM, there is currently no versatile tool that can be directly applied to CM's safety question , and that gives the user a choice among different algorithmic approaches to solve a concrete verification question.

We aim at filling this gap by presenting a *Model Checker for Systems of Communicating Machines* (McScM) that combines different algorithms for the safety verification problem of CM under the same roof and provides a ready-to-use front-end with the tool `verify`. McScM is available via our project's web page [14] either as a precompiled binary distribution (including man pages, a suite of examples, and some benchmarking scripts), or as source code release. McScM is programmed in OCaml and available under a BSD license. The development of McScM takes place in a software forge [14] that provides a wiki for documentation (including man pages and API), a bug and issue tracker, as well as our theoretical work [6, 7]. The following discussion refers to McScM's release 1.2.

In the following, we present our implemented algorithms (Section 2) and `verify`'s modular architecture (Section 3) before applying the tool in a small comparative benchmark in Section 4 that shows its capabilities. Finally, we change the focus to McScM as generic API/framework (Section 5) to implement novel algorithms and ideas and compare to existing tools and frameworks.

2 Safety Verification of Communicating Machines

In our setting, an instance of the safety verification problem is given by a textual representation of a CM (in a simple and intuitive automata-based language), and a set of bad states, i.e., a set of global control states together with a representation of the channel's contents by regular expressions (for details, see `scm(5)` man page). The tool `verify` allows the user to input this instance and to choose among a variety of verification techniques. After completion of the analysis, `verify` outputs either "model safe" if it finds an inductive invariant that proves the system safe, or returns a counterexample, i.e., a proof that the system is not safe. The tool aborts if it runs out of a resource that was a priori limited by the user, e.g., the number of analysis steps or the maximum precision allowed for abstraction. A closer look on the modular architecture of `verify` and the generic aspects underlying McScM is postponed to the next section.

McScM currently implements the following four different verification techniques:

absint: this abstract interpretation based approach [8] reduces verification to the calculation of a fixpoint in an abstract lattice, and terminates in a finite number of steps with either a positive answer (model safe), or aborts.

armc: the *Abstract Regular Model Checking* semi-algorithm [2] refines a global regular abstraction of the system by symbolic successor (or predecessor) calculation; we reimplemented the basic idea in our setting;

cegar: *Counterexample Guided Abstraction Refinement* is a semi-algorithmic approach that allows to start with a rough, safety-conservative abstraction that is refined along spurious counterexamples [4]; McScM started originally by porting this approach to CM relying on a novel notion of path invariant based refinement [6]; the implemented generic algorithm allows for a variety of parameterization (in particular, path invariant generation methods);

lart: we implemented the lazy abstraction approach [9] based on the construction of an abstract reachability tree; each vertex of the tree contains an abstract region, which may be refined with the help of path invariants when needed;

We can compare the algorithms on the same background as they share an underlying infrastructure implementing abstraction/extrapolating for CM, as well as a library of graph algorithms and (path) invariant generators. The first three techniques are semi-algorithms based on the *abstract-check-refine* paradigm. When the CM is not safe, they provide *counterexamples* that are an important feedback when using safety verification in practical (engineering) situations. Contrariwise, absint always terminates without guaranteeing a conclusive answer. A comparison of the four approaches with respect to a suite of example protocols derived from practice follows in Section 4.

Already revealing the benchmark's outcome, there is no silver bullet among the four techniques. Hence, to tackle a given instance of a CM safety verification problem, one has to choose among approaches and need fine-grained influence by additional parameterizations to the underlying algorithms (e.g., depth-first versus breadth-first exploration of the CM). This is exactly what `verify` offers: a "swiss army knife" for model checking systems of communicating machines.

In addition, McScM includes a supervisory control tool: `control`. If a CM system does not satisfy a safety property, `control` automatically computes a restriction of this system that assures safety by implementing the distributed control algorithm presented in [7] (see `control(1)` man page for details).

3   A Closer Look on `verify`'s Modular Architecture

Figure 1 shows the modular architecture of McScM's `verify` tool. The latter provides a common (command line based) interface and infrastructure for the implemented verification algorithms (absint, armc, cegar, lart), as well as allows to plug in a symbolic representation of the infinite data part of the CM, i.e., the queue contents. Currently, we only provide a wrapper for a library based on queue-content decision diagrams (QDD) [15]. The tool's input is an instance of the safety decision problem. Each algorithm accepts additional adaptations via command line parameters. The tool outputs either a counterexample, a positive result, or an abort message. McScM provides additional logging and profiling information that can be output by the tool, and helps to benchmark and compare algorithms, as seen in the next section.

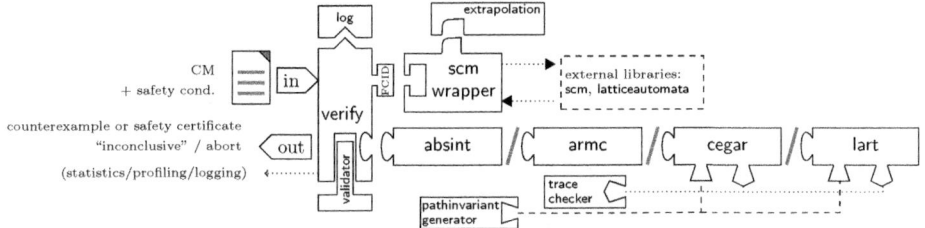

Fig. 1. Modular Architecture of McScM's Verification Tools

The cornerstone of McScM's development is *generic programming* which is supported by OCaml's modules and functors. This functional programming language based on type inference encloses the proof of behavioural guarantees at compile time with respect to our implemented algorithm's interfaces. In addition, we also provide means for the on-the-fly validation of both intermediary results (i.e., checking the result of a path invariant generation in cegar) and the inductive safety invariant. Both add an additional layer of *reliability*, especially when implementing new algorithms in McScM.

As our API is well specified and reasonably documented, it is relatively easy to implement also other algorithms for CMs. For example, the previously mentioned control tool is build of top of several OCaml modules of verify, and uses the same fixpoint computation as absint.

4   A Comparative Benchmark of Verification Algorithms

Figure 2 was generated by using verify to benchmark the included verification algorithms (on default parameters) on a suite of examples derived from practice. The latter includes the alternating bit protocol (ABP), a simplified version of TCP, and a distributed leader election algorithm ("Peterson"); the examples range from simple protocols with 5 global control states ("c/d") to around 10^4 for Peterson. The benchmark was run on an off-the-shelf computer (3.2GHz Intel i7-965, 64-bit Linux) and is contained as shell script in the latest McScM release.

In general, verify is able to give a solution for each example in a reasonable amount of time and memory. However, there is no algorithm that proves to be superior. absint provides a fast way of determining if a protocol is safe, however, it is not able to cope with unsafe examples. Due to its termination guarantee, it proves to be ideal as first line of attack when trying to verify an unknown protocol. The main difference between cegar and armc is their way of refining the abstraction, either locally and adapted, or globally for the whole model. This gives an advantage for cegar in the examples that require a "precise" abstraction only for a few control loops (like the Peterson algorithm, the erroneous load balancer, or the token ring example), and for armc in most other cases. However, our armc implementation is not able to cope with a simple non-regular protocol. As cegar allows a variety of additional parameters to the algorithm, we can fill the two gaps in the table by changing the underlying path invariant generation

	ABP	BRP	c/d	load-balancer	load-balancer nested c/d (err)	non-regular	Peterson	POP3	server/2 clients	sliding windows (simplified)	TCP (error)	TCP	token ring	
	ω	ω	ω	ω	ω	ω	b	b	b	ω	b	ω	ω	b	bounded
	✓	✓	↯	✓	↯	✓	✓	✓	✓	✓	✓	✓	↯	✓	safe
absint	0.05	1.70	0.02	0.00	1.72	0.00	0.07	85.00	1.18	298.6	6.53	0.10	0.16	27.43	time (s)
	2.97	5.88	2.97	2.75	4.91	2.97	2.97	54.31	6.84	5.58	7.81	3.94	3.94	14.44	mem (MiB)
armc	0.11	331.92	0.01	0.00	—	0.02	—	4.79	3.14	195.88	0.21	0.12	0.03	328.18	
	4.95	773.8	2.97	2.97		2.97		106.62	31.06	14.59	6.84	5.88	3.94	3143.52	
cegar	0.23	—	0.02	0.06	2.66	0.40	0.02	1.41	8.06	—	7.92	0.89	0.12	14.9	
	3.94		2.97	2.97	7.91	3.94	2.97	46.56	14.59		18.47	6.84	3.94	35.91	
lart	—	—	0.01	0.02	56.21	—	0.02	1.62	1184.4	—	437.81	—	0.01	—	
			2.97	2.97	16.53		2.97	41.56	18.47		73.69		2.97		

Fig. 2. Benchmarking `verify`'s Different Algorithms on a Suite of Examples (we denote an abort due to an $1h$ time limit by "—", and note for each example if it is safe ($\checkmark$) or not safe ($\not\downarrow$), as well as if queues are used in a bounded way (b) or without restriction (ω); inconclusive results of `absint` are marked gray)

(e.g., `-tc-engine apinv-fwd -k-min -1` leads to $13.48s/15.56MiB$ (BRP) and $5.67s/10.72MiB$ (server)); however, there is also no default parameterization for cegar that can be shown to be superior (see [6] for details).

To conclude, there is no silver bullet for the safety verification of CM among our algorithms; however, their common front-end via `verify` proves to be a flexible tool that can cope with all our examples.

5   The McScM Framework

McScM's generic approach is based on symbolic finite control infinite data transition systems (FCID) (a notion inspired by [5]); the latter are given by a finite transition system enhanced by an appropriate region algebra as symbolic representation of the infinite data. For CM, the region algebra is given by QDD, and we define a regular abstraction for our systems thereupon.

Thus, McScM provides a generic API for, on the one hand, implementing new algorithms on a high level of abstraction; and, on the other hand, allows to apply the implemented algorithms to other members of the FCID family, by supplying a fitting region algebra and a suited notion of abstraction.

Related Tools: McScM relates to other symbolic model checking tools that can verify CM or subclasses thereof. SPIN [16], for example, allows to verify only CM with a priori bounded channels (e.g., those marked b in Figure 2), but allows for deciding properties specified in linear temporal logic. The CADP [10] toolbox includes a μ-calculus model checking tool limited to finite labelled transition systems, i.e., CM with bounded channels only. The same restriction to finite transition systems holds for other tools, like LTSA [13]. TReX [18] analyzes infinite state systems: *lossy* channel systems with local timed/counter automata. LEVER [12] is a learning-based model checker that supported CM with regular channel languages in a previous, not further available version. So, McScM offers— to our knowledge—the only currently freely available tool that can directly verify CM with reliable, unbounded FIFO channels.

The LASH library [11] offers only symbolic data structures channels, but does not provide any model checking algorithm for CM. The LASH API permits to symbolically present several classes of FCID (e.g., by QDD, number decision diagrams (NDD), real vector automata (RVA)), and to implement algorithms for each. Modular front-ends like TaPAS [17] (for FCIDs based on Presburger arithmetic) even allow to implement for multiple FCID libraries at once. Even tough McScM can be used in the same spirit to implement and compare model checking algorithms for a given class of FCID, we are able to provide *generic algorithms* that can be parameterized by any FCID for which we can provide a symbolic representation. The latter must only be conform to the above mentioned region algebra and supply an appropriate notion of abstraction, e.g., a suitable wrapper for LASH's RVA would directly port cegar to FCID representable by real vector automata, e.g. timed or hybrid systems.

Future Work: McScM is a work in progress, hence, we are always optimizing internals and provide extensions that prove handy for practical verification tasks, like our planned direct support for PROMELA as input language. Our next big step will lead beyond CM by allowing the local machines to have infinite data (like counters or timers), which demands new insights and notions for abstractions and invariants for these systems, as well as practicable algorithmic data-structures for implementing a region algebra.

References

[1] Aziz Abdulla, P., Jonsson, B.: Undecidable verification problems for programs with unreliable channels. Information and Computation 130(1), 71–90 (1996)

[2] Bouajjani, A., Habermehl, P., Vojnar, T.: Abstract Regular Model Checking. In: Alur, R., Peled, D.A. (eds.) CAV 2004. LNCS, vol. 3114, pp. 372–386. Springer, Heidelberg (2004)

[3] Brand, D., Zafiropulo, P.: On Communicating Finite-State Machines. J. ACM 30(2), 323–342 (1983)

[4] Clarke, E., Grumberg, O., Jha, S., Lu, Y., Veith, H.: Counterexample-guided Abstraction Renement for Symbolic Model Checking. J. ACM 50(5), 752–794 (2003)

[5] Henzinger, T., Majumdar, R., Raskin, J.-F.: A classification of symbolic transition systems. ACM Transactions on Computational Logic 6, 1–32 (2005)

[6] Heußner, A., Le Gall, T., Sutre, G.: Extrapolation-Based Path Invariants for Abstraction Refinement of Fifo Systems. In: Păsăreanu, C.S. (ed.) SPIN 2009. LNCS, vol. 5578, pp. 107–124. Springer, Heidelberg (2009)

[7] Kalyon, G., Le Gall, T., Marchand, H., Massart, T.: Global State Estimates for Distributed Systems. In: Bruni, R., Dingel, J. (eds.) FORTE 2011 and FMOODS 2011. LNCS, vol. 6722, pp. 198–212. Springer, Heidelberg (2011)

[8] Le Gall, T., Jeannet, B., Jéron, T.: Verification of Communication Protocols Using Abstract Interpretation of FIFO Queues. In: Johnson, M., Vene, V. (eds.) AMAST 2006. LNCS, vol. 4019, pp. 204–219. Springer, Heidelberg (2006)

[9] McMillan, K.L.: Lazy Abstraction with Interpolants. In: Ball, T., Jones, R.B. (eds.) CAV 2006. LNCS, vol. 4144, pp. 123–136. Springer, Heidelberg (2006)

[10] CADP, http://www.inrialpes.fr/vasy/cadp/

[11] LASH, http://www.montefiore.ulg.ac.be/~boigelot/research/lash/

[12] LEVER, http://abhayspace.com/static/lever.html

[13] LTSA, http://www.doc.ic.ac.uk/ltsa/

[14] McScM, https://altarica.labri.fr/forge/projects/mcscm
[15] SCM, Lattice Automata, http://gforge.inria.fr/projects/bjeannet/
[16] SPIN, http://spinroot.com
[17] TaPAS, http://altarica.labri.fr/forge/projects/3/wiki/TaPAS/
[18] TReX, http://www.liafa.jussieu.fr/~sighirea/trex/

SLMC: A Tool for Model Checking Concurrent Systems against Dynamical Spatial Logic Specifications

Luís Caires and Hugo Torres Vieira

CITI and Departamento de Informática, Faculdade de Ciências e Tecnologia,
Universidade Nova de Lisboa, 2829-516 Caparica, Portugal

Abstract. The Spatial Logic Model Checker is a tool for verifying π-calculus systems against safety, liveness, and structural properties expressed in the spatial logic for concurrency of Caires and Cardelli. Model-checking is one of the most widely used techniques to check temporal properties of software systems. However, when the analysis focuses on properties related to resource usage, localities, interference, mobility, or topology, it is crucial to reason about spatial properties and structural dynamics. The SLMC is the only currently available tool that supports the combined analysis of behavioral and spatial properties of systems. The implementation, written in OCAML, is mature and robust, available in open source, and outperforms other tools for verifying systems modeled in π-calculus.

1 Introduction

Model-checking is one of the most widely used verification techniques in the analysis of software applications. The usual focus is on behavioral/temporal properties, which allow to check liveness and safety properties of systems, from the standpoint of their externally observable behavior. However, it is often the case that verification really needs to address on properties about (spatial) distribution, mobility, or resource usage. It is then crucial to be able to observe the structural/spatial configuration of systems. Examples of such properties include connectivity – there is always an access route between two sites – or unique handling – there is at most one server listening on a given channel name – or race absence – no simultaneous sends/writes to the same receiver/reader.

The Spatial Logic Model Checker [3] is a tool that allows the user to automatically verify behavioral and spatial properties of distributed and concurrent systems expressed in the π-calculus. Its base logic is a very rich dynamical spatial logic for concurrency, conveniently containing as a subset the logics supported by other model-checkers for π-calculi (e.g., [15,11,18]). The verification algorithm (using on-the-fly techniques) is provably correct for all expressible processes, and complete for the class of bounded processes [4], including the finite control π-calculus. In the next section, we present the SLMC by going through a simple example, which already illustrates the usefulness of the tool, briefly presenting the input languages in the meanwhile (see, e.g., [12,14] and [4,5,6,7] for background on π-calculus and on dynamic spatial logics, respectively).

2 Checking a Topological Property of a Distributed Protocol

The example we now discuss models a protocol which allows a set of nodes to organize itself into a ring like structure. The basic idea of the protocol is that in each step

C. Flanagan and B. König (Eds.): TACAS 2012, LNCS 7214, pp. 485–491, 2012.

two rings (which includes the case of the singleton ring, i.e., a ring with one node) are merged into a larger ring. Then, regardless of the intermediate configurations, a sequence of such steps leads to the point in which the whole set of nodes is included in the ring. The correctness of such protocol may be verified by our tool, since we are able to observe the topology of the system and check if the protocol yields a ring configuration.

We start by the specification of the three possible states of a node in our system: state *Node* represents the initial state of a node, which has no connections; state *Link* represents a node which is in a ring, hence connected to its left and to its right (we use left and right for the sake of illustration); state *Leader* also represents a node which is in a ring, but is the only node in its ring that is willing to connect to other nodes.

The specification of the *Node* in SLMC syntax is as follows:

```
defproc Node(com) =
    new link,chan in select {
         com!(link,chan).chan?(right).Leader(com,link,right);
         com?(right,newch).newch!(link).Link(link,right)};
```

The `defproc` introduces a π-calculus process definition in the system, named Node. The parameter com is the name of a public channel used by nodes to connect to each other. The process specified creates names link and chan (cf., π-calculus name restriction) and then may select one of two possible behaviors: either it outputs on channel com the freshly created names link and chan or it receives some names right and newch in channel com. In the former case, the process proceeds by receiving right in chan, after which becomes the Leader of the ring. In the latter case, the process proceeds by sending link in the received newch, and then becomes a Link node.

We then specify a Link node as a process that either inputs from the node on its left or outputs to the node on its right, and after which proceeds as a Link node:

```
defproc Link(left,right) =
    select { left?().Link(left,right);
             right!().Link(left,right)};
```

Like the Link, a Leader also receives from its left node and outputs to its right:

```
defproc Leader(com,left,right) =
    new chan in select {
        left?().Leader(com,left,right);
        right!().Leader(com,left,right);
        com?(newr,newch).newch!(right).Link(left,newr);
        com!(right,chan).chan?(newr).Leader(com,left,newr)};
```

Furthermore, a Leader node is willing to connect to another ring via channel com (and a freshly created chan). Intuitively, two Leaders connect by swapping their right links, in such way merging two rings into one. This is the case both when the Leader receives or outputs on com, the difference is that the former implies yielding the Leader status (proceeding as Link), while the latter does not (proceeding as Leader).

The system is specified as a set of (e.g., four) Nodes that share a public com channel:

```
defproc System = Node(com) | Node(com) | Node(com) | Node(com);
```

We may now present the spatial/behavioral properties that characterize the system. For starters, we describe a leader node:

```
defprop leader(a,b) =
    1 and (a != b) and (@com) and (<a?> true) and (<b!> true);
```

This `defprop` command defines property `leader` (with parameters a,b), which describes processes which are indivisible (1), that have `com` as a free name (`@com`) and that are able to input on a name (`<a?>` after which proceeding as processes that satisfy `true`, i.e., any) and output on another name (`<b!>` after which proceeding as any process). A link has a similar description, where `com` is not a free name:

```
defprop link(a,b) =
    1 and (a != b) and (not @com) and (<a?> true) and (<b!> true);
```

A node may be described as an indivisible process which is not a link nor a leader:

```
defprop node =
    1 and not exists a. exists b. (leader(a,b) or link(a,b));
```

Notice properties `link`, `leader` and `node` are specially suited for the node specification of this system in particular. However, testing for indivisibility (single-threaded) is a generic feature of a node, which is possible to observe thanks to the expressiveness of the logic. Property 1 may be taken as an abbreviation of "non-empty system which cannot be decomposed into two non-empty parts" – `not 0 and not (not 0 | not 0)`.

The separating composition A | B is a key operator of the dynamic spatial logic, characterizing systems that can be decomposed (via structural congruence) in two parts, one satisfying property A and the other satisfying property B. Using parallel composition, we may, e.g., specify the initial state of the system as a composition of four nodes:

```
defprop initial = inside ( node | node | node | node );
```

Property `inside` is used so as to *reveal* all name restrictions, i.e., open the scopes of all name restrictions, in such way allowing for spatial decomposition to split threads otherwise indivisible because of the sharing of some restricted name.

We now turn to the verification of the correctness of the protocol. In order to characterize rings, we first introduce the notion of a chain of connected link nodes:

```
defprop chain(c,d) = (minfix C(a,b).( link(a,b) or
                       (exists x. (link(a,x) | C(x,b))))))(c,d);
```

Intuitively, the least fixpoint (`minfix`), parameterized by a,b initially instantiated by c,d, characterizes a chain of linked nodes where the leftmost and rightmost links are a and b, respectively. Such chain may either be a single link node, or there `exists` (the existential quantifier) name x such that there is a `link(a,x)` in *parallel* with a chain from x to b. Then, a ring is a chain of `link`s in parallel with a `leader`:

```
defprop ring = exists a. exists b. (leader(a,b) | chain(b,a));
```

Notice the chain connects b to a, for some names b, a, which are the right and left link of the leader node, respectively. We may now ask the tool if all execution paths lead (`always` and `eventually`, defined as usual) to a ring configuration:

```
check System |= always (eventually (inside (ring)));
```

The success of this verification, which explores all possible execution paths of the system and exploits the unique combination of behavioral and spatial properties supported by the tool, guarantees the protocol always leads, regardless of intermediate steps, to a final configuration of a ring that connects all nodes.

3   Verification Algorithms and Implementation

In this section we discuss the verification algorithms, based on [4] and on a canonical representation of processes, and present some benchmark figures.

The model-checking procedure is based on an on-the-fly technique, which means the model state space is explored gradually, guided by the deconstruction of the formula. In our case, the verification comprises observing both structure and behavior of processes. Namely, model-checking relies on decompositions of processes – up to structural congruence – to check the composition formula $A \mid B$, on observing behaviors of processes – up to the labeled transition system which defines the operational semantics – to check action modalities, and, crucially to check fixpoints, on the ability to compare two processes – up to the identification of some *irrelevant* names for the purpose of the model-checking, i.e., names that are not referred by the formula [4]. A great deal of the reasoning performed by our algorithms is optimized by relying on equivariance [10] (working up to name permutations).

Processes are modeled by data structures representing sets of equations in a normal form. Each equation describes a *flat* state of the process, where only immediate actions are represented. Using $\mathcal{X}$ to range over equation identifiers and α to range over actions we then write $\mathcal{X}(\overline{x}) \mapsto (\boldsymbol{\nu}\overline{a}) \, (\alpha_1 \mid \ldots \mid \alpha_k)$ to represent the equation identified by variable $\mathcal{X}$, parameterized by the $\overline{x}$ variable set ($\overline{x}$ abbreviates $x_1, \ldots, x_j$), specifying a flat configuration with restricted names $\overline{a}$ and consists of the composition of k actions α_1 to α_k. Action prefixes, denoted by p, and actions, denoted by α, are given by:

$$p \; ::= \; n!(\overline{m}) \quad \text{(Output)} \quad \mid \; n?(\overline{x}) \; \text{(Input)} \qquad \alpha \; ::= \; \alpha + \alpha \quad \text{(Sum)}$$
$$\mid \; [n = m] \; \text{(Test)} \quad \mid \; \tau \qquad \text{(Internal)} \qquad \mid \; p.\,\mathcal{X}(\overline{n}) \; \text{(Prefix)}$$

An action prefix may be an output $n!(\overline{m})$ (read "send names $\overline{m}$ on channel n"), an input $n?(\overline{x})$ (read "instantiate variables $\overline{x}$ with the names received on channel n"), a test $[n = m]$ (read "if n is the same as m proceed") and τ which represents a process internal action. Actions are either the non-deterministic choice of two actions $\alpha + \alpha$, or a prefix and its respective continuation. Continuations in our setting are specified by the corresponding equation variable, hence $p.\,\mathcal{X}(\overline{n})$ represents a process which after p behaves as specified in the equation identified by $\mathcal{X}$, instantiating its parameters with $\overline{n}$.

Abstracting continuation states with equation variables is crucial to quickly verify if two processes are the same, since we only need to check if the immediate actions and their respective continuations are the same. This simplification is crucial for the verification of fixpoints, which rely on an approximation of the fixpoint which is updated and consulted throughout unfolding, via the process comparison mechanism.

A process model may then be represented by a set of equations, together with an entry point. To further optimize the verification we model active (top-level) processes considering the set of *connected components* [9]: two processes that share a restricted name and thus cannot be decomposed via π-calculus structural congruence. Hence, for the sake of verifying the composition formula, which involves exploring all possibilities for decomposing a process, it is vital that the representation clearly identifies which threads are not decomposable, identifying the basic units of decomposition. The SLMC top-level process representation is then given by:

$$(\boldsymbol{\nu}\overline{a}_1) \, (\alpha_1^1 \mid \ldots \mid \alpha_k^1) \mid \ldots \mid (\boldsymbol{\nu}\overline{a}_j) \, (\alpha_1^j \mid \ldots \mid \alpha_k^j)$$

where each $(\boldsymbol{\nu}\overline{a}_i) \, (\alpha_1^i \mid \ldots \mid \alpha_k^i)$ piece (for some i) is an indivisible process because the α_t^i actions share between them the $\overline{a}_i$ restricted names. This way, the several possi-

Table 1. Model-Checking Deadlock Absence (in seconds)

	SLMC	Petruchio	MWB (prove)	MWB (check)	MMC
Handover	0.0005	0.2	0.002	0.015	0.01
Arrow (a)	0.01	0.8	0.115	–	–
Arrow (b)	0.3	4.3	6.2	–	–

Table 2. Model-Checking Spatial properties (in seconds)

Ring	Handover	Arrow (a)	Arrow (b)
0.08	0.02	0.11	7.76

bilities for decomposing a process in two pieces are obtained by the possible combinations of gathering these basic indivisible blocks.

At the level of the optimizations, the main challenge we address is the expedite (re-)building of the process normal form, i.e., updating the top-level process representation as the consequence of observing an action/transition. For example, observing an output action entails updating the top-level process with the continuation of the output which then becomes active. So, actions and name restrictions specified in the continuation configuration (given by the equation identified by the variable prefixed by the output) must be integrated in the top-level representation, and the set of connected components must be updated considering the "new" actions and due to restricted names scope changes.

We now present some benchmark figures, comparing with other existing π-calculus model-checkers: the Petruchio tool [11], the Mobility Workbench (MWB) [15], and the Mobility Model Checker (MMC) [18]. The comparison is established for the verification of a fundamental behavioral property: *deadlock absence*. We consider two challenging systems: Milner's implementation of the Handover protocol [12] and a π-calculus implementation of the Arrow Distributed Directory Protocol [8], both available in the SLMC homepage [3].

The numbers shown in Table 1 list the amount of time needed for each tool to verify the systems are deadlock free, obtained running the tools on a Mac OS X 10.5.8, 2.4GHz Intel Core 2 Duo. For the Petruchio tool in particular, the figures indicate the time needed to translate π-calculus specifications into petri-nets [13], since Petruchio exports such petri-nets to external verification engines to carry out the model-checking. In the case of the MWB, we distinguish between the `prove` and `check` procedures. We consider the arrow system as available in [3] – (a) – and also a small variation obtained by adding one node to the system – (b). Notice that both the `check` procedure available in the MWB and the MMC did not provide results for checking deadlock absence for the arrow system, due to timeout and memory overflow, respectively. The comparison with Petruchio and the MWB `prove` procedure is favorable to the SLMC, where the figures obtained hint on the complexity of the arrow system itself.

Notice however none of the tools mentioned above supports the verification of structural properties, and there exists none, to the best of our knowledge, which allows for the combined analysis of behavioral and spatial specifications as the ones expressible in dynamic spatial logic. Table 2 reports on figures obtained for verifying spatial properties over the same systems, namely the verification shown in the previous section in the Ring system, race-freedom in the Handover system, and a complex correctness property of the Arrow system (see [3]): it is always the case that every node may eventually gain exclusive access to the shared object. As stated above, no comparison is possible in this case, since no other tool can handle spatial properties as the SLMC does.

4   Concluding Remarks

The SLMC is publicly available online, in open source, and is often downloaded. The tool, in development since 2004, has reached a very high maturity and robustness level, and is very fast in practical use. It has been routinely used for teaching purposes in our department, and we would like to further promote its use elsewhere. The development of the first version of the tool was supported by the FET Profundis project [1]. The development continued under the support of project IP Sensoria [2], where the tool was included in the Sensoria tool suite, and extended for the verification of service-oriented systems, as described in [17]. In particular, we have also concluded recently further extensions to the tool, namely an extension to the applied π-calculus (for security), and another for checking choreography conformance of service-oriented applications, based on an encoding of the Conversation Calculus [16]. Further information about the Spatial Logic Model Checker may be found in http://ctp.di.fct.unl.pt/SLMC/.

References

1. FET Profundis Project, http://www.it.uu.se/profundis/
2. IP Sensoria Project, http://www.sensoria-ist.eu/
3. Spatial Logic Model Checker, http://ctp.di.fct.unl.pt/SLMC/
4. Caires, L.: Behavioral and Spatial Observations in a Logic for the π-Calculus. In: Walukiewicz, I. (ed.) FOSSACS 2004. LNCS, vol. 2987, pp. 72–89. Springer, Heidelberg (2004)
5. Caires, L.: Dynamical Spatial logics: A Tutorial Survey. Bulletin of the EATCS (2008)
6. Caires, L., Cardelli, L.: A Spatial Logic for Concurrency (Part I). Information and Computation 186(2), 194–235 (2003)
7. Cardelli, L., Gordon, A.: Anytime, Anywhere: Modal Logics for Mobile Ambients. In: Proceedings of POPL 2000, pp. 365–377. ACM Press (2000)
8. Demmer, M.J., Herlihy, M.P.: The Arrow Distributed Directory Protocol. In: Kutten, S. (ed.) DISC 1998. LNCS, vol. 1499, pp. 119–133. Springer, Heidelberg (1998)
9. Engelfriet, J., Gelsema, T.: Multisets and Structural Congruence of the π-Calculus with Replication. Theor. Comput. Sci. 211(1-2), 311–337 (1999)
10. Gabbay, M., Pitts, A.: A New Approach to Abstract Syntax with Variable Binding. Formal Aspects of Computing 13(3-5), 341–363 (2002)
11. Meyer, R., Strazny, T.: Petruchio: From Dynamic Networks to Nets. In: Touili, T., Cook, B., Jackson, P. (eds.) CAV 2010. LNCS, vol. 6174, pp. 175–179. Springer, Heidelberg (2010)
12. Milner, R.: Communicating and Mobile Systems: the π-Calculus. CUP (1999)
13. Petri, C., Reisig, W.: Petri net. Scholarpedia 3(4), 6477 (2008)
14. Sangiorgi, D., Walker, D.: The π-Calculus: A Theory of Mobile Processes. CUP (2001)

15. Victor, B., Moller, F.: The Mobility Workbench - A Tool for the π-Calculus. In: Dill, D.L. (ed.) CAV 1994. LNCS, vol. 818, pp. 428–440. Springer, Heidelberg (1994)
16. Vieira, H.T., Caires, L., Seco, J.C.: The Conversation Calculus: A Model of Service-Oriented Computation. In: Gairing, M. (ed.) ESOP 2008. LNCS, vol. 4960, pp. 269–283. Springer, Heidelberg (2008)
17. Wirsing, M., Hölzl, M. (eds.): SENSORIA. LNCS, vol. 6582. Springer, Heidelberg (2011)
18. Yang, P., Ramakrishnan, C., Smolka, S.: A Logical Encoding of the π-Calculus: Model-Checking Mobile Processes Using Tabled Resolution. STTT 6(1), 38–66 (2004)

TAPAAL 2.0: Integrated Development Environment for Timed-Arc Petri Nets[*]

Alexandre David, Lasse Jacobsen, Morten Jacobsen, Kenneth Yrke Jørgensen,
Mikael H. Møller, and Jiří Srba

Department of Computer Science, Aalborg University,
Selma Lagerlöfs Vej 300, 9220 Aalborg Øst, Denmark

Abstract. TAPAAL 2.0 is a platform-independent modelling, simulation and verification tool for extended timed-arc Petri nets. The tool supports component-based modelling and offers an automated verification of the EF, AG, EG and AF fragments of TCTL via translations to UPPAAL timed automata and via its own dedicated verification engine. After more than three years of active development with a main focus on usability aspects and on the efficiency of the verification algorithms, we present the new version of TAPAAL 2.0 that has by now reached its maturity and offers the first publicly available tool supporting the analysis and verification of timed-arc Petri nets.

1 Introduction

Timed-arc Petri nets (TAPN) are a particular time extension of the classical Petri net model where the time information is attached to the tokens in the net, representing their ages, and arcs from places to transition contain time intervals that restrict the ages of tokens which are moved along the arcs. The model of TAPN was first studied by Bolognesi, Lucidi, Trigila and Hanisch [4,9] and it has proved particularly suitable for modelling of manufacturing systems, workflow management systems and other applications [1,2,14,15,16].

We present TAPAAL 2.0, an open source and platform-independent tool (available from www.tapaal.net) that allows users to edit, simulate and verify extended TAPN models in a component-based fashion through shared interfaces. There exist several other verification tools for timed models, like UPPAAL [19] for networks of timed automata, and Romeo [8] and Tina [3] for time Petri nets (where time intervals are associated with transitions, as opposed to tokens like in the TAPN model). However, all these models are rather different (and complementary) and even though translations among them are possible [18], their suitability from the modeller's point of view depends on the application domain. For example, time Petri nets use an a priory fixed number of clocks (one for each transition) while TAPN allow for dynamic creation of tokens that carry their own local clocks. Some criticism on the classical TAPNs mentions the lack of modelling features for ensuring urgent behaviour and lack of read-arcs. In

---

[*] The paper was supported by VKR Center of Excellence MT-LAB.

C. Flanagan and B. König (Eds.): TACAS 2012, LNCS 7214, pp. 492–497, 2012.

the extended TAPN model used in TAPAAL, the weak points are dealt with by enforcing urgency via age invariants, modelling read-arcs in a more general setting via transport arcs and introducing other features, like inhibitor arcs and components, facilitating a more convenient modelling.

We are not aware of other tools for the analysis of TAPN, except for two tool prototypes implementing a backward coverability algorithm based on the better-quasi-ordering technique [2] and a forward reachability algorithm presented in [1]. Both implementations consider only the basic TAPN model. The former tool allows to verify solely coverability queries (remarkably also for unbounded nets) while the latter one may not terminate as the reachability questions are in general undecidable. Neither of the tools supports a GUI interface and they do not seem to be maintained.

TAPAAL 1.1 was presented in [6] as a tool providing a TAPN editor, simulator and translator to UPPAAL timed automata. The two translations implemented in TAPAAL 1.1 preserve only safety properties but have showed the potential of the translation approach. In the present version of TAPAAL 2.0, the tool now has a completely new and dedicated verification engine (implemented in C++) and two novel translations preserving liveness. In addition, the GUI and the modelling features were significantly extended; notably we now support component-based model development, constants, inhibitor arcs, advanced query creation dialog, batch processing engine and numerous other features. The theory behind the tool has been published in [17,5,11,12].

2   Tool Description

The architecture of the tool is outlined in Figure 1. The GUI of the tool has originally been developed as an open source project PIPE 2.5 [10] but since its TAPAAL branch in 2008, it has been significantly extended with new features like a component-based editor that allows to describe component interfaces via shared places and transitions. The composed net can be simulated in a timed simulator, displaying traces with concrete delays, or verified either via TAPAAL's own engine or via automatic translations to the UPPAAL engine. Verification can be initiated from a user-friendly query dialog or as a batch job.

Verification Options. TAPAAL 2.0 adds two new translations from extended TAPNs to networks of timed automata that use the broadcast communication feature of UPPAAL [11] and contrary to the previous two translations that rely on handshake synchronization (already present in TAPAAL 1.1), they do not produce additional deadlocks and hence allow us to verify also liveness properties. The two new translations can additionally handle all the new extended features, including inhibitor arcs. The main extension of TAPAAL 2.0 is its own dedicated engine implementing an efficient forward reachability algorithm on abstract markings (ages of tokens represented via zones) while applying on-the-fly active clock reduction (resizing of zones to contain only clocks of the active tokens) and other optimizations. The monotonicity property (more tokens added

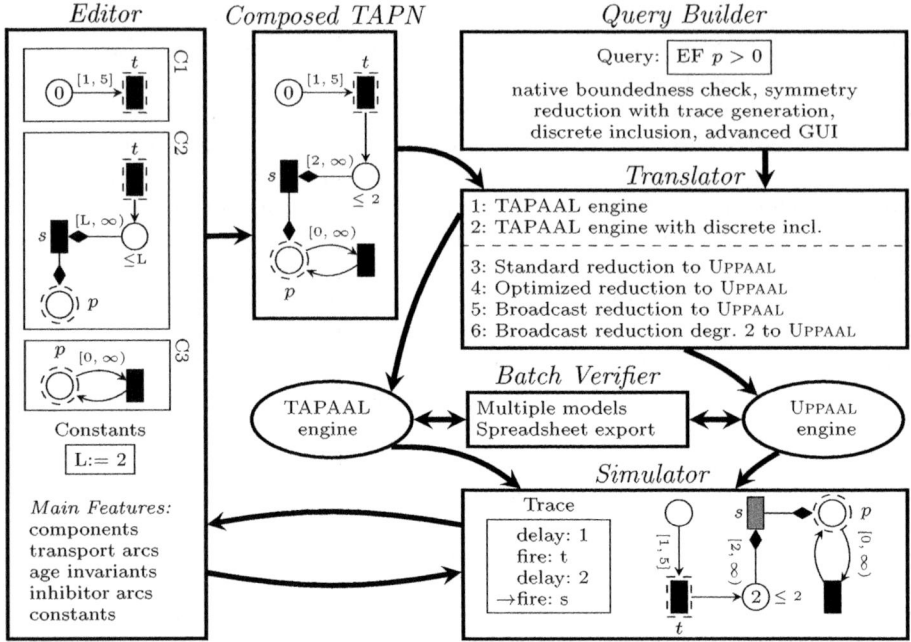

Fig. 1. Architecture of TAPAAL 2.0

to the net cause only more behaviour) that holds for the basic TAPNs, but breaks for the extended TAPNs due to the features like age invariants and inhibitor arcs, can be used to speed-up the verification. From the static analysis of the net, we define a novel ordering on markings of the net so that monotonicity is preserved even for the extended TAPN model and we exploit this in the reachability algorithm, still providing exact verification answers but often with considerable speed-up as demonstrated in Section 3. This technique, called *discrete inclusion*, can be further optimized by a manual intervention of the modeller. The verification engine also implements an automatic symmetry reduction technique and returns executable traces even if symmetry is activated (unlike e.g. UPPAAL or TAPAAL 1.1). Finally, the new engine implements a k-boundedness check of a given net. Even if the net is unbounded, the verification up to k tokens in the net is possible, providing a suitable under-approximation of the net behaviour.

Management of Tool Development. To facilitate easy collaboration between the TAPAAL tool contributors, we utilize launchpad.net/tapaal, a free tool-chain for collaboration in open source projects. Among others, all software bugs found in TAPAAL are registered and tracked using launchpad's bug management system. To this day, more than 20 individuals have contributed to the development of TAPAAL, working on more than 200 registered bugs and features, during over ten official releases of TAPAAL.

	UPPAAL engine				TAPAAL translations				TAPAAL engine			
	original		improved		original		improved		original		improved	
#	no	yes	no	yes	no	yes	no	yes	no	yes	no	yes
3	0.1	<0.1	<0.1	<0.1	0.4	0.2	<0.1	<0.1	0.2	<0.1	<0.1	<0.1
4	0.4	<0.1	<0.1	<0.1	16.8	0.3	<0.1	0.1	2.8	<0.1	<0.1	<0.1
5	5.3	0.1	<0.1	<0.1	—	0.6	<0.1	0.1	89.1	0.2	<0.1	<0.1
6	220.5	0.2	<0.1	<0.1	—	1.8	<0.1	0.1	—	0.9	<0.1	<0.1
7	—	1.1	0.1	<0.1	—	14.5	0.1	0.1	—	6.3	<0.1	<0.1
8	—	3.6	0.5	<0.1	—	104.8	0.1	0.1	—	48.9	<0.1	<0.1
9	—	20.7	3.1	<0.1	—	—	0.1	0.1	—	—	<0.1	<0.1
10	—	143.6	23.2	<0.1	—	—	0.1	0.1	—	—	<0.1	<0.1
11	—	—	148.0	<0.1	—	—	0.1	0.1	—	—	<0.1	<0.1
40	—	—	—	0.9	—	—	0.6	0.7	—	—	4.1	0.6
80	—	—	—	22.8	—	—	11.1	12.7	—	—	158.9	11.0
120	—	—	—	159.8	—	—	73.9	84.8	—	—	—	68.3
160	—	—	—	—	—	—	—	293.8	—	—	—	262.3

Fig. 2. Scheduling Feasibility of MPEG-2 Encoder (time in seconds)

3   Experiments

We present two new case studies in order to argue for the efficiency and applicability of the tool. More experimental results can be found e.g. in [5,12] and several TAPN models are available within the tool (under File/Example nets). All the models used in the following experiments can be obtained from the tool homepage (section Download). The experiments were carried out on a MacBook Pro equipped with a 2.7GHz Intel Core i7 and 8GB of RAM with a 300 seconds time limit. We used the 64-bit versions of TAPAAL 2.0.2 and UPPAAL 4.1.4.

MPEG-2 Case Study. We model the MPEG-2 algorithm that encodes a group of frames on a multiprocessor architecture. The algorithm treats one initial I-frame, a number of B-frames (we parameterize our model on this number), and a final P-frame. The TAPN model was taken directly from [14]. We recreated the UPPAAL model from the descriptions in [7] since their original UPPAAL model was not available anymore. The results are in Figure 2. The columns called *original* list the verification times for the model described in [14]; in the *improved* variant we employed several additional modelling optimizations (both in the timed automata and the TAPN model) via the use of invariants and symmetry reduction (features not available to the authors of [14]). The query asks whether the encoding can be performed within a given time bound (that we vary). In positive cases (*yes* columns) we used DFS, otherwise (*no* columns) we used BFS. This allows us to see how good the tools are to find a trace to a reachable state or to explore the whole state-space. The discrete inclusion technique does not improve the performance of the TAPAAL engine in this particular case.

Lynch-Shavit Protocol. The second case study is a timed-based mutual exclusion algorithm by Lynch and Shavit [13]. Both the TAPN and timed automata models

#	Uppaal engine	TAPAAL translations	TAPAAL engine	TAPAAL inclusion	TAPAAL M-incl.	TAPAAL M*-incl.
15	0.2	0.4	0.9	0.2	0.2	0.2
25	2.4	3.0	8.5	1.1	0.9	0.9
35	14.6	15.8	42.7	4.6	4.00	3.9
45	62.6	57.1	153.3	14.3	11.3	10.5
55	190.8	164.7	—	38.8	27.2	25.0
65	—	—	—	106.5	57.7	52.7
75	—	—	—	262.4	113.2	100.5
85	—	—	—	—	203.2	178.6
95	—	—	—	—	—	299.2

Fig. 3. Lynch-Shavit Protocol for Mutual Exclusion (time in seconds)

were taken from [1]. The column called *inclusion* refers to the generic application of the discrete inclusion technique, *M-incl.* refers to a manual optimization of the technique and *M*-incl.* shows a possible best performance of the technique (optimized by a brute force search with an automatic script). The results are presented in Figure 3 for a different number of processes participating in the protocol. The discrete inclusion technique is not an over-approximation of the behaviour and provides conclusive answers applied to any model, not only the protocol in this case study.

TAPAAL's performance is convincing in comparison with state-of-the-art model checkers like Uppaal and, in several cases, it provides a considerably faster verification. Due to space limitation, we present only two case studies but we also observed similar performance improvements for other models. In particular, if the net structure allows for more tokens in the same place, the generic discrete inclusion technique often gives a significant speed up and it can be further manually tuned up. Moreover, if symmetry reduction is applicable, it is often better exploited in the net models (where its detection is automatic on contrary to the user defined one in Uppaal models) and our translations create networks of timed automata that are significantly faster to verify than the native Uppaal models. Last but not least, TAPAAL allows for simulation of concrete error traces (even in case of symmetry reduction) while many other timed automata and Petri net tools display only the abstract ones, which makes them difficult to understand for the end-users.

4   Conclusion

TAPAAL 2.0 is an open source, platform-independent modelling and verification tool for extended timed-arc Petri nets. The tool has reached its maturity both in the GUI aspects as well as in the actual verification performance. The tool is becoming increasingly popular as documented by the total number of 2089 downloads (calculated in October 2011), out of which more than 650 downloads took part since April 2011.

References

1. Abdulla, P.A., Deneux, J., Mahata, P., Nylén, A.: Using forward reachability analysis for verification of timed Petri nets. Nordic J. of Computing 14, 1–42 (2007)
2. Abdulla, P.A., Nylén, A.: Timed Petri Nets and BQOs. In: Colom, J.-M., Koutny, M. (eds.) ICATPN 2001. LNCS, vol. 2075, pp. 53–70. Springer, Heidelberg (2001)
3. Berthomieu, B., Ribet, P.-O., Vernadat, F.: The tool TINA — construction of abstract state spaces for Petri nets and time Petri nets. International Journal of Production Research 42(14), 2741–2756 (2004)
4. Bolognesi, T., Lucidi, F., Trigila, S.: From timed Petri nets to timed LOTOS. In: Proc. of the IFIP WG 6.1 10th International Symposium on Protocol Specification, Testing and Verification, pp. 1–14. North-Holland, Amsterdam (1990)
5. Byg, J., Jørgensen, K.Y., Srba, J.: An Efficient Translation of Timed-Arc Petri Nets to Networks of Timed Automata. In: Breitman, K., Cavalcanti, A. (eds.) ICFEM 2009. LNCS, vol. 5885, pp. 698–716. Springer, Heidelberg (2009)
6. Byg, J., Jørgensen, K.Y., Srba, J.: TAPAAL: Editor, Simulator and Verifier of Timed-Arc Petri Nets. In: Liu, Z., Ravn, A.P. (eds.) ATVA 2009. LNCS, vol. 5799, pp. 84–89. Springer, Heidelberg (2009)
7. Cambronero, M.E., Ravn, A.P., Valero, V.: Using UPPAAL to analyze an mpeg-2 algorithm. In: Proc. of VII Workshop Brasileiro de Tempo Real, pp. 73–82 (2005)
8. Gardey, G., Lime, D., Magnin, M., Roux, O.H.: Romeo: A Tool for Analyzing Time Petri Nets. In: Etessami, K., Rajamani, S.K. (eds.) CAV 2005. LNCS, vol. 3576, pp. 418–423. Springer, Heidelberg (2005)
9. Hanisch, H.M.: Analysis of Place/Transition Nets with Timed-Arcs and Its Application to Batch Process Control. In: Ajmone Marsan, M. (ed.) ICATPN 1993. LNCS, vol. 691, pp. 282–299. Springer, Heidelberg (1993)
10. Platform Independent Petri net Editor 2.5, http://pipe2.sourceforge.net
11. Jacobsen, L., Jacobsen, M., Møller, M.H., Srba, J.: A Framework for Relating Timed Transition Systems and Preserving TCTL Model Checking. In: Aldini, A., Bernardo, M., Bononi, L., Cortellessa, V. (eds.) EPEW 2010. LNCS, vol. 6342, pp. 83–98. Springer, Heidelberg (2010)
12. Jacobsen, L., Jacobsen, M., Møller, M.H., Srba, J.: Verification of Timed-Arc Petri Nets. In: Černá, I., Gyimóthy, T., Hromkovič, J., Jefferey, K., Králović, R., Vukolić, M., Wolf, S. (eds.) SOFSEM 2011. LNCS, vol. 6543, pp. 46–72. Springer, Heidelberg (2011)
13. Lynch, N., Shavit, N.: Timing-based mutual exclusion. In: Proceedings of the 13th IEEE Real-Time Systems Symposium, pp. 2–11 (1992)
14. Pelayo, F.L., Cuartero, F., Valero, V., Macia, H., Pelayo, M.L.: Applying timed-arc Petri nets to improve the performance of the MPEG-2 encoding algorithm. In: Proc. of MMM 2004, pp. 49–56. IEEE (2004)
15. Valero, V., Pardo, J.-J., Cuartero, F.: Translating TPAL Specifications into Timed-Arc Petri Nets. In: Esparza, J., Lakos, C.A. (eds.) ICATPN 2002. LNCS, vol. 2360, pp. 414–433. Springer, Heidelberg (2002)
16. Ruiz, V.V., Pelayo, F.L., Cuartero, F., Cazorla, D.: Specification and analysis of the MPEG-2 video encoder with timed-arc Petri nets. ENTCS 66(2) (2002)
17. Srba, J.: Timed-Arc Petri Nets vs. Networks of Timed Automata. In: Ciardo, G., Darondeau, P. (eds.) ICATPN 2005. LNCS, vol. 3536, pp. 385–402. Springer, Heidelberg (2005)
18. Srba, J.: Comparing the Expressiveness of Timed Automata and Timed Extensions of Petri Nets. In: Cassez, F., Jard, C. (eds.) FORMATS 2008. LNCS, vol. 5215, pp. 15–32. Springer, Heidelberg (2008)
19. UPPAAL, http://uppaal.org

A Platform for High Performance Statistical Model Checking – PLASMA

Cyrille Jegourel, Axel Legay, and Sean Sedwards⋆

INRIA Rennes – Bretagne Atlantique
sean.sedwards@inria.fr

Abstract. Statistical model checking offers the potential to decide and quantify dynamical properties of models with intractably large state space, opening up the possibility to verify the performance of complex real-world systems. Rare properties and long simulations pose a challenge to this approach, so here we present a fast and compact statistical model checking platform, PLASMA, that incorporates an efficient simulation engine and uses importance sampling to reduce the number and length of simulations when properties are rare. For increased flexibility and efficiency PLASMA compiles both model and property into bytecode that is executed on an in-built memory-efficient virtual machine.

1 Introduction

The need to provide accurate predictions about the behaviour of complex systems is increasingly urgent. With computational power ever-more affordable and compact, man-made systems are inevitably becoming increasingly computerised, distributed and concurrent, creating a correspondingly increased burden to check that they function correctly. At the same time, following the success of the human genome project, there is an increased expectation that computers can provide answers to important questions raised by complex systems in the life sciences.

Complex systems tend to pose two particular challenges to formal verification: the non-determinism caused by concurrency and unpredictable environmental conditions and the size of the state space. Our focus here is *model checking*, that can verify the most intricate details of a system's dynamical behaviour and where non-determinism may be handled by assigning probabilistic distributions to unknowns and by quantifying results with a probability - *probabilistic model checking*. 'Exact' probabilistic model checking quantifies these probabilities to the limit of numerical precision by an exhaustive exploration of the state space, but is restricted in practise by what can be conveniently stored in memory. Techniques exist to work with a reduced state space (abstraction, lumping, etc.), but the state space of most real natural and man-made systems remain intractable.

Statistical model checking (SMC) avoids an explicit representation of the state space by building a statistical model of the executions of a system and estimating results within confidence bounds. An executable model of the system is run repeatedly and each simulation trace is verified against a property specified in

---

⋆ Corresponding author.

C. Flanagan and B. König (Eds.): TACAS 2012, LNCS 7214, pp. 498–503, 2012.

temporal logic. Examples of tools that have successfully applied this approach are [10,7]. Knowing a result with less than 100% confidence is often sufficient, since the confidence bounds may be made arbitrarily tight, however the key challenges of this approach are to reduce the length (simulation steps and cpu time) and number of simulation traces necessary to achieve a result with given confidence. The current proliferation of parallel computer architectures (multiple cpu cores, grids, clusters, clouds and general purpose computing on graphics processors, etc.) makes the production of multiple independent simulation runs relatively easy, but it is still necessary to make simulation as efficient as possible. Rare properties pose a particular problem for simulation-based approaches, since they are not only difficult to observe (by definition) but their probability is difficult to bound [2].

In what follows we present the prototype of a flexible SMC platform, PLASMA[1], that incorporates an in-built compiler and virtual machine to perform memory- and time-efficient simulations. PLASMA incorporates an efficient discrete event simulation algorithm and uses importance sampling to reduce the necessary number of simulation runs when properties are rare.

2   Software Architecture

PLASMA adopts a modular architecture to facilitate the extension of its features (Fig. 1). Models can already be specified using the PRISM reactive modules syntax [3] and biochemical syntax of the form $A + B \rightarrow C + D$, while the implementation of other modelling formalisms, such as timed automata and procedural programming languages such as C and Java, is in prospect. The input specification is translated into a notional common intermediate language based on elements (referred to as *simple commands* because they have no explicit synchronisation and no choice of actions) having the structure (*guard,rate,actions*), where *guard*, *rate* and *actions* are functions over the current state (constants, variables, clocks) of the system. The intermediate language thus expresses the semantics of a system that advances by discrete events: the *guard* enables the command, the *rate* resolves non-determinism between enabled commands (and controls the delay in continuous time systems) and the *actions* update the state of the system. Different input languages may be facilitated by implementing parsers that construct and fill data structures that reflect simple commands. Once the model is represented in the intermediate language it is compiled into an executable form (the *model program*).

PLASMA uses in-built compilers to create bytecode for execution on its own stack-based virtual machine (VM) that comprises a logic VM (LogicVM in Fig. 1) and a simulation VM (SimVM in Fig. 1). PLASMA's bytecode instructions constitute a domain-specific, low level, platform-independent language designed for efficient statistical model checking. This language contains standard low level instructions, such as *push, pop, add, sub, mul, div*, etc., as well as non-standard

---

[1] A demonstration version of PLASMA may be downloaded from
https://sites.google.com/site/plasmasmc

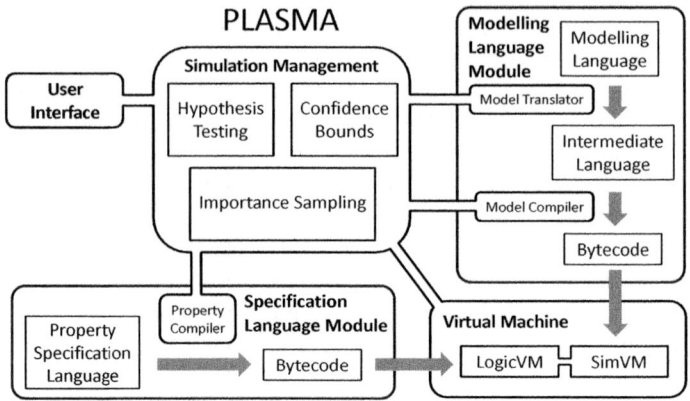

Fig. 1. The architecture of PLASMA

instructions to construct efficient model checking algorithms. The VM is implemented in a high level procedural programming language (currently Java, but the code uses no features that cannot easily be adapted to other languages) and is efficient because it is optimised for its domain of application: high level instructions are efficient sub-parts of model-checking algorithms and all instructions are optimised with respect to the hardware level. The compiler and VM are also sufficiently compact to allow PLASMA to be implemented as a browser application, a distributed component or in an embedded system etc.

PLASMA verifies properties specified in bounded temporal logic. Such properties are compiled into bytecode programs (*property programs*) and then executed on the logic VM. Our current focus is discrete time, however continuous time and other logics may be easily facilitated by implementing additional logic parser-compilers. Overall control of the verification process is maintained by the simulation management kernel (SMK) according to the options specified by the user. In general, the property program executes the model program until it has seen sufficient steps to decide a result and the SMK executes the property program until it has sufficient results to return an answer to the user. In this way, simulation traces contain the minimum number of states necessary to decide the property and the minimum number of simulations are generated. The logic accepts arbitrarily nested path formulae, however formulae that are not nested are particularly memory efficient: by employing a multivalued logic (*true, false, undecided*) PLASMA need only store the current state of the system. Nested formulae are also handled efficiently. In general, PLASMA stores only a subset of the full trace, having length equal to the maximum sum of the time bounds of any nested formulae.

2.1 Stochastic Simulation Algorithm

PLASMA performs discrete event simulation using the 'method of arbitrary partial propensities' (MAPP [6]). The MAPP is based on the Gillespie 'direct method' (DM [8]) but performs significantly better in large-scale practical applications than either the DM or the asymptotically better 'next reaction method' (NRM [1]). In a system of M simple commands, each step of the DM is $\mathcal{O}(M)$ because it iterates through all the commands in the system to find the command to execute and then again to update all the guards and rates following the execution of the chosen command. The MAPP divides the M commands into $\sqrt{M}$ subsets of $\sqrt{M}$ commands and thus divides choosing a command into two operations of $\mathcal{O}(\sqrt{M})$: choosing a subset and choosing a command within the subset. By performing an initial dependency analysis of the system the MAPP avoids updating commands whose guards and rates are not affected during the simulation. Since the average number of dependent commands, D, tends to be smaller than and independent of M, the overall complexity of the MAPP is $\mathcal{O}(\sqrt{M})$. The NRM achieves asymptotic complexity of $\mathcal{O}(\log M)$ by performing a similar dependency analysis and by arranging the commands in an ordered binary tree whose root always contains the next command to be executed. Choosing a reaction is $\mathcal{O}(1)$, but maintaining the invariant property of the tree is proportional to $D \log_2 M$. D is assumed constant, but the fact that it multiplies the complexity of the NRM tends to make the MAPP more efficient in most small to large-scale applications. See Figure 3.

2.2 Rare Properties and Importance Sampling

The process of statistical model checking estimates the probability of a property by verifying the execution paths of multiple independent simulation runs. If Ω is a probability space of traces ($\omega \in \Omega$), $f(\omega)$ the probability measure over Ω and $z(\omega) \in \{0, 1\}$ is a function indicating whether ω satisfies some property, then the expected probability of the property is given exactly by $\gamma = \int_\Omega z(\omega)f(\omega) \, d\omega$. This leads directly to the standard Monte Carlo estimator: $\tilde{\gamma} = \frac{1}{N} \sum_{i=1}^{N} z(\omega_i)$, where ω_i are simulation paths realised under distribution f.

As $\gamma \to 0$, however, it becomes increasingly difficult to bound the relative error in $\tilde{\gamma}$ and N becomes prohibitively large [2]. Importance sampling can be used to reduce N by performing simulations under a 'tilted' (importance sampling) distribution that produces the rare paths more frequently and by compensating for the tilt using the 'likelihood ratio'. If f' is the importance sampling distribution then $l(\omega) = \frac{f(\omega)}{f'(\omega)}$ is the likelihood ratio and $\gamma = \int_\Omega z(\omega) \frac{f(\omega)}{f'(\omega)} f'(\omega) \, d\omega$. This leads to the importance sampling estimator $\tilde{\gamma} = \frac{1}{N} \sum_{i=1}^{N} l(\omega_i')z(\omega_i')$, where ω_i' are simulation paths realised under the the importance sampling distribution and $l(\omega_i')$ is calculated on the fly.

By individually optimising all the probabilities in the transition system ('state dependent tilting') it is conceivable to create very good importance sampling distributions, however this is not in general tractable. PLASMA therefore adopts a parameterised (state *independent*) tilting scheme based on its intermediate language representation of the model. For each simple command in the system, an importance sampling parameter taking strictly positive values is introduced to bias the rates. To test the performance of PLASMA's paramterised importance sampling engine we applied it to repair models from [5] that have previously been considered in the context of state dependent tilting and which may be verified by PRISM's numerical algorithm. We have found that our state independent tilting scheme is nevertheless capable of achieving dramatic increases in performance. For instance, example 1 of [5] considers a property with probability 1.17×10^{-7}, requiring an expected 10^8 simulation runs to see a few examples. Using just six parameters PLASMA is able to make a 10^6-fold increase in the frequency of observing the rare event.

3   Results

Figure 2 illustrates typical performance scaling[2] of PLASMA's SMC engine relative to PRISM's numerical algorithm, applying them to increasing instance sizes of a classic probabilistic model checking problem (the randomised dining philosophers protocol of [4]). The state space increases exponentially with respect to instance size, hence PRISM's time scaling is exponential and its maximum instance size is here limited by available memory to about 35 philosophers. By estimating a result with (arbitrarily) bounded error, PLASMA can work with much larger models and its performance scales linearly in time proportional to the length of the property formula. Since PLASMA's memory requirement also scales linearly with instance size, its limit is much higher than the maximum shown in Figure 2.

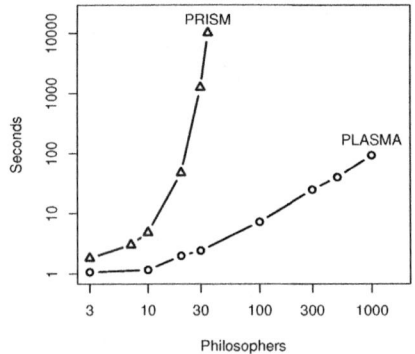

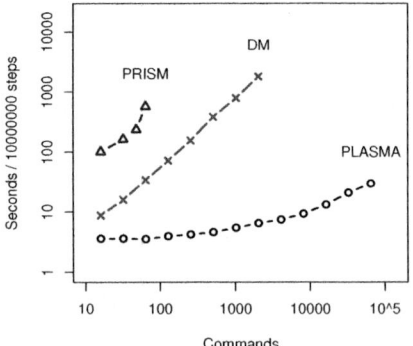

Fig. 2. Performance scaling of PLASMA's SMC engine vs. PRISM's numerical algorithm. PLASMA performed 118595 simulations per point.

Fig. 3. Simulation algorithm performance scaling using the genetic oscillator of [9] as a building block. The DM was implemented in PLASMA.

Figure 3 illustrates typical simulation performance scaling of PLASMA's MAPP algorithm in comparison to the direct method of [8] and PRISM's simulation en-

---

[2] All results generated under Windows 7 Enterprise 64-bit and Java 1.6.0_26 32-bit on Intel Core i7 CPU M640 @ 2.80Ghz with 4GB RAM. PRISM 4.0.1 was used.

gine. A stochastic oscillatory model from systems biology [9] was used as the building block to construct plausible biological models of increasing complexity. Using the DM's $\mathcal{O}(M)$ scaling as reference, the lower order scaling of PLASMA's MAPP algorithm is clear. The performance of PRISM's simulation algorithm is also here limited by memory, but in this case the limit is not related to the state space, which is intractable for even the smallest instance.

4   Conclusion and Future Challenges

PLASMA is a compact, efficient and flexible SMC platform that incorporates a novel importance sampling engine. Its broad goal is to take SMC beyond proof of concept and to tackle the analysis of real-world systems. Since such systems are usually not written in abstract modelling languages, we foresee a need to implement other input languages to avoid errors and make the process of verification more convenient. Of particular interest are timed and hybrid systems, since these are commonly used in industrial applications. Importance sampling constitutes a major thread of our research, as it has the potential to dramatically increase the performance of simulation-based techniques. A key challenge is the discovery of good parameterised importance sampling distributions and we are currently developing algorithms to infer these automatically.

References

1. Gibson, M., Bruck, J.: Efficient exact stochastic simulation of chemical systems with many species and many channels. J. of Physical Chemistry A 104, 1876 (2000)
2. Heidelberger, P.: Fast simulation of rare events in queueing and reliability models. ACM Trans. Model. Comput. Simul. 5, 43–85 (1995)
3. The PRISM manual,
   http://www.prismmodelchecker.org/manual/
4. Pnueli, A., Zuck, L.: Verification of multiprocess probabilistic protocols. Distributed Computing 1, 53–72 (1986)
5. Ridder, A.: Importance sampling simulations of markovian reliability systems using cross-entropy. Annals of Operations Research 134, 119–136 (2005)
6. Sedwards, S.: A Natural Computation Approach To Biology: Modelling Cellular Processes and Populations of Cells With Stochastic Models of P Systems. PhD thesis, University of Trento (2009)
7. Sen, K., Viswanathan, M., Agha, G.A.: Vesta: A statistical model-checker and analyzer for probabilistic systems. In: QEST, pp. 251–252. IEEE (2005)
8. Daniel, T., Gillespie: A general method for numerically simulating the stochastic time evolution of coupled chemical reactions. Journal of Computational Physics 22(4), 403–434 (1976)
9. Vilar, M.G., Kueh, H.Y., Barkai, N., Leibler, S.: Mechanisms of noise-resistance in genetic oscillators. Proceedings of the National Academy of Science 99 (2002)
10. Younes, H.L.S.: Ymer: A Statistical Model Checker. In: Etessami, K., Rajamani, S.K. (eds.) CAV 2005. LNCS, vol. 3576, pp. 429–433. Springer, Heidelberg (2005)

Competition on Software Verification[*]
(SV-COMP)

Dirk Beyer

University of Passau, Germany

Abstract. This report describes the definitions, rules, setup, procedure, and results of the 1st International Competition on Software Verification. The verification community has performed competitions in various areas in the past, and SV-COMP'12 is the first competition of verification tools that take software programs as input and run a fully automatic verification of a given safety property. This year's competition is organized as a satellite event of the International Conference on Tools and Algorithms for the Construction and Analysis of Systems (TACAS).

1 Introduction

The area of verification, in particular model checking, has grown to an own major research area within computer science, which is witnessed and acknowledged by a recent ACM Turing Award in the area and the growth of conferences in the field of verification to some of the top computer-science conferences with high impact on the research community. Model checking started to get adopted in software industry (e.g., Microsoft, NASA, NEC) about ten years ago, and major tool-development projects in software model checking began around that time (BLAST at UC Berkeley, SLAM at MSR, MAGIC at CMU).

Several new and powerful software-verification tools became available, but they have not been compared systematically in the past. The reason for this is that no widely distributed benchmark suite was available and most concepts were only validated in research prototypes. This can be changed by a competition. Comparison, and thus competition, is a driving force for the invention of new methods, technologies, and tools. This article describes the competition of software-verification tools, which this year is organized as a satellite event of the conference TACAS. SV-COMP'12 is the first competition that compares verification tools for software source code.

Only few research projects aim at producing stable tools that can be used by people outside the respective development groups, and the development of such tools is not continuous. PhD students and post-docs do not adequately benefit from tool development because theoretical papers are still considered more relevant than papers that present technical contributions, like tool papers. Through its visibility, the Competition on Software Verification changes this, by showing

---

[*] http://sv-comp.sosy-lab.org

C. Flanagan and B. König (Eds.): TACAS 2012, LNCS 7214, pp. 504–524, 2012.

off the latest implementation of the research results in our community, and giving credits and benefits to researchers and students who spend considerable amounts of time implementing verification algorithms in practical software packages (winning the verification competition in a category serves as acknowledgment). More discussion on problems and barriers in developing tools for software verification can be found in a position paper by Alglave et al. [1].

A competition event fosters the transfer of theoretical and conceptual advancements in software verification into practical tools. The main goals of the Competition on Software Verification are the following:

- Establish a set of benchmarks for software verification in the community, i.e., create and maintain a set of programs together with explicit properties to check, and make those publicly available for researchers to be used in performance comparisons when evaluating a new technique.
- Provide an overview of the state-of-the-art in software verification for the community, i.e., compare, independently from particular paper projects and specific techniques, different verification tools in terms of precision and performance.
- Increase the visibility and credits that tool developers receive, i.e., provide a forum for presentation of tools and discussion of the latest technologies, and give students the opportunity to publish about the implementation work that they have done.

Related Events. Competitions are widely acknowledged as a means to improve the available tools, the visibility of their strengths, and to establish a publicly available set of benchmark problems. In the formal-methods community (loosely interpreted), there are competitions on, e.g., SAT [1], SMT [2], Planning [3], QBF [4], HWMC [5], and Theorem Proving [6]. These events seem to have a positive impact on the development speed and the quality of the participating software tools; theoretical results are transferred to practical tools almost instantly.

2   Procedure and Schedule

The competition compared state-of-the-art software verifiers with respect to effectiveness and efficiency. The overall process was composed of several phases, as described in the following.

Announcement and Benchmark Submission. The competition was publicly announced on July 19, 2011 at the conference event CAV. During the preparation phase, calls for contributions were made in various mailings, the web page was set

---

[1] http://www.satcompetition.org

[2] http://www.smtcomp.org

[3] http://ipc.icaps-conference.org

[4] http://www.qbflib.org/competition.html

[5] http://fmv.jku.at/hwmcc11

[6] http://www.cs.miami.edu/~tptp/CASC

up, and benchmark verification tasks were collected and classified into competition categories. Since this was the first competition, all contributed benchmarks were initially accepted, and we only disqualified benchmark programs (after discussion) if they violated the requirements below.

Training Phase. The set of all benchmark verification tasks was finalized and made publicly available on September 14, 2011. During the training phase, the teams of the competition candidates were able to download the benchmarks in order to train their tools on the given verification tasks. At the end of this phase, the competition contributions (consisting of the software together with a three-page description of the competition candidate) were submitted. Also during the training phase, some benchmark programs were corrected (without changing the verification outcome), and some verification tasks were disqualified (by the rules below and after community discussion) and removed from the benchmark set.

Benchmark Evaluation Phase. The submission of competition contributions ended on October 14, 2011; all competition candidates were downloaded and installed on a competition machine, and the verification tools were applied to the sets of benchmark verification tasks. All submitted artifacts of the competition contribution (tool description and software archive files) were stamped with SHA hash values. The hash values were sent to all members of the program committee (= jury) of the competition, in order to eliminate the possibility of undue advantages of any tool.

Also in this phase, all descriptions of competition candidates (the three-page summary papers) were reviewed, each by several members of the program committee, in order to ensure the quality standards of the TACAS proceedings.

Approval of Verification Results. After the results were obtained on a competition machine [7] (the number of solved instances and the run time were measured), each participating team received the (preliminary) results that were obtained using their submitted competition candidate. This step gave the jury the opportunity to discuss some unexpected results with the corresponding authors of the competing tools. This approval phase was completed by December 9, 2011. By this time, a list of all participating teams was publicly announced.

Notification. On December 16, the notification of acceptance of the competition contribution, together with the reviews, were sent to all authors. All teams were informed of the results of all competition candidates, and tables with rankings were made available to all teams.

---

[7] One complete competition run of all candidates on all verification tasks required a total of 163 hours of non-stop machine time; several such competition runs were necessary.

3   Definitions and Rules

This section presents the definitions and rules that regulated the execution of the competition and how the results were evaluated towards a ranking.

Definition of Verification Task. A verification task consists of a C program and a safety property. For simplicity, the safety properties to be verified are reduced to reachability problems and encoded in the program source code (using the error label 'ERROR'). In other words, the competition candidate is asked, given a C program and the error label 'ERROR', whether there is a concrete execution path through the program such that the error label can be reached. A verification run is a non-interactive execution of a competition candidate on a single verification task. The result of a verification run is either

SAFE: there is no path that reaches the error location,
UNSAFE + Path: there exists a path that reaches the error location, or
UNKNOWN: the competition candidate does not succeed in computing an answer 'SAFE' or 'UNSAFE'.

There is no particular fixed format for the error path. The error path has to be written to a file or on stdout in a reasonable format to make it possible to manually check validity.

Benchmark Verification Tasks. All verification tasks were provided by the specified date on the competition web site [8]. Most programs were provided in CIL (C Intermediate Language). The programs were assumed to be written in GNU C (most of them adhere to ANSI C).

Potential competition participants were invited to submit benchmark verification tasks until the specified date. Programs had to fulfill two requirements to be eligible for the competition: (1) the program has to be written in GNU C or ANSI C, and can be successfully CIL-pre-processed [9] with the parameters `--dosimplify --printCilAsIs --save-temps --domakeCFG --no-convert-field-offsets --no-convert-direct-calls`, and (2) the property is instrumented into the program and is violated if the label 'ERROR' is reached.

As a further convention, a verification tool can assume that a function call `__VERIFIER_assume(expression)` has the following meaning: If `expression` is evaluated to '0', then the function loops forever, otherwise the function returns (no side effects). The verification tool can assume the following implementation:

```
void __VERIFIER_assume(int expression) {
  if (!expression) { LOOP: goto LOOP; }
  return;
}
```

---

[8] http://sv-comp.sosy-lab.org
[9] We used CIL version 1.3.7, from http://cil.sourceforge.net, with extensions.

Similarly, the following functions can be assumed to return an arbitrary value of the indicated type: `__VERIFIER_nondet_X()` (and `nondet_X()`, deprecated) with X being one of `int`, `float`, `char`, `short`, or `pointer` (no side effects, `pointer` refers to `void *`). The verification tool can assume that the functions are implemented according to the following template:

```
X __VERIFIER_nondet_X() {
  X val;
  return val;
}
```

Setup. The verification runs of the competition were (natively) executed on a dedicated unloaded compute server with a 3.4 GHz 64-bit Quad Core CPU (Intel i7-2600K) and a GNU/Linux operating system (x86_64-linux). The machine had 16 GB of RAM, of which exactly 15 GB were made available to the competition candidate. Every verification run had a run-time limit of 15 min. The run time was measured in seconds of CPU time.

The verification runs were started by a batch script that collects statistics and interprets the result of every competition candidate on every verification task as one of the following categories of verification results: SAFE (verifier states that the property holds), UNSAFE (verifier states that the property does not hold, an error path is reported), UNKNOWN (result does not fall into the other two categories: verification result not known, resources exhausted, verifier crashed).

Qualification. A verification tool was qualified to participate as competition candidate if the tool was publicly available (for the GNU/Linux platform, more specifically, it had to run on an x86_64 machine) and succeeded in more than 50 % of all training verification tasks to parse the input and start the verification process (a tool crash during the verification phase does not disqualify). A person (participant) was qualified as competition contributor for a competition candidate if the person was a contributing designer/developer of the submitted competition candidate (witnessed by occurrence of the person's name on the tool's project web page, a tool paper, or in the revision logs). A contribution paper was qualified if the quality of the description of the competition candidate sufficed to run the tool in the competition and was appropriate as competition-candidate representation for the TACAS proceedings.

A verification tool could participate several times as an independent competition candidate, if a significant difference of the conceptual or technological basis of the implementation is justified in the accompanying description paper. This applies to different versions as well as different configurations, in order to avoid forcing developers to create a new tool name for every new concept. Competition candidates were allowed to opt-out from certain categories.

Evaluation by Scores and Run Time. The scores were assigned according to the scoring schema in Table 1. Every verification task comes with an expected result, which was provided by the contributor of the verification task. The

Table 1. Scoring schema

Reported result	Points	Description
UNKNOWN	0	Failure to compute verification result, out of resources, program crash.
UNSAFE correct	+1	The error in the program was found and an error path was reported.
UNSAFE incorrect	−2	An error is reported for a program that fulfills the property (false alarm, imprecise analysis).
SAFE correct	+2	The program was analyzed to be free of errors.
SAFE incorrect	−4	The program had an error but the competition candidate did not find it (missed bug, unsound analysis).

interpretation of 'UNSAFE' is that a verification tool is supposed to find a path to the error label. The interpretation of 'SAFE' is that no executable path to the error label exists in the program, assuming the C semantics [2] and a standard POSIX run-time environment. The results of type 'SAFE' yield higher absolute score values compared to type 'UNSAFE', because it is expected to be heuristically easier to detect errors than it is to prove correctness. The absolute score values for incorrect results are higher compared to correct results, because a single correct answer should not be able to compensate for a wrong answer. This scoring schema ensures a disadvantage for (hypothetical) competition candidates that always return the same result or random results.

The participating competition candidates are ranked according to the sum of points. Competition candidates with the same sum of points are sub-ranked according to success run time. The success run time for a competition candidate is the total CPU time over all verification tasks for which the competition candidate reported a correct verification result.

The participants had the opportunity to check the verification results against their own expected results and discuss inconsistencies with the competition chair (cf. Sect. 2). A candidate that opted out from a category or obtained a negative total score in a category, was assigned zero points in that category as total score.

To ensure that no undue advantage occurs from knowing the benchmark programs beforehand, we obfuscated all benchmark programs (by renaming all variable and function names, as well as the file name) and ran the competition candidates on the obfuscated versions of the benchmark programs. All verification results obtained using obfuscated versions matched the verification results of the corresponding original program.

Publication and Presentation of the Competition Candidates. A description of every qualified competition candidate (contribution paper) was published in the LNCS proceedings of TACAS 2012. In addition, every qualified competition candidate was granted a demonstration slot in the TACAS program to present the competition candidate to the TACAS audience.

Competition Jury. The program committee that oversees the process of the competition consists of one member of each participating team. The tasks of this committee are to review the competition contribution papers and help the organizer to resolve any disputes that might occur. Deviation from the competition rules need to be approved by the committee. The 2012 competition jury consists of the following members:

> Dirk Beyer, University of Passau, Germany (Chair)
> Bernd Fischer, University of Southampton, UK
> Vadim Mutilin, Russian Academy of Sciences, Russia
> Andrey Rybalchenko, TU Munich, Germany
> Carsten Sinz, Karlsruhe Institute of Technology, Germany
> Michael Tautschnig, University of Oxford, UK
> Helmut Veith, TU Vienna, Austria
> Tomas Vojnar, Brno University of Technology, Czech Republic
> Georg Weissenbacher, Princeton University, USA
> Philipp Wendler, University of Passau, Germany
> Daniel Wonisch, University of Paderborn, Germany

The term of the jury is one year, and the next jury consists of the chair and one member of each participating team of the next competition.

4   Benchmark Verification Tasks

All verification tasks are available for browsing and download via the public SVN repository for the Competition on Software Verification [10]. The competition was organized in several categories of benchmark verification tasks, which are explained in the following.

The benchmark verification tasks were contributed by several research and development groups. After the submission deadline for benchmarks, a group of people (organizer and participants) were working on improving the quality of the verification tasks. This means that after the benchmark sets were made public, some programs were removed (not qualified, no property encoded, unknown architecture), and some programs were technically improved (CIL simplifications, compiler warnings, memory model). These changes have improved the overall quality of the final set of verification tasks for the competition, and have not changed the intended verification result; all changes are tracked in the public repository.

The expected verification result is encoded in the file name of each verification task: the sub-strings 'BUG' and 'unsafe' indicate that the program violates the property, i.e., the error label is reachable.

---

[10] https://svn.sosy-lab.org/software/sv-benchmarks/tags/svcomp12

Control Flow and Integer Variables. The first set of verification tasks consists of the programs in the set ControlFlowInteger:

```
ntdrivers-simplified/*_BUG.cil.c
ntdrivers-simplified/*[!G].cil.c
ntdrivers/*.BUG.i.cil.c
ntdrivers/*[!G].i.cil.c
ssh-simplified/*_BUG.cil.c
ssh-simplified/*[!G].cil.c
ssh/*.BUG.i.cil.c
ssh/*[!G].i.cil.c
locks/*.BUG.c
locks/*[!G].c
```

The programs and properties in this category use problems that relate mostly to control-flow structure and integer variables. There is no particular focus on pointers, data structures, and concurrency. The verification tasks were taken from the source-code repositories of the tools BLAST [6] and CPACHECKER [9].

The directories 'ntdrivers*' contain 19 verification tasks that were derived from (parts of) device drivers of the Windows NT kernel. The directories 'ssh*' contain 61 verification tasks s3_clnt* and s3_srvr*, which represent the subroutine for the connection handshake protocol (a state machine) of the SSH client and server. The different versions represent various protocol-specific safety properties (one program for each property). The directories with the suffix 'simplified' contain versions of the drivers and SSH programs that were manually pre-processed in order to remove heap access. The verification tasks with the suffix 'BUG' have artificial bugs injected, which cause the assertions to fail. The 13 verification tasks in directory 'locks' were taken from the CPACHECKER project, where they served the purpose of demonstrating the advantage of adjustable-block encoding [5, 10].

Linux Device Drivers 32-bit. This category consists of problems that require the analysis of pointer aliases and function pointers (32-bit machine model):

```
ldv-regression/*-unsafe*.cil.c
ldv-regression/*-safe*.cil.c
ddv-machzwd/*_BUG.cil.c
ddv-machzwd/*[!G].cil.c
```

The 46 verification tasks in directory 'ldv-regression' were contributed by the Linux Driver Verification (LDV) project [11]. The verification tasks are used in the LDV project as regression tests for BLAST and CPACHECKER. The benchmark set consists of small programs that check for features rather than imposing a high verification load; some of these tests are inspired by the problem patterns that were seen in real device-driver code.

The 13 verification tasks in the directory 'ddv-machzwd' were generated using DDVerify [30]. The main file ddv_machzwd_all contains several assertions. Then,

---

[11] http://linuxtesting.org/project/ldv

there is one separate file for each assertion in file `ddv_machzwd_all`; the file names of these separate files have a suffix that indicates the name of the function in which the assertion occurs.

Linux Device Drivers 64-bit. This category consists of problems that require the analysis of pointer aliases and function pointers (64-bit machine model):

```
ldv-drivers/*-unsafe*.cil.c
ldv-drivers/*-safe*.cil.c
```

The verification tasks in this category were contributed by the LDV project. The directory contains 41 recent (Sept. 2011) driver-verification tasks that were taken directly from the x86_64 Linux kernel. Among them are 16 programs with bugs, which are accompanied by sample error traces. Some of these are confirmed bugs that were reported by the LDV project to the kernel developers.

Heap Manipulation. The problems in this category require the analysis of data structures on the heap and consist of the programs in the set HeapManipulation:

```
heap-manipulation/*BUG.cil.c
heap-manipulation/*[!G].cil.c
list-properties/*.cil.c
```

The eight verification tasks in directory 'heap-manipulation' were provided by the PREDATOR project [12]. The program `bubble_sort_linux` is a bubble-sort implementation that operates on Linux lists. Verification tasks with the suffix 'BUG' have an artificial bug injected. The program `dll_of_dll` operates on a NULL-terminated doubly-linked list of doubly-linked lists. The program creates a doubly-linked list of doubly-linked lists, and then destroys the data structure in several phases. The program `merge_sort` is an implementation of the merge-sort algorithm that operates on two-level singly-linked lists. The program `sll_to_dll_rev` converts a singly-linked list to a doubly-linked list, then reverses the list, and coverts it back to a singly-linked list.

The six verification tasks in the directory 'list-properties' are taken from a supplementary web page of the BLAST 3.0 project [7]. This set contains several C programs that manipulate list data structures containing integers as data elements. The programs `simple` and `simple_built_from_end` both create a list that represents a sequence of integers that matches 1*0 (regular expression), i.e., an arbitrary number of list elements that are initialized with the data value 1 with the last element initialized with 0. Then, the programs traverse the list to check that every element is set to 1 and the last to 0. The difference between the two programs is the order in which the list elements are created. The program `list` creates a sequence that matches 1*2*3. The program `list_flag` creates a sequence that matches c*3, where c is a constant determined by a flag. Then, the program traverses the list to check that the integers occur in the correct order.

---

[12] http://www.fit.vutbr.cz/research/groups/verifit/tools/predator

The program `alternating` is similar to `list` except that the list begins with alternating 1s and 2s, and ends with a value 3, i.e., it creates a sequence that matches (12)*3. The program `splice` first builds the same list as `alternating`. Then, the list is split into two different lists: the first list contains the nodes at odd positions and the second list contains nodes at even positions of the original list, without the last value 3. Each new list is then traversed to check that all its elements have the same data value.

SystemC. This category contains SystemC-related problems:

```
systemc/*BUG.cil.c
systemc/*[!G].cil.c
```

This set of 62 verification tasks was provided by the SyCMC project [14]. The programs were transformed to sequential C programs by incorporating the scheduler into the C code. More details can be found in the research article that defines the benchmark [14].

Concurrency. Some concurrency problems are contained in this set:

```
pthread/*BUG.cil.c
pthread/*[!G].cil.c
```

This benchmark set of eight verification tasks was contributed by the Esbmc project [13]. The program `fib_bench` starts two threads, which are together computing a Fibonacci number, and then compares if the results of the two threads are smaller than an upper bound. The program `fib_bench_BUG` is a version which checks a wrong bound and thus is expected to yield an error. The programs `fib_bench_longer` and `fib_bench_longer_BUG` are using the same algorithm but a larger number of iterations.

The programs `queue_ok` and `queue_BUG` operate on a queue data structure, where the former is expected to work correctly and the second to reach the error label. Two threads are started, one trying to write to the queue and one trying to read from the queue, after acquiring a mutex lock, respectively, while the programs check for some properties to hold.

The program `reorder_5_BUG` lets a set of threads write values to two variables and another set of threads verify that the two values are either both untouched or both changed to the new values. Due to certain interleavings in the execution, a violation of the property is possible. The program `twostage_3_BUG` creates two sets of threads. One set of threads is writing a value to one global variable and an increased value to a second variable. The other set of threads verifies the success of the first set of threads. Again, due to certain interleavings, the property might be violated in some executions.

Overall. The category 'Overall' consists of the union of all above-mentioned sets of verification tasks.

[13] http://esbmc.org

5   Participating Teams

In the following, we briefly introduce the competition candidates, listed in alphabetical order. Table 2 gives an overview of the participating candidates. The top-three placements achieved in the competition for each category are given in the paragraph for the corresponding tool. The detailed summary of the results is presented in Sect. 6.

Table 3 provides an overview of the technologies and concepts used by the various competition candidates. The technique of counterexample-guided abstraction refinement (CEGAR) [15] is used by the majority of tools. Other techniques that are offered by the competition candidates are predicate abstraction [3,19], bounded model checking [12], shape analysis [23], construction of an abstract reachability tree (ART) as proof of correctness [6], lazy abstraction [21], and Craig interpolation for discovering new predicates to refine a predicate analysis [17,25]. Only three tools provide verification of concurrent programs.

BLAST 2.7 [26], submitted by *Pavel Shved, Vadim Mutilin, and Mikhail Mandrykin* (Institute for System Programming of the Russian Academy of Sciences, Russia), has achieved the following placements:

- *Winner* in DeviceDrivers64
- *Bronze* in DeviceDrivers

BLAST 2.7[14] is a model checker that is based on predicate abstraction, with a focus on verifying control-flow intensive programs such as device drivers and system programs. It is based on the CEGAR algorithm [15] and uses Craig interpolation [17] on infeasible error paths to discover new predicates for increasing the precision of the predicate abstraction. The tool was originally developed at the University of California at Berkeley and at EPFL Lausanne [6], but later significantly improved by the Linux Driver Verification group at the Institute for System Programming of the Russian Academy of Sciences in Moscow. The tool uses CVC 3 [28] as SMT solver, CSISAT [11] as interpolation procedure, and is implemented in OCaml.

CPACHECKER 1.0.10-ABE [24], submitted by *Stefan Löwe and Philipp Wendler* (University of Passau, Germany), has achieved the following placements:

- *Winner* in ControlFlowInteger
- *Silver* in Overall
- *Bronze* in SystemC
- *Bronze* in HeapManipulation

CPACHECKER 1.0.10-ABE is based on a predicate analysis, with an application focus similar to that of BLAST. CPACHECKER [9][15] is a flexible verification framework that implements the formalism of configurable program

---

[14] http://mtc.epfl.ch/software-tools/blast
[15] http://cpachecker.sosy-lab.org

Table 2. Competition candidates with their system-description references and representing jury members

Competition candidate	Ref.	Representing jury memb.	Affiliation
BLAST 2.7	[26]	Vadim Mutilin	Moscow, Russia
CPACHECKER 1.0.10-ABE	[24]	Philipp Wendler	Passau, Germany
CPACHECKER 1.0.10-MEMO	[31]	Daniel Wonisch	Paderborn, Germany
ESBMC 1.17	[16]	Bernd Fischer	Southampton, UK
FSHELL 1.3	[22]	Helmut Veith	Vienna, Austria
LLBMC 0.9	[27]	Carsten Sinz	Karlsruhe, Germany
PREDATOR 2011-10-11	[18]	Tomas Vojnar	Brno, Czech Republic
QARMC-HSF(C)	[20]	Andrey Rybalchenko	Munich, Germany
SATABS 3.0	[4]	Michael Tautschnig	Oxford, UK
WOLVERINE 0.5C	[29]	Georg Weissenbacher	Princeton, USA

Table 3. Technologies and features that the competition candidates offer

Competition candidate	CEGAR	Predicate Abstraction	Bounded Model Checking	Shape Analysis	ART-based Analysis	Lazy Abstraction	Interpolation	Concurrency Support
BLAST	✓	✓			✓	✓	✓	
CPA-ABE	✓	✓			✓	✓	✓	
CPA-MEMO	✓	✓			✓	✓	✓	
ESBMC			✓					✓
FSHELL			✓					
LLBMC			✓					
PREDATOR				✓				
QARMC-HSF(C)	✓	✓			✓		✓	✓
SATABS	✓	✓						✓
WOLVERINE	✓				✓	✓	✓	

analysis (CPA) [8]. The competition candidate CPACHECKER 1.0.10-ABE uses the concept of adjustable-block encoding [10], which is implemented as a CPA in the framework. The algorithm uses an interpolation-based refinement of the predicate precision and explores the abstract state space by building an abstract reachability graph. The framework currently uses MATHSAT [13] as SMT solver and interpolation procedure, and is implemented in Java.

CPACHECKER 1.0.10-MEMO [31], submitted by *Daniel Wonisch* (University of Paderborn, Germany), has achieved the following placements:

- *Winner* in Overall
- *Silver* in ControlFlowInteger
- *Silver* in DeviceDrivers64
- *Bronze* in HeapManipulation

CPACHECKER 1.0.10-MEMO is based the verification framework CPACHECKER, configured for large-block encoding [5] and boolean predicate abstraction. The novel feature of the competition candidate CPACHECKER 1.0.10-MEMO is the integration of the concept of block-abstraction memoization as a CPA. Intermediate analysis results of large blocks are cached in order to avoid repeated verification of similar program traces. This concept yields a significant improvement over the standard configuration of CPACHECKER in the category 'DeviceDrivers64', as shown in Table 4.

ESBMC 1.17 [16], submitted by *Lucas Cordeiro, Jeremy Morse, Denis Nicole, and Bernd Fischer* (University of Southampton, UK), has achieved the following placements:

- *Winner* in SystemC
- *Winner* in Concurrency
- *Bronze* in Overall

ESBMC 1.17[16] is a bounded model checker that is based on the concept of generating verification conditions for the program, which are then passed to an SMT solver for checking if a feasible error path exists. The focus of ESBMC is to provide a context-bounded verification of multi-threaded C programs, in addition to sequential C programs. The tool uses components of the CPROVER framework[17], the external solvers Z3[18] and BOOLECTOR[19], and is implemented in C++.

FSHELL 1.3 [22], submitted by *Andreas Holzer, Daniel Kröning, Christian Schallhart, Michael Tautschnig, and Helmut Veith* (TU Vienna, Austria), is a test-generation tool for C programs, which is based on bit-precise bounded model checking for identifying program paths that fulfill a given test-coverage

---

[16] http://esbmc.org

[17] http://www.cprover.org

[18] http://research.microsoft.com/projects/z3

[19] http://fmv.jku.at/boolector

criterion, for which test vectors can be derived using satisfying assignments. The tool FSHELL[20] is based on the CPROVER framework, uses the SAT solver MINISAT [21], and is implemented in C++.

LLBMC 0.9 [27], submitted by *Carsten Sinz, Stephan Falke, and Florian Merz* (Karlsruhe Institute of Technology, Germany), has achieved the following placements:

- *Winner* in DeviceDrivers
- *Silver* in HeapManipulation

LLBMC 0.9[22] is a bounded model checker that operates on LLVM's intermediate representation, with a focus on providing a bit-precise analysis of C code, in particular for detecting violations of safe memory usage. The tool is based on the LLVM compiler infrastructure, and passes the verification conditions to the SMT solver STP [23], which supports bit-vectors and arrays. LLBMC is implemented in C++.

PREDATOR 2011-10-11 [18], submitted by *Kamil Dudka, Petr Muller, Petr Peringer, and Tomas Vojnar* (Brno University of Technology, Czech Republic), has achieved the following placements:

- *Winner* in HeapManipulation
- *Silver* in DeviceDrivers

PREDATOR 2011-10-11[24] is a program analyzer that is based on separation logic, with a focus on verifying C programs with dynamically linked list data structures. The separation-logic formulas that describe (infinite) sets of heaps are internally represented as heap graphs. The main objective of the PREDATOR project is to support the verification of system code, which also uses low-level programming techniques like pointer arithmetics. The tool uses no external decision procedure, is designed as a plug-in for GCC, and is implemented in C++.

QARMC-HSF(C) [20], submitted by *Sergey Grebenshchikov, Ashutosh Gupta, Nuno P. Lopes, Corneliu Popeea, and Andrey Rybalchenko* (TU Munich, Germany), has achieved the following placements:

- *Bronze* in ControlFlowInteger

QARMC-HSF(C)[25,26] is a model checker that is based on predicate abstraction with a special focus on liveness properties in addition to being able to check safety

---

[20] http://code.forsyte.de/fshell

[21] http://minisat.se

[22] http://baldur.iti.uka.de/llbmc

[23] http://sites.google.com/site/stpfastprover

[24] http://www.fit.vutbr.cz/research/groups/verifit/tools/predator

[25] This tool participated in the competition as QARMC and was renamed to HSF(C) when the proceedings were due.

[26] http://www7.in.tum.de/tools/hsf

properties. The tool is based on the CEGAR algorithm, but instead of operating on transition systems, it operates directly on Horn-clause representations of the program and its proof rules. The tool is based on the ARMC infrastructure and the constraint solver CLP(Q). The frontend CIL is used as parser and for the transformation of C programs into the internal representation; the backend is implemented in Prolog and requires the SICStus compiler package.

SATAbs 3.0 [4], submitted by *Alastair Donaldson, Alexander Kaiser, Daniel Kröning, Michael Tautschnig, and Thomas Wahl* (Oxford University, UK), has achieved the following placements:

- *Silver* in SystemC
- *Silver* in Concurrency
- *Bronze* in DeviceDrivers64

SATAbs 3.0[27] is a model checker that is based on predicate abstraction with a focus on bit-precise analysis of program variables. The tool implements an explicit abstract-check-refine loop of the CEGAR algorithm for sequential and concurrent programs. In every iteration, an abstract (boolean) program is computed, then this abstract program is model-checked, then the error path is checked for feasibility, and predicates are discovered in order to compute a more precise abstract program in the next iteration. The tool uses components from the CPROVER framework, SMV or BOOM as model checkers, MINISAT as SAT solver, and is implemented in C++.

WOLVERINE 0.5c [29], submitted by *Georg Weissenbacher, Daniel Kröning, and Sharad Malik* (Princeton University, USA), is a model checker that is based on interpolation-based predicate analysis without computing predicate abstractions during single post-operations. Instead of discovering predicates and collecting them in a predicate precision, the interpolants from infeasible paths are directly used as part of the abstract states. WOLVERINE 0.5c[28] is based on an integrated interpolating decision procedure, uses components from the CPROVER framework, and is implemented in C++.

6 Results and Discussion

The results in this paper represent the state-of-the-art in software verification in terms of precision and performance, as available and participated, when the benchmark verification runs for the 1st Competition on Software Verification were performed. We sent all results to the participating competition teams for review; all results shown in this paper are approved by the competing teams. The total quantitative overview is provided in Table 4. The run time in the tables is given

---

[27] http://www.cprover.org/satabs
[28] http://www.cprover.org/wolverine

Table 4. Summary of all results. The tools are listed in alphabetical order. In every table cell for competition results, we list the points in the first row and the CPU time for successful runs in the second row (cf. Table 1 for the scoring schema). The top-three candidates have their score set in bold face and in larger font size. The entry '—' means that the competition candidate opted-out or obtained a total of less than 0 points in the category.

Competition candidate / Representing jury member / Affiliation	ControlFlowInteger (144 points max. / 93 verification tasks)	DeviceDrivers (103 points max. / 59 verification tasks)	DeviceDrivers64 (66 points max. / 41 verification tasks)	HeapManipulation (24 points max. / 14 verification tasks)	SystemC (87 points max. / 62 verification tasks)	Concurrency (11 points max. / 8 verification tasks)	Overall (435 points max. / 277 verification tasks)
BLAST							
Vadim Mutilin	71	**72**	**55**	—	33	—	231
Moscow, Russia	9900 s	30 s	1400 s		4000 s		15000 s
CPA-ABE							
Philipp Wendler	**141**	51	26	**4**	**45**	0	**267**
Passau, Germany	1000 s	97 s	1900 s	16 s	1100 s	0 s	4100 s
CPA-MEMO							
Daniel Wonisch	**140**	51	**49**	**4**	36	0	**280**
Paderborn, Germany	3200 s	93 s	500 s	16 s	450 s	0 s	4300 s
ESBMC							
Bernd Fischer	102	63	10	1	**67**	**6**	**249**
Southampton, UK	4500 s	160 s	870 s	220 s	760 s	270 s	6800 s
FSHELL							
Helmut Veith	28	20	0	—	—	0	48
Vienna, Austria	580 s	3.5 s	0 s			0 s	580 s
LLBMC							
Carsten Sinz	100	**80**	1	**17**	8	—	206
Karlsruhe, Germany	2400 s	1.6 s	110 s	210 s	2.4 s		2700 s
PREDATOR							
Tomas Vojnar	17	**80**	0	**20**	21	0	138
Brno, Czech Republic	1100 s	1.9 s	0 s	1.0 s	630 s	0 s	1700 s
QARMC-HSF(C)							
Andrey Rybalchenko	**140**	—	—	—	8	—	148
Munich, Germany	4800 s				820 s		5600 s
SATABS							
Michael Tautschnig	75	71	**32**	—	**57**	**1**	236
Oxford, UK	5400 s	140 s	3200 s		5000 s	1.4 s	14000 s
WOLVERINE							
Georg Weissenbacher	39	68	16	—	36	—	159
Princeton, USA	580 s	65 s	1300 s		1900 s		3800 s

Table 5. Overview of the top-five candidates for each category. The run time is given in seconds of CPU usage for the verification tasks that were successfully solved. The column 'False Alarms' indicates the number of verification tasks for which the tool reported an error but the program was safe (false positive), and column 'Missed Bugs' indicates the number of verification tasks for which the tool claims that the program is safe although it contains a bug (false negative).

Rank	Candidate	Score	Run Time	Solved Tasks	False Alarms	Missed Bugs
ControlFlowInteger						
1	**CPAchecker-abe**	**141**	1000	91		
2	CPAchecker-memo	140	3200	91		
3	QArmc-Hsf(c)	140	4800	91		
4	Esbmc 1.17	102	4500	70		4
5	Llbmc 0.9	100	2400	79	5	3
DeviceDrivers						
1	**Llbmc 0.9**	**80**	1.6	46		
2	Predator	80	1.9	46		
3	Blast 2.7	72	30	51	6	1
4	SATabs 3.0	71	140	43		1
5	Wolverine 0.5c	68	65	48	2	3
DeviceDrivers64						
1	**Blast 2.7**	**55**	1400	33		
2	CPAchecker-memo	49	500	33	2	
3	SATabs 3.0	32	3200	17		
4	CPAchecker-abe	26	1900	23	2	
5	Wolverine 0.5c	16	1300	12		
HeapManipulation						
1	**Predator**	**20**	1.0	12		
2	Llbmc 0.9	17	210	10		
3	CPAchecker-abe	4	16	9	5	
3	CPAchecker-memo	4	16	9	5	
5	Esbmc 1.17	1	220	6	3	1
SystemC						
1	**Esbmc 1.17**	**67**	760	58		4
2	SATabs 3.0	57	5000	40		
3	CPAchecker-abe	45	1100	34		
4	CPAchecker-memo	36	450	30		
5	Wolverine 0.5c	36	1900	25		
Concurrency						
1	**Esbmc 1.17**	**6**	270	7		1
2	SATabs 3.0	1	1.4	1		
Overall						
1	**CPAchecker-memo**	**280**	4300	209	20	
2	CPAchecker-abe	267	4100	203	20	
3	Esbmc 1.17	249	6800	191	9	11
4	SATabs 3.0	238	15000	149		1
5	Blast 2.7	231	15000	158	6	1

in seconds of CPU time. All measurement values are rounded to two significant digits. The points are calculated according to the scoring schema in Table 1. Some more details on the top-five tools for each category are given in Table 5.

The main result of this competition is that there is currently no single technique that is absolutely superior in comparison with the other tools. The competition candidates have scored differently in the various categories, with no single candidate being the absolute winner.

Towards Robustness. There is one single competition candidate that achieved positive scores in all categories: Esbmc 1.17. The following candidates participated in all categories, with a non-negative score in all categories: CPAchecker 1.0.10-abe, CPAchecker 1.0.10-memo, Esbmc 1.17, and Predator 2011-10-11.

Towards Soundness. There are four competition candidates that never reported the answer 'SAFE' for a benchmark program that actually contains a bug (missed a bug): CPAchecker 1.0.10-abe, CPAchecker 1.0.10-memo, Predator 2011-10-11, and QArmc-Hsf(c).

Towards Completeness. There are three competition candidates that never reported a bug for a safe program (false alarm): FShell 1.3, QArmc-Hsf(c), and SATabs 3.0.

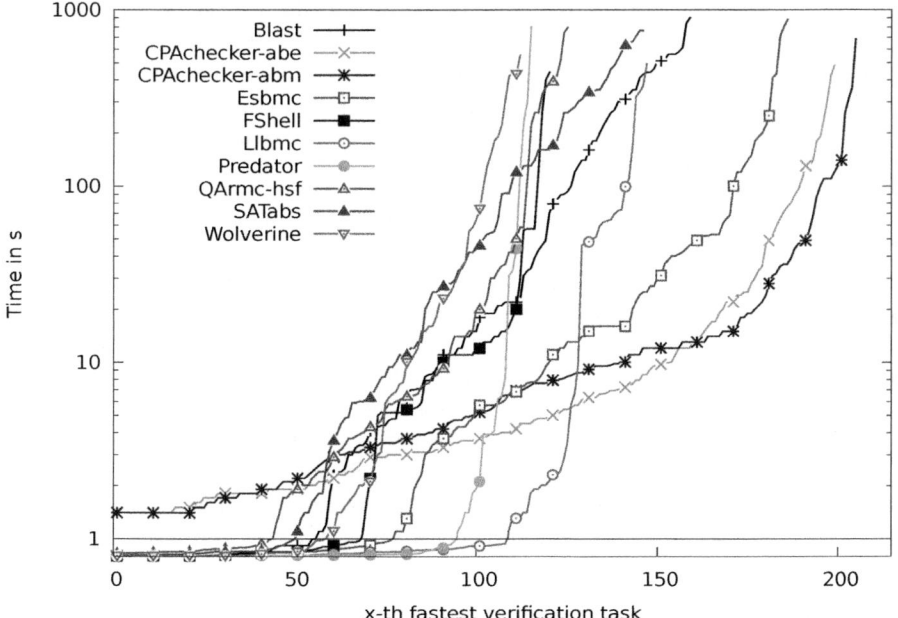

Fig. 1. Quantile functions: For each competition candidate, we plot all pairs (x, y) such that the maximum run time of the x fastest results is y. A logarithmic scale is used for the time range from 1 s to 1000 s, and a linear scale is used for the time range between 0 s and 1 s. The graphs are decorated with symbols at every tenth data point in order to make the graphs distinguishable on gray-scale prints.

About Solved Instances and Run Time. Figure 1 illustrates the competition results using the quantile functions over all benchmark verification tasks. The function graph for a competition candidate yields the maximum run time y for the x fastest computed correct results. On the left, the plot shows that two candidates need a few seconds of run time even for the simplest benchmark programs; this seems due to the setup time for the Java virtual machine that these two candidates are using. The right-most data point of each graph yields the number of successfully solved verification tasks by the corresponding competition candidate. The area below a graph (its integral) is the accumulated run time for all successfully solved verification tasks.

7 Summary and Future Plans

The competition on software verification was well received in the research community, and the participants were enthusiastic about the event. The participation of ten teams in the first competition, which exceeded the expectation, witnesses the need for such an event. The organizer and the jury were making sure that the competition follows the high quality standards of the TACAS conference, in particular to respect the important principles of fairness, community support, transparency, and technical accuracy. The conclusion is that the event shall be held annually from now on. One important objective for the next competition is to significantly extend the benchmark set, especially in the categories 'Heap-Manipulation' and 'Concurrency'. Since software verification becomes more and more relevant in practice, we are convinced that the pool of available benchmarks will considerably grow in the next few years. We also hope that the number of participants even increases in the next years, and that a wider range of verification technologies will be covered.

Acknowledgments. We thank the TACAS steering committee and the program chairs for hosting the Competition on Software Verification as satellite event of the conference TACAS, and for the encouragement and support during the design of the event. Most importantly, we thank the participating teams for contributing their tools and system descriptions. In particular, we want to thank (among others) Pavel Shved, Kamil Dudka, Georg Weissenbacher, Corneliu Popeea, Bernd Fischer, and Carsten Sinz for their help in preparing the benchmark verification tasks for the competition (contributing verification tasks, sending patches and comments). The biggest thanks goes to Karlheinz Friedberger, who programmed the benchmark processing script and helped with configuring the tools and infrastructure that we used for the competition.

References

1. Alglave, J., Donaldson, A.F., Kröning, D., Tautschnig, M.: Making Software Verification Tools Really Work. In: Bultan, T., Hsiung, P.-A. (eds.) ATVA 2011. LNCS, vol. 6996, pp. 28–42. Springer, Heidelberg (2011)

2. American National Standards Institute. ANSI/ISO/IEC 9899-1999: Programming Languages — C. American National Standards Institute, 1430 Broadway, New York, NY 10018, USA (1999)

3. Ball, T., Rajamani, S.K.: The SLAM Project: Debugging System Software via Static Analysis. In: Proc. POPL, pp. 1–3. ACM (2002)

4. Basler, G., Donaldson, A., Kaiser, A., Kröning, D., Tautschnig, M., Wahl, T.: SATABS: A Bit-Precise Verifier for C Programs. In: Flanagan, C., König, B. (eds.) TACAS 2012. LNCS, vol. 7214, pp. 551–554. Springer, Heidelberg (2012)

5. Beyer, D., Cimatti, A., Griggio, A., Keremoglu, M.E., Sebastiani, R.: Software Model Checking via Large-Block Encoding. In: Proc. FMCAD, pp. 25–32. IEEE (2009)

6. Beyer, D., Henzinger, T.A., Jhala, R., Majumdar, R.: The Software Model Checker BLAST. Int. J. Softw. Tools Technol. Transfer 9(5-6), 505–525 (2007)

7. Beyer, D., Henzinger, T.A., Théoduloz, G.: Lazy Shape Analysis. In: Ball, T., Jones, R.B. (eds.) CAV 2006. LNCS, vol. 4144, pp. 532–546. Springer, Heidelberg (2006)

8. Beyer, D., Henzinger, T.A., Théoduloz, G.: Program Analysis with Dynamic Precision Adjustment. In: Proc. ASE, pp. 29–38. IEEE (2008)

9. Beyer, D., Keremoglu, M.E.: CPACHECKER: A Tool for Configurable Software Verification. In: Gopalakrishnan, G., Qadeer, S. (eds.) CAV 2011. LNCS, vol. 6806, pp. 184–190. Springer, Heidelberg (2011)

10. Beyer, D., Keremoglu, M.E., Wendler, P.: Predicate Abstraction with Adjustable-Block Encoding. In: Proc. FMCAD, pp. 189–197. FMCAD (2010)

11. Beyer, D., Zufferey, D., Majumdar, R.: CSISAT: Interpolation for LA+EUF. In: Gupta, A., Malik, S. (eds.) CAV 2008. LNCS, vol. 5123, pp. 304–308. Springer, Heidelberg (2008)

12. Biere, A., Cimatti, A., Clarke, E., Zhu, Y.: Symbolic Model Checking without BDDs. In: Cleaveland, W.R. (ed.) TACAS 1999. LNCS, vol. 1579, pp. 193–207. Springer, Heidelberg (1999)

13. Bruttomesso, R., Cimatti, A., Franzén, A., Griggio, A., Sebastiani, R.: The MATHSAT 4 SMT Solver. In: Gupta, A., Malik, S. (eds.) CAV 2008. LNCS, vol. 5123, pp. 299–303. Springer, Heidelberg (2008)

14. Cimatti, A., Micheli, A., Narasamdya, I., Roveri, M.: Verifying SystemC: A Software Model Checking Approach. In: Proc. FMCAD, pp. 51–59. FMCAD Inc. (2010)

15. Clarke, E.M., Grumberg, O., Jha, S., Lu, Y., Veith, H.: Counterexample-Guided Abstraction Refinement for Symbolic Model Checking. J. ACM 50(5), 752–794 (2003)

16. Cordeiro, L., Morse, J., Nicole, D., Fischer, B.: Context-Bounded Model Checking with ESBMC. In: Flanagan, C., König, B. (eds.) TACAS 2012. LNCS, vol. 7214, pp. 534–537. Springer, Heidelberg (2012)

17. Craig, W.: Linear Reasoning. A New Form of the Herbrand-Gentzen Theorem. J. Symb. Log. 22(3), 250–268 (1957)

18. Dudka, K., Müller, P., Peringer, P., Vojnar, T.: PREDATOR: A Verification Tool for Programs with Dynamic Linked Data Structures. In: Flanagan, C., König, B. (eds.) TACAS 2012. LNCS, vol. 7214, pp. 544–547. Springer, Heidelberg (2012)

19. Graf, S., Saïdi, H.: Construction of Abstract State Graphs with PVS. In: Grumberg, O. (ed.) CAV 1997. LNCS, vol. 1254, pp. 72–83. Springer, Heidelberg (1997)

20. Grebenshchikov, S., Gupta, A., Lopes, N.P., Popeea, C., Rybalchenko, A.: HSF(C): A Software Verifier Based on Horn Clauses. In: Flanagan, C., König, B. (eds.) TACAS 2012. LNCS, vol. 7214, pp. 548–550. Springer, Heidelberg (2012)

21. Henzinger, T.A., Jhala, R., Majumdar, R., Sutre, G.: Lazy Abstraction. In: Proc. POPL, pp. 58–70. ACM (2002)
22. Holzer, A., Kröning, D., Schallhart, C., Tautschnig, M., Veith, H.: Proving Reachability Using FSHELL. In: Flanagan, C., König, B. (eds.) TACAS 2012. LNCS, vol. 7214, pp. 537–540. Springer, Heidelberg (2012)
23. Jones, N.D., Muchnick, S.S.: A Flexible Approach to Interprocedural Data-Flow Analysis and Programs with Recursive Data Structures. In: POPL, pp. 66–74 (1982)
24. Löwe, S., Wendler, P.: CPACHECKER with Adjustable Predicate Analysis. In: Flanagan, C., König, B. (eds.) TACAS 2012. LNCS, vol. 7214, pp. 527–529. Springer, Heidelberg (2012)
25. McMillan, K.L.: Interpolation and SAT-Based Model Checking. In: Hunt Jr., W.A., Somenzi, F. (eds.) CAV 2003. LNCS, vol. 2725, pp. 1–13. Springer, Heidelberg (2003)
26. Shved, P., Mandrykin, M., Mutilin, V.: Predicate Analysis with BLAST 2.7. In: Flanagan, C., König, B. (eds.) TACAS 2012. LNCS, vol. 7214, pp. 524–526. Springer, Heidelberg (2012)
27. Sinz, C., Merz, F., Falke, S.: LLBMC: A Bounded Model Checker for LLVMS Intermediate Representation. In: Flanagan, C., König, B. (eds.) TACAS 2012. LNCS, vol. 7214, pp. 541–543. Springer, Heidelberg (2012)
28. Stump, A., Barrett, C.W., Dill, D.L.: CVC: A Cooperating Validity Checker. In: Brinksma, E., Larsen, K.G. (eds.) CAV 2002. LNCS, vol. 2404, pp. 500–504. Springer, Heidelberg (2002)
29. Weissenbacher, G., Kröning, D., Malik, S.: WOLVERINE: Battling Bugs with Interpolants. In: Flanagan, C., König, B. (eds.) TACAS 2012. LNCS, vol. 7214, pp. 555–557. Springer, Heidelberg (2012)
30. Witkowski, T., Blanc, N., Kröning, D., Weissenbacher, G.: Model Checking Concurrent Linux Device Drivers. In: Proc. ASE, pp. 501–504. ACM (2007)
31. Wonisch, D.: Block Abstraction Memoization for CPACHECKER. In: Flanagan, C., König, B. (eds.) TACAS 2012. LNCS, vol. 7214, pp. 531–533. Springer, Heidelberg (2012)

Predicate Analysis with BLAST 2.7
(Competition Contribution)

Pavel Shved, Mikhail Mandrykin, and Vadim Mutilin

Institute for System Programming of the Russian Academy of Sciences
{shved,mandrykin,mutilin}@ispras.ru

Abstract. We present the software verification tool BLAST 2.7, which we submitted for the Competition on Software Verification. The tool is an improvement over BLAST 2.5, and its development is mostly targeted at its performance and usability in the Linux Driver Verification project. The paper overviews the tool and outlines our contribution to it.

1   Verification Approach

BLAST uses the CounterExample-Guided Abstraction Refinement approach (CE-GAR) with "lazy abstraction" [1], a decision procedure to explore all possible paths from the entry point, abstracting away from the realizable memory states as far as possible to prove the unreachability of the error label. BLAST marks each location with a conjunction of predicates over program variables in a path-sensitive way, such conjunctions being overapproximations of the set of feasible memory states at the locations. Path validity is checked with *formula satisfiability solvers*; *interpolating provers* automatically retrieve predicates to track. The concepts of BLAST are more thoroughly described in [1] and [2]. BLAST may also combine the predicate domain with lattice-based explicit-value dataflow analysis [3], which we used in the competition setup.

This has been implemented in BLAST 2.5, which was maintained by Dirk Beyer et al. [2]. In this paper, we present the improvements that we added to BLAST since the release of version 2.5 in 2008; we assigned version 2.7 to the competition release. Most of the improvements are merely more efficient implementations of already known algorithms.

The tool as of version 2.7 is capable to track states that may be expressed in terms of logical formulæ over atomic predicates that only contain linear (in)equalities over program variables, including aliases of pointers to scalar or structure variables. The analysis may be unsound if the unreachability proof requires reasoning about bit-vectors, bounded integers, or arrays.

2   Tool Architecture

The tool first converts the program into a set of per-function control flow automata with aid of the **CIL C frontend** (integrated into the tool). It converts

C. Flanagan and B. König (Eds.): TACAS 2012, LNCS 7214, pp. 525–527, 2012.

the structure of C source code directly into OCaml memory structures, and helps to perform transformations that simplify the semantics of individual operators.

Having built the CFAs, BLAST starts the abstract-check-refine loop, inlining each function call it encounters on demand, and skipping recursive calls. In the forward search phase, it uses an SMTlib solver to compute the abstract post-condition for predicates (**CVC3** is shipped alongside the tool, but any other decent SMTlib solver would work), and updates *symbolic execution lattice* elements [3]. The predicates are stored as BDDs over atomic predicate symbols. A potential error path is converted to a path formula (weakest precondition of each operator starting from the end with explicit substitution, see section 5.3 of [1]). It is checked for satisfiability with SMTlib solver, then filtered with multiple SMTlib solver calls to get unsatifiability cores, which then undergo Craig interpolation with **CSIsat** (or any other tool that supports the FOCI format) to get the predicates to track.

Both forward exploration and path analysis are supplemented with interprocedural points-to may-alias information provided by an subset-based Andersen analysis with BDDs as a storage and querying mechanism.

BLAST is implemented in OCaml.

3   Tool Improvements and Benchmarking

As BLAST 2.5 demonstrated some success in verifying drivers, we used it in the Linux Driver Verification (LDV) project [4], and improved it further to make the tool faster for Linux drivers. The rules instrumented into drivers used states expressed as simple integers, but the drivers themselves used structures and pointers to maintain data flow; therefore, we focused on improving implementations of the existing theoretical achievements, and did not try to extend the abstract domain.

In version 2.6, we improved folmulæ conversion between the internal OCaml representation and SMTlib format, making its overhead negligible, tuned the CVC3 solver to work faster for quantified formulæ, decreased the asymptotics of pre-interpolation trace filtering from $O(N)$ to $O(\log N)$ solver calls, and implemented $stop^{sep}$ and *merge-pred-join* for combining predicate analysis with lattices. We thorougly described these and many other improvements of 2.6 over 2.5 in [5]. The speedup we achieved on Linux drivers, compared with version 2.5, ranged from a factor of 5–8 on average to 30 on the most complex drivers. Our tool performed well on all driver-related benchmarks.

In the competition version 2.7, we also dramatically improved alias and structure analysis that are used to generate additional variable updates at assignment preconditions. We noticed that updates of indexed variables that are not used in the bottom part of the path formula were useless, and, while they would be ruled out by solvers anyway, BLAST spent a lot of time generating them. We revamped the generation algorithms so that only the variables that are in the already built part of a formula are considered as potential aliases or targets for structure updates; this made alias analysis overhead negligible. The new analysis is sound, but sometimes incomplete for variable-depth shapes. It improved the results on `list-properties` benchmarks, but the `heap-manipulation` benchmarks are analyzed with errors due to abuse of low-level-style accesses to structure fields via casts and raw pointer shifts.

Other benchmarks, such as `SystemC` or synthetic `locks`, involved state explosions that should be mitigated by verification algorithms that automatically merge states without loss of precision. BLAST does not merge paths with different predicate states assigned, so it times out on most of such benchmarks, which more recent tools should pass.

4   Downloading and Using BLAST

To use the tool, download binaries, unpack, add the `bin/` folder to your `PATH`, and run: `ocamltune pblast.opt -alias bdd -enable-recursion -noprofile -cref -sv-comp -lattice -include-lattice symb FILE_NAME.c`. Download: http://forge.ispras.ru/attachments/download/1157/blast-2.7-bin-x86_64.tgz.

Visit http://forge.ispras.ru/projects/blast/ to get the source code and 32-bit binaries. BLAST is licensed under Apache-2.0, and *all* external tools it relies on during compilation or at runtime are free software.

The verdict and the error trace, if any, are written to standard output. For more information on the tool usage, please, refer to the `README` file.

The binary distribution of the tool does not require external tools except for the Perl interpreter, C and C++ runtime libraries. BLAST is compatible with most modern Linux distributions, including Ubuntu 8.04 or newer.

Acknowledgements. BLAST 2.7 was prepared as a part of the Linux Driver Verification project with the help of our colleagues at ISPRAS. A number of people contributed to BLAST, including its former maintainers Dirk Beyer, Rupak Majumdar, Ranjit Jhala, and Thomas Henzinger, and the others mentioned in the `README` file.

References

1. Henzinger, T.A., Jhala, R., Majumdar, R.: Lazy abstraction. In: Symposium on Principles of Programming Languages, pp. 58–70. ACM Press (2002)
2. Beyer, D., Henzinger, T.A., Jhala, R., Majumdar, R.: The software model checker BLAST: Applications to software engineering. Int. J. Softw. Tools Technol. Transf. 9(5), 505–525 (2007)
3. Fischer, J., Jhala, R., Majumdar, R.: Joining dataflow with predicates. SIGSOFT Softw. Eng. Notes 30, 227–236 (2005)
4. Khoroshilov, A., Mutilin, V., Novikov, E., Shved, P., Strakh, A.: Towards an open framework for C verification tools benchmarking. In: Proceedings of PSI (2011)
5. Shved, P., Mutilin, V., Mandrykin, M.: Static verfication "under the hood": Implementation details and improvements of BLAST. In: Proceedings of SYRCoSE, vol. 1, pp. 54–60 (2011)

CPACHECKER with Adjustable Predicate Analysis
(Competition Contribution)

Stefan Löwe and Philipp Wendler

University of Passau, Germany

Abstract. CPACHECKER is a freely available software-verification framework, built on the concepts of CONFIGURABLE PROGRAM ANALYSIS (CPA). CPACHECKER integrates most of the state-of-the-art technologies for software model checking, such as counterexample-guided abstraction refinement (CEGAR), lazy predicate abstraction, interpolation-based refinement, and large-block encoding. The CPA for predicate analysis with adjustable-block encoding (ABE) is very promising in many categories, and thus, we submit a CPACHECKER configuration that uses this analysis approach to the competition.

1 Verification Approach

Predicate analysis is a common approach to software verification, and tools like BLAST and SLAM showed that it can be used effectively for software verification of medium sized programs. CPACHECKER [2] constructs —like BLAST— an abstract reachability graph (ARG) as a central data structure, by continuous successor computations along the edges of the control-flow automaton (CFA) of the program. The nodes of the ARG, representing sets of reachable program states, store relevant information like control-flow location, call stack, and, most importantly, the formulas that represent the abstract data states.

When single-block encoding (as implemented in BLAST) is used, abstractions are computed for every single edge in the CFA. The major drawback of this approach is the large number of successor computations, each requiring expensive calls to a theorem prover. Furthermore, boolean abstraction is prohibitive for such a large number of successor computations, and only the more imprecise cartesian abstraction can be used.

Therefore, CPACHECKER implements an approach called *adjustable-block encoding* [3], which completely separates the process of computing successors from the process of computing a predicate abstraction for a formula. The post operations in this approach (purely syntactically) assemble formulas for the strongest postcondition. Then, at certain points that can be chosen arbitrarily, the procedure applies an (expensive) computation of the predicate abstraction for a given abstract state. This method reduces the number of theorem-prover calls by effectively combining program blocks of arbitrary size into a single formula before computing an abstraction. Because the model checker now delegates much larger problems to the SMT solver (the formulas will contain a disjunction for each control-flow join point inside a block), this technique is able to leverage

C. Flanagan and B. König (Eds.): TACAS 2012, LNCS 7214, pp. 528–530, 2012.

the huge performance increase of SMT solvers being witnessed over the last decade. Experiments have shown that using adjustable blocks (e.g., loop-free blocks spanning across function calls) is orders of magnitudes faster than computing an abstraction for every single abstract state. Furthermore, the reduced number of abstractions (and refinements) makes it feasible to use the more expensive boolean abstraction, which makes the analysis more precise. This predicate analysis is wrapped in an algorithm for counterexample-guided abstraction refinement that uses Craig interpolation and lazy abstraction.

2 Software Architecture

CPACHECKER is designed as an extensible framework for software verification and is written in JAVA. The framework provides the parsing of the input program (by using the C parser from the Eclipse CDT project[1]), interfaces to the SMT solver and interpolation procedures (using the SMT solver MathSAT4[2]), and the central verification algorithms. In CPACHECKER, every analysis is implemented as a CONFIGURABLE PROGRAM ANALYSIS (CPA) [1], which makes it easier to implement new concepts (separation of concerns). Different CPAs can be flexibly combined on demand, enabling reuse of verification components. For the software verification competition, we use a configuration consisting of the CPAs for predicate analysis, program-counter tracking, call-stack analysis, and function-pointer analysis.

3 Strengths and Weaknesses

CPACHECKER is meant as an infrastructure for implementing and evaluating innovative verification algorithms. Due to that, the framework is not focused on optimizing as much as possible, but instead advocates a strong compliance of the theoretical concepts and its respective implementation, thus easing the integration of new algorithms and concepts. Furthermore, the use of CPAs provides a high degree of re-usability, which makes the tool kit highly interesting for other groups, some of which already use CPACHECKER to build their own extensions.

From a conceptional point of view, CPACHECKER, and the CPA for predicate analysis in particular, lack support for checking multi-threaded or recursive programs. Further areas of improvement, well documented by the false positives given in the categories DeviceDrivers and HeapManipulation, include a more complete handling of pointers as well as proper support for more advanced constructs of the C programming language, like structs and unions.

4 Setup and Configuration

The source code for CPACHECKER is released under the Apache 2.0 license and is available online at http://cpachecker.sosy-lab.org. Because the tool is written in JAVA, it runs on almost any platform. The predicate analysis currently works only under GNU/Linux because the MathSAT library is available only for this

---

[1] http://www.eclipse.org/cdt/
[2] http://mathsat4.disi.unitn.it/

platform. CPACHECKER requires a JAVA 1.6 compatible JDK (e.g., OpenJDK), Ant 1.7, and the GNU Multiprecision library for C++ (required by MathSAT). The build process is performed by calling ant from the CPACHECKER root directory. For the purpose of the software-verification competition, we use the trunk directory in revision 4569 and the configuration -sv-comp12. Thus the command line for running CPACHECKER is

```
./scripts/cpa.sh -sv-comp12 -heap 12500m path/to/sourcefile.cil.c
```

For C programs that assume a 64-bit environment (i.e., those in the category DeviceDrivers64) the below parameter needs to be added:

```
-setprop cpa.predicate.machineModel=64-Linux
```

The programs in the category DeviceDrivers need the following additional option, because they make heavy use of pointers:

```
-setprop cpa.predicate.handlePointerAliasing=true
```

For general purpose verification tasks (outside the competition), we recommend the configuration -predicateAnalysis instead. Also, the amount of memory given to the Java VM needs to be adjusted on machines with less RAM. CPACHECKER will print the verification result and the name of the output directory to the console. Additional information (such as the error path) will be written to files in this directory.

5 Project and Contributors

The CPACHECKER project was founded in 2007 by Dirk Beyer, and is hosted by the Software Systems Lab at the University of Passau. CPACHECKER is an international open-source project which is used and contributed to by several research groups, e.g., the Russian Academy of Science, the Technical University of Vienna, and the University of Paderborn.

We thank all contributors for their help and efforts spent on the CPACHECKER project. A complete list of contributors is provided on the project homepage at http://cpachecker.sosy-lab.org/. In particular, we would like to thank Dirk Beyer as the project leader and main architect, and Peter Häring, Michael Käufl, and Andreas Stahlbauer for their eager implementation work on CPACHECKER as student assistants.

References

1. Beyer, D., Henzinger, T.A., Théoduloz, G.: Configurable Software Verification: Concretizing the Convergence of Model Checking and Program Analysis. In: Damm, W., Hermanns, H. (eds.) CAV 2007. LNCS, vol. 4590, pp. 504–518. Springer, Heidelberg (2007)
2. Beyer, D., Keremoglu, M.E.: CPACHECKER: A Tool for Configurable Software Verification. In: Gopalakrishnan, G., Qadeer, S. (eds.) CAV 2011. LNCS, vol. 6806, pp. 184–190. Springer, Heidelberg (2011)
3. Beyer, D., Keremoglu, M.E., Wendler, P.: Predicate abstraction with adjustable-block encoding. In: Proc. FMCAD, pp. 189–197. FMCAD (2010)

Block Abstraction Memoization
for CPAchecker*
(Competition Contribution)

Daniel Wonisch

University of Paderborn, Germany
dwonisch@mail.upb.de

Abstract. Block Abstraction Memoization (ABM) is a technique in software model checking that exploits the modularity of programs during verification by *caching*. To this end, ABM records the results of block analyses and reuses them if possible when revisiting the same block again. In this paper we present an implementation of ABM into the predicate-analysis component of the software-verification framework CPACHECKER. With our participation at the Competition on Software Verification we aim at providing evidence that ABM can not only substantially increase the efficiency of predicate analysis but also enables verification of a wider range of programs.

1 Verification Approach

Currently, software model checking is getting more and more successful and is getting applied to industrial-size programs. Yet, scalability of the applied methods is still an issue. One approach to improve the scalability of model checking is *block abstraction memoization* (ABM). ABM exploits the modularity of programs by *caching* intermediate analyses of blocks. That is, ABM records the results of block analyses as for example analyses of loops or functions, and reuses them if possible when revisiting the same block again. We have implemented ABM into the predicate-analysis component of the software verification-framework CPACHECKER [3], including support for lazy refinements. It shows that ABM does not only increase the efficiency of the predicate analysis but also allows to successfully analyze programs that were not possible to analyze without.

As illustrative example, consider the C program (fragment) shown in the left of Figure 1. The program consists of three nested while loops, each incrementing a respective counter variable twice. After the execution of all loops it is asserted that the counting variables of the loops are indeed 2. On the right, Figure 1 shows a representation of this program as control-flow automaton (CFA). When analyzing the program using predicate abstraction we can prove the safety of the program with, e.g., the set of predicates $\{i = 0, i = 1, i = 2, j = 0, j = 1, j = 2, k = 0, k = 1, k = 2\}$. While doing so, CPACHECKER will visit, e.g.,

---

* This work was partially supported by the German Research Foundation (DFG) within the Collaborative Research Centre "On-The-Fly Computing" (SFB 901).

C. Flanagan and B. König (Eds.): TACAS 2012, LNCS 7214, pp. 531–533, 2012.

```
L0:   int i=2, j=2, k=2;
L1:   i = 0;
L2:   while(i < 2) {
L3:     j = 0;
L4:     while(j < 2) {
L5:       k = 0;
L6:       while(k < 2) {
          //no-op
L7:         k++;}
L8:       j++;}
L9:     i++;}
L10:  assert(i==2 && j==2 && k==2)
```

Fig. 1. Program NESTED and its control-flow automaton with 3 blocks

location L5 four times. At each visit the respective predicate abstractions only differ from each other in the valuation of the outer loop counter variables i and j. ABM allows to avoid this redundancy by considering the loop bodies as separate blocks as indicated in the right of Figure 1 by dotted rectangles. ABM initiates at each visit of a block a separate analysis whose result is *cached*. For example, if $L3$ is reached with an abstract state $i = 0 \land j = 2 \land k = 2$, ABM first recognizes that i is irrelevant for the block and thus analyzes the loop body starting with the *reduced* initial element $j = 2 \land k = 2$. After reaching the end of the block, the final abstract element $j = 2 \land k = 2$ is *expanded* to the full state space again, i.e., to $i = 0 \land j = 2 \land k = 2$ (i cannot change due an execution of this block), and the analysis of the program continues. In the next loop iteration, when reaching $L3$ again, the cached analysis can be reused and a re-analysis avoided. Similarly, with ABM, the most inner block starting with location $L5$ has to be analyzed only once. Hence, in this example, the amount of explored abstract elements for ABM-based model checking only grows linearly with the number of nested loops, leading to an exponential speed-up compared to the exponentially growing amount with classical model checking.

2   Implementation

We implemented our approach in the program-analysis tool CPACHECKER [3]. CPACHECKER is a framework for *configurable program analyses* (CPAs) [2]. CPAs allow users to specify different verification approaches in a uniform formalism. ABM is implemented as CPA in CPAchecker in order to benefit from existing verification components. Because ABM always functions as extension CPA (e.g., predicate analysis), it is technically implemented as a *WrapperCPA*. The wrapped CPA needs to comply with the *CPAWithABM* interface that basically requires the CPA to provide a *reduce* and *expand* operation for its abstract elements. So far, we have only implemented this interface in the *PredicateAnalysis* CPA. However, in principle, we could also easily enable ABM for other analyses (like e.g. shape analysis) by just implementing a *reduce* and *expand* operation for abstract elements of the respective domain.

In its supplied configuration, ABM considers loops and functions as blocks. Furthermore, it will consider all those predicate as relevant for a block for which a contained variable occurs in the block. As wrapped CPA we use predicate analysis with adjustable-block encoding (ABE) [1] configured to compute an abstraction at the start and end of a loop or function body. Math-SAT 4 (`http://mathsat4.disi.unitn.it`) is used as underlying SMT solver. Using this configuration, CPACHECKER with ABM performs very well on the *ControlFlowInteger* and reasonably well on the *SystemC* and *DeviceDrivers64* benchmark sets. With an incomplete analysis of pointer-aliases and heap structures, our approach is naturally rather unsuccessful for the *DeviceDrivers* and *HeapManipulation* sets. *Concurrency* (Pthreads) is currently not supported at all. Compared to CPACHECKER without ABM, our technique is especially beneficial on the *DeviceDrivers64* benchmark set.

3   Installation Instructions

ABM is fully integrated into the official source code of CPACHECKER. It can thus be downloaded from the official CPACHECKER webpage `http://cpachecker.sosy-lab.org` (Apache 2.0 license; Software Systems Lab, University of Passau). We use Revision 4573 for the competition. CPACHECKER can be compiled by executing `ant` in the checkout folder. To run CPA-CHECKER with the ABM configuration on a test file, execute `scripts/cpa.sh -sv-comp12-abm -heap 12500m source_file` in the checkout folder. For the *DeviceDrivers64* set, `-setprop cpa.predicate.machineModel=64-Linux` needs to be specified additionally. Contrary to this, for the *DeviceDrivers* set the script should be called with the arguments `-setprop cpa.predicate.handlePointerAliasing=true`, `-setprop cpa.abm.blockHeuristic=LoopPartitioning`, `-sv-comp12-abm-funpoint`, and `-heap 12500m`. Counterexamples, if found, are written to `test/output/ErrorPath.txt`.

Acknowledgement. We would like to thank Philipp Wendler for his extensive help with the integration of ABM into CPACHECKER.

References

1. Beyer, D., Keremoglu, M.E., Wendler, P.: Predicate Abstraction with Adjustable-Block Encoding. In: FMCAD 2010, pp. 189–197 (2010)
2. Beyer, D., Henzinger, T.A., Théoduloz, G.: Configurable Software Verification: Concretizing the Convergence of Model Checking and Program Analysis. In: Damm, W., Hermanns, H. (eds.) CAV 2007. LNCS, vol. 4590, pp. 504–518. Springer, Heidelberg (2007)
3. Beyer, D., Keremoglu, M.E.: CPACHECKER: A Tool for Configurable Software Verification. In: Gopalakrishnan, G., Qadeer, S. (eds.) CAV 2011. LNCS, vol. 6806, pp. 184–190. Springer, Heidelberg (2011)

Context-Bounded Model Checking with ESBMC 1.17
(Competition Contribution)

Lucas Cordeiro[1], Jeremy Morse[2], Denis Nicole[2], and Bernd Fischer[2]

[1] Electronic and Information Research Center, Federal University of Amazonas, Brazil
[2] Electronics and Computer Science, University of Southampton, UK
esbmc@ecs.soton.ac.uk

Abstract. ESBMC is a context-bounded symbolic model checker for single- and multi-threaded ANSI-C code. It converts the verification conditions using different background theories and passes them directly to an SMT solver.

1 Overview

ESBMC is a context-bounded symbolic model checker that allows the verification of single- and multi-threaded C code with shared variables and locks. ESBMC supports full ANSI-C (as defined in ISO/IEC 9899:1990), and can verify programs that make use of bit-level operations, arrays, pointers, structs, unions, memory allocation and floating-point arithmetic. It can reason about arithmetic under- and overflows, pointer safety, memory leaks, array bounds violations, atomicity and order violations, local and global deadlocks, data races, and user-specified assertions. However, as with other bounded model checkers, ESBMC is in general incomplete.

ESBMC uses the CBMC [2] frontend to generate the verification conditions (VCs) for a given program, but converts the VCs using different background theories and passes them to a Satisfiability Modulo Theories (SMT) solver. ESBMC natively supports Z3 and Boolector but can also output the VCs using the SMTLib format.

ESBMC supports the analysis of multi-threaded ANSI-C code that uses the synchronization primitives of the POSIX Pthread library. It traverses a reachability tree (RT) derived from the system in depth-first fashion, and calls the SMT solver whenever it reaches a leaf node. It stops when it finds a bug, or has explored all possible interleavings (i.e., the full RT). This combination of symbolic model checking with explicit state space exploration is similar to the ESST approach [1] for SystemC.

2 Verification Approach

We model a program as a transition system $M = (S, R, S_0)$, where S is the set of states, $S_0 \subseteq S$ the initial states, and $R \subseteq S \times S$ the transition relation. A state $s \in S$ consists of the values of all program variables, including the program counter pc. We use $I(s_0)$ to denote that $s_0 \in S_0$ and $\gamma(s_i, s_{i+1})$ to denote the constraints that correspond to a transition between two states s_i and s_{i+1}.

C. Flanagan and B. König (Eds.): TACAS 2012, LNCS 7214, pp. 534–537, 2012.

Given a transition system M, a safety property ϕ, a context switch bound C, and a bound k, ESBMC builds an RT that represents the program unfolding for C, k, and ϕ. For each interleaving π that passes through the RT nodes ν_1 to ν_k, it derives a formula $\psi_k^\pi = I(s_0) \wedge \bigvee_{i=0}^{k} \bigwedge_{j=0}^{i-1} \gamma(s_j, s_{j+1}) \wedge \neg\phi(s_i)$ which is satisfiable iff ϕ has a counterexample of depth k or less that is exhibited by π. Since $I(s_0) \wedge \bigwedge_{j=0}^{i-1} \gamma(s_j, s_{j+1})$ represents an execution of M of length i, ψ_k^π is satisfied iff for some time step $i \leq k$ there exists a reachable state along π at which ϕ is violated. The SMT solver then provides a satisfying assignment, from which we can extract a counterexample trace to the property violation. If ψ_k^π is unsatisfiable, we can conclude that no error state is reachable in k steps or less along π. Finally, we use $\psi_k = \bigvee_\pi \psi_k^\pi$ to check all interleavings.

ESBMC uses a quantifier-free fragment of a logic of linear integer arithmetics with arrays and uninterpreted functions to represent the VCs derived for a given ANSI-C program. Scalar datatypes can be represented either as bitvectors (i.e., using the SMTLib logic QF_AUFBV), or as abstract integers (i.e., QF_AUFLIA). Floating point arithmetic is approximated either by abstract reals (i.e., using the SMTLib logic AUFLIRA), by fixed-point arithmetic using bit vectors, or by rational arithmetic over abstract integers. Structures, unions, and pointer types are encoded using tuples [3,4]. ESBMC uses a simple but effective heuristic to select the best data representation and SMT solver.

ESBMC uses an instrumented model of the Pthread synchronization primitives to handle multi-threaded code [3]. The model allows an unbounded set of threads T, but ESBMC will only explore a finite number of context switches. It assumes that an enabled thread $t_j \in T$ can transition between statements to any enabled thread, and computes all states for which a transition exists to (implicitly) build the RT. ESBMC assumes sequential consistency, and by default assumes that variable accesses in individual statements are atomic.

3 Architecture, Implementation, and Availability

ESBMC is implemented in C++. It has been branched off CBMC v2.9, and still uses its parser, goto conversion, and core of the symbolic execution. The goto conversion replaces all control structures by (conditional) jumps, which simplifies the program representation. The symbolic execution of this goto representation converts the program into SSA form and unrolls loops and recursive functions on-the-fly, generating unwinding assertions that fail if the given bound is too small. It also generates the VCs for the safety properties and user-specified assertions.

Changes to the CBMC components include the addition of the new safety properties and more optimizations (e.g., better constant propagation), and the integration of native SMT backends. In order to support the analysis of multi-threaded code ESBMC implements a partial-order reduction (POR) [3,5] scheme to reduce the number of states that have to be explored. It first applies the visible instruction analysis POR, which removes the interleavings of instructions that do not affect the global variables, followed by the read-write analysis POR in which two (or more) independent interleavings can be safely merged into one. Additionally, it implements a two-level symbolic state hashing scheme [6] that represents a particular RT node and all constraints affecting a particular assignment to a variable separately. Since each new RT node can only change the (symbolic) value of at most one variable, this scheme reduces the computational effort, as it allows us to retain the hash values of the unchanged variables.

User Interface. ESBMC can be invoked through a standard command-line interface or configured through an Eclipse plug-in. When a property violation is detected, it produces a counterexample trace in the CBMC format. The plugin visualizes such traces and provides direct access to the corresponding code.

Availability and Installation. Self-contained binaries for 32-bit and 64-bit Linux environments are available at www.esbmc.org; versions for other operating systems are available on request. The competition version only uses the Z3 solver (V3.1). It assumes a 64-bit architecture and uses experimentally determined unwinding bounds; setting explicit context switch bounds is not required for the given concurrency benchmarks. It only checks for the reachability of the error label and ignores all other properties, including unwinding assertions. It is called as follows:

```
esbmc --timeout 15m --memlimit 15g --64 --unwind <n>
 --no-unwinding-assertions --no-assertions --error-label ERROR
 --no-bounds-check --no-div-by-zero-check --no-pointer-check <f>
```

4   Results

With unwinding assertions enabled, ESBMC proves 30 programs correct and finds errors in 27. However, it also claims errors in two correct programs and fails to find existing errors in another nine. ESBMC's performance is largely similar across all categories, although unwinding assertions and timeouts are, as expected, concentrated on the larger benchmarks.

With unwinding assertions *disabled*, a correctness *claim* is not a full correctness *proof*, because errors could occur for larger unwinding bounds. In fact, the number of false negatives/positives increases to ten and nine, resp., but their rate remains roughly the same, so that ESBMC's overall performance improves markedly: with 121 programs rightly claimed correct and 71 errors identified, it achieves a total score of 249. Four programs do not conform to the supported ANSI-C standard and cause parsing errors. The remaining programs time out during the symbolic execution (21) or fail with an internal error (41). These errors results are mostly caused by problems in ESBMC's pointer handling that are exposed by the excessive typecasts in the CIL-converted code.

Acknowledgments. The development of ESBMC is funded by EPSRC, EC FP7, the Royal Society, and INdT. Q. Li implemented the Eclipse plugin.

References

1. Cimatti, A., Micheli, A., Narasamdya, I., Roveri, M.: Verifying SystemC: a software model checking approach. In: FMCAD, pp. 121–128 (2010)
2. Clarke, E., Kroning, D., Lerda, F.: A Tool for Checking ANSI-C Programs. In: Jensen, K., Podelski, A. (eds.) TACAS 2004. LNCS, vol. 2988, pp. 168–176. Springer, Heidelberg (2004)

3. Cordeiro, L.: SMT-Based Bounded Model Checking of Multi-Threaded Software in Embedded Systems. PhD Thesis, U Southampton (2011)
4. Cordeiro, L., Fischer, B., Marques-Silva, J.: SMT-based bounded model checking for embedded ANSI-C software. In: ASE, pp. 137–148 (2009)
5. Cordeiro, L., Fischer, B.: Verifying Multi-Threaded Software using SMT-based Context-Bounded Model Checking. In: ICSE, pp. 331–340 (2011)
6. Morse, J., Cordeiro, L., Nicole, D., Fischer, B.: Context-Bounded Model Checking of LTL Properties for ANSI-C Software. In: Barthe, G., Pardo, A., Schneider, G. (eds.) SEFM 2011. LNCS, vol. 7041, pp. 302–317. Springer, Heidelberg (2011)

Proving Reachability Using FShell[*]
(Competition Contribution)

Andreas Holzer[1], Daniel Kroening[2], Christian Schallhart[2],
Michael Tautschnig[2], and Helmut Veith[1]

[1] Vienna University of Technology, Austria
[2] University of Oxford, United Kingdom

Abstract. FShell is an automated white-box test-input generator for
C programs, computing test data with respect to *user-specified* code
coverage criteria. The pillars of FShell are the declarative specification
language FQL (FShell Query Language), an efficient back end for computing test data, and a mathematical framework to reason about coverage criteria. To solve the reachability problem posed in SV-COMP we
specify coverage of ERROR labels. As back end, FShell uses bounded
model checking, building upon components of CBMC and leveraging the
power of SAT solvers for efficient enumeration of a full test suite.

1 Overview

FShell implements automatic white-box test-input generation according to a
user-defined coverage specification given in the declarative language FQL [5,6].
To resemble formal verification and solve the reachability problem presented in
the SW Verification Competition we specify coverage of all ERROR labels.

FQL is built on top of a concise mathematical framework for formalising
code coverage criteria. This framework enables automatic processing of FQL
queries and together with FQL makes our approach oblivious to the algorithmic
details of test-input generation. As this overall architecture is analogous to that
of databases, we refer to our approach as *query-driven program testing* [4].

As back end for solving FQL queries, i.e., computing test inputs, FShell
uses components of the C bounded model checker (CBMC) [1], which enables
support for full C syntax and semantics, and makes efficient use of SAT solving
(using MiniSat 2.2.0 [2]). An overview of the architecture is presented in the next
section. The technical approach was first sketched in [3], further refined in [4]
and full details of the current implementation can be found in [7].

2 Architecture

FShell comprises two main parts: The *front end* handles user interactions with
a command-line interface. There, control commands such as loading source files,

---

[*] This research is supported by the FWF research network RiSE, the WWTF grant
PROSEED and EPSRC project EP/H017585/1.

C. Flanagan and B. König (Eds.): TACAS 2012, LNCS 7214, pp. 538–541, 2012.

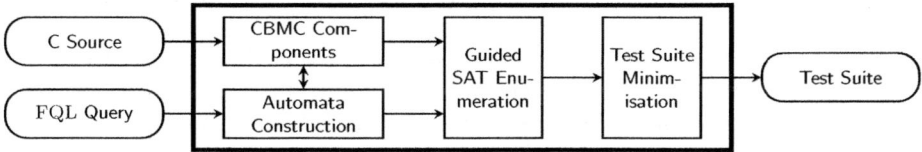

Fig. 1. Query processing

macro definitions, and FQL queries are entered by the user. The *back end* performs the actual test case generation. Figure 1 sketches the conceptual steps:

1. We use the code base of CBMC to first obtain an intermediate representation (GOTO functions) of the program under test.
2. We translate the given FQL query into automata over statements in the GOTO functions. We refer to these automata as *trace automata*.
3. CBMC builds a Boolean equation describing all states yielding a violation of a given property (assertion) of the input program. It is then able to produce an example of such a violation (counterexample).

   In test case generation, we use this scheme by adding the property stating that *no* program execution accepted by the trace automata exists. Any counterexample (i.e., test case) then describes a path that fulfills the query. We efficiently enumerate test cases using guided SAT enumeration [4]. The resulting test suite satisfies the given coverage specification by construction.
4. To remove redundant test cases, we perform test suite minimisation. This problem is an instance of the minimum set cover problem, which we reduce to a series of SAT instances.
5. We display the generated test suite as a list of initial values of variables.

3   Strengths and Weaknesses

The main advantage of FSHELL over its underlying back end CBMC is computing counterexamples for *all* reachable error locations. Apart from this fact, however, FSHELL mainly acts as bounded model checker in this competition. As such, FSHELL was successful as it was, together with SatAbs, one of only two competition participants that never reported a spurious "UNSAFE".

As bounded model checking is necessarily incomplete for all benchmarks with unbounded loops, a choice had to be made how to handle those. Without further options, FSHELL performs an additional step to prove given loop unwindings (see below) to be sufficient. This step would ensure that FSHELL does not return "SAFE" without having properly proved safety, but instead aborts early.

As consequence, however, FSHELL would not have scored any points on programs with unbounded loops. Therefore we decided to disable the early abort and instead perform verification under the assumption that the given loop bounds suffice. On the one hand, this is helpful in bug finding and permitted to prove unsafety on several benchmarks with unbounded loops. Whenever this loop bound is insufficient for finding paths to the error location, however, FSHELL wrongly reports "SAFE" and becomes unsound.

This is the classic bounded model checking setting, but proved to be even less successful under the presented scoring system: in the categories "Control-FlowInteger" and "DeviceDrivers" FSHELL correctly determined the result in more than 70% of the instances, but scored only 19% of the possible points. For all other categories (except for "Concurrency", which is presently unsupported) problems in the back end caused verification to fail; these have mostly been fixed by now and future versions are expected to perform much better.

4   Tool Setup

The competition participant is FSHELL version 1.3, which can be downloaded from http://code.forsyte.de/fshell. To avoid the interactive operation of FSHELL, a file "query" should first be set up containing the following statements:

```
cover @label(ERROR)
quit
```

Then, FSHELL can be run as

```
fshell --unwind 10 --no-unwinding-assertions --query-file query FOO.c
```

for a source file "FOO.c". By default, FSHELL will assume a 64-bit memory model as this is the competition platform. For those benchmarks written with a 32-bit memory model in mind, the option --32 must be given in addition.

By definition, FSHELL produces test inputs instead of full counterexample traces; each set of inputs uniquely determines a single execution, however. An instance is found to be "SAFE" if no test inputs are found. In this case, FSHELL prints Test cases: 0 – an "UNSAFE" instance yields a non-zero count.

Software Project. FSHELL is maintained by Michael Tautschnig as an extension of CBMC. FSHELL was jointly designed by the authors. FSHELL is released at the above web site as binary for several platforms under an Apache 2.0 license.

References

1. Clarke, E., Kroning, D., Lerda, F.: A Tool for Checking ANSI-C Programs. In: Jensen, K., Podelski, A. (eds.) TACAS 2004. LNCS, vol. 2988, pp. 168–176. Springer, Heidelberg (2004)
2. Eén, N., Sörensson, N.: An Extensible SAT-solver. In: Giunchiglia, E., Tacchella, A. (eds.) SAT 2003. LNCS, vol. 2919, pp. 502–518. Springer, Heidelberg (2004)
3. Holzer, A., Schallhart, C., Tautschnig, M., Veith, H.: FSHELL: Systematic Test Case Generation for Dynamic Analysis and Measurement. In: Gupta, A., Malik, S. (eds.) CAV 2008. LNCS, vol. 5123, pp. 209–213. Springer, Heidelberg (2008)
4. Holzer, A., Schallhart, C., Tautschnig, M., Veith, H.: Query-Driven Program Testing. In: Jones, N.D., Müller-Olm, M. (eds.) VMCAI 2009. LNCS, vol. 5403, pp. 151–166. Springer, Heidelberg (2009)

5. Holzer, A., Schallhart, C., Tautschnig, M., Veith, H.: How did you specify your test suite? In: ASE, pp. 407–416. ACM (2010)
6. Holzer, A., Tautschnig, M., Schallhart, C., Veith, H.: An Introduction to Test Specification in FQL. In: Raz, O. (ed.) HVC 2010. LNCS, vol. 6504, pp. 9–22. Springer, Heidelberg (2010)
7. Tautschnig, M.: Query-Driven Program Testing. Ph.D. thesis, Vienna University of Technology (2011)

LLBMC: A Bounded Model Checker for **LLVM**'s Intermediate Representation[*]
(Competition Contribution)

Carsten Sinz, Florian Merz, and Stephan Falke

Institute for Theoretical Computer Science
Karlsruhe Institute of Technology (KIT), Germany
{carsten.sinz,florian.merz,stephan.falke}kit.edu

Abstract. We present LLBMC, a bounded model checker for C programs. LLBMC uses the LLVM compiler framework in order to translate C programs into LLVM's intermediate representation (IR). The resulting code is then converted into a logical representation and simplified using rewrite rules. The simplified formula is finally passed to an SMT solver. In contrast to many other tools, LLBMC uses a flat, bit-precise memory model. It can thus precisely model, e.g., memory-based re-interpret casts.

1 Verification Approach

Bounded model checking (BMC) has proven to be a very successful technique in hardware verification. More recently, it has also been applied for verifying software written in C [1,4]. Applying BMC for verifying C programs, however, comes with many obstacles that have to be tackled. One of the most important differences is that the syntax and semantics of a programming language like C is much more complicated than a hardware description. One has to deal, e.g., with memory allocation and de-allocation, (function) pointers, complex data structures, and function calls.

LLBMC uses an approach which, instead of exploring the source code directly, makes use of existing compiler technology and performs the analysis on a compiler intermediate representation. Such an intermediate representation offers a much simpler syntax and semantics than a programming language like C, and thus eases a logical encoding of the verification problem considerably.

We have chosen the LLVM [5] compiler infrastructure and its assembler-like intermediate representation as the starting point for our approach, but the idea can also be applied to other low-level languages. LLVM is both a (GCC-compatible) C/C++/Objective-C compiler and a library of compiler technologies, providing, e.g., source- and target-independent optimizations.

Our primary goal is to detect memory errors in C code [7,2,6]. Memory errors include invalid memory accesses, heap and stack buffer overflows, and invalid frees (e.g., double frees).

---

[*] This work was supported in part by the "Concept for the Future" of Karlsruhe Institute of Technology within the framework of the German Excellence Initiative.

C. Flanagan and B. König (Eds.): TACAS 2012, LNCS 7214, pp. 542–544, 2012.
© Springer-Verlag Berlin Heidelberg 2012

2   Software Architecture

While LLBMC is designed for C programs, its input format is LLVM-IR, the intermediate representation of the LLVM compiler framework. LLVM-IR is an abstract assembler language that is programming-language-independent. This makes it easier to extend LLBMC to other languages supported by LLVM (like C++ or Objective-C). Furthermore, the challenges in parsing complex high-level language syntax, such as C++, are eliminated. Instead, only a limited instruction set needs to be supported. LLVM-IR is architecture-dependent in the sense that the compiler frontend selects, e.g., the bitwidth of pointers and integer data types.

After reading in the LLVM-IR code, LLBMC applies a number of transformations to it. In particular, loops are unrolled, functions are inlined, and the control flow graph is simplified. The transformed code is then converted to ILR, which is a representation of a program in the logic of bit-vectors and arrays plus some extensions, related to memory allocation. ILR provides an explicit state object for the memory content as well as for the state of the memory allocation system. These state objects encode the dependencies between memory accessing instructions in the ILR formula. Because of this, dependencies between instructions in LLVM, which were implicitly given by the ordering of the read and write operations are made explicit in the ILR formula. This change makes the expressions in an ILR formula ordering-independent. The ILR formula is then simplified using rewrite rules, and memory access correctness expressions are reduced to bit-vector formulas (see [2,7] for details). If no more rewrite rules can be applied, the formula is passed to the SMT solver STP [3].

3   Strengths and Weaknesses of the Approach

LLBMC is tailored towards finding bugs in C programs, especially memory-related ones (not so much towards proving their absence). Detectable errors include:

- arithmetic overflow and underflow, including shift overflow,
- invalid memory access operations,
- invalid memory allocation, including invalid `free`s, and
- overlapping memory regions in `memcpy`.

Furthermore, LLBMC supports checking of user assertions and reachability of labels named "ERROR" in the C-code. It can also detect whether the loop unrolling and function inlining bound was sufficient or has to be increased in order to achieve full coverage.

In the competition, LLBMC was used with a fixed unwinding bound of 7 and an automatically determined function inlining bound. It was not checked whether the unwinding bound is sufficient, but only whether the "ERROR" label was reachable within these bounds (as other comparable tools have chosen similar settings). If no error was found, the instance was considered safe. LLBMC was able to successfully handle 146 out of 269 benchmark instances (not participating in category "Concurrency", as this is not supported by LLBMC), resulting in a first place in category "Device Drivers" and a second place in category "Heap

Manipulation". Among the unsolved instances, 65 were due to time-outs, and 48 due to current restrictions of LLBMC (e.g., related to memcpy or inline assembly). LLBMC produced 7 false positives and 3 false negatives. The false negatives (i.e. where an error was missed) were due to an insufficient loop unrolling bound. Among the 7 false positives, one was due to an error in LLBMC related to detecting a malloc function. The other 6 were due to uninitialized pointer variables, by which other (e.g., global) variables could be overwritten and thus be modified, resulting in the "ERROR" label becoming reachable. We do not consider these errors as "false positives", but see here a special strength of LLBMC and its precise memory model, as such errors are very hard to detect and, in practice, result in non-deterministic program behavior.

4   Tool Setup and Configuration

The version of LLBMC (0.9) submitted to TACAS can be downloaded from

http://llbmc.org/llbmc-tacas12.zip.

LLBMC requires llvm-gcc (version 2.9) in order to convert C input files to LLVM's intermediate representation. For instructions on how to use LLBMC, just enter llbmc --help. The ZIP archive also contains two wrapper shell scripts to run LLBMC on individual C files. The first, llbmcc, iteratively increases the loop unwind bound and also checks whether the unwind bound is sufficient. The second, llbmcc2, which was used in the competition, also increases the unwind bound, but only up to a maximal value of 7, and does not perform unwind bound checks. Both shell scripts compile the C program, run LLBMC, and perform only a reachability check for a basic block labelled "ERROR", but no other checks, such as for invalid memory accesses. They output either SAFE, if the error label is unreachable (within the given bound for llbmcc2), or UNSAFE otherwise.

Further information on LLBMC is available on the web at http://llbmc.org.

References

1. Clarke, E., Kroning, D., Lerda, F.: A Tool for Checking ANSI-C Programs. In: Jensen, K., Podelski, A. (eds.) TACAS 2004. LNCS, vol. 2988, pp. 168–176. Springer, Heidelberg (2004)
2. Falke, S., Merz, F., Sinz, C.: A theory of C-style memory allocation. In: Proc. SMT 2011, pp. 71–80 (2011)
3. Ganesh, V., Dill, D.L.: A Decision Procedure for Bit-Vectors and Arrays. In: Damm, W., Hermanns, H. (eds.) CAV 2007. LNCS, vol. 4590, pp. 519–531. Springer, Heidelberg (2007)
4. Ivančić, F., Yang, Z., Ganai, M.K., Gupta, A., Ashar, P.: Efficient SAT-based bounded model checking for software verification. TCS 404(3), 256–274 (2008)
5. Lattner, C., Adve, V.S.: LLVM: A compilation framework for lifelong program analysis & transformation. In: Proc. CGO 2004, pp. 75–88 (2004)
6. Merz, F., Falke, S., Sinz, C.: LLBMC: Bounded Model Checking of C and C++ Programs Using a Compiler IR. In: Joshi, R., Müller, P., Podelski, A. (eds.) VSTTE 2012. LNCS, vol. 7152, pp. 146–161. Springer, Heidelberg (2012)
7. Sinz, C., Falke, S., Merz, F.: A precise memory model for low-level bounded model checking. In: Proc. SSV 2010 (2010)

Predator: A Verification Tool for Programs with Dynamic Linked Data Structures*
(Competition Contribution)

Kamil Dudka, Petr Müller, Petr Peringer, and Tomáš Vojnar

FIT, Brno University of Technology, IT4Innovations Centre of Excellence, Czech Republic

Abstract. Predator is a tool for automated formal verification of sequential C programs with dynamic linked data structures. It is in principle based on separation logic, but uses a graph-based heap representation. This paper first provides a brief overview of Predator and then discusses experience with its participation in the Software Verification Competition of TACAS'12.

1 Introduction

Predator is a tool for automated formal verification of sequential C programs with dynamic linked data structures. Currently, it supports verification of various linked list variants, including nested, cyclic, and/or shared lists, possibly using limited pointer arithmetics to navigate through list nodes as is usual in real-life implementations of list manipulating programs. Predator implicitly detects various memory-related errors and can also check for reachability of error labels, which made its participation in the TACAS'12 Software Verification Competition (SV-COMP'12) possible. However, most of the capabilities of Predator to detect memory-specific errors could not be applied in the competition. Predator is publicly available[1] as open-source under GPLv3.

This paper provides a brief overview of Predator's design principles and capabilities, and then discusses experiments with Predator on the benchmarks of SV-COMP'12. More details about Predator can be found in the tool paper [1].

2 Overview of Predator

Predator is conceptually based on *separation logic* with *higher-order inductive predicates*. It encodes infinite sets of heaps in a finite symbolic way using a *graph-based representation* of separation logic formulae. There are two kinds of nodes in the graphs: (1) possibly nested *objects* corresponding to statically and automatically allocated program variables, dynamically allocated storage, list segments, etc. and (2) *values* of the objects, e.g., addresses of objects and the special values undefined, deleted, and

---

* This work was supported by the Czech Science Foundation (project P103/10/0306), the Czech Ministry of Education (projects COST OC10009, MSM 0021630528), and the EU/Czech IT4Innovations Centre of Excellence project CZ.1.05/1.1.00/02.0070.

[1] http://www.fit.vutbr.cz/research/groups/verifit/tools/predator

C. Flanagan and B. König (Eds.): TACAS 2012, LNCS 7214, pp. 545–548, 2012.

null in the case of pointers and function pointers. This allows for dealing with pointers to any variable (not only dynamically allocated) and detection of some classes of memory-related bugs (stack smashing, buffer overrun, and the like). Predator also implicitly detects other memory-related errors like memory leaks, invalid dereferences, and double frees. Predator uses a specialised *join operator*, applicable both on entire symbolic heaps as well as on their parts. The latter functionality serves for discovering new list predicates used for a subsequent abstraction (summarisation) of list segments. The join algorithm is also used for checking entailment of symbolic heaps (by checking that the join of two symbolic heaps produces one of them). Hence, Predator does not use any off-the-shelf decision procedures.

The main goal of Predator is to verify real system code in a fully automated way. Since real-life implementations of lists often use limited *pointer arithmetics* (e.g., in Linux, the list header structure is embedded at any place in list nodes, and pointer arithmetics is used to move within the list nodes), Predator is capable of handling typical patterns of such low-level programming techniques. For this, *offsets* of sub-objects within their encapsulating objects are tracked.

Predator is implemented as a *GCC plug-in*[2], which brings significant advantages. For the many real-life programs whose production versions are built using GCC, the analysis is performed on the same program representation as the one used for producing the actual binary. GCC has also a very large coverage of what it can parse (standard C and GNU extensions), while other tools, like CIL, used by some other analysers, usually implement only a subset of the C language standard. Further, using GCC allows an easier integration with other commonly used development tools. In particular, Predator presents errors and warnings in the standard GCC format, which many development tools can handle by default.

Predator is written in C++ and uses Boost libraries. The only dependencies that need to be installed are GCC 4.4.6+ with C++ support and CMake. It is recommended to build Predator against a local build of GCC. This can be done in several steps, which are described in the README file[3] of the Predator's distribution. The local build of GCC is fully automated and the whole process takes up to 10 minutes assuming a fast enough Internet connection (needed for downloading GCC sources). The current version of Predator runs on Linux, but the code of the analyser itself is architecture-independent.

3   Experience with Predator

Predator is distributed with a collection of more than 200 test cases. The test cases include real-world code snippets as well as code focused on corner cases in the use of the dynamic linked data structures. This collection also serves as our internal benchmark for measuring the performance of the tool. The errors sought in these benchmarks are typical memory manipulation errors (memory leakage, double free, etc.) for which no user-provided specification is needed. More information on our test cases can be found

---

[2] Predator uses the low-level GIMPLE representation of the GCC's intermediate code.

[3] Instructions specific for the Software Verification Competition held at TACAS'12 are located in the file named README-sv-comp-TACAS-2012 available in the distribution of Predator.

in [1]. Four of our test cases were extended by explicit checks of shape properties and sent as a contribution to the SV-COMP'12 benchmarks.

Below, we summarise our experience with Predator from the training phase of SV-COMP'12, describe the results we reached on the competition benchmarks, and discuss problems encountered on particular test cases. We refrain from stating precise quantitative data about our results, which is a part of the presentation of SV-COMP'12 itself.

During the experiments with the training set of test cases of SV-COMP'12, we encountered problems resulting from Predator's implicit detection of memory-related errors (such as invalid dereferences or buffer overruns), which Predator never ignores. Some of the test cases in the benchmark were problematic from this point of view since they assumed idealised memory models. These test cases caused Predator to terminate prematurely and report memory errors unrelated to the reachability of the given error label. Hence, we needed to create a special layer on top of Predator that distinguishes between errors defined by the competition rules (where UNSAFE means that an error label is reachable) and errors caused by using memory in a wrong way.

From the SV-COMP'12 categories, we focused on those that Predator is designed to verify, especially on the Dynamic Data Structures category. This category contained test cases contributed by the Predator project itself (*heap-manipulation*) and test cases taken from the web page of the BLAST 3.0 project (*list-properties*). In this category, Predator succeeded in all but two test cases, which contained lists with alternating small integral numbers in their nodes. The current version of Predator cannot represent numbers in an abstract way (as, e.g., integral ranges), which prevented the list segment abstraction from being applied and consequently the analysis failed to terminate.

Verifying Linux drivers is one of the major goals for Predator, so we expected good results in this category too. On the *ldv-regression* benchmark, we ended up with the highest possible score (at least in our home environment). For that to happen, it was necessary to improve some test cases to make them allocate the memory they use and to write a few dummy models of external functions, but these changes were accepted by the competition organisers. Unfortunately, Predator was not successful in the *ddv-machzwd* benchmark due to the lack of abstraction over integral values as mentioned above.

Across all benchmark categories, Predator never returned SAFE for a test case declared UNSAFE, which confirms that the analysis done by Predator is sound. On the *pthread* benchmark, Predator instantly returned UNKNOWN for all test cases because of an unhandled call to pthread_create(). As Predator does currently not aim at the analysis of concurrent programs, this was an expected response. In the *locks* benchmark, Predator could solve all of the test cases, however, given time and space larger than allowed by the rules of SV-COMP'12. In the given limits, Predator managed to analyse only a few of these test cases. This is due to the analysis done by Predator is quite inefficient for such kind of programs since no (refinable) abstraction of non-pointer data is currently supported by Predator. Such data is either tracked precisely or completely discarded. Predator also succeeded on several test cases from the *systemc* benchmark, which were proven SAFE. Like in the *ddv-machzwd* benchmark, Predator lost many points here because of the lack of abstraction over integral values.

4   Conclusions and Future Work

We have briefly presented Predator and its participation in SV-COMP'12. Predator is regularly updated and enhanced. We plan to add support for further kinds of dynamic data structures (like trees), improve the support for non-pointer data, provide support for analysing C++ code, and possibly add techniques allowing one to analyse incomplete programs (e.g., using bi-abduction). SV-COMP'12 provides many interesting test cases, which represent a good motivation for further development of Predator.

Reference

1. Dudka, K., Peringer, P., Vojnar, T.: Predator: A Practical Tool for Checking Manipulation of Dynamic Data Structures Using Separation Logic. In: Gopalakrishnan, G., Qadeer, S. (eds.) CAV 2011. LNCS, vol. 6806, pp. 372–378. Springer, Heidelberg (2011)

HSF(C): A Software Verifier Based on Horn Clauses
(Competition Contribution)

Sergey Grebenshchikov[1], Ashutosh Gupta[2],
Nuno P. Lopes[3], Corneliu Popeea[1], and Andrey Rybalchenko[1]

[1] Technische Universität München
[2] IST Austria
[3] INESC-ID / IST - TU Lisbon

Abstract. HSF(C) is a tool that automates verification of safety and liveness properties for C programs. This paper describes the verification approach taken by HSF(C) and provides instructions on how to install and use the tool.

1   Verification Approach

HSF(C) is a tool for verification of C programs based on predicate abstraction and refinement following the counterexample-guided abstraction refinement (CEGAR) paradigm [4]. There are a number of successful tools [1,7,5,10,2] based on abstraction refinement. We give here a brief description of our verification algorithm; interested readers can find more details about the underlying theory behind our implementation in [10,6].

The algorithm used in HSF(C) is a generalization of the CEGAR scheme that deals with Horn-like clauses instead of transition systems/programs with procedures. We use Horn clauses to represent both the program to be verified and the proof rule used for verification, i.e., safety checking for programs with procedures. The proof rule lists premises for program safety and requires auxiliary assertions that represent inductive invariants. Given Horn clauses as input, our algorithm proceeds in three steps.

1. With a fixed set of predicates, initially empty, we find a solution for the auxiliary assertions. At this step we perform logical inference and rely on abstraction to ensure termination in the presence of recursion and to ensure efficiency in the presence of large sets of clauses.
2. We check whether the computed solution satisfies program safety. If so, the verification succeeds and the algorithm returns "safe".
3. We check whether the logical inference performed in the first step in the setting without any abstraction yields a solution that still violates program safety. If the violation is still present then we return "unsafe" and the inference tree as an error path that reaches the error location. Otherwise, we use the obtained solution to refine the abstraction function and go back to the first step.

C. Flanagan and B. König (Eds.): TACAS 2012, LNCS 7214, pp. 549–551, 2012.

2   Software Architecture

HSF(C) relies on the CIL library [9] as a frontend for our tool. An additional frontend step transforms the CIL abstract syntax tree representation in Horn clauses. The Horn clauses are generated automatically from a proof rule for safety checking with support for procedure summarization [11]. These Horn clauses are then solved with the CEGAR algorithm described in the previous section. Our solver is implemented in Prolog and compiled using the SICStus Prolog compiler [12]. Our implementation relies on the constraint solver for linear arithmetic CLP(Q) [8].

3   Discussion

In this section, we present our experience running HSF(C) on the benchmarks from the software competition.

ControlFlowInteger. We found the approach based on abstraction and refinement well suited for the benchmarks in the category. The proofs found by HSF(C) typically involve a small number of predicates. For handling the few pointers and heap-allocated structures that are used by these benchmarks, we found the use of the pointer analysis provided by CIL to be sufficiently precise.

SystemC. The benchmarks from this category encode concurrency from the program in finite-domain auxiliary variables. Some heuristics proposed in [3] combine explicit-state model checking to model the states of the SystemC scheduler with predicate abstraction. Such heuristics would have been more appropriate to handle these benchmarks than our current approach based only on predicate abstraction.

Other Categories. HSF(C) did not participate in the other four competition categories, i.e., Concurrency, DeviceDrivers, DeviceDrivers64, and HeapManipulation. Here we list the current limitations of our tool that we need to address to be able to handle these benchmarks:

- Concurrency: the frontend of HSF(C) needs a model for the functions from the Pthreads library.
- DeviceDrivers, DeviceDrivers64: some of the features of the C language that HSF(C) does not precisely model are fixed-size integers, union types, volatile variables, and bitwise operations.
- HeapManipulation: the pointer analysis that we use is not precise to handle the data structures from these benchmarks.

To summarize, we ran our tool on 158 benchmarks from two categories. HSF(C) obtained the following results:

- ControlFlowInteger (140/144 points): for all benchmarks where HSF(C) reported a result, the result was correct. HSF(C) ran out of time for two benchmarks.
- SystemC (8/87 points): for all benchmarks where HSF(C) reported a result, the result was correct. HSF(C) ran out of time or memory on 57 benchmarks.

4   Tool Setup

HSF(C) can be downloaded from the following webpage:

`http://www7.in.tum.de/tools/hsf/`

The HSF(C) distribution consists of three statically compiled binaries that correspond to the C frontend, a converter to Horn clauses, and the Horn clause solver. The distribution also contains a script that runs the three executables with appropriate parameters. The tool should be run as follows: `./qarmc.sh <file.c>`. The working directory (`PWD`) must be the directory where the HSF(C)'s files are located. The only required library is the standard glibc 32-bit.

References

1. Ball, T., Rajamani, S.K.: The SLAM project: debugging system software via static analysis. In: POPL (2002)
2. Beyer, D., Keremoglu, M.E.: CPACHECKER: A Tool for Configurable Software Verification. In: Gopalakrishnan, G., Qadeer, S. (eds.) CAV 2011. LNCS, vol. 6806, pp. 184–190. Springer, Heidelberg (2011)
3. Cimatti, A., Micheli, A., Narasamdya, I., Roveri, M.: Verifying SystemC: A software model checking approach. In: FMCAD, pp. 51–59 (2010)
4. Clarke, E.M., Grumberg, O., Jha, S., Lu, Y., Veith, H.: Counterexample-Guided Abstraction Refinement. In: Emerson, E.A., Sistla, A.P. (eds.) CAV 2000. LNCS, vol. 1855, pp. 154–169. Springer, Heidelberg (2000)
5. Clarke, E., Kroning, D., Sharygina, N., Yorav, K.: SATABS: SAT-Based Predicate Abstraction for ANSI-C. In: Halbwachs, N., Zuck, L.D. (eds.) TACAS 2005. LNCS, vol. 3440, pp. 570–574. Springer, Heidelberg (2005)
6. Gupta, A., Popeea, C., Rybalchenko, A.: Predicate abstraction and refinement for verifying multi-threaded programs. In: POPL, pp. 331–344 (2011)
7. Henzinger, T.A., Jhala, R., Majumdar, R., Sutre, G.: Lazy abstraction. In: POPL, pp. 58–70 (2002)
8. Holzbaur, C.: OFAI clp(q,r) Manual, Edition 1.3.3. Austrian Research Institute for Artificial Intelligence, Vienna, TR-95-09 (1995)
9. Necula, G.C., McPeak, S., Rahul, S.P., Weimer, W.: CIL: Intermediate Language and Tools for Analysis and Transformation of C Programs. In: Horspool, R.N. (ed.) CC 2002. LNCS, vol. 2304, pp. 213–228. Springer, Heidelberg (2002)
10. Podelski, A., Rybalchenko, A.: ARMC: The Logical Choice for Software Model Checking with Abstraction Refinement. In: Hanus, M. (ed.) PADL 2007. LNCS, vol. 4354, pp. 245–259. Springer, Heidelberg (2007)
11. Reps, T.W., Horwitz, S., Sagiv, S.: Precise interprocedural dataflow analysis via graph reachability. In: POPL, pp. 49–61 (1995)
12. The Intelligent Systems Laboratory. SICStus Prolog User's Manual. Swedish Institute of Computer Science, Release 4.2.0 (2011)

SatAbs: A Bit-Precise Verifier for C Programs[*]
(Competition Contribution)

Gérard Basler, Alastair Donaldson[1], Alexander Kaiser[2], Daniel Kroening[2], Michael Tautschnig[2], and Thomas Wahl[3]

[1] Imperial College, London, United Kingdom
[2] University of Oxford, United Kingdom
[3] Northeastern University, Boston, United States

Abstract. SATABS is a bit-precise software model checker for ANSI-C programs. It implements sound predicate-abstraction based algorithms for both sequential and concurrent software.

1 Verification Approach

SATABS [7] is a verifier for C programs that uses counterexample-guided abstraction refinement [8] (Fig. 1), based on predicate abstraction [12], as pioneered by SLAM [2]. By interpreting variables of the C program as bit-vectors, efficient SAT procedures are used for abstraction and simulation [6]. This renders the theorem prover calls that are made during abstraction decidable, and enables bit-precise verification, which is essential when analysing system-level software.

In [10] the first sound and symmetry-aware predicate abstraction based approach towards model checking multi-threaded programs was presented. These results have now been integrated with SATABS, allowing scalable verification of concurrent C programs comprised of replicated threads.

Efficient symmetry-aware predicate abstraction requires amendments in all four key components of Figure 1: (1) the Boolean program computed as abstraction will use *passive predicates and broadcasts* [10]; (2) the underlying model checker for Boolean programs must be able to make use of symmetry and support passive predicates, which presently only Boom [3] does; (3) as adding new predicates makes Boolean program model checking more expensive, we primarily rely on *transition refinement* (cf. [1]) –

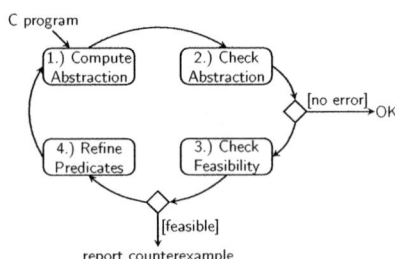

Fig. 1. Key components of SATABS

symmetry-aware analysis requires a particular variant that handles both active and passive threads; (4) adding new predicates in case of replicated threads requires extra care (cf. [10]).

[*] This research is supported by EPSRC projects EP/G026254/1, EP/G051100/1 and EP/H017585/1 and ERC project 280053.

C. Flanagan and B. König (Eds.): TACAS 2012, LNCS 7214, pp. 552–555, 2012.

2   Architecture

SatAbs shares large portions of the underlying C++ framework with CBMC [5], including the ANSI-C front end, the internal representation as GOTO programs, and interfaces to decision procedures. The competition candidate is linked with MiniSat 2.2.0 [11] as SAT solver.

Boolean Program Model Checking. Boolean program model checking is typically the bottleneck for a CEGAR-based verifier. SatAbs treats the Boolean program model checker as black box, and can be configured to use any suitable tool. For the competition, we use either Boom [4] together with a wrapper to analyse concurrent programs with as few threads as necessary, or Cadence SMV.

Refinement Strategy. SatAbs starts verification with a coarse abstraction, cheaply computed over predicates derived from assertions appearing in the program. The abstraction is refined in response to spurious counterexamples according to the following strategy: first, a spurious counterexample is checked for spurious transitions. If any exist, they are refined away using the technique of Das and Dill [9], following the approach of SLAM [1]. If no individual transition is spurious, weakest precondition calculations are used to derive new predicates from the counterexample, which are used to compute a more precise abstraction.

3   Strengths and Weaknesses

SatAbs supports all categories, including "Concurrency". In the presence of replicated concurrent programs, SatAbs exploits symmetry [13] to curb state explosion. Although the tool can also be applied to asymmetric concurrent programs, no scalability is expected for this case. As the focus on symmetric threads is not reflected in the benchmarks, SatAbs timed out on most of the benchmarks in the category "Concurrency". In the category "HeapManipulation", SatAbs failed because of a bug in the counterexample analysis; this has been fixed and in future we expect positive results there as well.

Overall, SatAbs proved to be reliable: bit-precise reasoning paired with some degree of maturity made SatAbs return only a single wrong result, which was due to bugs that have been fixed in the meantime. Yet we are aware of several limitations and weaknesses, which concern both sequential and concurrent code. Current technical limitations of predicate discovery may lead to SatAbs reporting "refinement failures". Furthermore overall efficiency and performance require closer inspection to reduce the number of timeouts that SatAbs had in the competition.

4   Tool Setup

SatAbs is hosted at http://www.cprover.org/satabs/ and is available both in binary form for popular platforms and as source code under a 4-clause BSD license. A C preprocessor is required (as provided by GCC on Unix-like platforms or

Visual Studio on Microsoft Windows). The model checkers Boom and Cadence SMV were used in the competition – SMV must be downloaded separately.[1]

The following command-line options were used for the competition, depending on category: 1) `--modelchecker boom`: Select Boom as model checker; without this option, Cadence SMV is used as default. 2) `--full-inlining`: Inline all functions. This is required for proper operation when using Boom. 3) `--error-label ERROR`: Instead of searching for violated assertions, prove (un)reachability of the label "ERROR" as specified in the competition rules. 4) `--32`: Select the basic bit-width of the architecture; by default, the bit-width of the execution platform is assumed, but some categories were designated to contain 32-bit benchmarks. 5) `--concurrency`: Enable use of passive threads and broadcast assignments, as described above. 6) `--max-threads 5`: Passed to Boom: only analyse executions involving no more than 5 concurrently running threads. 7) `--iterations 500`: Sets the upper bound on the number of CEGAR iterations to 500.

For categories "ControlFlowInteger" and "SystemC" we used SMV as model checker, i.e., option 1) was not given. Options 5) and 6) were only used for the category "Concurrency".

References

1. Ball, T., Cook, B., Das, S., Rajamani, S.K.: Refining Approximations in Software Predicate Abstraction. In: Jensen, K., Podelski, A. (eds.) TACAS 2004. LNCS, vol. 2988, pp. 388–403. Springer, Heidelberg (2004)
2. Ball, T., Majumdar, R., Millstein, T.D., Rajamani, S.K.: Automatic predicate abstraction of C programs. In: PLDI, pp. 203–213 (2001)
3. Basler, G., Hague, M., Kroening, D., Ong, C.-H.L., Wahl, T., Zhao, H.: BOOM: Taking Boolean Program Model Checking One Step Further. In: Esparza, J., Majumdar, R. (eds.) TACAS 2010. LNCS, vol. 6015, pp. 145–149. Springer, Heidelberg (2010)
4. Basler, G., Mazzucchi, M., Wahl, T., Kroening, D.: Context-aware counter abstraction. Formal Methods in System Design 36(3), 223–245 (2010)
5. Clarke, E., Kroning, D., Lerda, F.: A Tool for Checking ANSI-C Programs. In: Jensen, K., Podelski, A. (eds.) TACAS 2004. LNCS, vol. 2988, pp. 168–176. Springer, Heidelberg (2004)
6. Clarke, E., Kroening, D., Sharygina, N., Yorav, K.: Predicate abstraction of ANSI–C programs using SAT. Formal Methods in System Design (FMSD) 25, 105–127 (2004)
7. Clarke, E., Kroning, D., Sharygina, N., Yorav, K.: SATABS: SAT-Based Predicate Abstraction for ANSI-C. In: Halbwachs, N., Zuck, L.D. (eds.) TACAS 2005. LNCS, vol. 3440, pp. 570–574. Springer, Heidelberg (2005)
8. Clarke, E.M., Grumberg, O., Jha, S., Lu, Y., Veith, H.: Counterexample-guided abstraction refinement for symbolic model checking. J. ACM 50(5), 752–794 (2003)
9. Das, S., Dill, D.L.: Successive approximation of abstract transition relations. In: LICS (2001)

---

[1] Available at http://www.kenmcmil.com/

10. Donaldson, A., Kaiser, A., Kroening, D., Wahl, T.: Symmetry-Aware Predicate Abstraction for Shared-Variable Concurrent Programs. In: Gopalakrishnan, G., Qadeer, S. (eds.) CAV 2011. LNCS, vol. 6806, pp. 356–371. Springer, Heidelberg (2011)
11. Eén, N., Sörensson, N.: An Extensible SAT-solver. In: Giunchiglia, E., Tacchella, A. (eds.) SAT 2003. LNCS, vol. 2919, pp. 502–518. Springer, Heidelberg (2004)
12. Graf, S., Saïdi, H.: Construction of Abstract State Graphs with PVS. In: Grumberg, O. (ed.) CAV 1997. LNCS, vol. 1254, pp. 72–83. Springer, Heidelberg (1997)
13. Wahl, T., Donaldson, A.F.: Replication and abstraction: Symmetry in automated formal verification. Symmetry 2(2), 799–847 (2010)

WOLVERINE: Battling Bugs with Interpolants[*]
(Competition Contribution)

Georg Weissenbacher[1,**], Daniel Kroening[2], and Sharad Malik[1]

[1] Department of Electrical Engineering, Princeton University
[2] Department of Computer Science, Oxford University
(georg.weissenbacher@magd.oxon.org)

Abstract. WOLVERINE is a software verifier that checks safety properties of sequential ANSI-C and C++ programs, deploying Craig interpolation to derive program invariants. We describe the underlying approach and the architecture, and provide instructions for installation and usage.

1 Approach

WOLVERINE [1] is a software verification tool for ANSI-C and C++ programs that aims at finding either a Hoare-style correctness proof or a counterexample for a given reachability property. The tool is an implementation of the interpolation-based lazy abstraction algorithm [2] outlined in Figure 1. A description of the steps ① to ⑤ of Figure 1 is provided below.

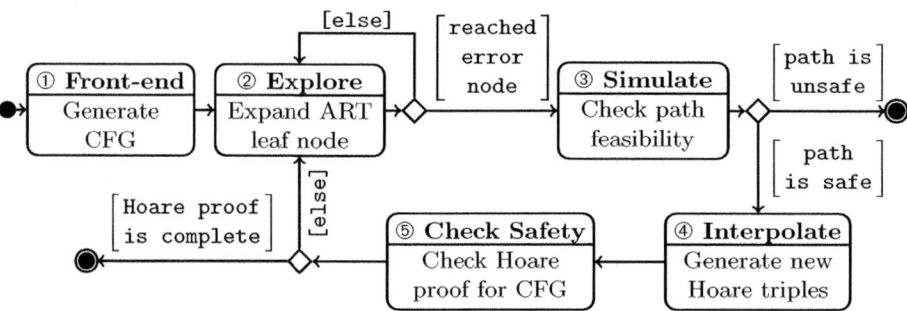

Fig. 1. UML activity diagram describing the work-flow of WOLVERINE

① WOLVERINE generates a control flow graph (CFG) representation of the program and encodes reachability properties using assertions/error nodes.
② Following the lazy abstraction paradigm established by [3], WOLVERINE constructs an abstract reachability tree (ART). To this end, it explores the paths of the CFG (in a depth first search manner) until it encounters an assertion.[1]

---

[*] Supported by a gift from the Intel Labs Academic Research Office.
[**] Corresponding author.
[1] The search algorithm of WOLVERINE 0.5c incorporates a constant propagation domain in order to enable early pruning of infeasible execution traces.

C. Flanagan and B. König (Eds.): TACAS 2012, LNCS 7214, pp. 556–558, 2012.

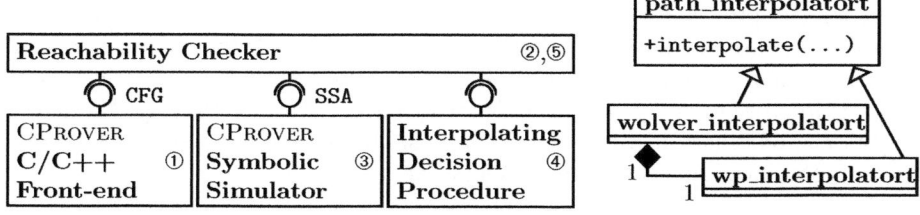

(a) UML component diagram for WOLVERINE   (b) Interpolator class hierarchy

Fig. 2. Software architecture of WOLVERINE

③ Given a path that reaches an assertion, WOLVERINE deploys symbolic simulation to determine whether the path corresponds to a feasible program execution violating the assertion; such unsafe executions are reported.

④ If the path is safe, WOLVERINE uses Craig interpolation to generate Hoare triples that prove that the assertion holds (c.f. [4]) and updates the edges and nodes of the ART accordingly: the spurious counterexample serves as a catalyst for refining the current approximation of safely reachable states [5].

⑤ If the Hoare triples of the ART are sufficient to prove the safety of all paths of the CFG, WOLVERINE concludes that the program is correct. Otherwise, the tool continues to expand paths that are not yet covered (step ②).

2   Software Architecture

Figure 2a shows the components and architecture of WOLVERINE. Our implementation uses the front-end (①) and the symbolic simulator (③) of the CPROVER framework (http://www.cprover.org). WOLVERINE uses an interpolating decision procedure (④) to extract Hoare triples from infeasible paths. To this end, the tool deploys its built-in decision procedure for equality logic with uninterpreted functions and limited support for bit-vectors [6,7,8] and falls back on the weakest precondition should this interpolator fail (see Figure 2b).

3   Tool Setup and Usage

Installation. Binaries for Linux, Windows, and MacOS X can be downloaded from the project website (http://www.cprover.org/wolverine) and should be deployed in a directory listed in the PATH environment variable. WOLVERINE requires a pre-processor (cl.exe, which is part of Visual Studio Express, on Windows and GNU's gcc on Unix-based platforms) and the header files typically packaged with it to be installed.

Usage. WOLVERINE must be executed from within the Visual Studio command prompt on Windows or a terminal on Linux and Mac OS X, and accepts options and source file names of the program to be verified as operands. By default, WOLVERINE scans the program for assertions and checks whether they hold. If executed with the option --error-label ERROR, WOLVERINE checks whether the label ERROR is reachable.

By default, WOLVERINE assumes that the host platform and the target platform for the program under test coincide. Therefore, in order to verify a Windows device driver on a Linux host, the options `--no-library --i386-win32` are recommended (but were not applied in the competition). The target processor architecture of the program under test has to be specified using the options `--32` or `--64` where it differs from the host. In the competition, these options were applied accordingly to all benchmarks.

4   Strengths and Limitations

While WOLVERINE shares many of the characteristics of predicate abstraction-based verifiers (most prominently, SLAM [9]), it avoids the computationally expensive image computation required to construct the abstraction (c.f. [2]), enabling the rapid detection of counterexamples (discussed in [1]).

WOLVERINE's performance is contingent on the Hoare triples that the interpolating decision procedure derives from spurious counterexamples. The inherent properties of interpolants typically enable concise abstractions. A "diverging" sequence of predicates, however, can result in a failed verification attempt. The built-in interpolator of WOLVERINE version 0.5c provides no support for linear arithmetic, quantified invariants, and heap models. In the competition, this led to a sub-optimal performance of WOLVERINE for benchmarks containing arithmetic expressions, unbounded arrays, or dynamic data structures. Moreover, WOLVERINE does currently not support the verification of concurrent programs.

References

1. Kroening, D., Weissenbacher, G.: Interpolation-Based Software Verification with WOLVERINE. In: Gopalakrishnan, G., Qadeer, S. (eds.) CAV 2011. LNCS, vol. 6806, pp. 573–578. Springer, Heidelberg (2011)
2. McMillan, K.L.: Lazy Abstraction with Interpolants. In: Ball, T., Jones, R.B. (eds.) CAV 2006. LNCS, vol. 4144, pp. 123–136. Springer, Heidelberg (2006)
3. Henzinger, T.A., Jhala, R., Majumdar, R., Sutre, G.: Lazy abstraction. In: POPL, pp. 58–70. ACM (2002)
4. Henzinger, T.A., Jhala, R., Majumdar, R., McMillan, K.L.: Abstractions from proofs. In: POPL, pp. 232–244. ACM (2004)
5. Clarke, E.M., Grumberg, O., Jha, S., Lu, Y., Veith, H.: Counterexample-Guided Abstraction Refinement. In: Emerson, E.A., Sistla, A.P. (eds.) CAV 2000. LNCS, vol. 1855, pp. 154–169. Springer, Heidelberg (2000)
6. D'Silva, V., Kroening, D., Purandare, M., Weissenbacher, G.: Interpolant Strength. In: Barthe, G., Hermenegildo, M. (eds.) VMCAI 2010. LNCS, vol. 5944, pp. 129–145. Springer, Heidelberg (2010)
7. Kroening, D., Weissenbacher, G.: Lifting propositional interpolants to the word-level. In: FMCAD, pp. 85–89. IEEE (2007)
8. Kroening, D., Weissenbacher, G.: An Interpolating Decision Procedure for Transitive Relations with Uninterpreted Functions. In: Namjoshi, K., Zeller, A., Ziv, A. (eds.) HVC 2009. LNCS, vol. 6405, pp. 150–168. Springer, Heidelberg (2011)
9. Ball, T., Rajamani, S.K.: The SLAM project: Debugging system software via static analysis. In: POPL, pp. 1–3. ACM (2002)

Author Index

Abdulla, Parosh Aziz   204
Ábrahám, Erika   299
Albarghouthi, Aws   157
Alur, Rajeev   188
Armando, Alessandro   267
Arsac, Wihem   267
Atig, Mohamed Faouzi   204
Avanesov, Tigran   267

Babiak, Tomáš   95
Barbot, Benoît   331
Barletta, Michele   267
Basler, Gérard   552
Becker, Bernd   299
Beyer, Dirk   504
Bjørner, Nikolaj   472
Bloem, Roderick   362
Bouajjani, Ahmed   451
Bozga, Marius   252

Caires, Luís   485
Calvi, Alberto   267
Cappai, Alessandro   267
Carbone, Roberto   267
Cassel, Sofia   466
Chadha, Rohit   437
Chang, Bor-Yuh Evan   33
Chechik, Marsha   157
Chen, Taolue   315
Chen, Yu-Fang   204
Chevalier, Yannick   267
Compagna, Luca   267
Cordeiro, Lucas   534
Cox, Arlen   33
Cuéllar, Jorge   267

David, Alexandre   492
Donaldson, Alastair   552
Drossopoulou, Sophia   407
D'Silva, Vijay   48
Dudka, Kamil   545

Eisenbach, Susan   407
El Ghazi, Aboubakr Achraf   422

Emmi, Michael   451
Erzse, Gabriel   267

Falke, Stephan   542
Fehnker, Ansgar   173
Finkbeiner, Bernd   392
Fischer, Bernd   534
Forejt, Vojtěch   315
Frau, Simone   267
Friedmann, Oliver   64

Geilmann, Ulrich   422
Gopinath, Divya   2
Grebenshchikov, Sergey   549
Gupta, Ashutosh   549
Gurfinkel, Arie   157
Gurney, Alexander J.T.   283

Haddad, Serge   331
Haller, Leopold   48
Hamlen, Kevin W.   126
Hardin, David   18
Hermanns, Holger   1
Heußner, Alexander   478
Höfner, Peter   173
Holzer, Andreas   538
Hölzl, Johannes   347
Howar, Falk   466
Huang, Chung-Yang (Ric)   377

Iosif, Radu   252

Jacobs, Swen   362
Jacobsen, Lasse   492
Jacobsen, Morten   492
Jansen, Nils   299
Jegourel, Cyrille   498
Jiang, Zhihao   188
Jin, Huafeng   220
Jørgensen, Kenneth Yrke   492
Jones, Micah M.   126
Jonsson, Bengt   466

Kaiser, Alexander   552
Katoen, Joost-Pieter   299

Khurshid, Sarfraz   2
Konečný, Filip   252
Křetínský, Mojmír   95
Kroening, Daniel   48, 538, 552, 556
Kwiatkowska, Marta   315

Lang, Frédéric   141
Lange, Martin   64
Le Gall, Tristan   478
Legay, Axel   498
Lengál, Ondřej   79
Leonardsson, Carl   204
Loo, Boon Thau   283
Lopes, Nuno P.   549
Löwe, Stefan   528

Madhusudan, P.   437
Malik, Sharad   556
Mandrykin, Mikhail   525
Mangharam, Rahul   188
Mateescu, Radu   141
McIver, Annabelle   173
McKinley, Kathryn S.   2
Merten, Maik   466
Merz, Florian   542
Minea, Marius   267
Moarref, Salar   188
Mödersheim, Sebastian   267
Møller, Mikael H.   492
Morse, Jeremy   534
Müller, Petr   545
Mutilin, Vadim   525

Nicole, Denis   534
Nipkow, Tobias   347
Nokhbeh Zaeem, Razieh   2

Pajic, Miroslav   188
Parker, David   315
Pellegrino, Giancarlo   267
Peringer, Petr   545
Peter, Hans-Jörg   392
Pham, Tuan-Hung   18
Picaronny, Claudine   331
Ponta, Serena Elisa   267
Popeea, Corneliu   237, 549
Portmann, Marius   173

Řehák, Vojtěch   95
Rezine, Ahmed   204
Rocchetto, Marco   267

Rusinowitch, Michael   267
Rybalchenko, Andrey   237, 549

Sanders, Beverly A.   220
Sankaranarayanan, Sriram   33
Scedrov, Andre   283
Schallhart, Christian   538
Sedwards, Sean   498
Shved, Pavel   525
Šimáček, Jiří   79
Simaitis, Aistis   315
Sinz, Carsten   542
Slind, Konrad   18
Song, Fu   110
Sonnex, William   407
Srba, Jiří   492
Sridhar, Meera   126
Steffen, Bernhard   466
Strejček, Jan   95
Sutre, Grégoire   478

Taghdiri, Mana   422
Talcott, Carolyn   283
Tan, Wee Lum   173
Tautschnig, Michael   48, 538, 552
Torabi Dashti, Mohammad   267
Touili, Tayssir   110
Turuani, Mathieu   267

Ulbrich, Mattias   422

van Glabbeek, Rob   173
Veanes, Margus   472
Veith, Helmut   538
Vieira, Hugo Torres   485
Viganò, Luca   267
Viswanathan, Mahesh   437
Vojnar, Tomáš   79, 545
von Oheimb, David   267

Wahl, Thomas   552
Wang, Anduo   283
Weissenbacher, Georg   556
Wendler, Philipp   528
Whalen, Michael   18
Wimmer, Ralf   299
Wonisch, Daniel   531
Wu, Cheng-Yin   377

Yavuz-Kahveci, Tuba   220
Yeh, Hu-Hsi   377